Broadband Wireless Access and Local Networks

Mobile WiMAX and WiFi

For a listing of recent titles in the
Artech House Mobile Communications Series,
turn to the back of this book.

Broadband Wireless Access and Local Networks

Mobile WiMAX and WiFi

Byeong Gi Lee
Sunghyun Choi

ARTECH HOUSE
BOSTON | LONDON
artechhouse.com

Library of Congress Cataloging-in-Publication Data
A catalog record for this book is available from the U.S. Library of Congress.

British Library Cataloguing in Publication Data
A catalogue record for this book is available from the British Library.

ISBN-13: 978-1-59693-293-7

Cover design by Yekaterina Ratner

685 Canton Street
Norwood, MA 02062

10 9 8 7 6 5 4 3 2 1

To our wives
Hyeon Soon Kang and Yoonjung Ryu

Contents

Preface

The recent trend of the convergence and diversification of data services and the growth of data traffic is truly phenomenal. The convergence of various different types of services (e.g., of voice, data, and video services), the convergence of conversational bidirectional services with distributive unidirectional services, the convergence of narrowband and broadband services, and the convergence of wireline and wireless services is now an established trend that is augmented by the convergence of user terminals. The growth of data traffic is exponential with the deep penetration of mobile wireless services, the popularization of music, video, and other forms of downloading and exchange, and the convergence of communications, entertainment, broadcasting, and financial services. The diversification of services has been driving a paradigm shift in communication services today toward *user-created content* (UCC). This grand trend has made the concept of *prosumers* (producers + consumers) a reality in the communication world and has brought other new terminologies into use as well, such as *motizens* (mobile citizens), *cyberlations* (cyber relations), and *digital natives*. In response, the Web has evolved to Web 2.0, with an increasing movement from Internet portal services to mobile Internet services.

Underlying the grand trend, and functioning as the enabler of the convergence, growth, and diversification of data services, is the *Internet protocol* (IP), which is a packet-mode technology designed to support the processing and transport of data in packets among different types of communication networks. Convergence of a diverse set of data services, for which circuit-mode technologies tried to offer a platform of service integration in the past, can now find a flexible and dependable platform in IP-based protocol stacks. The rapid growth of data services relies on the widely accepted means of processing and transport supported by IP-oriented technologies. The diversification of data services is also supported by IP technology with most emerging data services built upon IP.

The environment for data services now and in the near future is the Internet, or more generally, the networks using the IP. Originally developed for data communication among computers and terminals, IP has become a foundation of networking for all services in its short history of existence (only four decades). The key to today's success of IP technology is the omnipresence of Internet devices and IP's robust capability for realizing interoperability among many networks. The abundance of software that operates on top of IP is another strength of the IP technology. The secret behind the wide acceptance of IP was the simplicity of the IP. It was designed to support intermittent data communications among computers and terminals, based on *best effort* routing of packets in a variety of physical configurations, including bus, ring, and mesh. This simple protocol was generally considered not sophisticated enough to support real-time services. It targeted academic research networks, supporting the limited world of computer professionals. Even after opening to the commercial world, IP for a long time covered only the data portion of the business world, as the simple skeleton of IP had to be augmented with many other protocols to accommodate the real-time multimedia part of the commercial world.

IP-based networking, with its (originally) modest goal and simple architecture, has spread to encompass the last meters of every computer network. IETF *requests for proposals* (RFPs), together with IEEE 802 PHY/MAC standards, contributed significantly to the wide penetration of IP to the customer end, strengthening reliability and capacity and reducing costs. IEEE 802.3 Ethernet, as the name predicted, has become an ether-like network, existing everywhere computers meet users. Its predominance in the local area was followed by IEEE 802.11 WiFi when the wireless technology became practical for end-user services. Over two decades, *carrier-sense multiple access with collision avoidance* (CSMA/CA)-based wireless *local area network* (LAN) technology has frequently replaced *CSMA with collision detection* (CSMA/CD)-based wireline LANs, establishing a truly ether-like presence in the air. Today laptop computers, *personal data assistants* (PDAs), and other user-carried devices are mostly equipped with the WiFi capability from the manufacturing stage. For personal area networks, Bluetooth is often the choice because of its very low power and cost.

On the other side of the communication world, there still exists circuit-mode communication technology with the star-topology network architecture. It was born with the invention of the telephone. Over 125 years, the telephone network has spread all over the world, covering all inhabited areas down to almost every residence. Throughout this long period of development and deployment, the copper-based telephone network became omnipresent but, in contrast, the service concentrated mainly on voice. Technology was developed with the spread of the telephone network, but the technology development was focused more on providing high-quality voice services by expanding the transmission distance, increasing the transmission capacity, and accelerating the switching speed, than on developing new protocols and architecture to accommodate new types of service. As a consequence, the central offices in the telephone network were filled with intelligent transmission, switching, and signaling devices, but the subscriber's telephone set has been providing the same function for more than 100 years. The main reason for this voice-centric development was probably the lack of visible demand for data services during that long development period of telephone networks. Even after data services began to grow, telephone service providers were not successful in migrating into the data service market for various reasons. The long tradition of voice-centric telephone service made it difficult to admit the importance and the potential growth of data services. The circuit-switched telephone network, with dumb end-user devices and central operation, was not well suited to effectively accommodate data services. The growth of data services was not fast enough to justify big investments for renovating the huge established telephone network and installing new data-centric equipment. Lacking a competitive option in the switched services of the telephone network, new IP-based networks (using, however, physical lines provided by the telephone carriers) were readily established and widely accepted among end data users.

Throughout the long development period of the telephone network and the comparatively short development period of computer networks, the two were living in quite different worlds, though the computer networks such as ARPANet used the leased lines of the telephone network for wide area networking. The circuit-based telco network served the large commercial voice market using telephone, and the packet-based research networks served the small computer-communication community with noncommercial operation. They maintained these totally different identities until data services began to indicate some potential growth in the 1980s and the Internet was opened to the commercial world in the 1990s. Around that time, there appeared the first attempt to combine circuit-based voice services with packet-based data services, based on circuit-mode technology. It was the first encounter of the two different worlds, which was made in the context of wireline networks. The second encounter came later in the arena of wireless networks with an effort to harmonize circuit-mode and packet-mode wireless services based on packet-mode (or IP) technology.

It was the *International Telecommunication Union* (ITU) CCITT, later renamed ITU-T, that initiated the first encounter. It standardized the *integrated services digital network* (ISDN) with the goal of integrating voice and data services on a circuit-based platform, and it advanced digitization, which had been successfully completed in the core network, down to the access network by digitizing subscriber lines. This visionary project, started at the turn of the 1980s, was followed by the standardization of the *broadband ISDN* (BISDN), which progressed in harmony with the standardization of optical transmission in the *synchronous optical network* (SONET) and the *synchronous digital hierarchy* (SDH). BISDN introduced a new technology for integrating voice, data, and other broadband services in the *asynchronous transfer mode* (ATM), which gracefully combines circuit switching with the packet format. Moreover, it supports distribution services in addition to conversational services, and real-time services in addition to nonreal-time services. As a services integration strategy, ATM was ideal in theory but less successful in reality because of its relative complexity and high overhead cost. Moreover, the deployment of broadband optical networks was not done in a timely manner. The promotion of ATM technology in the 1990s, ironically, aroused a strong reaction in the Internet world, stimulating it to strengthen the competitiveness of the Internet. It is worth noting that the ATM concept lives on in IP networks in the form of *multiprotocol label switching* (MPLS).

The second encounter between circuit-mode and packet-mode networks began recently in wireless communications, encouraged by the booming success of mobile wireless businesses. The circuit-mode wireline telephone network was succeeded by cellular mobile communications in two major streams—the GSM/WCDMA family harmonized in the *Third Generation Partnership Project* (3GPP) and the IS-95/cdma2000 family harmonized in 3GPP2. Both were rooted in circuit switching, with packet-mode hybridization introduced in the course of evolution. The competing streams penetrated wide area networks within most countries and expanded coverage through international roaming services. They were very successful in providing voice services and began to provide high-quality data services to mobile users with comparatively low data rates and comparatively high service charges.

At the same time, the packet-mode Ethernet LAN was followed by the WiFi WLAN, using CSMA/CA instead of CSMA/CD. It has been very successful in providing data device users with the last 50-m access service into a wired LAN. It provides very high data rates to *hot spot* users at very low cost, but service quality is often unpredictable, and both mobility support and coverage are limited. The harmonization of those wireless extensions of the telephone and computer networks, constituting a second encounter of those communication worlds, is taking several paths, one of which is the handoff of multimode devices between cellular mobile and WLAN networks and a second, more intense path is the development of Mobile WiMAX networks. Whereas the first attempt of integration was made by the ITU, this second attempt was made by the IEEE, specifically the IEEE 802.16e standard working group. In contrast to the first attempt that adopted circuit-mode and then ATM technology, which is midway between circuit and packet modes, this second attempt employs IP-packet technology as the common vehicle for harmonization. Mobile WiMAX was designed on an IP foundation, maintaining the spirit of support for an IP network level seen in all IEEE 802 standards, thereby realizing efficient deployment of all types of data services. For effective provision of real-time multimedia services, it adopted a connection-oriented approach, not the connectionless approach of WiFi. It was designed to be capable of providing high-rate, high-quality data services to mobile users in medium to wide areas at very reasonable service charges.

Mobile WiMAX is very new, with the first IEEE 802.16e standard published in 2006 and the first system development and commercial service launched in 2007 in Korea. Commitments to Mobile WiMAX service are being made in a large number of countries, and allocations of frequency spectrum for Mobile WiMAX services have been announced in many countries. Furthermore, Mobile WiMAX has been accepted as a viable technology for the *fourth generation* (4G) mobile communications and was recently adopted as an IMT-2000 standard by ITU-R. Mobile WiMAX is now a reality. It incorporates many strong technologies, such as *orthogonal frequency division multiple access* (OFDMA), *time-division duplexing* (TDD), *multi-input multi-output* (MIMO), *adaptive modulation and coding* (AMC), IP, and security features, that can be combined to produce high spectral efficiency and resilient channels, resulting in high-rate, low-cost, wide-area, mobile multimedia services. Singling out OFDMA, Mobile WiMAX is the first mobile wireless specification to adopt this technology. Everything is ready to realize the second encounter of the descendents of the traditional communication and computer worlds. It is the investment made by network operators that will dictate the success of this second attempt for a harmonious services integration.

This book introduces the network technologies adopted by Mobile WiMAX for the implementation of IP-based broadband mobile wireless access and the WiFi technologies that have steadily evolved for the past 10 years, establishing a firm foundation for IP-based wireless local network access. These access and local technologies have many things in common, most prominently that both are oriented toward IP traffic and standardized by IEEE 802 working groups. The book is organized in two parts separately addressing Mobile WiMAX and WiFi, plus a preliminary chapter to provide a common ground of discussions for the two network technologies.

For the Mobile WiMAX part, we collected the most recent experience and knowledge of the design and field engineers, especially from Samsung Electronics and the Korea Telecom (KT) Corporation, who have been involved in the first development and deployment of Mobile WiMAX systems in Korea (with the nickname of "WiBro," an abbreviation for *wireless broadband*). The WiFi part is based on the extensive experience of one of the authors in IEEE 802.11 standards and on industry collaboration among Philips Electronics as a chip vendor, Samsung Electronics as a chip/system vendor, and KT as a service provider. The authors believe that understanding these two IP-oriented wireless network technologies will help readers deepen their insight into today's wireless networks and enhance their competence and competitiveness in the design of future wireless networks.

Acknowledgments

This book was made possible thanks to the contributions of many colleagues in industry and academia who accumulated the most up-to-date and practical knowledge of Mobile WiMAX and WiFi networks through direct involvement in standardization, system development, and network deployment. In the particular case of Mobile WiMAX, we gratefully acknowledge the contributions and assistance of standards, design, and field engineers in Samsung Electronics, KT (Korea Telecom), and other companies.

First, we would like to thank those colleagues who contributed by writing some chapters or sections: Hyunpo Kim at KT (Chapters 1 and 10); Hyeonwoo Lee (Chapter 2), Euiseok Hwang (Chapter 2), Hokyu Choi (Chapter 2), Dae Woo Lee (Chapter 2), Han-Seok Kim (Chapters 3 and 6), Jae Hwan Chang (Chapter 3), Seungjoo Maeng (Chapter 4), Myung-Kwang Byun (Chapter 4), Yonwoo Yoon (Chapter 4), Jaehee Cho (Chapter 4), Jinhan Song (Chapters 5 and 7), Inseok Hwang (Chapter 9), Eun Yong Kim (Chapter 9), and Jaekon Lee (Chapter 10), all at Samsung Electronics; Woojune Kim at Airvana (Chapter 3); Chung Gu Kang at Korea University (Chapters 5, 6, and 7); Pil Joong Lee at Pohang University of Science and Technology (POSTECH) (Chapter 8); and Hanbyul Seo (Chapters 2 and 7) and Hoojoong Kwon (Chapter 8 and 9) both at Seoul National University (SNU). We are especially indebted to those who dedicated a great deal of time and effort: Inseok Hwang, Chung Gu Kang, Han-Seok Kim, Woojune Kim, Seungjoo Maeng, and Jinhan Song.

We also thank the reviewers who helped improve the contents: Hyunpo Kim at KT (Chapters 1 and 10), Soon Young Yoon at Samsung Electronics (Chapters 1, 2, and 3), Chung Gu Kang at Korea University (Chapter 2), Dong Ho Cho at Korea Advanced Institute of Science and Technology (KAIST) (Chapters 3, 5, 6, and 7), Yong Hoon Lee at KAIST (Chapter 4), Jae Hyeong Kim at Posdata (Chapters 4, 9, and 10), Jae Hwan Chang at Samsung Electronics (Chapter 5), Saewoong Bahk at SNU (Chapters 5, 6, and 7), Dan Keun Sung at KAIST (Chapters 6, 7, 9, and 10), Seung Woo Seo at SNU (Chapter 8), Kwang Bok Lee at SNU (Chapter 9), Sunggeun Jin at SNU (Chapter 16), Youngsoo Kim at SNU (Chapter 18), Seongkwan Kim at SNU (Chapters 11, 16, and 18), Jeonggyun Yu at SNU (Chapters 11 and 14), Youngkyu Choi at SNU (Chapters 12 and 18), Hyewon Lee at SNU (Chapters 13 and 17), and Munhwan Choi at SNU (Chapters 13 and 15).

We are grateful to those who provided a comfortable environment and high-quality facilities to author the book: Seoul National University and the Research Institute of New Media and Communications at SNU. Byeong Gi Lee would like to thank Samsung Electronics, which offered an office for a year, and two Buddhist temples, Cheon Kwan Sa and Ssang Bong Sa, which offered quiet rooms for stays of several weeks.

We would like to thank those who encouraged and supported the writing of this book in various ways: Ki Tae Lee, vice chairman; Woon Seob Kim, executive vice president; Young Ky Kim, executive vice president; Sei Jei Cho, senior vice president; and Soon Young Yoon, senior manager, of Samsung Electronics. Byeong Gi Lee would like to give special thanks to Ki Tae Lee, who arranged accommodations at Samsung Electronics during a sabbatical leave and supported the project in every aspect.

We would like to thank our graduate students at SNU for supporting the writing of this book by drawing figures, correcting typos, and doing other miscellaneous work: Seo Shin Kwak, Soo Min Koh, Joon Ho Lim, Seon Wook Kim, and Seung Han Ryu in the Telecommunications and Signal Processing (TSP) Laboratory and Okhwan Lee, Youngwoo Hwang, Minsoo Na, Heeyoung Lee, Changyeon Yeo, and Seungmin Woo in the Multimedia and Wireless Networking Laboratory (MWNL). Byeong Gi Lee would like to thank Kyung Hee Choi at Samsung Electronics for help in drawing the figures and Hojoong Kwon at the TSP Laboratory for helping out all miscellaneous works throughout the authoring process. Sunghyun Choi would like to thank both Seongkwan Kim and Dongmyoung Kim at the MWNL for taking care of all miscellaneous works.

Last, but not least, we deeply thank our wives, Hyeon Soon Kang and Yoonjung Ryu, whose love and support enabled us to accomplish this heavy authoring task while conducting our regular jobs at school and performing other work in the community.

CHAPTER 1

Preliminaries

The concept of wireless access networks emerged in the late 1980s as a byproduct of cellular wireless technology. As the demand for cellular service exploded worldwide, the cost of wireless network components decreased, while the cost for deploying and maintaining the conventional copper-based subscriber network increased. The subscriber network, though it appears to be a small part of the overall telecommunications network, in reality occupies a considerably large portion of the overall network expenses, most of which is spent for deployment, operation, and maintenance of the subscriber lines. For this reason, the wireless subscriber network was first deployed in rural areas in the beginning where the initial cost is comparatively low. Later, it has become an effective alternative to the copper-based subscriber network in urban areas. It is very recent that the concept of mobile wireless access network was introduced. As an extension of a series of IEEE 802.16 standardization, whose commercial specification is known to be *worldwide interoperability for microwave access* (WiMAX) for fixed wireless access, a task group was formed in 2002 for enhancing the IEEE 802.16 standards by including mobility. Its first standard product, IEEE Std 802.16e, was published in 2006.

In contrast, the concept of the *wireless local area network* (WLAN), based on IEEE 802.11, was introduced much earlier, in early 1990s, in order to support IEEE 802.3 Ethernet-like best-effort packet access network without wires using unlicensed bands. As it operates at unlicensed bands, the protocols were designed to operate where other types of devices coexist. A distinctive feature of the WLAN is in that a user can purchase WLAN devices and then can enjoy high-speed wireless networking without permission from anybody. Today, many portable devices are being shipped with embedded WLAN radios, and the application space of WLAN is fast expanding from niche markets to general and wide markets.

This introductory chapter deals with the background Mobile WiMAX and WiFi. Noting that both of them operate in the wireless environment, we first introduce the wireless communication channel characteristics: we discuss the limitations of wireless channels and introduce the techniques developed to combat against the limitations. Second, we introduce the frequency spectrum used for wireless communications in general as well as the frequency bands allocated for Mobile WiMAX and WiFi communications. Third, we introduce the history of the IEEE 802.16 and IEEE 802.11 standardization for the generation of the Mobile WiMAX and WiFi standards, respectively, extending to the current status and future direction. Fourth, we provide a high-level sketch of the Mobile WiMAX and WiFi technologies in the aspects of *physical* (PHY) layer, *media access control* (MAC) layer, and network configuration, and, based on them, we compare the similarities and differences of the two network technologies.

1.1 Wireless Communication Channel Characteristics[1]

The most salient difference of wireless communications from wireline communications is in the propagation medium, or the channel for communication signal transmission. Whereas the media for wireline communications are mostly in protected form, for example, twisted-pair, coaxial cable, and fiber, the medium for wireless communications is unprotected and thus subject to various different kind of noises and external interferences. Such interfering factors cause attenuation on the received signal power and various different types of distortion in the received signal waveform. Moreover, in the case of mobile communications, user movement and environment dynamics make those impediments change over time in unpredictable ways. As a consequence, the radio propagation conditions and their statistical characteristics seriously affect the operation of wireless communication systems as well as their performances. Therefore, in preparation for the design or analysis of high-performance wireless communication systems like Mobile WiMAX, it is crucial to understand the characteristics of wireless communication channels and, further, the technologies to reinforce the wireless channels against those impediments.

1.1.1 Channel Gain

One of the most important characteristics of wireless channels is the *channel gain*, which determines how much portion of the transmission power is received at the receiver. The channel gain is defined by the ratio of the received power to the transmission power, so it dictates the level of the received signal power, which serves as the metric of link quality of the wireless channel. The transmitter-receiver pair, in general, gets a clearer channel as the channel gain gets higher.

Channel Gain Components

In general, the channel gain is determined by three different factors: path loss, shadowing, and multipath fading.

Path loss is the decay of the signal power dissipated due to the radiation on the wireless channels, so it is determined by the channel's physical characteristics of signal propagation. In general, path loss is modeled as a function of the distance between the transmitter and receiver on the assumption that signal loss is identical at a given distance.

Shadowing is caused by obstacles such as walls and trees located in the middle of the transmitter-receiver communication path. These obstacles absorb, diffract, and reflect the transmitted signal, thereby attenuating its received power. Since the location of the obstacles and their property are not predictable, shadowing is a random factor that can vary the received power even at the same distance.

1. The content of this section is generated out of the work of B. G. Lee, D. Park, and H. Seo. Refer to [1].

Multipath fading refers to the signal power variation caused by the constructive and destructive addition of signal components arriving at the receiver through multiple different paths between the transmitter and receiver. These multiple paths are generated by the scattering and reflecting objects located around the transmitter-receiver communication path. Like the case of shadowing, multipath fading is, in general, modeled as a statistical process.

Channel gain G is determined as the product of the power gains obtained from path loss, shadowing, and multipath fading. That is, for the transmission power P_T, the received power P_R is represented by

$$P_R = P_T G = P_T G_{PL} G_S G_{MF} \tag{1.1}$$

where G_{PL}, G_S, and G_{MF} denote the power gains determined by path loss, shadowing, and multipath fading, respectively.

Effects of Fading

Channel gain changes depend on the environmental dynamics, including the mobility of the transmitter-receiver pair. The three gain factors—path loss, shadowing, and multipath fading—are involved in the change but affect the change in different ways. The path loss factor is almost insensitive to the movement by several meters as the transmitter-receiver pair is usually assumed separated by hundreds of meters. The shadowing factor is affected by the length of the obstacles, which is in the range of 10–100m. As such, the two gain factors are affected by the environmental dynamics in the scale of meters, so those variations are called *large-scale fading*.

On the other hand, the multipath fading factor is strongly related to the wavelength of the propagating signal, which, in general, is in the scale of micrometers. Thus, even a small movement may affect the multipath fading factor, so this variation is called *small-scale fading*.

Figure 1.1 illustrates the characteristics of the channel fadings with respect to the distance between the transmitter and receiver. Both the channel gain and the distance are plotted in a log-scale.

1.1.2 Fading

Among the three channel gain factors, path loss and shadowing are affected by large-scale fading, while multipath fading is affected by small-scale fading.

Path Loss

Path loss is determined by how the signal power decays as it propagates through the wireless channel. The critical factor that affects path loss is the distance between the transmitter and receiver. It is known that the signal power decreases as the distance increases. The issue in modeling the path loss is how to determine the "rate" at which the signal power decays during propagation. There are several path loss models, including free space model, two-ray propagation model, and simplified path loss models.

In the case of the *free space model*, according to the propagation theory of electromagnetic wave, the signal power reduces inversely proportional to the square of

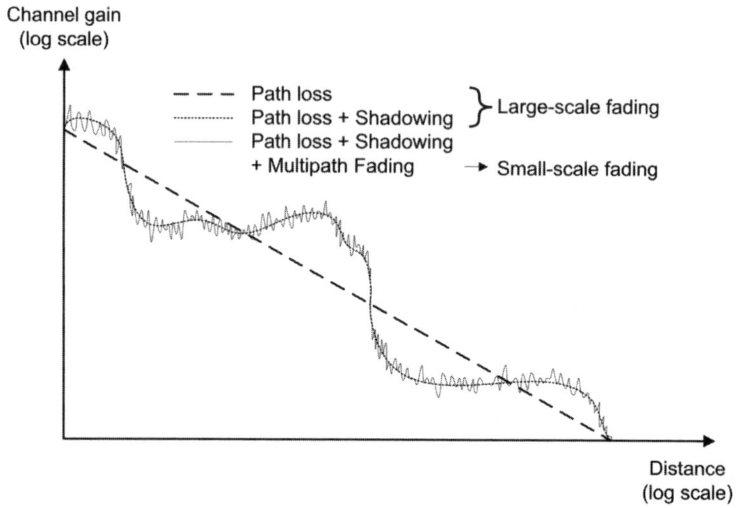

Figure 1.1 Illustration of the characteristics of channel fadings.

the distance in this free space, which equally applies to the power gain. To be specific, the power gain from path loss takes the expression

$$G_{PL} = \left(\frac{\lambda}{4\pi d}\right)^2 A \tag{1.2}$$

in the free space, for the transmitter-receiver distance d, the passband carrier wavelength λ, and the antenna power gain A.

In the case of the *two-ray model*, two propagation components—the *line-of-sight* (LOS) ray and the ground-reflected ray—are taken into account. It is the simplest model that considers the effects of reflection, diffraction, and scattering. In this case, the power gain from path loss is approximated to [2]:

$$G_{PL} = A\frac{(h_t h_r)^2}{d^4} \tag{1.3}$$

for the transmitter and receiver antenna heights h_t and h_r, and the transmitter-receiver distance d.

These two models may not be suitable for practical use due to their unrealistic assumptions, but they commonly established the fact that the power gain from path loss is proportional to $1/d^{\alpha}$ for a *path loss exponent* α, which is 2 and 4 for the two cases, respectively. A more practical variation of them is the *two-slope* path loss model, which employs two different path loss exponents α_1 and $\alpha_2 - \alpha_1(\alpha_1 \leq \alpha_2)$, according to the distance between the transmitter and receiver. To be specific [2],

$$G_{PL} = \begin{cases} A\left(\dfrac{d_0}{d}\right)^{\alpha_1}, d_0 \leq d \leq d_c \\ A\left(\dfrac{d_0}{d_c}\right)^{\alpha_1}\left(\dfrac{d_c}{d}\right)^{\alpha_2}, d > d_c \end{cases} \tag{1.4}$$

for the reference distance d_0 at which the channel gain becomes A and the critical distance d_c at which the path loss exponent changes. For example, the set of parameters determined for a microcellular systems with a carrier frequency of 1.9 GHz is $\alpha_1 = 2.07$, $\alpha_2 = 4.16$ and $d_c = 573$m for the antenna height 13.3m [3].

A more simplified path loss model is to use only a single path loss exponent α. This model is useful in evaluating macrocellular systems, since most of users are located further than the critical distance from the *base station* (BS). For example, the channel gain model with $\alpha = 3.76$, specifically,

$$G_{Pl} = \frac{0.0295}{d^{3.76}} \tag{1.5}$$

was derived for the macrocell systems with a carrier frequency at 2 GHz [4].

Shadowing

Shadowing refers to the additional attenuation of signal power caused by blocking objects in between the transmitter and receiver. Since the location and effects of those obstacles are unpredictable, the shadowing brings in variations of channel gain even at a fixed transmitter-receiver distance. For this reason, shadowing effect is formulated by statistical model.

The most common model of the shadowing effect is the *log-normal shadowing*. In this model, the channel gain determined by shadowing is assumed to be a log-normal distributed. That is, the logarithm of the channel gain is a zero-mean Gaussian random variable, which is given by

$$10\log_{10} G_s \sim N\left(0, \sigma_S^2\right) \tag{1.6}$$

for the standard deviation σ_S of the log-normal shadowing, whose value ranges from 6 to 10 dB, with a larger σ_S rendering more fluctuation in channel gain.

The shadowing effect is dictated by the obstacles such as walls and buildings. The shadowing effect is nearly the same at two closely located points but the signal power attenuation becomes more uncorrelated as the locations get further separated. In formulating this correlated property of the shadowing effect, it is sufficient to consider the covariance of the two different locations. It is because the joint distribution of any multivariate Gaussian random variable can be determined by its covariance. A simple method to formulate the correlation of shadowing effect is to assume that the correlation is proportional to the distance between two different locations. In this case, the correlation distance is defined by the distance at which the correlation becomes $1/e$ of the variance of the shadowing.

Multipath Fading

Multipath fading refers to the channel fluctuation caused by the superposition of multiple different propagation paths. In general, there exist a large number of objects between the transmitter and receiver, which create multiple propagation paths by reflecting and scattering the transmitted signal. The transmitted signal arrives at the receiver after undergoing multiple paths as illustrated in Figure 1.2. The multipath components may be added in a constructive or destructive manner,

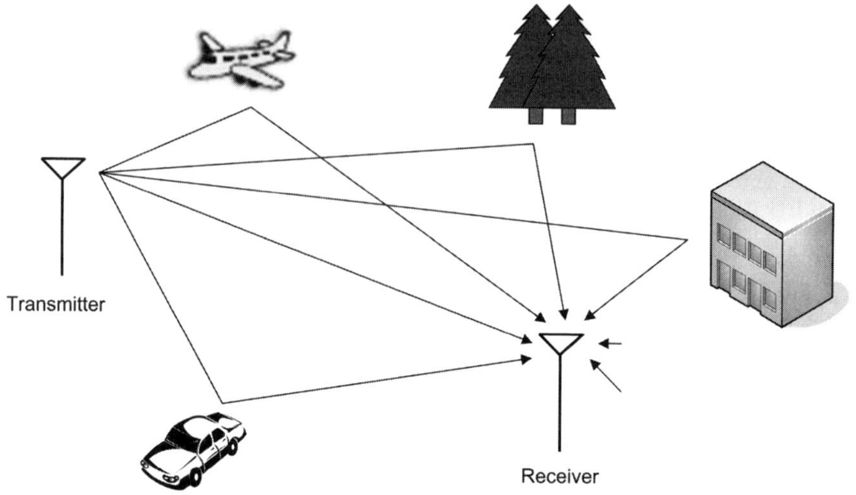

Figure 1.2 Illustration of multipath environment.

and a small phase shift caused by a wavelength-scale movement may yield a large variation in the aggregate channel characteristics (see Figure 1.1). Consequently, the multipath environment induces rapid channel fluctuations and, therefore, small-scale fading.

Those multiple paths have different attenuation property and add different phase shifts to the transmitted signal. Moreover, each multipath component of the same transmitted signal arrives at the receiver at different time instants since the lengths of multiple paths differ. As a result, even though the transmitter sends a single pulse with very short duration, the receiver sees a dispersed and distorted version of it. In terms of circuit theory, the channel is like a *time-varying*, *memory* channel. Figure 1.3 illustrates this by showing that the same impulse signal yields a different received waveform if it is transmitted at different time instance.

The channel impulse response in Figure 1.3 is, in fact, a superposition of multipath components that have different attenuations and phase shifts. This means that if we take a sample at a point of the channel response, it consists of multiple replicas of the transmitted signal that experienced different paths. Thus, the "effective"

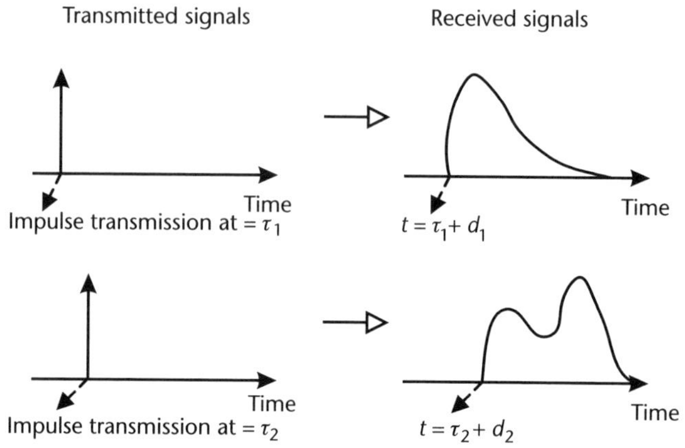

Figure 1.3 Illustration of multipath channel impulse response.

channel seen by the receiver at a time instance is determined by how the multipath components are added (i.e., constructively or destructively). Since the constructive/destructive addition is sensitive to a small phase shift caused by a wavelength-scale motion, it may yield a large variation of the wireless channel. It is referred to as fading in time domain. This time domain fading is usually modeled as a random variable such as Rayleigh and Rician random variables.

The fading channel in time domain can be characterized by the *Doppler power spectrum*, which basically describes how the bandwidth of the output signal is spread. The amount of bandwidth spread is determined by the Doppler frequency f_D, which refers to the variation of the carrier frequency observed at the receiver when the transmitter and/or receiver move. The range of the nonzero Doppler power spectrum, B_D $(= 2f_D)$, is called the *Doppler spread* of the channel. The Doppler spread renders a measure indicating how fast the channel varies in time domain. If we define by the *coherence time* T_C the difference of the observation times with which the channel becomes uncorrelated, it can be approximated by

$$T_C \approx \frac{1}{B_D} \tag{1.7}$$

The coherence time plays an important role in relation to the delay requirement of the traffic. If the coherence time is much shorter than the transmission time of a unit data, the transmission is likely to experience varying channel fading. If channel fading varies during a unit data transmission, it is called *fast fading*. On the other hand, if the coherence time is much longer than the delay requirement, the channel fading is not likely to change during a unit data transmission. It is called *slow fading*.

Multipath fading causes channel fluctuation not only in time domain but also in frequency domain. If there exists only one multipath component, yielding a delayed and scaled impulse as the impulse response at the receiver, then the channel frequency response will be flat. However, in reality, there exist multiple multipath components, so the channel impulse response becomes a sum of multiple delayed and scaled impulses, with each component having different delays and scaling factors, as shown in Figure 1.3. Consequently the resulting channel frequency response at each channel observation takes a nonflat shape. This is referred to as fading in frequency domain. Figure 1.4 illustrates fading in frequency domain, as well as in time domain.

The fading channel in frequency domain can be described by *multipath intensity profile*, which shows the average channel gain of the multipath component with respect to the multipath delay. The range of the multipath delay over which the multipath intensity profile is nonzero is denoted by T_M and called the *delay spread* of the channel. If we define by the *coherence bandwidth* B_C the difference of sinusoidal frequencies for which the channel becomes uncorrelated, then, similarly to the case of the Doppler spread and coherence time, the delay spread and coherence bandwidth take a reciprocal relation; that is,

$$B_C \approx \frac{1}{T_M} \tag{1.8}$$

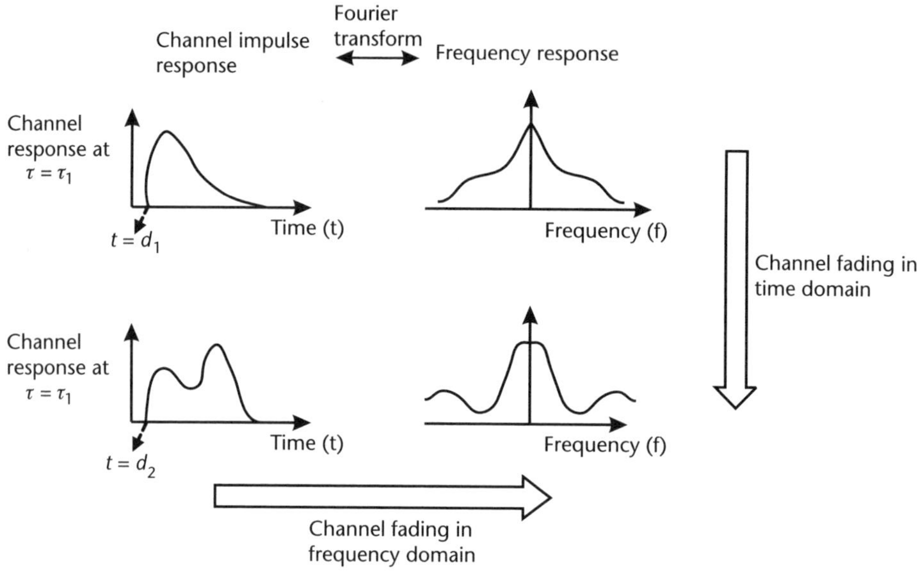

Figure 1.4 Illustration of fading in frequency and time domain.

The coherence bandwidth gives an important interpretation about the distortion of the transmitted signals. If the coherence bandwidth is small as compared with the bandwidth of the transmitted signal, some sinusoidal components of the signal may undergo channel fades differently from the others. As a result, the signal becomes severely distorted by the channel, so the receiver has difficulty in recovering the original signal. Such a channel is called a *frequency-selective fading* channel in the sense that the channel fluctuates over the bandwidth of the transmitted signal. On the other hand, if the coherence bandwidth is large, all the frequency components of the transmitted signal experience the same channel fading. In this case, the channel seen by the receiver is flat in frequency domain, and so is called a *frequency-flat fading* channel.

As noted earlier, the frequency selectivity of the channel distorts the transmitted signals. This distortion can be explained by introducing the *intersymbol interference* (ISI) as follows: from the reciprocal relation between the coherence bandwidth and the delay spread, the signal bandwidth larger than the coherence bandwidth (i.e., $B > B_C$) corresponds to the symbol time shorter than the delay spread (i.e., $T_S = 1/B < T_M$). Thus, in a frequency-selective fading channel, it happens that a new symbol is transmitted into the channel before the previously transmitted symbol finishes its arrival at the receiver. As a result, the previous symbol is interfered by the new symbol, or an ISI occurs. The effect of ISI becomes more severe for high-speed communication systems that employ wider bandwidth (i.e., shorter symbol time).

1.1.3 Channel Reinforcing Techniques

Various techniques have been introduced to overcome the unreliability and reinforce the performance of fading channels. Conventionally, equalizers have been used for eliminating or reducing the ISI effect by simply placing an equalizing filter at the receiver. Besides, there are system-level approaches that mitigate the negative effects of channel fading by implementing the channel-reinforcing techniques both at the transmitter and receiver. One is the diversity technique that transmits the information through multiple independent channels, thereby increasing the probability that the receiver detects the information correctly. *Orthogonal frequency division multiplexing* (OFDM), which divides the whole bandwidth into multiple narrow subcarriers, can be used as a diversity technique by associating multiple subcarriers in different frequency regions. The *hybrid automatic repeat request* (HARQ) technique combines *forward error correction* (FEC) and *automatic repeat request* (ARQ) techniques to strengthen the error correction capability. A third is the *adaptive modulation and coding* (AMC) technique, which adjusts the transmission rate by changing the modulation and coding techniques adaptively according to the state of the fading channel.

Diversity

Diversity techniques mitigate the effect of channel fading by using multiple independent channels. If several replicas of the same information signal are transmitted through independent fading channels, the probability that all the channels undergo deep fading at the same time decreases significantly. In this situation it is much more likely that at least one channel stays in a favorable condition, and the information can be delivered more reliably through that favorable channel. More specifically, the effect of diversity may be described by using probability as follows: If the probability that one channel suffers from deep fading is p, then the probability that the N independent channels provided by adopting the diversity technique go into deep fading all together drops to p^N. The number of independent channels, N, is called the *diversity order*.

There are several ways to implement the diversity technique. First, we may employ *time diversity*, where the same signal is transmitted on N different time slots. In order to secure the independency of each time slot, we should separate out two successive time slots to be more than the coherence time of the channel. Second, we may employ *frequency diversity*, where the same signal is transmitted over N (sub)carriers, with the separation of the carrier frequencies made larger than the coherence bandwidth of the channel. Third, we may use *space diversity*, which achieves the diversity effect by using multiple transmitter and/or receiver antennas. In this case we should position the multiple antennas sufficiently apart from each other (usually more than 10 wavelengths) such that the multipath fading in each antenna becomes nearly independent. There are other ways to achieve diversity, such as angle diversity and polarization diversity, but they are not used as widely as these three methods.

When diversity technique is used, the receiver combines the N independent diversity components in various ways to detect the transmitted signal. The nth diversity component takes the expression

$$r_n = h_n s + n_n \tag{1.9}$$

for the diversity channel function h_n and the additive noise n_n of the nth channel, both of which are complex variables. The noise term n_n may be assumed to be a zero mean white Gaussian process with an identical variance σ^2. So the signal generated at the receiver after weighting each component with a proper weighting factor w_n and then combining the n diversity components takes the form

$$r = \sum_{n=1}^{N} w_n r_n \tag{1.10}$$

Depending on how to choose the weighting factor, there exist three different ways of diversity combining, namely, selective combining, equal-gain combining, and maximal-ratio combining. First, *selection combining* is the simplest method that arranges w_n such that the sample with the largest channel gain is selected. Second, *equal-gain combining* (EGC) arranges w_n such that the phase differences of the samples are eliminated and thus all the diversity samples can be added. Third, *maximal-ratio combining* (MRC) arranges w_n such that the phase differences are eliminated and the magnitudes are enlarged in proportion to the amplitudes of the channels. As the MRC maximizes the *signal-to-noise ratio* (SNR), which governs the reliability of the transmission, it performs better than the other two combining techniques.

Hybrid ARQ

Hybrid ARQ is a technique that combines two representative transmission error recovery techniques—ARQ and FEC.

ARQ is a technique that recovers transmission error by requesting a retransmission of the same data to the transmitter. ARQ scheme requires an error-detecting mechanism such as the *cyclic redundancy check* (CRC) and a feedback channel that reports the successful/failed reception of a data block. If the reception is successful, the receiver sends an *acknowledgment* (ACK) message and, otherwise, a *negative acknowledgment* (NAK) to signal a retransmission. Since ARQ confirms whether or not a data block is reliably delivered, it is an essential technique in wireless data networks.

FEC is a technique that recovers bit error in the receiver by utilizing the redundancy carried over the *channel-coded* data. In essence, the channel coding technique helps to overcome the unreliability of wireless channels at the cost of added redundancy. The redundancy part, called *parity bits*, does not contain any new information but increases the dimension of the signal space and, as a result, increases the distance among different encoded sequences of the information bits. According to the information theory, the transmission errors can be corrected if the number of erred bits does not exceed half the minimum distance of the employed channel coding scheme. Such a coding gain is, therefore, achieved at the cost of lowered coding rate: coding gain increases as the coding rate decreases (i.e., as the number of added redundancy bits increases).

ARQ technique can operate more efficiently when used in conjunction with the FEC technique. This hybridization of FEC and ARQ is called the HARQ technique. HARQ combines FEC and ARQ by encoding the data block with a channel coding scheme prior to transmission. The HARQ technique can be better exploited by utilizing the information contained in the erroneously received block to assist the decoding of the retransmitted block. Depending on the methods of utilizing the previous information, HARQ is categorized into *chase combining* and *incremental redundancy* schemes (refer to Sections 2.1.6 and 4.2.3 for details).

HARQ technique may be interpreted as an *implicit* channel adaptation method. The unreliability of poor-conditioned channels can be overcome by using a stronger channel coding with a lower coding rate. When the channel suffers from deep fading, HARQ protocol reduces the coding rate by requesting additional information to be delivered via retransmissions. Through repeated retransmissions, the receiver obtains a stronger codeword, which can be successfully decoded against high-level noises and interferences. The transmitter-receiver pair equipped with an HARQ adjusts the transmission rate adaptively to the level that the current channel condition can accommodate. This adaptation is performed without exchanging any explicit message but by utilizing the implicit message contained in the ACK/NAK message.

Adaptive Modulation and Coding

AMC refers to the transmission technique that adjusts the transmission rate based on the current channel condition. If channel is in good condition, AMC increases the transmission rate by using a higher order modulation and/or a higher rate channel coding. Otherwise, it operates to the opposite direction. This enables a robust and spectrally efficient transmission over time-varying fading channels: it brings in higher throughput when the channel is in good state and brings in reliable communications when the channel suffers from deep fading.

AMC technique inherently requires a channel estimation process at the receiver and a feedback mechanism to report the estimated channel condition to the transmitter. So in implementing AMC, it is important to arrange such that the channel information is kept as accurate as possible. This requires making the delay caused by the channel-estimating and reporting process less than the coherent time of the channel. Once the transmitter receives the channel status information, it selects the *modulation and coding scheme* (MCS) best fitted to the received channel condition. Since AMC operates based on the reported channel condition, it is a kind of *explicit* channel adaptation method, in contrast to the HARQ case. (Refer to Section 2.1.3 for more discussions on AMC.)

1.2 Frequency Spectrum for Wireless Communications

Frequency spectrum is the unique resources for wireless communications that should be shared among different services and different network operators. Every wireless communication network requires a portion of the radio spectrum for operation. The radio spectrum is infinitely large in theory but, in practice, only a very limited part of spectrum is available for commercial access networks. This limited spectrum is divided into a diverse set of access and distribution networks, or communication and broadcasting services, including terrestrial access, satellite communication, television broadcasting, mobile communication, *industrial scientific medical* (ISM) usage, and others. Therefore, a wireless network operator must acquire suitable frequency spectrum first to establish the intended access network. Then it should come up with technical/engineering solutions to the challenges of the wireless environment of the acquired frequency band.

Tables 1.1 to 1.3 illustrate the usage of frequency spectrum for a diversified set of communication services in Korea, the United States, and Europe, respectively. We observe a similar trend in the usage of the frequency bands in all cases. The frequency bands in the 800–960-MHz and 1,700–2,500-MHz ranges are used for cellular mobile communications in all three cases. A number of large bands in the 24–27-GHz, 27–32-GHz, and 24–33-GHz ranges are allocated for *local multipoint distribution service* (LMDS) in the United States, *broadband wireless local loop* (B-WLL) in Korea, and the point-to-multipoint services in Europe, respectively. In the case of the Mobile WiMAX services, Korea allocates the 100-MHz band in the 2.3–2.4-GHz frequency region; the United States allocates bandwidths in the 2.3-GHz, 2.5-GHz, and other frequency bands; and Europe allocates 3.5-GHz and other frequency bands (see Table 1.5). In the case of the ISM bands, Korea allocates some frequency bands in the 2.4-GHz and 5.7-GHz regions and the United States allocates a large number of frequency bands stretched in the 6-MHz–244-GHz range. For the wireless LAN services, several bands in the vicinity of 2.4 GHz and 5.15 GHz are allocated in all three cases (see Tables 1.6–1.8).

1.2.1 Frequency Spectrum for WiMAX

The IEEE 802.16 standard system operates in two different frequency bands: one is in the 10–66-GHz band, and the other is the "below 11 GHz" band, or the 2–11-GHz band, specifically.

Table 1.1 Frequency Spectrum Allocation for Wireless Communication Services in Korea

Services		Frequency Band(MHz)	Bandwidth	Operators
Mobile Phone		US 824~849 DS 869~894	25MHz x 2	1, Nationwide
Wireless Data		US 898~900 DS 938~940	2MHz x 2	3, Nationwide
Trunked Radio System (TRS)		US 811~821 DS 856~866	10MHz x 2	1, Nationwide 5, Local
Personal Communication Service (PCS)		US 1,750~1,780 DS 1,840~1,870	30MHz x 2	3, Nationwide
Wireless Paging		161.2~169.0 322.0~328.6	7.8MHz 6.6MHz	1, Nationwide 8, Local
Bidirectionl Wireless Paging		US 923.55~924.45 DS 317.9875~320.9875	900kHz 2MHz	1, Greater Capital Area
WiBro*		2,300~2,400	9MHz x 9	2, Nationwide
Broadband Wireless Local Loop (B-WLL)		US 24,250~24,750 DS 25,500~26,700	500MHz 1,200MHz	3, Nationwide
IMT - 2000	Terrestrial	US 1,920~1,980(FDD) DS 2,110~2,170(FDD) US/DS 1,885~1,920(TDD) 2,010~2,025(TDD)	60MHz 60MHz 35MHz 15MHz	2, Nationwide
	Satellite	US 1,980~2,010 DS 2,170~2,200	30MHz 30MHz	
ISM Band		2,400~2,483.5 5,725~5,850	83.5MHz 125MHz	

Table 1.2 Frequency Spectrum Allocation for Wireless Communication Services in the United States

Services	Frequency Band (MHz)	Bandwidth	Operators
Wireless Services 700MHz	698~704 / 728~734 704~710 / 734~740 722~728 746~757 / 776~787	6MHz x 2 6MHz x 2 6MHz x 1 11MHz x 2	
Cellular	US 824~849 DS 869~894	25MHz x 2	
Trunked Radio System (TRS)	US 806~824 DS 851~869 US 896~901 DS 935~940	18MHz x 2 5MHz x 2	2, Nationwide 5, Local
Narrowband PCS spectrum for two-way paging systems	901~902 930~931 940~941	1MHz x 3	1, Nationwide 8, Local
Broadband PCS spectrum for cellular-like systems	1,850~1,910 1,910~1,930 (unlicensed) 1,930~1,990	60MHz x 2	3, Nationwide
Advanced Wireless Services (AWS)	1,710~1,720 / 2,110~2120 1,720~1730 / 2,120~2,130 1,730~1735 / 2,130~2,135 1,735~1740 / 2,135~2,140 1,740~1745 / 2,140~2,145 1,745~1755 / 2,145~2,155	10MHz x 2 10MHz x 2 5MHz x 2 5MHz x 2 5MHz x 2 10MHz x 2	1, Greater Capital Area
Multi-channel Multi-point Distribution Services (MMDS)	2,150~2,162 2,500~2,596 2,596~2,644 2,644~2,686		
Wireless Communications Service (WCS)	2,305~2,310 / 2,350~2,355 2.310~2,315 / 2,355~2,360 2,315~2,320 2,345~2,350	5MHz x 2 or unpaired 5MHz x 2 or unpaired 5MHz x 1 5MHz x 1	
Unlicensed National Information Infrastructure (U-NII)	5,150~5,350 5,725~5,825		
Local Multi-point Distribution Services (LMDS)	(Block A) 27,500~28,350 (Block A) 29,100~29,250 (Block B) 31,000~31,075 (Block A) 31,075~31,225 (Block B) 31,225~31,300		
ISM Band	6.765~6.795 13.553~13.567 26.957~27.283 40.66~40.70 902~928 2,400~2,483.5 5,725~5,875 24~24.25GHz 61~61.5GHz 122~123GHz 244~246GHz	30kHz 14kHz 308kHz 40kHz 26MHz 100MHz 150MHz 250MHz 500MHz 1GHz 2GHz	3, Nationwide

The 10–66-GHz band provides a physical environment where, due to the short wavelength, *line-of-sight* (LOS) transmission is required and multipath effect is negligible. The channels used in this band are large (e.g., typically 25 or 28 MHz) and the raw data rates could be over 120 Mbps. With such a large bandwidth, this band is well suited for *point-to-multipoint* (PMP) access serving applications from *small-office home-office* (SOHO) through medium to large office applications. *Single-carrier* (SC) modulation is used for the air interface in the 10–66-GHz band, which is referred to as wireless MAN-SC in Table 1.4.

Table 1.3 Frequency Spectrum Allocation for Wireless Communication Services in Europe

Services	Frequency Band (MHz)	Bandwidth
GSM 900	US 890~915 (G) DS 935~960 (G)	25MHz x 2
Extended GSM 900	880~915 / 925~960 (F, U, I)	35MHz x 2
GSM 1800	US 1,725~1,780 / DS 1,820~1,875 (G), US 1,710~1,785 / DS 1,805~1,880 (F, U) US 1,735~1,780 / DS 1,830~1,880 (I)	55MHz x 2 75MHz x 2 50MHz x 2
DECT	1,880~1,900 (F, G) 1,835~1900 (I)	20MHz 65MHz
IMT – 2000 / UMTS	US 1,920~1,980 (FDD) DS 2,110~2,170 (FDD) 1,900~1,920 (TDD) 2,010~2,025 (TDD)	60MHz 60MHz 20MHz 15MHz
Radio LANs	2,300~2,483.5 (I), 2,400~2,483.5 (G, U) 2,445~2,455 (U), 5,150~5,725 (I) 5,150~5,350 (G), 5,470~5,725 (G) 5,150~5,875 (U), 10,675~10,699 (U)	183.5MHz, 83.5MHz 10MHz, 575MHz 200MHz, 255MHz 725MHz, 24MHz
HIPERLAN	5,150~5,255 (F) 5,150~5,350 (I) 5,460~5,725 (I) 17,100~17,300 (I)	105MHz 200MHz 265MHz 200MHz
Fixed links	(F) 1,375~1,400, 1,427~1,460, 1,484~1,492, 2,300~2,310, 3,400~4,200, 5,925~7,250 7,375~7,890, 8,025~8,500, 10,500~10,680 10,700~11,700, 12,750~13,250, 14,250~14,500 15,250~15,350, 17,700~19,700, 21,200~23,600 24,250~26,500, 27,500~27,940.5, 28,192.5~28,450 29,200.5~29,460, 31,000~31,300, 31,800~33,400 37,268~38,220, 38,528~39,480, 40,500~43,500 47,200~50,200, 51,400~52,600, 55,780~66,000	25MHz, 33MHz, 8MHz 10MHz, 800MHz, 1325MHz 515MHz, 475MHz, 180MHz 1GHz, 500MHz, 250MHz 100MHz, 2GHz, 2.4GHz 2.25GHz, 440.5MHz, 257.5MHz 259.5MHz, 300MHz, 1.6GHz 952MHz, 952MHz, 3GHz 3GHz, 1.2GHz, 10.22GHz
	(G) 14,250~14,500, 27,828.5~28,052.5, 28,836.5~29,060.5 48,500~50,200, 51,400~52,600, 55,780~66,000	250MHz, 224MHz, 224MHz 2GHz, 1.2GHz, 10.22GHz
	(I) 1,427~1,530, 2,040~2,110, 2,215~2,300 2,440~2,450, 2,468~2,483.5, 3,600~4,200 5,250~5,450, 5,925~7,075, 7,125~7,750 10,000~10,680, 10,700~11,700, 14,250~14,620 15,230~15,350, 17,700~19,700, 22,000~22,330 22,768~23,380, 24,250~24,450, 29,100~29,500 31,983~32,599, 32,795~33,400, 37,338~38,300 38,598~39,500, 51,400~52,600, 55,780~59,000 64,000~66,000	103MHz, 70MHz, 85MHz 10MHz, 15.5MHz, 600MHz 200MHz, 1.15GHz, 625MHz 680MHz, 1GHz, 370MHz 100MHz, 2GHz, 330MHz 612MHz, 200MHz, 400MHz 616MHz, 605MHz, 962MHz 902MHz, 1.2GHz, 3.22GHz 2GHz
	(U) 3,480~3,500, 3,580~3,600, 3,605~3,689 3,925~4,009, 28,052.5~28,445, 29,060.5~29,452.5	20MHz, 20MHz, 84MHz 84MHz, 392.5MHz, 392MHz
Point-to-multipoint	(G) 2,540~2,670, 3,410~3,594, 24,500~25,165 25,500~26,173, 28,052.5~28,444.5, 29,060.5~29,452.5 31,000~31,101, 32,300~32,600, 33,131~33,400	130MHz, 184MHz, 665MHz 673MHz, 392MHz, 392MHz 101MHz, 300MHz, 269MHz
	(I) 24,450~25,109, 25,445~26,117, 27,500~29,500	659MHz, 672MHz, 2GHz
	(U) 31,000~31,800, 57,000~59,000	800MHz, 2GHz

*F: France G: Germany I: Italy U: United Kingdom

In the IEEE 802.16 standards there are specified four different types of physical layers for operation in different frequency bands, based on different multiple access technologies—namely, wireless MAN-SC, MAN-SCa, MAN-OFDM, and MAN-OFDMA. In addition, wireless *high-speed unlicensed metropolitan area net-*

Table 1.4 Air Interface Nomenclature

Designation	Applicability	Options	Dulex
WirelessMAN-SC	10 ~ 66 GHz		TDD, FDD
WirelessMAN-SCa	2 ~ 11 GHz licensed bands	AAS ARQ STC mobile	TDD, FDD
WirelessMAN-OFDM	2 ~ 11 GHz licensed bands	AAS ARQ Mesh STC mobile	TDD, FDD
WirelessMAN-OFDMA	2 ~ 11 GHz licensed bands	AAS ARQ HARQ STC mobile	TDD, FDD
WirelessHUMAN	2 ~ 11 GHz licensed-exempt bands	AAS ARQ Mesh STC	TDD

Source: [5].

work (HUMAN) PHY is additionally specified for use in the license-exempt bands. Specifically, wireless MAN-SC PHY is designed for operation on the 10–66-GHz frequency band, based on a combination of TDMA and *demand assigned multiple access* (DAMA) in the uplink and TDM in the downlink. Wireless MAN-SCa PHY is for use on the 2–11-GHz frequency band based on single carrier operation. Both wireless MAN-OFDM PHY and wireless MAN-OFDMA PHY are designed for operation on the 2–11-GHz frequency band based on the OFDM and the *orthogonal frequency division multiple access* (OFDMA) technologies, respectively. They are most commonly used in the fixed and mobile wireless access networks, respectively.

Table 1.4 lists a summary of the nomenclature for various air interface specifications in IEEE 802.16 standards. It shows five different types of physical layer designations in conjunction with the applicable frequency ranges.

The 2–11-GHz band provides a physical environment where, due to the longer wavelength, LOS is not necessarily the case and multipath effect may be significant. The ability to support *nonline of sight* (NLOS) scenarios requires additional physical layer functionality, such as the support of advanced power management techniques, interference mitigation, and multiple antennas. Additional MAC features, such as ARQ and the support of mesh topology, are also introduced. For the air interface of the licensed frequencies in the 2–11-GHz band, three different types of physical layers are used, namely single-carrier, OFDM, and OFDMA, which are referred to as wireless MAN-SCa, wireless MAN-OFDM, and wireless MAN-OFDMA, respectively, in Table 1.4.

The 2–11-GHz license-exempt (or unlicensed) bands are similar to that of 2–11-GHz licensed bands in physical environment, so their physical layer complies with the three PHY features specified for the 2–11-GHz licensed frequencies (i.e., wireless MAN-SCa, wireless MAN-OFDM, and wireless MAN-OFDMA). However, it has some additional issues caused by the license-exempt nature, such as interference, coexistence, and power radiation. So the license-exempt bands adopt the physical and MAC layer mechanisms that facilitate the detection and avoidance of interference and the prevention of harmful interference into other users.

Geographical distribution of potential WiMAX frequency allocation as well as the license status is listed in Table 1.5.

Table 1.5 Frequency Allocation for WiMAX in Some Selected Countries

Country	Frequency allowed/ Considered [GHz]	Licenses
Argentina	2.5, 3.5, 5.8	
Australia	2.3, 3.5, 5.8 / 2.5	2.3: Austar, Unwired
Brazil	2.5 / 2.3, 3.3	
Canada	2.3, 2.5, 3.5, 5.8 / 3.3	
China	3.4~3.6 / 2.3, 2.5	3.4~3.6: China Mobile, CECT Chinacomm, etc.
France	3.4~3.6 / 2.3, 2.5, 5.8	3.4~3.6: Altitudes Telecom , Maxtel, etc.
Germany	2.5~2.69, 3.4~3.6 / 2.3, 5.8	2.5~2.69: Airdata 3.4~3.6: Arcor, German Networks, etc.
India	/ 2.3, 2.5, 3.3, 3.4, 5.8	
Italy	3.4 / 2.3, 2.5, 3.3, 3.6, 5.8	3.4: ARIADSL, AFT, Telecom Italia , etc.
Japan	2.5 / 2.3, 3.4, 3.6	2.5: WBPK, Wilcom
Korea	2.3 / 2.5, 3.4, 5.8	2.3: KT, SKT
Malaysia	2.3~2.4, 2.5~2.69, 3.4~3.6	2.3~2.4: Bizsurf, MIB Comm, etc. 2.5~2.69: EB Technologies , TT Dotcom, etc. 3.4~3.6: Arized Broadband, Atlasone, etc.
New Zealand	2.3~2.4, 2.5~2.69, 3.4~3.6	2.3~2.4: Kordia, BCL, Telecom, etc. 2.5~2.69: Telecom Leasing , Vodafone, etc. 3.4~3.6: TelstraClear, Vodafone, etc.
Russia	2.5, 3.5, 5.8 / 2.3	Summa Telecom , Start Telecom , etc.
Singapore	2.3~2.4, 2.5~2.69 / 3.5	2.3~2.4: Qala, Inter-touch 2.5~2.69: Mobile One, Pacific Internet, etc.
Spain	3.4~3.6 / 2.3, 2.5, 5.8	3.4~4.6: Iberbanda, Neo-Sky, etc.
Taiwan	2.5~2.69 / 3.3, 3.4	2.5~2.69: FET, Tatung, Vastar, etc.
Venezuela	2.5~2.69, 3.4~3.6 / 2.3, 3.3	2.5~2.69: Omnivision 3.4~3.6: Telcel, Genesis, etc.
U.K.	3.4~3.6 / 2.3, 2.5, 3.3, 5.8	3.4~3.6: UK Broadband, Pipex
U.S.A.	2.3, 2.5, 3.6, 5.8 / 3.3	2.5: Sprint-Nextel

(Note) In the WRC-07 meeting, the two frequency bands, 450-470 MHz and 2.3-2.4 GHz, were selected as the world's common 4G bands, which may possibly include the Mobile WiMAX.

1.2.2 Frequency Spectrum for WiFi

WiFi devices operate at the unlicensed bands in the neighborhood of 2.4 GHz and 5 GHz. Specifically, the WiFi devices based on IEEE 802.11b/g operate at the 2.4-GHz bands while those based on IEEE 802.11a operate at the 5-GHz bands. The particular unlicensed bands differ from region to region. In this section, we discussed the available bands in Korea, thc United States, and Europe.

Unlicensed Bands at 2.4 GHz and 5 GHz

According to a rule published in September 2007, by the Ministry of Information and Communications (MIC) of the Korean government, the unlicensed bands in Korea, which can be used by WLAN devices, include 2.400–2.4835 GHz, 5.150–5.350 GHz, 5.470–5.650 GHz, and 5.725–5.825 GHz. Note that 2.400–2.4835 GHz and 5.725–5.825 GHz are ISM bands.

For each subband, specific rules are defined, including the maximum transmission power density (in mW/MHz) and the maximum antenna gain (in dBi). Additional rules are defined for subbands of 5.250–5.350 and 5.470–5.650 GHz, including *transmit power control* (TPC) and *dynamic frequency selection* (DFS). (Refer to Chapter 17 as to the detailed operations of TPC and DFS according to IEEE 802.11h.) Table 1.6 lists a summary of the unlicensed bands along with the corresponding maximum transmission power density and the maximum antenna gain.

We consider the case of the 5.725–5.825-GHz band, for example, assuming the transmission bandwidth of 20 MHz. The maximum power can be 200 mW (or 23 dBm) ideally, but depending on the spectral mask of the signal, the total power level may be quite smaller than 23 dBm due to the constraint of power density per MHz. Accordingly, if the total transmission power of 20 dBm is used, the *equivalent isotropically radiated power* (EIRP) can be up to 26 dBm (= 20 dBm + 6 dBi) including the antenna gain.

In the United States, the Federal Communications Commission (FCC) regulated by FCC CFR47 [6], Part 15 and Subpart E,[2] that 2.400–2.4835-GHz, 5.15–5.35-GHz, 5.47–5.725-GHz, and 5.725–5.850-GHz bands are available for WLAN operations as summarized in Table 1.7 along with the maximum transmission power and requirements. Among those bands, the 2.400–2.4835-GHz and 5.725–5.850-GHz bands are ISM bands, and 5.15–5.35-GHz, 5.47–5.725-GHz, and 5.725–5.825-GHz bands are called *unlicensed national information infrastructure* (U-NII) bands. Note that the 5.725–5.825-GHz band belongs to both ISM and U-NII bands, while the 5.825–5.850-GHz band belongs only to the ISM.

The maximum allowed transmission power differs depending on the subbands. The maximum power for the ISM bands is 1W, while that of U-NII bands depend

Table 1.6 Unlicensed Bands Useful for WLANs in Korea

Bands (GHz)	Max. Tx power density (mW/MHz)	Max. antenna gain (dBi)	Remark
2.4~2.4835	10	6	ISM
5.15~5.25	2.5	6	
5.25~5.35	10	7	TPC/DFS
5.47~5.65	10	7	TPC/DFS
5.725~5.825	10	6	ISM

2. More specifically, FCC CFR47, Part 15, Sections 15.205, 15.209, and 15.247; and Subpart E, Sections 15.401–15.407, Section 90.210, and Section 90.1201–90.1217.

Table 1.7 Unlicensed Bands Useful for WLANs in the United States

Bands (GHz)	Max. Tx RF Power (with up to 6 dBi antenna gain)	Remark
2.4~2.4835	1W	ISM
5.15~5.25	40 mW	Indoor only
5.25~5.35	200 mW	TPC & DFS
5.47~5.725	200 mW	TPC & DFS
5.725~5.825	800 mW	ISM

on each subband. Actually, the power limit for U-NII supersedes that for ISM, and hence the limit for the 5.725–5.825-GHz band becomes 800 mW, instead of 1W, as shown in Table 1.7. A transmission antenna with up to 6-dBi gain is basically allowed, and, for certain cases, an antenna with higher antenna gain can be used with the reduced transmission power limit. The 5.15–5.25-GHz band is limited for only indoor usage. For 5.25–5.35-GHz and 5.47–5.725-GHz bands, both TPC and DFS are required. (Refer to Chapter 17 for the detailed operations for TPC and DFS according to IEEE 802.11h.)

In Europe, European Conference of Postal and Telecommunications (CEPT) defines the available spectrum for WLANs, and rules are specified in ETSI EN 301 389 [7]. The unlicensed bands useful for WLANs are summarized in Table 1.8 along with the corresponding maximum transmission power and requirements. Note that for 5 GHz bands, the power limit is specified in terms of EIRP without considering the maximum antenna gain. As is the case in the United States, both TPC and DFS are required for 5.25–5.35-GHz and 5.47–5.725-GHz bands. Differently from the United States, TPC is also required for the 5.15–5.25-GHz bands, which is reserved for indoor usage. Note that the 5-GHz ISM band is not available in Europe for the WLANs.

Policy and Issues for Unlicensed Bands

As discussed earlier, the 2.4–2.4835-GHz and 5.725–5.85-GHz bands are the ISM bands, which were originally reserved internationally by ITU-R for the use of RF

Table 1.8 Unlicensed Bands Useful for WLANs in Europe

Bands (GHz)	Max. Tx Power	Remark
2.4~2.4835	100 mW	ISM
5.15~5.25	200 mW EIRP	Indoor only
5.25~5.35	200 mW EIRP	TPC & DFS
5.47~5.725	1 W EIRP	TPC & DFS

electromagnetic fields for industrial, scientific, and medical purposes other than communications. There are other ISM bands, including those at 902–928 MHz and 61–61.5 GHz. The most commonly encountered ISM device for many people is probably the microwave oven operating at 2.45 GHz. It is known that the WiFi operating at 2.4 GHz (especially near 2.45 GHz) is severely affected by the microwave oven operating at the proximity.

Generally, communication devices should be able to live with any interference generated by ISM equipment. As communication devices using the ISM bands must tolerate any interference from the ISM equipments, these bands are typically given over to the usage intended for unlicensed operations. Note that unlicensed operations typically need to be tolerant of interference from other communication devices anyway. The first generation WiFi devices were defined to operate at the 2.4-GHz ISM bands. Then, when a new PHY of IEEE 802.11 (i.e., IEEE 802.11a) for 5-GHz operations was defined, new bands (e.g., U-NII bands in the United States), in addition to the 5-GHz ISM bands were given over for the 802.11a operations. Today, IEEE 802.11a is there in the market. However, the most widely used type of WiFi, which is based on IEEE 802.11g PHY, operates at the 2.4-GHz band due to various reasons, including lower cost, longer transmission range, wider deployment, and others, even if the 2.4-GHz band is more crowded due to the wide deployment of many other types of devices operating at this band (e.g., microwave ovens, Bluetooth devices, and cordless phones).

The rule for the unlicensed communication operations at the ISM bands has been evolving over time. For the unlicensed operations, it is necessary to have a limit on the transmission power level for the reason of coexistence among multiple devices. In addition, the usage of spread spectrum was once mandated to reduce the interference to other devices. That is the reason why the first generation 802.11 PHYs defined for the 2.4-GHz operations employed spread spectrum technologies including *direct-sequence spread spectrum* (DSSS) and *frequency-hopping spread spectrum* (FHSS). Moreover, there were also specific constraints for the spread spectrum technologies as well, such as the number of frequency slots and the width of the frequency slots in the case of FHSS. However, as the technologies, based on different transmission schemes, arise over time, the rules for the unlicensed operations have been modified accordingly. For example, the *complementary code keying* (CCK) of IEEE 802.11b and the *orthogonal frequency division multiplexing* (OFDM) of IEEE 802.11g are not spread spectrum technologies, and they are now allowed to be used at the 2.4-GHz ISM bands.

As discussed earlier, some 5-GHz bands (e.g., U-NII bands) not included in the ISM require the WiFi devices to be capable of TPC and DFS. Note that TPC allows the adjustment of the transmission power and DFS allows the WiFi devices to jump to another frequency channel once a primary user is detected. This is due to the fact that there are other types of devices operating in these bands. Those include radar and satellite systems, which are called primary users of these bands. Note that the WiFi devices are secondary users, and hence the signals from the secondary users should not disturb the operations of the primary users. By utilizing TPC and DFS, this constraint can be achieved.

1.3 Standardization History

Standardization of Mobile WiMAX and WiFi was dictated by IEEE 802.16 and 802.11 *Working Groups* (WGs), respectively, both under the IEEE 802 LAN/MAN committee. The IEEE 802.16 standardization dates back to 1999, with the first IEEE 802.16 standard published in 2002. A series of IEEE 802.16 standards soon followed, with the 802.16a, 802.16d, and 802.16e standards published in 2003, 2004, and 2006, respectively. In contrast, the IEEE 802.11 standardization was started much earlier, with the first baseline standard of IEEE 802.11 published in 1997. Subsequently, the IEEE 802.11a and 802.11b standards were published in 1999 and 802.11g in 2003.

1.3.1 IEEE 802.16/WiMAX Standardization

The IEEE 802.16 standard has been developed by the IEEE 802.16 WG on *broadband wireless access* (BWA) since 1999. This standard was initially designed for fixed wireless services and was expanded to support mobility feature in IEEE 802.16e. In parallel with that, the *Telecommunications Technology Association* (TTA) of Korea worked on the standardization of WiBro system in harmonization with IEEE 802.16e, thereby positioning it as a special profile of the IEEE 802.16e standard. WiMAX system, the commercial profile of IEEE 802.16 standard, has been developed by WiMAX Forum. More detailed history of standardization is described in the following.

IEEE 802.16 Standardization

Among IEEE 802 standard committees responsible for standardization of PHY and MAC layers of LAN/MAN, IEEE 802.16 WG has developed BWA standards since it was formed in 1999.

This standard was initially designed to support fixed BWA service in LOS environment of 10–66-GHz band and IEEE 802.16-2001 [8] was approved by the Standard Board of the *IEEE Standard Association* (IEEE-SA) in 2001. Later, in the NLOS environment of the 2–11-GHz band, IEEE 802.16a standard [9], which includes three types of physical layers, SCa, OFDM, and OFDMA was developed in 2003. These standards were later revised and consolidated by IEEE 802.16 *Task Group d* (TGd) and its final version, IEEE 802.16-2004 [10], was approved.

On the other hand, IEEE 802.TGe was organized to enhance the standards by including mobility in 2002. In the beginning, it started with incorporating a limited mobility to the existing OFDMA specification with 1,024-point *fast Fourier transform* (FFT) and 5-MHz bandwidth. Later, it was expanded to encompass full mobility.

In May 2004, harmonization work was started between TGe and TTA. In the beginning TGe took the results of TTA's study for WiBro but later TTA also adopted the TGe's specifications to its system called WiBro. In September 2004, TGe started the sponsor ballot process (or the specification review process among the IEEE SA members), during which process a significant number of improvements were made following some heated discussions among the large group of participants. Throughout the sponsor ballot process, the specifications on HARQ, MIMO,

AAS, and PKM v2 were refined and matured. In parallel with the TGe standardization process, the TG Cor1 worked on correcting the errors in the 802.16-2004 standards document. After the completion of the two task groups, the standards document 802.16e-2005 [5] was approved in December 2005. Table 1.9 provides a comparison among various IEEE 802.16 standards.

In May 2005, IEEE 802.16 WG launched a new standardization activity on *mobile multihop relay* (MMR). MMR is intended to adopt relay stations to provide multihop communication to the *mobile station* (MS) that follows the IEEE 802.16e standards. It then would enable to expand the service coverage in fast-speed and low-cost manner and also helps to increase the system capacity.

TTA Standardization of WiBro

In Korea, the standardization activity of WiBro, which is a 2.3 GHz–based Mobile WiMAX, was led by the TTA. It started the standardization in June 2003 and took some features as the basic requirements from the beginning, which include the frequency reuse factor of 1, the use of *time-division duplex* (TDD), and the handover time of less than 150 ms. The *Elecronics and Telecommunications Research Institute* (ETRI) and Samsung Electronics submitted a joint contribution in June 2004, which was later taken as the WiBro phase 1 standards. TTA started amending the phase 1 standards for the harmonization with IEEE 802.16, and the amended phase 1 standard, which was made compatible with IEEE 802.16 TGe draft version 5, was approved as of December 2004. TTA started standardization of the WiBro phase 2 standards in 2005, including the topics on MIMO technology support and others. The topics were later added to the IEEE 802.16 standards too.

The WiBro profile task force team of TTA performed a study on the profile of WiBro specification. As to the functionalities of the PHY and MAC layers, the task force team classified them into basic and extended items and decided the items to implement in the BS and the user terminal, respectively. The result is that the BS must implement all the basic and extended items but the user terminal may implement all the basic items and some limited number of extended items determined by network operators. In October 2005, TTA launched a task force team to work on *interoperability test* (IOT) and *protocol implementation conformance statement* (PICS). In December 2005, TTA performed another stage of harmonization process

Table 1.9 Comparison of IEEE 802.16 Standards

Standards	802.16-2001	802.16a	802.16-2004, 16d	802.16e
Frequency band	10~66GHz,LOS	2~11 GHz,NLOS 10~66 GHz, LOS	2~11 GHz,NLOS 10~66 GHz, LOS	2~11 GHz,NLOS
PHY layer	SC	SCa, OFDM, OFDMA	SC, SCa, OFDM, OFDMA	SCa, OFDM, OFDMA
Duplex	TDD, FDD	TDD, FDD	TDD, FDD	TDD, FDD
Mobility	Fixed	Fixed	Fixed	Mobile
Release date	Apr. 2002	Apr. 2003	Oct. 2004	Feb. 2006

with the WiMAX Forum, which studied the profile and IOT of IEEE 802.16 standards and amended the standards harmoniously [11].

WiMAX Standardization Activities

The WiMAX Forum was organized by manufacturers and service providers for the commercialization of a BWA system based on IEEE 802.16 standards in 2001. The WiMAX Forum worked on regulating the specifications on profile, IOT, and others for the realization of the Mobile WiMAX system. Major participants of the WiMAX Forum are the companies supporting BWA and the companies supporting mobile systems.

Standardization works on the profiles and IOT for the radio access specification of Mobile WiMAX were done by the technical working group (TWG). In February 2006, the TWG selected the mobile WiMAX system profile of the functions based on the IEEE 802.16-2004 and TGe specifications [12]. This profile is shown in Table 1.10.

For the product certification, they worked on the development of *protocol implementation conference statement* (PICS) and *test suite structure* (TSS)/*test purposes* (TP), defining the band class groups listed in Table 1.11 as of 2007.

Standardization of the end-to-end network architecture for application services was performed by the network working group (NWG) of the WiMAX Forum. Among the release 1 standards to support the basic functions of IEEE 802.16, stages 2 and 3 specifying the network architecture and message flow have been completed by the NWG of the WiMAX Forum.

IEEE 802.16/WiMAX Evolution Standardization

In December 2006, IEEE 802.16 TGm was organized to develop an evolution version of IEEE 802.16. This project aimed at the amendment of the IEEE 802.16 standard to meet the requirements of the IMT-Advanced radio interface, which will be developed by the ITU-R. Major system requirements agreed within 802.16 TGm meeting in September 2007 are as listed in Table 1.12. They expect to complete 802.16m standardization by 2009 [13].

The WiMAX Forum also defined a Mobile WiMAX evolution roadmap as shown in Figure 1.5. In conformance with 802.16m standardization, they plan to develop the mobile WiMAX system profile 2.0. In addition to this profile, WiMAX

Table 1.10 Mobile WiMAX System Profile

Layer	Items
PHY	OFDMA, TDD, channel bandwidth (7, 8.75, 5, 10 MHz) DL-PUSC, DL-FUSC, UL-PUSC ,DL/UL B-AMC All 4 rangings, 6-bit CQI, TB-CC, CTC (DIUC), H-ARQ Modulation: (DL) 4,16,64-QAM, (UL) 4,16-QAM BS/MS synchronization, open-loop and closed-loop power control RSSI, CINR measurement, ECINR, normal MAP, compressed MAP MIMO & BF package(UL sounding, DL MIMO, UL C-SM, dedicated pilots)
MAC	PHS, ROHC, ARQ, H-ARQ MAC support QoS (BS-initiated), QoS (MS-initiated) BE, rtPS, nrtPS, ertPS, UGS, IPv4 CS, IPv6 CS scanning, PKMv2 Sleep & idle mode, OH-HO, MBS

Table 1.11 Band Class for Mobile WiMAX Product Certification

Band class	Frequency band (GHz)	Bandwidth (MHZ)	Regulatory Readiness
1.A	2.3-2.4	8.75	Korea
1.B		5 & 10	Singapore
2.A	2.305-2.320, 2.345-2.360	5	WCS spectrum in U.S.A. and Canada
2.B		10	
3.A	2.496-2.690	5 & 10	U.S.A., Europe
4.A	3.3-3.4	5	China
4.B		7	India, China, Europe
4.C		10	China
5.A	3.4-3.8	5	Europe
5.B		7	Europe
5.C		10	

Table 1.12 IEEE 802.16m Requirements

Item	Requirements
Operating bandwidth	Scalable bandwidth between 5~20 MHz
Duplex	Full-duplex FDD, Half-duplex FDD, TDD
Normalized peak data rate	Downlink: > 8.0 bps/Hz, Uplink: > 2.8 bps/Hz
Handover interruption time	Intra-frequency: max 30 ms, Inter-frequency: max 100 ms
User throughput	Downlink: > 2 x 802.16e, Uplink: > 2 x 802.16e
Mobility	Up to 350 km/h

network requirement specification, Release 2.0, developed by WiMAX Forum, will be the baseline of WiMAX certification, Release 2.0 [14].

1.3.2 IEEE 802.11/WiFi Standardization

The IEEE standard 802.11 has been developed by the IEEE 802.11 WG on WLAN since 1991. The first standard was published in 1997, and since then, the 802.11 WG has been developing many amendments to enhance this technology in various ways, including higher speed, QoS support, and security enhancement. The WiFi Alliance, which started in 1999, has been testing and certifying the interoperability of IEEE 802.11-based WLAN products. More detailed history of standardization is described in the following.

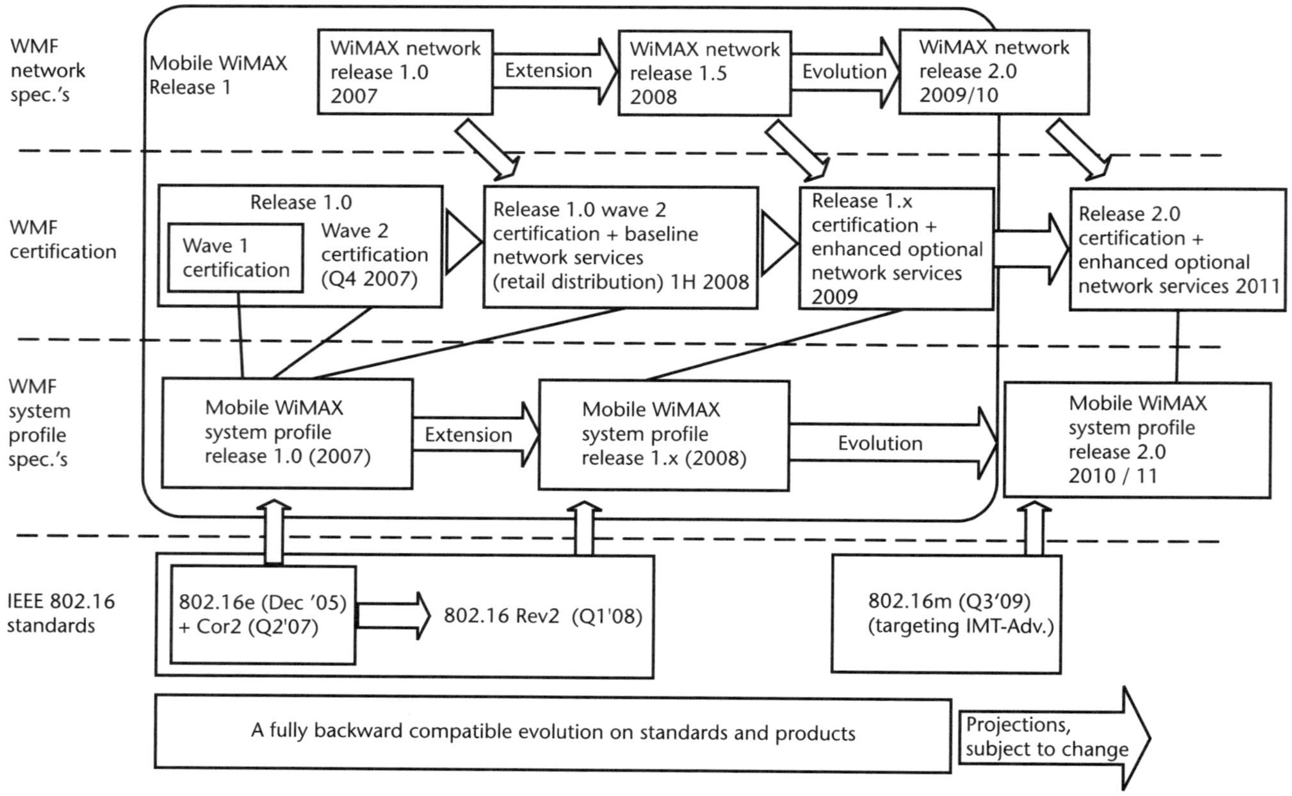

Figure 1.5 Mobile WiMAX technology evolution vision. (*After:* [14].)

IEEE 802.11 Standardization

As other IEEE 802 standards do, the 802.11 standards specify the protocols for both the MAC sublayer and PHY layer. The IEEE 802.11 WG started its standardization activities in 1991, and published the first standard specification, referred to as IEEE 802.11-1997, in 1997, and then a revision, referred to as IEEE 802.11-1999 [15], in 1999. In fact, the 802.11-1999 reflects mostly minor changes compared with the 802.11-1997. Since then, many extensions (officially called amendments) were generated in order to enhance the performance of the 802.11 in various aspects.

This standard was initially designed to support an Ethernet-like service without wires in local areas. For this reason, IEEE 802.11 was often referred to as "wireless Ethernet." The first generation 802.11 standard, IEEE 802.11-1997, and its revision IEEE 802.11-1999 (published in 1997 and 1999, respectively) include a single connectionless MAC and three PHYs, namely, DSSS, FHSS, and *infrared* (IR). All three PHYs support only 1- and 2-Mbps transmission rates, and both DSSS and FHSS PHYs were defined to operate at the 2.4-GHz ISM bands. We refer to the original protocols found at IEEE 802.11-1999 as the baseline protocols (i.e., baseline MAC and baseline PHY, respectively) for the rest of this book.

After the completion of the base protocols, the 802.11 WG started developing higher speed PHYs. Whenever a new project for the extension or amendment of the existing standard is planned, a new *task group* (TG) is established, and the people involved in the TG basically generate a new amendment. A TG is assigned an alphabet, and the name of the newly generated amendment is coined according to the alphabet. For example, IEEE 802.11a-1999 [16] was generated in 1999 out of the IEEE 802.11 *Task Group a* (TGa) activity. The first two amendments of the base 802.11 protocol, namely, 802.11a and 802.11b [17], were in the PHY. While TGa started its standardization before TGb, both 802.11a and 802.11b were published in 1999. IEEE 802.11a-1999 defined a new PHY based on OFDM to operate at the 5-GHz bands, while IEEE 802.11b-1999 defined a new PHY based on CCK to operate at the 2.4-GHz ISM bands.

The 802.11b is backward compatible with the DSSS PHY in the 802.11-1999 so that an 802.11b device can communicate with a DSSS PHY-based device. That is, the 802.11b is defined as an extension of the DSSS PHY. Accordingly, while the newly defined transmission rates by the 802.11b are only 5.5 and 11 Mbps, the 802.11b is actually meant to support 1, 2, 5.5, and 11 Mbps, including the baseline DSSS rates. In the market of the first generation 802.11 devices, both DSSS- and FHSS-based 802.11 devices were competing with each other since both of them had their own advantages compared with the other. However, along with the emergence of the 802.11b, the first generation technologies started disappearing very fast. The 802.11b products were introduced in the market around the completion of the standardization. Thanks to the fast transmission rate of the 802.11b, which is comparable with that of the earlier Ethernet (supporting only 10 Mbps), the 802.11b WLAN started booming. The standardization of the 802.11a was also completed in 1999, but its market introduction was only made in 2002 due to the relative difficulty of the implementation and so on. In 2002, the 802.11b-based WLAN APs had been already deployed widely, especially, in the hot-spot environments. Moreover, the 802.11a devices were at that time more expensive than the 802.11b devices.

Accordingly, for many people, the 802.11a, which was not compatible with the existing WLAN APs, was not very attractive even if its transmission speed was expected to be much faster.

More bad news for the 802.11a was the introduction of IEEE 802.11g-2003 [18], which was defined as an extension of the 802.11b for the 2.4-GHz operations. Again, the 802.11g is backward compatible with the 802.11b by supporting all the 802.11b rates. The newly defined rates by the 802.11g are exactly the same as those of the 802.11a since the 802.11g uses the exact same transmissions schemes as the 802.11a. Due to the backward compatibility requirements, the performance of the 802.11g in ideal environments was worse than that of the 802.11a, but for many people, the 802.11g devices, which were lower cost and backward compatible with the widely deployed 802.11b, were much more attractive. In fact, the 802.11g products were introduced in 2003 around the completion of the 802.11g standardization. Today, IEEE 802.11g is still the most popular PHY in the market.

The 802.11 MAC has been evolving as well. IEEE 802.11e-2005 [19] was defined to support QoS for multimedia applications over WLAN. The baseline MAC defined for wireless Ethernet was not capable of supporting multimedia applications including *voice over Internet protocol* (VoIP) and video streaming. The 802.11e MAC defines new QoS-supporting MAC, scheduling and admission control mechanisms, and other new features, which enhance the efficiency of the 802.11 network. In early 2000, a big hurdle against wide acceptance of the 802.11 products was the security threats. At that time, a number of papers reporting the security threats in the 802.11 products were published, and some tools to break the legacy security mechanisms became available in the public domain. To overcome such threats, IEEE 802.11i-2004 [20] was introduced by defining new encryption schemes, new authentication and key management schemes, and so on.

Another amendment is IEEE 802.11h-2003 [21] for spectrum and transmit power management extensions. This amendment defines DFS and TPC for the operations of the 802.11a-based WLAN devices in Europe. As presented in Section 1.2.2, many countries now require the DFS and TPC mechanisms at some 5-GHz bands, but when the 802.11h was being defined, these mechanisms were required only in Europe, which made the 802.11h constrained for European 5-GHz bands. However, technically the 802.11h can be used for the DFS and TPC operations in other countries as well.

There are some other amendments including IEEE 802.11d-2001 [22] for operations in various countries and IEEE 802.11j-2004 [23] for 4.9–5-GHz operations in Japan. Moreover, IEEE 802.11F-2003 [24], which is a recommended practice for *interaccess point protocol* (IAPP), was also specified. The IAPP is for the communications among APs to support handoff within a WLAN. This protocol defines the operations above the MAC, and it is a major reason why the 802.11F was defined as a recommended practice, since the 802.11 standards are meant for the MAC and PHY.

In fact, the IEEE published a new standard specification, called IEEE 802.11-2007 [25], in 2007. IEEE 802.11-2007 revision describes the IEEE 802.11 standard for WLANs with all the amendments that have been published until June 2007. The amendments, which were standardized after the publication of the baseline protocols, and were then rolled into IEEE 802.11-2007, include the following. Note that some projects took more time than others. For example, IEEE

802.11e-2005 is "amendment 8" after IEEE 802.11j-2004, even if its project alphabet (i.e., "e") is much earlier than that of the 802.11j.

- IEEE Std 802.11a-1999 (Amendment 1) for "High-Speed Physical Layer in the 5 GHz Band";
- IEEE Std 802.11b-1999 (Amendment 2) for "Higher-Speed Physical Layer Extension in the 2.4 GHz Band";
- IEEE Std 802.11b-1999/Corrigendum 1-2001 for "Higher-Speed Physical Layer (PHY) Extension in the 2.4 GHz Band—Corrigendum1";
- IEEE Std 802.11d-2001 (Amendment 3) for "Specification for Operation in Additional Regulatory Domains";
- IEEE Std 802.11g-2003 (Amendment 4) for "Further Higher Data Rate Extension in the 2.4 GHz Band";
- IEEE Std 802.11h-2003 (Amendment 5) for "Spectrum and Transmit Power Management Extensions in the 5 GHz Band in Europe";
- IEEE Std 802.11i-2004 (Amendment 6) for "Medium Access Control (MAC) Security Enhancements";
- IEEE Std 802.11j-2004 (Amendment 7) for "4.9 GHz–5 GHz Operation in Japan";
- IEEE Std 802.11e-2005 (Amendment 8) for "Medium Access Control (MAC) Enhancements for Quality of Service (QoS)."

WiFi Activities

In 1999, several companies came together to form the WiFi Alliance with the goal of driving the adoption of a single worldwide-accepted standard for high-speed WLAN. Through comprehensive interoperability testing, the WiFi Alliance certification programs ensure that WLAN products from multiple manufacturers work with each other. As a result, the WiFi Alliance certification programs have been a catalyst for the rapid adoption of WLAN products at homes, offices, and public access environments around the world. The WiFi Alliance is not intended to do any standardization, and relies on IEEE 802.11 WG for the standard generations. However, it should be also noted that a WiFi certification does not imply standard compliance, as their interoperability tests do not actually check such compliance. Note that a standard specifies a set of mandatory operations and a set of optional operations, and in order to be compliant with the standard, all the mandatory features must be implemented. In fact, the interoperability test for a certification program considers the operations of only a subset of the protocols in a specification. Such a subset might not include all the mandatory operations. For example, a certification program might check the interoperability based on only some optional operations without worrying about mandatory operations. That is the case with many certification programs, including *WiFi protected access* (WPA), *WiFi multimedia* (WMM), and *WMM power save*, which are explained next.

The WiFi certification programs cover the following categories. First, the mandatory programs include the following:

- WiFi products based on IEEE PHY standards: These are 802.11a, 802.11b, and 802.11g in single, dual mode (802.11b and 802.11g) or multiband (2.4 GHz and 5 GHz) products.
- WiFi wireless network security: WPA and WPA2 offer government-grade security mechanisms for personal and enterprise. WPA and WPA2 are based on *temporal key integrity protocol* (TKIP) and *CTR with CBC-MAC protocol*[3] (CCMP) defined in IEEE 802.11i.
- *Extensible authentication protocol* (EAP): An authentication mechanism used to validate the identity of a network device (for enterprise devices) is certified.

Second, the optional programs include:

- Next generation WiFi: Support for the IEEE 802.11n draft 2.0 standard is certified.
- Setup of security features: "WiFi Protected Setup" facilitates an easy setup of security using a *personal identification number* (PIN) or a button located on the WiFi device.
- Support for multimedia applications over WiFi networks: WMM enables WiFi networks to prioritize traffic generated by various applications, using QoS mechanisms of IEEE 802.11e, specifically, *enhanced distributed channel access* (EDCA).
- Power savings for multimedia content over WiFi networks: WMM Power Save, based on IEEE 802.11e, helps conserve battery life while running voice and multimedia applications by intelligently managing the time during which the device spends in a doze state.
- Devices equipped with both WiFi and cellular technologies: This provides detailed information about the performance of the WiFi radio in such a converged handset, as well as how the cellular and WiFi radios interact with one another.

Evolution of IEEE 802.11 Standards

There are a number of ongoing projects within IEEE 802.11 WG as summarized in Table 1.13.

A brief description of each project is as follows:

- IEEE 802.11k for radio resource measurement enhancements is to provide mechanisms to higher layers for radio and network measurements. It provides knowledge about the radio environment to improve performance and reliability in unlicensed radio environments.
- IEEE 802.11n for higher throughput will amend and extend the 802.11 WLAN protocol to incorporate new technologies for increasing the throughput of WLANs. The amended standard would specify mechanisms to increase transmission rates up to 600 Mbps.

3. CTR and CBC-MAC stand for counter mode and cipher-block chaining message mode authentication code, respectively.

Table 1.13 Ongoing Standardization of IEEE 802.11

Documents	Task Group	Type	Project name
802.11k	TGk	Amendment	Radio resource measurement ehnancement
802.11n	TGn	Amendment	High data rate
802.11p	TGp	Amendment	Wireless access for vehicle environment
802.11r	TGr	Amendment	Fast roaming
802.11s	TGs	Amendment	ESS mesh networking
802.11.2	TGT	Recommended practice	Wireless performance prediction
802.11u	TGu	Amendment	Wireless interworking with external networks
802.11v	TGv	Amendment	Wireless networking management
802.11w	TGw	Amendment	Protected management frames
802.11y	TGy	Amendment	Contention based protocol
802.11z	TGz	Amendment	Direct link setup

- IEEE 802.11p for *wireless access in vehicular environment* (WAVE) is to define enhancements to the 802.11 required to support *intelligent transportation system* (ITS) applications in vehicular environments. New licensed bands at 5.9 GHz were defined for this purpose.
- IEEE 802.11r for fast roaming is to define a protocol supporting fast handoff operations in WLANs without sacrificing QoS and security. Such a demand is coming from the emergence of *VoIP over WLAN* (VoWLAN) devices.
- IEEE 802.11s for *extended service set* (ESS) mesh networking is to define the protocols for the 802.11 APs to establish peer-to-peer wireless links with neighboring APs in order to establish a wireless mesh backhaul infrastructure.
- IEEE 802.11.2 (generated by Task Group T) will be a recommended practice for evaluation of the 802.11 WLAN performances by defining a set of performance metrics, measurement methodologies, and test conditions in order to enable such measurements and permit the prediction of the performance of installed WLAN devices and networks.
- IEEE 802.11u for interworking with external networks is to define a protocol specifying interworking with external networks, as typically found in hotspots. In this case, interworking refers to MAC layer enhancements that allow higher layer functionality to provide the overall end-to-end solution.
- IEEE 802.11v for wireless network management is to define extensions to the 802.11 MAC/PHY to provide network management for non-AP stations, including: (a) *basic service set* (BSS) transition management, (b) colocated

interference reporting, (c) diagnostic and event reporting, and (d) a traffic filtering service.

- IEEE 802.11w for protected management frames is developing enhancements to the 802.11 MAC layer to provide mechanisms that enable data integrity, data origin authenticity, replay protection, and data confidentiality for selected 802.11 management frames. Note that management frames are not protected under IEEE 802.11i.
- IEEE 802.11y for contention-based protocol will be an amendment for the operation of the WLAN in the United States 3.65–3.7 GHz, newly allocated for broadband wireless services.
- IEEE 802.11z for *direct link setup* (DLS) enhancement will be an amendment for a new DLS mechanism to allow operation with non-DLS capable access points and allow stations with an active DLS session to enter power save mode by enhancing the existing DLS mechanism defined in IEEE 802.11e.

1.4 Mobile WiMAX Versus WiFi

As discussed earlier, Mobile WiMAX and WiFi were generated independently each other by different standards groups and for different target frequency bands. In addition, the goals were different: Mobile WiMAX was intended to offer a wireless access means to wide area networks, whereas WiFi was intended to function as a wireless extension to the existing local area networks. Consequently, the employed PHY and MAC technologies were different from each other as each was optimized to its own targeted goals. Nevertheless, they possess an important commonality: both standard groups are under the same umbrella of IEEE 802 LAN/MAN committee, which has developed various access and connectivity protocols on the foundation of the same packet-mode (or IP-mode) concept, as opposed to other existing cellular wireless systems that employ the circuit-mode concept.

1.4.1 Mobile WiMAX: Broadband Wireless Access Networks

As discussed in Section 1.3.1, IEEE 802.16 generated a family of standards for BWA, among which the IEEE 802.16e standard includes the mobility feature that yields the Mobile WiMAX. Mobile WiMAX supports roaming service in metropolitan and regional networks, so allows mobile connectivity to mobile users. The target mobility is 120 km/h and the peak throughput is 18.7-Mbps downlink and 5.0-Mbps uplink in the case of DL/UL ratio = 29:18 and 10-MHz bandwidth. It utilizes the cell concept and the coverage of a cell is in the range a few kilometers. Equipped with such features, Mobile WiMAX is advantageous in supporting low-latency data, video, and real-time voice services for mobile users at high speed.

The protocol layering of the IEEE 802.16 system consists of a MAC layer and a physical layer, with the MAC layer divided into three sublayers, namely, service-specific *convergence sublayer* (CS), MAC *common part sublayer* (CPS), and security sublayer. The service-specific CS performs functions of converging user services to MAC CPS. There are two CS specifications, namely ATM CS and packet CS, but the packet CS is more commonly used for transporting all packet-based protocols such as IP, *point-to-point protocol* (PPP), and Ethernet. Among these CSs, only IP CS is included in the WiMAX profile. MAC CPS is the main body of the MAC layer, which supports all different types of service-specific CSs in common. It provides a mechanism that enables all the users to share the wireless medium effectively. Specifically, it provides the core MAC functionality such as system access, bandwidth allocation, connection establishment, and connection maintenance. The security sublayer (or privacy sublayer) provides authentication, privacy key exchange, and encryption functions. The security function is supported by an authenticated client/server *key management protocol* (KMP) in which the BS controls the distribution of the keying material to mobile stations.

Physical Layer

As discussed in Section 1.2.1, there are specified, in the IEEE 802.16 standards, four different types of physical layers for operation in different frequency bands, based on different multiple access technologies—namely, Wireless MAN-SC, Wireless-SCa, Wireless-OFDM, and Wireless-OFDMA. In addition, Wireless HUMAN PHY is specified for use in the license-exempt bands. The most commonly used among the four are the Wireless MAN-OFDM and Wireless MAN-OFDMA, which are used in the fixed and mobile BWA networks, respectively.

The OFDM/OFDMA-based IEEE 802.16 WiMAX system has several distinctive features in employing advanced technologies: First, it adopts the *time division duplex* (TDD) scheme for sharing communication channels between uplink and downlink, in addition to the *frequency division duplex* (FDD) that has been widely adopted in the existing circuit-mode mobile wireless systems. Second, it adopts the OFDMA scheme for sharing the communication link among multiple users, whereas the existing mobile wireless systems adopted TDMA or CDMA schemes. Third, it uses AMC technology for an efficient modulation, demodulation, coding, and decoding of communication signals. AMC dynamically changes the modulation and coding techniques depending on the channel status, thereby enhancing the system efficiency in varying wireless channel conditions. Fourth, it employs multiple antenna technologies so that it can significantly enhance the system performance and increase transmission capacity by taking advantage of the space diversity, spatial multiplexing, and beamforming with interference nulling effects. In addition, it takes a larger channel bandwidth (e.g., 10 MHz) for operator allocation than the existing mobile system did, within which the operator can actively apply these technologies.

In support of the mobility, which is a very important feature of the Mobile WiMAX system, it adopts efficient technologies for battery power saving and IP-based mobility. For battery power saving, the Mobile WiMAX system adopts the sleep/idle mode terminal operation. When each MS is not in awake mode, it goes into the sleep mode, and for further power saving, it can go into the idle mode, in which case it does not register to any BS but only receives the downlink paging messages periodically. For mobility, the Mobile WiMAX can use mobile IP, which manages the location information by home and foreign agents. It realizes terminal mobility through the handover function among the neighboring BSs and it basically supports the *hard handover* scheme.

Supported by these advanced technologies, the Mobile WiMAX system sets aggressive requirements on system performance and data services. It supports the data transmission rate of 18.7 Mbps downstream and 5.0 Mbps upstream in case of DL/UL ratio = 29:18 and 10 MHz bandwidth for non-MIMO case. The peak data rates are doubled if MIMO technology is applied. The Mobile WiMAX supports the *frequency reuse factor* (FRF) of 1 for all cells, the user mobility of 120 km/h, and the handover latency of 150 ms. Table 1.14 lists the features of Mobile WiMAX system in 10 MHz bandwidth with 29:18 DL/UL ratio.

MAC Layer

Mobile WiMAX is connection-oriented system, which enables it to tightly control the resource allocation and QoS, as well as the security function, needed for broadband wireless access. The MAC function of Mobile WiMAX is divided into three sublayers, namely service-specific CS, CPS, and security sublayer. The service specific CS performs the functions needed for converging user services to MAC CPS, including the reception, classification, and processing of the higher layer PDUs, the delivery of CS PDUs to the appropriate MAC SAP, and the receiving of CS PDUs from the peer entity. The MAC CPS performs the core MAC functionality including system access, bandwidth allocation, connection establishment, and connection maintenance, as well as effective user sharing of the wireless medium.

For the enhancement of reliability of data transmission, Mobile WiMAX adopts both ARQ and HARQ mechanisms. ARQ is a primitive form of error recovery technique that totally relies on the retransmission of the erred packets, and HARQ is an enhanced form of ARQ that utilizes FEC for the improvement of detection capability. Specifically, HARQ exploits the information in the original message to aid the decoding of the retransmitted messages. ARQ in Mobile WiMAX is enabled on a per-connection basis and is specified and negotiated during connection setup. Mobile WiMAX defines four ARQ feedback types to signal ACK/NAK, namely, selective ACK, cumulative ACK, cumulative with selective ACK, and cumulative ACK with block sequence ACK. In the case of HARQ, both Chase combining and *incremental redundancy* (IR) HARQ methods are defined in the IEEE 802.16e standards but only Chase combining HARQ is included in the WiMAX profile.

Table 1.14 System Feature of the Mobile WiMAX System (10-MHz BW)

Parameters	Value or technology
Duplex	TDD
Multiple Access	OFDMA
Frequency reuse factor	1
Peak data rate (non-MIMO)	18.7 Mbps/sector DL 5.0 Mbps/sector UL
Peak data rate (MIMO)	37.4 Mbps/sector DL 10.0 Mbps/sector UL
Mobility	120 km/h
Handoff	150 ms

The principal mechanism of Mobile WiMAX for providing QoS is to associate packets traversing the MAC interface into a service flow. The MS and BS provide the QoS according to the QoS parameter set defined for the service flow. As the mechanisms of providing QoS services, Mobile WiMAX defines several bandwidth allocation types, which reflect the delay requirements and traffic characteristics, and their corresponding data delivery services, such as *unsolicited grant service* (UGS), *extended real-time variable-rate* (ERT-VR) service, *real-time variable-rate* (RT-VR) service, *nonreal-time variable-rate* (NRT-VR) service, and *best effort* (BE) service. In order to enhance the efficiency of the bandwidth usage, the Mobile WiMAX adopts well-organized bandwidth request, grant, and polling mechanisms, which are supported by these five different types of delivery services. The downlink bandwidth is solely managed by the downlink scheduler at the BS, but the uplink bandwidth is allocated by BS to MSs through the resource request and grant process. For the implementation of QoS services, Mobile WiMAX employs the enforcement functions such as scheduling, CAC, and policing. These functions are designed to maximize the QoS satisfaction, minimize the QoS violation, and protect the QoS of the contract-conforming connections.

For the mobility, Mobile WiMAX basically supports the hard handover scheme, as it is optimized for IP data traffic, but it also supports soft handover (in the standard, but not in the WiMAX profile). Handover is performed in two main processes, namely, network topology acquisition process and handover execution process: The network topology acquisition periodically updates the parameter values needed for making handover decision, and the handover execution practically executes the handover through a series of processes such as neighbor scanning, handover capability negotiation, MS release, and network re-entry. Power saving is crucial to terminal mobility, and Mobile WiMAX supports both sleep mode and idle mode operations: the sleep mode allows MS to be absent from the serving BS air interface while not in use, and the idle mode allows MS to be mostly idle and only listen to the paging messages periodically.

Network Configuration

The WiMAX system is originally designed to be a data-centric network based on the IP technology, different from the existing voice-centric mobile communication networks that used circuit-mode technology. It adopts an all-IP network structure tailored for Internet service provision, so the network structure is simple and is adequate for provision of diverse set of services.

Figure 1.6 illustrates the configuration of Mobile WiMAX network. As the network is designed based on the all-IP network concept, the network configuration is very simple and the network construction cost is low. The network has a star architecture, with the *mobile stations* located at the end of the branches. The IP packets sent by MSs get accessed to the Internet via the BS to *access service network gateway* (ASN-GW) path. This demonstrates how simple it is to provide Internet services over the WiMAX network. Consequently, a diverse set of services can be provided over the WiMAX network at low cost, with the voice service provided in VoIP form.

As shown in Figure 1.6, overall Mobile WiMAX network consists of *access service network* (ASN) and *connectivity service network* (CSN). ASN consists of three basic building blocks, namely, MS, BS, and ASN-GW, and the CSN consists of vari-

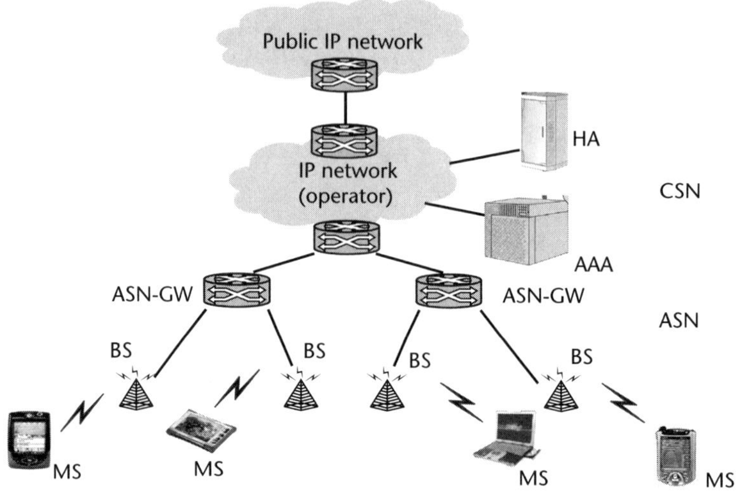

Figure 1.6 Illustration of Mobile WiMAX network configuration.

ous servers and core routers/switches. So the Mobile WiMAX network configuration is much simpler than existing circuit-based mobile communication networks such as the IS-95/EV-DO family system, which includes *base station controller* (BSC), *mobile switching center* (MSC), or the GSM/WCDMA family system, which includes *radio network controller* (RNC), *serving GPRS support node* (SGSN), and *gateway GPRS support node* (GGSN), in place of ASN-GW (see Section 2.4).

To be more specific, the BS collects user terminal data via wireless path, passes it to the ASN-GW in the upstream, and distributes the data received from the ASN-GW to the MSs in the downstream. The functions of BS include wireless access processing, radio resources management and control, mobility support for seamless services while moving, QoS support for stable service quality, and overall equipment control and management. On the other hand, the ASN-GW connects the BS with the various servers and core routers/switches in the CSN. It performs the routing function transferring data between the BS and the CSN and the control function controlling the MSs, services, and mobility.

1.4.2 WiFi: Wireless Local Area Networks

IEEE 802.11 WLAN or WiFi is probably the most widely accepted broadband wireless networking technology, providing the highest transmission rate among standard-based wireless networking technologies. Today's WiFi devices based on IEEE 802.11a and 802.11g provide transmission rates up to 54 Mbps and, further, a new standard IEEE 802.11n, which supports up to 600 Mbps, is being standardized. The transmission range of a typical WiFi device is up to 100m, where its exact range can vary depending on the transmission power, the surrounding environments, and others. The 802.11 devices operate in unlicensed bands at 2.4 and 5 GHz, where the exact available bands depend on each county.

Most of today's laptop computers as well as many PDAs and smart phones are shipped with embedded WLAN interfaces. Moreover, many electronic devices including VoIP phones, personal gaming devices, MP3 players, digital cameras, and camcorders are being equipped with WLAN interfaces as well. The most typical applications of the 802.11 WLAN should be the Internet access of portable devices in various networking environments including campus, enterprise, home, and hot-spot environments, where one or more *access points* (APs) are deployed to provide the Internet service in a given area. The 802.11 can be used for a peer-to-peer communication among devices where APs are not deployed. For examples, laptops and PDAs in proximity can use the 802.11 to share their local files. Also, people in proximity can do networked gaming using their gaming devices with the 802.11 interface. It is primarily being used for the indoor purpose. However, it can be also used in outdoor environments, and some level of mobility (e.g., the walking speed) can be also supported.

As discussed in Section 1.3.2, IEEE 802.11 WG has generated a family of standards for WLAN. The 802.11 specifications are limited to PHY and MAC layers, and the existing higher layer protocols, which were originally developed for wireline networking technologies, can work on top of the 802.11 since it was basically developed to provide the service similar to the 802.3 Ethernet. At one point, this technology was referred to as "Wireless Ethernet." In typical 802.11 devices, the 802.2 LLC protocol sits on top of the 802.11 MAC, where IP sits on top of the LLC. Through its evolution, the 802.11 is becoming much more than Ethernet. For example, the 802.11e MAC enables multimedia applications such as *voice over IP* (VoIP) *over WLAN* (or simply VoWLAN). The protocols for seamless mobility are being developed since the support of seamless mobility became quite critical along with the emergence of WLAN-based VoIP phones (or VoWLAN phones). In fact, people are also trying to use this technology for vehicular networking (e.g., car-to-car and car-to-roadside) as well.

Physical Layer

IEEE 802.11 PHYs have been evolving dramatically. The baseline standard of IEEE 802.11 (published in 1997) defined three different PHY protocols, namely, *direct-sequence spread-spectrum* (DSSS), *frequency-hopping spread-spectrum* (FHSS), and IR, where all three PHYs supported only the transmission rates of 1 and 2 Mbps. The extensions of the 802.11 PHY include the 802.11a (published in 1999) supporting up to 54 Mbps based on the OFDM, the 802.11b (published in 1999) supporting up to 11 Mbps based on the *complementary code keying* (CCK), and the 802.11g (published in 2003) again based on OFDM to support up to 54-Mbps transmission rates.

The 802.11 PHYs operate in unlicensed bands at 2.4 GHz and 5 GHz. While most of other PHYs, including DSSS, FHSS, 802.11b, and 802.11g operate at the 2.4-GHz bands, the 802.11a operates at the 5-GHz bands. The 802.11g, in fact, includes the mandatory transmission schemes of the 802.11b, while the 802.11b includes the baseline DSSS PHY. That is, the 802.11g is backward compatible with the 802.11b, while the 802.11b is backward compatible with the baseline DSSS PHY. This implies that an 802.11g device can communicate with an 802.11b device using the transmission schemes of the 802.11b. Today, the most popular 802.11 PHY is the 802.11g, thanks to its fast transmission rate as well as low-cost chipset availability even though the 2.4-GHz bands, where the 802.11g operate, are much more crowded than the 5-GHz bands of the 802.11a.

The 802.11 basically operates with a *time division duplexing* (TDD) scheme for the sharing between uplink and downlink transmissions. That is, a single frequency channel is used for all the transmissions in a *basic service set* (BSS), which is a similar concept as a cell in typical cellular networks. The transmission bandwidth depends on the PHY as well. For example, the 802.11a and 802.11g signals occupy a 20 MHz band while the 802.11b signals occupy a 22-MHz band.

The 802.11 PHYs support multiple transmission rates by using different combinations of *modulation and coding schemes* (MCSs). Both the 802.11a and 802.11g support up to 54 Mbps, which make the 802.11 the fastest standards-based wireless technology as of today. In fact, as discussed in Section 1.3.2, the emerging 802.11n PHY will support up to 600 Mbps by utilizing multiple antenna technologies (i.e., MIMO schemes) and channel bonding (i.e., using 40-MHz bandwidth instead of 20 MHz). As 802.11 PHYs support multiple transmission rates, selecting a rate for a given packet (or frame in the 802.11 term) transmission is a very important issue for the performance optimization of the network. In general, the higher the transmission rate, the shorter the transmission range is since high-order modulation schemes require higher *signal-to-interference-and-noise ratio* (SINR) for successful transmissions.

Table 1.15 lists a summary of various PHYs of the 802.11 along with their transmission schemes, frequency bands, and supported transmission rates.

The transmission power level for the 802.11 PHY depends on the regulation of the corresponding country. Each country defines the upper limit of the transmission power at particular unlicensed bands. Typical 802.11 devices emit the power up to 20 dBm (or 100 mW).

MAC Layer

The 802.11 baseline standard defines connectionless MAC for the best-effort service. The baseline MAC is composed of two coordination functions, namely, the mandatory contention-based *distributed coordination function* (DCF) and the optional contention-free *point coordination function* (PCF). The DCF is based on *carrier-sense multiple access with collision avoidance* (CSMA/CA) and the PCF is a poll-and-response MAC. In fact, the PCF was rarely implemented in real products due to its complexity, the lack of needs, the lack of desirable operational functions, and others. Under the DCF, which was employed by most, if not all, WiFi devices, a

Table 1.15 Various PHYs of IEEE 802.11

PHY	Transmission schemes	Frequency bands	Transmission rates (Mbps) supported
Baseline	DSSS, FHSS and IR	DSSS, FHSS – 2.4 GHz IR – 850~950 nm	1, 2
802.11a	OFDM	5 GHz	6, 9, 12, 18, 24, 36, 48, 54
802.11b	CCK	2.4 GHz	5.5, 11 + DSSS rates
802.11g	OFDM	2.4 GHz	6,9, 12, 18, 24, 36, 48, 54 + 802.11b rates
802.11n	OFDM, MIMO	2.4 GHz, 5 GHz	Up to 600

station transmits only when it determines that the channel is not occupied by other transmissions, and this makes this MAC a perfect fit to the operation at unlicensed bands, at which various types of devices should coexist with some etiquette.

The baseline MAC is enhanced by the 802.11e to support *quality-of-service* (QoS) for multimedia applications such VoWLAN, video streaming, and so forth. The 802.11e MAC is called *hybrid coordination function* (HCF), which comprises the contention-based *enhanced distributed channel access* (EDCA) and the poll-and-response *HCF controlled channel access* (HCCA). EDCA and HCCA enhance DCF and PCF, respectively. EDCA provides prioritized channel access to frames with different priorities, where lower priority frame might be transmitted before higher priority frames due to the contentious nature of the EDCA. HCCA relies on the polling and downlink frame scheduling of the AP to meet the QoS described by a set of parameters. The 802.11e also defines various features needed for QoS provisioning, including the means for admission control of QoS streams.

Thanks to the carrier-sensing feature of the MAC, the 802.11 inherently supports FRF of one. That is, even if the neighboring cells (or BSSs) use the same frequency channel, the performance degradation due to the cochannel interference will be minimal since stations transmit frames only when they determine other neighboring stations are not transmitting. Apparently, depending on how the cells are deployed and which frequency channels are used for cells, there is room to improve the networking performance. That is, it is the best if neighboring cells can operate at nonoverlapping channels. However, the number of available nonoverlapping channels varies depending on the countries. The number of nonoverlapping channels at the 2.4-GHz bands is only three in most countries, and hence it is almost impossible to allocate nonoverlapping channels to all neighboring cells. This is particularly true in multistory building environments, since the cell structure is three-dimensional.

The 802.11 MAC supports reliable transmission of frames using ARQ. The baseline MAC defines a stop-and-wait ARQ, for which a receiver of a data frame responds with an ACK frame immediately after a successful reception. The 802.11e MAC then defines an enhanced ARQ scheme (i.e., selective repeat ARQ) using a mechanism called *block ACK*, in which a control frame called block ACK is transmitted by the receiver after the transmission of a number of data frames. A block ACK includes a bit map indicating which of the previous transmitted frames were successfully received and which were not.

The mobility support has not been a major concern of the WiFi since people rarely use their laptop or PDA to access the Internet via WLAN while they are moving around. However, some level of mobility is supported by the 802.11. For example, the walking speed mobility is surely supported. The 802.11 allows a station to be associated with a single AP at a given time. That is, a hard handoff is supported. Along with the emergence of the VoWLAN applications, supporting seamless and smooth handoffs in the 802.11 WLAN becomes a hot topic.

Power saving is one of the major concerns for portable mobile communication devices. The 802.11 MAC defines *power-saving mode* (PSM) operation, in which a station switches back and forth between the active and the doze states, where the station consumes minimal energy in the doze state since it can neither transmit nor receive frames while staying in the doze state. The 802.11e further enhances the power-saving scheme, thus defining a scheme called *automatic power-save delivery*

(APSD), which allows a station to save some power even during a QoS stream operation (e.g., VoWLAN operation).

The baseline MAC of the 802.11 had security mechanisms for confidentiality (via encryption) and authentication, but these schemes were found to be too weak to protect the security of the WiFi users. The problems included the cryptographic weakness of the encryption scheme (called RC4), the lack of key management, and so on. For example, under the legacy security mechanism, the same security key is basically used for every station in the network, while the key is rarely changed over time. Such a security hole of the 802.11 was a big hurdle for the wide acceptance of WiFi at one point. Especially for enterprise networking, a strong security support was a mandatory requirement. Then, IEEE 802.11i enhanced security features by defining the *robust security network* (RSN), which is composed of stronger encryption schemes, per-frame authentication, per-station key management, and so on.

IEEE 802.11h defines mechanisms for spectrum managements including *dynamic frequency selection* (DFS) and *transmit power control* (TPC). While the 5-GHz bands, where the 802.11a operates, are unlicensed bands, there are in fact primary users who also use these bands. Those primary users are satellite and radar systems. The regulatory body in Europe required a WLAN device to have both DFS and TPC functions to minimize the interference of the WLAN to these primary users. That is, when a radar system is detected, the WLAN devices should leave the current channel to switch to another channel, and when a satellite system is detected, the WLAN devices have to limit their transmission power to the regulatory maximum minus 3 dB.

Network Configuration

The basic form of the 802.11 network is called a *basic service set* (BSS), which comprises a number of stations. IEEE 802.11 supports two types of network configurations, namely, infrastructure and ad hoc modes. An infrastructure BSS is composed of an AP and a number of stations that are associated with the AP [see Figure 1.7(a)]. A station in an infrastructure BSS communicates with other stations or nodes outside the WLAN through its AP. IEEE 802.11 AP contains all the functions for a sta-

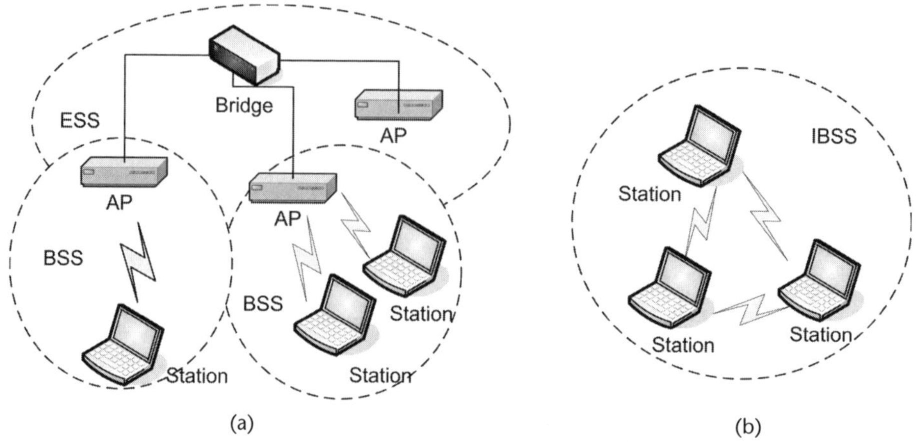

Figure 1.7 Illustration of WiFi network configuration: (a) ESS composed of infrastructure BSS, and (b) IBSS.

tion, and also provides various services, including the routing of the frames from and to its stations. APs are connected through the backbone, called *distribution system* (DS), to form an *extended service set* (ESS), which can provide a seamless WLAN service to a given area. One can understand an infrastructure BSS as a cell in a cellular network. An 802.11 station can hand off from an AP to another AP while it moves around within an ESS.

Independent BSS (IBSS) is the other type of BSS, which is used for the ad hoc mode. An IBSS is composed of a number of stations that can communicate directly each other [see Figure 1.7(b)]. The 802.11 does not support wireless multihop communications. In order to support wireless multihop communications, the stations should implement a layer-3 routing function, such as by employing a *mobile ad hoc network* (MANET) routing protocol.

The 802.11 standards do not define how to implement the DS. That is, how to connect multiple APs is not specified in the standard. There are different ways to construct a DS. In typical deployments of the 802.11 WLAN, APs are connected via Ethernet. However, the standard also allows them to be connected wirelessly (i.e., using the 802.11 links). Another issue is whether an AP is a layer-2 or layer-3 device. By default, an AP is a layer-2 bridging (or switching) device, and all the APs are connected via layer-2 bridges. In such a case, all the APs along with the associated stations are within the same subnet. However, an AP can be implemented as a layer-3 device (or router) such that the frame (or packet) forwarding is made based on the layer-3 IP address. The 802.11 MAC can be actually divided into a time-critical lower MAC, including the frame transmission/reception, and a less time-critical upper MAC related with the network management. In fact, an AP can be also implemented as a lower layer-2 device. That is, an AP might include only the lower MAC functions, and then less time-critical upper MAC functions are implemented in a so-called WLAN switch, which connects multiple APs with only lower MAC functions.

1.4.3 Similarities and Differences

Mobile WiMAX and WiFi are access technologies developed by IEEE 802.16 and 802.11 WGs, respectively, where both 802.16 and 802.11 WGs are under the umbrella of IEEE 802 LAN/MAN committee. Various access and connectivity protocols in the IEEE 802 family are developed for the packet-switched networking, and both IEEE 802.16 and 802.11 are also along the same line. This is quite different from other cellular technologies (e.g., those developed by 3GPP and 3GPP2, which evolved from voice communication-oriented circuit-switched networking). While both 802.16 and 802.11 define peer-to-peer, mesh, or ad hoc modes of operation, their primary network configuration is a star topology, where a user station communicates though its AP or BS to connect to the rest of the world.

It should be also mentioned that many people envision that these two technologies are quite complementary in that WiFi is better for lower-mobility networking while Mobile WiMAX is better for higher-mobility networking. Portable devices supporting both technologies are emerging today, and the protocols for interworking of heterogeneous access technologies like Mobile WiMAX and WiFi are being developed today (e.g., IEEE 802.21).

There are a number of differences between Mobile WiMAX and WiFi. First of all, Mobile WiMAX is developed for *wireless metropolitan area network* (WMAN), providing the transmission range of a few kilometers, while WiFi is for *wireless local area network* (WLAN) with the transmission range up to 100m. Mobile WiMAX is also mostly for commercial networks operated by service providers. However, WiFi is mainly for noncommercial networks deployed and maintained by an individual or a company. Home and enterprise networking are good examples of WiFi. In their typical commercial deployment scenarios, Mobile WiMAX is meant for the seamless service coverage in a city or even a whole country, while WiFi is for spotty coverage provisioning at hot-spot areas, such as airports, coffee shops, and shopping malls, where many people gather. Mobile WiMAX was developed to support high mobility so that users can use this technology even inside a moving car or a train, but WiFi is mainly for nomadic users, who use this technology while mostly staying at a given place. WiFi can also support some low mobility (e.g., walking speed) but most of WiFi devices are not optimized for mobility support since people rarely use WiFi devices while moving around.

In terms of their technical operations, there are a number of differences as well. First of all, the MAC protocols are very different. The baseline MAC for WiFi relies on CSMA/CA, which is connectionless and contention-based. As WiFi operates at unlicensed bands, where various heterogeneous devices have to smoothly coexist, the adoption of CSMA/CA, which allows a device to transmit only when the channel is deemed to be free, seems a very natural and perfect choice. On the other hand, Mobile WiMAX employs a connection-oriented bandwidth request and allocation MAC. As Mobile WiMAX operates at licensed bands, for better QoS support, this type of centralized MAC seems to be a good choice. While QoS provisioning in wireless networks is always challenging due to time-varying nature of the network, it should be more feasible to provide proper QoS by using licensed bands. While Wi-Fi only supports TDD, Mobile WiMAX supports both TDD and FDD.[4] As discussed in the previous sections, WiFi supports various kinds of PHYs such as 802.11 and 802.11a/b/g, whereas the Mobile WiMAX supports OFDMA PHY based on 802.16e.[5] The OFDMA PHY allows multiple users to receive/transmit simultaneously by using different subcarriers. IEEE 802.11a and 802.11g are OFDM PHYs, but not OFDMA. That is, all the subcarriers are used for the transmission to a single receiver at a given time.

References

[1] Lee, B. G., D. Park, and H. Seo, *Wireless Communications Resource Management*, New York: Wiley-IEEE, 2008.

[2] Goldsmith, A., *Wireless Communications*, Cambridge, U.K.: Cambridge University Press, 2005.

4. The FDD profile is being defined, and the specification was scheduled to be completed by the first half of 2008 in the WiMAX Forum.
5. The 802.16m, the next version of OFDMA PHY specification, was scheduled to be completed by the end of 2009.

[3] Feuerstein, M. J., et al., "Path Loss, Delay Spread, and Outage Models as Functions of Antenna Height for Microcellular System Design," *IEEE Trans. on Vehicular Technology*, Vol. 43, No. 3, August 1994, pp. 487–498.

[4] 3GPP TR 25.814, Physical Layer Aspects for Evolved Universal Terrestrial Radio Access (UTRA), v. 7.1.0, September 2006.

[5] IEEE Std 802.16e-2005 and IEEE Std 802.16-2004/Cor1-2005, Part 16: Air Interface for Fixed and Mobile Broadband Wireless Access Systems, February 2006.

[6] FCC CFR47, Title 47 of the Code of Federal Regulations; Part 15: Radio Frequency Devices, Federal Communication Commission, September 2007.

[7] ETSI EN 301 893, Broadband Radio Access Networks (BRAN); 5 GHz High Performance RLAN; Part 2: Harmonized EN Covering Essential Requirements of Article 3.2 of the R&TTE Directive, v. 1.3.1 August 2005.

[8] IEEE Std 802.16-2001, Part 16: Air Interface for Fixed Broadband Wireless Access Systems, April 2002.

[9] IEEE Std 802.16a-2003, Amendment 2 to Part 16: Air Interface for Fixed Broadband Wireless Access Systems: Medium Access Control Modifications and Additional Physical Layer Specifications for 2–11 GHz, April 2003.

[10] IEEE Std 802.16-2004, Part 16: Air Interface for Fixed Broadband Wireless Access Systems, revision of IEEE Std 802.16-2001, October 2004.

[11] WiMAX Forum, "Relationship Between WiBro and Mobile WiMAX," white paper, October 2006.

[12] WiMAX Forum, Mobile System Profile, Release 1.0, May 2007. For the latest release, refer to http://www.wimaxforum.org

[13] IEEE 802.16m-07/001r1, "Work Plan for Development of IEEE P802.16m Draft Standard & IMT-Advanced Submission," July 2007.

[14] WiMAX Forum, "Wireless Technology Roadmap Strategy," June 2007.

[15] IEEE 802.11-1999, Part 11: Wireless LAN Medium Access Control (MAC) and Physical Layer (PHY) Specifications, June 1999.

[16] IEEE 802.11a-1999, Amendment 1 to Part 11: Wireless LAN Medium Access Control (MAC) and Physical Layer (PHY) Specifications: High-Speed Physical Layer in the 5 GHz Band, September 1999.

[17] IEEE 802.11b-1999, Supplement to Part 11: Wireless LAN Medium Access Control (MAC) and Physical Layer (PHY) Specifications: Higher-Speed Physical Layer Extension in the 2.4 GHz Band, September 1999.

[18] IEEE 802.11g-2003, Amendment 4 to Part 11: Wireless LAN Medium Access Control (MAC) and Physical Layer (PHY) Specifications: Further Higher Data Rate Extension in the 2.4 GHz Band, June 2003.

[19] IEEE 802.11e-2005, Amendment 8 to Part 11: Wireless LAN Medium Access Control (MAC) and Physical Layer (PHY) Specifications: Medium Access Control (MAC) Enhancements for Quality of Service (QoS), November 2005.

[20] IEEE 802.11i-2004, Amendment 6 to Part 11: Wireless LAN Medium Access Control (MAC) and Physical Layer (PHY) Specifications: Medium Access Control (MAC) Security Enhancements, July 2004.

[21] IEEE 802.11h-2003, Amendment 5 to Part 11: Wireless LAN Medium Access Control (MAC) and Physical Layer (PHY) Specifications: Spectrum and Transmit Power Management Extensions in the 5GHz Band in Europe, October 2003.

[22] IEEE 802.11d-2001, Amendment 3 to Part 11: Wireless LAN Medium Access Control (MAC) and Physical Layer (PHY) Specifications: Specification for Operation in Additional Regulatory Domains, July 2001.

[23] IEEE 802.11F-2003, IEEE Recommended Practice for Multi-Vendor Access Point Interoperability via an Inter-Access Point Protocol Across Distribution Systems Supporting IEEE 802.11 Operation, July 2003.

[24] IEEE 802.11j-2004, Amendment 7 to Part 11: Wireless LAN Medium Access Control (MAC) and Physical Layer (PHY) Specifications: 4.9 GHz–5 GHz Operation in Japan, October 2004.

[25] IEEE 802.11-2007, Part 11: Wireless LAN Medium Access Control (MAC) and Physical Layer (PHY) Specifications, revision of IEEE Std 802.11-1999, June 2007.

Selected Bibliography

Biglieri, E., J. Proakis, and S. Shamai, "Fading Channels: Information-Theoretic and Communications Aspects," *IEEE Trans. on Information Theory*, Vol. 44, No. 6, October 1998, pp. 2619–2692.

Eklund, C., et al., "IEEE Standard 802.16: A Technical Overview of the WirelessMAN Air Interface for Broadband Wireless Access," *IEEE Communications Magazine*, Vol. 40, No. 6, June 2002, pp. 98–107.

Gesbert, D., et al., "Technologies and Performance for Non-Line-of-Sight Broadband Wireless Access Networks," *IEEE Communications Magazine*, Vol. 40, No. 4, April 2002, pp. 86–95.

Ghosh, A., et al., "Broadband Wireless Access with WiMAX/8O2.16: Current Performance Benchmarks and Future Potential," *IEEE Communications Magazine*, Vol. 43, No. 2, February 2005, pp. 129–136.

Koffman, I., and V. Roman, "Broadband Wireless Access Solutions Based on OFDM Access in IEEE 802.16," *IEEE Communications Magazine*, Vol. 40, No. 4, April 2002, pp. 96–103.

Sklar, B., "Rayleigh Fading Channels in Mobile Digital Communication Systems Part I: Characterization," *IEEE Communications Magazine*, Vol. 35, No. 7, July 1997, pp. 90–100.

Sklar, B., "Rayleigh Fading Channels in Mobile Digital Communication Systems Part II: Mitigation," *IEEE Communications Magazine*, Vol. 35, No. 7, July 1997, pp. 102–109.

WiMAX Forum, "A Review of Spectrum Requirements for Mobile WiMAX Equipment to Support Wireless Personal Broadband Services," white paper, September 2007.

WiMAX Forum, "Fixed, Nomadic, Portable and Mobile Applications for 802.16-2004 and 802.16e WiMAX Networks," white paper, November 2005.

WiMAX Forum, "Mobile WiMAX—Part I: A Technical Overview and Performance Evaluation," white paper, August 2006.

WiMAX Forum, "WiMAX and IEEE 802.16a—Igniting BWA," white paper, May 2004.

PART I

Mobile WiMAX: Broadband Wireless Access Network

Mobile WiMAX is a new technology rooted on the IEEE 802.16e standard, published in 2006, with the first commercial service commenced in 2007. The IEEE 802.16e standard distinguishes itself from its predecessor IEEE 802.16 standards with its mobility feature. It is designed to adopt the cellular concept and support roaming service with the coverage of a few kilometers in metropolitan and regional networks, allowing mobile connectivity to mobile users.

Mobile WiMAX is equipped with novel technological tools, such as *orthogonal frequency division multiplexing* (OFDMA), *time division duplexing* (TDD), *multi-input multi-output* (MIMO), *adaptive modulation and coding* (AMC), *Internet protocol* (IP), security, and others, which are combined together to offer high-rate, low-cost, wide-area, secured mobile multimedia services. In particular, Mobile WiMAX is the first mobile wireless system that has adopted the OFDMA technology for multiple access. Even the MIMO, TDD, and AMC technologies find their first combined implementation in the Mobile WiMAX system to realize enhanced spectal efficiency and system flexibility.

The protocol layering of the Mobile WiMAX system, which is common to all IEEE 802.16 systems, consists of *medium access control* (MAC) layer and physical layer, with the MAC layer divided into service-specific *convergence sublayer* (CS), MAC *common part sublayer* (CPS), and security sublayer. Another distinctive feature of Mobile WiMAX is that it defines the security sublayer as a formal layer in the system protocol architecture. The security sublayer provides authentication, privacy key exchange, and encryption functions that are necessary to realize secured communications.

This part is intended to describe the Mobile WiMAX network by introducing each constituent technological component comprehensively but concisely. To this end, we arranged the chapters such that a more essential component comes earlier than others, with an initial overview of the Mobile WiMAX system and a description of the overall network operation. The ordering of the subsequent chapters is as follows: After the discussion of the network operation issues such as network initialization and maintenance, we discuss the frameworks of the OFDMA-based physical layer and MAC layer, bandwidth management and *quality of service* (QoS) issues, mobility support issue, security control issue, and multiple antenna technology. In the last chapter, we discuss the system design, network deployment, and service provision issues of the WiBro system, which is the first Mobile WiMAX system implemented based on its 2.3-GHz profiles.

Specifically, Chapter 2 discusses the key technologies adopted by the Mobile WiMAX network, such as TDD, OFDMA, AMC, MIMO, bandwidth management, *hybrid automatic repeat request* (HARQ), mobility management, and security management. Then it describes the protocol layering and network architecture of Mobile WiMAX. At the end, it explains the evolutionary process of the circuit-mode cellular mobile systems, such as GSM/WCDMA and IS-95/EV-DO, and compares them with the Mobile WiMAX system, finally discussing how to interwork them with the Mobile WiMAX system.

Chapter 3 examines the overall system operation of the Mobile WiMAX system by discussing the operational procedure that happens when the Mobile WiMAX system is turned on. The procedure includes network discovery, network initialization, connection setup, and connection maintenance. Mobility-related procedures are involved in support of user mobility, in addition to the operation at the nonconnected state and paging procedures.

Chapter 4 describes the OFDMA-based Mobile WiMAX physical layer framework. It first investigates OFDMA-related communication signal processings, including channel coding, HARQ, modulation, OFDMA mapping, and DFT transform. Then it describes the OFDMA frame structuring issues, introducing OFDMA slots, bursts, OFDMA frame, *downlink/uplink* (DL/UL) MAPs, and others. On that basis, it discusses the subchannelization issues, including the description of four different types of subchannlizing methods, namely, DL *partial usage subchannel* (PUSC), DL *full usage subchannel* (FUSC), UL PUSC, and DL/UL AMC.

Chapter 5 deals with the MAC framework of the Mobile WiMAX system. It first describes the service specific sublayer, focusing on the packet CS, then discusses the MAC common part sublayer, describing the concept of connection in detail and explaining the MAC management messages. In addition, it discusses how to arrange MAC *protocol data unit* (PDU) formats in conjunction with the fragmentation, packing, and concatenation processes. As an addendum, it discusses the *automatic repeat request* (ARQ) issues at the end.

Chapter 6 discusses the issues of bandwidth management and QoS. It first describes the scheduling and data delivery services such as *unsolicited grant service* (UGS), *real-time polling service* (rtPS), *extended rtPS* (ertPS), *nonreal-time polling service* (nrtPS), and best-effort service, and, based on this, it discusses the bandwidth request and allocation mechanisms. Then it discusses various issues related to the QoS, including the concepts of service flow and service class; QoS messages and parameters; service flow setup and release procedures; and scheduling, connection admission control, and policing issues.

Chapter 7 discusses the mobility support issues of the Mobile WiMAX system. It starts the discussion with the introduction of the cellular concept and the methods of intercell interference management. Then it discusses the handover management issues, including network topology acquisition, handover execution, and soft handover. In addition, it discusses the power-saving methods such as sleep mode and idle mode, which are closely related to the mobility issue.

Chapter 8 deals with a rather unique issue of security control. To begin with, it provides a comprehensive introduction to the fundamentals of cryptography and information security, providing an overview of the Mobile WiMAX security system at the end. Based on this, it describes the security system architecture of the Mobile

WiMAX system, in terms of security association, encapsulation, authentication, and key management. At the end, it makes a more detailed description on the *privacy key management* (PKM) techniques, namely PKMv1 and PKMv2, exemplifying them with state machines.

Chapter 9 handles a rather independent topic of multiple antenna technology. It begins with an introduction of fundamental multiple antenna technology, introducing the concepts of space diversity, spatial multiplexing, and beamforming. It then divides the technology in two categories—open-loop technology and closed-loop technology—and discusses each of them independently. The discussions are comprehensive, using an adequate amount of mathematics. The last section focuses on the MIMO receiver algorithm and discusses different methods of receiver design.

Chapter 10 introduces the WiBro system, the 2.3 GHz–based first Mobile WiMAX system, which was designed by Samsung Electronics and deployed by KT, with the first commercial service having commenced in June 2007. It first dicusses the WiBro network configuration and the WiBro system requirements, which are followed by the descriptions of the design issues of WiBro *access control router* (ACR), or *base station* (BS), system and WiBro *radio access station* (RAS), or *access service network–gateway* (ASN-GW), system. It then addresses the deployment issues of WiBro access network and WiBro core network. It finally describes the WiBro services that KT provided in its first commercialization or plans to provide in the future.

CHAPTER 2

Introduction to Mobile WiMAX Networks

Mobile WiMAX network is rooted on the IEEE 802.16e standard, which is a mobile version of the IEEE 802.16d standard for fixed broadband wireless access services. Mobile WiMAX network has the mobility feature in addition to the broadband capability and the IP-based framework that its predecessors possessed. Those three attributes—*broadband*, *IP-based*, and *mobile*—were the design goal of the Mobile WiMAX standardization from the beginning and now have become the distinctive features of the Mobile WiMAX network. The three attributes render a perfect means to accommodate the requirements of the ever-evolving communication services: The broadband capability can support the high data rate for downloading/uploading multimedia services and the convergence of multiple services. The IP-based framework makes Mobile WiMAX compatible with the omnipresent Internet world and yields a simple network architecture to support a diverse set of data services. The mobility nature enables mobile Internet service, satisfying the desire for unrestricted and unlimited communications of customers. The three attributes, in a combined form, enables the provision of bidirectional, high-data rate, user-participated, mobile, broadband, triple services, including data services such as Web access, Web search, and *user-created contents* (UCC) uploading; communication services such as video telephony, chatting, mobile VoIP, and multimedia message service; and media services such as live TV and streaming media.

These three attributes enable the Mobile WiMAX network to be equipped with a technological competence that is more effective in providing multimedia data services than the existing cellular wireless networks and *wireless LAN* (WLAN) networks. The cellular wireless networks such as GSM/WCDMA family or IS-95/EV-DO family, which were originated from the circuit-mode technology, are efficient in providing voice services or mobile, high-quality data services in very wide area, but the data rate is not high enough and the service charge is comparatively high. On the other hand, the WLAN, or WiFi network, which employs packet-mode (or IP-mode) technology, is efficient in providing IP services and high-data-rate services in small areas at very low cost, but the service quality is low, the coverage is small, and the mobility is not supported. In contrast, the Mobile WiMAX network can provide mobile, high-quality, high-data-rate services in wide areas at low cost. Whereas WiFi network targets small hot-spot services, Mobile WiMAX network may target medium metro-zone services in the beginning but can expand services to wide area as well.

Broadband

From technological aspects, Mobile WiMAX tries to achieve the broadband feature by allocating large bandwidths to the network operators and by adopting advanced multiple access technologies that can enhance the spectral efficiency. The bandwidth allocated in Mobile WiMAX goes up to 10 MHz, which is much larger than that allocated to cellular mobile networks, even if not as large as that of the WiFi network. As a data rate increases with the large bandwidth, however, *an intersymbol interference* (ISI) problem becomes critical over the multipath fading channel, which is equivalently translated to the frequency-selective fading phenomenon in a frequency domain. So Mobile WiMAX adopts OFDMA as the multiple access technology for a broadband system, which is more robust to the frequency-selective fading channel than TDMA technology. As compared to CDMA technology, which employs much larger spreading bandwidth as a data rate increases for the given processing gain, in turn requiring a unrealistically high complexity, it is resorting to a practically simpler transceiver architecture. In addition, OFDMA enables to take advantage of frequency diversity or channel averaging effects as well.

Due to the scarcity of wireless frequency bandwidth and the power limitation of user devices, Mobile WiMAX does apply the cellular concept as other existing cellular mobile networks do. To achieve the high spectral efficiency goal, Mobile WiMAX recommends using the *frequency reuse factor* (FRF) of 1. Maintaining FRF = 1 at the boundary of a cell is a very challenging task, as the *intercell interference* will be very strong at the boundary. Mobile WiMAX solves the problem by adopting the *adaptive modulation and coding* (AMC) technology, which enables *mobile station* (MS) to survive at the cell boundary with FRF = 1 by employing a robust modulation scheme with a heavy channel coding (e.g., QPSK at a coding rate of 1/2 with a repetition factor of 6). On the other hand, AMC helps to enhance the frequency spectral efficiency significantly in the vicinity of the *base station* (BS), where the *carrier-to-interference and noise ratio* (CINR) is very high, by taking a high-efficiency modulation such as 64-QAM with less heavy channel coding. The AMC technology will be effective for wireless data network, in which a higher possible data rate is always better for higher data throughput. It is different from the design principle in the conventional circuit-mode cellular network (e.g., IS-95 CDMA cellular network), which was originally designed to support a fixed data rate for warranting a uniform voice quality throughout the coverage area of each cell. In addition, Mobile WiMAX can adopt the *multiple-input multiple-output* (MIMO) technology, which again helps to increase the data rate substantially.

IP-Based

IP-based design and operation of the Mobile WiMAX network makes packet-mode data processing and transport very efficient. It also renders an easy and simple means to interwork with the existing Internet and other IP-based networks. The resulting Mobile WiMAX network architecture is much simpler than the existing cellular mobile networks: The four-step processing of the GSM/WCDMA family network (i.e., BTS/RNC/SGSN/GGSN) reduces to two-step processing in the Mobile WiMAX network (i.e., BS/ASN-GW) (see Section 2.4.2). The *time-division duplex* (TDD) technology adopted by Mobile WiMAX enables flexible bandwidth allocation between the uplink and downlink, reflecting the asymmetric nature of the uplink-downlink traffic in general multimedia data transport, which was not possible in the *frequency-division duplex* (FDD) technology of the existing voice-centric cellular mobile networks. Furthermore, TDD does not waste extra bandwidth for a guard band, which is always required between uplink and downlink band in FDD. Even though designed based on the IP technology, the Mobile WiMAX technology adopts the connection-oriented operation and well-developed bandwidth management technologies so that it can provide high-quality services demanded by real-time multimedia services. In addition, the Mobile WiMAX network is equipped with a strong security technology, which is built in as a sublayer below the MAC layer so that it can guarantee secured communications between the BS and MSs.

Mobile

As the cellular concept sectorizes the network into cells, mobility of the Mobile WiMAX network can be achieved through handover mechanism between two adjacent cells. Though the cellular concept, as well as the handover mechanisms, is readily established in the existing cellular mobile networks, the implementation is a different issue for the Mobile WiMAX network because it uses OFDMA, not CDMA: Mobile WiMAX supports only hard handover, whereas IEEE 802.16e defines both hard hadover and soft handover. Another important issue that is accompanied by the cellular concept is the power-saving issue: Mobile WiMAX is offering two battery power-saving modes—sleep mode and idle mode.

This introductory chapter is organized as follows: To begin with, we will discuss the technological aspect of Mobile WiMAX first, as an extension of this discussion. Among the various technologies that Mobile WiMAX adopts, we will discuss the following eight items—TDD, OFDMA, AMC, MIMO, QoS, HARQ, mobility, and security. Then we will describe the protocol layering of Mobile WiMAX, which is composed of four sublayers, and the network architecture of the Mobile WiMAX networks in terms of network reference model, functional entities, and reference points. Finally, we will examine the relations of the Mobile WiMAX network and the cellular mobile networks in the aspects of interworking and comparison of functionalities.

2.1 Key Network Technologies

As briefly discussed earlier, there are a large number of advanced technologies that are involved in the Mobile WiMAX system. Various different types of technologies that were developed independently are combined together to build up the Mobile WiMAX system. They include the conventional radio interface technologies such as duplexing and multiple access; newly emerging radio technologies such as MIMO and other multiple antenna technologies; communication system technologies such as AMC; mobility support technologies such as power saving and handover; bandwidth management and QoS technologies; and security technologies. Among the multitude of the available technologies in each category, the Mobile WiMAX system selects the most advanced ones, some of which have been long considered to be employed for 4G mobile system in the future, as will be exemplified in the following.

2.1.1 Duplexing: TDD

Duplexing refers to the mechanism of sharing a communication link for two-way communications. There are two typical duplexing techniques—*frequency division duplexing* (FDD) and *time division duplexing* (TDD). FDD divides the given frequency band into two bands—one for the uplink transmission and the other for the downlink transmission, respectively. In contrast, TDD uses the frequency band as a whole but divides the time slots into two groups for uplink and downlink transmissions, respectively.

FDD, when the two bands are equally divided, is more adequate for symmetric traffic like voices than for asymmetric traffic like Internet services. So FDD has been traditionally used in mobile cellular communication systems, such as GSM, IS-95, WCDMA, and cdma2000. In contrast, TDD enables asymmetric allocation to uplink and downlink interval while their interval can be dynamically configured as suited to the traffic demand, so is adequate for asymmetric services like Internet services. Due to this dynamic load-balancing feature, WiBro has adopted TDD as the preferred duplexing technology.

The operations of FDD and TDD may be well distinguished by the illustrations in Figure 2.1. FDD system contains a duplexer, consisting of two bandpass filters, which filter out the uplink and the downlink (i.e., transmit and receive) frequency bands, respectively. In contrast, TDD system contains a time division switch in front of the antenna to switch the connection of the antenna to the transmit circuitry and the receive circuitry alternately. Whereas FDD requires a guard band between the uplink and the downlink frequency bands, TDD requires guard time in between the transmit time interval and the receive time interval.

TDD may require synchronization in transmit/receive timing between channels or between operators so as to minimize the interference from the adjacent frequency channels or the neighboring operators. TDD is more adequate for applying antenna technologies than FDD. TDD RF switch has less insertion loss and lower cost than FDD duplexer. TDD has advantages over FDD in that the transmitter and receiver can share some devices like filter and oscillator. However, TDD has the disadvantages that the equalizer length is twice, the DAC/ADC speed is twice, and the RF switch is expensive.

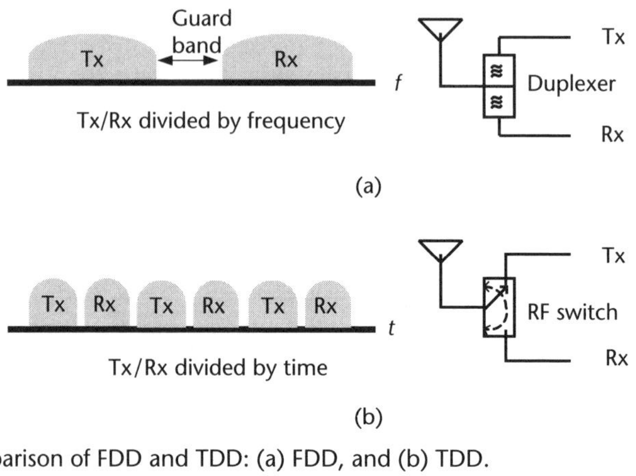

Figure 2.1 Comparison of FDD and TDD: (a) FDD, and (b) TDD.

Mibile WiMAX system adopts TDD profile, while IEEE 802.16e standard defines FDD operation as well.[1] In the circuit-mode cellular systems, which was originally designed for voice services, FDD was an adequate choice with the given frequency spectrum equally divided between uplink and downlink. However, the IP-mode Mobile WiMAX system was designed for high-speed data services from the beginning, so adopted TDD to be capable of asymmetric and dynamic bandwidth allocation.

2.1.2 Multiple Access: OFDMA

Multiple access refers to the mechanism of sharing a communication link among multiple users. For multiple access among different users, the three techniques—*frequency division multiple access* (FDMA), *time division multiple access* (TDMA), and *code division multiple access* (CDMA)—have been widely used in the past. However, WiBro adopts *orthogonal frequency division multiple access* (OFDMA), which falls within the category of FDMA in wide sense but incorporates the orthogonal characteristic as the basic feature. (Refer to Section 4.1 for more detailed discussions on OFDMA communication signal processing.)

Figure 2.2 compares the three multiple access techniques pictorially, on the two-dimensional basis of frequency and time. FDMA allows multiple access among multiple users by allocating different frequency bands to different users, but TDMA

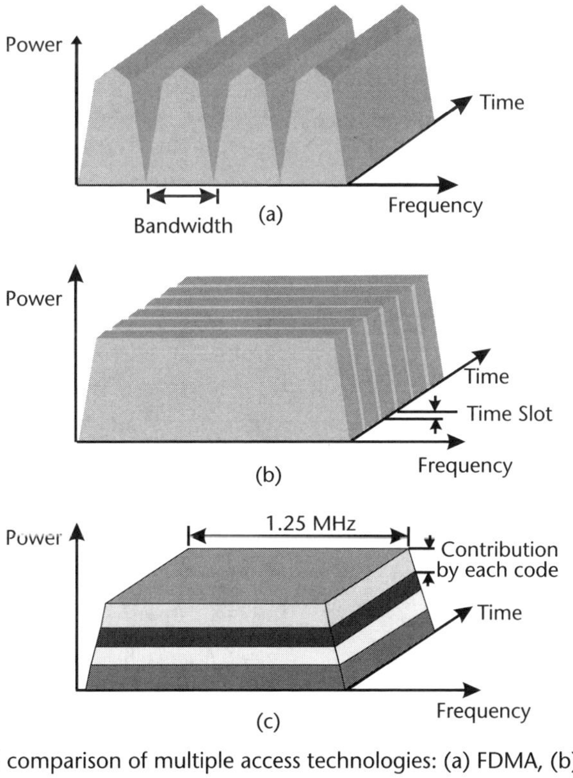

Figure 2.2 Pictorial comparison of multiple access technologies: (a) FDMA, (b) TDMA, and (c) CDMA.

1. FDD profile is being defined in the WiMAX Forum, with the expected completion in 2008.

allocates different time slots to different users. In contrast, CDMA provides multiple access, not by allocating frequency or time, but by allocating different codes to different users. Differently from those three techniques, OFDMA seeks for multiple access by dividing the frequency band into a large number of frequency components (commonly known as subcarriers), which are shared by one or more users in the same symbol. OFDMA is different from OFDM, in which all subcarriers are assigned to a single user rather than shared by a multiple number of users in each symbol.

Basically, OFDMA falls within the category of FDMA in wide sense but incorporates the orthogonal characteristic as the basic feature. Existing FDMA was inefficient in using spectrum, since overlapping in spectrum bands was not allowed. OFDMA resolved this problem by securing orthogonality among the constituent subcarriers. OFDMA divides the given frequency band into multiple subcarriers, each of which is equally spaced, and modulates the user data on the subcarriers. Orthogonality implies that a subcarrier is not affected by another subcarrier, which is guaranteed because all the other subcarriers take value 0 when any of the subcarriers takes the peak value. Figure 2.3 illustrates this property in the frequency band.

OFDMA has various advantages over other multiple access techniques. First of all, frequency band utilization becomes very efficient due to the division of the band into a large number of subcarriers: In case narrowband interference signal exists, we can either eliminate the corresponding subcarrier components from service or apply a more robust modulation such as BPSK or QPSK, thereby blocking the spread of the narrowband interference into other users. As OFDMA is a just variant of OFDM transmission as a multiple access technology, it inherits a broadband transmission feature of OFDM with robustness against multipath fading. More specifically, it is attributed to the long symbol period that is yielded by the division of frequency band into multiple subcarriers. For example, if the frequency band is divided into N, the symbol length would become N times as long as that of the TDMA case, so the multipath fading can be effectively absorbed even if a small guard time is used.

As a multiple access technology, OFDMA allows the subcarriers within the same OFDM symbol to be allocated to one or more users, which facilitates a multiuser diversity in a frequency domain. In other words, some good subcarriers for an individual user can be preferentially sorted out, rather than all subcarriers allocated to a single user in each OFDM symbol. When there are a sufficient number of active users in the cell, many users can be selected in such a way that their total throughput becomes maximized in each symbol, which typically is referred to as a *multiuser diversity*. Note that the multiuser diversity will be effective only if the

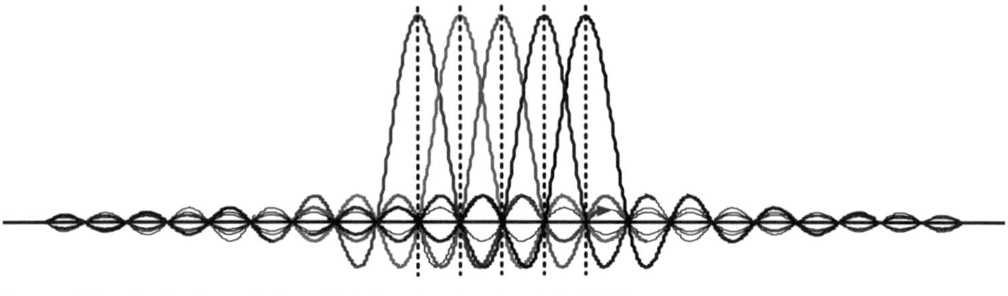

Figure 2.3 Illustration of the orthogonal property of OFDMA.

channel condition of each subcarrier for every user is instantaneously known to the scheduler. Otherwise, a subset of subcarriers is randomly selected for each user, achieving the averaging effect in a frequency domain, which is known as *frequency diversity*.

On the other hand, OFDMA has disadvantages in that it requires much computation and keen synchronization. Frequency offset or phase noise could seriously affect the maintaining of the orthogonality among different subcarrier components. In addition, *peak-to-average power ratio* (PAPR) could grow high when multiple subcarriers add coherently, consequently decreasing the efficiency of the power amplifier.

In practice, FDMA was used in the *first generation* (1G) wireless communication systems (e.g., AMPS system). TDMA was adopted in the GSM system, the 2G system widely used in Europe and other countries. CDMA was adopted in the IS-95 system, the other 2G system used in Korea, the United States, and other countries. Also CDMA was the ground technology for the 3G wireless communication systems such as WCDMA and cdma2000. However, the Mobile WiMAX system adopts OFDMA, and, further, it is expected that the *fourth generation* (4G) systems yet to come will also adopt OFDMA.

2.1.3 Coding and Modulation

The wireless communication environment is filled with various sources of noises and interferences. The communication channel state varies in time even in fixed wireless communications, and the variation becomes much more severe in mobile communications. In order to combat against the fluctuation of channel state, a diverse set of modulation and coding techniques can be employed. Among the multitude of modulation techniques, BPSK, QPSK, 16-QAM, and 64-QAM are used in Mobile WiMAX: BPSK and QPSK are the phase-shift keying techniques that map binary data to the subcarriers by taking the phase shifts of 180 and 90 degrees, respectively, whereas 16-QAM and 64-QAM are the quadrature-amplitude modulation techniques that map binary data to the subcarriers by taking different amplitudes and phases that constitute 16 and 64 constellation points, respectively. (Refer to Sections 4.1.1 and 4.2 for more discussions on channel coding and modulation.)

FEC

Channel coding is a technique that intends to correct bit errors occurred during transmission by utilizing the redundant bits added to the information bits before transmission. For this reason, channel coding is also called *forward error-correction coding* (FEC). The additional bits do not contain any new information, as they are determined solely by the transmitting information bits. However, they increase the dimension of the signal space and, as a result, increase the distance among different encoded sequences of the information bits. Thus, all transmission errors can be corrected as long as the number of bits in error does not exceed a half of the minimum distance of the employed channel coding scheme. This *coding gain* is achieved at the cost of a lowered coding rate, and more coding gain can be exploited as the coding rate decreases (i.e., as the number of additional bits increases). The channel coding is a major technique to overcome the unreliability of wireless channels and to achieve robust communications.

There have been developed various channel coding schemes, namely, linear block code, convolutional code, and concatenated code. *Linear block code* uses several parity bits in a block of data bits to detect and correct transmission errors. The parity bits are determined by linear combinations of the data bits in a finite field. By multiplying a proper parity-check matrix to the received bits, linear block codes can identify the error pattern as long as it is detectable. This property enables the coding scheme to correct the transmission errors. *Convolutional code* generates the coded bit sequence by passing the data bits through a linear finite-state shift register. Since each bit of the coded sequence is a linear combination of the data bits in the shift register, the original data bits can be recovered by tracing which state of the shift register generates the received sequence. This operation can be easily done by tracing the trellis diagram, which illustrates the relation between the shift register state and the generated sequence, and searching for the minimum likelihood sequence with the well-known detection algorithms (e.g., Viterbi algorithm). The convolution code is adopted for Mobile WiMAX. *Concatenated code* uses two levels of channel coding usually separated by an interleaver used for the randomization of the coded sequence. One advantage of the concatenated code is that it can exploit the merit of iterative decoding, where a decoder can utilize the output of the other decoder as a preknowledge. Owing to this merit, concatenated codes such as turbo codes are also adopted in Mobile WiMAX, generally achieving very low error probability at a reasonable level of complexity.

AMC

In order to strengthen the robustness of communications in the mobile wireless environment, channel coding techniques may be employed in conjunction with the modulation techniques. As a means for increasing the system performance in varying channel condition, Mobile WiMAX uses a combined form of modulation and channel coding techniques, which is called adaptive modulation and coding (AMC). AMC dynamically changes the modulation and coding techniques depending on the channel status. When the channel condition is good, it selects a high-efficiency modulation (i.e., 64-QAM) and coding technique, but when the channel condition is poor it selects a low-efficiency modulation (i.e., BPSK or QPSK) and coding technique. OFDM technique is adequate to use in conjunction with AMC, as it can use different modulation and coding techniques for different groups of subcarriers (or subchannels).

For an effective operation of AMC, each MS reports its channel status to the BS, which is mainly done in terms of *signal-to-noise ratio* (SNR) or *carrier-to-interference ratio* (CIR or C/I). On receiving this channel status report, the BS decides which modulation and coding techniques to use. Therefore, the AMC technique inherently requires a channel estimation process at the receiver and a mechanism to feed the estimated channel condition back to the transmitter. It is important in implementing AMC to report the current channel condition to the transmitter as accurately as possible. So the delay caused in estimating and delivering the channel condition should be maintained below the coherent time of the channel, as the AMC technique will perform poorly if the channel changes faster than this delay. The transmitter regards the fed-back channel condition as the current one and selects the *modulation and coding scheme* (MCS) that is the most appropriate under the current channel condition. Figure 2.4 illustrates the overall operation of AMC technique.

The CIR value increases as the mobile terminal moves close to the BS and decreases as it moves away from it, so the CIR value becomes small when the mobile terminal approaches the cell boundary. Mobile terminal periodically reports such channel status information to the BS through the *channel quality indicator* (CQI) channel, so that the BS can change the modulation and coding dynamically to the most appropriate ones. By taking such adaptive operation, the Mobile WiMAX system can maintain the transmission capacity at a maximum level. Table 2.1 illustrates the MCS set and the corresponding data rates for the Mobile WiMAX (10 MHz)/WiBro (8.75 MHz) system.[2] The minimum required SNR for each MCS level is as summarized in Table 2.2 [1].[3]

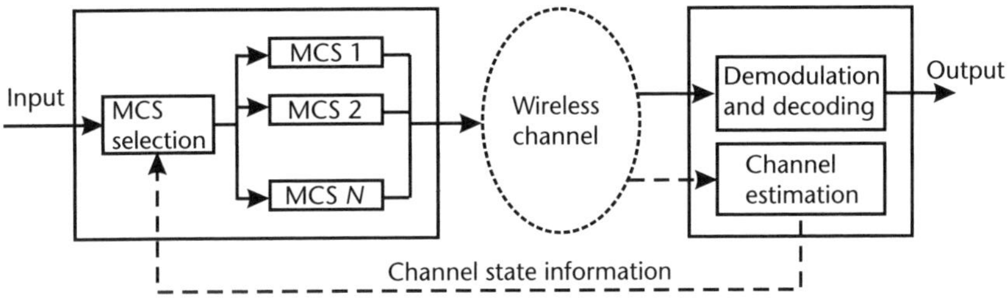

Figure 2.4 Illustration of AMC operation.

Table 2.1 Illustration of AMC Sets and the Corresponding Data Rates for the Mobile WiMAX/WiBro System: (a) Downlink and (b) Uplink

(a)

Data rate (kbps) WiMAX	Data rate (kbps) WiBro	FEC rate	Modul-ation
624	576	1/12	QPSK
936	864	1/8	QPSK
1872	1728	1/4	QPSK
3744	3456	1/2	QPSK
5616	5184	3/4	QPSK
7488	6912	1/2	16QAM
11232	10368	3/4	16QAM
11232	10368	1/2	64QAM
14976	13824	2/3	64QAM
16848	15552	3/4	64QAM
18720	17280	5/6	64QAM

(b)

Data rate (kbps) WiMAX	Data rate (kbps) WiBro	FEC rate	Modul-ation
280	224	1/12	QPSK
420	336	1/8	QPSK
840	672	1/4	QPSK
1680	1344	1/2	QPSK
2520	2016	3/4	QPSK
3360	2688	1/2	16QAM
5040	4032	3/4	16QAM
5040	4032	1/2	64QAM
6720	5376	2/3	64QAM
7560	6048	3/4	64QAM
8400	6720	5/6	64QAM

2. The data rates are the values obtained by filling in all the subchannels with a particular MCS level, assuming single antenna link in PUSC zone (see Section 4.3.2). Two symbols (used for MAP signaling) are excluded in the DL frame and three control symbols are excluded in the UL frame. The TDD DL to UL symbol ratio is set to 29:18 and 27:15 for WiMAX and WiBro, respectively. The cases of 64-QAM modulation in UL (the shaded region in Table 2.1) are out of system profile, even though they are included in the IEEE 802.16 standards.
3. This operation guarantees 10^{-6} BER in AWGN channels under single antenna reception in PUSC zone. The number of packets per UL frame is determined assuming the TDD DL to UL symbol ratio of 26:21 and 10-MHz bandwidth.

Table 2.2 Minimum Required SNR Specified by WiMAX Forum: (a) Downlink and (b) Uplink

FEC rate	Modulation	Packet length (bytes)	Packets per DL	SNR (dB)
1/2	QPSK	540	1	2.9
3/4	QPSK	540	1	6.3
1/2	16QAM	540	1	8.6
3/4	16QAM	540	1	12.7
1/2	64QAM	540	1	13.8
2/3	64QAM	540	1	16.9
3/4	64QAM	540	1	18.0
5/6	64QAM	540	1	19.9

(a)

FEC rate	Modulation	Packet length (bytes)	Packets per UL	SNR (dB)
1/2	QPSK	60	21	2.9
3/4	QPSK	54	35	6.3
1/2	16QAM	60	42	8.6
3/4	16QAM	54	70	12.7

(b)

Source: [1], Tables 28 and 113, modified.

2.1.4 Multiple Antennas

Single antenna has long been perceived as the natural way of building a wireless communication system, but recent studies unveiled that the use of multiple antennas can significantly enhance the reliability and/or increase the capacity substantially. The reliability originates from the beamforming and spatial diversity effects, while the high data rate originates from the spatial multiplexing effect of the multiple antenna system. Multiantenna technologies may be categorized into *adaptive antenna system* (AAS) and *multi-input multi-output* (MIMO) technologies: AAS technology is intended to take advantage of the *beamforming* (BF) effect, and the MIMO technology is intended to take advantage of *space diversity* (SD) or *spatial multiplexing* (SM). (Refer to Chapter 9 for a detailed description of the multiple antenna technology.)

AAS

AAS provides spatial division access by utilizing multiple antennas in array. By utilizing highly directional antennas or arrayed antennas, it becomes possible to optimize the usage and minimize the system cost of the given radio resources. With the usage of highly directional antennas, the AAS would yield good transmission quality but mobility would be restricted. The multiple antennas used in AAS bring forth spatial processing gain, and the resulting antenna diversity decreases the multipath interference as well as adjacent cell interference. In addition, it is possible to provide a stable transmission rate of services to the *nonline-of-sight* (NLOS) users as well, by taking advantage of the beams reflected by the neighboring buildings or objects. However, mobility is rather limited due to the highly directional nature of the antenna beam characteristics, since it is difficult to steer beams fast enough according to the high mobility of the user terminal.

As such, AAS can help to maximize the power of the desired signals while minimizing the power of the interfering signals by the directional beamforming effect and possibly with a capability of nulling the interference. Different forms of beams can be shaped by controlling the relative magnitudes and phases of the signals transmitted/received by the antenna elements. Figure 2.5(a) illustrates the structure of the AAS, which consists of an array of antennas. The arrayed antennas control the

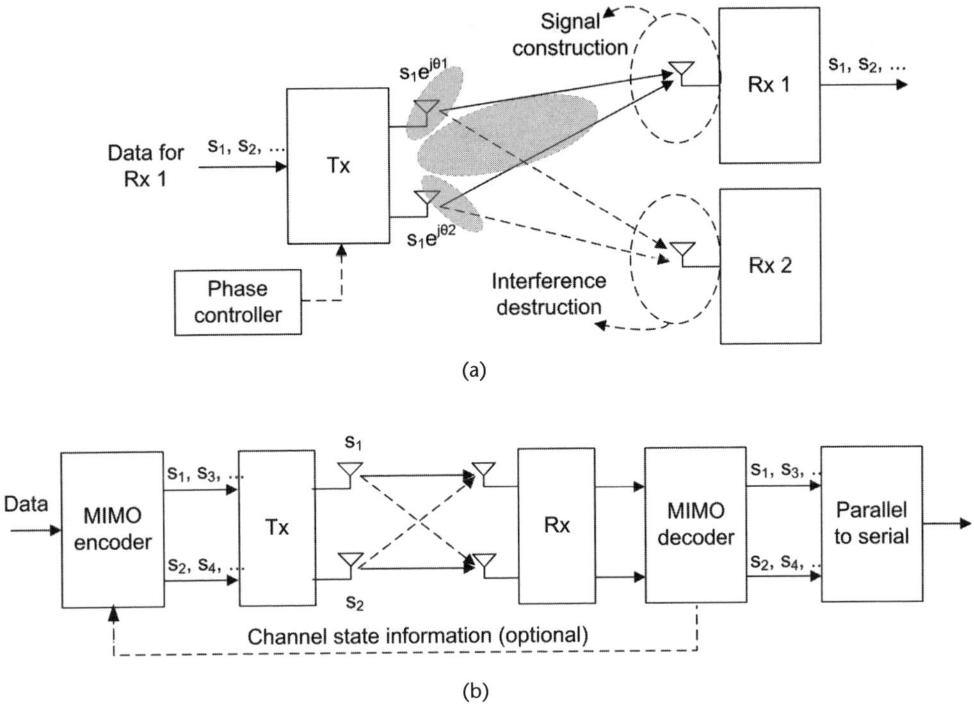

Figure 2.5 Illustration of AAS and MIMO operations: (a) AAS, and (b) MIMO.

phases of the signals to transmit the signal toward some desired particular direction selectively in such a way that the effects of interference are minimized and the signal components combine in constructive manner at the desired destination (i.e., Rx1), whereas the signal components combine in destructive manner at the other destinations (i.e., Rx2). This enables expanding the coverage or increasing the throughput of the system to the level that amounts to the decreased interference or the increased CINR.

MIMO

MIMO systems typically consist of multiple antennas in both the transmitter and the receiver. The multiple antennas may be arranged to increase the reliability by taking the space diversity effect or to increase the capacity by taking the spatial multiplexing effect.

Space diversity is intended to combine multiple signals that were transmitted from the same source but traveled through statistically independent channels. It is possible to send the same signal through an array of transmit antennas or receive/combine multiple signals obtained through an array of receive antennas.

Spatial multiplexing is intended to transmit multiple independent signals over the same frequency at the same time by employing multiple transmit and multiple receive antennas. To take advantage of spatial multiplexing, both transmitter and receiver should be equipped with multiple antennas. The original signal may be split into multiple streams before transmission at each antenna or the multiple antennas may carry different signals, thereby increasing the per-user transmission rate significantly. The receive antennas receive a combination of the data streams transmitted through the multiple transmit antennas and the MIMO demultiplexer decodes the original data streams correctly. If the *channel state information* (CSI) is available at the transmitter, it can help to enhance the capacity further by enabling to choose the best subset of transmit antennas or to precode the input signal so that the total throughout can be maximized.

2.1.5 Bandwidth Management

Providing quality data services by satisfying subscribers' QoS requirements is an important goal of the Mobile WiMAX system. The principal mechanism for providing QoS in Mobile WiMAX is to associate packets traversing the MAC interface into a service flow. So MS and BS provide the QoS according to the QoS parameter set defined for the service flow. The MAC in the Mobile WiMAX is connection-oriented, and the mappings to the services on MSs, as well as their association with varying levels of QoSs, are all done in the context of connection. Note that a connection defines the mapping between peer convergence processes that utilize the MAC and the relevant service flow. (Refer to Chapter 6 for the details of QoS and bandwidth management technology.)

As to the implementation of the QoS, there are three QoS enforcement functions involved, namely, scheduling, *connection admission control* (CAC), and policing. Scheduling deals with how to maximize the system throughout while supporting the degree of QoS satisfaction for the admitted connections; CAC deals with how to minimize the chance of unnecessary blocking of connection requests and the chance of QoS violation due to excessive admitted connections; and policing deals with how to protect the QoS of the contract-conforming connections against malicious connections.

Among various QoS parameters, bandwidth and delay play the most important role. These two QoS parameters are directly related to how much and how fast the required bandwidth is allocated. In fact, bandwidth is a very precious resource in wireless communication so its efficient use is very crucial. In order to enhance the efficiency of the bandwidth usage, the Mobile WiMAX system adopts well-organized bandwidth request, grant, and polling mechanisms. The downlink bandwidth is solely managed by the downlink scheduler at the BS, but the uplink bandwidth is allocated by BS to MSs through the resource request and grant process.

Bandwidth management, or bandwidth request and allocation, in Mobile WiMAX is supported by the five different types of scheduling service categories, namely, *unsolicited grant service* (UGS), *real-time polling service* (rtPS), *extended rtPS* (ertPS), *nonreal-time polling service* (nrtPS), and *best effort* (BE) service. If the BS receives a particular category among the five in request, it can anticipate the throughput and latency needed for the corresponding uplink traffic, so can apply polls and/or grants accordingly.

The UGS supports real-time service flows that generate fixed-size data packets on a periodic basis, whose typical examples are T1/E1 and VoIP without silence suppression. So, resource allocation is guaranteed to the UGS traffic without requiring contention or request. The rtPS supports real-time service flows that generate variable-size data packets on a periodic basis, so it offers real-time, periodic, unicast request opportunities which meet the flow's real-time needs and allow the MS to specify the desired grant size. The ertPS is similar to rtPS but makes requests only when change occurs in the desired transmission bandwidth, thereby reducing the request overhead. The nrtPS supports nonreal-time service flows that require variable size data grant burst on a regular basis, so it offers unicast polls regularly. The BE service is intended to serve the best-effort traffic.

The five scheduling services are closely associated with the five data delivery services in mobile networks: The five data delivery services are *unsolicited grant service* (UGS), *real-time variable-rate* (RT-VR) service, *nonreal-time variable-rate* (NRT-VR) service, *best effort* (BE) service, and *extended real-time variable-rate* (ERT-VR) service. Table 2.3 lists the five data delivery services in association with the five scheduling services, including their application examples.

2.1.6 Retransmission: HARQ

In order to further increase the robustness of the communications in the mobile wireless environment, it is desirable to employ the retransmission capability. For the retransmission of erred data, the *automatic repeat request* (ARQ) technique, which has been widely used for data transmission in the MAC layer, is applied. If an error is detected, ARQ processor requests the sender to retransmit the data block that is erred and then corrects the error. The ARQ technique can render more efficient operations when used in conjunction with the FEC technique in the physical layer. The hybridization of channel coding and ARQ, which is called *hybrid ARQ* (HARQ), enables us to exploit the coding gain in the retransmissions and thus enhance the overall transmission rate and error correction capability simultaneously. (Refer to Sections 1.1.3, 4.2.3, and 5.3 for more discussions on HARQ and ARQ.)

Table 2.3 Data Delivery Services and Scheduling Services

Service type	Scheduling type	Example
UGS (unsolicited grant service)	UGS (unsolicited grant service)	T1/E1 leased line, VoIP without silence suppression
ERT-VR (extended real-time variable-rate service)	ertPS (extended real-time polling service)	VoIP with silence suppression
RT-VR (real-time variable-rate service)	rtPS (real-time polling service)	MPEG video
NRT-VR (non-real-time variable-rate service)	nrtPS (Non-real-time polling service)	FTP
BE (best effort service)	BE (best effort service)	HTTP

Chase Combining HARQ

Whereas the simple hybridization of FEC and ARQ brings forth performance improvement to some extent, a major performance gain can be achieved by taking advantage of the *packet combining* technique, which utilizes the information contained in the erroneously received block. In this packet combining function, the erred data blocks are stored at the receiver and are combined with the retransmitted block before being fed to the decoder of the error-correction code. When the transmitter repeats sending the same coded data block in retransmissions, a newly received block can be combined with the previous ones by applying the *maximal ratio combining* (MRC) to their soft value (i.e., the amplitude and phase of each modulated symbol) before entering the hard decision process. This type of data block combining is called the *Chase combining*. As it is likely that an erroneously received block contains relatively small number of erred bits, the Chase combining technique can take advantage of the useful information remaining in the erred data block to improve the data detection performance. Notice that this improvement is achieved at the cost of additional memory deployment at the receiver.

Incremental Redundancy HARQ

The performance of the packet combining function can be further improved if different codes are used at different (re)transmissions, in contrast to the Chase combining case in which the same codes are simply repeated. The received data blocks are combined into a single codeword, and this combined codeword can be decoded more reliably since coding is done effectively across retransmissions. Specifically, this idea can be implemented as follows: Initially, the information bits are encoded by a low rate channel coder. At the first transmission, the information bits and a selected number of parity bits are transmitted. If the transmission is not successful, the transmitter sends additional selected parity bits in the next retransmissions. The receiver puts together the newly received parity bits with those previously received. This operation produces a new codeword with more parity bits (i.e., lower code rate). Thus, the receiver can decode a stronger codeword as the number of retransmissions increases. As the total number of parity bits is incremented in each retransmission, this scheme is called the *incremental redundancy*. Usually, the frame used in each (re)transmission is obtained by puncturing the output of the mother code for a rate-compatible channel encoder. The punctuation pattern used during each (re)transmission is different, so different coded bits are sent at each time.

2.1.7 Mobility Management

The mobility, in general, is realized through the handover function among the neighboring BSs. Handover refers to the operation of converting the wireless link connecting an MS to the BS in service to another wireless link connecting the MS to another BS in such a way that the communication connection is continuously maintained without degrading the QoS while an MS moves from a cell to another. There are two different types of handover, hard handover and soft handover. Hard handover disconnects the existing link before making a new connection to another BS. Mobile WiMAX supports only hard handover, whereas IEEE 802.16e defines both hard handover and soft handover. (Refer to Chapter 7 for detailed discussions of the mobility management technology. Also refer to Section 3.6 for viewing the mobility issue from the network initialization's point of view.)

Handover Process

Hard handover is performed in two steps: One is the network topology acquisition process and the other is handover execution process. Topology acquisition refers to the process of periodically updating the parameter values that are needed to make handover decision between MS and BS. Handover execution refers to the process of actually executing the handover by performing handover decision and initiation, synchronization to the target BS downlink, ranging, and termination of the MS context processes.

MS periodically receives the channel parameter information of the neighboring BS via the serving BS. The MS sends a scanning request message to request parameters needed to perform the scanning process in the case of interfrequency handover. For intrafrequency handover, scanning requests are not needed and the MS modems are designed in such a way that scanning is done as a background procedure during normal transmission. The BS sends back a scanning response to the MS, which includes the information on the permitted time and duration for measuring the signal quality of the neighboring BSs. Then during the allowed scanning period, the MS acquires synchronization with each neighboring BS, measures the CINR and other parameters, and finally determines whether or not each neighboring BS is adequate as the target handover BS. In addition, the MS performs an association process to get the ranging information that helps to select the target handover BS and to expedite the handover.

If the MS judges that the CINR values of the neighboring BSs is good enough to conduct a handover, it requests to start the handover to the serving BS. The serving BS notifies the neighboring BSs of the MAC address, the requested resources, and the QoS level of the MS and then receives the QoS value that the neighboring BSs can support. Using the received data, the serving BS selects the most appropriate handover target BS, and then notifies it to the target BS and to the MS. Finally the MS notifies to the serving BS with its final decision on disconnecting the link. Then the newly selected target BS offers a noncontention-based fast-ranging opportunity to the MS so that the MS can join the new BS quickly. Aside from this MS-initiated handover process, another option is that the serving BS can also initiate the handover process utilizing the scanning results periodically reported by the MSs.

In order to increase the cell coverage and improve the QoS performance at the cell boundary, soft handover techniques may be optionally used. There are two defined soft handover techniques used in the IEEE 802.16e standard, namely the *macro diversity handover* (MDHO) technique and the *fast BS switching* (FBSS) technique. In the case of the MDHO, the MS communicates simultaneously with a collection of BSs within the diversity set that allocates wireless resources to the MS. In the case of FBSS, the MS communicates only with the *anchor BS* that the MS was initially registered with and synchronized to. Enabling or disabling those optional soft handover techniques is determined through the exchange of the REG-REQ/RSP message and, in addition, each soft handover process starts with the renewal of the diversity set and the anchor BS by conducting the network topology acquisition and handover execution processes discussed earlier. (Refer to Section 7.2.3 for more discussions on soft handover.)

Power Saving

Since the mobile devices are likely to be compact and portable, the battery size is likely to be limited, so the power saving is an important design issue to the Mobile WiMAX system. In support of power savings, Mobile WiMAX supports sleep mode and idle mode of operations in such a way that the MS can operate in those power-saving modes if not in use but can return to the normal operation mode whenever needed. The sleep mode operation is designed to save power by allowing the MS to be absent from the serving BS air interface while not in use and the idle mode operation is designed to save power by allowing the MS to be mostly idle and only listen to the broadcast messages periodically.

In the case of the sleep mode, each MS and the BS exchange the sleep request and response messages for the transition to the sleep mode. The messages include the time to start transition to the sleep mode, the minimum and the maximum length of the duration of sleep mode, and the time period to wake up to listen to the signals from the BS. During the listen period, the BS sends a traffic indication message to the MS to notify whether or not new traffic appeared to the MS. Depending on the message, the MS decides whether to move to the normal operation mode or to return back to the sleep mode.

In the case of the idle mode, the MS does not register to a particular BS while moving over cells but only receives the downlink broadcast traffic periodically. The idle mode does not require performing any functions for the activation and operation of mobile communications, such as handover, but requires only the scanning operation for some discrete time period. This limited operation helps to save the terminal power and the operation resources further. When the BS receives packets to forward to a particular MS in idle mode it broadcasts a paging message to access the MS. (Refer to Section 7.3 for more detailed discussions on power saving. Also refer to Section 3.4 for more discussions on sleep and idle modes.)

2.1.8 Security Management

The Mobile WiMAX system offers a strong security function by installing a dedicated security sublayer between the MAC layer and physical layer. This structured security capability distinguishes the Mobile WiMAX system from the existing cellular mobile and WiFi systems. Cellular mobile systems did not take the security issue seriously because the user channels were protected by the circuit-mode operation, and the WiFi system did not seriously consider the security issue even if it was packet-mode-based (i.e., IP-based) because it was designed for LAN operation. In the case of the Mobile WiMAX system, however, security is a very important issue as it is IP-based and provides access to *wide area networks* (WANs). (Refer to Chapter 8 for detailed discussions on security management technology.)

The security sublayer in the Mobile WiMAX system is designed to provide users with privacy, authentication, or confidentiality across the fixed and mobile broadband wireless network. It also provides operators with strong protection against any unauthorized access to the data transport services by securing the associated service flows across the network. Further, it employs an authenticated client/server key management protocol in which the BS (i.e., the server) controls distribution of keying material to the MS (i.e., the client).

The security function has two component protocols, namely, an *encapsulation protocol* and a *key management protocol* (KMP). The encapsulation protocol is intended to secure packet data across the fixed or Mobile WiMAX network. It defines a set of supported data encryption and authentication algorithms as well as the rules for applying those algorithms to MAC PDU payload. The KMP is intended to provide a secure distribution of the keying data from the BS to the MS. It enables the MS to synchronize the keying data with the BS and enables the BS to enforce conditional access to network services.

Specifically, the Mobile WiMAX security system is designed to provide secured communications of the data traffic by encrypting the data traffic using the *traffic encryption key* (TEK), so its function is centered on generating and distributing the TEK between the BS and the MS in secured manner. The overall operations for the generation of the TEK and the distribution of the TEKs take place in the following procedure:

1. To begin with, the MS sends an authentication information to the BS so that the BS can authenticate the MS. Then BS performs authentication (i.e., entity identification) on the MS using the received authentication information.
2. Soon after sending the authentication information, the MS sends an authorization request message to the BS to request an *authorization key* (AK). On receiving the request, the BS generates an AK and sends it to the MS.
3. Once a common AK is shared between the BS and the MS, each station derives, independently, two additional keys, namely *key encryption key* (KEK) and HMAC key, out of the AK. KEK is used for encrypting the TEK, and HMAC key is used for protecting the TEK request and reply messages.
4. Then the MS transmits a TEK request message to the BS to request a TEK, by appending the HMAC value calculated using the HMAC key to the TEK request message. Receiving the message, the BS verifies the HMAC value using its own generated HMAC key. Next, the BS generates a TEK, encrypts it using a symmetric-key cipher, and then distributes it to the MS.
5. On receiving the encrypted TEK, the MS decrypts it using the same symmetric code and keeps the decrypted TEK for use in data traffic encryption.
6. From then on, the MS uses the TEK as the key for encrypting data traffic using a symmetric-key cipher.

2.2 Protocol Layering

The protocol layering of IEEE 802.16 standards is shown is Figure 2.6. This protocol layering equally applies to IEEE 802.16e mobile broadband access system and its 2.3 GHz–based implementation (i.e., the WiBro system in Korea).

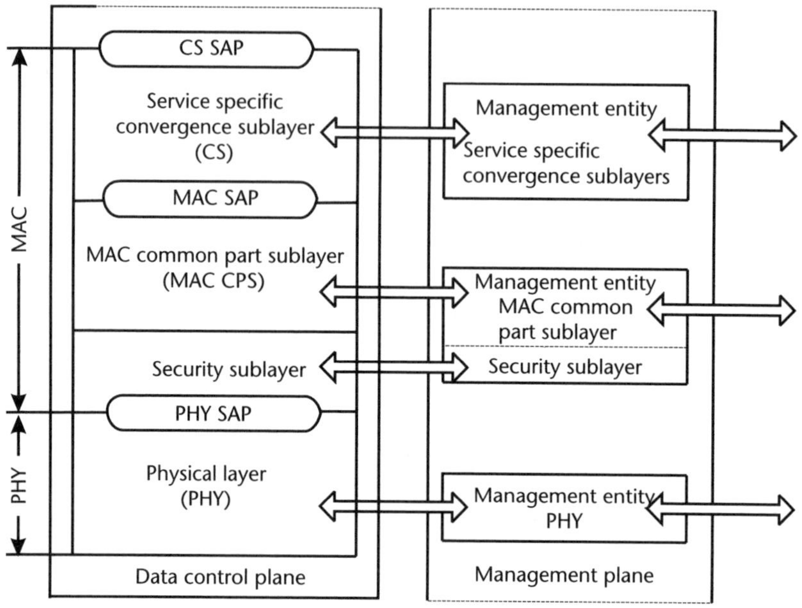

Figure 2.6 IEEE 802.16 protocol layering. (*Source:* [2].)

The IEEE 802.16 protocol layering consists of a MAC layer and a physical layer. The MAC comprises three sublayers, namely, the service-specific *convergence sublayer* (CS), the *common part sublayer* (CPS), and the security sublayer (or privacy sublayer). We briefly discuss the functions of the three sublayers and the physical layer in the following.

2.2.1 Service-Specific Convergence Sublayer

Basically, the service-specific CS performs the function of converging user services to MAC CPS. Specifically, the CS accepts higher layer PDUs from the higher layer, classifies them into the appropriate transport connections, processes them based on the classification, and then delivers the CS PDUs to the lower layer.

More specifically, the CS provides a transformation or mapping function on the external network data, received through the CS *service access point* (SAP), into the MAC *service data units* (SDUs), which are then sent to the MAC CPS through the MAC SAP. This includes classifying external network SDUs and associating them to the proper MAC *service flow identifier* (SFID) and *connection identifier* (CID). It may also include such functions as *payload header suppression* (PHS).

In principle, two CS specifications are provided, namely, the *asynchronous transfer mode* (ATM) CS and the packet CS.[4]

The packet CS is used to transport all packet-based protocols such as IPv4 and IPv6. The packet CS performs classification of the higher layer protocol PDU into the appropriate connection. When performing PHS, the sender and receiver exchange the PHS rules at connection creation or modification, even for the case of

4. The ATM CS is a logical interface that associates different ATM services with the MAC CPS *service access point* (SAP). We omit the discussion of ATM CS as the Mobile WiMAX profile excludes it, even though the IEEE 802.16d standard defined it.

connectionless higher-layer services such as IP. The internal format of the CS payload is unique to the CS, and the MAC CPS is not required to understand the format of, or to parse any information from, the CS payload.

2.2.2 MAC Common Part Sublayer

The MAC CPS is the main body of the MAC layer. It supports all different types of service-specific CSs in common. It provides the core MAC functionalites, such as system access, bandwidth allocation, connection establishment, and connection maintenance.

The MAC CPS processes the MAC SDUs received from the CS through MAC SAP and constructs a MAC PDU by putting a header and CRC. More specifically, the user data packet PDU, which is mapped into the payload of MAC SDU, is prefixed by a generic MAC header and postfixed by a CRC code to yield a MAC PDU. Note that MAC PDU may also be constructed to carry MAC management messages. In general, the MAC PDU is of variable length, which depends on the length of the carried payload. So, in the MAC PDU constructing process, fragmentation, packing, and concatenation processing are applied appropriately: to be specific, a MAC SDU or a MAC management message may be divided into multiple MAC PDUs if its length is long (*fragmentation*), and multiple MAC SDUs or MAC management messages may be combined into a MAC PDU if their lengths are short (*packing*). It is also possible to combine multiple MAC PDUs into a single serial transmission in the transmission process of MAC PDU (*concatenation*).

As the IEEE 802.16 network utilizes a shared medium to provide services to multiple users, MAC CPS provides a mechanism that enables all the users to share the wireless medium effectively. Two typical examples of the shared wireless media are two-way PMP and mesh topology wireless networks.

The PMP network has a star architecture, with the BS located at the center and MSs at the end of the branches. The PMP wireless link operates with a central BS and a sectorized antenna that is capable of handling multiple independent sectors simultaneously. Within a given frequency channel and antenna sector, all MSs receive the same transmission. As the BS is the only transmitter operating in the downlink direction, it may transmit, without having to coordinate with other stations, within the given TDD time period. The PDUs that each MS receives may contain an individually addressed message or a multicast or broadcast message. Every MS listens to the broadcast message and checks the *connection identifiers* (CIDs) in the received PDUs, then retains only those PDUs addressed to it. The uplink of PMP network is shared among all the MSs in the same cell or sector on demand basis. Depending on the class of service in subscription or in service, MS may be given an unsolicited grant to transmit, or may be granted by the BS with a right to transmit by polling or by contention procedures after making a request.

The mesh network differs from the PMP network in that traffic can be routed though other MSs and can occur directly between MSs, as opposed to the PMP network where traffic occurs only between the BS and MSs. Depending on the transmission protocol algorithm used, medium access can be done on the basis of distributed scheduling, centralized scheduling, or a combination of them.

2.2.3 Security Sublayer

The security sublayer in the Mobile WiMAX system is intended to provide users with privacy, authentication, or confidentiality across the fixed and mobile broadband wireless network. The security function also provides with strong protection from unauthorized access to the data transport service operators, thereby securing the service flows across the network. In support of such secured communications, the security layer defines two major component protocols, namely, encapsulation protocol and key management protocol. The *encapsulation protocol* is for securing packet data across the network and the *key management protocol* is for providing a secure distribution of the keying data from BS to MS.

The security layer acquires secured transmission of data traffic by encrypting the data traffic using the *traffic encryption key* (TEK). So it puts emphasis on generating and distributing the TEK between BS and MS in secured manner. The overall procedure of security control in the Mobile WiMAX system may be summarized as follows.

In the beginning stage, the BS authenticates the MS using the authentication information sent by the MS. Once authenticated, the MS requests an AK and then the BS generates and sends an AK to the MS. Out of this AK, commonly shared by the MS and the BS, both stations derive a KEK and a HMAC key, independently. Next, the MS transmits a TEK request message to the BS and then the BS generates and sends a TEK to the MS. In this process, each transmitted message is protected by a HMAC key or a KEK. In the last stage, the MS decrypts the received TEK and stores it for use in encrypting the data traffic to transmit. (Refer to Section 2.1.8 for a more detailed summary and Chapter 8 for full description of the security layer function.)

2.2.4 Physical Layer

As discussed in Section 1.4.1, the physical layer of IEEE 802.16e standard has multiple specifications, namely, WirelessMAN-SC, -SCa, -OFDM, and -OFDMA, each appropriate to a particular frequency range and applications. Among them, WirelessMAN-OFDM PHY and WirelessMAN-OFDMA PHY, which are designed based on the OFDM and the OFDMA technologies, respectively, for operation on the NLOS 2-11GHz frequency band, are opted for use in the fixed and mobile wireless access networks, respectively. From the Mobile WiMAX aspect, we focus on the WirelessMAN-OFDMA case in the following.

The WirelessMAN-OFDMA Mobile WiMAX system adopts most state-of-the-art technologies that are assembled together to enhance system performance, including throughput and QoS. From the communication structural aspect, it adopts the *time division duplex* (TDD) scheme for sharing communication channels between the uplink and downlink, adopts the *orthogonal frequency division multiple access* (OFDMA) scheme for sharing the communication link among multiple users, and takes a large channel bandwidth (e.g., 10 MHz) for operator allocation. From the functional block aspects, it uses *adaptive modulation and coding* (AMC) technology for an efficient modulation/demodulation and coding/decoding of communication signals and employs multiple antenna technologies such as the *beamforming* (BF) function of *adaptive antenna system* (AAS) and *multi-input multi-output* (MIMO). In addition, it takes efficient technologies for battery power saving and IP-based mobility.

The signal processing and frame structuring of the physical transmission data are basically tailored for the OFDMA technology. In the transmit direction, the input data signal is first encoded, modulated, and mapped to multiple subcarriers (or an OFDMA symbol) and then randomized. The resulting parallel signal is inverse-transformed via *inverse discrete Fourier transform* (IDFT), lowpass filtered, and finally converted to analog signal for upconversion to RF frequency and transmission. Conversely, in the receive direction, the received RF signal is downconverted to a baseband signal, sampled to digital signal, and lowpass filtered, before passing through the *discrete Fourier transform* (DFT) process. The transformed parallel streams, which are carried over the multiple subcarriers, are derandomized and demapped, and finally demodulated and decoded to yield the output signal.

The OFDMA symbol consists of the orthogonal subcarriers, which are classified into data subcarriers, pilot subcarriers, and null subcarriers. The data and pilot subcarriers are grouped in multiple to form *subchannels*. In this case, the subcarriers are selected according to a predetermined pattern that considers the effects of frequency diversity and other factors. The concept of subchannel is effectively used in allocating subcarriers to the users in the cells or sectors, enabling multiple users to share an OFDM symbol aiming at the advantage of multiuser diversity gain in the frequency domain.

In Mobile WiMAX system, transmission data allocation is *OFDMA slot* based. The OFDMA slot is defined in two dimensions, one in time (i.e., the OFDMA symbol number) and the other in frequency (i.e., the subchannel number). The size and the shape of the OFDMA slot varies depending on the transmit direction and channel type.

Finally, the subchannels are mapped into the frame structure in the unit of OFDMA slots. The mapping is done according to the TDD scheme. In the *downlink* (DL) part, the first symbol is allocated to the preamble, which is followed by DL-MAP and DL bursts which carry UL-MAP and UL bursts. In the *uplink* (UL) part, the channels in the first three symbols are usually allocated for ranging/ACK/CQI channels and then OFDMA data slots follow. A *Tx/Rx transition gap* (TTG) is inserted between the downlink and uplink parts in the same frame, and an *Rx/Tx transition gap* (RTG) is inserted between the end of a frame and the start of the next frame.

AMC refers to a combined modulation and channel coding technique intended to aid the system performance in the varying wireless mobile environment: a high-efficiency modulation (i.e., 16-QAM or 64-QAM) and coding technique is selected when the channel condition is good, and a low-efficiency modulation (i.e., BPSK or QPSK) and coding technique is used otherwise. AMC is an adequate technique to use in conjunction with OFDMA, as it can use different modulation and coding techniques for different group of subcarriers (i.e., subchannels), whose channel condition may be varying. In order to further enhance the system performance, the Mobile WiBro system adopts the ARQ technique in the MAC layer and the *hybrid ARQ* (HARQ) technique, which is a hybridization of channel coding and ARQ, in the PHY layer.

The Mobile WiMAX system adopts both BF and MIMO technologies. BF and MIMO technologies are intended to increase the system performance and enhance the system capacity by employing and properly controlling multiple antennas. With the BF technique, the multiple antennas are controlled in such a way that the terminal interference decreases and the received signal strength increases, which leads to an expanded coverage or increased throughput. With the MIMO technique, the multiple antennas in the transmitter (in the receiver) are arranged to transmit (receive) multiple different signal streams simultaneously, thereby increasing the per-user transmission rate significantly.

2.3 Network Architecture

Whereas the protocol layering offers an insight into the internal architecture of the Mobile WiMAX network, the network architecture offers an insight to view the external, physical configuration of the network. The network architecture may be described in several components, including network reference model, functional entities, and reference points.

Figure 2.7 illustrates the network architecture of the Mobile WiMAX. The Mobile WiMAX network is divided into *access service network* (ASN) and *connectivity service network* (CSN). The ASN contains BSs and *ASN gateways* (ASN-GWs). The CSN is composed of routers/switches and various servers, such as *authentication, authorization, and accounting* (AAA) server, *home agent* (HA), *dynamic host configuration protocol* (DHCP) server, *domain name service* (DNS) server, and *policy and charging rules function* (PCRF) server. The ASN is connected with the CSN via router and switch.

2.3.1 Network Reference Model

The *network reference model* (NRM) of IEEE 802.16 standards is as depicted in Figure 2.8. NRM refers to a logical representation of the network architecture. NRM, in general, identifies functional entities and reference points over which

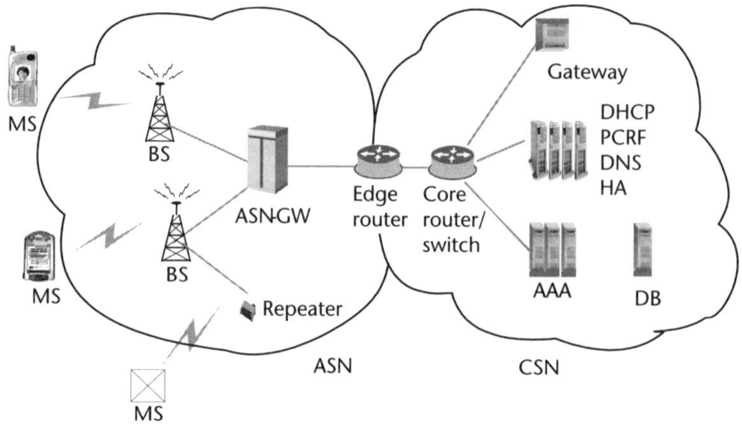

Figure 2.7 Mobile WiMAX network architecture.

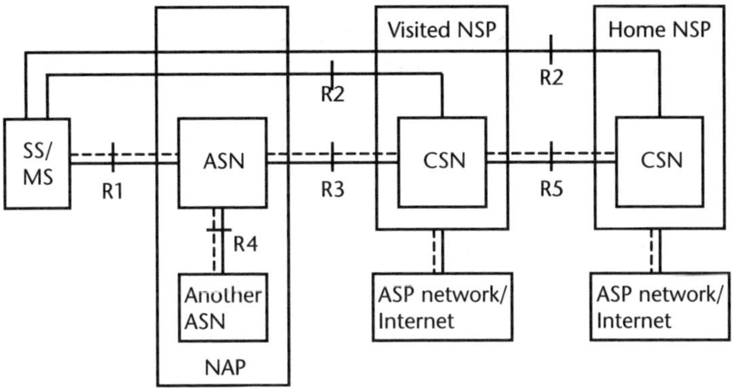

Figure 2.8 Network reference model of IEEE 802.16 system. (*After:* [3], Figure 6-1, redrawn.)

interoperability between functional entities is achieved. The NRM of the Mobile WiMAX system consists of the functional entities, MS, *access service network* (ASN), and *connectivity service network* (CSN), as well as the *reference points* (RPs) R1 through R5.

In implementation, each functional entity may be realized in different ways. It may be implemented in a single physical device or may be distributed over multiple physical devices. How to group and distribute the functions into physical devices is the implementer's choice. NRM enables multiple implementation options for a given functional entity on one hand, and achieves interoperability among different forms of implementations on the other.

Interoperability is intended to achieve an overall end-to-end function by properly defining communication protocols and data plane treatment between functional entities. As mentioned earlier, interoperability is referenced on the reference points. The functional entities on either side of a reference point may be implemented in different ways, but the protocol exposed at the reference point should meet the interoperability requirements.

2.3.2 Functional Entities

The functional entities in the Mobile WiMAX system may be categorized into three groups—MS, ASN, and CSN. MS is the end terminal of the network, ASN is the main access network where BSs and ASN/GWs reside, and CSN is the core network that provides various supporting functions and servers.

MS

Mobile station or *subscriber station* (SS) is a generalized mobile equipment set providing connectivity between subscriber equipment and a BS. Apparently, the term *subscriber* in SS may see to indicate a *stationary* or *fixed* station in contrast to the term *mobile*, which clearly indicates mobile station. In reality, however, subscriber indicates a station for the access network as opposed to a station in the local area network.[5] So the two terms, SS and MS, may be intermixed in the Mobile WiMAX,

5. Note that the term *subscriber* has been widely used in the public-switched telephone network in the form of *subscriber loop*, which corresponds to the access network.

with SS having a broader coverage, including both fixed and mobile stations. MS/SS may be one host or may support multiple hosts.

ASN

ASN is a complete set of network functions needed to provide wireless access to Mobile WiMAX subscribers. It consists of network elements such as one or more *base stations* (BSs) and one or more *ASN gateways* (ASN-GWs). In addition to the network reference model in Figure 2.8, Figure 2.9 illustrates the ASN reference model for the case containing multiple ASN-GWs. As shown in the two figures, ASN shares RP R1 with an MS, RP R3 with a CSN, and RP R4 with another ASN. Internally, RP R6 is shared between a BS and an ASN-GW, and RP R8 is shared between two different BSs.

Among the ASN components, BS is a *logical* (not physical) entity that embodies a full instance of the WiMAX MAC and physical layers according to IEEE 802.16 standards, where a BS instance represents one sector with one frequency assignment. It also incorporates scheduler functions for uplink and downlink resource management. For every MS, a BS is associated with exactly one ASN-GW, but each BS is required to be connected to two or more ASN-GWs for load balancing or redundancy purposes. On the other hand, ASN-GW is a logical entity that represents an aggregation of control plane functions. It also performs bearer plane routing or bridging function. ASN-GW functions may be viewed as consisting of two groups of functions, namely, the *decision point* (DP) and the *enforcement point* (EP), which are interfaced by RP R7.

ASN basically provides layer-2 (L2) connectivity with WiMAX MS; transfers the *authentication, authorization, and accounting* (AAA) messages to *home network service provider* (H-NSP); helps subscriber's network to discover and select the preferred NSP; helps to establish layer-3 (L3) connectivity with a WiMAX MS; and conducts radio resource management. H-NSP is the operator or service provider that provides AAA service to the subscriber according to the precontracted *service level agreements* (SLAs). In the mobile environment, ASN additionally supports the following functions: ASN anchor mobility, CSN anchor mobility, paging and location management, and ASN-CSN tunneling.

Here, ASN anchor mobility means enabling the handover of an MS from the *serving BS* to the *target BS* without changing the traffic anchor point in the serving

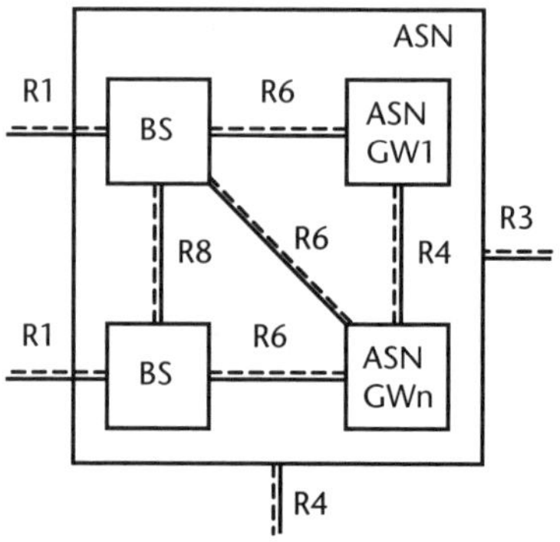

Figure 2.9 Illustration of ASN reference model containing multiple ASN-GWs. (*After:* [3], Figure 6-3, redrawn.)

(or anchor) ASN. (Refer to Chapter 7 for more details of the mobility support functions.) CSN anchor mobility means the changing of the traffic anchor point, for the MS, from one anchor point to another one in the ASN without changing the CSN anchor. ASN-CSN tunneling refers to the capability that enables ASN and CSN to exchange IP data via intermediate networks, while hiding the protocol details from the intermediate networks.

CSN

CSN is a set of network functions that provide IP connectivity services to the subscribers. To this end, CSN provides various functions, including MS IP address and endpoint parameter allocation; Internet access; AAA proxy or server; policy and admission control; ASN-CSN tunneling support; subscriber billing and interoperator settlement; inter-CSN tunneling for roaming; and inter-ASN mobility [3].

In order to conduct this diverse set of functions, CSN contains routers/switches, AAA and other servers, user databases, and interworking gateways. CSN servers include HA for the management of home address, the AAA server for security and accounting functions, the DNS server for conversion of IP addresses and system names, the DHCP server for dynamic allocation of IP, and the PCRF server for managing the service policy and for sending QoS setting and accounting rule information.

2.3.3 Reference Points

A *reference point* (RP) is a conceptual point between two groups of functions that reside in different functional entities on either side of it. These functions expose various protocols associated with an RP. All protocols associated with an RP may not always terminate in the same functional entity. That is, two protocols associated with a RP may originate and terminate in different functional entities. The normative RPs between the major functional entities are R1-R5 and the inter-ASN informative RPs between ASN-internal functional entities are R6-R8 (refer to Figures 2.8 and 2.9).

RP R1 consists of the protocols and procedures between MS and ASN as per the air interface (PHY and MAC) specifications. It may additionally include management plane-related protocols.

RP R2 consists of protocols and procedures between the MS and CSN associated with authentication, services authorization, and IP host configuration management. This RP is logical in that it does not reflect a direct protocol interface between MS and CSN.

RP R3 consists of the set of control plane protocols between the ASN and the CSN to support AAA, policy enforcement, and mobility management capabilities. It also encompasses tunneling to transfer user data between the ASN and the CSN.

RP R4 consists of the set of control and bearer plane protocols originating/terminating in various functional entities of an ASN that coordinate MS mobility between ASNs and ASN-GWs. R4 is the only interoperable RP between similar or heterogeneous ASNs.

RP R5 consists of the set of control plane and bearer plane protocols for internetworking between the CSN operated by the home NSP and that operated by a visited NSP.

RP R6 consists of the set of control and bearer plane protocols for communication between the BS and the ASN-GW. The bearer plane consists of intra-ASN data path between the BS and ASN-GW. The control plane includes protocols for datapath establishment, modification, and release control in accordance with the MS mobility events.

RP R7 consists of the optional set of control plane protocols, such as for AAA and policy coordination in the ASN gateway as well as other protocols for coordination between the two groups of functions identified in R6 (not shown in Figure 2.9). The decomposition of the ASN functions using the R7 protocols is optional.

RP R8 consists of the set of control plane message flows and optionally bearer plane data flows between the BSs to ensure fast and seamless handover. The bearer plane consists of protocols that allow the data transfer between the BS involved in handover of a certain MS [3].

2.4 Mobile WiMAX Versus Cellular Mobile Networks

Mobile WiMAX network stands between WiFi network and cellular mobile networks: Mobile WiMAX network, as its early name WirelessMAN indicates, was originally intended to cover the metropolitan area, even though it was later converted to a mobile access network covering a wide area, whereas WiFi and cellular mobile networks were intended to cover the local area and the wide area, respectively. Mobile WiMAX is similar to WiFi in that it is based on the packet mode, or IP technology, and is similar to cellular mobile networks in that it is based on the cellular concept, supporting mobility and high-quality mobile services. Mobile WiMAX can go down to the WiFi area for high-speed local access on one side and can reach out to the wide area to provide large bandwidth to mobile users in competition with the cellular mobile networks on the other. Therefore, in order to better understand Mobile WiMAX, it is important to view it in comparison with the cellular mobile networks, as we viewed it in comparison with WiFi in Section 1.4. So, in this section, we review the evolution process of the cellular mobile networks first in terms of two different technical families—the GSM/WCDMA family and the IS-95/cdma2000 family—and then compare the Mobile WiMAX system with those cellular mobile systems, finally discussing how to do interworking between them.

2.4.1 Evolution of Cellular Mobile Networks

Mobile wireless networks that provide narrowband voice and data access services to subscribers take the form of cellular network, as it can achieve high spectral efficiency. Similar to the use of analog modems in wireline telephony networks, wireless modems in the laptop and hand-held devices find applications in remote access to corporate LANs and the Internet.

Mobile wireless access networks in cellular network configurations enable frequency reuse, thereby enhancing spectrum efficiency significantly. Cellular mobile wireless access is basically a circuit-mode technology, so its services have evolved from pure voice services to combined voice, data, and other multimedia services. In some advanced mobile wireless systems, data-only services are also provided.

The *first generation* (1G) mobile access systems, launched in the early 1980s, were analog systems whose typical example is the *advanced mobile phone service* (AMPS) system. Those 1G systems used FDMA technology and worked mainly in the 800–900-MHz frequency bands. They were limited to voice services and had very low transmission rates (typically a few Kbps), unreliable handover, poor voice quality, and poor security.

After the first generation, the mobile access system evolved into digital systems, but the evolution path was divided into two different streams—one by TDMA technology and the other by CDMA technology. The TDMA-based *global system for mobile communications* (GSM) system [4] was mainly developed in Europe and diffused into a large number of countries. As the GSM system evolved to *general packet radio services* (GPRS), *enhanced data-rate GSM evolution* (EDGE), *wideband CDMA* (WCDMA), and HSDPA/HSUPA systems [5–9], however, it accommodated the CDMA technology as well. On the other hand, the CDMA-based system was actively studied in the United States and the system implementation and commercialization were first realized in Korea. The CDMA-based system evolved from IS-95 to cdma2000 1x, EV-DV, and EV-DO systems [10–13].

Figure 2.10 shows the evolutionary path of the circuit-mode cellular mobile wireless access networks and services.

The cellular mobile services initially started with voice-only services but evolved to include data services and other multimedia services, and the provided data rate increased significantly from generation to generation. For example, the *second generation* (2G) cellular mobile access networks such as GSM, GPRS, EDGE, and CDMA IS-95 were more voice-centric than data-centric, but the *third generation* (3G) cellular mobile networks such as WCDMA and cdma2000 may be said to be data-centric.

In the case of the GSM/WCDMA family, the phase 2+ GSM system provided low-rate data services of 9.6–14.4 Kbps but GPRS increased it up to 115.2 Kbps, and EDGE further increased it to 384 Kbps. The WCDMA system has increased the data service rate significantly up to 2 Mbps, providing both high-rate packet data and high-rate circuit-switched data services. The HSDPA specification, released by the *Third Generation Partnership Project* (3GPP), increases the best-effort packet data rates dramatically, up to 8–10 Mbps. To make it better, HSDPA is specified in such a way that an operator can deploy it at low incremental cost of implementation, mainly through a straightforward software upgrade.

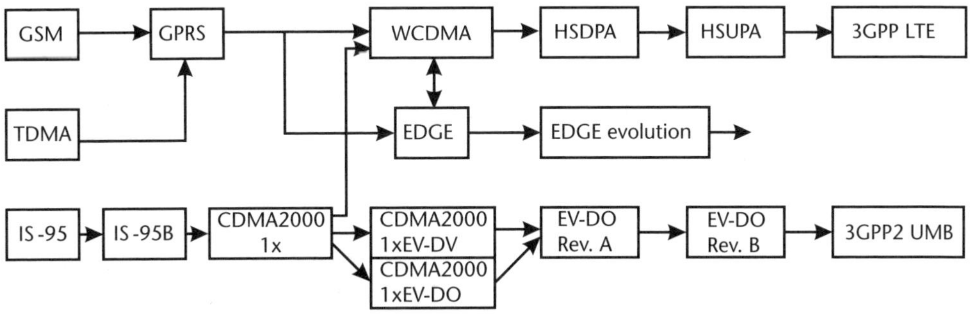

Figure 2.10 Evolution path of cellular mobile wireless access networks and services.

On the other hand, in the case of the IS-95/cdma2000 family, IS-95 supported a low data rate of 9.6–14.4 Kbps, but IS-95B supported an increased data rate of up to 64–115 Kbps in both directions and a burst mode packet data service as well. The 3G cdma2000 system has increased the data rate dramatically, up to 2 Mbps, and also provides multiple concurrent data and voice services. The 1xEV-DO system can support even higher data rates of 2.4 Mbps. In fact, the 1xEV-DO system uses a separate carrier dedicated to data-only services, whereas the 1xEV-DV system can provide both high-speed data and voice services using one carrier (i.e., can support real-time packet-data services as well). The 1xEV-DO *Revision A* (Rev.A) system [14] provides the better support of VoIP service and higher data rate for uplink than 1xEV-DO. The 1xEV-DO *Revision B* (Rev.B) system [15] is the *N* times multicarrier version of the EV-DO Rev.A system. Table 2.4 compares the two families of mobile data services.

For further enhancement of mobile data services, *fourth generation* (4G) mobile systems are being researched for the goal of attaining a much higher data rate, such as 100 Mbps in mobile state and 1 Gbps in nomadic state. In order to make this really happen, various different types of new physical layer technologies, such as *orthogonal FDM* (OFDM), *adaptive modulation and coding* (AMC), *multi-input multi-output* (MIMO), and intelligent radio resource management technologies have to be fully exploited. The *long-term evolution* (LTE) system [16] pursued by 3GPP and the *ultra mobile broadband* (UMB) system [17] pursued by 3GPP2 are both expected to render such 4G mobile systems. Note that the IEEE 802.16e Mobile WiMAX system, whose first implementation by Korea named WiBro (refer to Chapter 10), is perceived as an interim solution toward the 4G mobile data services, and its subsequent versions are admitted to yield another 4G mobile system. Table 2.4 includes the expected data rates of the LTE and UMB systems.

The network architecture of cellular mobile networks is different for each technical family and for each generation. Figure 2.11 illustrates the configuration of the WCDMA and the cdma2000/1xEV-DO mobile communication networks in one network in mixed form. On the far left, GSM network elements are shown, including MS, *base transceiver station* (BTS), *base station controller* (BSC), *serving GPRS support node* (SGSN), and *gateway GPRS support node* (GGSN). On its right are shown the WCDMA network elements, including *user equipment* (UE), nodeB,

Table 2.4 Comparison of Data Rates of CDMA IS-95 and GSM Families

GSM Family		CDMA IS-95 Family	
GSM phase 2+	9.6 ~14.4 kbps	IS-95 (CDMA)	9.6 ~ 14.4 kbps
GPRS	115.2 kbps	IS-95	64 ~ 115 kbps
EDGE	384 kbps	cdma2000 1xEV	2 Mbps
WCDMA	2 Mbps	cdma 1xEV-DO	2.4 Mbps
HSDPA	7 ~ 14 Mbps	EV-DO Rev.A, Rev.B	3.1~4.9 Mbps
LTE	172~326 Mbps	UMB	288 Mbps

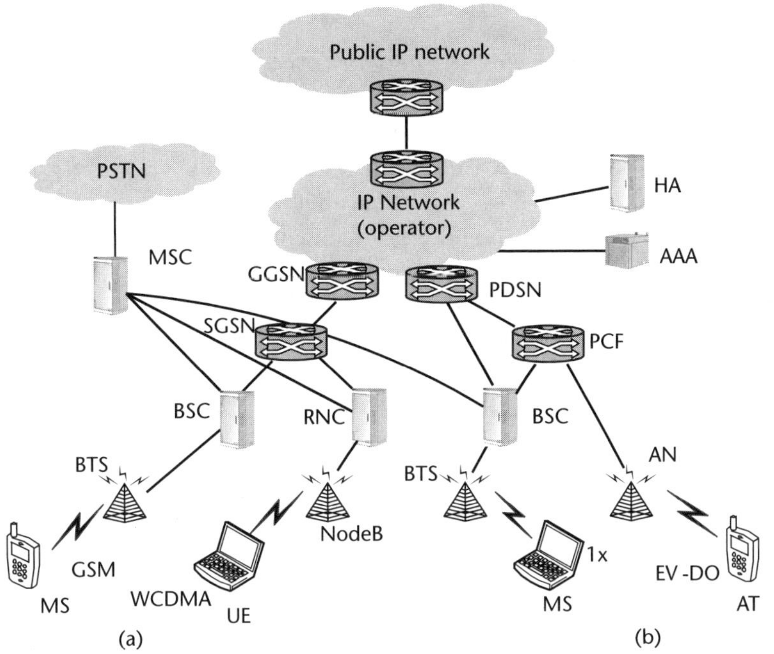

Figure 2.11 Conceptual configuration of (a) GSM/WCDMA network, and (b) IS-95/EV-DO network drawn in mixed form.

radio network controller (RNC), SGSN, and GGSN. Shown on its right are the cdma2000 1x network elements, including MS, BTS, BSC, and *packet data serving node* (PDSN). On the far right are the EV-DO network elements, including *access node* (AN), *packet controller function* (PCF), and PDSN.

2.4.2 Comparison of Mobile WiMAX and Cellular Mobile Networks

GSM/UMTS family or CDMA IS-95/cdma2000 family cellular wireless networks differ from the Mobile WiMAX network in various aspects. First of all, the cellular networks are founded on the CDMA technology, whereas the Mobile WiMAX network is founded on the OFDMA technology. This difference in multiple access technology leads to rather big differences in network performance and complexity issues. Second, the cellular networks were initially based on the circuit-mode technology and later migrated to a hybrid technology by augmenting the packet mode for data services, whereas the Mobile WiMAX network is solely based on the IP-mode (or packet mode) based technology. Such a difference in traffic mode leads to the difference in network architecture, which is magnified in practical network implementation as the resulting network configurations differ significantly. In addition to those two salient differences, there are several other aspects that distinguish the cellular networks from the Mobile WiMAX network, including the coverage (wide area versus metro zone) and the incumbency (widely deployed versus newly emerging). In the following, however, we will discuss the comparisons focusing mainly on the evolutionary aspects that the Mobile WiMAX network has over the existing cellular networks.

Network Architecture

The existing cellular wireless networks are initially designed on the circuit-mode basis, and the packet-mode part was augmented for data services. In contrast, the Mobile WiMAX network was designed for IP service and IP traffic from the beginning. The Mobile WiMAX network consists of only two layers: one is the radio unit called *base station* (BS), and the other is the *access service network gateway* (ASN-GW). This is all the network elements needed for the access.

Figure 2.12 compares the access networks of the existing cellular wireless network (a) and the Mobile WiMAX network (b) (refer to Figure 2.11 and Figure 1.6). In Figure 2.12(a), the left-hand-side branch is the circuit-mode network and the right-hand side is the augmented packet-mode part network, which consists of SGSN and GGSN in addition to RNC and BTS. We observe that the existing cellular network has a hybrid configuration comprised of multiple network elements in four stacks of access network components, whereas the IP-based Mobile WiMAX network consists of two stacks only. Due to this simplicity of network configuration, the Mobile WiMAX can be applied to many applications economically, and it can support IP services much faster and more efficiently than the existing cellular networks.

CDMA Versus OFDMA

All the existing cellular networks currently in use, both the GSM/UMTS family and the IS-95/cdma2000 family, adopt CDMA as their multiple access technologies. Throughout the past practical services, CDMA proved itself to be a great technology for the voice and other circuit-mode services, as it can provide very reliable and high-quality voice service as well as wide coverage.

However, the effectiveness of the cellular networks does not extend well to IP traffic and IP services, including interactive multimedia services. The reason is that the required data rates in those cases are much higher than those for voice services, for example, in the range of 1 Mbps, 10 Mbps, or even 100 Mbps. As a consequence, the mobile systems need bandwidths that are much wider than is the case of voice services. Under the current cellular network environment, in which the cell radius ranges from a few hundred meters to a few kilometers, CDMA requires modems

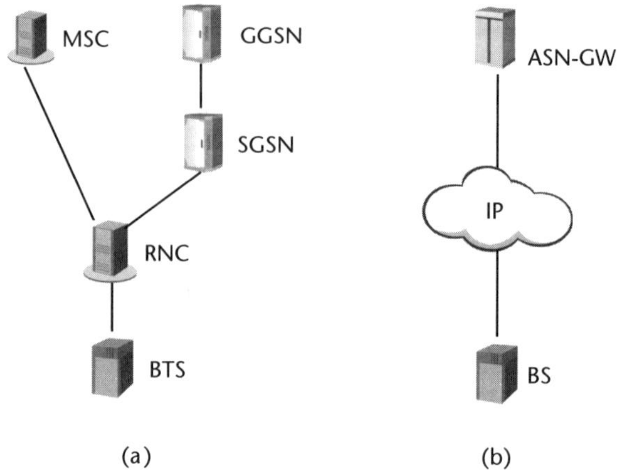

Figure 2.12 Access network architecture of (a) WCDMA-HSDPA and (b) Mobile WiMAX.

whose complexity increase exponentially with respect to the increase of the bandwidth. In contrast, the OFDM-based technology requires wider bandwidth modems with linearly increasing complexity with respect to the bandwidth increase. In addition to wireless broadband transmission capability, furthermore, OFDMA is a variant of OFDM as a multiple access technology, which attempts to fully exploit the multiuser diversity gain in a frequency domain.

Therefore, in serving very high data rate IP services, OFDMA turns out to be the right choice of multiple access technology, and therefore the OFDMA-based Mobile WiMAX renders a much more economic solution to the very high data rate environment. For the same reason, the latest evolution technologies such as LTE and UMB also adopt OFDMA.

CIR Performance

The reason why OFDMA provides better throughput than CDMA does may be explained by comparing the *carrier-to-interference ratio* (CIR) performance. CDMA, as the name indicates, is a code division–based multiple access technique, so there always exist interference among the users even in the same cell, in both uplink and downlink under realistic environmental delay spread. In contrast, OFDMA provides pure orthogonality both in uplink and in downlink, even for reasonable delay spread, and there exists no intracell interferences among the users within the same cell. Due to this orthogonality and the immunity to intracell interferences, OFDMA provides much better CIR performance than CDMA does. This results in higher data rates for the OFDMA-based Mobile WiMAX system, even if processing gain is considered in the CDMA system.

Interference Cancellation

We now consider some more advanced features, such as *interference cancellation* (IC) and *multi-input multi-output* (MIMO) technology. In these aspects, CDMA is very limited, since multiple users share the same resource. In this case the interference originates from the users sharing the same resource, and the interference is spread all over the shared spectrum. However, in the case of the OFDMA, even though there are interference signals, the interference is injected mostly by one or two dominant interferers. So when interference cancellation technology is applied to OFDMA, it is much easier to cancel out the interference than in the case of CDMA. As a consequence, OFDMA ends up with better performance and reasonable complexity for realizing the advanced features such as IC and MIMO.

Connection Setup

Another benefit of Mobile WiMAX lies in its quick connection capability. For the voice service, quick connection is not a fundamentally important issue, because, once it is connected, the session is maintained for a while. In contrast, in the case of the IP traffic and IP services, the traffic has a bursty nature, which means that the connection and disconnection occur more often than in the case of voice traffic. So the connection time becomes an important element for IP traffic and IP services. With the CDMA network, where the resource is shared among all the users, the connection should be established very carefully so as not to cause much interference to other users in the same cell and other cells. In the case of OFDMA networks like Mobile WiMAX, however, the caused interference is limited to certain time and frequency zone in the other cells, while there is no intracell interference, as all resources are orthogonally shared by the users in the same cell. Therefore we may send the first initial setup message with much stronger power, which leads to fast connection time. The dedicated control capability for ranging in the Mobile WiMAX networks also contributes to fast connection time.

2.4.3 Mobile WiMAX to Cellular Mobile Network Interworking

The Mobile WiMAX network will interwork with the cellular mobile networks and other pre-WiMAX wireline and wireless networks such as *public-switched telephone network* (PSTN) and the Internet. Figure 2.13 illustrates the overall network architecture, including the Mobile WiMAX network and the existing pre-WiMAX wireline and wireless networks. Basically, the Mobile WiMAX network is an IP-based network. So it supports free communications with the existing IP-based wireline and wireless networks without any particular requirements. However, the communications with the circuit-based PSTN or cellular mobile networks can only be done via the *media gateway* (MGW) in the *IP multimedia subsystem* (IMS) domain.

When a network operator deploys an access network containing multiple access technologies, it usually segregates the access networt depending on each access technology. However, when deploying a service network, composed of core service network and IMS, the operator usually arranges the service network to be shared by multiple access networks in common. In Figure 2.13, the IMS and the operator's IP service network are not necessarily the required part of the network. Depending on the service policy of the operator, it is also possible to arrange the network such that the Internet connction is made directly at the core router in the access network.

Standardization for interworking among heterogeneous networks is currently under processing in standard committees such as the WiMax Forum, 3GPP, and 3GPP2. The main purpose of interworking is two-fold: One is to guarantee a *dual-band dual-mode* (DBDM) terminal equipped with two or more different access technologies to be provided with continuous session and service after handover between different networks deployed by a common service provider. The other is to provide roaming service among different service providers using the access technologies that are different from each other. One example is the roaming service between a WiMAX-only network and a 3GPP network. In this section, we only deal with the first category—the interworking developed to support handover among heterogeneous networks deployed by a common service provider.

Each access network (3GPP, 3GPP2, WiMAX) has its own session management and authentication procedure. Thus, in order to support session/service continuity after handover among heterogeneous networks, it is necessary to define how to share the required information across the involved networks:

1. *Authentication information* is required for the recognition of the handover terminal's identity in both networks.

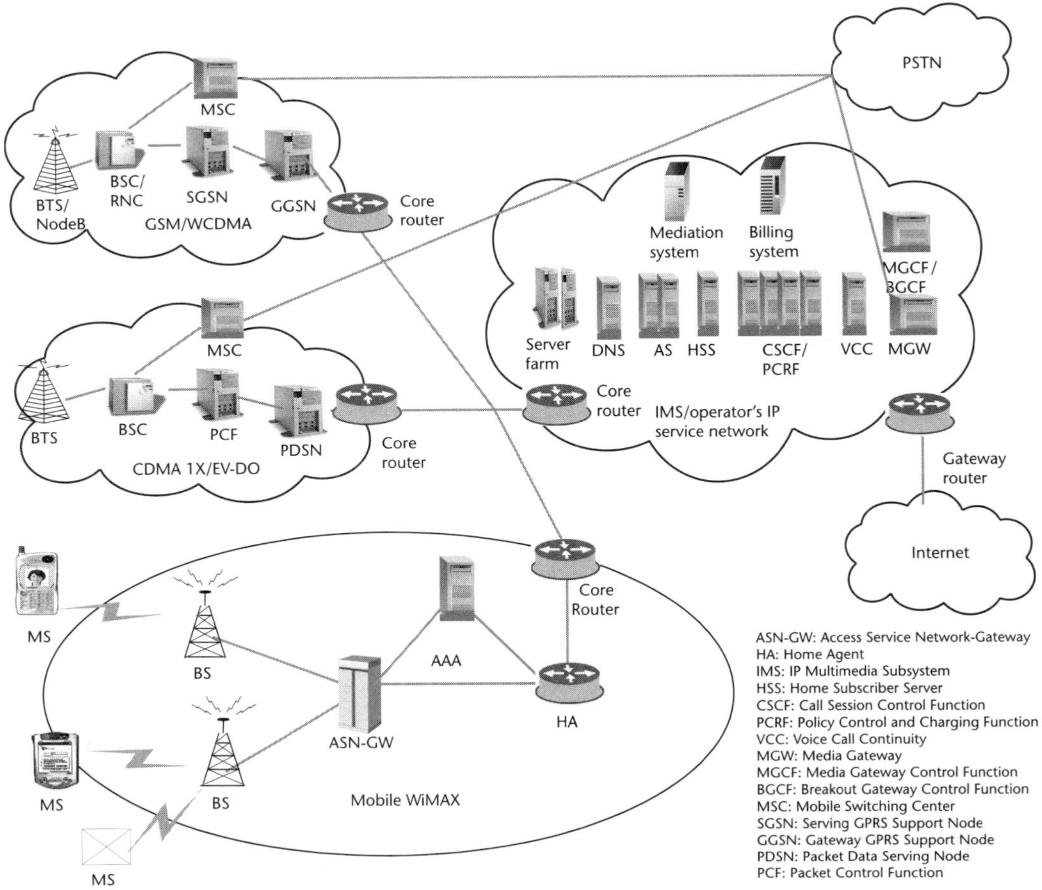

Figure 2.13 Overall network architecture, including Mobile WiMAX and pre-WiMAX wireline and wireless networks.

2. *Session information* is required to support session/service continuity.[6]
3. *Handover optimization* is required for the minimization of handover break time.[7]

In the following, we will first describe how handover is performed among heterogeneous networks deployed by a common service provider and then will briefly examine how interworking can be done between the Mobile WiMAX network and the 3GPP and 3GPP2 networks, respectively.

Handover Among Heterogeneous Networks

In conducting internetwork handover among heterogeneous networks, there are two different mechanisms—the nonoptimized handover and optimized handover. *Nonoptimized handover* refers to the mechanism that begins the handover procedure without any coordination between the handover terminal and the target access network. The terminal performs the initial network entry procedure only after moving into the target access network. The IP address of the terminal is maintained by MIP. In contrast, *optimized handover* refers to the mechanism that exchanges some information required for the access to the target network (i.e., authentication, session parameter, radio parameter) via the radio link of the source access network before moving to the target access network. This information exchange helps to define the tunneling method in addition to MIP. In the tunneling method, the terminal and the target access network obtain the information needed for the session setup via a signaling path or traffic path of the source network.[8] The information helps to simplify the network entry procedure in the target access network, and, as a result, the handover break time reduces.

When performing handover among heterogeneous networks, the traffic path switching is done in the following way: In the case of nonoptimized handover, the traffic path is switched from the source access network to the target network by using MIP. After moving into the target access network, the terminal performs binding update of MIP to the HA, and the HA switches the traffic path from the source access network to the target access network according to the MIP binding update. In this process, the packets that were buffered in the source acces network without being delivered to the terminal are dropped. In the case of the dual-radio terminal, it can minimize packet loss by delivering the traffic data to two access networks simultaneously by using the simultaneous binding of MIP.[9]

In the case of the optimized HA, no specific methods are standardized so far, but the following two methods are possible: One is that the terminal switches the traffic path of the HA to the target access network after moving into the target access network, similar to the nonoptimized handover mechanism. The other is that the termi-

6. IP address is enough as the session information in the best-effort network, but the session information in a QoS network includes various QoS information (i.e., packet classification rule, traffic parameter, scheduling parameter, and so on) in addition to IP address.
7. This is not a required function in a terminal equipped with dual radio.
8. Both paths are considered at this time but only one of them will be selected as the standard method eventually. The method using the signaling path requires an additional definition on the delivery of the tunneling information to MAC signaling layer. The method using the bearer path may minimize the change of MAC layer protocol, but it requires the definition on the delivery of the tunneling information in IP layer in turn.
9. Simultaneous binding has the drawback of consuming more radio resources.

nal stops forwarding traffic from the HA to the source access network before entering the target access network. In this case it is necessary to consider whether or not it is best to arrange it such that the data buffered at the source access network, without being delivered to the terminal, can be delivered to the target network.[10]

It is not necessary to perform the optimized handover on a terminal equipped with dual radio, as it can access both networks simultaneously. In the single radio case, however, a terminal can access only one network at a time. Therefore, unless the single-radio terminal supports the optimized handover, handover break time may increase to several seconds while performing the initial network entry procedure in the target access network after terminating the connection to the source access network.

Interworking with 3GPP

We consider only the nonoptimized handover method, as the standardization of the optimized handover method is currently underway [18]. The legacy 2G/3G packet data network does not support the MIP, while Mobile WiMAX network supports it. So it is impossible to implement seamless handover between the two networks. The 3GPP LTE network, however, is defined to support the MIP; thus, it becomes possible to support seamless handover to/from the Mobile WiMAX network. Figure 2.14 depicts the nonroaming architecture for the nonoptimized handover between the 3GPP network and the Mobile WiMAX network. In the figure, S2c, S5, and S2a are the reference points using MIP defined for the purpose of supporting an MS to access the same PDN gateway and maintain the same IP address after hand-

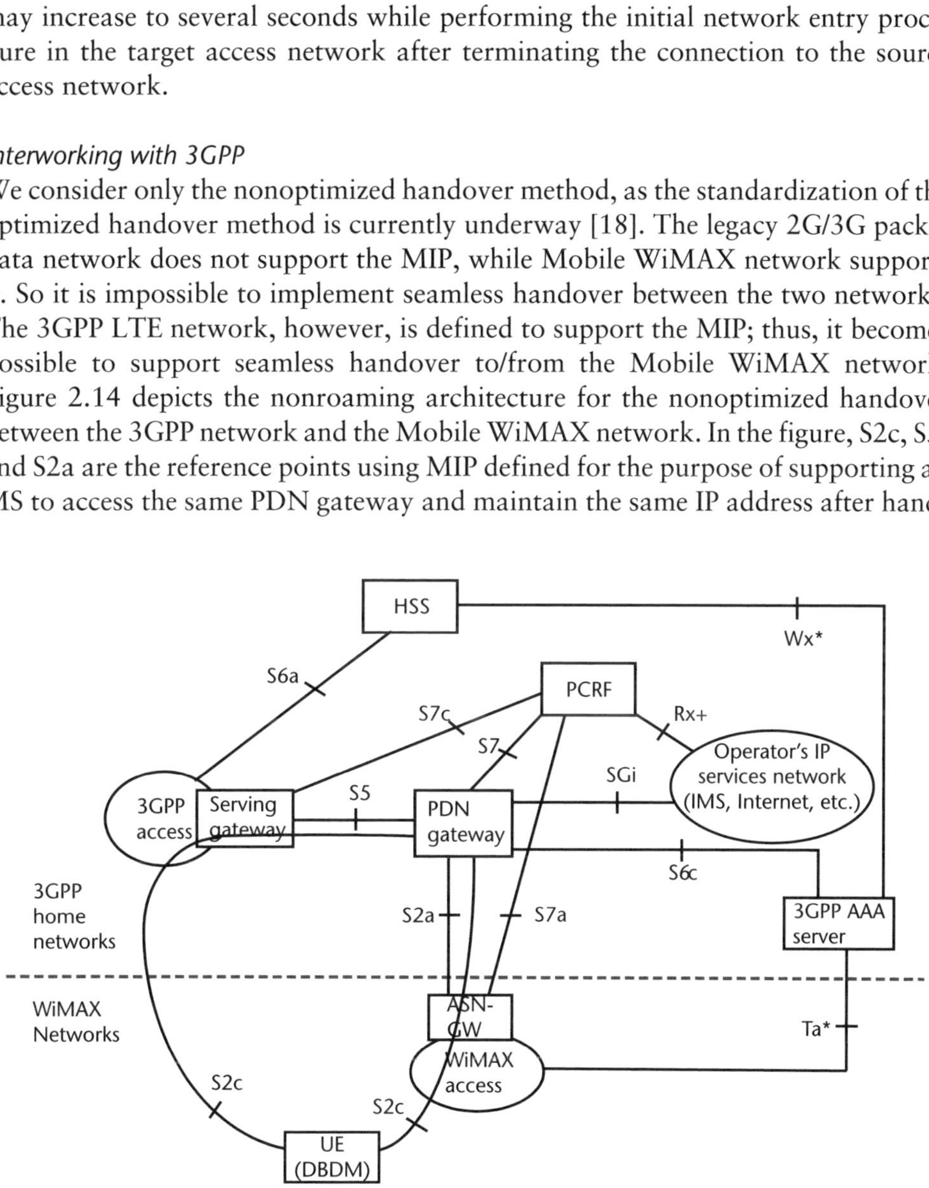

Figure 2.14 Nonroaming architecture for interworking between 3GPP and Mobile WiMAX.

10. It is desirable to minimize the communications among the heterogeneous access networks. In this context, the approach of delivering the buffered data from the source access network directly to the target access network may not be selected in the standard.

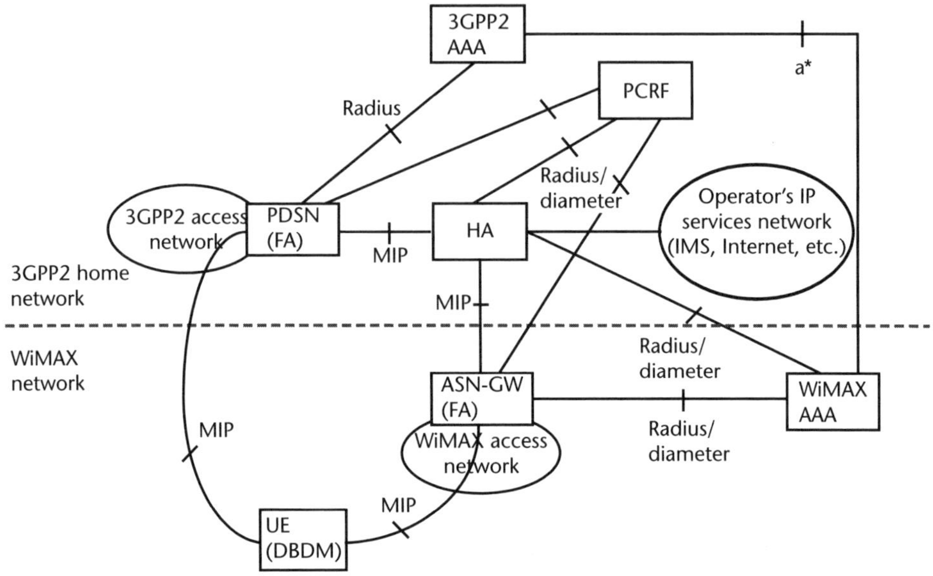

Figure 2.15 Nonroaming architecture for interworking between 3GPP2 and Mobile WiMAX.

over [19–21]. Reference points S7, S7a, and S7c use the DIAMETER protocol. They are associated with a single PCRF and are utilized to provide the two networks with the same session QoS policy information of the MS. Reference points S6a, S6c, and Ta* are utilized to perform the authentication procedures that are defined for each access network. Reference point Wx* is used in sharing the authentication information of MS between the 3GPP AAA server and HSS.

Interworking with 3GPP2

We consider only the nonoptimized handover method, as the standardization of the optimized handover method is currently underway. As both the 3GPP2 packet data network and the Mobile WiMAX network support the MIP, the nonoptimized handover between the two networks may be implemented based on the MIP [22, 23]. Figure 2.15 depicts the nonroaming architecture for nonoptimized handover between the 3GPP2 network and the Mobile WiMAX network. We omit the detailed explanation of this 3GPP2-Mobile WiMAX architecture, as it is similar to that of the 3GPP-Mobile WiMAX architecture in Figure 2.14. Note that MIP or RADIUS/DIAMETER protocols are used in most interfaces connecting different network elements.

References

[1] WiMAX Forum, Mobile Radio Conformance Tests (MRCT), Revision 1.1.0, 2007. For the latest release, refer to http://www.wimaxforum.org.

[2] IEEE Std 802.16e-2005 and IEEE Std 802.16-2004/Cor1-2005, Part 16: Air Interface for Fixed and Mobile Broadband Wireless Access Systems, February 2006.

[3] WiMAX Forum, Network Architecture–Stage 2: Architecture Tenets, Reference Model and Reference Points [Part 1], Release 1.1.0, July 2007.

[4] 3GPP TR 01.01, Digital Cellular Telecommunications System (Phase 2+); GSM Release 1999 Specifications, v. 8.8.0, December 2002.

[5] 3GPP TR 03.64, General Packet Radio Service (GPRS); Overall Description of the GPRS Radio Interface; Stage 2, v. 8.12.0, April 2004.

[6] 3GPP TR 43.051, GSM/EDGE Radio Access Network (GERAN); Overall Description; Stage 2, v. 7.0.0, August 2007.

[7] 3GPP TR 25.201, Physical Layer—Gernal Description, v. 7.4.0, September 2007.

[8] 3GPP TR 25.308, High Speed Downlink Packet Access (HSDPA); Overall Description; Stage 2, v. 7.3.0, June 2007.

[9] 3GPP TR 25.319, Enhanced Uplink; Overall Description; Stage 2, v. 7.2.0, March 2007.

[10] TIA/EIA IS-95, Mobile Station–Base Station Compatibility Standard for Wideband Spread Spectrum Cellular Systems, July 1993.

[11] 3GPP2 C.S0002-B, Physical Layer Standard for CDMA2000 Spread Spectrum Systems—Release B, v. 1.0, April 2002.

[12] 3GPP2 C.S0002-D, Physical Layer Standard for CDMA2000 Spread Spectrum Systems—Revision D, v. 2.0, September 2005.

[13] 3GPP2 C.S0024, CDMA2000 High Rate Packet Data Air Interface Specification, v. 4.0, October 2004.

[14] 3GPP2 C.S0024-A, CDMA2000 High Rate Packet Data Air Interface Specification, v. 3.0, September 2006.

[15] 3GPP2 C.S0024-B, CDMA2000 High Rate Packet Data Air Interface Specification, v. 2.0, May 2006.

[16] 3GPP TS 25.814, Physical Layer Aspects for Evolved Universal Terrestrial Radio Access (UTRA), v. 7.1.0, September 2006.

[17] 3GPP2 C.S0084, Overview for Ultra Mobile Broadband (UMB) Air Interface Specification, v. 2.0, August 2007.

[18] 3GPP TR 23.402, Architecture Enhancements for Non-3GPP Accesses, v. 1.3.0, September 2007.

[19] IETF RFC 3344, Mobility Support for IPv4, August 2004.

[20] IETF RFC 3775, Mobility Support in IPv6, June 2004.

[21] IETF Internet-Draft, IPv4 Support for Proxy Mobile IPv6, draft-ietf-netlmm-pmip6-ipv4-support-00.txt, July 2007.

[22] WiMAX Forum, Network Architecture—Stage 2: Architecture Tenets, Reference Model and Reference Points [3GPP2–WiMAX Interworking], Release 1.1.0, July 2007.

[23] WiMAX Forum, Network Architecture—Stage 2: Architecture Tenets, Reference Model and Reference Points [Annx: WiMAX–3GPP2 Interworking], Release 1.1.0, July 2007.

Selected Bibliography

Bhushan, N., et al., "CDMA2000 1xEV-DO Revision A: A Physical Layer and MAC Layer Overview," *IEEE Communications Magazine*, Vol. 44, No. 2, February 2006, pp. 75–87.

Drewes, C., W. Aicher, and J. Hausner, "The Wireless Art and the Wired Force of Aubscriber Access," *IEEE Communication Magazine*, Vol. 39, No. 5, May 2001, pp. 118–124.

Honkasalo, H., et al., "WCDMA and WLAN for 3G and Beyond," *IEEE Wireless Communications*, Vol. 9, No. 2, April 2002, pp. 14–18.

Knisely, D. N., et al., "Evolution of Wireless Data Services: IS-95 to cdma2000," *IEEE Communications Magazine*, Vol. 36, No. 10, October 1998, pp. 140–149.

Milstein, L. B., "Wideband Code Division Multiple Access," *IEEE Journal on Selected Areas in Communications*, Vol. 18, No. 8, August 2000, pp. 1344–1354.

Nanda, S., K. Balachandran, and S. Kumar, "Adaptation Techniques in Wireless Packet Data Services," *IEEE Communications Magazine*, Vol. 38, No. 1, January 2000, pp. 54–64.

Recommendation ITU-R M.1645, "Framework and Overall Objectives of the Future Development of IMT-2000 and Systems beyond IMT-2000," January 2003.

Sarikaya, B., "Packet Mode in Wireless Networks: Overview of Transition to Third Generation," *IEEE Communications Magazine*, Vol. 38, No. 9, September 2000, pp. 164–172.

Smith, C., and J. Meyer, *3G Wireless with 802.16 and 802.11*, New York: McGraw-Hill, 2004.

Willenegger, S., "CDMA2000 Physical Layer: An Overview," *Journal of Communications and Networks*, Vol. 2, No. 1, March 2000, pp. 5–17.

WiMAX Forum, "Mobile WiMAX—A Comparative Analysis of Mobile WiMAX Deployment Alternatives in the Access," white paper, May 2007.

WiMAX Forum, "Mobile WiMAX—Part II: A Comparative Analysis," white paper, May 2006.

WiMAX Forum, "WiMAX and IMT-2000," white paper, January 2007.

CHAPTER 3

Network Initialization and Maintenance

All mobile wireless systems carry out similar activities at a high level. From a mobile system's perspective, these steps consist of the following:

1. *Network discovery:* Finding a network system using the same radio characteristics with which the MS can try to establish connectivity.
2. *Network initialization:* Establishing some sort of air link and network context so as to enable more connections for transmitting and receiving data.
3. *Connection setup:* Ability to establish connections for transmitting and receiving data.
4. *Handover (or handoff):* Ability to carry out handover between neighboring BSs of the network to enable continuous service.
5. *Nonconnected state:* Ability to support the ability to go into nonconnected mode for power-saving purposes.
6. *Paging:* Ability to be alerted by the network to set up a connection if the network has data to send to the MS.

Note that these aspects are from the perspective of a mobile wireless network. If the system does not support mobility, it will not need to support handover. In addition, the need to support efficient battery usage drives the need for a nonconnected mode and support for paging. In fact, efficient battery usage may be useful even for nonmobile systems.

Although there are differences in the specific methods used to support these abilities among different wireless systems, most systems can be understood within this framework. An easy way to understand the specifics of any cellular wireless system is to examine it from these perspectives.

Example of cdma2000 Network

For example, in a traditional cellular system such as the cdma2000 network, the issues are resolved in the following manner.

- *Network discovery*: All MSs know which radio band class they are assigned to and within that band class the standard has determined a priori specific CDMA channels and PN codes to search for.
- *Network initialization*: Once a cdma2000 1xRTT handset discovers a network, the handset registers with the network by initializing key context for the handset within the network. This usually consists of updating various

location servers and also downloading some key information (such as the allowed services) from its home network down to the visited network.

- *Connection Setup*: cdma2000 handsets use CDMA.
- *Handover*: cdma2000 handsets use the CDMA technology to carry out handover between neighboring BSs. This requires the maintenance of air link connections between the handset and multiple BSs simultaneously. This is enabled in an easy manner by the basic CDMA technology, and it is the reason CDMA technology is so closely tied to the wireless system.
- *Nonconnected state*: All handsets automatically go into an idle state once communication has stopped and only wake up periodically to see if they are paged. The handsets can also quit (or also escape) the nonconnected state if they wish to open a connection.
- *Paging*: cdma2000 handsets are paged by the network based on the unique identifier. Such messages are sent over a broadcast channel that all handsets must listen to.

By tying these key aspects together, one can understand how a cellular wireless system such as cdma2000 operates. In addition, by understanding weaknesses in the approach to these problems and also how a new technology such as OFDMA physical layer affects these problems, one can see what must be invented or modified for new wireless systems such as Mobile WiMAX.

Bearing these aspects in mind, in this chapter we will try to understand the WiMAX 802.16e standard within this structure. The aim is not in giving a detailed description of each aspect, which will be done in the subsequent chapters, but in providing the reader with a basic idea of how the overall system works together to achieve these aims.

Key Mobile WiMAX PHY Characteristics

One key aspect to understand in the following is that at the physical layer Mobile WiMAX is a multiple access point-to-point communication system that relies on a type of joint TDM and FDM data transmission on both forward and reverse links. It is multiple access because multiple systems can use the network. It is point-to-point because at any single time, and via any single frequency subchannel, a BS receives data from only one MS. Due to the characteristics of radio communications, transmission can be a mix of sending messages to a single MS and to multiple MSs simultaneously. But even when sending data to multiple users, it is the same data that is being sent—not multiple streams of data being sent simultaneously. Finally, the specific method for sharing the resources is *orthogonal frequency division multiple access* (OFDMA), which could be considered as a type of joint time and frequency division multiplexing—each MS is given a number of time and frequency slots to transmit or receive data (refer to Figure 4.26).

These aspects drive many of the choices and specific functions in the Mobile WiMAX system, including various aspects of network initialization, such as ranging, scheduling of a packet transmission and reception, and inter-BS handover.

It should be noticed that these choices are not the only possible methods. For example, in certain CDMA networks such as the cdma2000 system, due to the CDMA-based physical layers, it is possible for multiple users to transmit simultaneously and receive data in an asynchronous fashion. Consequently there is no need to try to allocate MSs specific time slots for transmitting/receiving data.

Basic Steps in Network Initialization

The overall procedure for such network entry and initialization may be summarized into the 10 phases listed in Table 3.1. Among them phases e, g, h, and i are optional.[1] We discuss these steps in the following sections.

Note that each MS device has a unique physical identifier and security identifier from the time it is shipped out of the manufacturer: Specifically, the 48-bit universal MAC address is used to identify the MS to the various provisioning servers during initiation, and the security information is used to authenticate the MS to the security server and to authenticate the responses from the security and provisioning servers. This identity is assumed to be tied to the X.509 certificate that every mobile device is supposed to be supplied with so that when authentication is done using the PKM procedures, it can be tied to the mobile device.

3.1 Network Discovery

Network discovery entails how an MS decides what radio frequencies to scan/search through to discover the messages being broadcast by a BS. Once the MS handset has decided what channels it should scan through, it will scan through every channel for a certain period of time to select the frequency channel to use. Then the MS gets its physical time synchronized using the preamble in the DL frame transmitted by the BS. Once the physical time is synchronized, the MS acquires the protocol information needed for initial ranging from the UCD or DCD information

Table 3.1 A Summary of Network Entry and Initialization Procedures [1]

Phase	Process
a	Scan downlink channel, Synchronize with BS
b	Obtain transmit parameters
c	Perform ranging
d	Negotiate basic capabilites
e*	Authorize MS, perform key exchange
f	Perform registration
g*	Establish IP connectivity
h*	Establish time of day
i*	Transfer operational parameters
j	Set up connections

* optional

1. Phase e is performed if both the MS and the BS support authentication policy. Phases g, h, and i are performed only when the MS is a managed station.

message and then accesses the initial random access field from DL-MAP and UL-MAP. We describe these steps in more detail next.

3.1.1 Scanning

The first action an MS takes to get connected to the network is to acquire a downlink channel. For the channel acquisition, it scans all the possible channels in the downlink frequency band until it finds a valid downlink signal. The same processing is needed when acquiring a downlink channel after signal loss. In this case, however, the MS first tries to reconstruct the previous operational parameters stored in the nonvolatile storage that is supposed to be equipped in the MS device. If it cannot reconstruct the parameters, it takes the same action that a newly entering MS does.

3.1.2 Synchronization

As soon as it finds a valid downlink signal, an MS tries to find the frame boundary by locating the downlink preamble, an OFDM symbol that appears at the beginning of each downlink subframe. Immediately after the MS finds the downlink preamble, it identifies its BS by searching for the preamble index of the strongest received signal strength. It is at this point that the MS is said to have acquired initial synchronization in the PHY.

3.1.3 Parameter Acquisition

Once the MS has acquired synchronization in the PHY, then it attempts to acquire the channel control parameters for the downlink and uplink in the MAC. The main aim of these procedures is to acquire the DL-MAP, *downlink channel descriptor* (DCD), UL-MAP, and *uplink channel descriptor* (UCD).

The DL-MAP and DCD are critical parameters to understand, as they tell the MS how the downlink channel can be decoded and which timeslots are being used for which sort of data and/or user. DCD may be regarded as a note from the BS to the MS telling what modulation and encoding schemes are used in the OFDMA frame indicated by the DL-MAP whose DCD count is the same as the value of the configuration change count of the DCD. In that way the MS can demodulate and decode the messages. The DL-MAP describes to the MS many aspects of the system, including basic system parameters. But one of the key functions of the DL-MAP is that it notifies the MS how the slots in the downlink are allocated—specifically, to which MS or application (broadcast/multicast, for example) the data in the slots are to be sent.

The UL-MAP and UCD serve similar purposes as the DL-MAP and DCD but for the uplink channel. The UCD notifies the MS of the modulation/encoding to use for the uplink when the MS is to transmit. Similarly, the UL-MAP tells the MS which slots it may use when it attempts to transmit data. The UL-MAP indicates which slots are dedicated to individual MSs that have connections allocated and which slots are open to contention usage by the MSs that do not have connections allocated. This is an important role, as the contention-based slots are used for activities such as ranging by the MS.

Before the MS can decide to transmit, it must therefore understand the downlink messages first. Consequently it always looks for the DL-MAP and DCD before it looks for the UL-MAP and UCD. It must have both before it can attempt *any* transmission of data.

Downlink Parameters

Once the MS has acquired a downlink channel, it then searches for the DL-MAP MAC management message from the channel. If the message is detected at least once, then MAC synchronization is achieved, so the MS can decode the DL-burst profiles contained in the downlink frame structure to determine the downlink channel parameters. An MS MAC is considered in synchronization if it keeps receiving the DL-MAP and DCD messages successfully. If a valid DL-MAP message or a valid DCD message is not received until due time limit, the MS is considered out of synchronization, so it has to establish synchronization again.[2]

Uplink Parameters

After acquiring synchronization in the MAC, the MS scans through the uplink channels transmitted by the BS to detect an UCD message. Once the UCD message is detected, it retrieves a set of transmission parameters for possible uplink channels. From the parameters, the MS can determine whether or not it may use the uplink channel. If the channel turns out suitable, the MS then extracts the uplink channel parameters of that selected channel from the UCD. Otherwise, the MS continues scanning to find another suitable uplink channel. After getting the uplink parameters, the MS waits for a bandwidth allocation map for that selected channel and then begins transmitting on the uplink in accordance with the MAC operation and the bandwidth allocation mechanism.

3.2 Network Initialization

Once network discovery is done and all the key parameters for the initial transmission and reception of data with the network are known, the MS tries to proceed with network initialization to establish a context in the network so that it can open and close connections for transmission of user and application data streams. The basic steps are shown in Figure 3.1. As indicated in the figure, the network initialization consists of the following eight steps: (1) initial ranging (refer to Figure 3.2 for the details of the initial ranging procedure), (2) basic capabilities negotiation, (3) authentication, (4) security association, (5) key exchange, (6) registration, (7) transport connection setup, and (8) IP address allocation.[3] Note that the interaction with PCRF is not yet standardized and may be skipped if only non-QoS connection setup follows. Each step is explained in more detail in the next sections. In Figure 3.1, the

2. The due time limit is the lost DL-MAP interval for a valid DL-MAP message and the T1 interval for a valid DCD message. Refer to Table 342 in [1] for the details of parameter T1.
3. Note that at least one set of uplink and downlink transport connections should be established for the delivery of the following DHCP messages, which are just data packets to the BS and the ASN-GW. (Since the ASN-GW is the first hop router from the DCHP client (i.e., the MS), it works as a DHCP relay agent.)

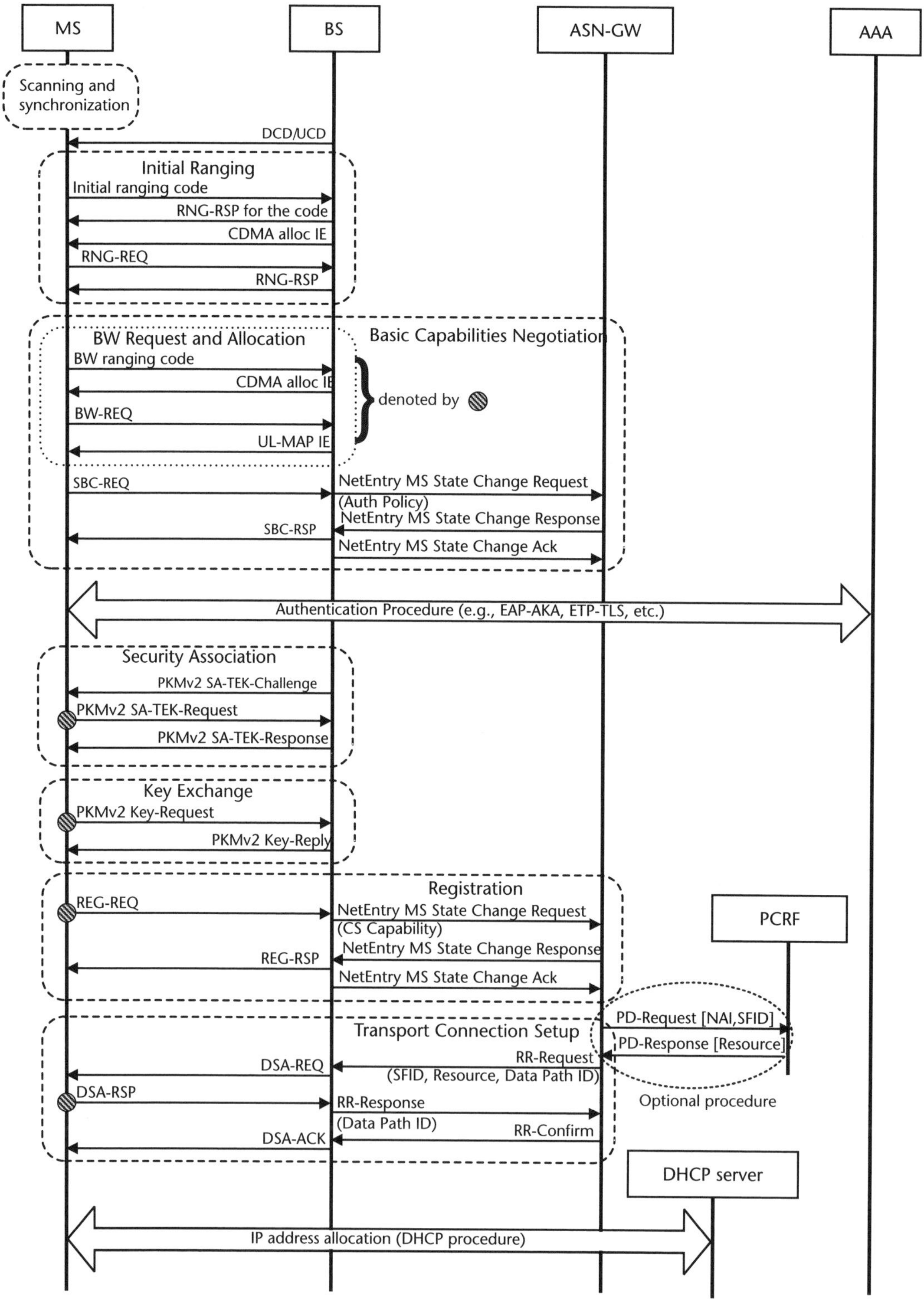

Figure 3.1 Network initialization call flows.

message arrow that originates from the dashed circle denotes that the corresponding message transmission follows the BW request and allocation procedure.

A high-level summary of the actions that occur in these steps is as follows.

After completing a series of acquisition processes, the MS transmits a ranging request message or ranging code to the BS through the initial random access channel. In the initial stage, the MS uses a minimum level of power for transmission and gradually increases the power. Ranging, or the process that adjusts the timing and power such that the frames of all terminals get synchronized at the BS, is necessary because the difference in the terminal-to-BS distance among multiple terminals in the same cell causes mismatch in frame synchronization.

In response to the received ranging request message or ranging code, the BS sends out a ranging response message. It then commands to control the reception power and timing synchronization of the MS, and allocates a basic *channel identification* (CID) and a primary management CID to the MS. The MS then performs association, authorization, and encryption processes using the allocated primary management CID. During the association process, it gets allocated with a secondary CID from the BS.

After the session is established, the BS and the MS conduct the ranging request and response mechanism periodically in order to proactively react to the rapid change of wireless link that may happen during the data transmission and reception. This process is called *periodic ranging*. In addition, if abrupt changes occur in channel environment, they exchange the physical layer parameters of the burst to *dynamic burst profile change* (DBPC) request and response messages. On the other hand, in case service traffic attributes of the connection change, they exchange *dynamic service change* (DSC) messages so that the relevant changes in the QoS requirements can be duly reflected in the appropriate parameters.

3.2.1 Initial Ranging

Initial ranging is the process of acquiring the correct timing offset and power adjustments of the MS such that the signal of the MS is aligned in timing and strength with the transceiver of the BS. Once the ranging process is complete, the timing delay through the PHY is maintained within the tolerance range, which can be accommodated within the guard time of the uplink PHY overhead. Aside from the initial ranging for the network entry and initialization stage, there is another type of ranging, called *periodic ranging*, which maintains the quality of the wireless communication link between the MS and the BS while in service (see Section 3.7.2).

Initial Ranging Procedure

The procedure of contention-based initial ranging is as follows.

We consider the MS side first. While the MS gets synchronized to the BS and learns the uplink channel characteristics through the UCD MAC management message, it scans the UL-MAP message to randomly choose a ranging slot using a binary truncated exponential algorithm to avoid possible collisions while performing the ranging. Then the MS randomly chooses a ranging code from the initial ranging domain and sends it to the BS as a CDMA ranging code in the case of the OFDMA PHY. The power level used in sending the CDMA code should be below the maximum transmission strength calculated through a power control algorithm. (Refer to Section 6.3.9.5.1 of [1] for the details of the power control algorithm.) If the MS does not receive a response from the BS until the due timeout, it sends a new CDMA ranging code at the next available initial ranging transmission opportunity with an adjusted power level.[4] If it receives a *ranging response* (RNG-RSP) message from the BS that contains the CDMA ranging code it has transmitted and a *continue* status, it makes the power level and timing offset correction specified in the RNG-RSP and continues the ranging process as done on the first entry with a ranging code randomly chosen from the initial ranging domain sent on the periodic ranging region. If it receives an UL-MAP containing a CDMA_Allocation_IE and the parameters of the code it has transmitted, then the RNG-RSP reception is successful. Initial ranging process is completed after receiving the RNG-RSP message, which includes a valid basic CID.

Now we consider the BS side. When the BS receives a CDMA ranging code, it cannot tell which MS has sent the CDMA ranging request. So upon receiving a CDMA ranging code, it broadcasts a ranging response message that advertises the received ranging code as well as the ranging slot (OFDMA symbol number, subchannel, and so on) where the CDMA ranging code has been identified. The ranging response message contains all the parameters to adjust, such as time, power, and possibly frequency corrections, and a status notification. When the BS receives an initial-ranging CDMA code that complies with the required adjustments, it sends an RNG-RSP message with *success* status and provides bandwidth allocation for the MS to send a *ranging request* (RNG-REQ) message. The ranging request/response step is repeated between the MS and the BS until fine tuning is completed.

Figure 3.2 shows the initial ranging procedure: First, the BS sends a UL-MAP containing the ranging channel information (UIUC=12) with the broadcast CID. After an MS acquires the UL parameters from the UL-MAP, it transmits a randomly selected initial ranging code in a randomly selected ranging slot from the initial ranging region. If the code has failed to be detected at BS so the timer T3 (which is the timeout required for MS to wait for RNG-RSP) expires, the MS randomly chooses a ranging slot (after truncated binary exponential backoff) and transmits an initial ranging code in the initial ranging region. If the code is detected outside the tolerable limits of the BS, the BS sends RNG-RSP message with timing and power corrections and ranging code attributes with the status of *continue*. Then the MS adjusts timing and power accordingly and transmits a randomly selected initial ranging code in a periodic ranging region. If the code is detected inside the tolerable limits of the BS, the BS sends an RNG-RSP message with timing and power corrections and ranging code attributes with the status of success. Then, the BS sends CDMA_allocation_IE in the UL-MAP with the broadcast CID, so that bandwidth allocation for the RNG-REQ message is performed. The MS transmits an RNG-REQ message with its own MAC address. Finally, the BS sends an RNG-RSP message with basic CID, primary CID, and MAC address, so that the basic and the primary CID allocation for the MS is performed.

Ranging Parameter Adjustment

In support of the initial ranging operation described here, ranging parameters are adjusted in the following manner: All parameters are adjusted to stay within the

4. The timeout for ranging response reception is called T3, and its typical value is 200 ms.

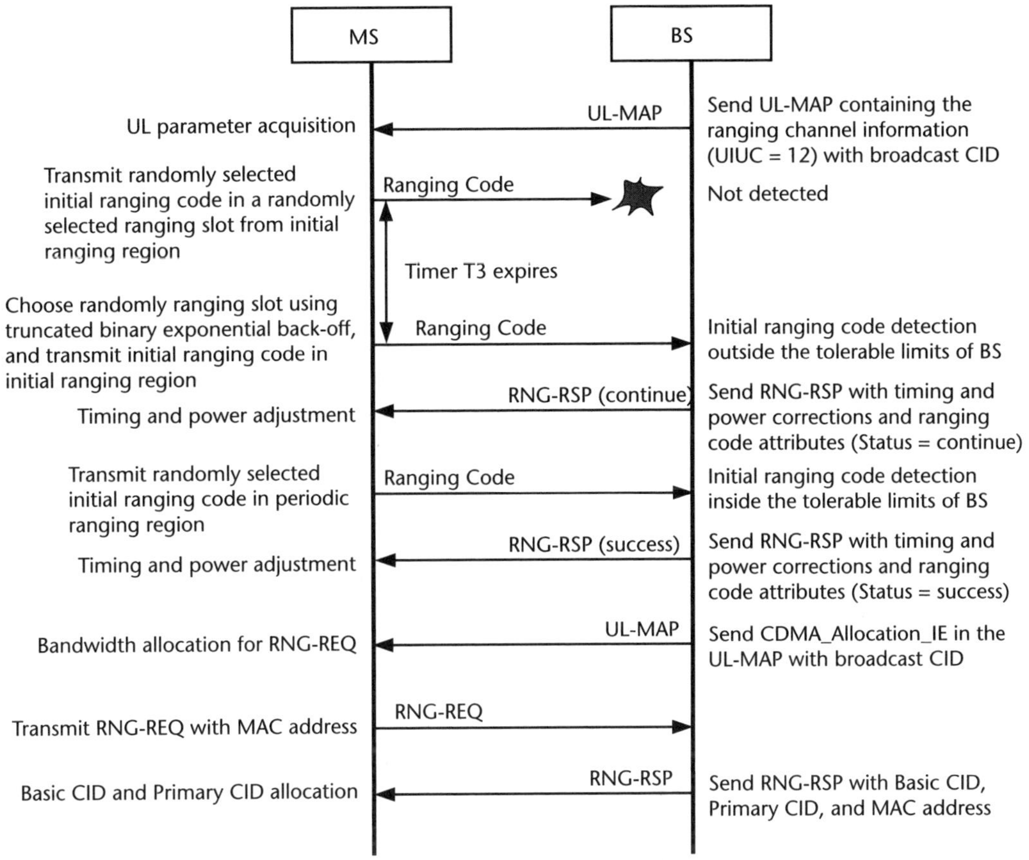

Figure 3.2 Initial ranging procedure.

approved range at all times. Power adjustment starts from the initial minimum value selected with the aforementioned algorithm unless a valid power setting is readily made in nonvolatile storage. Power adjustment is done, in increment or decrement, by the amount specified in the RNG-RSP message. If the power increases to the maximum value during the initialization procedure, the BS has to pull it down to the minimum. The MS has to adjust the RF signal in accordance with the RNG-RSP message and check that it is stabilized before making transmissions.

Contention Resolution

Collisions may occur during initial ranging (and request) intervals. In order to resolve the contention problem, the Mobile WiMAX system uses a contention resolution method that is based on a truncated binary exponential backoff, with the initial and maximum backoff windows controlled by the BS. The relevant parameter values are specified as part of the UCD message and represent a power-of-two value. If an MS wants to send information, it enters the contention process, sets its interval backoff window equal to the ranging (or request) backoff start defined in the UCD message, and randomly selects a number within the backoff window. This selected random value indicates the number of contention transmission opportunities that the MS has to defer before transmitting. If the MS receives an RNG-RSP message (or a data grant burst type IE) after the contention transmission, the contention resolution is completed. If no such message is received within due timeout time [i.e., T3 (or the contention-based reservation timeout)], the MS may regard the contention transmission as lost. So it increases the backoff window by a factor of two, unless it does not exceed the maximum backoff window, and repeats the deferring process described earlier.

3.2.2 Basic Capabilities Negotiation

Immediately after completing the ranging procedure, negotiation of basic capabilities follows. The capabilities negotiated at this point are based on the capabilities supported by the MS, such as the physical parameters and bandwidth allocation support. The *subscriber-station basic capabilities request* (SBC-REQ) contains information on the capabilities of the MS, while the *subscriber-station basic capabilities response* (SBC-RSP) contains information on what subset of those capabilities the BS can support.

Specifically the MS informs the BS of its basic capabilities by transmitting an SBC-REQ message with its capabilities set to *on*. Then the BS returns an SBC-RSP message with the intersection of the MS's and BS's capabilities set to on.

3.2.3 Authorization and Key Exchange

For the cases where the *privacy key management* (PKM) function is enabled, the authorization and key exchange between the MS and BS take place as described later. For more details, refer to Section 8.3.

MS authorization is the process of the BS authenticating the identity of client MS. The BS and MS first establish a shared *authorization key* (AK) by RSA, from which a *key encryption key* (KEK) and message authentication keys are derived. Then the BS provides the authenticated MS with the identities [i.e., *security association identifiers* (SAIDs)] and properties of the primary and static *security associations* (SAs).

To begin with, the MS sends to the BS an authentication information message, containing the MS manufacture's X.509 certificate. Right after that, it sends an authorization request message to the BS to request an AK and the SAIDs, which identifies any static security SAs that the MS is authorized to participate in. The authorization request message includes the manufacturer-issued X.509 certificate, a description of the cryptographic algorithms that the requesting MS supports, and the MS's basic CID. The basic CID is the first static CID that was assigned by the BS during the initial ranging stage.

In reply to this authorization request, the BS takes the following actions: it validates the requesting MS's identity, determines the encryption algorithm and protocol support, activates an AK for the MS, encrypts it with the MS's public key, and then sends it back to the MS in the authorization reply message. The authorization reply message includes an AK encrypted with the MS's public key, a 4-bit key sequence number to use to distinguish between successive generations of AKs, a key lifetime, identities (i.e., SAIDs), and the properties of the authorized SAs.

While in service, the MS periodically refreshes the AK by reissuing an authorization request to the BS. The procedure for reauthorization is identical to that for authorization except that the MS does not send authentication information massages. To avoid service interruptions during reauthorization, the BS and MS both are given the capability to support up to two simultaneously active AKs during the transition period.

3.2.4 Registration

Once the basic capabilities are negotiated successfully or MS authorization and key exchange process is completed, the MS registers to the network. In the case of managed MSs,[5] there follows additional stages, such as establishment of IP connectivity, establishment of time of day, and transfer of operational parameters, before setting up connections finally.

Registration

Registration refers to the process where an MS gains entry to the network and a managed MS gets a secondary management CID and thus becomes manageable. In order to register with a BS, the MS sends a REG-REQ message to the BS. Then the BS responds with a REG-RSP message. In the case of the managed MS, the REG-RSP message includes the secondary management CID.

If the MS chooses to support IPv6 on the secondary management connection, it should indicate it in the REG-REQ. If no specific indication is made, the BS interprets it as IPv4 support. On detecting a specific IP version number on the REG-REQ, the BS includes the IP version parameter in the REG-RSP to use on the secondary management connection. If the IP version number is omitted in the REG-RSP, it means that only IPv4 should be used.

Establishing IP Connectivity

An MS can establish IP connectivity using the secondary management connection either through mobile IP or DHCP. If the MS uses mobile IP, it may secure its address on the secondary management connection using mobile IP. Otherwise, it invokes DHCP mechanisms to get an IP address and other parameters needed to establish IP connectivity. If the MS has a configuration file, the DHCP response should contain the name of a file that contains further configuration parameters. If the MS uses IPv6, it invokes either DHCPv6 or IPv6 stateless address auto-configuration.

Establishing Time of Day

Acquirement of the time of day is not mandatory for registration, but the current date and time is required for time-stamping logged events, though the accuracy is not stringent. The time of day can be retrieved using the protocol described in IETF RFC 868, with the request and response messages transferred using UDP. The current local time can be determined by combining the time of day retrieved from the server with the time offset received from the DHCP response. The time of day is established using the MS's secondary management connection, as was the case for establishing IP connectivity.

5. Managed MS refers to the user terminals that the operator can manage using SNMP or other means.

Transferring Operational Parameters

The MS downloads the MS configuration file using *trivial file transfer protocol* (TFTP) on its secondary management connection, if it is specified in the DHCP response. When the downloading is completed, the MS notifies the BS by transmitting a *TFTP complete* (TFTP_CPLT) message on its primary management connection. The BS continues the transmission periodically until it receives a *TFTP response* (TFTP-RSP) message with OK response from the MS or it terminates retransmission due to retry exhaustion.

3.2.5 Establishing Connections

Once the transfer of operational parameters is completed for managed MSs or registration is completed for unmanaged MSs, the BS sends *dynamic service addition request* (DSA-REQ) messages to the MS to set up connections for the provisioned service flows. Then the MS sends back *dynamic service addition response* (DSA-RSP) messages in response. Unmanaged MSs that do not have the secondary management connection can now establish IP connectivity using the transport connection either through mobile IP or DHCP.

3.3 Connection Setup

Once the network initialization is done the MS should be able to set up connections. This step actually consists of two resource allocations—the allocation of connections with *connection IDs* (CIDs) and service flows with *service flow IDs* (SFIDs). A service flow is a unidirectional flow of packets that is provided with a particular QoS. A service flow is identified by the SFID. An active service flow is also tied with a CID.

Note the concept here of a nonconnected state. A service flow may be defined for a user and be maintained for many hours. But when the user does not have any active connections (i.e., an active WiMAX air link connection), a CID will not be assigned to that service flow. So the MS and the network will be aware of the SFID and the relevant packet flows that are tied to it but no connection (and hence no CID) will be tied to it. In essence, this is the nonconnected state (refer to Section 3.4 for further discussions).

Once this is done, there is one more key aspect that must be addressed, which is the allocation of bandwidth to each service flow. Conceptually the IDs are used only to identify flows and connections, but not for QoS purposes. That is, IDs do not indicate how much bandwidth has actually been allocated to the user.

3.3.1 Basic Connection Setup

Once the context for the MS is set up in the Mobile WiMAX network, the setting up of connections is relatively straightforward. Specifically, it is done by the exchange of DSx-REQ/DSx-RSP messages as follows:

- DSA-REQ—*Dynamic service addition request* to add an SF;

- DSA-RSP—*Dynamic service addition response* to add an SF;
- DSA-ACK—*Dynamic service addition acknowledgment* to acknowledge the addition;
- DSC-REQ—*Dynamic service change request* to change an SF's parameters;
- DSC-RSP—*Dynamic service change response* to change an SF's parameters;
- DSC-ACK—*Dynamic service change acknowledgment* to acknowledge the change;
- DSD-REQ—*Dynamic service delete request* to delete an SF;
- DSD-RSP—*Dynamic service delete response* to delete an SF.

Creation of a service flow may be initiated by an MS or by an *access service network* (ASN). This is done through the exchange of the DSA-REQ/DSA-RSP/DSA-ACK messages. For provisioned service flows, the BS will initiate it after the MS registration.

Specifically, the BS assigns a new CID in the DSA-RSP/REQ messages when adding a new service flow and in the DSC-REQ/RSP messages when activating an existing service flow. The CID identifies the MAC packets assigned to a user in the various MAC protocols. For example, when time slots for packet transmission are assigned to users, this is indicated in the DL-MAP (for downlink packets) and in the UL-MAP (for uplink packets). The CID is unique per user per BS. Whenever the MS moves its connection (i.e., does a handover to another BS) the CID must be reallocated by the new BS to which the MS moves. The CID is also used for bandwidth requests.

The SFID is the identifier used by network to deal with service flows, and is used in subsequent DSx signaling. As with the CID, the network assigns the SFID. When an SF is no longer admitted or active, its CID may be reassigned by the network. Note that on handover, the SFID is maintained because SFID should be unique across multiple BSs. The SFID differs from the CID in this aspect.

Finally, a critical aspect of the service allocation is that, in addition, per service flow security associations (SAs) may also be setup. This is based on the security association done during the network initialization phase.

Figure 3.3 shows the connection setup procedure initiated by the MS and the bandwidth allocation status. The procedure in the figure consists of three steps in the viewpoint of bandwidth allocation status: DSA, DSC, and DSD. The first step of DSA is to create a connection and reserve the bandwidth. After a connection is created with *admitted* status, bandwidth is reserved so that *connection admission control* (CAC) excludes such amount from the available bandwidth. From the scheduling point of view, however, all the bandwidth resource can be used by the existing *active* connections. The second step of DSC is to activate the bandwidth allocation. The connection becomes active and the scheduler starts to allocate bandwidth to the connection. The third step of DSD is to release the bandwidth.

3.3.2 QoS and Bandwidth Allocation

Once a connection or service flow is set up the user/application should be able to transmit and receive packets. But one question remains on how much the user is

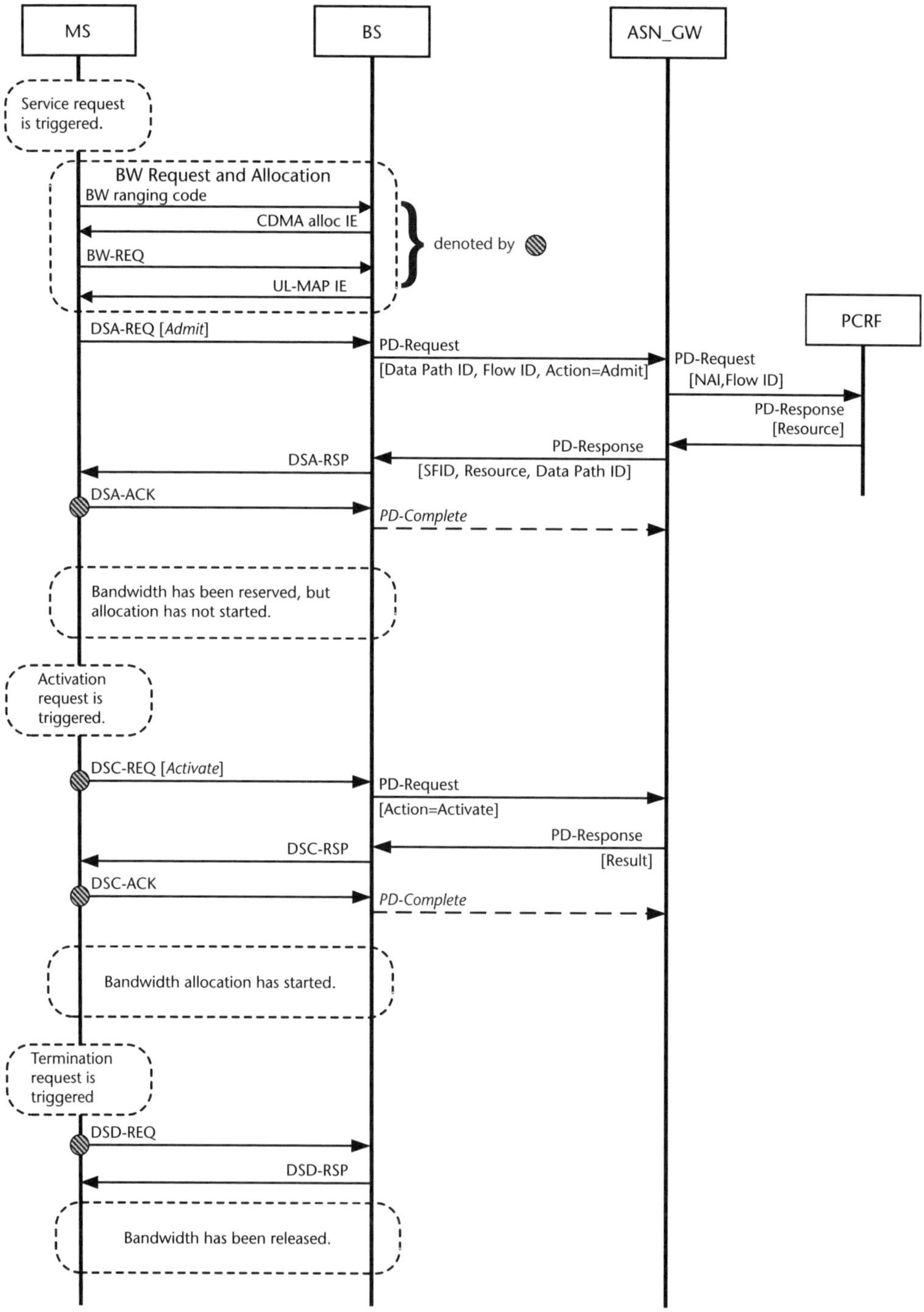

Figure 3.3 Connection setup procedure and bandwidth allocation status.

able to send or receive—what bandwidth is allowed to the user? A related question is the QoS applied to the user packets.

In Mobile WiMAX, the bandwidth that is tied to connections and service flows is not fixed. Instead, it may be adjusted based on the QoS requirements and the needs of the user/application.

Mobile WiMAX enables this by the use of requests that can be standalone requests or piggybacked on top of other messages. These may be done based on the QoS class of the user in autonomous ways or by polling by the network. Once requests are received by the network, it sends out grants to the MS, which indicate how much bandwidth it may consume. In addition, based on this information, the network sets the DL-MAP and UL-MAPs accordingly. This is needed due to the TDM nature of the OFDMA physical layer.

Multiple QoS classes are available in Mobile WiMAX. This subject is dealt with in more detail in Chapter 6.

3.4 Nonconnected State

A critical concept in all wireless systems is the nonconnected state. In this state, differently from the connected operational state, the MS will usually not have an air link resource assigned to it. One of the biggest changes that have occurred in Mobile WiMAX is the introduction and extension of this conception. There are two main aims that support such two-fold state arrangement.

The first aim is not to use too much of the potentially constrained air link resources. For example, in CDMA systems, the biggest constraint was the number of Walsh codes available. In the Mobile WiMAX system, this is less of a constraint given the TDM nature of the system and its use of long IDs, both for connections and service flows. However, due to implementation constraints, many systems will still try to reuse CIDs and SFIDs as much as possible.

The second aim is to conserve the battery life of mobile systems. By making the MS not have to keep air link resources such as modems operational when it has no specific data to send or receive, it is possible to significantly conserve battery life.

In Mobile WiMAX, there are two nonconnected state–related concepts: one is sleep mode and other is idle mode. The main difference is that in the sleep mode the MS maintains a relationship with a specific BS, whereas in the idle mode the MS essentially deregisters from the network and does not have a relationship with any specific BS.

3.4.1 Sleep Mode

In *sleep mode*, the MS and BS have basically agreed that the MS will be unreachable for periodic amounts of time. Essentially, it enables the MS to conserve battery life by turning off its modem during these stretches of time. There are three types of power-saving classes that the MS can enter into (refer to Section 7.3.1). Sleep mode can be initiated by the BS or the MS depending on the requirements. The MS will then wake up periodically.

Another use of the sleep mode may be for the MS to indicate that it will not be available for communication with its current BS because it is trying to communicate with another BS. This may be used when the MS is trying to handover to another BS and is trying to scan for the BS during which time it will be unavailable on the current BS.

3.4.2 Idle Mode

In *idle mode*, the MS essentially deregisters from the network. This is opposite to the registration stage discussed in the network initialization. In this state the MS essentially deregisters and only periodically wakes up to check whether there are any pages for it. (See Sections 3.5 and 7.3.2 for more information on paging.) In essence, the MS does not try to maintain a tight link with any specific BS, so can be accessed only by paging.

During this stage all BSs, including the last one that served the MS, will essentially forget about the MS. The network will keep information regarding the idle MS in a paging controller or some other centralized entity. The network will keep track of the "rough" location of the MS by waiting for location updates from the MS. As the MS moves through the network, it sends location updates into the network so that it may be paged.

One interesting aspect of the concept of a nonconnected state and its consequences was discussed in Section 3.3, as a part of the discussion on service flows and connections. There, we discussed that when an MS is in the idle state, it essentially loses any CID but maintains its SFIDs with the network.

3.5 Paging

As mentioned earlier, once the MS enters the idle mode, it can only be reached by the network through paging. Paging is the concept of sending out a message on a general broadcast channel that all MSs are expected to be listening to. General broadcast messages indicating the general system information are usually sent over such broadcast channels, but, in addition, MS specific messages are also sent. Usually these MS-specific messages are known as pages.

Pages are sent using the broadcast paging message by the paging controller over a large area based on the last reported location of the MS. This ties us to the concept of paging groups. When an MS is in the idle mode, it is essentially deregistered from the network and not tied to any specific BS. Consequently when the network wishes to send some message to the MS it does not have a specific BS through which it can reach the MS. While presumably the MS is reachable through some BSs, the network does not know specifically which it is.

This is where the concept of paging groups and location updates come in. The MS is assumed to send periodic location updates based on where it is. Paging groups are the groups of BSs that can be used to send page messages simultaneously. So whenever an MS moves into a BS that is a member of a different paging group from the last BS it was at, the MS must send a location update. This location update information is sent to the paging controller to keep track of the location of the MS. The paging mechanism after this is dependent somewhat on the implementation. One possible way of doing it is the following: When a packet for the user arrives at the *access service network gateway* (ASN-GW) or the last known BS, it will query the paging controller to page the MS through the last known paging group. The paging controller will page for the MS through the last paging group as indicated by the last location update of the MS. When the MS responds, a connection will be set up on the BS through which the MS responded, and the user data will be delivered.

Figure 3.4 illustrates the paging operartion. An MS first goes into idle mode through deregistration procedure. After the MS moves into a new BS in different paging group, it detects paging group change from the broadcast paging advertisement message. It then performs a location update. (Refer to Section 7.3.2 for more discussions on paging and paging groups.)

3.6 Mobility

The general concept of mobility in wireless systems relates to the fact that MSs can move between different wireless BSs. This can be done while the MS has a connection and also when it does not have a connection. When an MS has a connection, it is important that the MS is able to connect with the new target BS as quickly as possible to minimize any disruption of traffic that is flowing to or from the MS. When the MS does not have a connection, the aim is to maintain reachability to the MS so that it can be paged by the network when the network receives traffic for it.

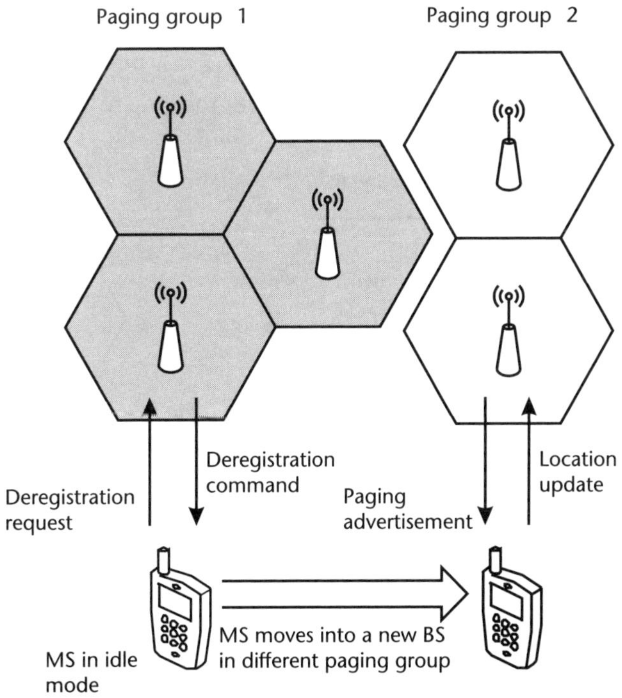

Figure 3.4 Illustration of paging operation.

3.6.1 Nonconnected-State Mobility

Nonconnected-state mobility is related to the question of what happens when an MS moves between BSs while in sleep mode or idle mode.

When the MS is in sleep mode it must register again with its new BS. So it is in essence very similar to the connected state mobility.

When the MS is in idle mode, it does not have to do anything unless the MS sees from the broadcast messages of the new BS that the paging group has changed. If it sees that it is beyond the boundaries of its old paging group, the MS will send up a location update after connecting to the new BS. This will result in the location information of the MS being sent to the paging controller in the core of the network that will keep track of it. Note that location update may be done for other reasons, such as timers on the MS.

3.6.2 Connected-State Mobility—Handover

When an MS is in the connected state, it is in essence in constant direct communication with a specific BS. The BS knows of the air link context (i.e., the information configured as part of the network initialization process in Section 3.2) and also specific air link resources (e.g., the transmit time offset from ranging, CID, SFID, transmit power) allocated to the MS.

Connected-state mobility, or *handover*, means moving from the current BS to a new BS. This essentially means recreating this information on the new BS so that the MS may maintain connectivity.

Handover is very complex subject, so here we will make an overview of how the overall handover structure works and why it is made that way. More discussions on handover may be found in Sections 7.1 and 7.2.

Figure 3.5 illustrates the handover operation. It shows that an MS communicates with the serving BS to determine the target BS. After moving into the cell of the target BS, the MS then performs handover ranging to get connected to the target BS.

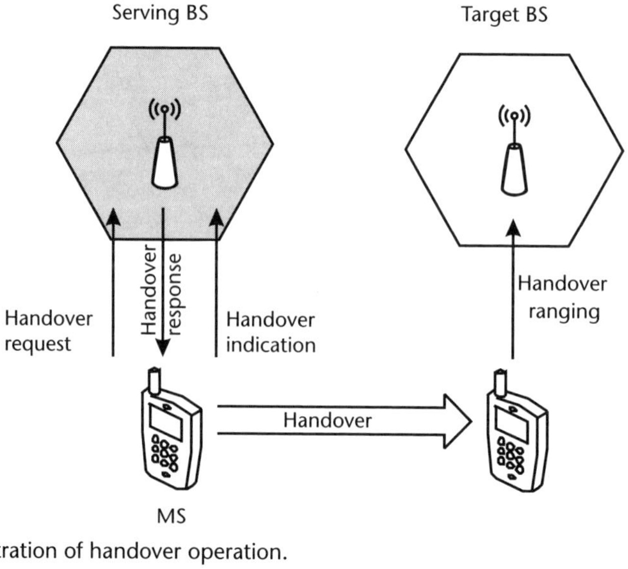

Figure 3.5 Illustration of handover operation.

Discovering Other BSs

The first step in carrying out handover is for the MS to discover candidate "target BS" to which it can do handover. The current "serving BS" helps this process by sending out *neighbor advertisements* in its broadcast messages. Essentially this tells the MS what BSs are nearby and roughly what their main characteristics are from a radio point of view. The MS can then initiate scanning for these BSs. During this time, the MS indicates to the BS that it cannot receive any data. Essentially it is entering a sleep mode–like state.

Preparing for Handover

Once the MS is aware of a neighboring BS, it can prepare for handover by setting up associations with the neighboring BS. Associations can be of three levels. The base level, or level 0, is when the MS tries to scan for the neighboring BS without any real coordination. In levels 1 and 2, the MS is able to do scanning with the target BS with preformed coordination. Effectively this relies on the MS using the serving BS to communicate with the target BS before trying to scan for the signals of the target BS. Specifically, one method is to get from the target BS the preallocated times for the MS to switch over to the target BS. By getting such preallocated times, the MS is able to time exactly when it should switch its radio to the new target BS and thereby gains fast network entry (i.e., downlink synchronization and ranging).

Handover Initiation

The actual decision to do handover is done either by the BS or the MS. Once the decision is made, the MS moves to the target BS and performs network entry. If preassociations have been done, the network entry into the new target BS can be done very efficiently.

Initial Entry upon Handover

At this stage, the MS aims to achieve downlink synchronization with the BS and acquires the important downlink and uplink parameters of the BS as quickly as possible. Only upon achieving downlink synchronization and getting these parameters can the MS start transmitting and receiving data with the BS. If the MS were to set up some sort of preassociation with the BS before physically moving to the target BS, then this process of initial entry after handover could be done much more efficiently. Once initial entry is done, the MS does a new registration with the BS and network.

As part of the handover preparation, it is possible that much of the previous network context for the MS may have been moved from the serving BS to the target BS. If not done at this stage, such information may be moved over as part of the handover process. This is an aspect that is discussed in much detail in the WiMAX Forum NWG specifications [2].

3.7 Maintenance

Once the network is initialized and reaches normal operation state, maintenance processes are needed to ensure that the MS is able to connect to the network when it wishes to transmit/receive data and also to maintain the quality of a connection when it is connected. Among the maintenance mechanisms, the most representative ones are synchronization, periodic ranging, and power control.

3.7.1 Synchronization

Synchronization in the aspect of maintenance is the process in which an MS maintains its timing and frequency such that it can track the timing and frequency of its BS. By using downlink signals, the MS measures and corrects its timing offset and frequency offset.

These offsets in timing and frequency in the mobile systems are caused partly by the fact that its local oscillator frequency cannot be exactly identical to that of the corresponding BS. In addition, these offsets are caused by the mobility of the MS. Timing offset occurs because of the variations in multipath environment as well as the changes in the distance and hence the propagation delay between the BS and the MS, while frequency offset can be a result of the Doppler frequency shift.

The accuracy and stability of the downlink synchronization is crucial to the receiver performance. Downlink channel estimation and channel decoding performance may be adversely affected by these offsets in timing and frequency, especially in the OFDM communication systems. It should be noted that the downlink synchronization not only affects the downlink receiver performance but also the uplink receiver performance, since it serves as the basis for the uplink synchronization.

3.7.2 Periodic Ranging

Ranging, in general, is the process of resolving the problems caused by the unknown spatial separation of the MS and the BS, which could be timing mismatch and the power mismatch. As noted in the beginning of this chapter, this is critical for the Mobile WiMAX system due to the TDM nature of its physical layer. MSs may transmit to the BS from many different locations, resulting in very different propagation times for the data transmitted by different MSs. By using ranging, the different propagation times are taken into account and each MS is delayed individually so that when the data from each MS arrives at the BS, the boundaries fall on the exact time slots allocated to the user. Simply speaking, ranging helps to adjust the MS's timing offset in such a way that it appears, after the offset, as if it were collocated with the BS.[6]

Along with ensuring that received packets fall within their allocated time slot boundaries, one must ensure that the power levels are also adjusted to be within the same level when the packets arrive at the BS. This is also achieved as part of the ranging procedure.

Ranging consists of two procedures—initial ranging and periodic ranging. Initial ranging is intended to allow the MS entering the network to acquire correct transmission parameters, such as time offset and transmission power level, so that the MS can start communicating with the BS as if it were collocated with the BS. On the other hand, periodic ranging is intended to allow the MS in service to adjust

6. This is a well-known problem in multiple access systems that use TDM on the reverse link. Examples include various *passive optical network* (PON) systems, cable modem systems, and wireless systems such as GSM.

transmission parameters so that the MS can maintain uplink communications with the BS. In the following, we deal only with periodic ranging to complement the discussions of initial ranging given in Section 3.2.1.

The procedure of periodic ranging is as follows: An MS that wants to perform periodic ranging randomly chooses a ranging slot using a binary truncated exponent algorithm to avoid possible recollisions at the time of performing the ranging. Then it randomly chooses a periodic ranging code and sends it to the BS as a CDMA code in the case of OFDMA PHY. If the MS does not receive a response, it sends a new CDMA code at the next appropriate periodic ranging transmission opportunity and adjusts its power level up to the allowed maximum level. As the BS does not recognize which MS has sent the CDMA ranging request, it broadcasts a ranging response message that advertises the received periodic ranging code as well as the ranging slot where the CDMA periodic ranging code has been identified. The ranging response message contains all the needed adjustment (e.g., time, power, and possibly frequency corrections) and a status notification. Examining the information, the MS identifies that the ranging response message corresponds to its ranging request. If the received ranging response message indicates a *continue* status, the MS continues the ranging process with further periodic ranging codes randomly chosen. The BS may send an unsolicited RNG-RSP in response to a CDMA-based bandwidth request or any other data transmission from the MS. Note that it is the MS, not the BS, that controls the periodic ranging timer when using the OFDMA ranging mechanism.

In case an MS wants to perform handover ranging, it takes a process similar to that for the initial ranging with the following modification. In the case of OFDMA PHY, the BS uses the CDMA handover ranging code selected from the handover-ranging domain, instead of the initial ranging code. Alternatively, the BS, when prenotified of the upcoming MS handover, provides bandwidth allocation information to the MS using Fast_Ranging_IE to send an RNG-REQ message.

3.7.3 Power Control

Power control in cellular or mobile systems usually refers to the transmit power control of MSs, and the Mobile WiMAX system is not an exception. Its main objective is to make the received power at the BS from multiple MSs located in different distances and channel conditions become similar to one another while being kept at an optimum level for the receiver performance as well as for the battery lifetime of the MSs.

Power control is categorized into two different but possibly coexisting modes: closed loop and open loop. *Closed-loop* power control means that there is a closed loop in the power control mechanism between the transmitter and the receiver. From the transmitter to the receiver is the transmitted signal itself and from the receiver to the transmitter is some kind of feedback or command that controls the transmit power level. On the contrary, *open-loop* power control does not require any feedback information from the receiver in determining the proper transmit power level so that the control mechanism does not form a closed loop. Instead of relying on the power control commands from the BS, an MS calculates its transmit power based on the assumption that the path losses of downlink and uplink are the same, which is considered valid in wireless TDD systems.

Power control is applied in every kind of uplink transmissions (i.e., ranging codes, channel quality information feedbacks, data or management messages, and so on). The required signal quality of these signals at the BS receiver is different from one another, and it depends on the modulation scheme and coding rate used in the case of data or management message transmission. This has to be taken into account when determining the transmit power level regardless of the power control mode.

References

[1] IEEE Std 802.16e-2005 and IEEE Std 802.16-2004/Cor1-2005, "Part 16: Air Interface for Fixed and Mobile Broadband Wireless Access Systems," February 2006.

[2] WiMAX Forum, Network Architecture—Stage 3: Detailed Protocols and Procedures, Release 1.1.0, July 2007. For the latest release, refer to http://www.wimaxforum.org.

Selected Bibliography

Fu X., Y. Li, and H. Minn, "A New Ranging Method for OFDMA Systems," *IEEE Trans. on Wireless Communications*, Vol. 6, No. 2, February 2007, pp. 659–669.

Nuaymi, L., *WiMAX Technology for Broadband Wireless Access*, New York: Wiley, 2007.

Morelli, M., and U. Mengali, "An Improved Frequency Offset Estimator for OFDM Applications," *IEEE Communications Letters*, Vol. 3, No. 3, March 1999, pp. 106–109.

Morelli, M., C.-C. J. Kuo, and M.-O. Pun, "Synchronization Techniques for Orthogonal Frequency Division Multiple Access (OFDMA): A Tutorial Review," *Proc. of the IEEE*, Vol. 95, No. 7, July 2007, pp. 1394–1427.

Schmidl, T. M., and D. C. Cox, "Robust Frequency and Timing Synchronization for OFDM," *IEEE Trans. on Communications*, Vol. 45, No. 12, December 1997, pp. 1613–1621.

CHAPTER 4

OFDMA PHY Framework

As discussed earlier in Section 2.2.4, IEEE 802.16 WiMAX system deals with four different types of physical layers that are specified for operation in different frequency bands, based on different multiple access technologies. They are WirelessMAN-SC, WirelessMAN-SCa, WirelessMAN-OFDM, and WirelessMAN-OFDMA. In addition, there is another physical layer, WirelessHUMAN PHY, which is specified for the license-exempt bands. In particular, WirelessMAN-OFDM PHY and WirelessMAN-OFDMA PHY are designed for operation below the 11-GHz licensed band, based on the OFDM and the OFDMA technologies, respectively. Among the four different types of physical layer technologies, mobile WiMAX adopts WirelessMAN-OFDMA PHY only. So the discussions in this chapter are all concentrated on this OFDMA system.

OFDMA PHY distinguishes itself from the physical layer of the existing CDMA- or TDMA-based wireless communication systems in that it adopts OFDMA technology for the multiple access and transmission of communication signals. So in discussing the mobile WiMAX system, it is very important to make a detailed description about the OFDMA communication processing. The OFDMA communication processing is composed of channel coding, modulation, subcarrier mapping, *discrete Fourier transform* (DFT), and other relevant signal processings (refer to Figure 4.1). *Hybrid ARQ* (HARQ) technique is additionally employed to strengthen the error recovery capability of the channel coding. The OFDMA PHY also includes some additional processings, such as OFDMA frame structuring and subchannelization. Besides, multiple antenna technology such as *adaptive antenna system* (AAS) and *multi-input multi-output* (MIMO) system has become an important PHY component for the enhancement of channel efficiency.

Among all the OFDMA PHY components listed here, we discuss the OFDMA communication processing in the first two sections—overall OFDMA communication signal processing in one section and a detailed discussion on channel coding and HARQ in another. Then we deal with the OFDMA frame structuring and the subchannelization in the subsequent two sections. As to the discussion of the multiple antenna technology, we allocate a full chapter, Chapter 9, in consideration of its differentiated and optional characteristics.

4.1 OFDMA Communication Signal Processing

The OFDMA-based communication basically divides the available frequency spectrum into a large number of subcarriers and transmits the input data by mapping it into the subcarriers. So the communication signal processing involved in the

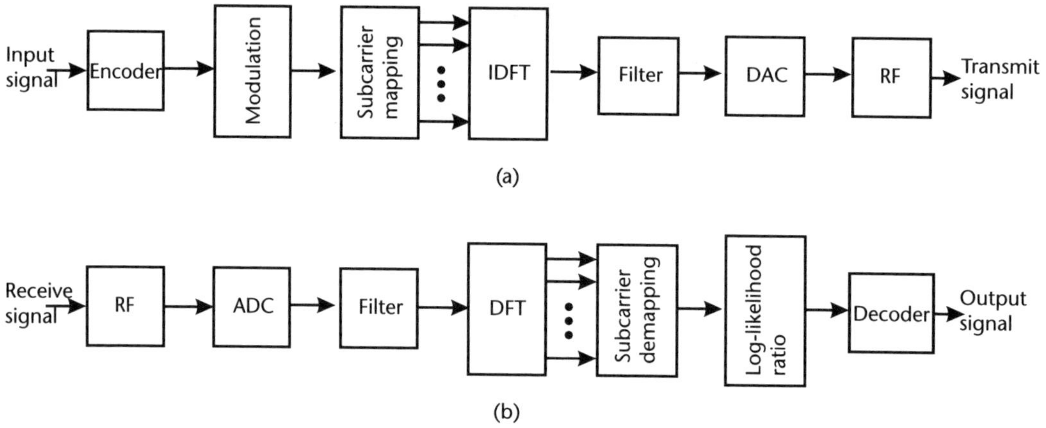

Figure 4.1 OFDMA communication processing: (a) transmitter, and (b) receiver.

OFDMA system takes the procedure shown in Figure 4.1. In the transmit direction, the input digital signal, or MAC data, is first encoded and modulated, then mapped to multiple OFDMA subcarriers. The resulting parallel signals are then inverse-transformed via IDFT and filtered and converted to analog signal for final transmission in RF frequency. Conversely, in the receive direction, the received RF signal is sampled to digital signal and filtered, before passing through the DFT process. The transformed parallel streams, which were allocated to multiple subcarriers, are demapped, demodulated, and finally decoded to yield the output signal.

4.1.1 Encoding and Modulation

As the first signal processing of the OFDMA communication processing in the transmit direction, MAC data undergo the channel coding and modulation process. This includes randomization, *forward error correction* (FEC) encoding, bit-interleaving, repetition, and modulation, as depicted in Figure 4.2. Among those components, FEC and bit-interleaving processes in the dashed box will be discussed in Section 4.2, so we deal only with the other components. Note that repetition is applied only to the case of QPSK modulation.

Randomizing/Scrambling

MAC data input injected to the physical layer may include a long string of the same bits (0s or 1s) or a repetition of an identical bit pattern. So the data stream is randomized first by employing a *shift register generator* (SRG), or a *pseudo-random binary sequence* (PRBS) generator, which is composed of shift registers and *exclusive-OR* (XOR) gates. The SRG has the characteristic polynomial $1 + x^{14} + x^{15}$, which generates a pseudo-random bit stream of period $2^{15}-1$ (see Figure 4.3). The

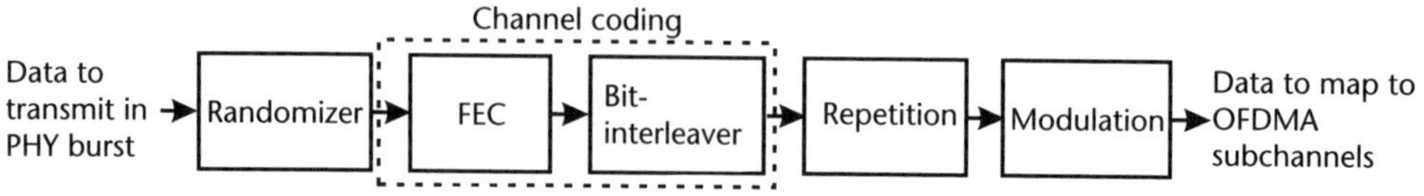

Figure 4.2 Channel coding and modulation process for OFDMA transmission. (*After:* [1, 2].)

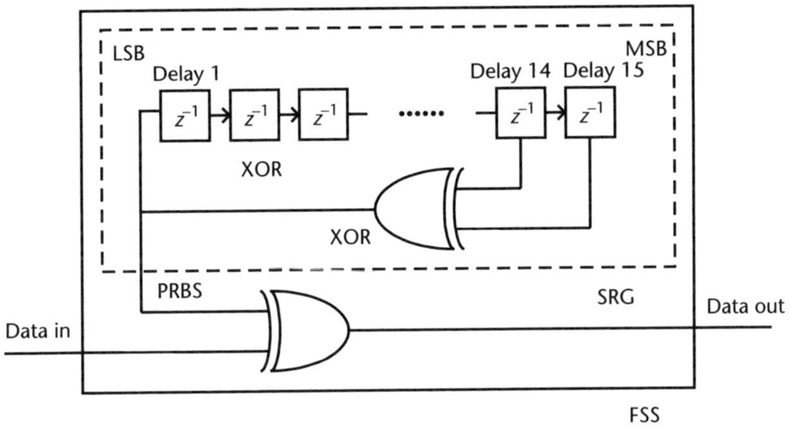

Figure 4.3 PRBS generator for data randomization. (*After:* [1, 2].)

pseudo-random bit stream is then combined with the MAC data stream through an XOR gate to yield a randomized MAC data stream. The randomization process is done in a frame synchronous fashion—the randomizer is a *frame-synchronous scrambler* (FSS). The SRG is reset to a predefined initial state at the start of each *forward error correction* (FEC) block. The predefined initial state is [LSB] 011011100010101 [MSB]. If the amount of data to transmit does not fit exactly the amount of data allocated or the number of slots allocated for the data burst, N_s, it is padded by the data stream 0×FF (1 only) up to the end of the transmission block.

FEC and Bit-Interleaving

Refer to the discussions in Sections 4.2.1 and 4.2.2.

Repetition

Repetition coding is the process of repeating the identical slot of bits by R times to further increase the signal margin over what is obtained by applying only FEC and modulation processes. R is called the repetition factor. The data obtained after passing through the FEC and bit-interleaving processes is segmented into multiple slots, and each group of bits designated to fit in a slot is repeated R times, thereby forming R contiguous slots with the bit ordering following the normal slot ordering used for data mapping. So when repetition coding with repetition factor R is applied, the number of binary data that fills in a data burst reduces by a factor of R. In the case $R = 2, 4$, or 6, the number of allocated slots, N_s, is a multiple of R for the uplink, and is a number in the range $[RK, RK+(R-1)]$ for the downlink, where K is the number of the required slots before applying repetition. For example, when $K = 10$ and $R = 6$ for a burst transmission, N_s, or the number of allocated slots for the burst, may be a number between 60 and 65 slots. This repetition scheme applies only to QPSK modulation.

Modulation

After the repetition process, the modulation process follows, as shown in Figure 4.2. Data bits that have gone through the repetition process are entered serially to the constellation mapper for the modulation process. Figure 4.4 shows the constella-

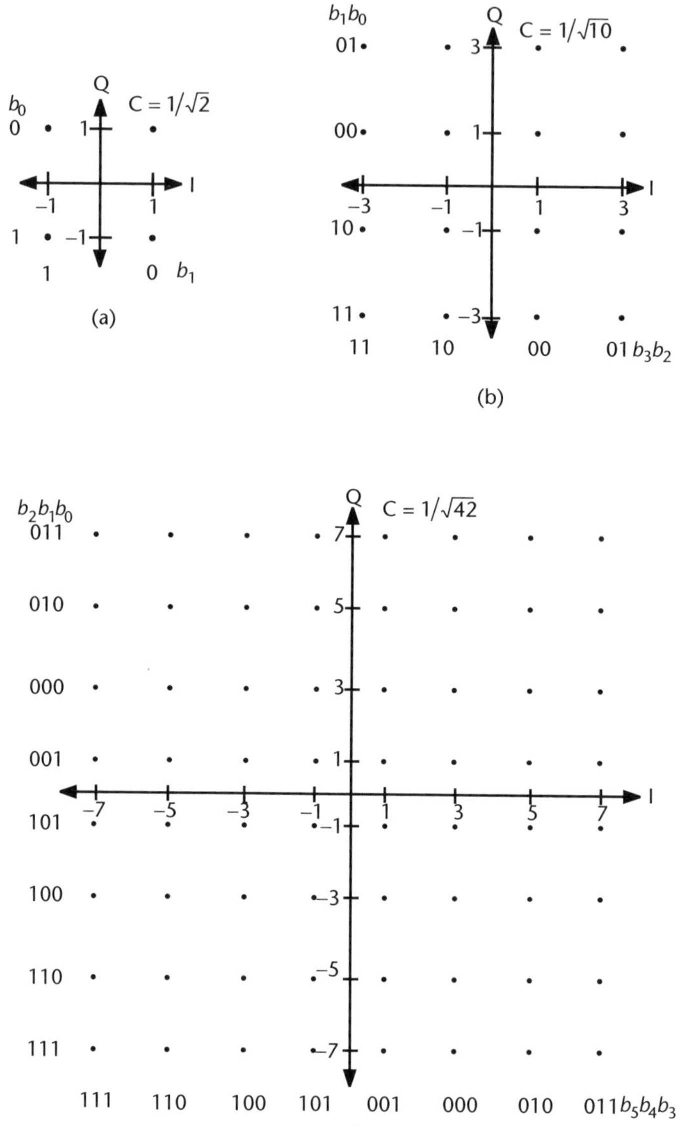

Figure 4.4 Constellations for modulation: (a) QPSK, (b) 16-QAM, and (c) 64-QAM. (*After:* [1, 2].)

tions for the three modulation schemes: (gray-mapped) QPSK, 16-QAM, and 64-QAM, respectively. In order to achieve equal average power, the three constellations are normalized by a factor c, which is $1/\sqrt{2}$, $1/\sqrt{10}$, and $1/\sqrt{42}$, respectively for the three modulation schemes.

Modulation applies differently to data and pilot. In the case of data, per-allocation adaptive modulation and coding is performed in the downlink, with different modulation and coding schemes applied depending on the channel state of the subcarriers that are allocated to carry the data.[1] In the uplink, different modulation schemes are applied for each MS based on the MAC burst configuration messages

1. In the case of pilot, modulation is done differently depending on the OFDMA slot structure. Refer to [1, Section 8.4.9.4.3], for a detailed description.

coming from the BS. All interleaved and repeated bits are mapped to the constellation bits $b_{M-1} - b_0$ in the MSB first order, where $M = 2, 4$, and 6 for QPSK, 16-QAM, and 64-QAM, respectively, and the constellation-mapped data is subsequently modulated onto the allocated data subcarriers. Before mapping the data to the physical subcarriers, each subcarrier is multiplied by the factor $2 * (1/2 - w_k)$ according to the subcarrier physical index, k, where w_k is the sequence generated by the PRBS generator whose generator polynomial is $1 + x^9 + x^{11}$. (Refer to [1, Section 8.4.9.4.1], for more details.)

4.1.2 Subcarrier Mapping and Transform

According to Figure 4.1, once the encoding and modulation processing is done, the output data is mapped into OFDMA subcarriers and then put through the IDFT processing. This means that the original input data is regarded as frequency domain signal, or frequency component, in the OFDMA-based communications. Consequently, the data generated after the IDFT processing—or the *inverse fast Fourier transform* (IFFT) processing—becomes a time domain signal.

Figure 4.5 depicts the block diagram for the subcarrier mapping and IDFT transform processing. The encoded and modulated signals are mapped into OFDMA subcarriers in the unit of OFDMA slot, which is a two-dimensional data region formed by the OFDMA symbols and subchannels (or a group of subcarriers).

We focus the discussions on the subcarrier mapping and the IDFT signal processing in this subsection, deferring all the discussions related to OFDMA frame structuring to Section 4.3.

Subcarriers Mapping

In the OFDMA systems, the encoded and modulated data is mapped into multiple frequency components called *subcarriers*. Subcarriers are classified into *data subcarriers* for carrying user data, *pilot subcarriers* for channel estimation and other functions, and *null subcarriers* for guard bands and DC carriers, which are not transmitted. The frequency band available for the OFDMA communications is divided into N_{FFT} subcarriers,[2] among which N_{used} subcarriers are actually "used" for carrying data and pilot tones, with the remainder allocated for guard bands.

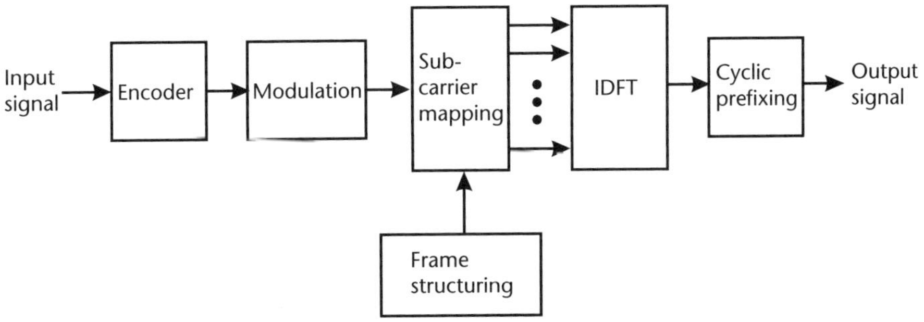

Figure 4.5 Block diagram of subcarrier mapping and IDFT processing.

2. We use the term FFT in the number N_{FFT}, as the number of subcarriers is determined in relation to the subsequent IDFT processing, for which the well-known *fast Fourier transform* (FFT) technique is used.

Those "used" subcarriers, excluding the null subcarriers, are the *active* subcarriers in the OFDMA systems.

As each subcarrier is confined by the time duration of an OFDMA symbol, T_b, in the time domain, it takes the form of a *sinc* function in the frequency domain, as shown in Figure 4.6. To be more specific, the inverse Fourier transform of the time duration pulse function $u(t) - u(t - T_b)$ is $\text{sinc}(\pi(f - k\Delta f)T_b)$, for the unit step function $u(t)$, the subcarrier spacing Δf $(=1/T_b)$, and the subcarrier number k. The *sinc* function has a long tail, so it can cause interference to the neighboring frequency band. In order to mitigate this interference, a guard band is appended at the both ends of the data and pilot subcarrier band, as shown in the figure. The guard bands are composed of null subcarriers which make the transmit signal naturally decay, thereby mitigating the interferences to the neighboring channels.

Subchannel Grouping

The active subcarriers are grouped in multiple to form a *subchannel*, as illustrated in Figure 4.7. The figure illustrates three subchannels formed by grouping two subcarriers each. It also shows the guard bands at both ends and the DC subcarrier in the middle. As is the case in the figure, the subcarriers selected for forming a subchannel are not necessarily adjacently located. They are selected according to a predetermined pattern that considers the effects of frequency diversity and other factors. The concept of subchannel can be effectively used in allocating subcarriers to the users in the neighboring cells or neighboring sectors. This enables multiple users to share an OFDM channel by taking advantage of the orthogonality in the frequency domain, as is the case of the FDMA.

In practice, different types of subchannels are formed depending on the method of grouping of the subcarriers. One type is the *distributed permutation*–based subchannel in which subcarriers are grouped by taking permutation over a wide range of subcarriers, thereby taking the effects of diversity. The other type is the *adjacent permutation*–based subchannel in which the subcarriers that are physically adjacent are grouped together. The former is further divided into the *partial usage*

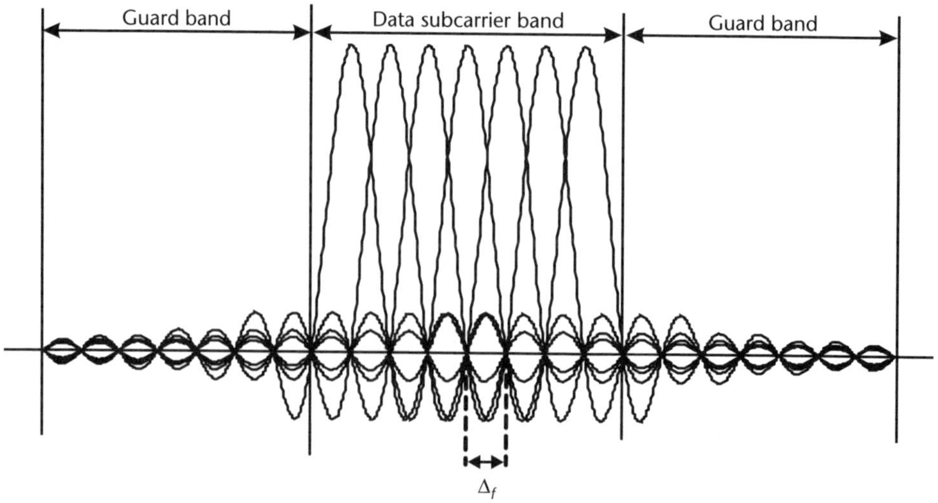

Figure 4.6 Illustration of guard bands in OFDMA.

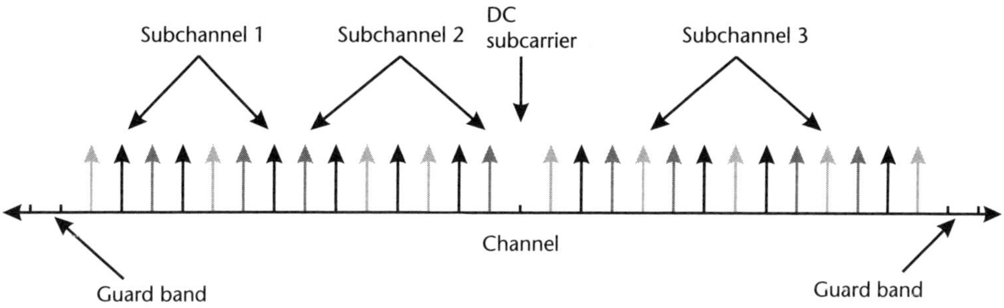

Figure 4.7 Illustration of subchannels in OFDMA. (*After:* [1, 2].)

subchannel (PUSC) and *full usage subchannel* (FUSC). In the case of the PUSC, subchannels are designed to be partially allocated to multiple transmitters (i.e., into three sectors), but a single transmitter operation is also possible. In contrast, in the case of FUSC, all subchannels are fully allocated to one transmitter. The latter type of subchannel (i.e., adjacent permutation–based subchannel) is called the *band AMC subchannel.* It enhances the band efficiency by allocating a group of bands in good channel state to the users, thereby taking advantage of the AMC effect (Refer to Section 4.4 for detailed discussions on the subcarrier allocation).

The number and position of the data and pilot subcarriers differ for PUSC, FUSC, and band AMC subchannels. Table 4.1 compares the subcarrier grouping to guard bands, data, pilot tones, and others for *downlink* (DL) and *uplink* (UL) PUSC, DL FUSC, and DL/UL band AMC subchannels in the case of the 1,024-FFT OFDMA system. (Refer to Section 4.4 for a more detailed description on subchannels.[3])

IDFT Processing

Once the data mapping process is completed, each subcarrier is allocated with an encoded and modulated data. Then for each full set of subcarriers there follows the IDFT (or IFFT) transform process, which converts the N_{FFT} OFDMA subcarriers, or the frequency domain signal, into the time-domain counterpart.[4] For the IDFT conversion, the two-dimemsional data space, which consists of the horizontal OFDMA symbol axis and the vertical subchannel (or subcarrier) axis, is processed columnwise. That is, each OFDMA symbol of length N_{FFT}, $X(k)$, $k = 0, 1, \ldots, N_{FFT} - 1$, is converted into the time-domain signal $x(n)$, $n = 0, 1, \ldots, N_{FFT} - 1$. The N_{FFT} subcarriers includes the N_{used} "used" subcarriers that carry data and pilot tones and guard subcarriers, whose values are specified in Table 4.1.

3. In the case of UL PUSC, "number of pilot subcarriers = 420/0" and "number of data subcarriers = 420/840" indicate that in case the OFDMA symbol carries pilot subcarriers, 420 subcarriers are allocated to pilot, with the other 420 subcarriers allocated to data; and in case the OFDMA symbol does not carry pilot subcarriers, whole 840 subcarriers are allocated to data. (Refer to Figure 4.40.) In addition, in the case of DL/UL AMC, "number of subchannels = 48" and "number of data subcarriers in each symbol per subchannel = 16" assume that type 2 × 3 AMC subchannel is used—that is, each OFDMA slot is composed of 2 bins and 3 OFDMA symbols (refer to Table 4.10 and Figure 4.42).
4. *Discrete Fourier transform* (DFT) is a generic term of a transform, and *fast Fourier transform* (FFT) refers to a special method of calculating DFT. A similar relation holds between IDFT and IFFT. As such, the two terminologies are different but are used intermixed in the book as every radix-2 DFT (i.e., a DFT whose length is in power of 2, such as 128, 512, 1,024, or 2,048) is calculated by FFT in practice.

Table 4.1 Subchannel Grouping in 1,024-FFT OFDMA System

Parameters	DL PUSC	DL FUSC	UL PUSC	DL/UL AMC
Number of null subcarriers	184	174	184	160
Number of DC subcarriers	1	1	1	1
Number of guard subcarriers, left	92	87	92	80
Number of guard subcarriers, right	91	86	91	79
Number of used subcarriers (*Nused*) excluding null subcarriers	840	850	840	864
Number of pilot subcarriers	120	82	420/0	96
Number of data subcarriers	720	768	420/840	768
Number of subchannels	30	16	35	48
Number of data subcarriers in each symbol per subchannel	24	48	12/24	16

Source: [1, 2].

We consider the continuous-time signal $x(t)$ in the right-hand side of Figure 4.8(a), whose Fourier transform $X(j\Omega)$ is as shown on the left. If we sample $x(t)$ at the sampling period ΔT, or with the sampling frequency F_S ($=1/\Delta T$), then the sampled waveform $x_s(t)$ takes the shape on the right-hand side of Figure 4.8(b) and its frequency spectrum $X_s(j\Omega)$ is as on its left, which is a continuous and periodic signal with the period $2\pi/\Delta T$. Now we convert the continuous-time signal into discrete-time signal $x(n)$ shown in Figure 4.8(c), whose frequency spectrum $X(e^{j\omega})$ is as shown on the left. Then we sample $X(e^{j\omega})$ at the spacing $2\pi/N_{FFT}$ to get the periodic waveform $\tilde{X}(k)$ with period N_{FFT}, as shown on the left-hand side of Figure 4.8(d). Then its discrete-time signal $\tilde{x}(n)$ on its right is also periodic with period N_{FFT}. Since ω is the normalized frequency $\Omega \cdot \Delta T$, the sampling spacing of $2\pi/N_{FFT}$ in ω axis corresponds to $2\pi/(N_{FFT}\Delta T)$ in Ω axis, and to $1/(N_{FFT}\Delta T)$ in the frequency domain (i.e., $\Delta f = F_s/N_{FFT}$). They form a *discrete Fourier series* (DFS) pair. Finally we take one period of the periodic waveforms to get $x(n)$ and $X(k)$ shown in Figure 4.8(e). Then they are the *discrete Fourier transform* (DFT) pair, which are both defined in the range $0, 1, ..., N_{FFT} - 1$.

Once the encoded and modulated symbols are mapped, together with the related pilot tones, to the N_{FFT} subcarriers, with only the N_{used} subcarriers actually filled with the symbols and pilot tones, we get the OFDMA symbol $X(k)$, $k = 0, 1, .., N_{FFT} -1$ on the left-hand side of Figure 4.8(e). If we take IDFT of it, taking advantage of the fast computation capability of the FFT, then we get $x(n)$, $n = 0, 1, .., N_{FFT} -1$ on the right-hand side, which corresponds to the sampled and discretized waveform of the analog signal $x(t)$ of length T_b shown in Figure 4.8(a). Note that the symbol time T_b is the inverse of the frequency sampling spacing Δf, or the subcarrier spacing (i.e., $T_b = N_{FFT}\Delta_T = N_{FFT}/F_S = 1/\Delta f$).

In mathematical expression, $\tilde{X}(k)$ is obtained by sampling $X(e^{j\omega})$ at frequencies $\omega_k = 2\pi k/N_{FFT}$, $k = 0, 1, ..., N_{FFT} - 1$. So we get

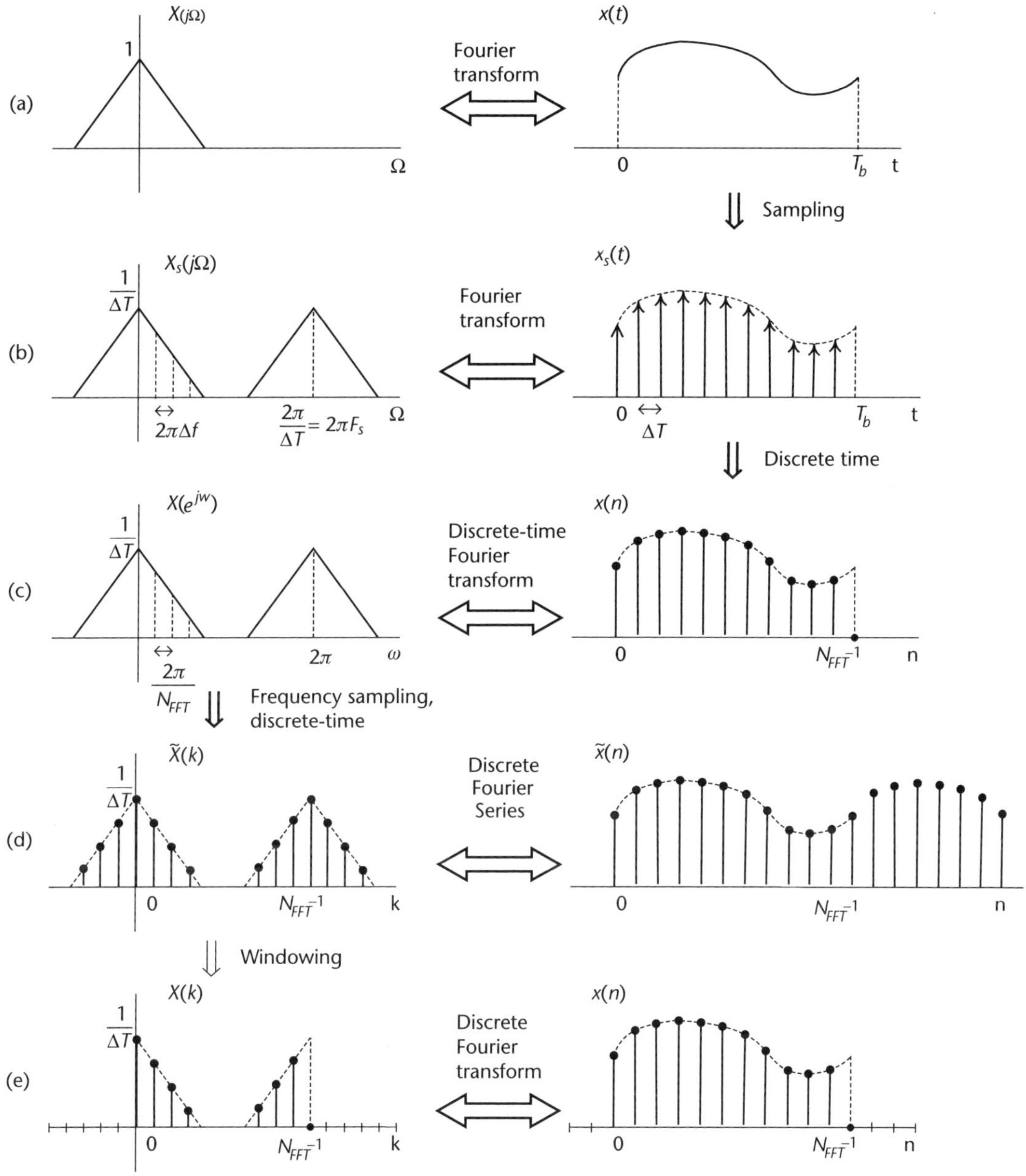

Figure 4.8 (a–e) Illustration of time-frequency waveforms for IDFT processing.

$$
\begin{aligned}
\tilde{X}(k) &= X\left(e^{j2\pi k/N_{FFT}}\right) \\
X(k) &= \begin{cases} \tilde{X}(k), & 0 \le k \le N_{FFT} - 1 \\ 0, & \text{otherwise} \end{cases}
\end{aligned}
\tag{4.1}
$$

Taking the discrete-frequency to continuous-time inverse Fourier transform of $X(k)$, we get the finite duration signal $x(t)$ in the form

$$x(t) = \sum_{k=0}^{N_{FFT}-1} X(k) e^{j2\pi k\Delta f t}, \quad 0 \le t \le T_b \tag{4.2}$$

Then we take the continuous-time Fourier transform of $x(t)$ to get the continuous-time frequency response $X(f)$ in the following manner.

$$\begin{aligned} X(f) &= \int_{-\infty}^{\infty} x(t) e^{-j2\pi ft} dt \\ &= \int_{-\infty}^{\infty} \sum_{k=0}^{N_{FFT}-1} X(k) e^{j2\pi k\Delta f t} \{u(t) - u(t - T_b)\} e^{-j2\pi ft} dt \\ &= \sum_{k=0}^{N_{FFT}-1} X(k) \int_{-\infty}^{\infty} \{u(t) - u(t - T_b)\} e^{-j2\pi (f - k\Delta f)t} dt \\ &= T_b \sum_{k=0}^{N_{FFT}-1} X(k) e^{-j\pi (f - k\Delta f) T_b} \operatorname{sinc}(\pi (f - k\Delta f) T_b) \end{aligned} \tag{4.3}$$

The waveform in Figure 4.6 corresponds to this expression for the case $X(k) = 1$ all k in the active band (i.e., data and pilot tones) and $X(k) = 0$ for all k in the guard band.

Cyclic Prefixing

In order to combat against the multipath fading, the IDFT transformed time-domain signal $x(n)$ is protected by a *cyclic prefix* (CP). Let $x(n)$ have the effective symbol length T_b, as shown in Figure 4.9.[5] We copy a portion of length T_g at the tail of the symbol, and paste it at the front end of the symbol, naming it *cyclic prefix* (CP). Then, after prefixing the CP part, the overall OFDMA symbol length increases to $T_s = T_g + T_b$.

The CP contributes to protecting the OFDMA symbol from multipath interference, while maintaining the orthogonality among the constituent subcarriers. Note that any contiguous T_b portion of the cyclic-prefixed OFDMA symbol of length T_s in time domain yields a phase-rotated version of the same frequency domain signal, and a multipath interference results in multiple of such phase-rotated versions of the same frequency domain signal, which yields only a scaling effect on the final received signal.

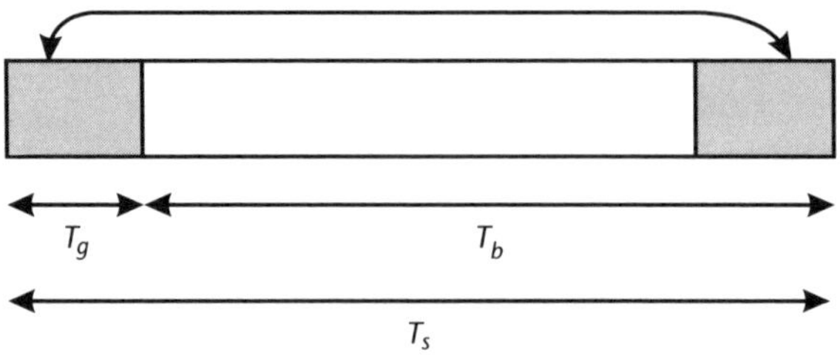

Figure 4.9 OFDMA symbol structure. (*After:* [1, 2].)

5. Rigorously speaking, T_b is the length of $x(t)$, the continuous-time counter part of $x(n)$. When sampled at the sampling interval ΔT, or at the sampling frequency F_s, the resulting number of points of $x(n)$ that corresponds to the length T_b is N_{FFT}. Therefore, the number of points that corresponds to the length T_g is $N_{FFT} \cdot T_g/T_b$.

More rigorously, the robustness of OFDMA to the multipath interference may be explained mathematically as follows: Let $x(n)$, $n = 0, 1, \ldots, N_{FFT} - 1$, denote the original time-domain signal obtained by taking IDFT in the transmitter on the input frequency-domain (or subcarrier domain) signal $X(k)$, $k = 0, 1, \ldots, N_{FFT} - 1$. Then we get the expression

$$x(n) = IDFT[X(k)] = \frac{1}{N_{FFT}} \sum_{k=0}^{N_{FFT}-1} X(k) e^{j\frac{2\pi kn}{N_{FFT}}} \tag{4.4}$$

Let $x_i(n)$, $n = 0, 1, \ldots, N_{FFT} - 1$, denote a delayed signal of $x(n)$, for the delay d_i; that is,

$$x_i(n) = x(n - d_i) \tag{4.5}$$

Then its DFT signal, $X_i(k)$, takes the expression

$$\begin{aligned} X_i(k) &= DFT[x_i(n)] \\ &= \sum_{n=0}^{N_{FFT}-1} x((n - d_i))_{N_{FFT}} e^{-j\frac{2\pi kn}{N_{FFT}}} \\ &= \sum_{m=0}^{N_{FFT}-1} x((m))_{N_{FFT}} e^{-j\frac{2\pi km}{N_{FFT}}} e^{-j\frac{2\pi kd_i}{N_{FFT}}} \\ &= e^{-j\frac{2\pi kd_i}{N_{FFT}}} DFT[x(n)] \end{aligned} \tag{4.6}$$

where $x((n - d_i))_{N_{FFT}}$ denotes the circular-shifted signal of $x(n - d_i)$ with modulo N_{FFT}. The multipath signal received by the receiver may be expressed by a scaled sum of multiple (or M) delayed signals; that is,

$$\hat{x}(n) = \sum_{i=0}^{M-1} \alpha_i x(n - d_i) \tag{4.7}$$

for the scaling factor α_i. So the final output signal of the receiver takes the expression

$$\begin{aligned} \hat{x}_i(k) &= DFT[\hat{x}_i(N)] \\ &= \sum_{n=0}^{N_{FFT}-1} \sum_{i=0}^{M-1} \alpha_i x((n \quad d_i))_{N_{FFT}} e^{-j\frac{2\pi kn}{N_{FFT}}} \\ &= \sum_{i=0}^{M-1} \alpha_i \sum_{n=0}^{N_{FFT}-1} x((n - d_i))_{N_{FFT}} e^{-j\frac{2\pi kn}{N_{FFT}}} \\ &= \left(\sum_{i=0}^{M-1} \alpha_i e^{-j\frac{2\pi kn}{N_{FFT}}} \right) DFT[x(n)] \\ &= \alpha DFT[x(n)] \end{aligned} \tag{4.8}$$

This demonstrates that the multipath interference yields only a scaling effect on the original signal in the case of OFDMA.

4.1.3 Transmit Processing

The converted time-domain signal $x(n)$ undergoes the final three processes in Figure 4.1, namely, lowpass filtering, *digital-to-analog conversion* (DAC), and *radio-frequency* (RF) transmission. The three stages of signal processing are done in such a way that the frequency spectrum of the signal does not interfere with the frequency band of other operators when modulated to the given carrier frequency band. The lowpass filtered signal is converted to analog signal, $x(t)$, and modulated up to the allocated carrier frequency f_c for transmission.

Actual implementation of this transmit processing, however, is the choice of the system designer. For example, in designing the analog filter, one can relax its specification by employing the interpolation technique in conjunction with a steep digital filter in the digital domain. Figure 4.10 shows one of the most practical ways of implementing the transmit signal processing, by employing the interpolation technique.

Interpolation and Digital LPF

In principle, the input signal $x(n)$, which is the IDFT signal of the original OFDMA symbol $X(k)$, needs to be filtered by a digital *lowpass filter* (LPF) so that the input signal to the DAC converter can be confined within the angular frequency $\omega=\pi$. After the DAC process, we place an analog LPF filter to compensate for the distorting effect of the zero-order holder circuit within the DAC. Unfortunately, it is difficult or highly costly to implement this analog compensation filter. In order to get around this difficulty, we apply the interpolation technique, which enables the use of a low-order analog compensation filter by adopting an interpolator and a high-order digital LPF, which is comparatively easy and cost-effective to implement.

The input discrete-time signal $x(n)$ is first interpolated by $L-1$ zeros (i.e., padded by $L-1$ zeros) to increase the sampling rate to L times the original rate. Then for the resulting signal $x_L(n)$, we get the expressions

$$x_L(n) = \begin{cases} x(n/L), & n = 0, L, \ldots \\ 0, & \text{otherwise} \end{cases} \tag{4.9}$$

$$X_L\left(e^{j\omega}\right) = \sum_n x_L(n)e^{-j\omega n} = \sum_n x(n)e^{-j\omega nL} = X\left(e^{j\omega L}\right) \tag{4.10}$$

For the input signal $x(n)$ whose time and frequency domain waveforms are as given in Figure 4.11(a), Figure 4.11(b) illustrates the waveforms of the interpolated

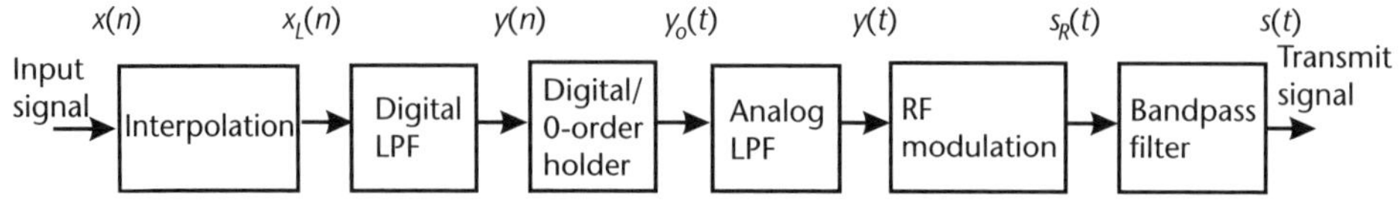

Figure 4.10 An implementation of the transmit signal processing.

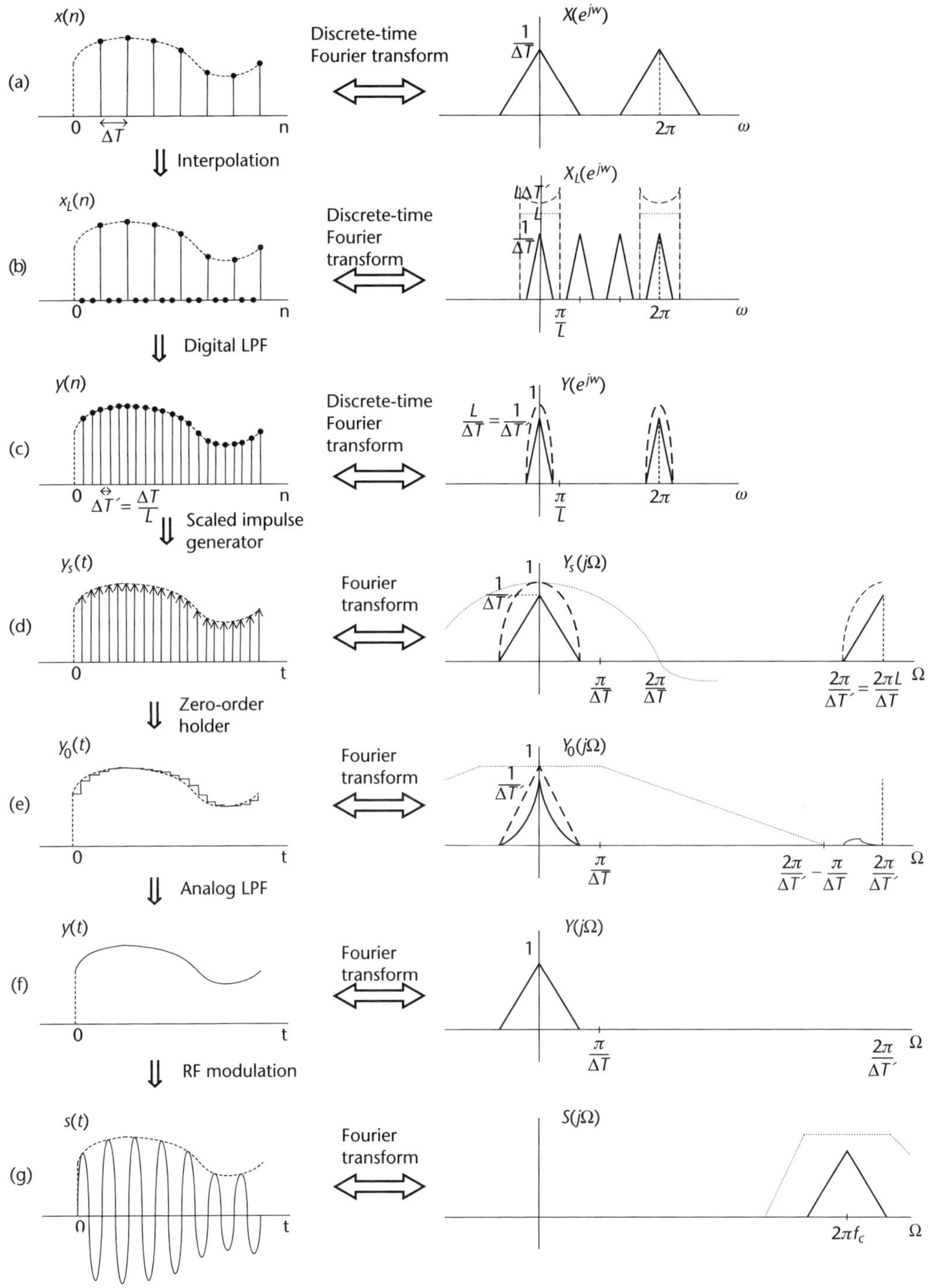

Figure 4.11 (a–g) Time and frequency domain waveforms.

(or zero-padded) signal, $x_L(n)$ and $X_L(e^{j\omega})$, for the case of $L = 3$ (solid line). Let the digital LPF $H_L(e^{j\omega})$ have the ideal filter characteristic shown in the dotted line of Figure 4.11(b) with the specification

$$H_L\left(e^{j\omega}\right)=\begin{cases}L, & |\omega|<\dfrac{\pi}{L},\\ 0, & \text{otherwise}\end{cases} \tag{4.11}$$

Then the filtered signal $y(n)$ and its frequency characteristic have the waveforms shown in Figure 4.11(c) (solid line). Notice that $y(n)$ has the sampling period $\Delta T' = \Delta T/L$, so the cutoff frequency of $Y(e^{j\omega})$ drops below the angular frequency π/L.

In practice, however, the LPF characteristic has to be predistorted in such a way that it compensates for the distortion to be made by the zero-order holder circuit in the DAC (see Figure 4.12). Noting that the zero-order holder has the effect of placing a LPF with the characteristics $H_0(e^{j\omega})$ in (4.14), we modify $H_L(e^{j\omega})$ to

$$\hat{H}_L\left(e^{j\omega}\right)=\begin{cases}L\dfrac{\omega/2}{\sin(\omega/2)}, & |\omega|<\dfrac{\pi}{L}\\ 0, & \text{otherwise}\end{cases} \tag{4.12}$$

Notice that the upscaling (by $\Delta T'$) needed to reconstruct the original analog signal from the discrete-time signal—see (4.17)—is also incorporated in the equation. This new filter is indicated by dashed line in Figure 4.11(b). Then the spectrum of the filtered output $y(n)$ changes to the shape shown in dashed line in Figure 4.11(c).

DAC and Analog LPF

We model the DAC as a cascade of a scaled impulse generator, a zero-order holder circuit having the impulse response $h_0(t)$, and an analog LPF, as depicted in Figure 4.12.

Then we get the impulse response

$$h_0(t)=\begin{cases}1, & 0<t<\Delta T'\\ 0, & \text{otherwise}\end{cases} \tag{4.13}$$

and the frequency characteristic

$$H_0(j\Omega)=\frac{2\sin(\Omega\Delta T'/2)}{\Omega}e^{-j\Omega\Delta T'/2} \tag{4.14}$$

whose magnitude characteristic is given in the dotted line in Figure 4.11(d). If we pass the scaled impulse signal through the zero-order holder circuit, the output signal takes the expressions

$$y_0(t)=\sum_n y(n)h_0(t-n\Delta T') \tag{4.15}$$

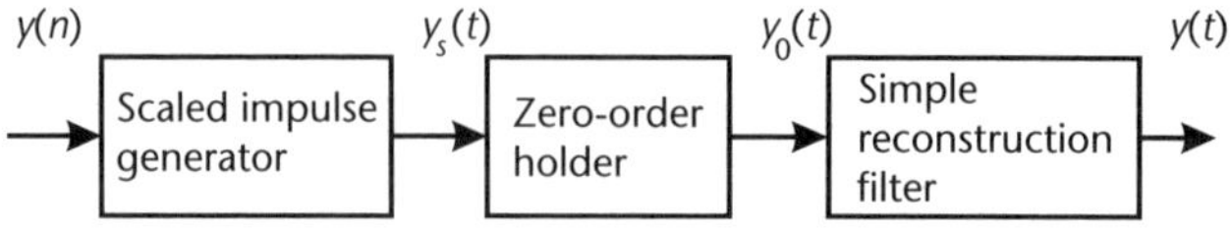

Figure 4.12 Internal signal processing for DAC.

$$
\begin{aligned}
Y_0(j\Omega) &= \int_{-\infty}^{\infty} \sum_n y(n) h_0(t - n\Delta T') e^{-j\Omega t}\, dt \\
&= Y\left(e^{j\Omega \Delta T'}\right) H_0(j\Omega)
\end{aligned}
\tag{4.16}
$$

where

$$
Y\left(e^{j\Omega\Delta T'}\right) = Y\left(e^{j\omega}\right)\Big|_{\omega=\Omega\Delta T'} = \frac{1}{\Delta T'} \sum_k Y\left(j\Omega + j\frac{2\pi k}{\Delta T'}\right) \tag{4.17}
$$

In order to reconstruct the original analog signal, we must take only the baseband component $Y(j\Omega)$ in (4.17) and scale it up by $\Delta T'$, eliminating all the high-order harmonic components.

In practice, however, when designing the digital LPF that follows the interpolator circuit, we have already reflected the distortion effect of the zero-order holder circuit, as well as the upscaling process, thereby getting the predistorted spectrum $Y(e^{j\omega})$ in Figure 4.11(c) (dashed line). Passing through the scaled impulse generator, the spectrum changes to $Y_s(j\Omega)$ in Figure 4.11(d) (dashed line); and again by passing through the zero-order holder circuit, it takes the shape in Figure 4.11(e) (dashed line). So we put a very simple analog LPF after the DAC, having the specification

$$
H_c(j\omega) = \begin{cases} 1, & |\omega| < \dfrac{2\pi}{\Delta T'} - \dfrac{\pi}{\Delta T} \\ 0, & \text{otherwise} \end{cases} \tag{4.18}
$$

Then we finally get the reconstructed signal $y(t)$, which takes the frequency spectrum shown in Figure 4.11(f).

RF Modulation and Analog BPF

As the final stage of signal processing, we modulate $y(t)$ up to the carrier frequency f_c. In the process of this modulation and the followed power amplification process, nonlinear signal processing intervenes, so the frequency spectrum of the modulated signal gets unwanted harmonic components. So we put an analog BPF in the final stage in order to filter out such harmonics.

If we assume that all the signal processing in Figure 4.10 is done ideally, the output signal $y(t)$ is supposed to be identical to the original signal $x(t)$. Therefore, considering the cyclic prefix T_g, we may express the transmitted signal $s(t)$ as follows:

$$
\begin{aligned}
s(t) &= \mathrm{Re}\left\{e^{j2\pi f_c t}\, x\left(t \quad T_g\right)\right\} \\
&= \mathrm{Re}\left\{e^{j2\pi f_c t} \sum_{k=0}^{N_{FFT}-1} X(k) e^{j2\pi\Delta f (t - T_g)}\right\}
\end{aligned}
\tag{4.19}
$$

Note that among the N_{FFT} subcarriers $X(k)$s in the summation, only N_{used} components are the actual data symbols and pilot tones and all the other components for the guard band and DC are zero.

Transmit Signal Waveform

As a summary of the discussions on OFDMA signal processing, Figure 4.13 illustrates the relations of the four continuous and discrete waveforms $X(k)$, $x(n)$, $x(t)$, and $X(f)$, for the case $N_{FFT} = 4$.

Figure 4.13(a) is the encoded and modulated symbols of $X(k)$, $k = 1, 2, 3, 4$, allocated to four different frequencies (e.g., ω_0, $2\omega_0$, $3\omega_0$, and $4\omega_0$). Figure 4.13(b) shows $x(n)$, $n = 1, 2, 3, 4$, the IDFT of $X(k)$, in vertical lines and shows the four frequency components of $X(k)$ in time domain, which are combined to yield the analog signal $x(t)$ shown in Figure 4.13(d). Note that $x(t)$ is the signal that is obtained after ideal LPF and DAC processing, which is $y(t)$ in Figure 4.11. It is $x(t)$ that is modulated up to the carrier frequency f_c to yield the transmission signal $s(t)$.

If we take the Fourier transform of $x(t)$ we get the continuous signal $X(f)$ in Figure 4.13(c), which corresponds to the sum of the four *sinc* signals at the four different frequencies. Note that if we sample the analog signal $X(f)$ at the four frequencies, then the original discrete signal $X(k)$, $k = 0, 1, 2, 3$ result. In practice, the original data $X(k)$ is reconstructed by taking DFT on $x(n)$, the A-to-D converted signal of $x(t)$, which is the demodulated baseband signal of the received RF signal $s(t)$.

4.1.4 OFDMA System Parameters

While the WirelessMAN-OFDMA PHY is defined for operation below 11 GHz licensed bands, the mobile WiMAX system, in practice, takes the frequency bands in 2.3, 2.5, or 3.5 GHz. It takes the bandwidth (or channel spacing) of 5, 7, 8.75, or 10 MHz by adopting the TDD scheme, but FDD is also allowed. FDD MSs may be half-duplex FDD (H-FDD). For multiple access, mobile WiMAX takes OFDMA for the subcarrier number (or FFT size) of 128, 512, 1,024, or 2,048, and for the TDD

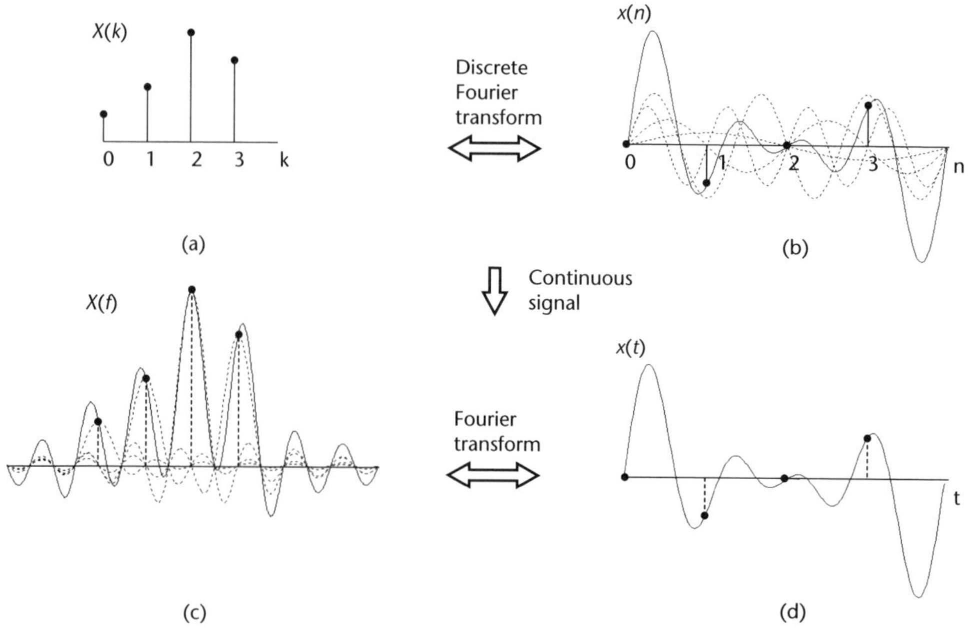

Figure 4.13 (a–d) Example of the four discrete and continuous signals for $N_{FFT} = 4$.

frame length of 5 ms. In addition, it adopts BPSK, QPSK, 16-QAM, and 64-QAM for modulation, convolutional turbo code for channel coding, and hybrid ARQ for data retransmission. Table 4.2 lists a summary of the main parameters of mobile WiMAX.

The characteristics of an OFDMA symbol are determined by the following basic parameters—the channel bandwidth BW; the sampling frequency F_s; the sampling factor (or the sampling frequency to bandwidth ratio) n; the number of DFT points N_{FFT}; and the guard time ratio (i.e., the CP time to "useful time" ratio) G. The channel bandwidth is the nominal channel bandwidth, not the band spacing of 9 MHz in the case of the 2.3-GHz WiBro system. Band spacing is the frequency band separation, which is determined in consideration of the interference among the neighboring bands, so the nominal channel bandwidth should be contained within this band. In OFDMA, the sampling frequency is the bandwidth multiplied by the oversampling rate (i.e., $F_s = n^*BW$). Once those basic parameters are given, the subcarrier spacing Δf, the effective symbol length T_b, and the OFDM symbol length T_s are determined by the relations $\Delta f = F_s/N_{FFT}$, $T_b = 1/\Delta f$, and $T_s = T_b + T_g$.

According to the standards, sampling factor n is set as follows: For channel bandwidths that are a multiple of 1.75 MHz, it is set to 8/7, and for channel bandwidths that are a multiple of 1.25, 1.5, 2, or 2.75 MHz, it is set to 28/25. For the channel bandwidths that are not otherwise specified, it is set to 8/7. Besides, for the guard time ratio G, the values 1/32, 1/16, 1/8, and 1/4 are supported in general.

For example, in the case of the 2.3-GHz band WiBro system, we are given BW = 8.75 MHz, $n = 8/7$, $N_{FFT} = 1{,}024$, and $G = 1/8$. The sampling frequency is $F_s = 10$ MHz, and the subcarrier spacing is $\Delta f = F_s / N_{FFT} = 9.765625$ kHz and, consequently, the effective symbol length becomes $T_b = 1/\Delta f = 102.4\ \mu s$. Besides, guard time is set to be 12.5 percent (i.e., $G = T_g/T_b = 1/8$) and the OFDM symbol length is $T_s = T_b + T_g = 115.2\ \mu s$. Table 4.3 lists a summary of these OFDMA signal parameters adopted for the design of WiBro, a 2.3-GHz Mobile WiMAX system.

Table 4.2 Main Parameters of Mobile WiMAX

Parameters	Values
Frequency band	2.3/2.5/3.5 GHz
Bandwidth	5/7/8.75/10 MHz
Duplexing	TDD, FDD
Multiple access	OFDMA
TDD frame length	5 ms
FFT size	128/512/1024/2048
Modulation	BPSK, QPSK, 16-QAM, 64-QAM
Channel coding	Convolutional turbo code
ARQ	Hybrid ARQ

Table 4.3 OFDMA Signal Parameters of WiBro

Parameters	Values	Remarks
Channel bandwidth (BW)	8.75 MHz	< Band spacing
FFT size (N_{FTT})	1,024	
Sampling factor (n)	8/7	$n = F_s / BW$
Sampling frequency (F_s)	10 MHz	
Sampling period (ΔT)	100 ns	$\Delta T = 1 / F_s$
Subcarrier spacing (Δf)	9.765625 KHz	$\Delta f = F_s / N_{FFT}$
Guard time ratio (G)	1/8	$G = T_g / T_b$
Effective symbol length (T_b)	102.4 μs	$T_b = 1 / \Delta f$
Cyclic prefix length (T_g)	12.8 μs	$T_g = G \cdot T_b$
OFDM symbol length (T_s)	115.2 μs	$T_s = T_b + T_g$

4.2 Channel Coding and HARQ[6]

Channel coding is a popular method of improving communication performance at the cost of reduced bandwidth and delay. Various *forward error correction* (FEC) encoding methods are employed in the IEEE 802.16e standards as mandatory requirements or as options. The mandatory coding method is the tail-biting *convolutional coding* (CC), and the optional coding methods are *convolutional turbo coding* (CTC), *block turbo coding* (BTC), zero-tailed CC, and *low density parity coding* (LDPC). So we discuss CC and CTC channel coding techniques in this section.

4.2.1 Convolutional Code

The encoding process is composed of the following three parts:

1. Making up the encoding blocks of adequate lengths;
2. Actual FEC encoding with puncturing;
3. Interleaving, as depicted in Figure 4.14.

We discuss the three subjects in terms of subchannel concatenation, convolutional encoder, puncturing, and interleaver, in the following.

6. Among the various topics in OFDMA communication signal processing, we deal with channel coding separately in this section. Readers familiar with this topic may skip Sections 4.2.1 and 4.2.2.

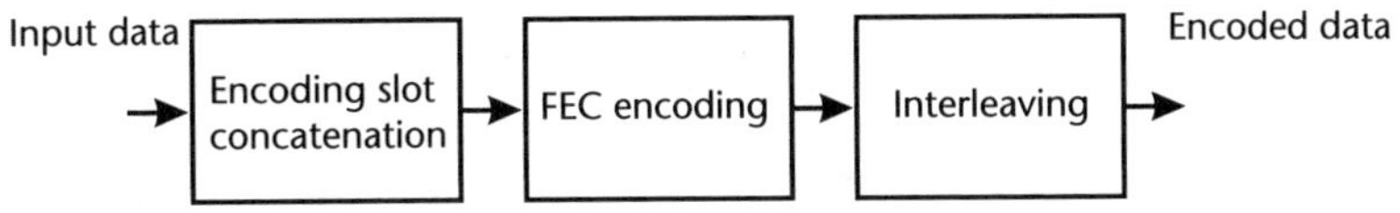

Figure 4.14 Block diagram of the encoding process.

Encoding Slot Concatenation in Convolutional Code

After the channel coding and modulation processes, information data are segmented in multiple blocks to fit into an OFDMA slot, consisting of 48 data subcarriers. Multiple OFDMA slots are concatenated to enable larger blocks to be encoded. The larger blocks may result in additional coding gain and, at the same time, larger latency as well. Thus, the encoding block size should be determined considering the tradeoff relation of coding gain and latency. When determining the encoding block size, the following three factors are taken into account: modulation type, code rate, and concatenation rule.

The maximum number of concatenated slots differs depending on the modulation type and code rate. For example, when QPSK with code rate 1/2 is used, the maximum number of concatenated slots is 6. Table 4.4 shows the maximum number of concatenated slots for different modulation types and code rates. Then, the concatenation rule applies as follows: Let n denote the *floor* of the number of allocated slots per repetition factor. Let k and m be the quotient and the remainder of n/j such that $n = k \cdot j + m$, where j denotes the maximum number of concatenated slots: (1) when $n \leq j$, 1 block of n slots is concatenated; and (2) when $n > j$, if n is a multiple of j, then k blocks of j slots are concatenated—otherwise, $k - 1$ blocks of j slots, 1 block of $ceil((m + j)/2)$ slots, and 1 block of $floor((m + j)/2)$ slots are concatenated ([1], Table 317; [2], Table 561). For example, (1) when $n = 4$ and $j = 6$, 1 block of 4 slots is concatenated; (2) when $n = 12$ and $j = 6$, 2 blocks of 6 slots are concatenated; and (3) when $n = 13$ and $j = 6$, 3 blocks of 6, 4, 3 slots are concatenated.

For a given number of concatenated slots, the encoding block size is determined to be the number of symbols contained in the concatenated slots. Table 4.5 lists the sizes of the data payloads to be encoded in relation to the modulation type, code rate, and concatenation number.

Table 4.4 Maximum Number of Concatenated Slots for Convolutional Coding

Modulation, code rate	j
QPSK 1/2	6
QPSK 3/4	4
16-QAM 1/2	3
16-QAM 3/4	2
64-QAM 1/2	2
64-QAM 2/3	1
64-QAM 3/4	1

Source: [1, 2].

Table 4.5 Data Payloads for Different Modulation and Concatenation

Modulation	QPSK		16-QAM		64-QAM		
Encoding rate	R=1/2	R=3/4	R=1/2	R=3/4	R=1/2	R=2/3	R=3/4
Concatenation	6	4	3	2	2	1	1
Data payload (bytes)	6						
		9					
	12		12				
	18	18		18	18		
	24		24			24	
		27					27
	30						
	36	36	36	36	36		

Source: [1, 2].

Convolutional Encoder

In mandatory binary convolutional encoding, we use the encoder of code rate 1/2 and constraint length 7 as shown in Figure 4.15, whose generator polynomials are $1 + x + x^2 + x^3 + x^6$ for the output X and $1 + x^2 + x^3 + x^5 + x^6$ for the output Y.

In the case of the zero tailing CC, a single 0×00 tail byte is appended at the end of each burst, which is needed for decoding operation. On the other side, the tail-biting convolutional encoder memories are initialized by the 6 last data bits of the FEC block being encoded.

Puncturing

The code rate of convolutional code can be varied by not transmitting certain code symbols, or *puncturing* the original code. In this case, one encoder and decoder can be used with variable code rates. In general, a rate p/q punctured convolutional code can be constructed from a $(n, 1)$ convolutional code by deleting $np - q$ code symbols from every np code symbols corresponding to q input information symbols. The $(n, 1)$ code is called the *mother code.*

As discussed earlier, the rate 1/2 convolutional code is used as the mother code. Table 4.6 describes the puncturing patterns and serialization order for different code rates, such as 2/3 and 3/4 with respect to the convolutional coder in Figure 4.15. Note that the "1" and "0" in X or Y read "transmit" and "remove," respectively. In the table, d_{free} denotes the free distance of the punctured code.

Interleaving

Interleaving is the process of shuffling the encoded data bits in a predetermined manner. Interleaving is used to protect the transmission data from long consecutive errors. The transmitted signal, in general, is subject to burst error as well as random error while being transmitted, and the burst error is hard to detect and correct in the

Table 4.6 Convolutional Code with Puncturing Configuration

	Code rates		
Rate	1/2	2/3	3/4
d_{free}	10	6	5
X	1	10	101
Y	1	11	110
XY	X_1Y_1	$X_1Y_1Y_2$	$X_1Y_1Y_2X_3$

Source: [1, 2].

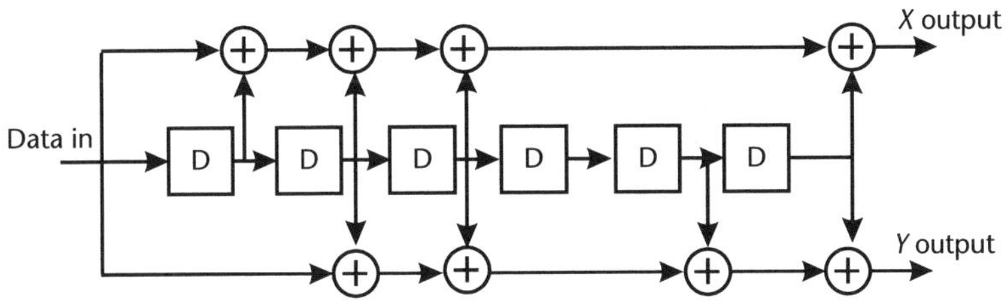

Figure 4.15 Convolutional encoder of code rate 1/2. (*After:* [1, 2].)

receiver. If the encoded data is interleaved before transmission, the burst error that might have occurred during transmission is converted into a sort of random error when deinterleaved in the receiver, which can be detected and possibly corrected after the decoding process. Specifically, the encoded data bits are interleaved by a block interleaver whose block size corresponds to the number of coded bits per encoded block. For example, in the case of the 16-QAM with code rate 1/2, the maximum number of concatenated slots is 3, as shown in Table 4.4. If we take the concatenated slot number of 2, then the encoded block size becomes 24 bytes, so the corresponding interleaver block size is 192 bits. The interleaving process is done by a two-step permutation: First, the adjacent coded bits are mapped onto nonadjacent subcarriers and the first permutation is defined by

$$m_k = \left(N_{cbps}/d\right)\cdot k_{\text{mod}(d)} + floor\left(k/d\right) \quad \text{for } k = 0, 1, \ldots, N_{cbps} - 1 \text{ and } d = 16 \tag{4.20}$$

where N_{cbps} is the number of coded bits per encoded block size. Second, the originally adjacent coded bits are mapped onto more or less significant bits of the constellation, alternatively, by

$$j_k = s\cdot floor(m_k/s) + \left(m_k + N_{cbps} - floor\left(d\cdot m_k/N_{cbps}\right)\right)_{\text{mod}(s)} \tag{4.21}$$

where $s = 1$, 2, and 3, respectively, for QPSK, 16-QAM and 64-QAM. Figure 4.16 illustrates this two-step permutation process for the case of 16-QAM with code rate 1/2.

Conversely, the deinterleaver performs the inverse operation of the interleaver, which can also be done in two-step permutation. In this case, the first permutation

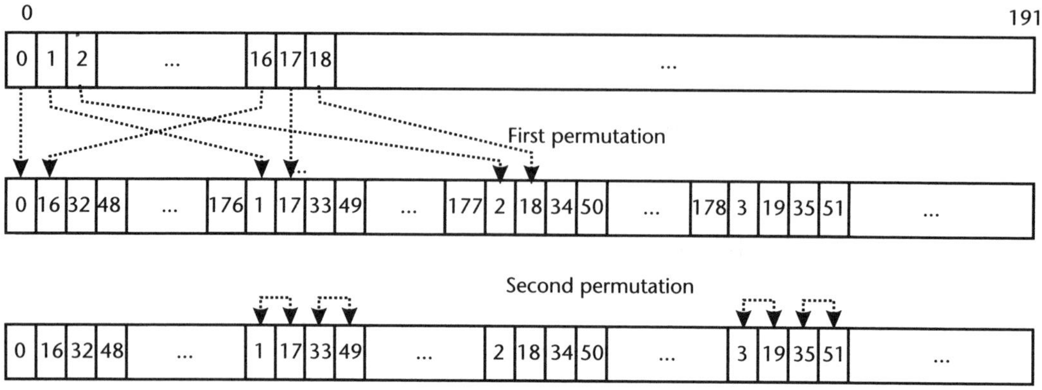

Figure 4.16 Illustration of the two-step permutation based interleaving process for 16-QAM with code rate 1/2.

in the deinterleaver is the inverse of the second permutation in the interleaver, and the same relation holds for the second permutation.

4.2.2 Convolutional Turbo Code (CTC)

Turbo code is one of the few FEC codes to come close to the Shannon limit, the theoretical limit of the maximum information transfer rate over a noisy channel. The turbo code [3] was proposed by Berrou and Glavieux in 1993. The main feature of turbo code that distinguishes it from the traditional FEC codes is the use of two error-correcting codes and an interleaver. Decoding is then done iteratively by taking advantage of the two sources of information.

Figure 4.17 depicts the block diagram of classical turbo encoder [4]. There are three blocks of bits that are multiplexed together: the first block is the m-bit block of uncoded data; the second block is $n/2$ parity bits added in sequence to the payload data and computed using a convolutional code; and the third block is another $n/2$ parity bit added in sequence to a known permutation of the payload data and computed using a convolutional code. Consequently, two different redundant blocks of parity bits are added to the originally sent payload. The resulting complete block contains $m + n$ bits of data with the code rate of $m/(m + n)$.

In the IEEE 802.16e standards, the *double binary turbo code* (DBTC) [5] is adopted as an optional *convolutional turbo code* (CTC). The DBTC uses double-input *linear feedback shift registers* (LFSRs), which enable several information bits to be encoded or decoded at the same time. This arrangement brings forth several

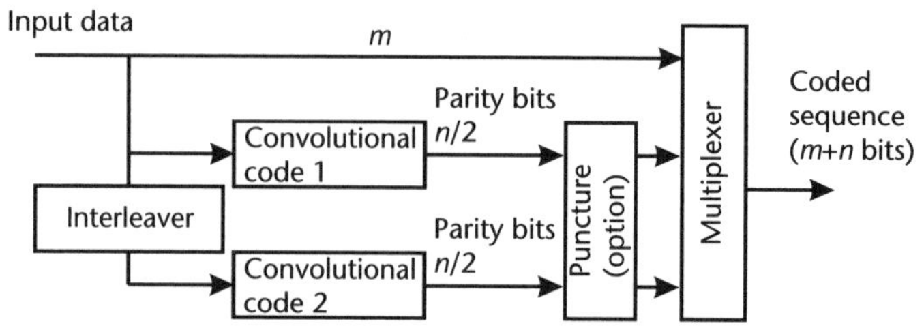

Figure 4.17 Turbo coded sequence generation.

advantages over the classical turbo code. A detailed description of the structure of DBTC and its advantages are provided separately in a later subsection.

Figure 4.18 depicts the block diagram of the encoding process of CTC. The input data is first concatenated and then CTC encoded. The CTC encoded data goes through an interleaving process and finally a symbol selection (or puncturing) process. The mother codeword is punctured to generate subpackets with various coding rates, such as 1/2, 2/3, 3/4, and 5/6.

Slot Concatenation in Convolutional Turbo Code

As in the case of the convolutional code, multiple slots are concatenated in order to fit information data into an OFDMA slot consisting of 48 data subcarriers and to enable larger blocks to be encoded. In determining the encoding block size, two considerations are taken into account: First, convolutional turbo encoding blocks are made larger in size than the convolutional encoding blocks so as not to lose the coding gain. Table 4.7 shows the maximum number of concatenated slots used in convolutional turbo encoding for different modulation types and code rates.

Second, convolutional turbo encoding block size is determined to not be a multiple of 7. If the number of symbols in a block is a multiple of 7 in the encoding process, there does not exist the initial state equal to the final state after encoding, therefore failing to satisfy the tail-biting condition. Therefore, the concatenation rule avoids making the length of blocks a multiple of 7.

The concatenation rule of CTC applies as follows: Let n denote the *floor* of the number of allocated slots per repetition factor. Let k and m be the quotient and the remainder of n/j such that $n = kj + m$. (1) When $n \leq j$ and $n \neq 7$, 1 block of n slots is

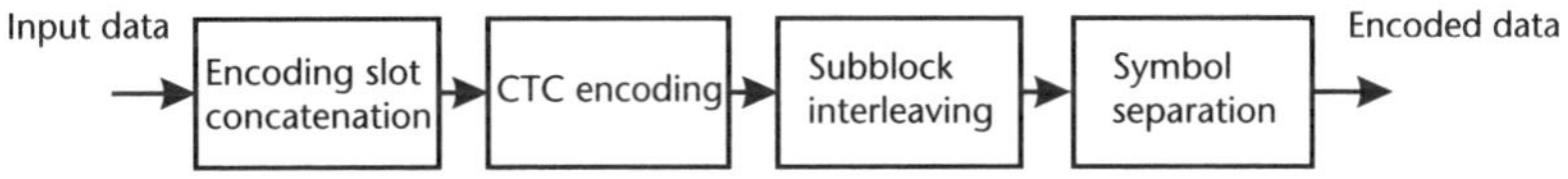

Figure 4.18 Block diagram of the convolutional turbo encoding process.

Table 4.7 Maximum Number of Concatenated Slots for Convolutional Turbo Encoding

Modulation, rate	j
QPSK 1/2	10
QPSK 3/4	6
16-QAM 1/2	5
16-QAM 3/4	3
64-QAM 1/2	3
64-QAM 2/3	2
64-QAM 3/4	2
64-QAM 5/6	2

Source: [1, 2].

concatenated. (2) When $n = 7$, 1 block of 4 slots and 1 block of 3 slots are concatenated. (3) When $n > j$, if n is a multiple of j, then k blocks of j slots are concatenated; otherwise, $k - 1$ blocks of j slots, 1 block of $ceil((m + j)/2)$ slots, and 1 block of $floor((m + j)/2)$ slots are concatenated. In this case if the *ceil* or the *floor* of $(m + j)/2$ is a multiple of 7, the ceil value is increased by 1 and the floor value is decreased by 1 ([1], Table 324; [2], Table 569).

CTC Encoder

Figure 4.19 depicts the configuration of the CTC encoder, which consists of a CTC interleaver and the constituent CTC encoder. The CTC can be used for the support of the optional *hybrid ARQ* (HARQ). The CTC uses a double binary circular recursive systematic convolutional code. The bits of the data to be encoded are alternatively fed to the inputs A and B, with the MSB of the first byte fed to input A. The data to the encoder is fed by blocks of k (a multiple of 8) bits or N (a multiple of 4) couples ($k = 2N$ bits), for a value N lying in the range $8 \leq N/4 \leq 1024$.

The encoded bits C_1 and C_2 are generated in two rounds as follows: In the first round, the encoder (after initialization) is fed by the input sequences A and B in the natural order (with the switch at position 1). This first-round encoding is called *C_1 encoding*. In the second round, the encoder (after initialization) is fed by the CTC interleaved sequence (with the switch at position 2). This second-round encoding is called *C_2 encoding*. The two sets of parity bits, Y_1, Y_2 and W_1, W_2, are generated as shown in Figure 4.19, with Y_1 and W_1 obtained for C_1 encoding and Y_2 and W_2 obtained for C_2 encoding, respectively. The generator polynomials are $1 + x^2 + x^3$ for the output Y and $1 + x^3$ for the output W.

Advantages of DBTC

The CTC encoder including its constituent encoder uses a double binary circular *recursive systematic convolutional* (RSC) code. Parallel concatenation of double binary RSC codes offers several advantages in comparison with the classical one input turbo codes.

1. *Better convergence of the iterative process*: The bidimensional iterative process converges better, since the density of the erroneous paths is lower in

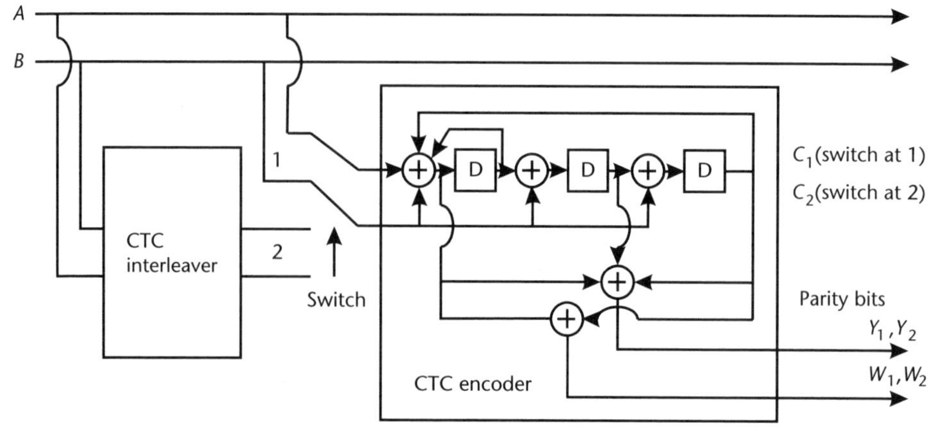

Figure 4.19 Convolutional turbo code encoder. (*After:* [1, 2].)

each dimension, which reduces the correlation effects between the component decoders.

2. *Larger minimum distances*: The minimum distances may be increased by one more degree in the design of permutation. A typical dilemma in the design of a good permutation lies in the need to obtain a large minimum distance for two distinct classes of codewords, which requires conflicting treatment. If some disorder may be instilled without altering the regular character of the classical permutation, it is possible to define very robust permutations toward all codewords.
3. *Less puncturing for a given rate*: In order to obtain a coding rate higher than 1/2 from the RSC encoder, it discards fewer numbers of redundant symbols, as compared with the binary TC. Consequently, the correcting ability of the constituent code is less degraded.
4. *Higher throughput and lower latency*: The decoder of DBTC provides 2 bits at each decoding step. Thus, once the data block is received, and for a given processing clock, the decoding throughput of hardware decoder is doubled. As a consequence, the latency (i.e., the number of clock periods required to decode a data block) is then halved, as compared with the classical binary TC.
5. *Robustness of the decoder*: The performance gap between the full algorithm and its simplified version, or between the *maximum a posteriori* (MAP) algorithm and the *soft-output Viterbi algorithm* (SOVA), lies in the 0.2–0.6 dB range, depending on the block size, in the case of the binary turbo codes. This gap halves if DBTC is used. So DBTC decoding may be done without using the full MAP decoder.

Tail-Biting Technique

From the strict definition of convolutional codes, it is clear that convolutional codes can be applied only to semi-infinite sequences. However, almost any communication system is block-oriented due to practical constraints. There are several techniques of transforming a convolutional code into a block code. While it is plausible to stop the encoding process after the information block length, it actually leads to weak error protection for the last codeword bits. The typical solution that avoids such performance degradation is to force the encoder back to the all-zero state by appending a block of the tail bits to the information vector. However, this method may become inefficient due to the rate loss caused by the termination process, especially when the codewords are very short. An alternative way to transform a convolutional code into a block code is to make the initial state identical to the final state, allowing the initial state to be any state of the encoder. The code trellis in this case can be viewed as a circle, without any state discontinuity. This termination technique is called *tail biting* [6]. The tail-biting technique is advantageous over the classical trellis-termination technique, which drives the encoder to the all-zero state. As no extra bits are added or transmitted, there is no rate loss and no reduction in the spectral efficiency of transmission. A weak point of the tail-biting technique is in the double encoding process, which finds the initial state equal to the final state for the given information sequence, but it is not a critical limitation, as the encoding process is a low-complexity process.

We approach the tail-biting encoder issue more rigorously: We consider a convolutional encoder that generates the n-tuple $\mathbf{v}_t$ of code bits at time t, given the k-tuple $\mathbf{u}_t$ of information bits, where $\mathbf{v}_t \in GF(2^n)$, $\mathbf{u}_t \in GF(2^k)$, and $t \geq 0$. The state of the encoder at time t is denoted by m-tuple $\mathbf{x}_t$, where m is the memory of the encoder. For encoders without feedback it is easy to fulfill the tail-biting boundary condition, $\mathbf{x}_0 = \mathbf{x}_N$, since the ending state $\mathbf{x}_N$ depends only on the last m inputs $\mathbf{u}_{N-m}, \ldots, \mathbf{u}_{N-1}$ to the encoder, where N is the number of information blocks. For feedback encoders, the situation is complicated because the ending state $\mathbf{x}_N$ depends on the entire information vector $\mathbf{u} = (\mathbf{u}_0, \ldots, \mathbf{u}_{N-1})$. Thus, for a given information vector $\mathbf{u}$, we must calculate the initial state $\mathbf{x}_0$ that will lead to the same state after N cycles.

We formulate the problem in state space representation as follows:

$$\begin{aligned} \mathbf{x}_{t+1} &= \mathbf{A}\mathbf{x}_t + \mathbf{B}\mathbf{u}_t^T \\ \mathbf{v}_t^T &= \mathbf{C}\mathbf{x}_t + \mathbf{D}\mathbf{u}_t^T \end{aligned} \tag{4.22}$$

where $\mathbf{A}$ is an $m \times m$ state transition matrix and $\mathbf{C}$ is an $n \times m$ measurement matrix. We express the complete solution of the state equation by the superposition of the zero-input solution $\mathbf{x}_t^{[zi]}$ and the zero-state solution $\mathbf{x}_t^{[zs]}$:

$$\mathbf{x}_t = \mathbf{x}_t^{[zi]} + \mathbf{x}_t^{[zs]} \tag{4.23}$$

As we require that the state at time $t = N$ is equal to the initial state $\mathbf{x}_0$, we obtain, by (4.23) and the relation $\mathbf{x}_t^{[zi]} = \mathbf{A}^t \mathbf{x}_0$, that

$$\mathbf{x}_0 = \left(\mathbf{I}_m + \mathbf{A}^N\right)^{-1} \mathbf{x}_N^{[zs]} \tag{4.24}$$

where $\mathbf{I}_m$ is the $m \times m$ identity matrix. Provided the matrix $(\mathbf{I}_m + \mathbf{A}^N)$ is invertible in binary sense, it is possible to calculate the correct initial state $\mathbf{x}_0$ if the zero-state response $\mathbf{x}_N^{[zs]}$ is known.

According to the previous discussion, the encoding process is divided into two steps:

1. Determine the zero-state response $\mathbf{x}_N^{[zs]}$ for the given information vector $\mathbf{u}$: We start the first step with the encoder set at the all-zero state, $\mathbf{x}_0 = 0$, feed the input information vector $\mathbf{u}$ to the encoder, but discard the output bits. After N cycles, the encoder reaches the state $\mathbf{x}_N^{[zs]}$. So we can calculate the corresponding initial state $\mathbf{x}_0$ using (4.24), and can initialize the encoder accordingly. Then we store the precomputed solutions of (4.24) for various $\mathbf{x}_N^{[zs]}$ in a lookup table.
2. Perform the actual encoding process: We start the second step with the encoder set at the correct initial state $\mathbf{x}_0$, and feed the input information vector $\mathbf{u}$. Then we can get the valid codeword vector $\mathbf{v}$.

For example, we derive the lookup table in IEEE 802.16e standards. From Figure 4.19, we get the state transition matrix in the form

$$\mathbf{A} = \begin{bmatrix} 1 & 0 & 1 \\ 1 & 0 & 0 \\ 0 & 1 & 0 \end{bmatrix} \tag{4.25}$$

and $m = 3$. By applying it to (4.24) for $N = 1, ..., 6$, we obtain the lookup table in Table 4.8. In case N is 7 or a multiple of it, $(\mathbf{I}_m + \mathbf{A}^N)$ in (4.24) becomes a singular matrix, so it is not possible to determine the initial state that is identical to the final state of the given information sequence.[7]

CTC Interleaver (Inner Interleaver)

Inside the overall CTC encoder block, a CTC interleaver is included (see Figure 4.19), which rearranges the input sequences before the C_2 encoding process. This rearrangement, or permutation, introduces randomness to the encoding process. The permutation may be done in numerous different forms but, among them, the permutation adopted in the IEEE 802.16e standards is given by the form

$$i = \prod(j) = Pj + Q(j) + i_0 \tag{4.26}$$

where $Q(j)$ is an integer whose value is taken from the set $\{0, Q_1, Q_2, ..., Q_{C-1}\}$ in cyclic way; P is an integer that is relatively prime to the length of the encoding blocks, N_{EP}; and i_0 is the starting index. The cycle of the permutation, C, must be a divisor of N and is typically chosen to be 4 or 8. For instance, in the case $C = 4$, the permutation law takes the expressions

$$\begin{aligned} &\text{if } j = 0 \bmod 4, \quad i = \prod(j) = Pj + 0 + i_0 \bmod N \\ &\text{if } j = 1 \bmod 4, \quad i = \prod(j) = Pj + Q_1 + i_0 \bmod N \\ &\text{if } j = 2 \bmod 4, \quad i = \prod(j) = Pj + Q_2 + i_0 \bmod N \\ &\text{if } j = 3 \bmod 4, \quad i = \prod(j) = Pj + Q_3 + i_0 \bmod N \end{aligned} \tag{4.27}$$

Table 4.8 Lookup Table for the Initial State of IEEE 802.16e Standards

N mod 7	**Final State from Zero State**							
	0	1	2	3	4	5	6	7
1	0	6	4	2	7	1	3	5
2	0	3	7	4	5	6	2	1
3	0	5	3	6	2	2	1	4
4	0	4	1	5	6	7	7	3
5	0	2	5	7	1	3	4	6
6	0	7	6	1	3	4	5	2

Source: [1, 2].

7. Since the encoder state circulates from the initial state to the final state, the initial state is also referred to as the *circulation state*.

where N is a multiple of 4.

Subblock Interleaver (Outer Interleaver)

In double binary CTC encoding, the encoded data form six subblocks, namely, A, B, Y_1, Y_2, W_1, and W_2 subblocks (see Figure 4.19). Two of them are systematic parts and four of them are parity parts. The six subblocks are interleaved separately, and the interleaving is done in the unit of symbols. The subblock interleaving basically uses the *bit reverse order* (BRO) interleaving technique. The BRO interleaving provides almost uniform puncturing patterns with the pruned symbols. The BRO interleaver is applicable only to the encoding blocks whose length is a power of 2. To overcome this demerit, the encoding blocks are first fragmented into multiple parts whose length is a power of 2, and then BRO interleaving is applied to each fragment. This interleaving scheme is called the *partial BRO* (PBRO) [7], and is used in the IEEE 802.16e system.

The PBRO operation is performed in such a way that the input address k is converted to the tentative output address T_k through the operation (see Figure 4.20)

$$T_k = 2^m (k \bmod J) + BRO_m(\lfloor k/J \rfloor) \tag{4.28}$$

where m is the exponent indicating the size of fragement, 2^m, and J is the number of fragments. The term $BRO_m(x)$ indicates the bit reversed m-bit value of x. Table 4.9 lists the subblock interleaver parameters for the case of the CTC encoder in Figure 4.19, with the code rate 3 and the number of subblocks 6.

After the subblock interleaving is completed, two sets of parity subblocks, (Y_1, Y_2) and (W_1, W_2), are interleaved in pairs, respectively. Figure 4.20 shows the block diagram of the overall interleaving process.

The interleaved symbols are selected to generate the subpacket. The length of subpacket is determined to support the various coding rates and modulations. Let k be the subpacket index when HARQ is used, which starts from 0 and increases by one in the subsequent subpackets. It is set to 0 when HARQ is not used. Let N_{EP} be the number of bits in the information block before encoding. Then the encoded bits are $3N_{EP}$ after CTC encoding because the mother code rate is 1/3. Among the $3N_{EP}$ encoded bits, $L_k = 48 \cdot N_{SCHk} \cdot m_k$ bits are needed to generate a subpacket, where N_{SCHk} is the number of the concatenated slots for the subpackets and m_k is the modulation order for the kth subpacket, (e.g., $m_k = 2$ for QPSK, 4 for 16-QAM, and 6 for 64-QAM). Noting that the encoded data symbols are carried by N_{SCHk} slots of 48 symbols each, we get the relation

$$N_{EP}/m_k \times \frac{1}{R} = N_{SCHk} \cdot 48 \tag{4.29}$$

where R is the code rate of the encoding. We define

$$F_k = (\mathrm{SPID}_k \cdot L_k) \bmod (3N_{EP}) \tag{4.30}$$

for the N_{EP} determined by (4.29), where SPID_k is the *subpacket ID* for the kth subpacket. Then the index of the ith symbol for the kth subpacket is

Table 4.9 Subblock Interleaver Parameters

Block size N_{EP}	Subblock size N	Subblock interleaver parameters	
		m	*J*
48	24	3	3
72	36	4	3
96	48	4	3
144	72	5	3
192	96	5	3
216	108	5	4
240	120	6	2
288	144	6	3
360	180	6	3
384	192	6	3
432	216	6	4
480	240	7	2

Source: [1, 2].

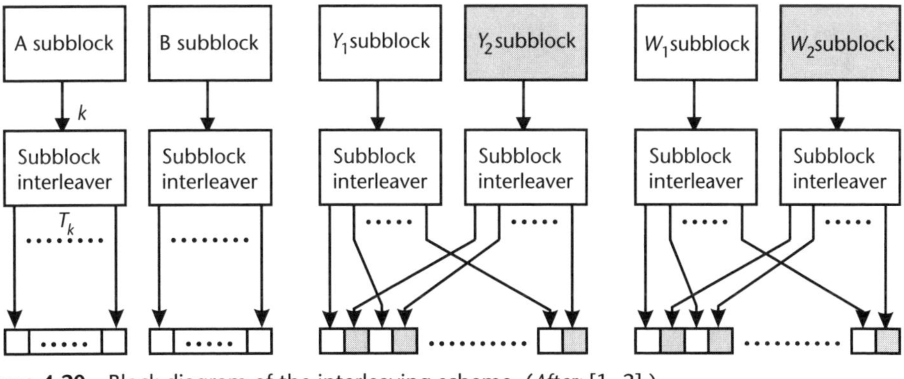

Figure 4.20 Block diagram of the interleaving scheme. (*After:* [1, 2].)

$$S_{k,i} = (F_k + i)\,\text{mod}(3N_{EP}) \tag{4.31}$$

For example, we consider the case of QPSK 1/2. Let N_{SCHk} be 2 for all subpackets. Then $L_k = 48 \cdot 2 \cdot 2 = 192$ for all subpackets and $3N_{EP} = 288$ by (4.29). In addition, $F_0 = 0$, $F_1 = 192$, $F_2 = 96$, and $F_3 = 0$ by (4.30). Then by (4.31) we get the following subpacket arrangement: the zeroth subpacket spreads over 0–191; the first subpacket, over 192–287 and 0–95; the second subpacket, over 96–287; and the third subpacket, over 0–191. Figure 4.21 illustrates the generation of those four subpackets.

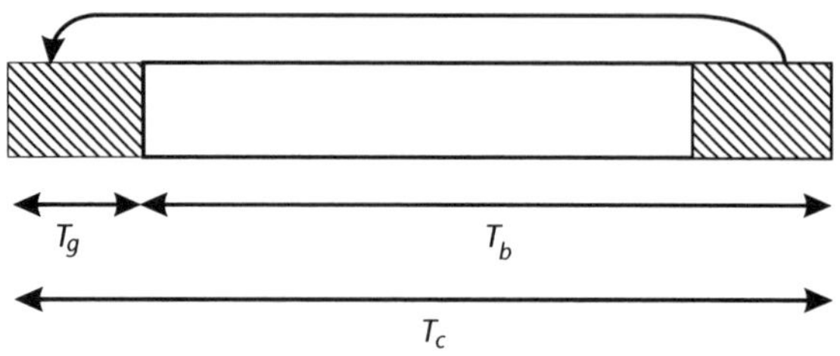

Figure 4.21 Subpacket generation through symbol selection.

4.2.3 Hybrid ARQ

Automatic repeat request (ARQ) and *hybrid automatic repeat request* (HARQ) are the mechanisms that are intended to enhance the reliability of the transmitted information by retransmitting the undelivered or incorrectly delivered messages. HARQ is a variation of the ARQ error control method supplied by the FEC subsystem. Whereas ARQ disregards the original incorrect messages and totally relies on the retransmitted messages, HARQ exploits the information in the original message to aid the decoding of the retransmitted messages. So HARQ yields better performance than ordinary ARQ does, particularly over the wireless channels, but it costs increased implementation complexity [8]. (Refer to Sections 1.1.3 and 2.1.6 for more details of HARQ.)

HARQ techniques are classified in terms of the combining method (chase or incremental redundancy) and protocol timing (synchronous or asynchronous). In the IEEE 802.16e standards, both Chase combining and *incremental redundancy* (IR) HARQ methods are adopted, but in the WiMAX profile only Chase combining HARQ is included. In addition, asynchronous HARQ is standardized in IEEE 802.16e.

Chase Combining and Incremental Redundancy HARQ

The simplest version of HARQ may combine FEC and ARQ by encoding the data block plus error-detection information with an error-correction code prior to transmission. However, the performance can be enhanced if the incorrectly received coded data blocks are stored at the receiver and used in combination with the next retransmitted data blocks for the error correction. This type of HARQ scheme is the *Chase combining* technique. In contrast, the IR technique takes different encoding for different (re)transmissions, instead of simply repeating the same codes. Then this increased redundancy results in an increased likelihood of error-recovery capability (refer to Section 2.1.6 for more details).

Synchronous and Asynchronous HARQ

The HARQ techniques that are classified according to time intervals between initial transmission and retransmission are *synchronous HARQ* (SHARQ) and *asynchronous HARQ* (AHARQ). Figure 4.22(a) illustrates the timing that occurs during SHARQ packet transmission/reception. In the figure, the horizontal axis indicates the time axis, and the upper part indicates the transmitter and the lower part the receiver. The arrow indicates the propagation process from transmitter to receiver and vice versa. In SHARQ, packet transmission/reception occurs in regular period, so no control overhead is needed.

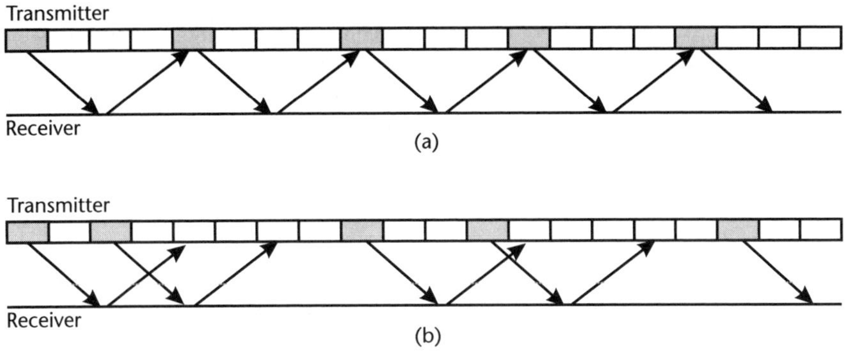

Figure 4.22 Illustration of timing for HARQ: (a) SHARQ (synchronous), and (b) AHARQ (asynchronous) [9].

In contrast, in the case of the AHARQ scheme, packet transmission timing is controlled by the transmitter. According to the channel quality information, the transmission timing is adjusted such that data transmission occurs when the channel is in good state. Figure 4.22(b) illustrates the timing of packet transmission/reception in AHARQ.

PHY Support of HARQ

HARQ is an optional function of the MAC layer and is supported only for the OFDMA PHY. One or more MAC PDUs are concatenated to form a PHY burst, and a CRC and parity fields are added to the PHY burst to construct a HARQ encoder packet. Figure 4.23 illustrates this process. Note that the PHY burst does not allow a mixture of HARQ and non-HARQ traffic.

The size of a HARQ encoder packet is 3,000 bytes at maximum. From a HARQ encoder packet, four subpackets are generated, which may differ at each retransmission. The four subpackets are identified by a *subpacket identifier* (SPID), such as 00, 01, 10, or 11. For the generation of subpackets, two main variants of HARQ, Chase combining and *incremental redundancy* (IR), are supported. In the case of the Chase combining, the PHY layer encodes the HARQ encoder packet, generating only one version of encoded subpacket, so SPID is not necessary. In the case of the IR, the PHY layer encodes the HARQ encoder packet to generate several versions of encoded subpackets, which are uniquely identified by SPID.

HARQ and its associated parameters are specified and negotiated during network entry or reentry procedure. Basically, HARQ operates on per-connection

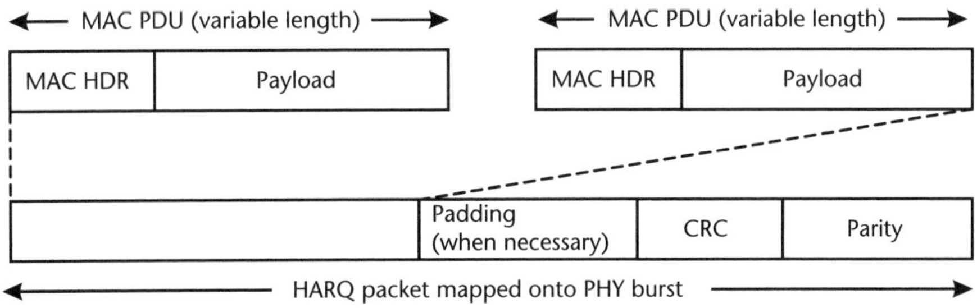

Figure 4.23 Illustration of HARQ encoder packet construction. (*After:* [1, 2].)

basis (i.e., it is enabled on a per CID basis, but it can be implemented in two different ways—per-terminal basis or per-connection basis). Specifically, HARQ may be enabled either on per-terminal basis for all active CIDs of a terminal, or on per-CID basis using the DSA/DSC messages. If HARQ is supported, MS normally supports per-terminal implementation, whereas BS supports per-connection implementation.

HARQ Operation

When an MS receives a subpacket, it tries to decode the encoded packet by comparing it with the previously received subpackets. It can achieve some coding gain by applying the Chase combining technique in case identical subpackets were retransmitted, and by applying the IR technique in case different redundancy information was transmitted. If the MS can decode the encoded packet, it sends an ACK message to the BS and, on receiving it, the BS terminates the HARQ transmission for the encoded packet. If the MS cannot decode the encoded packet, it sends a NACK message to the BS and the BS performs retransmission process. This procedure continues until the MS successfully decodes the packet and sends back acknowledgment.

For IR, each HARQ attempt has a uniquely encoded subpacket. When transmitting such subpackets, the following rule applies: At the first transmission, BS sends the subpacket labeled "00"; the BS may then send any numbered subpacket in any order. In addition, the BS may send more than one copy of any subpacket and may omit any subpacket except for those labeled "00."

Specifically, subpacket transmission follows the procedure given here:

1. In the initial transmission, the BS sends the subpacket whose SPID is 00.
2. The BS transmits the subpackets whose SPID is 00, 01, 10, 11 as long as the number of the transmitted packets does not exceed the maximum HARQ retransmission number indicated in the *channel descriptor* (CD) message. The retransmission follows the HARQ scheme. Note that the identical subpacket having the SPID 00 is retransmitted in the case of the Chase combining scheme, whereas the retransmission is done in the order of SPID in the case of the CTC incremental redundancy scheme.
3. The BS may send one or more copies of the subpackets or may omit transmitting the subpackets other than those having the SPID 00.

In order to indicate new subpacket transmission, we toggle the 1-bit HARQ identifier sequence number (AI_SN) whenever an encoded packet is successfully transmitted. Noticing that the AI_SN value is changed, the receiver recognizes that the corresponding subpacket is obtained from a new encoded packet, so it discards the previously received subpackets. The HARQ scheme basically operates based on the *stop-and-wait* protocol. The terminal sends, after a predefined fixed delay time, the ACK message for the HARQ burst on the HARQ_ACK_DELAY field in the CD message. The time for retransmission corresponds to the asynchronous part of HARQ and is not fixed. HARQ scheme supports multiple HARQ channels for each terminal and may do channel-based encoded packet transaction. The number of HARQ channels is determined by the BS, and each HARQ channel is distinguished by the *HARQ channel identifier* (ACID). The ACID for subchannels is determined by the control information carried on the MAP.

4.3 OFDMA Frame Structuring

As discussed in Section 4.1.2, modulated data symbols and pilot tones are mapped onto subcarriers to form an OFDMA symbol, which is then IDFT processed and transmission processed. Multiple subcarriers are bundled into a subchannel in different ways depending on the type of subchannel. The subchannel then serves as the basic unit in organizing the OFDMA slot, which is again the basic building block to organize the DL/UL data burst. The DL/UL bursts for different users are mapped into the OFDMA frame structure together with the relevant mapping information overheads. The OFDMA frame structure is equipped with separate time slots to accommodate both DL and UL traffic. All these functions are taken care of by the frame structuring block (seen later in Figure 4.26).

4.3.1 OFDMA Slots and Bursts

OFDMA slot and burst are both intermediate building blocks of the OFDMA frame structure. OFDMA slot is a two-dimensional (in time and frequency) basic building block to organize the DL/UL data burst. Conversely, data burst is a two-dimensional data block into which multiple OFDMA slots that belong to each individual user are mapped. Consequently, the OFDMA symbols generated out of a user station are mapped into a burst in the OFDMA frame in multiple units of OFDMA slots.

OFDMA Slots

In the mobile WiMAX specification, *OFDMA slot* is the minimum possible data allocation unit. It is defined in two dimensions, one in time (i.e., the OFDMA symbol number) and the other in frequency (i.e., the subchannel number). The definition of the OFDMA slot depends on the OFDMA symbol structure, which differs for DL and UL, and for PUSC, FUSC, and band AMC subchannels. (Refer to Section 4.4 for the detailed descriptions of PUSC, FUSC, band AMC, and other subchannels.)

The OFDMA slots for PUSC, FUSC, and band AMC subchannels are defined as follows: In the case of the DL FUSC using the distributed subcarrier permutation, each slot is composed of a subchannel and an OFDMA symbol (i.e., 1 subchannel × 1 OFDMA symbol). For the DL PUSC using distributed subcarrier permutation, one slot is one subchannel by two OFDMA symbols (i.e., 1 subchannel × 2 OFDMA symbols). For the UL PUSC each slot is composed of a subchannel by three OFDMA symbols (i.e., 1 subchannel × 3 OFDMA symbols) with each subchannel consisting of six tiles. However, in the case of the DL- and UL-band AMC subchannels, which are based on adjacent subcarrier permutation, one slot is one subchannel by three OFDMA symbols (i.e., 1 subchannel × 3 OFDMA symbols). In all the cases, an OFDMA slots contains 48 data subcarriers. Table 4.10 lists a summary of this arrangement.

Data Bursts

Multiple OFDMA slots that contain the OFDMA symbols generated out of a user terminal are mapped in bulk into a two-dimensional data block called *burst*. Burst, in general terms, is a *data region* that is a two-dimensional data block consisting of a group of contiguous subchannels in a group of contiguous OFDMA symbols. The constituent OFDMA symbols are allocated through the resource allocation message (or *MAP* message) that a base station generates at the MAC layer. Figure 4.24 illustrates a data region.

When the data stream coming out of the encoding and modulation block is mapped to the DL/UL bursts, the mapping is OFDMA slot based, according to the procedures described next.

In the downlink, the modulated data stream is segmented into blocks that are sized to fit into one OFDMA slot in such a way that each slot spans one subchannel in the frequency axis and one or more OFDMA symbols in the time axis, in line with the definition of the OFDMA slots given in Table 4.10. In mapping the slots, the lowest numbered slot is put at the lowest numbered subchannel in the lowest numbered OFDMA symbol. The mapping is done in the order of increasing OFDMA subchannel index. When reaching the edge of the data region, the mapping is continued from the lowest numbered OFDMA subchannel in the next available OFDMA symbol. Figure 4.25(a) illustrates the order of mapping OFDMA slots to subchannels and symbols in the downlink for the case of PUSC. Notice that the mapping in the vertical direction is double OFDMA symbol based, as defined in Table 4.10.

In the uplink, the data mapping is done in two steps: The first is to select the OFDMA slots to allocate to each burst, and the second is to map the allocated slots to data region.

For the allocation of OFDMA slots to bursts, the data is segmented into blocks that are sized to fit into one OFDMA slot such that each slot spans one or more subchannels in the frequency axis and one or more OFDMA symbols in the time axis, according to the definition of the OFDMA slots given in Table 4.10. In mapping the slots, the lowest numbered slot is put at the lowest numbered OFDMA symbol in the lowest numbered subchannel. The mapping is continued in the order of increasing OFDMA symbol index. When reaching the edge of the UL zone, the mapping is continued from the lowest numbered OFDMA symbol in the next available subchannel. The UL allocation is done in the increasing fashion by selecting an integer number of contiguous slots according to this ordering. Figure 4.25(b) illustrates the burst structure (shaded area) thus obtained for the case of uplink PUSC. Notice that the mapping in the vertical direction is triple OFDMA symbol based, as defined in Table 4.10.

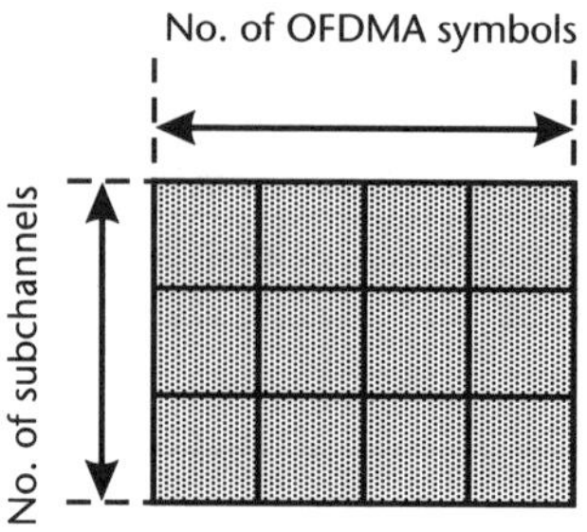

Figure 4.24 Illustration of data region. (*After:* [1, 2].)

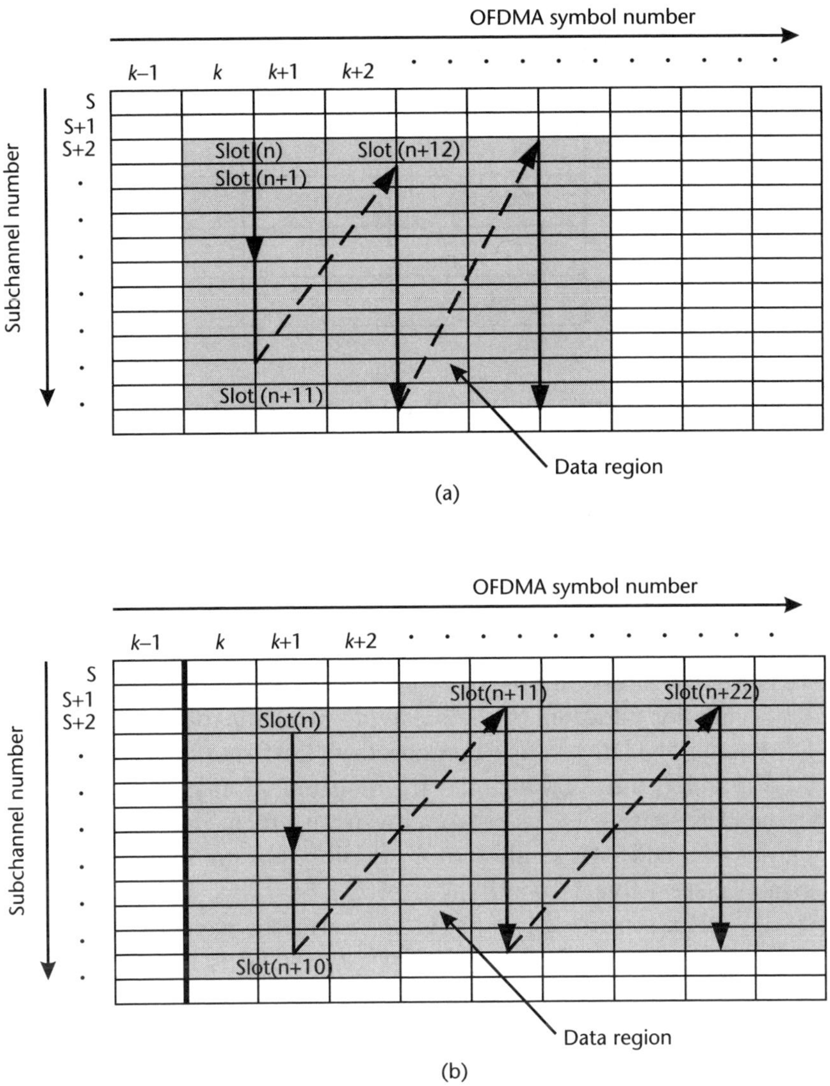

Figure 4.25 Illustration of data mapping to data region in PUSC mode: (a) downlink, and (b) uplink. (*After:* [1, 2].)

For the mapping of OFDMA slots within the UL burst, the slots are mapped in such a way that the lowest numbered slot occupies the lowest numbered subchannel in the lowest numbered OFDMA symbol. The mapping is continued in the order of increasing subchannel index. When reaching the last subchannel, the mapping is continued from the lowest numbered subchannel in the next OFDMA symbol that belongs to the UL burst. The resulting order is indicated by arrows in Figure 4.25(b).

4.3.2 OFDMA Frame

The data bursts are finally mapped into the OFDMA frame structure together with the relevant mapping description (or MAPs) and other overheads. OFDMA frame

Table 4.10 Definition of OFDMA Slot

Subchannelization scheme	Number of Subchannels per slot	Number of Symbols per slot
DL PUSC	1	2
DL FUSC	1	1
UL PUSC	1	3
DL/UL AMC	1	3

structure, in general, consists of a transmission period for DL transmission and another transmission period for UL transmission, which are interlaced each other with a gap intervening in between. OFDMA frame may also be comprised of multiple zones with each zone accommodating PUSC, FUSC, or AMC traffic.

Frame Structure

Figure 4.26 shows an example of the OFDMA frame structure of the Mobile WiMAX system employing the TDD scheme. It consists of a DL transmission period and a UL transmission period, with a gap following each period. In the DL part, the first symbol is allocated to the preamble, which is followed by DL-MAP and other DL-bursts. The DL part also contains the UL-MAP. In the UL part, ranging channel, CQI channel, and ACK channel are allocated. Notice the OFDMA frame structure in the PUSC case—it is clear from the arrangement that the DL bursts are in multiples of two OFDMA symbols and the UL bursts are in multiples of three OFDMA symbols (see Table 4.10 and Figure 4.25).

A *Tx/Rx transition gap* (TTG) is inserted between the DL and UL parts in the same frame and an *Rx/Tx transition gap* (RTG) is inserted between the end of a

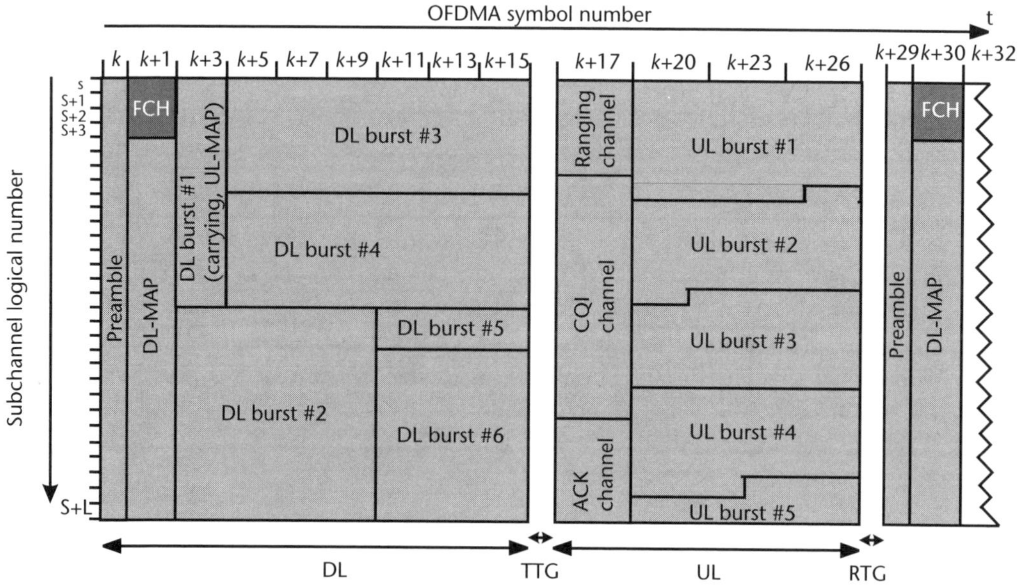

Figure 4.26 Example of an OFDMA frame structure in TDD mode. (*After:* [1, 2].)

frame and the start of the next frame. Those values of TTG and RTG are determined in consideration of the transmission delay of the user terminals located near to and far from the base station of a cell when adjusting timing of the user terminals through initial ranging. It also considers the coverage of the relay stations, the scheduling/processing time at the BS, and the processing time at the MS.

In the case of the 2.3-GHz-based WiBro system, for example, the values of 87.2 μs and 74.4 μs are selected for TTG and RTG, respectively. So, in this case, the OFDMA TDD frame structure has the length of 5 ms, which is composed of 42 OFDM symbols and TTG/RTG (i.e., 42 OFDM symbols * 115.2 μs/symbol + TTG 87.2 μs + RTG 74.4 μs = 5 ms).

Downlink transmission begins with one preamble symbol, followed by a *frame control header* (FCH), DL-MAP, and DL data bursts. The two symbols coming after the preamble are always allocated to PUSC subchannel, which contains FCH to transmit the information on the frame organization. Among the DL bursts, the first one contains the UL-MAP for the uplink. The UL zone allocates space for the ranging channel, CQI channel, and ACK channel, as illustrated in Figure 4.26.

Multiple Zone Frame Structure

An OFDMA frame may be comprised of multiple zones (namely, PUSC, FUSC, AMC), as illustrated in Figure 4.27. The downlink frame may contain the subchannels of PUSC, FUSC, AMC type, but, in contrast, the uplink frame may contain the subchannels of only PUSC and AMC type.

Among the plurality of the zones in the DL and UL subframes, only the preamble and the first PUSC in the DL subframe are mandatory—all the others are optional. This first PUSC contains the FCH and DL-MAP fields. Transition from one region to another is indicated by STC_DL_Zone_IE in the DL-MAP. DL-MAP and UL-MAP allocations cannot span over multiple zones. The physical parameters such as channel state and interference levels may change from zone to zone.

In allocating the downlink subframes, the maximum number of downlink zones in one downlink subframe is limited to five (by the WiMAX Forum). For each MS, the maximum number of bursts to decode in one downlink frame is limited to 64.

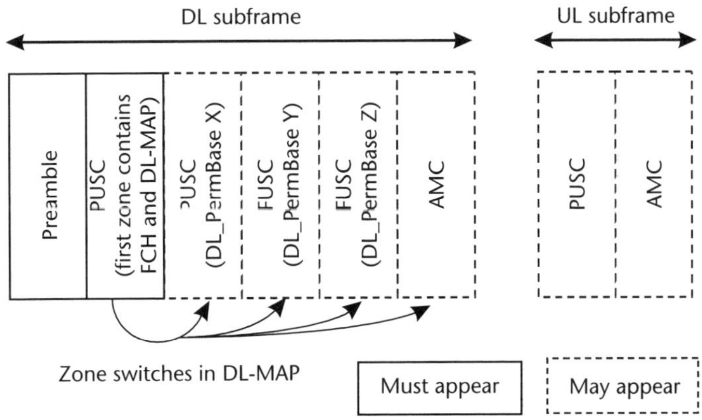

Figure 4.27 Frame structure comprised of multiple allocation zones. (*After:* [1, 2].)

For each MS, the maximum number of bursts transmitted concurrently[8] and directed to the MS is limited by the value specified in Max_Num_Bursts TLV. In the case of Mobile WiMAX, this value is 10.

4.3.3 FCH and DL/UL MAPs

The OFDMA frame structure is composed of DL/UL data bursts to carry user traffic and overhead fields to carry preamble, FCH, DL/UL MAPs, CQICH, ACK channel, and ranging subchannel. The preamble enables frame synchronization and equalization in the receiver, and DL/UL MAPs help to figure out how to access the DL/UL bursts.

Preamble

Preamble is a standard-defined sequence of symbols known by the receiver. The preamble is used by the PHY layer for synchronization and equalization. The preamble is the first symbol in the downlink transmission frame. There are three types of preamble carrier sets, which are defined by allocating different subcarriers for each one of them. Those subcarriers are modulated using a boosted BPSK modulation with a specific PN code.

The preamble is composed of each third subcarrier, and the preamble carrier sets are defined using the following equation:

$$PreambleCarrierSet_n = n + 3k \tag{4.32}$$

where $PreambleCarrierSet_n$ specifies all subcarriers allocated to the specific preamble for segment index $n = 0, 1, 2$; and k is a running index 0, 1, ..., 283. The total number of PN series modulating the preamble carrier set is 114, and the PN series modulated depends on the segment used and the IDCell parameter.

FCH

FCH is for the transmission of DL_Frame_Prefix. DL_Frame_Prefix is a data structure that is transmitted at the beginning of each frame. It contains information on the current frame such as the length of the subsequent DL-MAP message, the repetition encoding, and the channel coding that are applied to the DL-MAP.

The DL_Frame_Prefix has the format shown in Figure 4.28, which applies to all FFT sizes other than 128. It consists of 24 bits and is repeated twice to form a 48-bit FEC block. This 48-bit FEC block is repeated four times, and allocated to the first 4 slots as shown in Figure 4.26. In the case of the FFT size of 128, the format reduces to 12 bits, with each of the six fields reduced to 1, 1, 2, 3, 5, and 0 bits, respectively.

6 bits	1 bit	2 bits	3 bits	8 bits	4 bits
Used subchannel bitmap	Res	Rep. cod. Ind.	Cod. Ind.	DL-MAP length	Res.

Figure 4.28 OFDMA DL_Frame_Prefix format. (*After:* [1, 2].)

8. The concurrently transmitted bursts mean the bursts that share the same OFDMA symbol.

This 12-bit DL_Frame_Prefix is repeated four times to form a 48-bit block before being mapped to the FCH. Repetition is not applied in the case of 128 FFT size, so only the first slot is dedicated to FCH.

In the DL_Frame_Prefix format, the "used subchannel bitmap" field indicates which groups of subchannels are used on the first PUSC zone and on the PUSC zones with "use all SC=0" in the STC_DL_Zone_IE(). Value 1 means that it is used by this segment, and 0 means that it is not used by this segment. (Refer to Section 4.4.1 for details.) The 2-bit "repetition coding indication" field indicates the repetition code used for the DL-MAP. It may be 0, 1, 2, or 3, respectively, meaning that repetition coding of 0, 2, 4, or 6 is used on the DL-MAP. The "coding indication" field indicates the FEC code used for the DL-MAP. It could be CC, BTC, CTC, ZT CC, CC+interleaver, or LDPC encoding. In the case of Mobile WiMAX, CTC is used for DL-MAP encoding. The DL-MAP is transmitted with QPSK modulation at FEC rate 1/2. The "DL-MAP length" field defines the length (in slots) of the DL-MAP message, after repetition coding.

Location of FCH is dependent on the segment index. FCH subchannel starts from subchannel 0 in segment 0, 10 in segment 1, and 20 in segment 2, respectively, in the case of the 1.024 FFT OFDMA. After decoding the DL_Frame_Prefix message within the FCH, the MS knows how many and which subchannels are allocated to the PUSC segment. In order to observe the allocation of the subchannels in the downlink as a contiguous allocation block, the subchannels are renumbered. The renumbering for the first PUSC zone is performed in a cyclic manner such that the subchannel number starts from the FCH subchannels to the last allocated subchannel, then continues from the first allocated subchannel to the subchannel right before the FCH subchannels.

DL-MAP

DL-MAP is a MAC management message that defines the access to the downlink information. The DL-MAP message is mapped to the slots starting after the FCH field and continues to the next PUSC symbols if necessary. The length of DL-MAP message varies depending on the PHY-dependent information length. If the length of the DL-MAP message is not a multiple of a byte, then it is padded up to the next integral number of bytes, but the padded bits are disregarded by the MS. The DL-MAP message is broadcast downlink by the BS.

Figure 4.29 depicts the DL-MAP message format, which is assigned with the MAC management message type number 2 (see Table 5.3), and a detailed view of the DL-MAP_IE().

The "PHY synchronization" field depends on the PHY specification used. In the case of the OFDMA PHY, it consists of an 8-bit frame duration code and a 21 bit frame number field. For example, the frame duration code is 4 for the case of 5-ms OFDMA frame. The frame number of 24 bits increments by one modulo 2^{24} at each frame count.

The "*DL channel descriptor* (DCD) count" field matches the value of the "configuration change count" field in the DCD message, specified under MAC management message type 1 (see Section 4.3.4 and Figure 4.35).

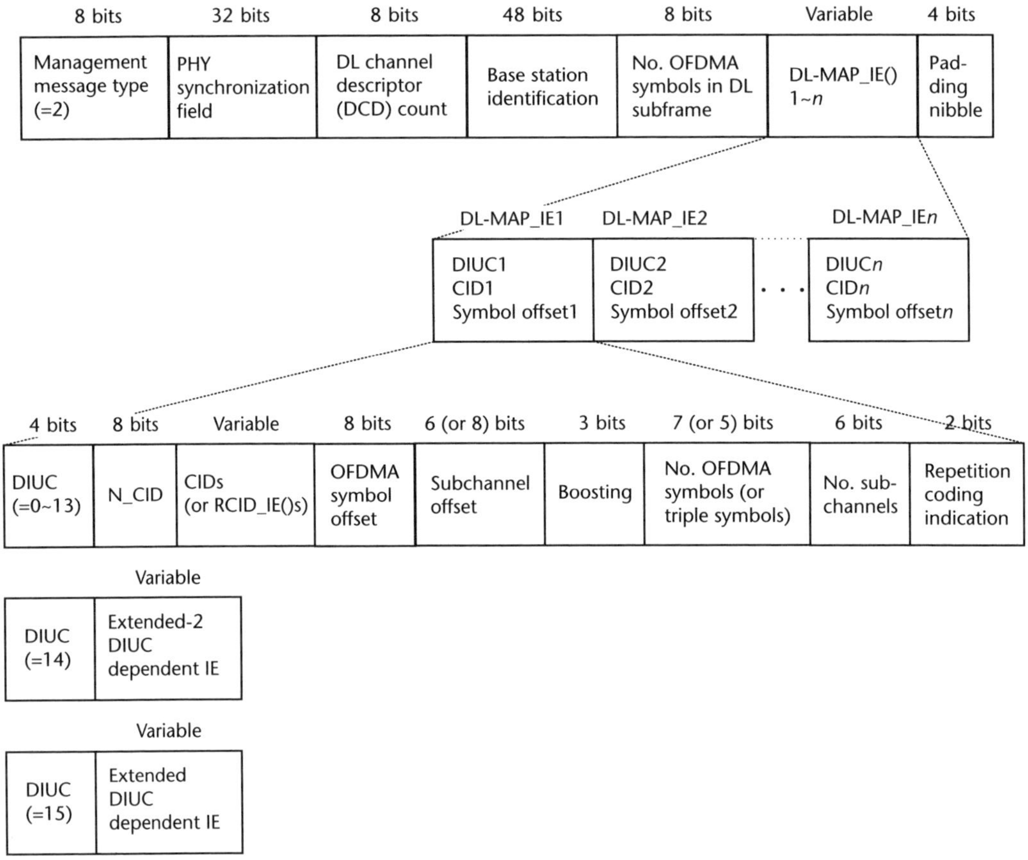

Figure 4.29 DL-MAP message format. (*After:* [1, 2].)

The "BS ID" field is to identify the BS. The most significant 24 bits among the 48 allocated bits are categorized as the operator ID, which is used to identify specific service provider.

The "DL-MAP_IE()" field is a PHY-dependent field that defines the downlink allocation patterns. It consists of various parameters including the "*downlink interval usage code* (DIUC)," which classifies/describes the usage of the burst;[9] the "*connection identifier* (CID)," which represents the MS to which the IE is assigned; the "OFDMA symbol offset," which is the offset of the OFDMA symbol in which the burst starts, measured from the downlink symbol in which the preamble is transmitted and counted in the number of OFDMA symbols; the "subchannel offset," which is the lowest index OFDMA subchannel used for carrying the burst, starting from subchannel 0; the "boosting," which indicates whether or not the subcarriers for this allocation are power-boosted; the "number of OFDMA symbols," which indicates the number of OFDMA symbols that are used to carry the downlink PHY burst; the "number of subchannels," which is the number of subchannles with subsequent indices used to carry the burst; and the "repetition coding indication,"

9. The DL-MAP_IE() format changes in the case of the extended DIUC and the extended-2 DIUC, which are the extensions of the DIUC with the DIUC values of 15 and 14, respectively. Refer to Section 4.3.4 for more details.

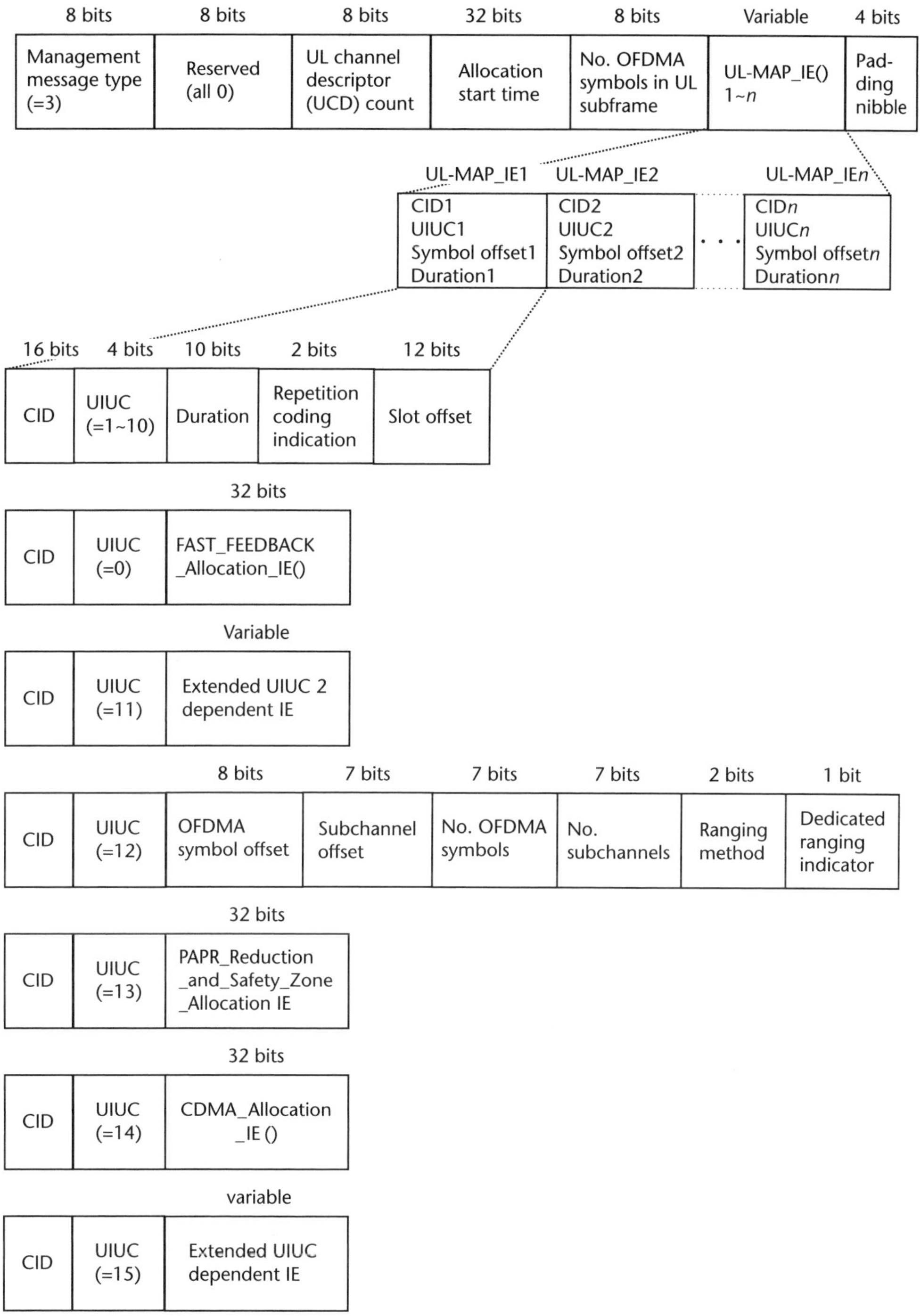

Figure 4.30 UL-MAP message format. (*After:* [1, 2].)

which applies only when the data in the burst are QPSK modulated (refer to [1, Table 275]; [2, Table 389]).

UL-MAP

UL-MAP is a MAC management message that defines the access to the entire uplink for all MSs during a scheduling interval. The UL-MAP message is mapped on the burst described by the first DL_MAP_IE of the DL-MAP message. In case there are multiple PDUs in the burst described by the first DL_MAP_IE, the UL-MAP message comes first. The length of UL-MAP message varies depending on the PHY-dependent information, UL-MAP_IE(). As in the case of the DL-MAP, the length of the UL-MAP message is padded up to the next integral number of bytes if it is not a multiple of a byte. The UL-MAP message is broadcast downlink by the BS.

Figure 4.30 depicts the UL-MAP message format, which is assigned with the MAC management message type number 3 (see Table 5.3), and a detailed view of the UL-MAP_IE().

The "*UL channel descriptor* (UCD) count" field matches the value of the "configuration change count" field in the UCD message, as specified under MAC management message type 0, which describes the uplink burst profile that applies to this map (see Section 4.3.4 and Figure 4.36).

The "allocation start time" field indicates the effective start time of the uplink allocation defined by the UL-MAP.

The "UL-MAP_IE()" field is a PHY-dependent field that defines the uplink bandwidth allocations. It consists of various parameters including the "*uplink interval usage code* (UIUC)," which defines the type of uplink access and the uplink burst profile associated with the access; the extended and extended-2 UIUCs, which are the extensions of the UIUC; the "*connection identifier* (CID)," which represents the MS to which the IE is assigned; ranging region allocation (in case UIUC=12); fast feedback channel region allocation (in case UIUC=0); UL allocation for CDMA bandwidth request (in case UIUC=14); the "duration," which indicates the duration of the UL burst allocation (in units of OFDMA slots); and the "repetition coding indication" for the bursts whose data are QPSK modulated (refer to [1, Table 287]; [2, Table 431]).

Compressed MAPs

The compressed DL-MAP format is a simplified version of the original DL-MAP format in which the 48-bit BS ID field is replaced with a subset of it, with the full 48-bit BS ID published in the DCD.

Figure 4.31 shows the compressed DL-MAP message format. In the figure, the "compressed map indicator" field is set to 110 to indicate a compressed map format and the "UL-MAP appended" field is set to 1 if a compressed UL-MAP is appended to the current compressed DL-MAP data structure. The "MAP message length" field specifies the length of the compressed map messages from the byte containing the "compressed map indicator" to the last byte of the compressed DL-MAP message, including the computed 32-bit CRC value. In case the "UL-MAP appended bit" is set, the "MAP message length" field specifies the extended length up to the last byte of the UL-MAP compressed message. The "DL IE count" field holds the number of the IE entries in the list of the DL-MAP_IEs that follow. The CRC is calculated in the same way as for the standard MAC messages across all the bytes of the compressed maps starting from the byte containing the "compressed map indicator" to the last byte of the maps as specified by the "MAP message length" field.

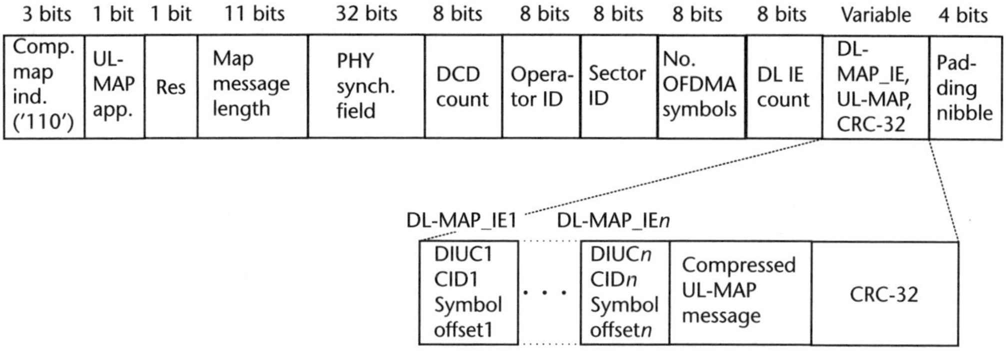

Figure 4.31 Compressed DL-MAP message format. (*After:* [1, 2].)

The presence of the compressed DL-MAP format is indicated by the contents of the most significant two bits of the first data byte, which overlay the HT and EC bits of the *generic MAC header* (GMH). When these bits are both set to 1, which is an invalid combination for a standard header, it indicates that the compressed DL-MAP format is present.

In the case of the compressed UL-MAP message, its format is identical to the standard UL-MAP message format, except that the GMH and the reserved fields are omitted. A compressed UL-MAP appears only after a compressed DL-MAP. The presence of a compressed UL-MAP is indicated by a bit in the compressed DL-MAP data structure.

Ranging Channel

A ranging channel is composed of a group of six adjacent UL PUSC subchannels.[10] This ranging channel is allocated to MSs for initial ranging, periodic ranging, handover ranging, and bandwidth request. (Refer to Sections 3.2.1 and 3.7.2 for details.) The binary ranging code of length 144 is transmitted on the 144 subcarriers of the ranging channel. The binary codes are the pseudo-noise codes produced by the PRBS shown in Figure 4.32, whose polynomial generator is $1 + x^1 + x^4 + x^7 + x^{15}$. The PRBS generator is initialized by the seed b14...b0 = 0, 0, 1, 0, 1, 0, 1, 1, s0, s1, s2, s3, s4, s5, s6, where s6 is the LSB of the PRBS seed, and s6:s0 = UL_PermBase, with s6 being the MSB of the UL_PermBase.

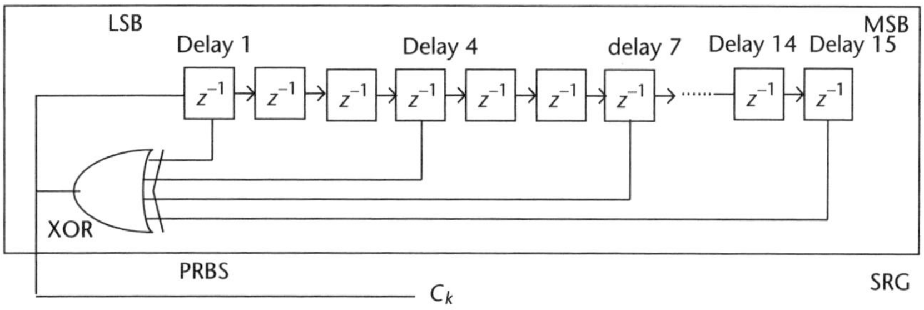

Figure 4.32 PRBS generator for ranging code generation. (*After:* [1, 2].)

10. Subchannels are considered adjacent if they have successive logical subchannel numbers.

The duration of initial ranging and handover ranging transmissions is two consecutive OFDMA symbols, since these transmissions occur before the MS acquires the correct timing offset of the system. The same ranging code is transmitted on the ranging channel during each symbol, with no phase discontinuity between the two symbols. Figure 4.33 illustrates the initial ranging or handover ranging transmission in time domain.

The duration of periodic ranging and bandwidth request transmissions is one OFDMA symbol, since these transmissions occur after MS is synchronized to the system. The indices of the subchannels that compose the ranging channel are specified in the UL-MAP message through UIUC = 12.

CQI Channel

The UL *channel quality information channel* (CQICH), or the fast feedback channel in synonym, is used for DL CINR report and MIMO mode selection feedback. One CQICH slot occupies one UL PUSC slot, and each tile of a CQICH slot carries an orthogonally modulated 8-ary alphabet such that a length 6 codeword over 8-ary alphabet of a CQICH slot can carry a data payload of 6 bits.[11]

Let $M_{n,8m+k}$ $(0 \le k \le 7)$ be the modulation symbol index of the kth modulation symbol in the mth UL PUSC tile of the nth CQICH. Then the possible modulation patterns composed of $M_{n,8m}$, $M_{n,8m+1}$, . . . , $M_{n,8m+7}$ are as defined in Table 4.11 for $P0 = \exp(j \cdot \frac{\pi}{4})$, $P1 = \exp(j \cdot \frac{3\pi}{4})$, $P2 = \exp(-j \cdot \frac{3\pi}{4})$, $P3 = \exp(-j \cdot \frac{\pi}{4})$, and $M_{n,8m+k}$ is mapped to a CQICH tile as shown in Figure 4.34.

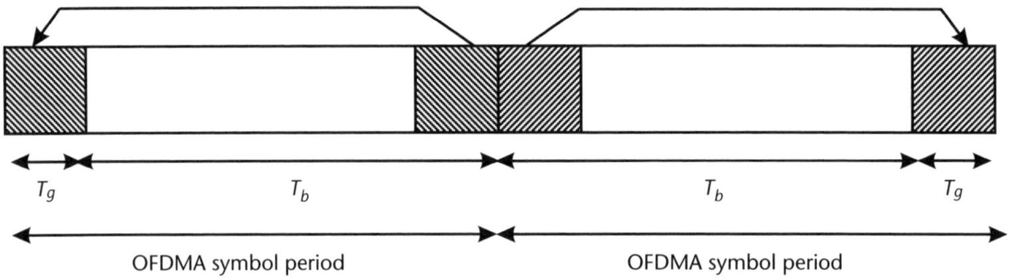

Figure 4.33 Initial ranging or handover ranging transmission. (*After:* [1, 2].)

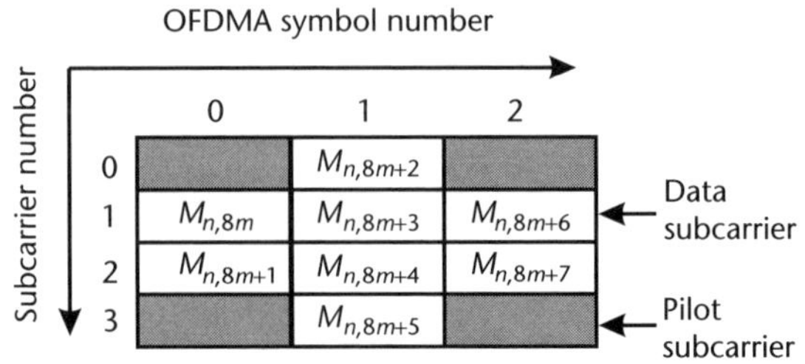

Figure 4.34 Subcarrier mapping of CQI modulation symbols. (*After:* [1, 2].)

11. It follows the Mobile WiMAX profile.

Table 4.11 Orthogonal Modulation Patterns of CQICH

Vector index	$M_{n,8m}, M_{n,8m+1}, \ldots, M_{n,8m+7}$
0	P0,P1,P2,P3,P0,P1,P2,P3
1	P0,P3,P2,P1,P0,P3,P2,P1
2	P0,P0,P1,P1,P2,P2,P3,P3
3	P0,P0,P3,P3,P2,P2,P1,P1
4	P0,P0,P0,P0,P0,P0,P0,P0
5	P0,P2,P0,P2,P0,P2,P0,P2
6	P0,P2,P0,P2,P2,P0,P2,P0
7	P0,P2,P2,P0,P2,P0,P0,P2

Source: [1, 2].

Among 64 CQICH codewords, the first 32 codewords are used for DL CINR report. The MS reports the DL CINR using the mapping rule in (4.33). The remaining CQICH codewords are used for MIMO mode selection feedback. CQICH region is allocated by UIUC=0 in the UL-MAP, and CQICH allocation to individual MS is done by CQICH allocation IE.

$$Payload\ bits = \begin{cases} 0, & CINR \le -3 \\ n, & (n-4) < CINR \le (n-3),\quad 0 < n < 31 \\ 31, & CINR > 27 \end{cases} \tag{4.33}$$

ACK Channel

The UL ACK channel provides feedback for DL HARQ. One ACK channel occupies a half subchannel, which is three pieces of UL PUSC tile. The even half subchannel consists of Tile(0), Tile(2), and Tile(4); and the odd half subchannel consists of Tile(1), Tile(3), and Tile(5). The Acknowledgment bit of the nth ACK channel is 0 (ACK) if the corresponding DL packet has been successfully received; otherwise, it is 1 (NAK). This 1 bit is encoded into a length 3 codeword over 8-ary alphabet for the error protection as described in Table 4.12.

Modulation and subcarrier mapping of UL ACK channel are the same as those of UL CQICH. ACK channel region is allocated by HARQ ACKCH region allocation IE. The HARQ-enabled MS that receives HARQ DL burst at frame i transmits the ACK signal through the ACK channel in the ACKCH region at frame $(i+1)$. The

Table 4.12 ACK Channel Subcarrier Modulation

ACK 1-bit symbol	Vector indices per tiles Tile(0),Tile(2),Tile(4) for even half subchannel Tile(1),Tile(3),Tile(5) for odd half subchannel
0	0,0,0
1	4,7,2

Source: [1, 2].

half-subchannel offset in the ACKCH region is determined by the order of HARQ-enabled DL burst in the DL MAP. For example, when an MS receives a HARQ-enabled burst at frame i, and the burst is the nth HARQ-enabled burst among the HARQ-related IEs, the MS transmits HARQ ACK at nth half-subchannel in ACKCH region that is allocated by the BS at frame $(i+1)$.

4.3.4 Burst Profiles

As shown in Figure 4.26, data burst refers to a contiguous portion of the data region that uses the same PHY parameters. A burst is called a DL burst if it belongs to the DL subframe and a UL burst if it belongs to the UL subframe. A burst is distinguished by the PHY parameters such as coding type and the DIUC or UIUC values, which are known as the burst profile. So each burst profile is dictated by DIUC and UIUC, respectively, in the downlink and uplink.

DL Burst Profile

Figure 4.35 shows the organization of the *type-length-value* (TLV) format of the DL_Burst _Profile and the DCD message into which the DL_Burst_Profiles of multiple DL bursts are mapped. The DCD message is a MAC management message classified as type 1, which is transmitted by the BS at a periodic interval (whose maximum value is 10 seconds) to define the characteristics of a downlink physical channel. The BS generates DCDs in the format shown in Figure 4.35(a). Among the constituent parameters, the "configuration change count" increments by one modulo 256 if any value in the DCD changes, and thereby enables the MS to quickly determine whether or not the remaining fields have changed and then properly react. TLV-encoded information such as Tx power of BS, TTG/RTG value, and handover-related parameters are also included in the DCD. At the end of the DCD message format, there follows a series of TLV-encoded information that describes the DL_Burst_Profiles of n DL bursts—see Figure 4.35(b). The TLV format of the DL burst profile, which is defined under the type number 1, has the structure shown in

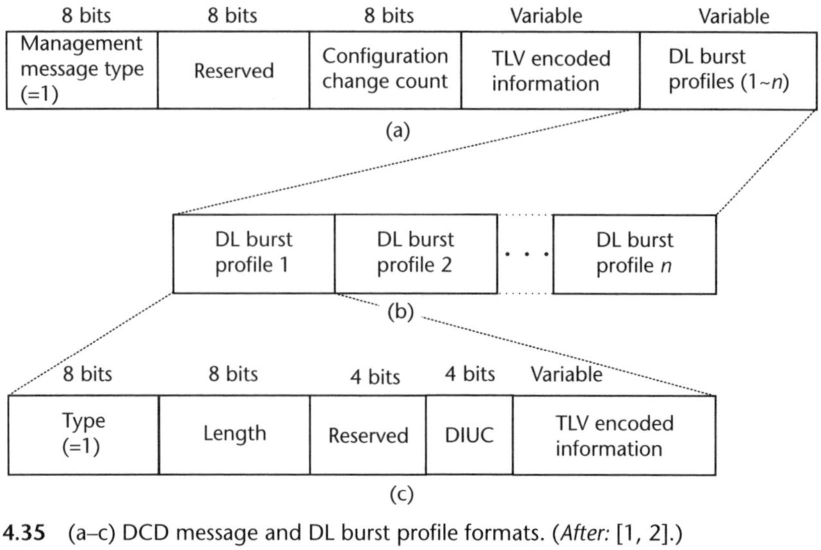

Figure 4.35 (a–c) DCD message and DL burst profile formats. (*After:* [1, 2].)

Figure 4.35(c). The DL_Burst_Profile is encoded with type 1 and a 4-bit DIUC. The DIUC value in the DL-MAP message is to specify the DL_Burst_Profile types for a specific DL burst.

DIUC

The DIUC field is associated with the DL_Burst_Profile and the DIUC values are used in the DL-MAP message to specify the Burst_Profile to be used for a specific DL burst. Table 4.13 lists the 16 DIUC values and their usage. In short, DIUC = 0 to 12 are allocated to different burst profiles; DIUC=13 is used for gap/PAPR reduction; and DIUC = 14 and 15 are both used for DIUC extensions.

To be more specific, DIUC = 0 has the burst profile parameters that are the same as those used for transmission of the DL-MAP message (i.e., the modulation and code rate of DIUC 0 is QPSK 1/2, and the coding type is determined by "coding indication" field of FCH). DIUC = 13 may be used for allocating subchannles to the *peak-to-average power ratio* (PAPR) reduction schemes.

The two DIUC extensions are intended to extend the coverage of the 4-bit DIUC. In the case of DIUC = 15, the DL-MAP extended IE format is comprised of a 4-bit extended DIUC indication, 4-bit length field, and a variable-sized unspecified data field. Similarly, in the case of DIUC = 14, the DL-MAP extended-2 IE format is comprised of a 4-bit extended-2 DIUC indication, 8-bit length field, and a variable-sized unspecified data field. The extended DIUC and extended-2 DIUC are used for allocation of subchannles to carry various different information elements as listed in Table 4.14.

UL Burst Profile

Figure 4.36 shows the organization of the TLV format of the UL_Burst _Profile and the UCD message into which the UL_Burst_Profiles of multiple UL bursts are mapped. The UCD message is a MAC management message classified as type 0, which is transmitted by the BS at a periodic interval (whose maximum value is 10 seconds) to define the characteristics of a uplink physical channel. The BS generates UCDs in the format shown in Figure 4.36(a). Among the constituent parameters, the "configuration change count" increments by one modulo 256 if any value in the UCD changes, and thereby enables the MS to quickly determine whether or not the remaining fields have changed and then properly react. The "ranging backoff start" field indicates the initial backoff window size for initial ranging contention, the "ranging backoff end" field indicates the final backoff window size for initial rang-

Table 4.13 DIUC Values and Usage

DIUC	Usage
0-12	Different burst profiles
13	Gap/PAPR reduction
14	Extended-2 DIUC IE
15	Extended DIUC

Source: [1, 2].

ing contention, the "request backoff start" field indicates the initial backoff window size for contention bandwidth requests, and the "request backoff end" field indicates the final backoff window size for contention bandwidth requests. In all cases, the window size is expressed as a power of 2, with the values of n ranging 0 to 15. TLV-encoded information such as the number of the ranging codes for each ranging type, Tx power report control, and power control–related parameters are also included in the UCD. At the end of the UCD message format, there follows a series of TLV encoded information that describes the UL_Burst_Profiles of n UL bursts—see Figure 4.36(b). The TLV format of the UL burst profile, which is defined under the type number 1, has the structure shown in Figure 4.36(c). The UL_Burst_Profile is encoded with type 1 and a 4-bit UIUC. The UIUC value used in the UL-MAP message is to specify the UL_Burst_Profiles for a specific UL burst.

UIUC

The UIUC field is associated with the UL_Burst_Profile, and the UIUC values are used in the UL-MAP message to specify the Burst_Profile to be used for a specific UL burst. Table 4.15 lists the 16 UIUC values and their usage. In short, UIUC = 0 is used for fast feedback channel. UIUC = 1 to 10 are allocated to different burst profiles;

Table 4.14 Code Assignment for Extended DIUC and Extended-2 DIUC

(DIUC=15) Extended DIUC	Usage
00	Channel_Measurement_IE
01	STC_Zone_IE
02	AAS_DL_IE
03	Data_location_in_another_BS_IE
04	CID_Switch_IE
05	*Reserved*
06	*Reserved*
07	HARQ_Map_Pointer_IE
08	PHYMOD_DL_IE
09	*Reserved*
0A	Broadcast Control Pointer IE
0B	DL PUSC Burst allocation in Other Segment IE
0C	PUSC ASCA Alloc IE
0D-0E	*Reserved*
0F	UL_interference_and_noise_level_IE

(DIUC=14) Extended -2 DIUC	Usage
00	MBX_MAP_IE
01	HO-Anchor_Active_DL_MAP_IE
02	HO_Active_Anchor_DL_MAP_IE
03	HO_CID_Translation_MAP_IE
04	MIMO_in_another_BS_IE
05	Macro-MIMO_DL_Basic_IE
06	Skip_IE
07	HARQ_DL_MAP_IE
08	HARQ_ACK_IE
09	Enhanced_DL_MAP_IE
0A	Closed-loop_MIMO_DL_Enhanced_IE
0B	MIMO_DL_Basic_IE
0C	MIMO_DL_Enhanced_IE
0D	*Reserved*
0E	AAS_SDMA_DL_IE
0F	*Reserved*

Source: [1, 2].

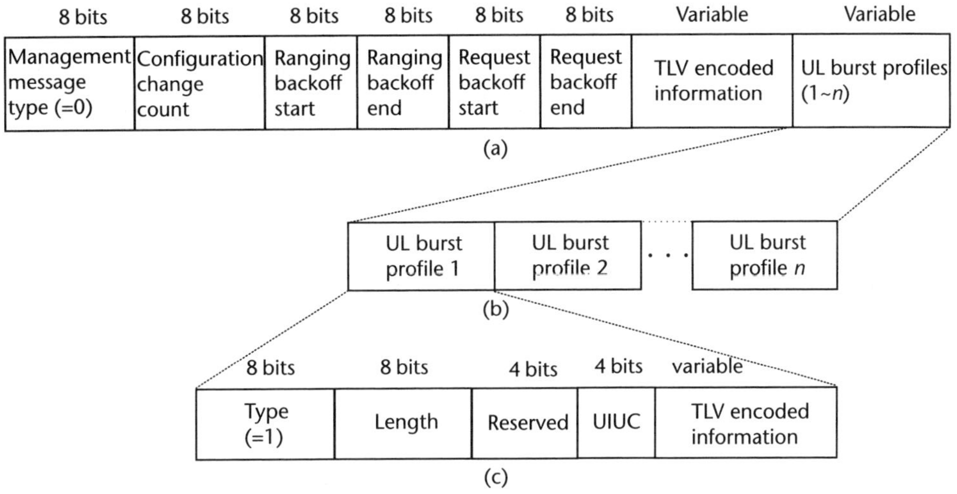

Figure 4.36 (a–c) UCD message and UL burst profile formats. (*After:* [1, 2].)

Table 4.15 UIUC Values and Usage

UIUC	Usage
0	Fast feedback channel
1-10	Different burst profiles (data grant burst type)
11	Extended-2 UIUC IE
12	CDMA bandwidth request, CDMA ranging
13	PAPR reduction allocation, Safety zone
14	CDMA allocation IE
15	Extended UIUC

Source: [1, 2].

UIUC = 12 to CDMA bandwidth request and CDMA ranging; UIUC = 13 to PAPR reduction allocation and safety zone; and UIUC = 11 and 15 are both used for UIUC extensions.

To be more specific, UIUC = 0 is used for allocation of fast-feedback channel region, and there is only one such UL-MAP_IE for a UL frame. UIUC = 1 to 10 are used for different burst profile parameters and are the same as those used in the DL-MAP message. The BS does not allocate to any MS more than one UL-MAP_IE with data burst profile UIUC = 1 through 10 in a single frame. UIUC = 12 is used for allocation of CDMA bandwidth request region and CDMA ranging region, and UIUC = 14 for UL allocation according to the CDMA bandwidth request. UIUC = 13 may be used for allocating subchannles to the PAPR reduction schemes. In this case the subchannels may be used by all MSs to reduce PAPR of their transmissions.

As in the case of DIUC, there are two UIUC extensions, which are intended to extend the coverage of the 4-bit UIUC. In the case of UIUC = 15, the UL-MAP extended IE format is comprised of a 4-bit extended UIUC indication, 4-bit length field, and a variable-sized unspecified data field. Similarly, in the case of UIUC = 11, the UL-MAP extended-2 IE format is comprised of a 4-bit extended-2 UIUC indication, 8-bit length field, and a variable-sized unspecified data field. The extended UIUC and extended-2 UIUC are used for allocation of subchannles to carry various different information elements as listed in Table 4.16.

4.4 Subchannelization

As discussed earlier, there are two different types of subcarrier permutations in subchannelization (i.e., grouping multiple subcarriers to form a subchannel—*distributed permutation*, or *diversity permutation*, and *adjacent permutation*). The diversity permutation draws subcarriers pseudorandomly by taking permutation over a wide range of subcarriers, thereby taking the effects of frequency diversity and intercell interference averaging. The subchannelizations based on the diversity permutation include DL PUSC, DL FUSC, and UL PUSC. The adjacent permutation

Table 4.16 Code Assignment for (a) Extended UIUC and (b) Extended-2 UIUC

(UIUC=15) Extended UIUC	Usage
00	Power_control_IE
01	*Reserved*
02	AAS_UL_IE
03	CQICH_Alloc_IE
04	UL Zone IE
05	PHYMOD_UL_IE
06	*Reserved*
07	UL-MAP_Fast_Tracking_IE
08	UL-PUSC_Burst allocation in other segment IE
09	Fast_Ranging_IE
0A	UL_Allocation_Start_IE
0B-0F	*Reserved*

(a)

(UIUC=11) Extended -2 UIUC	Usage
00	CQICH_Enhanced_Allocation_IE
01	HO-Anchor_Active_UL_MAP_IE
02	HO_Active_Anchor_UL_MAP_IE
03	Anchor_BS_switch_IE
04	UL_Sounding_Command_IE
05	*reserved*
06	MIMO_UL_Enhanced_IE
07	HARQ_UL_MAP_IE
08	HARQ_ACKCH_Region_Alloc_IE
09	MIMO_UL_Basic_IE
0A	Mini-subchannel_Allocation_IE
0B-0D	*Reserved*
0E	AAS_SDMA_UL_IE
0F	Feedback_Polling_IE

(b)

Source: [1, 2].

draws the subcarriers that are physically adjacent, thereby enhancing the band efficiency by taking advantage of the AMC effect. The subchannelization based on the adjacent permutation includes DL AMC and UL AMC subchannels, which take the same structure.

The OFDMA symbol structure is composed of data, pilot tones, and guard subcarriers. The data and pilot carriers are located in the middle part of the full spectrum, centered by the DC subcarrier, and the guard subcarriers are divided into the left- and the right-hand sides of the pilot and data subcarriers. The number and position of the subcarriers allocated to the null (i.e., DC and guard subcarriers), used (for data and pilot) subcarriers differ among DL PUSC, DL FUSC, UL PUSC, and DL/UL AMC. The subcarrier allocation in the case of the 1,024-FFT system is as shown in Table 4.17,[12] which provides a more detailed view on the subcarrier allocation in Table 4.1. We observe that among 1,024 subcarriers, only 840, 850, 840, and 864 subcarriers are respectively allocated to carry the data and pilot sig-

Table 4.17 Subcarrier Allocation in the 1,024-FFT OFDMA System

Parameters	DL PUSC	DL FUSC	UL PUSC	DL/UL AMC
Number of DC subcarriers	1	1	1	1
Number of guard subcarriers, left	92	87	92	80
Number of guard subcarriers, right	91	86	91	79
Number of pilot subcarriers	120	82	420/0	96
Number of data subcarriers	720	768	420/840	768
Number of subchannels	30	16	35	48
Number of data subcarriers in each symbol per subchannel	24	48	12/24	16
Number of clusters	60			
Number of subcarriers per clusters	14			
Number of clusters per subchannel	2			
Number of tiles			210	
Number of subcarriers per tile			4	
Number of tiles per subchannel			6	
Number of bins				96
Number of subcarriers per bin				9
Number of bins per subchannel				2

Source: [1, 2].

12. In the case of DL/UL AMC, "number of bins per subchannel = 2" applies to the case of type 2 × 3 AMC subchannel, with each OFDMA slot composed of 2 bins and 3 OFDMA symbols (refer to Table 4.10 and Figure 4.42).

nals in the four different cases, and the numbers of subcarriers for data carry reduce further.

Once the used subcarriers are determined, for downlink and uplink, they are allocated to pilot tones and data subcarriers. The method of allocating pilot tones and data subcarriers differs for the subcarrier permutation schemes. In the case of DL PUSC and DL FUSC, the pilot tones are allocated first, and the remainders are used as data subcarriers, which are then divided into subchannels. So there is one set of common pilot tones in each major group of DL PUSC and there is one set of common pilot tones overall in DL FUSC. In the case of UL PUSC and DL/UL AMC, however, the used subcarriers are partitioned into subchannels first, and some subcarriers within each subchannel are designated as the pilot subcarriers. So each subchannel contains its own pilot tones. The set of available subchannels is grouped as a *segment*, which may include all available subchannels.

Note that the method of allocating pilot tones differs for different subcarrier permutation schemes. In the case of DL PUSC and DL FUSC, the pilot tones are spread all over the frequency band in the form of common pilot tones, whereas the pilot tones are embedded in the subchannels in the case of UL PUSC and DL/UL AMC. In essence, it is a matter of the target channel state to estimate out of the received pilot tones: in the former case, the receiver needs to estimate the average channel state of the whole frequency band; in the latter case the receiver needs to estimate the channel state of each individual subchannel.

4.4.1 DL PUSC

In the case of the 1,024-FFT system, for example, the 840 active subcarriers used for DL PUSC are divided into 720 data subcarriers and 120 pilot tones. The symbol structure is constructed using those pilot tones and data subcarriers. The distributed permutation concept is incorporated in the form of cluster-based permutation in the cluster renumbering stage of the following multistage subcarrier allocation process.

First, the 840 subcarriers are divided into 60 *physical clusters* (i.e., $N_{clusters} = 60$, C_0, .., C_{59}), with each physical cluster containing 14 adjacent subcarriers. Among the 14 subcarriers in each physical cluster, 2 subcarriers are defined as pilot tones and 12 subcarriers as data carriers. Figure 4.37 shows the structure of the physical cluster, with the subcarriers allocated from top to bottom in the order of increasing subcarrier index. The two pilot tones are positioned at the inner part of the cluster for the odd-indexed OFDMA symbols and at the outer part of the cluster in the even-indexed OFDMA symbols.

Second, the 60 physical clusters are relocated into logical clusters according to a cluster relocation procedure. Specifically, they are renumbered into *logical clusters* in the following way: In the first DL PUSC zone (i.e., first downlink zone) and in the PUSC zone defined by STC_DL_ZONE_IE with "Use All SC Indicator = 0," the default renumbering sequence, or *RenumberingSequence*(*PhysicalCluster*), is used. For all the other cases, the DL_PermBase parameter in the STC_DL_Zone_IE is additionally involved in the renumbering as follows:

$$RenumberingSequence\{(PhysicalCluster + 13 \cdot DL\_PermBase) \bmod N_{clusters}\} \quad (4.34)$$

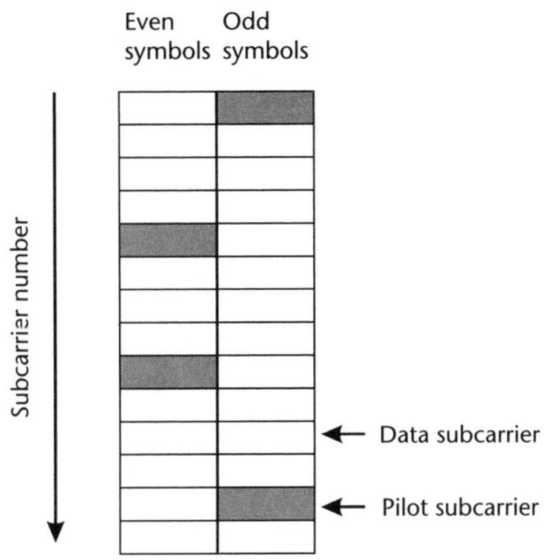

Figure 4.37 Cluster structure for odd and even symbols. (*After:* [1, 2].)

where *RenumberingSequence* is a predefined set of numbers in {0,1,..,59} (refer to [1, Tables 310, 310a, 310b, and 310c]; [2, Tables 511, 512, 513, and 514]). Table 4.18 shows the renumbering sequence for the case of 1,024-FFT system.

Third, the logical clusters are mapped into the six major groups in the following way: clusters 0–11 are mapped into group 0, clusters 12–19 into group 1, clusters 20–31 into group 2, clusters 32–39 into group 3, clusters 40–51 into group 4, and clusters 52–59 into group 5. Those groups may be allocated to three different segments, if a segment is being used, with at least one group allocated to each segment. By default, group 0 is allocated to segment 0, group 2 to segment 1, and group 4 to segment 2.

Fourth, the six major groups are regrouped into 6 sets of 24 groups, with the pilot tones in each constituent logical cluster excluded. Note that each of the even-numbered major groups (i.e., groups 0, 2, and 4) contains 12 logical clusters and each of the odd-numbered major groups (i.e., groups 1, 3, and 5) contains 8 logical clusters, with each cluster carrying 12 data subcarriers and 2 pilot tones. For the even-numbered major groups, each cluster is divided into 2 groups with 6 data subcarriers each; for the odd-numbered major groups, each cluster is divided into 3 groups with 4 data subcarriers each.

Table 4.18 DL PUSC Renumbering Sequence—1,024-FFT

Parameter	Value	Comments
Renumbering Sequence	6, 48, 37, 21, 31, 40, 42, 56, 32, 47, 30, 33, 54, 18, 10, 15, 50, 51, 58, 46, 23, 45, 16, 57, 39, 35, 7, 55, 25, 59, 53, 11, 22, 38, 28, 19, 17, 3, 27, 12, 29, 26, 5, 41, 49, 44, 9, 8, 1, 13, 36, 14, 43, 2, 20, 24, 52, 4, 34, 0	Used to renumber clusters before allocation to subchannels
Basic Permutation Sequence for 6 Subchannels	3, 2, 0, 4, 5, 1	—
Basic Permutation Sequence for 4 Subchannels	3, 0, 2, 1	—

Source: [1, 2].

Fifth, the subcarriers in each of the 24 groups are mapped into 6 subchannels or 4 subchannels as follows: in the case of the 24 groups made out of the even-numbered major groups, having 6 data subcarriers each, 1 subcarrier is pulled out of each of the 24 groups to form a subchannel. In this case, 6 subchannels are generated, with each subchannel composed of 24 subcarriers. In the case of the 24 groups made out of the odd-numbered major groups, having 4 data subcarriers each, 1 subcarrier is pulled out of each of the 24 groups to form a subchannel. In this case, 4 subchannels are generated, with each subchannel composed of 24 subcarriers. So the resulting subchannels have 24 subcarriers each, and the total number of subchannels is 30. Figure 4.38 illustrates this process.

The allocation of subcarriers to subchannels in the last two steps can be expressed by the following equation, called *permutation formula*.

$$subcarrier(k, s) = N_{subchannels} \cdot n_k + \{p_s[n_k \bmod N_{subchannels}] + DL\_PermBase\} \bmod N_{subchannels} \quad (4.35)$$

where *subcarrier*(k, s) denotes the subcarrier index of subcarrier k in subchannel s; s is the subchannel index in {0, 1, .., $N_{subchannels} - 1$}; k is the subcarrier-in-subchannel index in {0, 1, .., $N_{subcarriers} - 1$}; $n_k = (k + 13 \cdot s) \bmod N_{subcarriers}$; $p_s[j]$ is the series obtained by rotating the basic permutation sequence of Table 4.18 cyclically to the left by s times; and *DL_PermBase* is an integer ranging from 0 to 31, which is the preamble IDCell in the first zone and determined by the DL_PermBase field in the STC_DL_Zone_IE (transmitted through DL-MAP) for other zones. In the case of the 1,024-FFT system, $N_{subchannels} = 6$ (for even-numbered major groups) or 4 (for odd-numbered major groups), $N_{subcarriers} = 24$.

4.4.2 DL FUSC

In the case of the 1,024-FFT system, for example, the 850 active subcarriers used for DL FUSC are divided into 768 data subcarriers and 82 pilot tones. The 768 data subcarriers are divided into 16 subchannels of 48 subcarriers each. The distributed permutation concept is incorporated in the form of full-spectrum diversity in the process of dividing the 768 subcarriers into 16 subchannels in two steps as follows: The 768 data subcarriers are first partitioned into 48 groups, with each group consisting of 16 contiguous subcarriers. Then each subchannel is formed by taking one subcarrier from each of those 48 groups. Specifically, the *i*th group G_i, $i = 0, 1, ..., 47$, is formed by taking the *i*th 16 contiguous subcarriers from the beginning, and the *s*th subchannel Ss, $s = 0, 1, ..., 15$, is formed by taking the *s*th subcarrier in each group, sequentially. In general, the number of groups is equal to the number of subcarriers in each subchannel, $N_{subcarriers}$, and the number of subchannels in each group is equal to the number of subchannels, $N_{subchannels}$, which are 48 and 16, respectively, in the case of the 1,024-FFT system. This process is illustrated in Figure 4.39.

There are two variable pilot-sets and two constant pilot-sets ([1], Tables 311, 311a, 311b, and 311c; [2], Tables 515, 516, 517, and 518). In FUSC, each segment uses both sets of variable/constant pilot-sets. Table 4.19 summarizes the pilot subcarrier indices. The variable set of pilots embedded within the symbol of each segment follows the rule:

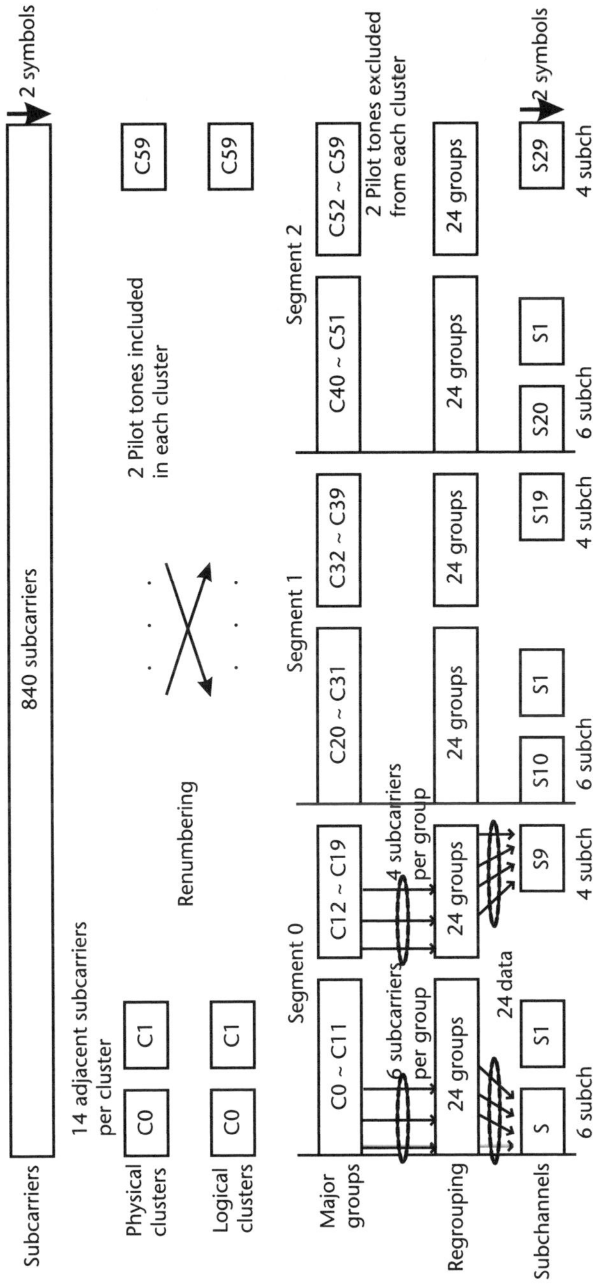

Figure 4.38 Illustration of DL PUSC mapping for 1,024-FFT system.

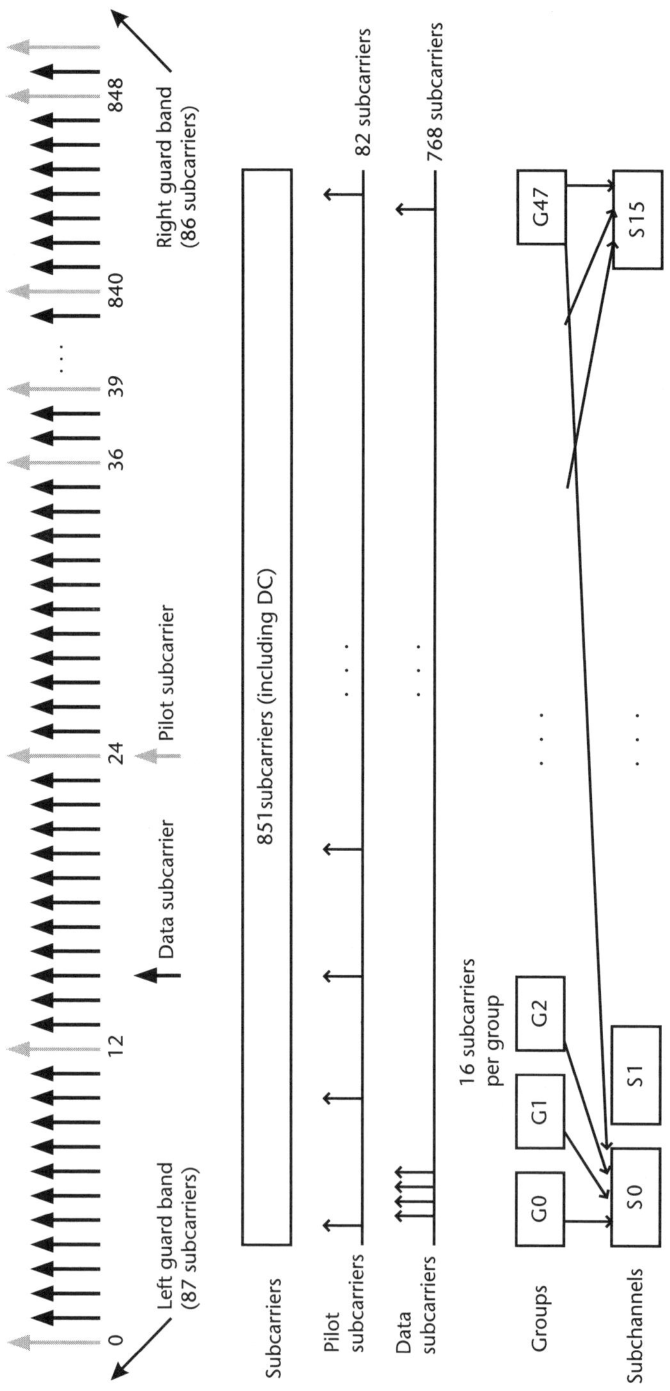

Figure 4.39 Illustration of DL FUSC mapping for 1,024-FFT system.

Table 4.19 DL FUSC Pilot Subcarrier Indices—1,024-FFT

Parameter	Value	Comments
Pilot Subcarrier Index : VariableSet #0	0, 24, 48, 72, 96, 120, 144, 168, 192, 216, 240, 264, 288, 312, 336, 360, 384, 408, 432, 456, 480, 504, 528, 552, 576, 600, 624, 648, 672, 696, 720, 744, 768, 792, 816, 840	36 Subcarriers
Pilot Subcarrier Index : ConstantSet #0	$72\times(2\times n+k) + 9$ when $k = 0$ and $n = 0,\dots,5$	DC subcarrier shall be included when the pilot subcarrier index is calculated by the equation.
Pilot Subcarrier Index : VariableSet #1	36, 108, 180, 252, 324, 396, 468, 540, 612, 684, 756, 828, 12, 84, 156, 228, 300, 372, 444, 516, 588, 660, 732, 804, 60, 132, 204, 276, 348, 420, 492, 564, 636, 708, 780	35 Subcarriers
Pilot Subcarrier Index : ConstantSet #1	$72\times(2\times n+k) + 9$ when $k = 1$ and $n = 0,\dots,4$	DC subcarrier shall be included when the pilot subcarrier index is calculated by the equation.
Basic Permutation Sequence	6, 14, 2, 3, 10, 8, 11, 15, 9, 1, 13, 12, 5, 7, 4, 0	—

Source: [1, 2].

$$PilotsLocation = VariableSet\#\,x + 6\cdot(FUSC\_SymbolNumber \bmod 2) \tag{4.36}$$

where *FUSC_SymbolNumber* counts the FUSC symbols used in the current zone starting from 0.

The partitioning of subcarriers into subchannels is done according to (4.35) in Section 4.4.1. In the case of the 1,024-FFT system, $N_{subchannels} = 16$ and $N_{subcarriers} = 48$ in (4.35).

4.4.3 UL PUSC

In the UL PUSC, a slot is composed of 1 subchannel and 3 OFDMA symbols (i.e., 1 subchannel × 3 OFDMA symbols). Each subchannel is composed of 24 subcarriers, so in each slot, there are 72 subcarriers, which are divided into 48 data subcarriers and 24 pilots. The distributed permutation concept is incorporated in the form of tile-based diversity permutation, with each tile containing some dedicated pilots. *Tile* is the basic unit that constitutes the UL PUSC. It is composed of 4 adjacent subcarriers taken out of 3 consecutive OFDMA symbols (i.e., 4 subcarriers × 3 OFDMA symbols), with 4 pilot tones located at the 4 corner points, as shown in Figure 4.40. Therefore, a UL PUSC slot is constructed out of 6 tiles. In the case of the 1,024-FFT system, for example, there are 840 active subcarriers, which are divided into 35 subchannels with 24 subcarriers each, so the total number of tiles is 210. These numbers are listed in Table 4.17. In the table, the number of pilot subcarriers is indicated as 420/0, which, according to Figure 4.40, means that the number is 420 (=2 subcarriers × 210 tiles) in the case of the first and the third symbols in each tile and 0 (=0 subcarriers × 210 tiles) in the case of the second symbol in each tile. Likewise, the number of data subcarrier indicated as 420/840 means that the number is 420 (=2 subcarriers × 210 tiles) in the case of the first and the third symbols and 840 (=4 subcarriers × 210 tiles) in the case of the second symbol.

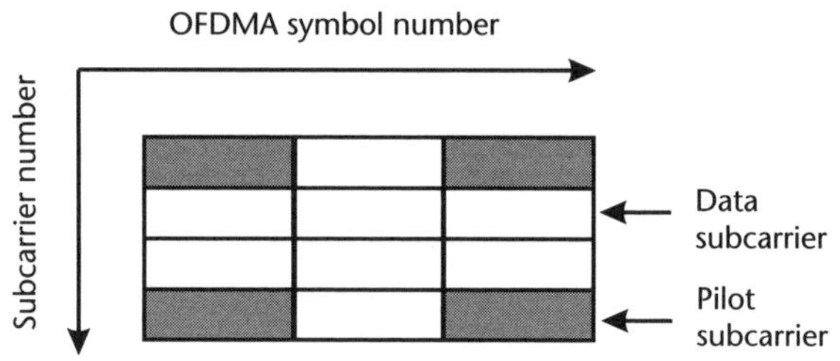

Figure 4.40 Tile structure for UL PUSC. (*After:* [1].)

In the UL PUSC, subcarrier allocation is done in the following procedure: The usable subcarriers in the allocated frequency band are divided into N_{tiles} physical tiles. The physical tiles are then allocated to logical tiles according the following equation:

$$\begin{aligned} &Tiles(s,n) = \\ &N_{subchannels} \cdot n + \{Pt[s + n \bmod N_{subchannels}] + UL\_PermBase\} \bmod N_{subchannels} \end{aligned} \tag{4.37}$$

where *Tiles* (*s*, *n*) is the physical tile index of the logical tile at (*n*,*s*) for the tile index *n* in {0, ..., 5} and the subchannel number *s* in {0, 1, ..., $N_{subchannels}$ – 1}; *Pt* is the tile permutation sequence ([1, Tables 313, 313a, 313b, and 313c]; [2, Tables 524, 525, 526, and 527]); *UL_PermBase* is an integer value in {0, 1, ..., 69}, which is assigned by UCD in the first zone and determined by the UL_PermBase field in the UL_Zone_IE (transmitted through the UL-MAP) for other zones; and $N_{subchannels}$ is the number of subchannels. In the case of the 1,024-FFT system, $N_{subchannels}$ = 35. Table 4.20 shows the tile permutation sequence for the case of 1,024-FFT system.

After mapping of the physical tiles to logical tiles is completed for each subchannel, the data subcarriers are enumerated per slot in the following procedure.

First, the pilot subcarriers are allocated within each tile, and the data subcarriers are indexed within each slot. The indexing begins with the first symbol at the lowest indexed subcarrier of the lowest indexed tile. It continues in an ascending order through the subcarriers in the same symbol. It then moves to the next symbol at the lowest indexed data subcarrier, and continues in that manner, with the data subcarrier increasing from 0 to 47.

Second, data is mapped onto the subcarriers according to

$$Subcarrier(n, s) = (n + 13 \cdot s) \bmod N_{subcarriers} \tag{4.38}$$

Table 4.20 UL PUSC Tile Permutation Sequence—1024-FFT

Parameter	Value	Comments
Tile Permutation Sequence	11, 19, 12, 32, 33, 9, 30, 7, 4, 2, 13, 8, 17, 23, 27, 5, 15, 34, 22, 14, 21, 1, 0, 24, 3, 26, 29, 31, 20, 25, 16, 10, 6, 28, 18	Used to allocate tiles to subchannels

Source: [1, 2].

where subcarrier(n,s) is the permutated subcarrier index corresponding to the data subcarrier located at (n,s) for n in {0, ..., 47} and the subchannel number s; $N_{subcarriers}$ is the number of subcarriers per slot.

Figure 4.41 illustrates the UL PUSC mapping process for the 1,024 OFDMA system.

Data subchannel rotation scheme is applied for the purpose of enhancing frequency diversity. A rotation scheme is applied per each OFDMA slot-duration in any zone, except for the zones marked as AMC zone. For each slot-duration, the rotation scheme is applied to all UL subchannels that belong to the segment, except for the UL subchannels allocated to the UL control channels, such as ranging, CQICH, and ACKCH. The rotation scheme for slot duration with no UL control channels is defined by

$$subchannel_{rotated}(s) = (s + 13 \cdot Slot_{index}) \bmod N_{subchannels} \quad (4.39)$$

where $subchannel_{rotated}(s)$ is the rotated subchannel number corresponding to the subchannel number s; $Slot_{index}$ is the slot index, numbered from 0 for each permutation zone; and $N_{subchannels}$ is the number of subchannels per symbol. In the case of the 1,024 FFT system, $N_{subchannels}$ = 35. Note that the rotated subchannel number replaces the original subchannel number for UL allocation.

The rotation scheme for slot duration, where part of the subchannels is allocated to the UL control channels, is defined by the following procedure [1].

For each OFDMA slot duration, we pick only the subchannels that are not allocated to the UL control channels and then renumber these subchannels contiguously from 0 to $N_{subchan}$, which is the total number of subchannels less the total number of control channels, such that the lowest numbered physical subchannel is renumbered with 0. We denote by f the mapping function for this renumbering—that is, *temp1_subchannel_number* = f(*original_subchannel_number*).

Then we calculate the rotated subchannel number for *temp1_subchannel_number* by applying (4.39) with $N_{subchannel}$ replaced by $N_{subchan}$. We apply the inverse mapping f^{-1} to the *rotated_subchannel_number* to obtain a *new_subchannel_number*—that is, *new_subchannel_number* = f^{-1}(*rotated_subchannel_number*).

For the subchannels that are allocated to the UL control channels, the original subchannel number is not changed.

4.4.4 DL/UL AMC

In the case of AMC, the basic allocation unit is a *bin*, which is a set of 9 contiguous subcarriers within an OFDMA symbol, both in downlink and uplink. Among the 9 subcarriers in each bin, 1 subcarrier is defined as pilot tone and 8 subcarriers as data carriers. Figure 4.42(a) depicts the bin structure.

An AMC slot is composed of 2 bins by 3 OFDMA symbols, as shown in Figure 4.42(b). The pilot tones located in each OFDMA symbol in a slot are put skewed as shown in Figure 4.42(b) to enhance the performance of the channel estimation and compensation.

For each slot data, subcarriers are allocated in the following procedure. First, the pilot subcarriers are allocated within each bin, and the data subcarriers are

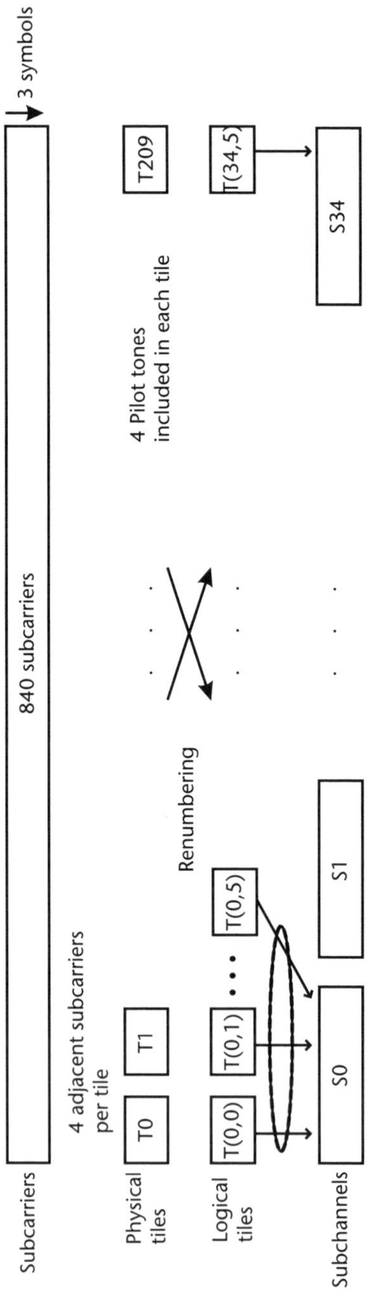

Figure 4.41 Illustration of UL PUSC mapping for 1,024-FFT system.

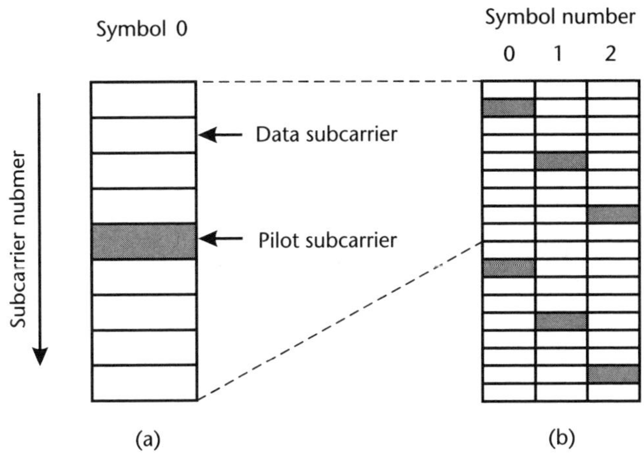

Figure 4.42 DL/UL AMC structure: (a) bin; and (b) slot. (*After:* [1, 2].)

indexed within each slot. The indexing begins with the first symbol at the lowest indexed subcarrier of the lowest indexed bin. It continues in an ascending order through the subcarriers in the same symbol. It then moves to the next symbol at the lowest indexed data subcarrier, and continues in that manner, with the data subcarrier increasing from 0 to 47.

Second, data is mapped onto the subcarriers according to the following equation:

$$subcarriers(k, s) = \begin{cases} \left(p_m[k] + p_{off}\right)_{GF(7^2)} - 1, & \text{if } \left(p_m[k] + p_{off}\right)_{GF(7^2)} \neq 0 \\ p_{off} - 1, & \text{otherwise} \end{cases} \tag{4.40}$$

where *subcarriers*(k,s) is the permutated subcarrier index of subcarrier k in slot s; $p_m[k]$ is the series obtained by rotating the basic permutation sequence defined in Galois field $GF(7^2)$ of Table 4.21 cyclically to the left by m times ([1], Tables 316, 316a, 316b, and 316c; [2], Tables 536, 537, 538, and 539); $m = PermBase \bmod 48$; and $p_{off} = (\lfloor PermBase/48 \rfloor) \bmod 49$ and is an element of $GF(7^2)$. Note that $(n_1 + n_2)_{GF(7^2)}$ denotes the addition of two element in $GF(7^2)$ [i.e., a compo-

Table 4.21 DL/UL AMC Permutation Sequence—1,024-FFT

Parameter	Value	Comments
Pilot Subcarrier Index	$9k+3m+1$, for k=0,1...95, and m=[symbol index] mod 3	Symbol of index 0 in pilot subcarrier index should be the first symbol of the current zone. DC subcarrier is excluded when the pilot subcarrier index is calculated by the equation.
Basic Permutation Sequence	01, 22, 46, 52, 42, 41, 26, 50, 05, 33, 62, 43, 63, 65, 32, 40, 04, 11, 23, 61, 21, 24, 13, 60, 06, 55, 31, 25, 35, 36, 51, 20, 02, 44, 15, 34, 14, 12, 45, 30, 03, 66, 54, 16, 56, 53, 64, 10	hepta-notation

Source: [1, 2].

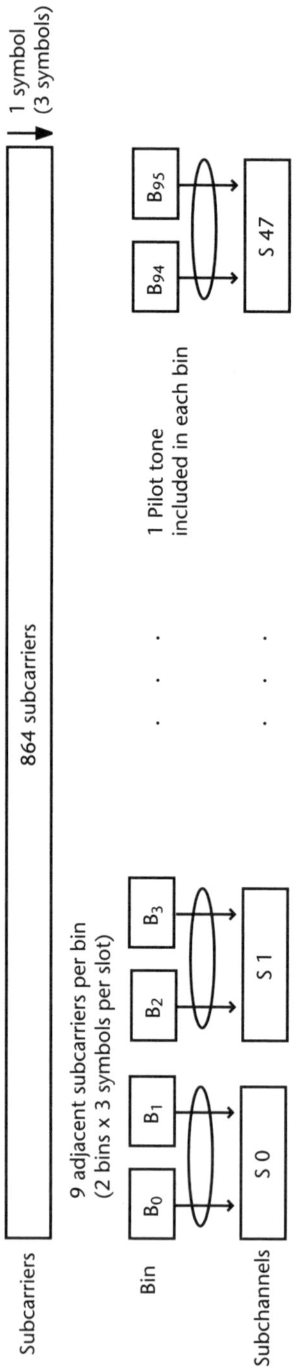

Figure 4.43 Illustration of DL/UL AMC mapping for 1,024-FFT system.

nent-wise addition modulo 7 of two representations, such as $(54 + 34)_{GF(7^2)} = 13$]. In the downlink, *PermBase* is set to *DL_PermBase* specified in the preceding STC_DL_Zone_IE, and, in the uplink, it is set to *UL_PermBase* specified in the preceding UL_Zone_IE.

The procedure of constructing subchannels from subcarriers is as follows: First, the 864 subcarriers are divided into 96 bins, B_0, ..., B_{95}, with each bin containing 9 contiguous subcarriers. Second, each two adjacent bins construct a subchannel, so the resulting subchannel contains 18 subcarriers, and, consequently, the total number of subchannels is 48. Note that an AMC subchannel is composed of 2 bins for each OFDMA symbol since an AMC slot is composed of 2 bins by 3 OFDMA symbols. Figure 4.43 illustrates the DL/UL AMC mapping process for the 1,024-FFT OFDMA system.

References

[1] IEEE Std 802.16e-2005 and IEEE Std 802.16-2004/Cor1-2005, Part 16: Air Interface for Fixed and Mobile Broadband Wireless Access Systems, February 2006.

[2] IEEE Std P802.16 Rev2/D1, Part 16: Air Interface for Fixed and Mobile Broadband Wireless Access Systems, October 2007.

[3] Berrou, C., and A. Glavieux, "Near Optimum Error Correcting Coding and Decoding: Turbo-Codes," *IEEE Trans. on Communications*, Vol. 44, No. 10, October 1996, pp. 1261–1271.

[4] Vucetic, B., and J. Yuan, *Turbo Codes: Principles and Applications,* Boston, MA: Kluwer, 2000.

[5] Doulillard, C., and C. Berrou, "Turbo Codes with Rate-*m*/(*m*+1) Constituent Convolutional Codes," *IEEE Trans. on Communications*, Vol. 53, No. 10, October 2005, pp. 1630–1638.

[6] Weiss, C., et al., "Turbo Decoding with Tail-Biting Trellises," *Proc. of IEEE International Symposium Signals, Systems, and Electronics*, 1998, pp. 343–348.

[7] Kim, M. G., and S. H. Ha, "Quasi-Complementary Turbo Codes (QCTC) for Applications in High-Data-Rate Systems," *Proc. of IEEE Vehicular Technology Conference*, 2003, pp. 2381–2385.

[8] Lin, S., and D. J. Costello, *Error Control Coding: Fundamentals and Applications,* Englewood Cliffs, NJ: Prentice-Hall, 1983.

[9] Das, A., et al., "Adaptive, Asynchronous Incremental Redundancy (A^2IR) with Fixed Transmission Time Intervals (TTI) for HSDPA," *Proc. of IEEE International Symposium on Personal, Indoor and Mobile Radio Communications*, 2002, pp. 1083–1087.

Selected Bibliography

Andrews, J. G., A. Ghosh, and R. Muhamed, *Fundamentals of WiMAX: Understanding Broadband Wireless Networking*, Englewood Cliffs, NJ: Prentice-Hall, 2007.

Chase, D., "Code Combining—A Maximum Likelihood Decoding Approach for Combining an Arbitrary Number of Noisy Packets," *IEEE Communications Magazine*, Vol. 33, No. 5, May 1985, pp. 385–393.

Cimini, L. J., "Analysis and Simulation of a Digital Mobile Channel Using Orthogonal Frequency Division Multiplexing," *IEEE Trans. on Communications*, Vol. 33, No. 7, July 1985, pp. 665–675.

Eklund, C., et al., "IEEE Standard 802.16: A Technical Overview of the WirelessMAN Air Interface for Broadband Wireless Access," *IEEE Communications Magazine*, Vol. 40, No. 6, June 2002, pp. 98–107.

Hagenauer, J., "Rate-Compatible Punctured Convolutional Codes (RCPC Codes) and Their Applications," *IEEE Trans. on Communications*, Vol. 36, No. 4, April 1988, pp. 389–400.

Han, S. H., and J. H. Lee, "Peak-to-Average Power Ratio Reduction in Multicarrier Communicationsystems," *IEEE Wireless Communications*, Vol. 12, No. 2, April 2005, pp. 56–65.

Kallel, S., "Complementary Punctured Convolutional (CPC) Codes and Their Applications," *IEEE Trans. on Communications*, Vol. 43, No. 6, June 1995, pp. 2005–2009.

Koffman, I., and V. Roman, "Broadband Wireless Access Solutions Based on OFDM Access in IEEE 802.16," *IEEE Communications Magazine*, Vol. 40, No. 4, April 2002, pp. 96–103.

Kwon, T., et al., "Design and Implementation of a Simulator Based on a Cross-Layer Protocol Between MAC and PHY Layers in a WiBro Compatible IEEE 802.16e OFDMA System," *IEEE Communications Magazine*, Vol. 43, No. 12, December 2005, pp. 136–146.

Ozdemir, M. K., and H. Arslan, "Channel Estimation for Wireless OFDM Systems," *IEEE Communications Surveys & Tutorials*, Vol. 9, No. 2, 2nd Quarter 2007, pp. 18–48.

Weinstein, S., and P. Ebert, "Data Transmission by Frequency-Division Multiplexing Using the Discrete Fourier Transform," *IEEE Trans. on Communications*, Vol. 19, No. 5, October 1971, pp. 628–634.

CHAPTER 5

MAC Framework

In Mobile WiMAX, the OFDM-based multiple access (referred to as OFDMA) has been employed to support the simultaneous user connections. As these connections must be sharing the same radio resource in a dynamic manner, there must be a *medium access control* (MAC) mechanism to manage the bandwidth and QoS for individual user applications while processing the data and control packets with appropriate header information. In particular, it must be elaborated to match with a connection-oriented feature of Mobile WiMAX. Unlike WiFi operating under the connectionless mechanism, therefore, Mobile WiMAX can provide a tight control of resource allocation and QoS (see Section 6.3). In particular, bandwidth can be reserved on an on-demand basis in a dynamic manner. In order to support both *real-time* (RT) and *nonreal-time* (NRT) services, a packet scheduler is employed in BS so that it can maximize the overall system throughput while supporting the QoS of individual connections. In fact, the different scheduling service types are defined to support the various data delivery services with their own attributes in bandwidth request/grant and QoS parameters (see Section 6.1). The MAC layer in Mobile WiMAX is closely associated with such a resource management feature, which is not explicitly available in WiFi.

The MAC function of Mobile WiMAX is divided into three sublayers, namely, *service-specific convergence sublayer* (CS), *common part sublayer* (CPS), and *security sublayer*, as shown in Figure 2.6. Among them, we discuss the first two sublayers in this section, separating out the discussions on security sublayer in Chapter 8. The sending CS in the transmitter delivers the MAC SDU down to the MAC SAP according to the classification process, which associates each application stream with a particular connection. Subsequently, the MAC CPS delivers the MAC SDU to the peer MAC SAP in the receiver according to the QoS, fragmentation, concatenation, and other transport functions associated with the QoS constraints of the given particular connection. The receiving CS accepts the MAC SDU from the peer MAC SAP and delivers it up to a higher layer entity. We discuss the service-specific CS in Section 5.1 and discuss the organizations and operations of the MAC CPS in Section 5.2. As a supplement, we discuss the ARQ operation in Section 5.3.

5.1 MAC Service-Specific Convergence Sublayer

The service-specific CS performs functions of converging user services to MAC CPS. Specifically, the functions include accepting PDUs from the higher layer, performing classification of higher-layer PDUs into the appropriate transport connection, processing the higher-layer PDUs based on the classification, delivering CS PDUs to the appropriate MAC SAP, and receiving CS PDUs from the peer entity. There are two service-specific CSs specified in the IEEE 802.16 standards, namely, ATM CS and packet CS. The ATM CS is a logical interface that associates different ATM services with the MAC CPS SAP. The ATM CS is supposed to accept ATM cells from the ATM layer, perform classification, and deliver CS PDUs to the appropriate MAC SAP. However, since ATM is not prevalent in mobile communications, the Mobile WiMAX profile excluded it. So in this section we briefly discuss the packet CS in the capacity of packet CS functions, classification functions, *payload header suppression* (PHS) functions, and MAC SDU and CS PDU formats.

The packet CS performs the following functions by utilizing the services of the MAC:

1. Classifying the higher layer PDU (e.g., IP packet) into the appropriate connection;
2. Delivering the resulting CS PDU to the MAC SAP associated with the service flow (i.e., a MAC transport service to provide unidirectional transport of packets, characterized by a set of QoS parameters—see Section 6.3.1) for transporting to the peer MAC SAP;
3. Receiving the CS PDU from the peer MAC SAP. Optionally, it performs PHS-related functions, such as suppressing payload header information and rebuilding any suppressed payload header information.

The packet CS is used for the transport of all packet protocols, such as Internet Protocol (IP CS), IEEE 802.3 LAN/MAN CSMA/CD access method (Ethernet CS), and IEEE 802.1Q (VLAN CS). IP CS enables transport of IPv4 and IPv6 packets directly over the MAC. In addition, packet CS supports *robust header compression* (ROHC) and *enhanced compressed real-time transport protocol* (ECRTP) header compression. As a result, packet CS defines multiple versions with different classifiers of IP, Ethernet, and VLAN CS, as listed in Table 5.1. Among them, Mobile WiMAX specification mainly uses IP CS as default.

5.1.1 Classification Functions

In order to support the connection-oriented mechanism with a tight control of resource allocation and QoS, MAC SDUs must be mapped onto a particular transport connection for transmission between MAC peers, a process called *classification*. Transport connection refers to a connection used to transport user data. The mapping process associates a MAC PDU with a transport connection, which is identifiable by a *connection ID* (CID) in the header of each MAC PDU, and with the service flow characteristics of that connection. This association process facilitates the delivery of MAC SDUs with the appropriate QoS constraints. In other words, a process of classification is to map a stream of user data to a particular connection associated with the service flow characterized by a set of QoS parameters to support the corresponding application service. Figure 5.1(a) illustrates the downlink classification process applied by the BS to the packets it is transmitting to the MS, and Figure 5.1(b) illustrates the uplink classification process applied by the MS to the MAC SDUs that it is transmitting to the BS.

Table 5.1 Types of Packet CS and ATM CS

Value	CS
0	Generic packet convergence sublayer (GPCS)
1	Packet, IPv4
2	Packet, IPv6
3	Packet, IEEE 802.3/Ethernet
4	Packet. IEEE 802.1Q VLAN
5	Packet, IPv4 over IEEE 802.3/Ethernet
6	Packet, IPv6 over IEEE 802.3/Ethernet
7	Packet, IPv4 over IEEE 802.1Q VLAN
8	Packet, IPv6 over IEEE 802.1Q VLAN
9	ATM
10	Packet, IEEE 802.3/Ethernet with ROHC header compression
11	Packet, IEEE 802.3/Ethernet with ECRTP header compression
12	Packet, IP with ROHC header compression
13	Packet, IP with ECRTP header compression

Source: [1].

A *classifier* is a set of matching criteria applied to each packet entering the WiMAX network. It consists of some protocol-specific packet matching criteria (e.g., destination IP address), a classifier priority, and a reference to a CID. If a packet matches with the specified packet matching criteria, it is then delivered to the MAC SAP for delivery on the connection specified by the CID. For example, a stream of IP packets from a particular application service with the same destination can be classified into the same connection by referencing to the same connection ID. Classifier priority is needed for ordering the application of classifiers to packets. The service flow characteristics of the transport connection provide the QoS for the associated packet.

5.1.2 MAC SDU and CS PDU Formats

Once classified and associated with a specific MAC connection, higher-layer PDUs (i.e., packet PDUs) are encapsulated in the MAC SDU format shown in Figure 5.2. In the figure, *payload header suppression index* (PHSI) is an 8-bit optional field, which indicates which bytes in the suppression target field to suppress and which bytes not to suppress. PHS is optional, so PHSI is included only when a PHS rule is defined for the associated connection. The details of PHS functions are described in Section 5.1.3. The CS PDU constructed as a MAC SDU of a specific MAC connection in the sending CS will be appended with a MAC header, which includes a CID

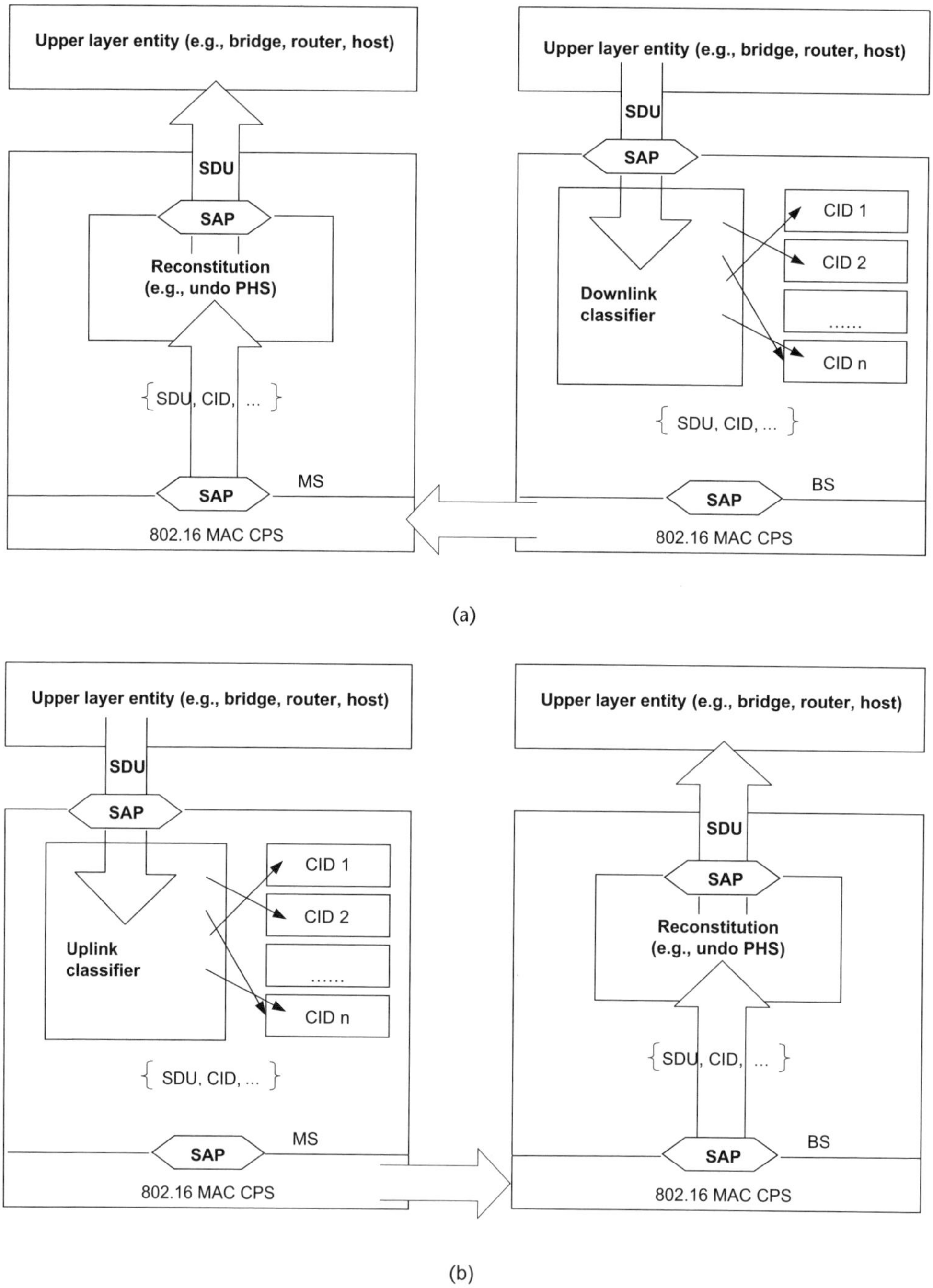

Figure 5.1 Classification and CID mappings: (a) BS to MS; and (b) MS to BS. (*After:* [1].)

to identify the corresponding connection in the CPS (see Figure 5.8, later in this chapter).

Mobile WiMAX specifies the CS PDU formats for IEEE 803.3 Ethernet, IEEE 802.1Q VLAN, and IP packets (IETF RFC 791, 2460). The formats are basically the same as MAC SDU format in Figure 5.2, with each packet PDU field replaced with an Ethernet packet, a VLAN packet, and an IP packet, respectively. As stated earlier,

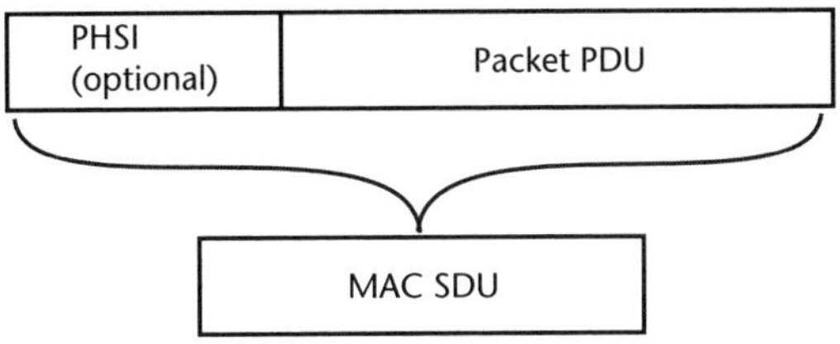

Figure 5.2 MAC SDU format. (*After:* [1].)

the PHSI field is omitted (i.e., PHSI=0) when PHS is not applied. In the case of Ethernet, the Ethernet FCS field is excluded when it is carried over the Ethernet CS PDU.

5.1.3 PHS Functions

PHS is to suppress a repetitive portion of the payload headers of the higher layer in the MAC SDU, which is done by the sending entity and restored by the receiving entity. The sending entity is the MS and the receiving entity is the BS on the uplink, whereas the sending entity is the BS and the receiving entity is the MS on the downlink. If PHS function is enabled at the MAC connection, each MAC SDU is prefixed with a *payload header suppression index* (PHSI), which references the *payload header suppression field* (PHSF). PHSF refers to a string of bytes representing the header portion of a PDU in which one or more bytes are to be suppressed.

The sending entity uses classifiers to map packets into a service flow. The classifier uniquely maps packets to its associated PHS rule. The receiving entity uses the CID and the PHSI to restore the PHSF. Once a PHSF is assigned to a PHSI, it cannot be changed. If the sending entity wants to change the PHSF value on a service flow, it has to define a new PHS rule first and then remove the old rule from the service flow. If it deletes a classifier, it has to delete any associated PHS rules as well.

PHS has options such as *payload header suppression valid* (PHSV) and *payload header suppression mask* (PHSM). The PHSV option is to validate the payload header before suppressing it, and the PHSM option is to indicate which bytes are to be suppressed. PHSM option facilitates the suppression of the header fields that remain static within a higher-layer session (e.g., IP addresses) by masking the fields that change from packet to packet (e.g., IP total length).

The BS assigns all PHSI values just as it assigns all CID values. Either the sending or the receiving entity specifies the PHSF and the *payload header suppression size* (PHSS) or the length of the suppressed field in bytes. The PHS rule that uniquely identifies the suppressed header within the service flow is generated by a higher-layer service entity. This higher-layer entity should guarantee that the byte strings that are being suppressed are kept constant from packet to packet for the duration of the active service flow.

Figure 5.3 illustrates the processes of packet suppression in the sender and packet restoration in the receiver when using PHS masking. The PHSM indicates that the first (A), the third (C), and the fifth (E) bytes in the packet header are to be suppressed, so the sender verifies whether or not they match the corresponding bytes (A', C', and E') in the PHSF. If the verification succeeds, the sender suppresses

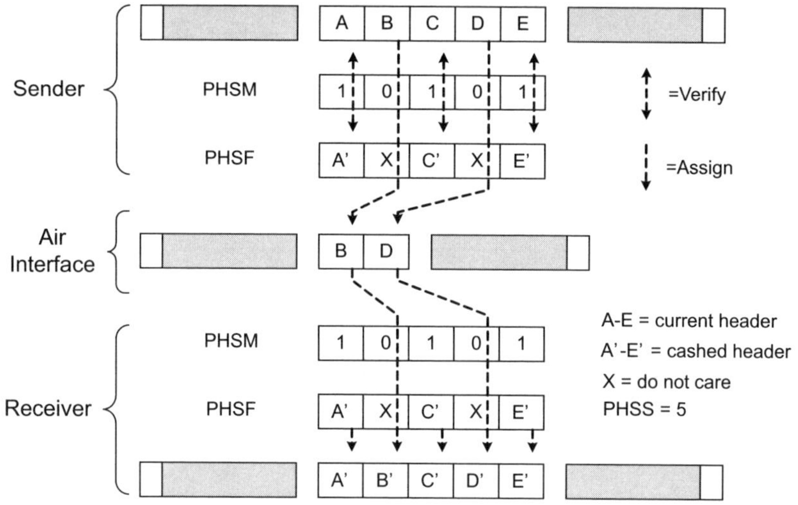

Figure 5.3 Illustration of PHS with masking. (*After:* [1].)

them and transmits the remaining bytes (B and D) to the receiver. Then the receiver restores the original packet header by filling the suppressed bytes with the cached bytes (A', C', and E') in the PHSF.

Figure 5.4 illustrates the operation of the PHS function. It describes the procedure to take on the uplink for the sender MS and the receiver BS. A similar procedure applies on the downlink. On the sender side, when a packet arrives from an upper-layer entity to the packet CS, the MS classifies the packet according to its classifier rule. If the rule matches, it generates an uplink service flow, CID, and a PHS rule. The PHS rule provides PHSF, PHSI, PHSM, PHSS, and PHSV. If PHSV is set or not present, the MS compares the bytes in the packet header with the bytes in the PHSF that are to be suppressed as indicated by the PHSM. If they match, the MS suppresses all the bytes in the uplink PHSF except for the bytes masked by PHSM. The MS then prefixes the PDU with the PHSI and presents the entire MAC SDU to the MAC SAP for transport on the uplink.

In the receiver side, when the MAC PDU arrives from the air interface to the BS, the BS MAC layer determines the associated CID by examining the generic MAC header. The BS MAC layer sends the PDU to the MAC SAP associated with that CID. The receiving packet CS uses the CID and the PHSI to look up PHSF, PHSM, and PHSS. The BS reassembles the packet and then proceeds with normal packet processing, presenting the packet to the CS SAP.

5.2 MAC Common Part Sublayer

The MAC CPS renders a common substrate to all different types of service-specific CSs in service. It provides the core MAC functionality, including system access, bandwidth allocation, connection establishment, and connection maintenance. As the IEEE 802.16 network operates on a shared wireless medium, or the air space through which the radio wave propagates, MAC CPS provides a mechanism that enables all the users to share the wireless medium effectively.

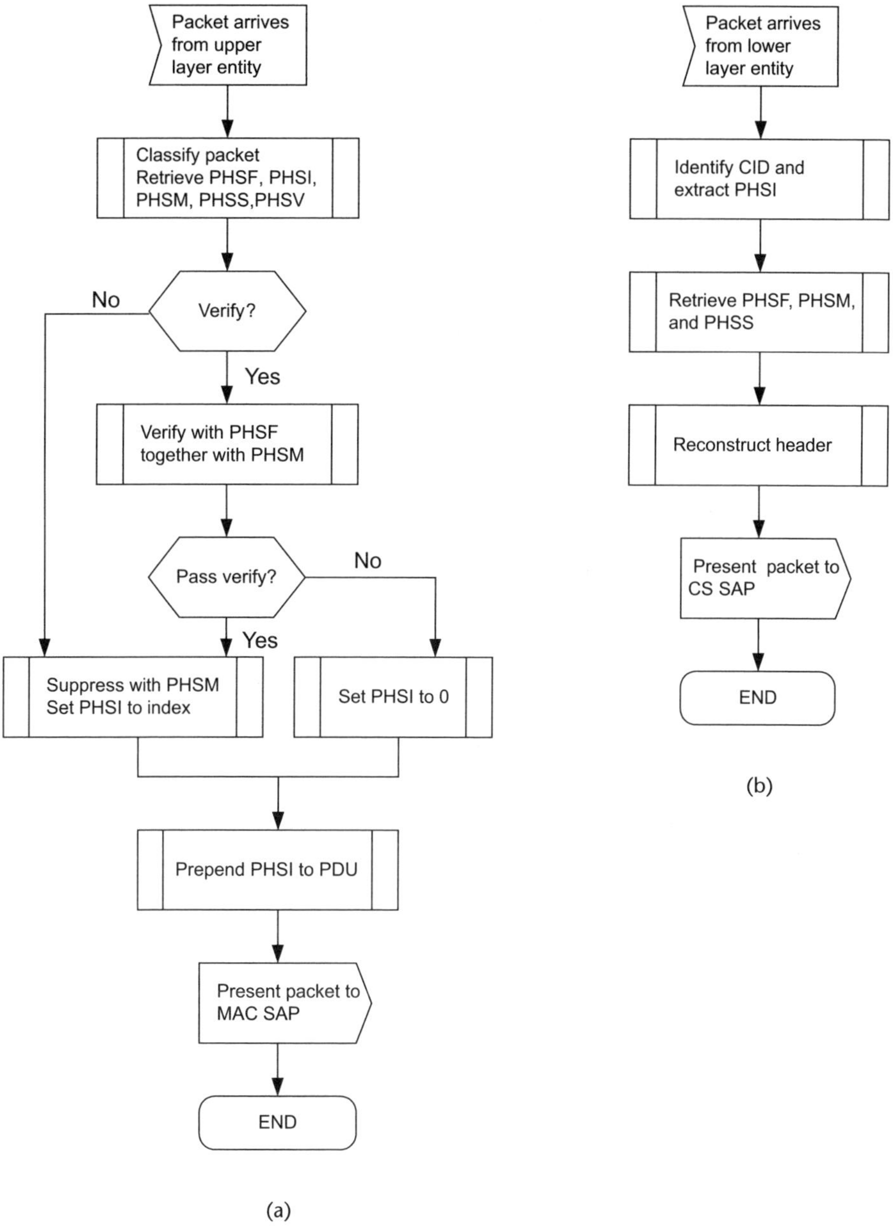

Figure 5.4 Illustration of PHS operation on the uplink: (a) sending by MS; and (b) receiving by BS. (*After:* [1].)

5.2.1 MAC CPS Functions

Before going into the details of the MAC CPS functions, we will briefly discuss the main principles of CPS design for the network architecture of Mobile WiMAX.

Network Architecture

The Mobile WiMAX network has a star architecture, with the BS located at the center and MSs at the end of the branches. The Mobile WiMAX wireless link operates with a central BS and a sectorized antenna that can handle multiple independent sectors simultaneously. All the MSs within a given frequency channel and antenna sector receive the same transmission.

Whereas the fixed WiMAX includes two-way *point-to-multipoint* (PMP) and mesh topology wireless networks for the shared wireless media, the 802.16e-2005 Mobile WiMAX standard considers only PMP architecture and does not specify the mobile extension of mesh network architecture. Instead, the 802.16j Relay Task Group discusses *mobile multihop relay* (MMR) architecture for application to mesh network. So we deal only with the PMP architecture in this section.

Connection-Oriented Communication

The MAC of the Mobile WiMAX operates in a connection-oriented manner. Note that *connection* refers to a unidirectional mapping between the BS and MS MAC peers, which is identified by a CID, 16-bit identifier that differentiates the connections for user data transport. All data communications are made in the context of a *transport connection* (or a connection used for user data transport) for the purpose of mapping to services on MSs and associating with varying levels of QoSs. Once an MS is registered, transport connections are associated with the *service flows* that were provisioned when installing the MS. A service flow refers to a unidirectional flow of MAC SDUs on a connection that provides a particular QoS in terms of a set of QoS parameters. Service flow can be dynamically configured by exchanging the MAC management messages (e.g., DSA, DSC, and DSD messages in Table 5.3) between MS and BS, which specify the QoS parameters (e.g., traffic priority, traffic rate, and service scheduling type) associated with the corresponding connection. Service flows provide a mechanism for uplink and downlink QoS management, so are important in the bandwidth allocation process.

Bandwidth Allocation

Basically, in Mobile WiMAX, the bandwidth allocation mechanism differs for the downlink and uplink. The downlink part of the Mobile WiMAX network is governed by the BS, with the downlink bandwidth managed by the downlink scheduler at the BS. As the BS is the only transmitter operating in the downlink direction, it can transmit, without having to coordinate with other stations, within the given TDD time period. Each MS receives the PDUs that may contain an individually addressed message or a multicast or broadcast message. The portion of downlink bandwidth (referred to as a *data region* in the context of OFDMA systems), scheduled for the individual MS is informed in one of the MAC management messages, referred to as *downlink map* (DL-MAP, see Table 5.3 and Figure 4.29). So unless the DL-MAP explicitly indicates that a portion of the downlink subframe is for a particular MS, every MS listens to it, checks the CIDs in the received PDUs, and retains only those PDUs addressed to it.

On the other hand, the uplink part of the Mobile WiMAX network is shared among all the MSs in the same cell or sector, so the bandwidth is allocated by the BS to MSs on a demand basis. Within each sector, MSs need a transmission protocol that can control contention among MSs and adjust the service to the required QoS (i.e., the delay and bandwidth requirements) of each user application. An MS requests an uplink bandwidth from the BS on a per-connection basis, implicitly identifying the associated service flow. In response to this per-connection request of the

MS, the BS grants the bandwidth to the MS as an aggregate of grants. In other words, whereas the bandwidth request by each MS references individual connections, the grant to the bandwidth request is addressed to the MS as an aggregate, not to individual CIDs. Once a transport connection is established, it requires active maintenance (refer to Section 6.2 for a description of the bandwidth request and allocation processes).

To provide QoS for each connection in the uplink, Mobile WiMAX employs five different types of scheduling mechanisms, which are implemented using unsolicited bandwidth grants, polling, and contention procedures. Unsolicited grants may be used for leased-line types of services, polling for the delay-sensitive applications like voice and video that demand services on a deterministic and periodic basis, and contention for the occasion when the individual polling service continues to be inactive for a long period of time. (Refer to Section 6.1.1 for a detailed discussion of scheduling mechanisms.)

5.2.2 Addressing and Connections

There are three different types of identifiers used in the Mobile WiMAX network: MAC address, service flow identifier, and connection identifier.

Each MS has a unique 48-bit IEEE MAC address, which differentiates it from all others within the set of all possible manufacturers and equipment types. This MAC address is used during the initial ranging process when each MS establishes a connection, and is also used in the authentication process for identifying the BS and MS.

Each MAC PDU is mapped to the packet flow in the uplink or downlink connection offering a particular type of QoS between BS and MS, which is referred to as a *service flow* (SF). Service flows are individually identified by a 32-bit *service flow identifier* (SFID).

Particular service flow traffic is transmitted by the unidirectional mapping between the MAC peers of BS and MS, referred to as a *connection*. Connections are identified by a 16-bit CID, which is used as the MAC layer address. The MAC SDUs associated with a particular service flow are sorted out by examining a CID within the MAC header of each MAC PDU in the receiving CPS. The 16-bit CID permits a total of approximately 64,000 connections within each downlink and uplink channel. So, all communications in the Mobile WiMAX system are identified by the CID in the MAC PDU header. All traffic is delivered over a connection. The service flow implementing the nominally connectionless traffic like IP is also delivered over a connection. CID may be considered a connection identifier even for such connectionless traffic, as it offers a pointer to the destination and the context information. Each CID is matched with the SFID that identifies the QoS parameters associated with the service flow that the particular connection belongs to.

There are three different types of management connections—basic, primary, and secondary. The *basic connection* is used by the BS MAC and MS MAC to exchange short, time-urgent MAC management messages; the *primary connection* to exchange longer, more delay-tolerant MAC management messages; and the *secondary connection* to transfer delay-tolerant, standards-based messages. The secondary management connection is required only for managed MS. In the initiation stage, while each MS performs the ranging and registration process, the BS allocates to the MS the management connections with basic CID, primary CID, and (optional) secondary CID. Whenever a new session begins, the BS allocates to the MS a unidirectional CID through a series of management messages and channel access mechanism. The BS withdraws the CID when the connection terminates. The CIDs for multicasting and broadcasting are defined separately. (Refer to Chapter 3 for a detailed discussion of network initiation and ranging.)

There are other types of CIDs, including initial ranging CID, multicast CID, and broadcast CID. Table 5.2 lists all the CIDs, in conjunction with their values and descriptions.

5.2.3 MAC Management Messages

Unlike the air interfaces of the existing cellular systems, there is no explicit channelization for control signals in the Mobile WiMAX system. To support the

Table 5.2 List of CIDs

CID	Values	Description
Initial ranging	0x0000	Used by SS and BS during initial ranging process.
Basic ID	0x0001 - m	The same value is assigned to both the DL and UL connection.
Primary management	m+1 – 2m	The same value is assigned to both the DL and UL connection.
Transport IDs, Secondary mgmt CIDs	2m+1 – 0xFE9F	For the secondary management connection, the same value is assigned to both the DL and UL connection.
Multicast CIDs	0xFEA0 – 0xFEFE	For the downlink multicast service, the same value is assigned to all MSs on the same channel that participate in this connection.
AAS initial ranging CID	0xFEFF	A BS supporting AAS shall use this CID when allocating an AAS ranging period (using AAS Ranging Allocation IE).
Multicast polling CIDs	0xFF00 – 0xFFF9	A BS may be included in one or more multicast polling groups for the purposes of obtaining bandwidth via polling. These connections have no associated service flow.
Normal mode multicast CID	0xFFFA	Used in DL-MAP to denote bursts for transmission of DL broadcast information to normal mode MS.
Sleep mode multicast CID	0xFFFB	Used in DL-MAP to denote bursts for transmission of DL broadcast information to Sleep mode MS. May also be used in MOB_TRF-IND messages.
Idle mode multicast CID	0xFFFC	Used in DL-MAP to denote bursts for transmission of DL broadcast information to Idle mode MS. May also be used in MOB_PAG-ADV messages.
Fragmentable broadcast CID	0xFFFD	Used by the BS for transmission of management broadcast information with fragmentation. The fragment subheader shall use 11-bit long FSN on this connection.
Padding CID	0xFFFE	Used for transmission of padding information by SS and BS.
Broadcast CID	0xFFFF	Used for broadcast information that is transmitted on a downlink to all SS.

Source: [1].

Mobile WiMAX network-specific operations (e.g., bandwidth management, channel description, sleep mode, and handover), predefined management messages are exchanged between the MS and BS as a MAC layer PDU. Table 5.3 lists the MAC management messages, including their connection types (e.g., basic, primary, secondary, or broadcast). There are 64 different types of messages readily defined, with the other types reserved for future definition. Among the MAC management mes-

Table 5.3 MAC Management Messages

Type	Message name	Message description	Connection	Type	Message name	Message description	Connection
0	UCD	Uplink Channel Descriptor	Fragmentable Broadcast	34	ARQ-Discard	ARQ Discard message	Basic
1	DCD	Downlink Channel Descriptor	Fragmentable Broadcast	35	ARQ-Reset	ARQ Reset message	Basic
2	DL-MAP	Downlink Access Definition	Broadcast	36	REP-REQ	Channel measurement Report Request	Basic
3	UL-MAP	Uplink Access Definition	Broadcast	37	REP-RSP	Channel measurement Report Response	Basic
4	RNG-REQ	Ranging Request	Initial Ranging or Basic	38	FPC	Fast Power Control	Broadcast
5	RNG-RSP	Ranging Response	Initial Ranging or Basic	39	MSH-NCFG	Mesh Network Configuration	Broadcast
6	REG-REQ	Registration Request	Primary Mgt.	40	MSH-NENT	Mesh Network Entry	Basic
7	REG-RSP	Registration Response	Primary Mgt.	41	MSH-DSCH	Mesh Distributed Schedule	Broadcast
8		Reserved		42	MSH-CSCH	Mesh Centralized Schedule	Broadcast
9	PKM-REQ	Privacy Key Management Request	Primary Mgt.	43	MSH-CSCF	Mesh Centralized Schedule Configuration	Broadcast
10	PKM-RSP	Privacy Key Management Response	Primary Mgt. or Broadcast	44	AAS-FBCK-REQ	AAS Feedback Request	Basic
11	DSA-REQ	Dynamic Service Addition Request	Primary Mgt.	45	AAS-FBCK-RSP	AAS Feedback Response	Basic
12	DSA-RSP	Dynamic Service Addition Response	Primary Mgt.	46	AAS-BEAM-Select	AAS Beam Select message	Basic
13	DSA-ACK	Dynamic Service Addition Acknowledge	Primary Mgt.	47	AAS-BEAM-REQ	AAS Beam Request message	Basic
14	DSC-REQ	Dynamic Service Change Request	Primary Mgt.	48	AAS-BEAM-RSP	AAS Beam Response message	Basic
15	DSC-RSP	Dynamic Service Change Response	Primary Mgt,	49	DREG-REQ	SS De-registration message	Basic
16	DSC-ACK	Dynamic Service Change Acknowledge	Primary Mgt.	50	MOB-SLP-REQ	Sleep request message	Basic
17	DSD-REQ	Dynamic Service Deletion Request	Primary Mgt.	51	MOB-SLP-RSP	Sleep response message	Basic
18	DSD-RSP	Dynamic Service Deletion Response	Primary Mgt.	52	MOB-TRF-IND	Traffic indication message	Broadcast
19		Reserved		53	MOB-NBR-ADV	Neighbor advertisement message	Broadcast, Primary Mgt.
20		Reserved		54	MOB-SCN-REQ	Scanning interval allocation request	Basic
21	MCA-REQ	Multicast Assignment Request	Primary Mgt.	55	MOB-SCN-RSP	Scanning interval allocation response	Basic
22	MCA-RSP	Multicast Assignment Response	Primary Mgt.	56	MOB- BSHO-REQ	BS HO request message	Basic
23	DBPC-REQ	Downlink Burst Profile Change Request	Basic	57	MOB-MSHO-REQ	MS HO request message	Basic
24	DBPC-RSP	Downlink Burst Profile Change Response	Basic	58	MOB-BSHO-RSP	BS HO response message	Basic
25	RES-CMD	Reset Command	Basic	59	MOB-HO-IND	HO indication message	Basic
26	SBC-REQ	SS Basic Capability Request	Basic	60	MOB-SCN-REP	Scanning result report message	Primary Mgt.
27	SBC-RSP	SS Basic Capability Response	Basic	61	MOB-PAG-ADV	BS broadcast paging message	Broadcast
28	CLK-CMP	SS network clock comparison	Broadcast	62	MBS-MAP	MBS MAP message	
29	DREG-CMD	De/Re-register Command	Basic	63	PMC-REQ	Power control mode change request message	Basic
30	DSx-RVD	DSx Received Message	Primary Mgt.	64	PMC-RSP	Power control mode change response message	Basic
31	TFTP-CPLT	Config File TFTP Complete Message	Primary Mgt.	65	PRC-LT-CTRL	Setup/Tear-down of long term MIMO precoding	Basic
32	TFTP-RSP	Config File TFTP Complete Response	Primary Mgt.	66	MOB-ASC-REP	Association result report message	Primary Mgt.
33	ARQ-Feedback	Standalone ARQ Feedback	Basic	67-255		Reserved	

Source: [1].

sages in the table UCD, DCD, DL-MAP, and UL-MAP are the representative examples of management messages that control the physical layer parameters directly.

MAC management messages are carried in the payload of the MAC PDU. Their format consists of an 8-bit management message type field and a variable-size management message payload field. The MAC management messages are broadcast or sent on three CIDs in each direction, such as basic, primary, and secondary, and three management connections in each direction are established between the MSs and the BS. MAC management messages are not to be carried on transport connections.

The MAC management messages on the basic, broadcast, and initial ranging management connections are not fragmented and not packed. (Refer to Section 5.2.5 for a detailed discussion of fragmentation and packing.) However, the MAC management messages on the primary management connection may be packed or fragmented, and the MAC management messages on the fragmentable broadcast connection may be fragmented too. For the OFDM or OFDMA-based physical layer, the management messages carried on the initial ranging, broadcast, fragmentable broadcast, basic, and primary management connections are used with the CRC function enabled.

As an example, we examine the DL-MAP message (message type 2), which defines the access to the downlink information. (Refer to Section 4.3.3 for a more detailed description of DL-MAP as well as its uplink counterpart, UL-MAP.) All other messages can be referred to the full specification, including the syntax and functional description, in [1]. Among the large number of MAC management messages, MAP messages are very important ones, especially in the MAC layer of the OFDMA-based Mobile WiMAX system, because they notify the resource allocation results to the MSs dynamically in every frame. To ensure its robustness to the varying channel condition, the DL-MAP message is protected with a most reliable modulation and coding scheme (refer to Section 2.1.3). It is located at the front-most position of each frame (see Figure 4.26) and performs the function of notifying the data region, specified in terms of time-domain and frequency-domain location and the number of the OFDMA symbols, as well as the number of subchannels, in each frame to MSs in the downlink. The DL-MAP message has the format illustrated in Table 5.4 (also in Figure 4.29) for the case of WiressMAN-OFDMA, or Mobile

Table 5.4 DL-MAP Message Format for Mobile WiMAX

Syntax	Size	Notes
Management message type	8 bits	Prespecified to 2
PHY synchronization field	variable	Depends on PHY specifications
DCD count	8 bits	Configuration change count of DCD
Base station ID	48 bits	Front 24 bits used as operator ID
Number of OFDMA symbols	8 bits	All OFDMA symbols in the DL subframe
DL-MAP_IE(), i=1~n	variable	For each DL-MAP element, 1~n (see the relevant PHY specification)
Padding nibble	4 bits	Padding to reach byte boundary

Source: [1].

WiMAX. The description of each field in the message format may be found in Section 4.3.3, in relation with Figure 4.29.

5.2.4 MAC PDU Formats

Basically, a MAC PDU takes the format shown in Figure 5.5. It begins with a 48-bit fixed-length generic MAC header, which may be followed by the payload of the MAC PDU and CRC. The payload consists of subheaders and/or MAC SDUs if it carries user data and consists of MAC management messages if it carries management information. As the length of the payload information may vary, MAC PDU usually represents a variable number of bytes. The only exception is the MAC PDU for MAC signaling header (see the following for MAC header format), which does not carry any data payload. In other words, MAC signaling header is a header-only PDU and used only for uplink. Meanwhile, note that a CRC field at the end of payload is optional in the sense that it is not required when ARQ protocol is not implemented.

MAC Header Formats

There are two different categories of MAC header: One is the *generic MAC header*, which begins each MAC PDU containing either MAC management messages or CS data. The other is the *MAC signaling header*, which carries only signaling information without any MAC PDU payload or CRC following. The reason for using the MAC signaling header without a payload is that short signaling information (e.g., bandwidth request or feedback information) can be accommodated within a MAC header without incurring the overhead associated with payload. One particular example is a MAC signaling header with a *bandwidth request* (BR) field to specify how many bytes an MS expects to reserve in the uplink bandwidth. As bandwidth must be frequently requested and granted on a demand basis, a short BR field within the MAC header will be beneficial for saving bandwidth. The generic MAC header is defined for both the downlink and uplink, whereas the MAC signaling header is defined for the uplink only. The MAC signaling header is further divided into types I and II, with type I used for bandwidth request and other signaling and type II for feedback information.

The various MAC header formats described here are differentiated by the bit fields *header type* (HT) and *encryption control* (EC): HT=0 for the generic MAC header for both DL and UL; and HT=1 for the MAC signaling header. Further, EC=0 if not encrypted and EC=1 if encrypted in the case of the generic MAC header; EC=0 for type I and EC=1 for type II in the case of the MAC signaling header. Note that the EC field is used for the distinction of types in the case of the MAC signaling header, as encryption is not applied in this case. Table 5.5 lists a summary of this categorization. (Note that compressed DL-MAP and reduced private MAP do not use MAC headers as defined in this table. For details, refer to Section 6.3.2 of [1].)

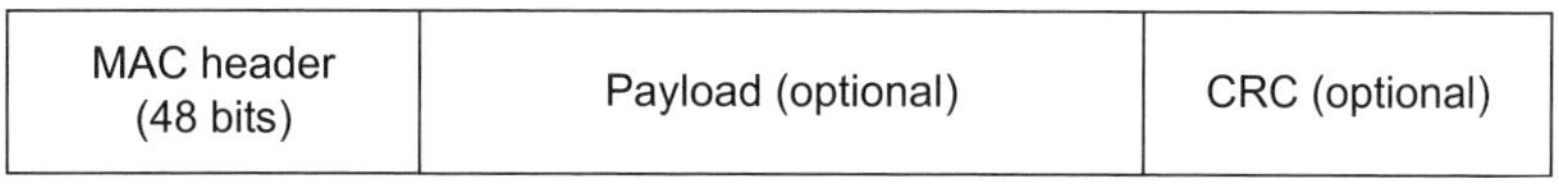

Figure 5.5 MAC PDU format. (*After:* [1].)

Table 5.5 Categorization of MAC Headers

	HT=0	HT=1
EC=0	Generic MAC header for DL and UL, MAC PDU with data payload, no encryption	MAC signaling header type I for UL, MAC PDU without data payload
EC=1	Generic MAC header for DL and UL, MAC PDU with data payload, with encryption	MAC signaling header type II for UL, MAC PDU without data payload

Generic MAC Headers

Generic MAC header is the header of ordinary MAC PDUs that carries payload and CRC fields as described in Figure 5.5. The payload contains either MAC management messages or the user packets in CS PDU format. It can contain six different types of MAC subheaders in the position immediately following the generic MAC header.

Figure 5.6 shows the structures of the generic MAC header format. The use of the various fields in the generic MAC header format is as follows: *Type* field indicates the presence of various subheaders as discussed in the next subsection (see Table 5.6). *Extended subheader field* (ESF) indicates the presence of the extended subheader, which follows the *generic MAC header* (GMH) immediately if present. *CRC indicator* (CI) field indicates the presence of CRC. *Encryption key sequence* (EKS) field contains the index of the traffic encryption key and initialization vector used for encryption. As the length of payload varies, *length* (LEN) field is required for indicating the length (in bytes) of the MAC PDU, including the MAC header and the CRC. CID field contains the CID value, which is represented in 16 bits. *Header check sum* (HCS) is the 8-bit CRC for error-checking of the header. In case that CRC does not check, the corresponding header must be discarded.

MAC Signaling Headers

Type I and type II MAC signaling headers are both header-only headers without payload and CRC field, and are both applied to the uplink only. Figure 5.7 shows the structures of the both MAC signaling header formats. The *header content* field in the figure is the space to put different parameters depending on the content that the MAC signaling header carries. Usage of the other fields is as described previously.

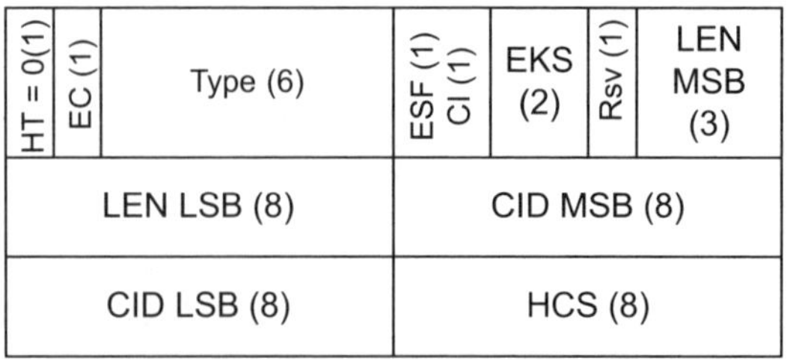

Figure 5.6 Generic MAC header format. (*After:* [1].)

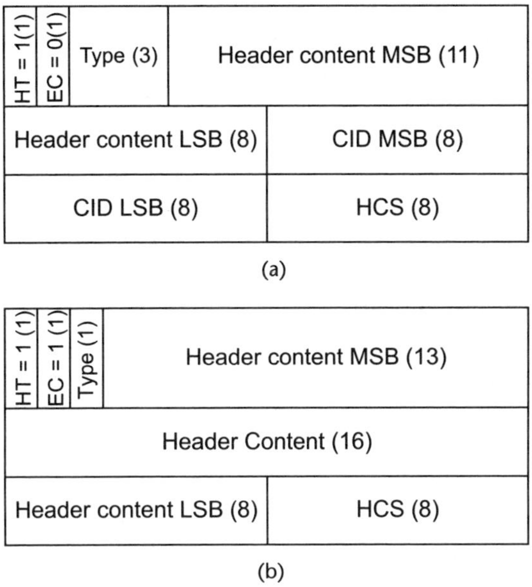

Figure 5.7 MAC signaling header formats: (a) type I; and (b) type II. (*After:* [1].)

Type I MAC signaling header is used as bandwidth request header for requesting uplink bandwidth (incremental and aggregate), bandwidth request and uplink transmit power report header, bandwidth request and *carrier-to-interference-and-noise ratio* (CINR) report header, *channel quality indicator* (CQI) channel allocation request header, physical channel report header, bandwidth request and uplink sleep control header, and *sequence number* (SN) report header.

For example, when the type I signaling header is used as the bandwidth request header, both header content fields turn into the *band request* (BR) field, which indicates the number of bytes requested, and the CID field, which indicates the connection for which uplink bandwidth is requested. Similarly, when the type I signaling header is used as the bandwidth request and uplink transmit power report header, the header content MSB field turns into the BR field and the header content LSB field into uplink transmit power field. (Refer to Section 6.3.2.1.2.1 of [1] for the details of the type I signaling headers.)

Type II MAC signaling header is used as feedback header to form feedback PDU. The first 5 bits in the header content MSB field in Figure 5.7(b) are used for *CID inclusion indication* (CII) (1 bit) and feedback type (4 bits). The 1-bit type field, if the value is 0, indicates that the 4-bit feedback type field describes the feedback contents. The case of type value 1 is reserved for later use. The feedback contents include the CQI and MIMO channel feedback, DL average CINR of the serving or anchor BS, MIMO coefficients feedback, UL transmit power, PHY channel feedback, and others. (Refer to Section 6.3.2.1.2.2 of [1] for the details of feedback contents.) Note that all these contents are frequently required but are short enough to be carried within the MAC header.

MAC Subheaders

In the MAC PDU with generic MAC header, there exist six different types of subheaders, including per-PDU subheaders, per-SDU subheader, ARQ feedback payload, and extended subheaders. There are four per-PDU subheaders—mesh, fragmentation, grant management, and fast-feedback allocation subheaders. In addition, there is one per-SDU subheader, which is the packing subheader.

The mesh subheader is used to put the node IDs when the network operates in mesh mode; the fragmentation subheader is to indicate the fragmentation state of the payload; the grant management subheader is used by the MS to convey bandwidth request messages to the BS. The packing subheader is to put multiple SDUs into a single MAC PDU when packing is used, and the fast-feedback allocation subheader always appears as the last per-PDU subheader. The support of the fast-feedback allocation subheader is PHY specification specific. Table 5.6 lists a summary of the six MAC subheaders in relation to the type field values.[1]

The per-PDU subheaders may be inserted in MAC PDUs immediately following the *generic MAC header* (GMH), preceding all per-SDU subheaders. If both the fragmentation subheader and grant management subheader are present, the grant management subheader is put first. If the mesh subheader is present, it is put ahead of all other subheaders. The fast-feedback allocation subheader is always put at the last position. The packing and fragmentation subheaders are mutually exclusive, so are not simultaneously present within the same MAC PDU.

Extended subheaders start with an 8-bit length field, which specifies the total length of the subheader group, including all the extended subheaders and the length byte. Each extended subheader consists of a 7-bit extended subheader type field, a variable-size extended subheader body, and a reserved bit. The extended subheaders are used for various purposes, including to carry the last virtual MAC SDU sequence number, sleep control message, fast-feedback request, MIMO mode feedback, uplink transmit power report, mini-feedback, sequence number report header, and PDU sequence number.

Table 5.6 MAC Subheaders

Type Bit	MAC Subheader	Usage
5	Mesh subheader	Mesh network mode
4	ARQ feedback payload	Presence
3	Extended type	Extension of present packing/fragment. subheader
2	Fragmentation subheader	Presence
1	Packing subheader	Presence
0	DL: fast feedback allocation UL: grant management	Presence

1. The ARQ feedback payload (for type bit 4) is transported if the type bit 4 is set. The extended type (for type bit 3) indicates whether or not the present packing or fragmentation subheader is extended to non-ARQ-enabled connections.

5.2.5 Construction and Transmission of MAC PDU

MAC PDU may be constructed out of user data packet PDU or out of MAC management messages. In the case of the user data packet PDU, the MAC PDU is generated via MAC SDU. That is, the packet PDU is first mapped as MAC SDU (see Figure 5.2), and a generic MAC header and CRC are added to it to build a MAC PDU (see Figure 5.5). In this MAC PDU constructing process, a MAC SDU or a MAC management message may be divided into multiple MAC PDUs if its length is long, and multiple MAC SDUs or MAC management messages may be combined into a MAC PDU if their lengths are short. The former process is called *fragmentation* and the latter *packing*. In the transmission process of MAC PDU, it is possible to combine multiple MAC PDUs into a single burst. It is called *concatenation.*

Fragmentation

Fragmentation is intended to enhance the efficiency of the air interface. Fragmentation is required on both BS and MS, and may be initiated either by the BS (if it is for downlink connection) or by the MS (if it is for uplink connection). The authority to fragment traffic on a connection is defined when the connection is created by the MAC SAP. For each connection, the size of the *fragment sequence number* (FSN) in fragmentation subheaders is fixed to 3 or 11 bits. Both BS and MS must support 11-bit FSN and may support 3-bit FSN.

Each fragment is tagged with an indication of its position in the parent SDU (e.g., 10 if the first segment, 11 if a continuing fragment, and 01 if the last fragment). For non-ARQ connections, fragments are transmitted in sequence, and the receiver reassembles them according to their sequence numbers. If any fragment belonging to the parent SDU is lost, the receiver discards all the MAC PDUs on the connection until a new first fragment is detected. This inefficiency is improved in the ARQ-enabled connections. In this case, fragments are formed for each transmission by concatenating sets of ARQ blocks having adjacent sequence numbers.

Packing

Packing is also intended to enhance the efficiency of air interface, by condensing multiple MAC SDUs into a single MAC PDU. The transmitter decides whether or not to do packing, and the receiver is mandated to be capable of unpacking. It is also possible to partition a MAC SDU into multiple fragments and then pack them again into a MAC PDU, or to construct a MAC PDU with the fragments from different MAC SDUs or from a mixture of first transmissions and retransmissions.

The packing mechanism differs between the non-ARQ and the ARQ-enabled connections. Even among the non-ARQ connections, the mechanism differs depending on whether the MAC SDUs to pack are of fixed length or variable length. If they are of fixed length, packing can be done without adding *packing subheaders* (PSHs); otherwise, a PSH is necessary to help delineate the original MAC SDU blocks in the receiver. Besides, in the fixed-length case, fragmentation is not allowed, while packing but in the variable-length case fragmentation is allowed in conjunction with packing. Efficiency increases by the simultaneous use of packing and fragmentation, but, as the price, the handling of packing and fragmentation indication information is complicated.

In the case of the ARQ-enabled connections, the use of PSH is similar to that for non-ARQ connections except that ARQ-enabled connections set the *extended type bit* (i.e., bit 3 of the type field) in the generic MAC header to 1 (see Figure 5.6). Packing of variable-length MAC SDUs for the ARQ-enabled connections is similar to that of non-ARQ connections. The *block sequence number* (BSN) of the PSH is used by the ARQ protocol to identify and retransmit ARQ blocks.

Figure 5.8 depicts the procedure of constructing MAC PDUs applying fragmentation and packing. It contains all possible variations of the construction. Figure 5.9 illustrates the construction of MAC PDUs: (a) without fragmentation or packing, (b) with fragmentation, and (c) with packing in one figure.

Concatenation

Combining of multiple MAC PDUs is possible in the transmission stage. They may be *concatenated* into a single transmission in the uplink or downlink direction. Since each constituent MAC PDU can be identified by its own CID, the receiving MAC entity can deliver the MAC SDU to the correct instance of the MAC SAP. Concatenation applies not only to the same type of MAC PDUs but also to different types of MAC PDUs, including user MAC PDU, MAC management messages, and bandwidth request MAC PDU. Figure 5.10 illustrates how to concatenate multiple MAC PDUs of different types into an uplink burst. Such a concatenation capability is an efficient feature that enables putting user data and management messages together whenever needed. This feature distinguishes the WiMAX system from other conventional systems that require differentiating data and control channels physically or logically.

5.3 ARQ

The ARQ mechanism is intended to recover transmission errors by retransmission of the erred packets. As it is a part of the MAC, it must be differentiated from the HARQ (see Sections 2.1.6 and 4.2.3), which handles retransmission in association with channel coding in the physical layer. As long as a reliable link can be provided with HARQ, the MAC-layer ARQ might be redundant and thus is optional for implementation. In support of the ARQ operation, an error-detecting mechanism is necessary, as well as a feedback channel reporting the successful/failed reception (i.e., ACK/NAK) of data blocks. ARQ in Mobile WiMAX is enabled on a per-connection basis and is specified and negotiated during connection setup. Once a connection is set up with ARQ enabled, only ARQ-enabled traffic is allowed on the connection. In other words, ARQ and non-ARQ traffic cannot be mixed on the same connection.

5.3.1 ARQ Block Processing

ARQ operates in the unit of *ARQ block*. ARQ block is formed in different ways depending on whether or not the fragmentation is enabled. When fragmentation is enabled, a MAC SDU is logically partitioned into ARQ blocks with the length specified by a connection TLV parameter—see Figure 5.11(a)—which are further divided

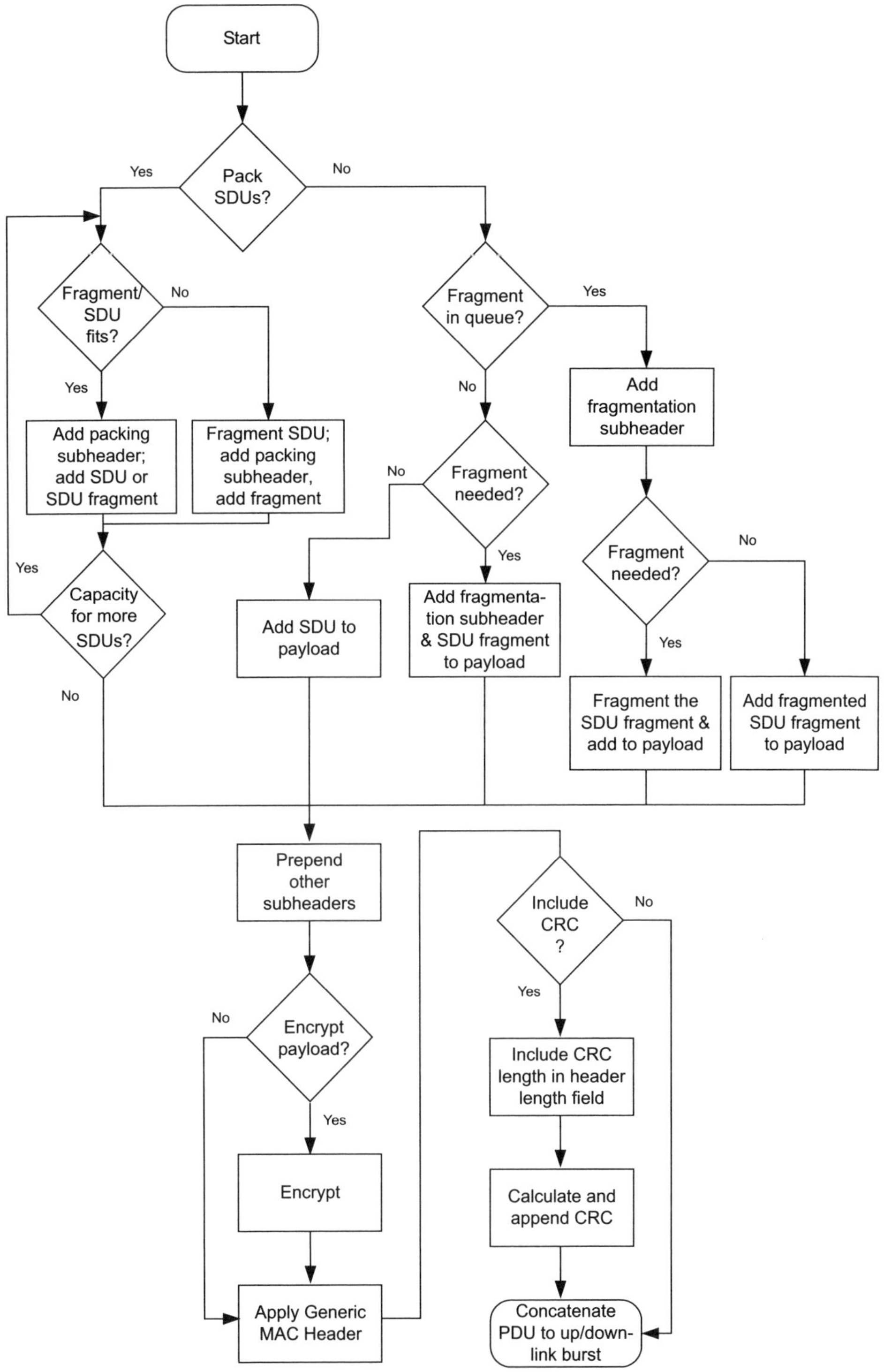

Figure 5.8 Procedure of constructing a MAC PDU. (*After:* [1].)

into two different fragments. The ARQ block is used as a basic unit to keep track in the course of MAC operation and to rearrange a subset of blocks in the course of

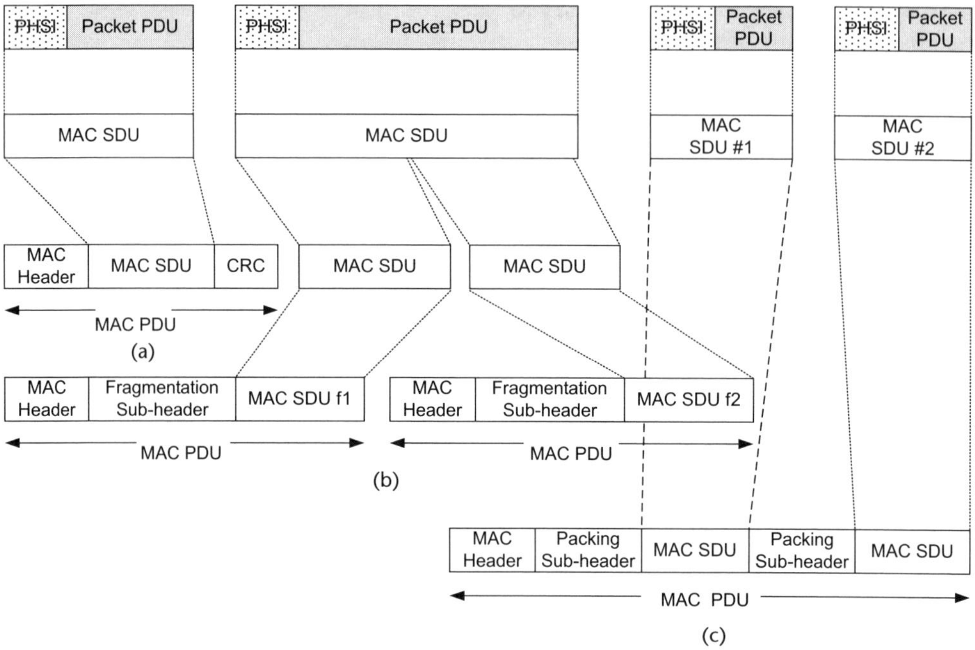

Figure 5.9 Illustration of MAC PDU construction: (a) regular; (b) fragmentation; and (c) packing.

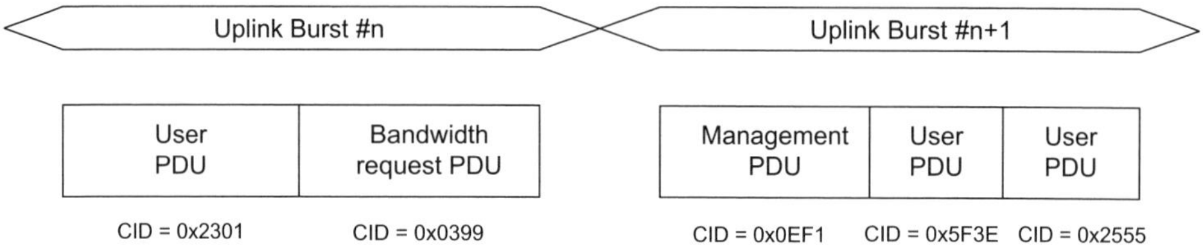

Figure 5.10 Illustration of MAC PDU concatenation. (*After:* [1].)

retransmission (see Figure 5.11). If the length of the SDU is not an integer multiple of the block length, the last block of the SDU is formed by the remaining SDU bytes. When fragmentation is not enabled, the connection is managed as if fragmentation were enabled but, regardless of the negotiated block size, each fragment contains all the blocks of data associated with the parent SDU.

The ARQ blocks selected for transmission or retransmission are encapsulated into a PDU. Either fragmentation or packing is applied to the ARQ blocks. Fragmentation or packing subheaders contain a *block sequence number* (BSN), which is the sequence number of the first ARQ block in the sequence of the blocks following the subheader. For fragmented PDU, all the blocks in the PDU have contiguous block numbers. For packed PDU, the sequence of blocks immediately between MAC subheaders and the sequence of blocks after the last packing subheader have contiguous block numbers.

In Figure 5.11, two options for retransmission are presented—with and without rearrangements of blocks. Figure 5.11(c) illustrates the case when PDU #2 has not been successfully received so is rearranged into two different PDUs. Note that a PDU

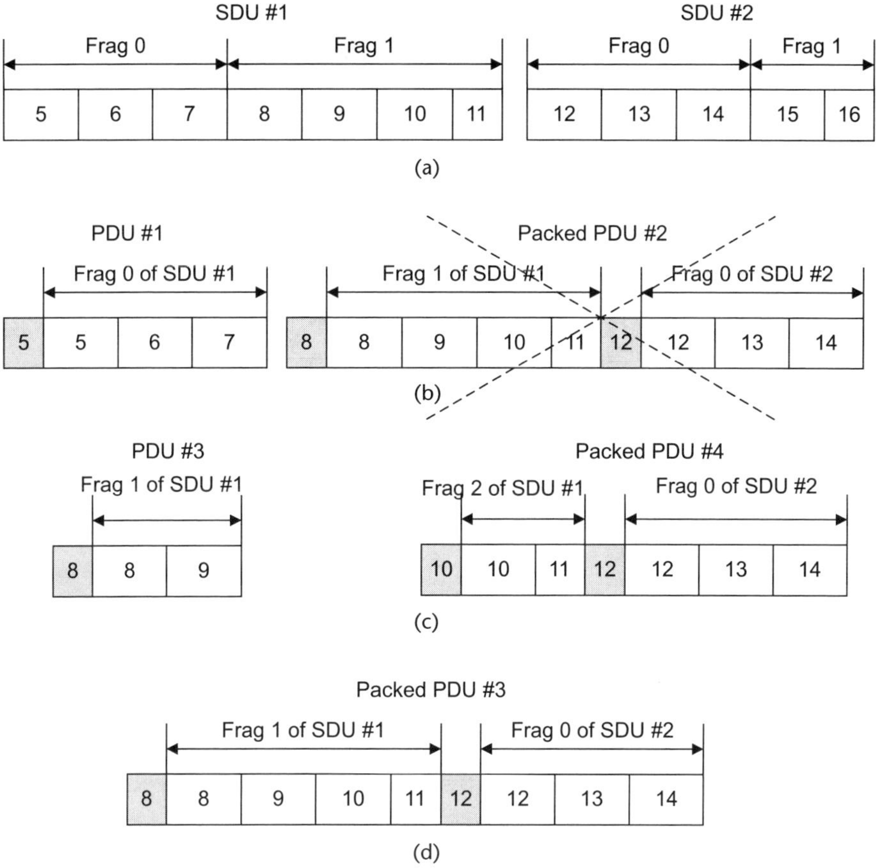

Figure 5.11 Illustration of block usage for ARQ: (a) two consecutive SDUs presented to MAC; (b) original transmission with PDU #2 failed; (c) retransmission of PDU #2 with rearrangement; and (d) retransmission of PDU #2 without rearrangement. (*After:* [1].)

may contain blocks that are transmitted for the first time as well as those being retransmitted.

5.3.2 ARQ Feedback

In order to signal positive or negative acknowledgments to each block as the ARQ feedback information, the ARQ feedback IE is used by the receiver. It can be sent as a standalone MAC management message on the appropriate basic management connection or piggybacked on an existing connection. ARQ feedback cannot be fragmented. Figure 5.12 depicts the ARQ feedback IE format used by the receiver. In the figure, CID field denotes the ID of the connection being referenced; LAST indicates whether or not the ARQ feedback IE is the last one; ACK type indicates the type of ACK; BSN has different meaning depending on the type of ACK.

There are four different types of ARQ feedback, as is identified by ACK type field: selective ACK (type 0), cumulative ACK (type 1), cumulative with selective ACK (type 2), and cumulative ACK with block sequence ACK (type 3). The size and

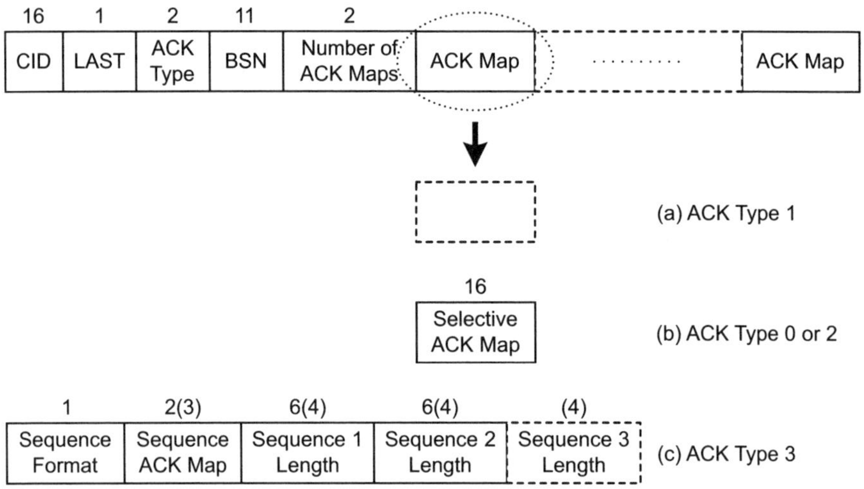

Figure 5.12 ARQ feedback IE format.

format of ACK map as well as the usage of BSN field differ for different ACK types—see (a)–(c) in Figure 5.12. ACK map is used to indicate which ARQ blocks have been received without error.

A set of IEs of this format may be transported either as a packed payload (i.e., piggybacked) within a packed MAC PDU or as a payload of a standalone MAC PDU. Depending on the ACK type, the length of ARQ feedback IE is variable. In order to see the difference between the types of ARQ feedback, we take an example in Figure 5.13.

In the *selective ACK* type (ACK type 0), each bit set to 1 in ACK map indicates the corresponding ARQ block has been received without errors. As there are 16 bits in the ACK map, 16 blocks can be acknowledged at most. Meanwhile, a value in the BSN field corresponds to the most significant bit of the first 16-bit ARQ ACK map. In the example of Figure 5.13(a), 4 out of 16 ARQ blocks are received with errors (those marked with X at positions of bit 4, bit 7, bit 8, and bit 12). Identifying the erred blocks by the positions of 0s, the ACK map is encoded as 1110110011101111.

In the *cumulative ACK* type (ACK type 1), no ACK map is used. Instead, only BSN is used to indicate that its corresponding block and all the blocks with smaller values within the transmission window have been successfully received. In the example of Figure 5.13(b), BSN = 3 indicates that the first three consecutive blocks have been received without error.

The *cumulative with selective ACK* type (ACK type 2) combines the functionality of selective ACK and cumulative ACK. In the example of Figure 5.13(c), BSN = 3 indicates that the first three blocks have been successfully received. The rest of the bit map is interpreted similar to selective ACK. In other words, the bit sequence in the ACK map represents the positions of bits that have been received with errors as in the selective ACK map, except for the most significant bit of the first map entry, which is set to one, and the IE is interpreted as a cumulative ACK for the BSN value in the IE. The rest of the bit map is interpreted similar to selective ACK.

In the *cumulative ACK with bock sequence ACK* (ACK type 3), two or three ARQ block sequences can be defined. Each bit set to 1 in sequence ACK map indicates that the corresponding sequence of ARQ blocks has been received without

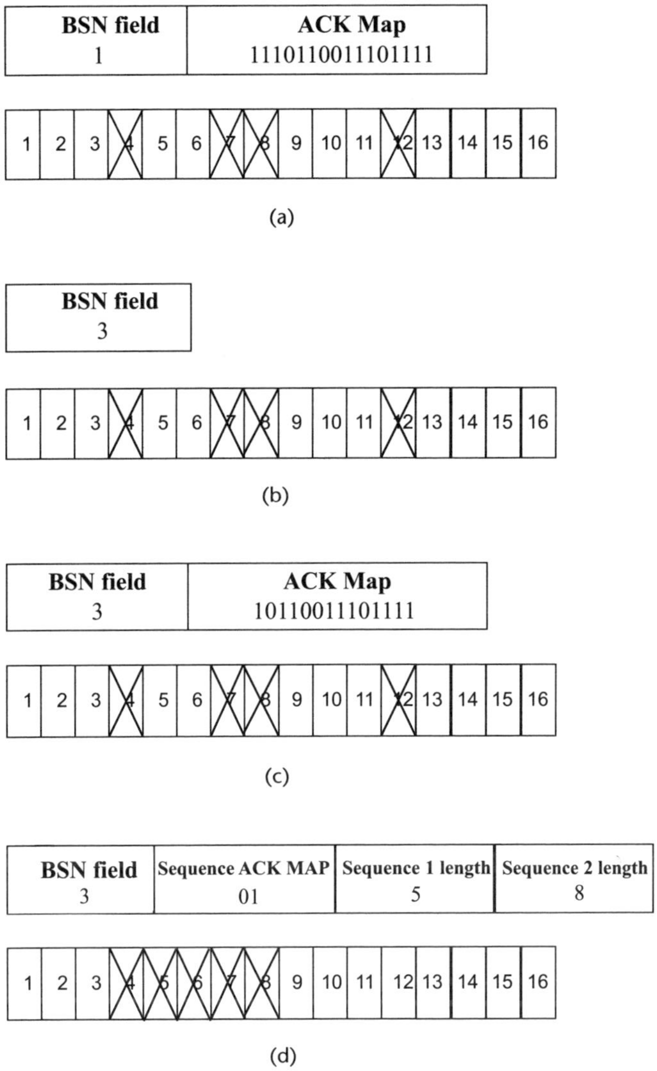

Figure 5.13 ARQ operation of four ACK types: (a) selective ACK; (b) cumulative ACK; (c) cumulative with selective ACK; and (d) cumulative ACK with block sequence ACK.

errors. In the example of Figure 5.13(d), two block sequences, one with 5 blocks and the other with 8 blocks, are defined. As the second bit of sequence ACK map is set to 1, it is known that only the second sequence has been successfully received.

5.3.3 ARQ Operation

ARQ operation is based on sliding windows. Windows indicate the number of unACKed ARQ blocks that can be sent. Figure 5.14 illustrates the operation of the sliding windows and the relevant ARQ parameters managed by the transmitter (a) and receiver (b).

In the case of the receive window in Figure 5.14(b), the window management at the receiver side is as follows: The sliding window is maintained such that the ARQ_RX_WINDOW_START variable always points to the lowest numbered ARQ block that has not been received or has been received with errors. It implies

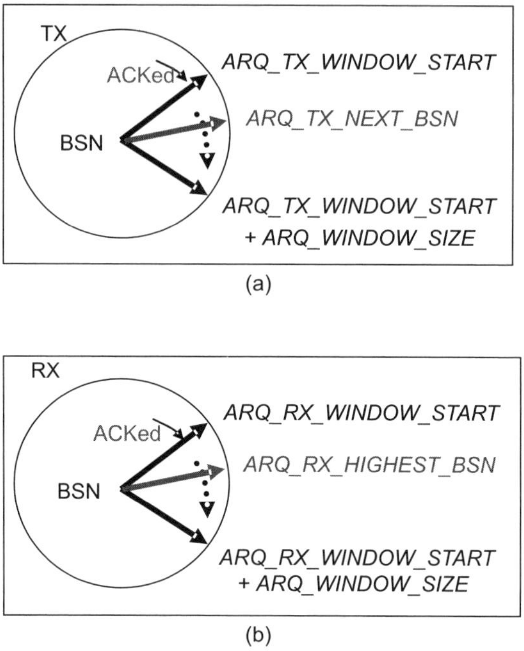

Figure 5.14 Sliding window and ARQ parameters: (a) transmitter; and (b) receiver.

that all BSNs up to (ARQ_RX_WINDOW_START—1) have been acknowledged. When an ARQ block with a BSN corresponding to the value of ARQ_RX_WINDOW_START is received, the window is advanced.

Every time a new ARQ block is received, the receiver first updates ARQ_RX_HIGHEST_BSN to be the BSN of the highest block received plus one. Since the maximum number of unacknowledged ARQ blocks at any given time is given by ARQ_WINDOW_SIZE, ARQ_RX_HIGHEST_BSN will be in the interval ARQ_RX_WINDOW_START to (ARQ_RX_WINDOW_START + ARQ_WINDOW_SIZE). Meanwhile, if the BSN of the new ARQ block is equal to ARQ_RX_WINDOW_START, it advances ARQ_RX_WINDOW_START to the BSN of the next ARQ block not yet received. Then, the timer for ARQ_SYNC_LOSS_TIMEOUT is reset. In case that ARQ_RX_WINDOW_START remains at the same value for ARQ_SYNC_LOSS_TIMEOUT, then loss of synchronization of receiver state machine is declared as long as data transfer is known to be active. For ARQ blocks not resulting in an advancement of ARQ_RX_WINDOW_START, it stores them after setting the timer for ARQ_RX_PURGE_TIMEOUT. When the timer for ARQ_RX_PURGE_TIMEOUT expires, it advances the ARQ_RX_WINDOW_START to the BSN of the next ARQ block not yet received after the ARQ block associated with the timer.

The sliding windows managed by the transmitter in Figure 5.14(a) operate in a similar fashion. ARQ_TX_WINDOW_START variable points to the lowest ARQ block to be acknowledged next. In other words, all BSNs up to (ARQ_TX_WINDOW_START –1) have been acknowledged. Meanwhile ARQ_TX_NEXT_BSN variable points to the BSN of the next block to send, which is in the interval of ARQ_TX_WINDOW_START to (ARQ_TX_WINDOW_START + ARQ_WINDOW_SIZE). When an acknowledged block with a BSN corresponding to the value of ARQ_TX_WINDOW_START is received, the window is advanced. If transmission of an ARQ block is not acknowledged by the receiver before the ARQ_BLOCK_LIFETIME is reached, the block is discarded. In order to retransmit an unacknowledged block for retransmission, a transmitter must wait for the minimum time interval of ARQ_RETRY_TIMEOUT.

Reference

[1] IEEE Std 802.16e-2005 and IEEE Std 802.16-2004/Cor1-2005, Part 16: Air Interface for Fixed and Mobile Broadband Wireless Access Systems, February 2006.

Selected Bibliography

Eklund, C., et al., "IEEE Standard 802.16: A Technical Overview of the WirelessMAN Air Interface for Broadband Wireless Access," *IEEE Communications Magazine*, Vol. 40, No. 6, June 2002, pp. 98–107.

Eklund, C., et al., *Wieless MAN: Inside the IEEE 802.16 Standard for Wireless Metropolitan Networks*, New York: IEEE Press, 2006.

Kuran, M. S., and T. Tugcu, "A Survey on Emerging Broadband Wireless Access Technologies," *Computer Networks*, Vol. 51, No. 11, August 2007, pp. 3013–3046.

Kwon, T., et al., "Design and Implementation of a Simulator Based on a Cross-Layer Protocol Between MAC and PHY Layers in a WiBro Compatible IEEE 802.16e OFDMA System," *IEEE Communications Magazine*, Vol. 43, No. 12, December 2005, pp. 136–146.

Lin, S., D. Costello, and M. Miller, "Automatic-Repeat-Request Error-Control Schemes," *IEEE Communications Magazine*, Vol. 22, No. 12, December 1984, pp. 5–18.

Ohrtman, F., *WiMAX Handbook: Building 802.16 Wireless Networks*, New York: McGraw-Hill, 2005.

CHAPTER 6

Bandwidth Management and QoS

In cellular networks, in general, the wireless bandwidth is shared among the mobile users according to the service requests. As the service request deals not only with the bandwidth but also with delay and other QoS parameters, the network conducts scheduling to decide the bandwidth to allocate and the type of the bandwidth allocation. The scheduling algorithm essentially arranges such that the network resources can be shared fairly among the users in consideration of the requested QoS. In addition, admission control gets involved to control the admission of new mobile users to the network without disruptiong the ongoing services to the existing users.

The method of bandwidth allocation differs depending on the networks in service: for example, in order to satisfy the given delay requirement, the cdma2000 system allocates dedicated channels (i.e., CDMA *codes*) and the CDMA 1xEV-DO system dynamically allocates *slots*. The efficiency of those bandwidth allocation schemes varies depending on the traffic characteristics and the QoS requirements. For example, a dedicated bandwidth allocation method is not suitable for handling the unpredictable bandwidth demand of a bursty traffic. In the case of Mobile WiMAX, bandwidth allocation is done according to the bandwidth allocation types, which reflect the delay requirements and traffic characteristics of the requested services.

In this chapter we first introduce five different types of bandwidth allocation (i.e., scheduling services) together with the relevant data delivery services and then discuss the bandwidth request and allocation mechanisms. On this foundation, we investigate the QoS issues of Mobile WiMAX, including service flows and classes, QoS messages and parameters, QoS-related network elements, service flow setup and release procedures, and other issues.

6.1 Scheduling and Data Delivery Services

Scheduling services represent the data-handling mechanisms supported by the MAC schedulers for data transport on a connection associated with a single scheduling service, whereas data delivery services are defined to support various user application services in association with the QoS-related parameters. These two services are tightly coupled together in connection with the traffic patterns and the related QoS requirements.

6.1.1 Scheduling Services

Scheduling services are designed to improve the efficiency of the polling[1]/granting process for requesting bandwidth in the uplink.[2] The BS can anticipate the throughput and latency required by the uplink traffic and can provide polls and/or grants at the appropriate times by specifying a scheduling service and its associated QoS parameters. Rigorously speaking, a scheduling service is determined by a set of QoS parameters that quantify the aspects of its behavior. (For more discussions on the QoS parameters, refer to Section 6.3.)

Depending on the traffic characteristics of user applications (e.g., constant bit rate or variable bit rate) and their associated individual QoS requirements (e.g., latency), scheduling services are divided into five different categories. The five scheduling services specified in the mobile WiMAX standards are *unsolicited grant service* (UGS), *real-time polling service* (rtPS), *nonreal-time polling service* (nrtPS), *best effort* (BE), and *extended rtPS* (ertPS). The *UGS* supports real-time service flows that generate fixed-size data packets on a periodic basis; the *rtPS* supports real-time service flows that generate variable-size data packets on a periodic basis; the *extended rtPS* is a variation of rtPS; the *nrtPS* supports nonreal-time service flows that require variable-size data grant burst on a regular basis; and the *BE service* supports best-effort traffic.

In order to maximize the efficiency of dynamic bandwidth allocation schemes, these scheduling services are mainly characterized by their uplink bandwidth request and grant processes, which vary with traffic characteristics and delay requirements. For example, the request opportunities can be either a periodic or on-demand basis, depending on the traffic characteristics. Once a specific means of bandwidth request is prescribed with a type of scheduling service, it is implemented by a procedure explained in Section 6.2. Upon the bandwidth requests from all active MSs, on the other hand, the uplink scheduler in the BS determines how much bandwidth to allocate to each of them, subject to their individual delay requirement. In other words, a packet-scheduling algorithm must be implemented for determining a share of uplink bandwidth to individual MSs, so as to meet their QoS requirement while maximizing the overall system throughput.

Table 6.1 lists a summary of the five different categories of scheduling services, including their individual application examples and attributes of bandwidth management. In the table, *piggyback request* refers to one of two special bandwidth request options, in which a bandwidth request is carried by a MS along with its uplink data transmission without any explicit bandwidth reservation for bandwidth request. Meanwhile, *bandwidth stealing* refers to another special option, which uses the granted bandwidth for sending another bandwidth request rather than sending data. Depending on the scheduling type, both of these special options for bandwidth requests can be applicable. In the following subsections, the detailed usage and the underlying key QoS parameters are described for each scheduling type.

1. Polling refers to the process that the BS allocates to MSs the bandwidth to use when making bandwidth requests; for more discussions on polling service, refer to Section 6.2.
2. Scheduling gets involved in handling downlink traffic too, but it is an independent function of the BS that does not require the involvement of MSs.

Table 6.1 Classification of Scheduling Services

Scheduling type	Example	Piggyback request	Bandwidth stealing	Polling method
Unsolicited Grant Service (UGS)	T1/E1 leased line, VoIP without silence suppression	Not allowed	Not allowed	PM bit used to request unicast poll for bandwidth needs on non-UGS connections.
Real-time Polling Service (rtPS)	MPEG video	Allowed	Allowed	Only allows unicast polling
Non-real-time Polling Service (nrtPS)	FTP	Allowed	Allowed	May restrict service flow to unicast polling via transmission / request policy: Otherwise all forms of polling are allowed.
Best Effort (BE) Service	HTTP	Allowed	Allowed	All forms of polling allowed.
Extended rtPS (ertPS)	VoIP with silence suppression	Allowed	Allowed	Uses unicast polling. BS offers unicast grant like unrequested UGS.

Source: [1].

UGS

The UGS supports real-time uplink service flows that generate fixed-size data packets on a periodic basis, whose typical examples are T1/E1 and *voice-over-IP* (VoIP) without silence suppression. The UGS offers fixed-size grants on a real-time periodic basis so that it can eliminate the overhead and latency of the MS requests and meet the real-time requirement of the service flow. Specifically, the BS provides data grants that can guarantee the maximum sustained traffic rate to the MS at periodic intervals so that the MS can send data without request or contention. The size of the grant should be large enough to service the requested service flow and may be made larger at the discretion of the BS scheduler.

For a proper operation of the UGS, the request/transmission policy should be set such that the MS cannot use any contention request opportunities for the connection. The key QoS parameters for the UGS, which are mandatory, are the maximum sustained traffic rate, maximum latency, tolerated jitter, uplink grant scheduling type, and request/transmission policy.

The UGS has a mechanism to provide long-term compensation for bandwidth fluctuation caused by lost MAP or clock rate mismatch, by acquiring additional bandwidth. In support of this, the MS passes the status information on the UGS service flow to the BS over the *slip indicator* (SI) bit in the grant management subheader. The SI bit is set to 1 if the service flow is determined to exceed the predefined transmit queue depth and is cleared when the transmit queue returns back within the limits. For acquiring the additional bandwidth, the *poll-me* (PM) bit may be used to request to be polled for an additional, non-UGS connection. In case the SI bit is set to 1, the BS may grant up to 1 percent additional bandwidth for the compensation of clock rate mismatch.[3]

3. Other than that, the BS does not allocate more bandwidth than the maximum sustained traffic rate parameter in the active QoS parameter set.

The *frame latency* (FL) and the *frame latency indication* (FLI) fields in the grant management subheader may be used to provide the BS with the synchronization information of the MS application that generates periodic data for the UGS (or the extended rtPS) service flows. Those fields may also be used to monitor if the latency of those service flows exceed a predetermined threshold. If the FL increases beyond the threshold, the BS may start bandwidth allocation to the corresponding service flows early.

rtPS

The rtPS supports real-time uplink service flows that generate variable-size data packets on a periodic basis, whose typical example is *moving pictures experts group* (MPEG) video. It offers real-time, periodic, unicast request opportunities that meet the real-time requirements of the service flows and allow the MS to specify the size of the desired grants. The overhead for making service requests is larger for the rtPS than for the UGS, but data transport efficiency is higher for the rtPS as it supports variable grant sizes.

With the rtPS, the MS is provided with periodic unicast request opportunities. Thus the request/transmission policy should be set to not allow the MS to use any additional contention request opportunities on the rtPS connection. Instead, the BS has to issue unicast request opportunities as prescribed by the rtPS even if the prior requests are not fulfilled.

The key QoS parameters, which are mandatory, are the minimum reserved traffic rate, maximum sustained traffic rate, maximum latency, uplink grant scheduling type, and request/transmission policy.

nrtPS

The nrtPS supports nonreal-time service flows that require variable-size data grant bursts on a regular basis, such as high-bandwidth FTP. It offers unicast polls on a regular basis to assure that the uplink service flow receives request opportunities even during network congestion. The interval that the BS polls the connections of nrtPS type is typically on the order of 1 second or less.

The BS should provide timely unicast request opportunities. For a proper operation of the nrtPS service, the request/transmission policy is set such that the MS is allowed to use contention request opportunities. Then, the MS can use the contention request opportunities as well as the unicast request opportunities and the data transmission opportunities.

The mandatory key QoS parameters are the minimum reserved traffic rate, maximum sustained traffic rate, traffic priority, uplink grant scheduling type, and request/transmission policy.

BE

The BE grant scheduling type provides efficient service to the best-effort traffic in the uplink. For a proper operation of this BE scheduling service, the request/transmission policy is set such that the MS is allowed to use contention request opportunities. Then the MS can use the contention request opportunities as well as the unicast request opportunities and the data transmission opportunities.

ertPS

The *extended rtPS* (ertPS) supports real-time service flows that generate variable-size data packets on a periodic basis, as was the case for the rtPS, but it adopts unsolicited method of bandwidth request, as was the case for the UGS. Whereas the rtPS repeats bandwidth requests whenever needs arise, the ertPS makes request only when changes occur in the requested bandwidth. In other words, it maintains fixed allocation while the transmission rate is constant but makes changes in request only when changes occur in the transmission rate. Thus, the BS provides unicast grants in an unsolicited manner, as for the UGS case, and thereby reduces the latency in making bandwidth requests. This helps mitigate the inefficiency, or bandwidth consumption, of the rtPS that is incurred by repeated requests. In contrast to the UGS whose allocations are fixed in size, however, the ertPS allocations are dynamic. A typical example of service amenable to the ertPS is the VoIP service with silence suppression.

The BS provides periodic uplink allocations that may be used for requesting the bandwidth as well as for data transfer. The size of uplink allocations normally corresponds to the current maximum sustained traffic rate at the connection. This allocation size is maintained until another bandwidth change request is made by the MS. For uplink bandwidth size change, the MS may send a request to the BS by either using an extended piggyback request field of the grant management subheader, or using the *bandwidth request* (BR) field in the MAC signaling headers, or sending a codeword over CQICH. Likewise, the MS, if it has data to send but cannot find any available unicast bandwidth request opportunities, may use contention request opportunities to acquire a grant for an ertPS connection, or send the CQICH codeword to the BS. Then the BS will start allocating the uplink grant that corresponds to the current maximum sustained traffic rate.

The mandatory key QoS parameters are the maximum sustained traffic rate, minimum reserved traffic rate, maximum latency, and request/transmission policy.

Figure 6.1 illustrates the difference of the bandwidth request and allocation process among UGS, rtPS, and ertPS scheduling services. In the case of the UGS, no separate request process is necessary—see Figure 6.1(a)—whereas a bandwidth request may be made at every given polling period in the case of the rtPS—see Figure 6.1(b). As such, the rtPS may make bandwidth requests periodically, but such periodic bandwidth requests may degrade the bandwidth efficiency, as the request is made using a separate bandwidth. If there is no user data to send even when the bandwidth request opportunity is given, the MS sends bandwidth requests with the size zero (i.e., zero-BR), which ends up wasting the extra bandwidth used for the request. The polling period may be determined by the unsolicited polling interval. In the case of the ertPS, a fixed size of grant allocation is maintained just as the UGS case while the transmission rate is constant, and the rate change is informed by a piggyback bandwidth request—see Figure 6.1(c). While the transmission rate is zero, the BS may continue or stop periodic polling to avoid the inefficiency problem that the rtPS had—Figure 6.1(c) shows continuous polling. In case that polling is stopped, the MS may request the maximum sustained rate by sending the CQICH codeword (refer to Section 4.3.3).

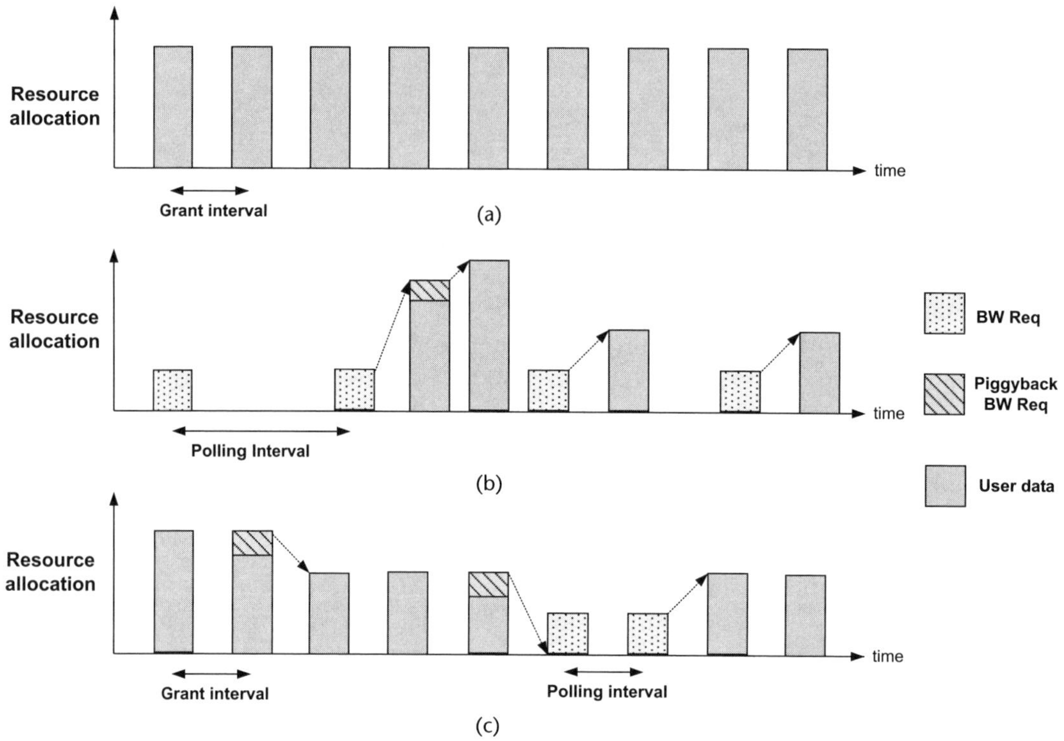

Figure 6.1 Illustration of bandwidth request and allocation: (a) UGS; (b) rtPS; and (c) ertPS.

6.1.2 Data Delivery Services

For data delivery in mobile networks, five different types of data delivery services are defined to support various user application services in association with the predefined sets of QoS-related service flow parameters: *unsolicited grant service* (UGS), *real-time variable-rate* (RT-VR) service, *nonreal-time variable-rate* (NRT-VR) service, *best-effort* (BE) service, and *extended real-time variable-rate* (ERT-VR) service. In particular, for uplink connections, the five types of data delivery services are tightly related to the scheduling services discussed in Section 6.1.1 and are as listed in Table 6.2. In other words, each type of delivery service is supported by one of the scheduling services.

The UGS is to support real-time applications generating fixed-rate data. This data can be provided as either fixed- or variable-length PDUs. The parameters specified for this service are the tolerated jitter, SDU size, minimum reserved traffic rate, maximum latency, request/transmission policy, and unsolicited grant interval.

The RT-VR service is to support real-time data applications with variable bit rates which require guaranteed data rate and delay. The parameters specified for this service are the maximum latency, minimum reserved traffic rate, maximum sustained traffic rate, traffic priority, request/transmission policy, and unsolicited polling interval.

The NRT-VR service is to support nonreal-time data applications that require a guaranteed data rate but are insensitive to delays. It is desirable in certain cases to limit the data rate of these services to some maximum rate. The parameters defined

Table 6.2 Types of Data Delivery Services

Type	Service type	Relation with scheduling services (for UL connections)
0	UGS (unsolicited grant service)	Supported by UGS scheduling service
1	RT-VR (real-time variable-rate service)	Supported by rtPS scheduling service
2	NRT-VR (non-real-time variable-rate service)	Supported by nrtPS scheduling service
3	BE (best effort service)	Supported by BE scheduling service
4	ERT-VR (extended real-time variable-rate service)	Supported by ertPS scheduling service

Source: [1].

for this service are the minimum reserved traffic rate, maximum sustained traffic rate, traffic policy, and request/transmission policy.

The BE service is for applications without rate or delay requirements. The parameters specified for this service are the maximum sustained traffic rate, traffic priority, and request/transmission policy.

The ERT-VR service is to support real-time applications with variable data rates, which require guaranteed data rate and delay. The parameters required for this service are the maximum latency, minimum reserved traffic rate, maximum sustained traffic rate, traffic priority, request/transmission policy, and unsolicited grant interval.

6.2 Bandwidth Request and Allocation

Among various resources for wireless communications, bandwidth is the most precious resource due to its scarcity and medium-sharing properties. In order to enhance the efficiency of the bandwidth usage while supporting an individual QoS requirement per connection basis, the mobile WiMAX system adopts well-organized bandwidth request, grant, and polling mechanisms, which are supported by the five different types of scheduling services as discussed in Section 6.1. Bandwidth allocation is governed by the BS: the downlink bandwidth is solely managed by the downlink scheduler at the BS, but the uplink bandwidth is allocated by BS to MSs through the resource request and grant process. In this subsection, we present the notions and the detailed procedures of the request and grant process for bandwidth request and allocation.

6.2.1 Requests

Request refers to the mechanism in which the MS informs the BS that it needs uplink bandwidth allocation. Request is normally made on a stand-alone bandwidth request signaling but, optionally, it may be made as a piggyback on an uplink data burst. As uplink burst profile changes dynamically, request is made in terms of the number of bytes needed to carry the MAC header and payload, excluding the PHY overhead. The bandwidth request message may be transmitted during any uplink allocation, except for the initial ranging interval. Recall that the bandwidth request is made in a type-I MAC signaling header (see Section 5.2.4).

Request may be incremental or aggregate. If it is an incremental request, the BS adds the requested bandwidth to the current bandwidth of the connection; and if it is aggregate request, the BS replaces the current bandwidth of the connection with the newly requested bandwidth. The request type is indicated on the type field in the bandwidth request header. In the case of the piggybacked bandwidth request, which does not have the type field, bandwidth request is always made incremental. The capability of incremental request is optional for MS and mandatory for BS, whereas the capability of aggregate request is mandatory for both MS and BS.

6.2.2 Grants

Whereas each bandwidth request of the MSs reference individual connections, each bandwidth grant is addressed to the MSs' basic CID, not to individual CIDs. Usually bandwidth grant takes a nondeterministic pattern, so some MSs may happen to get less frequent grants than expected. Thus the MSs should be prepared to perform backoff and repeat requests based on the latest information received from the BS.

The procedure of requests and grants taken by the MS local scheduler when deciding which connections would get the granted bandwidth is as follows: On arrival of SDUs on a connection, the scheduler makes incremental bandwidth requests for the particular CID and then processes the relevant UL-MAP IEs received in the later frame. Once the MS finds that the request is granted on the corresponding basic CID by looking into UL-MAP, it assigns bandwidth to the outstanding requests in a specific data region prescribed by UL-MAP IE (see Figure 4.30). Then it checks if the requests are fully satisfied and sends out data if satisfied. Otherwise, it makes either aggregate requests or incremental requests depending whether or not the timer for the aggregate requests expires and then sends out the data.

6.2.3 Polling

Polling refers to the process that the BS allocates to MSs the bandwidth to use when making bandwidth requests. Such an allocation may be done to individual MSs or to a group of MSs. In the latter case, bandwidth request contention gets involved among the multiple MSs in the groups. The allocation is done not in an explicit message form but as a series of IEs within the UL-MAP. Note that polling is done on an MS basis as was the case for bandwidth allocation, while bandwidth request is done on a CID basis.

When unicast polling is made on an MS individually, no explicit message is needed to poll the MS. The MS is allocated with a bandwidth sufficient to respond to a bandwidth request. Unicast polling is not made on the MSs having an active UGS connection unless it signals by setting the PM bit in the header to request additional non-UGS connection (see Section 6.1.1). Unicast polling is normally done on an MS basis, as is the case for other types of polling.

If the available bandwidth is not sufficient to poll inactive MSs individually, multicast or broadcast polling may be made to multicast groups or as a whole. In support of this, some CIDs are reserved for multicast and broadcast polling (see Table 5.2). As for the unicast polling case, the multicast or broadcast poll is not an explicit message but bandwidth allocated in the UL-MAP. However, as opposed to the unicast polling that associated the allocated bandwidth with an MS's basic CID, the multicast or broadcast poll associates the allocated bandwidth with a multicast or broadcast CID. Once the poll is directed at a multicast or broadcast CID, the actual bandwidth allocation to individual MSs is done on a contention basis. An MS belonging to the polled group may request bandwidth during any request interval allocated to that CID.

The MSs having an active UGS connection may signal to the BS by setting the PM bit in the *grant management* (GM) subheader in a MAC packet of the UGS connection to indicate that they want to be polled to request additional bandwidth over non-UGS connections. The BS, once it detects the PM bit set for polling, makes individual polling in reply to the request. As the BS may possibly miss detecting the signal on the PM bit, the MS may set the PM bit in all the UGS MAC grant management subheaders in the uplink scheduling interval to minimize the possibility of missing.

Figure 6.2 illustrates the bandwidth request and allocation processes of the three different types—*piggyback request* (PBR), polling, and PM bit–based requests. The PBR bandwidth request applies to rtPS, ertPS, and nrtPS services—see Figure 6.2(a). The three services allow PBR and bandwidth stealing for bandwidth requests and also get the request opportunity by polling. In the cases of rtPS and ertPS, bandwidth requests should be made possible only through unicast polling, without contention, so that it can get guarantee on the QoS of real-time traffic. On the other hand, in the case of nrtPS, polling may be done both by unicast and by contention. Whereas the bandwidth request by PBR is made through the GM subheader, the bandwidth request by polling scheme is made through the *bandwidth request* (BR) MAC header—see Figure 6.2(b). The unit of bandwidth request is a byte, so the required bandwidth is converted to the number of bytes in advance. In case the user terminal has both UGS and non-UGS connections, bandwidth requests may be made by setting the PM bit in the GM subheader in the MAC packet of the UGS connection—see Figure 6.2(c).

6.3 QoS

Since the MAC in the Mobile WiMAX is connection oriented, QoS is supported on a per-connection basis. The principal mechanism for providing QoS is to associate packets traversing the MAC interface into a service flow. The MS and BS provide the QoS according to the QoS parameter set defined for the service flow. For the purpose of mapping to services on MSs and associating them with varying levels of QoS, all data communications are done in the context of a connection. A connection defines both the mapping between peer convergence processes that utilize the MAC and a service flow.

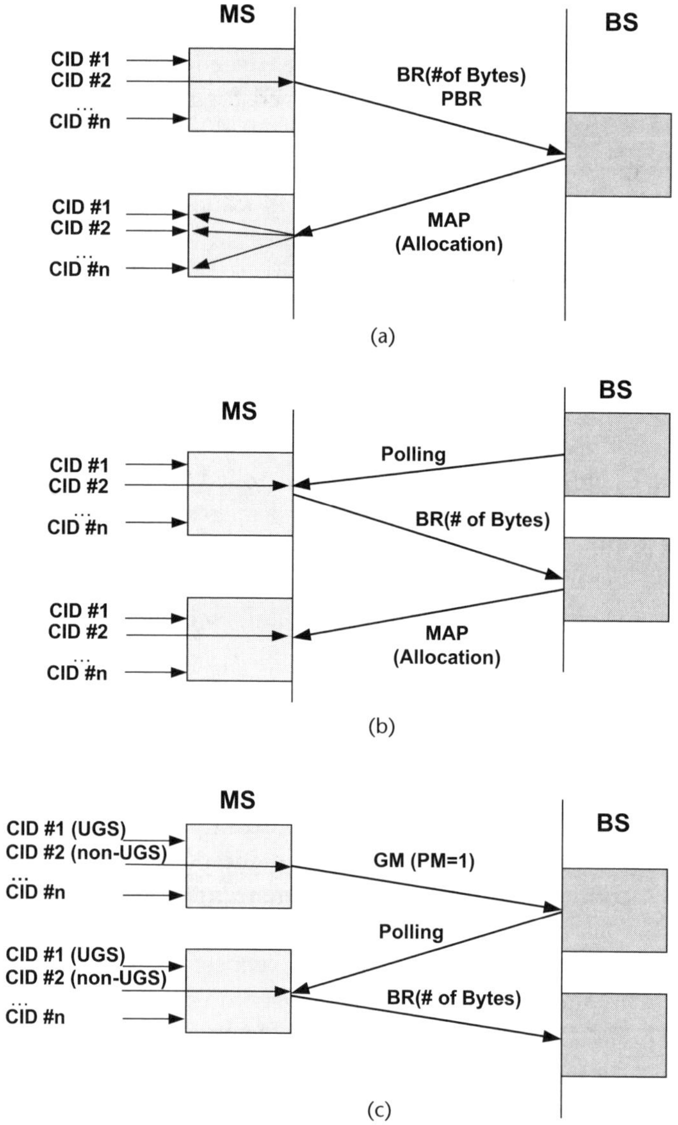

Figure 6.2 Bandwidth request and allocation process: (a) piggyback request; (b) polling; and (c) PM bit.

In Mobile WiMAX networks, every service flow is associated with a separate MAC connection according to IEEE 802.16e. User traffic is subject to traffic classification and conditioning at the convergence sublayer of the MS and ASN-GW. Then, the corresponding QoS policy is applied to the MAC connection. Traffic may be classified by many different parameters, such as source address, destination address, and source port. For example, VoIP traffic can get a higher priority treatment if the destination address of the VoIP server is available at the ASN-GW. The ASN-GW obtains classification information from the *policy and charging rule function* (PCRF) server.

QoS-related functions include QoS profile authorization, QoS admission control, *policy enforcement point* (PEP), PCRF, policing and monitoring, QoS parameter mapping across different QoS domains, and so on.[4] These functions may reside within a network or distributed across networks comprising MS, ASN, CSN, and networks interconnecting CSN. However, the discussions in this section pertain mostly to ASN of the Mobile WiMAX network.

To define the QoS functionalities in the Mobile WiMAX network, the following three elements are required: (1) the language to express QoS requirements—specifically, the DSA/DSC/DSD messages over air interface between BS and MS, and the RR/PD messages over the network interfaces among BS, ASN-GW, and PCRF server (see Section 6.3.2); (2) the talkers to have a dialogue in such language—specifically, the MS, BS, ASN-GW, and PCRF server; and (3) the procedures to define the order of sentences and actions in such dialogues—specifically, MS initiated, network initiated, static provisioning, and dynamic provisioning. This section deals with those elements.

6.3.1 Service Flows and Classes

A *service flow* refers to a unidirectional flow of packets that is associated with a particular QoS, whereas a *service class* is an identifier for a specific set of QoS parameter values.

Service Flows

A service flow is a MAC transport service that provides unidirectional transport of packets either to uplink packets or to downlink packets. It is characterized by a set of QoS parameters such as latency, jitter, and throughput. In order to standardize the operation between the MS and BS, these attributes include the details of how the MS requests uplink bandwidth allocations and the expected behavior of the BS uplink scheduler.

A service flow is partially characterized by the following attributes:

1. *Service flow ID* (SFID): The identification of service flow that is assigned to each existing service flow. It serves as the principal identifier for the service flow between a BS and an MS. Each service flow has a 32-bit SFID.
2. *Connection ID* (CID): The identification of a transport connection that exists only when the service flow is admitted or active. The relationship between SFID and transport CID is unique in that an SFID is never associated with more than one transport CID and a transport CID is never associated with more than one SFID. Each admitted or active service flow has a 16-bit CID.
3. *Provisioned QoS parameter set*: A QoS parameter set provisioned via, for example, network management system.
4. *Admitted QoS parameter set*: A QoS parameters set for which the BS (and possibly the MS) is reserving resources. The resources include bandwidth primarily, and other memory or time-based resources needed to activate the flow. The admitted QoS parameter set is a subset of the authorized module.
5. *Active QoS parameter set*: A set of QoS parameters defining the service actually being provided to the service flow, which can forward packets. The

4. All the necessary QoS functionalities are not yet fully defined in [2].

active QoS parameter set is always a subset of the admitted QoS parameter set.

6. *Authorized module*: A logical function within the BS that approves or denies every change to QoS parameters and classifiers associated with a service flow. In effect, it defines an "envelope" that limits the possible values of the admitted and active QoS parameter sets.

Service Classes

A service class is an optional identifier that identifies a specific set of QoS parameter values. It helps operators to move the burden of configuring service flows from the provisioning server to the BS. Operators can provision the MSs with the service class name, and the implementation of the name can be configured at the BS. In addition, service class allows higher-layer protocols to create a service flow by its service class name.

For example, when a service flow is created for VoIP, the PCRF sends RR_REQ only with a service class of "VoIP G.729," instead of defining all the necessary parameters such as minimum reserved rate and grant interval in the RR_REQ message. Then the BS refers to the corresponding set of QoS parameters that are predefined in their own database. If RR_REQ includes both a service class and some QoS parameters, parameter values in RR_REQ will override the corresponding parameter values in the predefined database.

With the service classes made available optionally, a service flow may have its QoS parameter set specified by explicitly including all traffic parameters, by indirectly referring to a set of traffic parameters by specifying a service class name, or by specifying a service class name along with modifying parameters.

In order to facilitate operation across a distributed topology, it is required to use common definitions of service class names and the associated authorized QoS parameter sets over the distributed mobile networks. To enable operation in this context, *global service class* names are defined. Global service class names are a rules-based, composite naming system parsed in eight information fields as listed in Table 6.3. They are employed as a baseline convention for communicating authorized or admitted QoS parameter sets.

Global service class name is similar in function to service class name except that: (1) the use of global service class name may not be modified by a BS, (2) the use of global service class names remain consistent among all BSs, and (3) global service class names are a rules-based naming system whereby the global service class name itself contains the referential QoS parameter codes. In practice, the global service class names are accompanied by extended or modified QoS parameter sets defining parameters, as needed, thereby providing a complete and expedited method for transferring authorized or admitted QoS parameter set information.

6.3.2 QoS Messages and Parameters

Protocol messages are defined between network entities to create/delete a service flow, and to exchange QoS information. When a service flow is created, its status becomes either *provisioned* (or *admitted*) or *activated*, in line with the three types of QoS parameter sets aforementioned:

Table 6.3 Global Service Class Names

Position	Name	Size (bits)	Value
I	UL/DL indicator	1	0/1 (UL/DL)
S	Max sustained rate	6	Extensible lookup table
B	Max traffic burst	6	Extensible lookup table
R	Min reserved traffic rate	6	Extensible lookup table
L	Max latency	6	Extensible lookup table
S	Variable/fixed-length SDU indicator	1	0/1 (variable/fixed-length)
P	Paging preference	1	0/1 (no/with paging generation)
S1	Request/Transmission policy	8	bit 0: broadcast polling/BW-REQ; bit 1: multicast polling/ BW-REQ; bit 2: piggyback BW-REQ; bit 3: fragmentation; bit 4: header suppression; bit 5: packing; bit 6: CRC; bit 7: reserved.
S2	Uplink grant scheduling type	3	1 to 6 (included only when I=0)
L1	Tolerated jitter	6	Extensible lookup table
S3	Traffic priority	3	0 to 7 (included only when I=0)
S4	Unsolicited grant interval	6	Extensible lookup table
S5	Unsolicited polling interval	6	Extensible lookup table
R	Padding	variable	For byte aligned. Set to zero

Source: [1].

1. *Provisioned service flow*: A type of service flow known via provisioning by, for example, the network management system. Status is that a service flow has been created but resource has not been reserved, and it has not been tested by *call admission control* (CAC).
2. *Admitted service flow*: A type of service flow that has resources reserved by the BS for the relevant admitted QoS parameter set, which is not yet active. Status is that resource has been reserved after passing CAC but resource allocation has not started yet.
3. *Active service flow*: A type of service flow that has resources committed by the BS for the relevant active QoS parameter set. Status is that resource allocation has started. For example, the UL grant is periodically allocated for an activated UGS flow.

DSA/DSC/DSD Messages

As to the QoS messages, it is important to examine the messages on the following three interfaces: between BS and MS, between ASN-GW and BS, and between ASN-GW and PCRF server.

IEEE 802.16e defines *dynamic service addition* (DSA), *dynamic service change* (DSC), and *dynamic service deletion* (DSD) messages (collectively called DSx) between BS and MS, which are used for exchanging QoS information for a MAC connection. The roles of those messages are as follows:

1. DSA_REQ/RSP/ACK is to create a service flow. It can be issued either by ASN-GW or MS. The former case is called network-initiated mode or push mode, and the latter MS-initiated or pull mode. For more details, refer to Section 6.3.4.
2. DSC_REQ/RSP/ACK is to change the status and the QoS requirements of an existing service flow. One example is that a VoIP service flow in *admitted* status during ring-back tone becomes *active* by the DSC message, which is triggered by picking up.
3. DSD_REQ/RSP is to delete a service flow. All resources are released. ACK message is not necessary for DSD.

The QoS parameters delivered by those three messages are as summarized in Table 6.4.

RR/PD Messages

Whereas DSx messages are defined over the air interface for BSs to interact with MSs, *resource reservation* (RR) and *policy decision* (PD) messages are defined over the network interface for ASN-GWs to interact with BSs and PCRF servers. The roles of those messages are as follows:

1. RR_REQ/RSP/CFM (confirmation) is to create a path—or *generic routing encapsulation* (GRE) tunnel—between BS and ASN-GW, and to ask the BS to reserve resources. It contains the same QoS parameters as in DSA.
2. PD_REQ/RSP is to request an authenticator in AAA or PCRF server to decide QoS policy for a service flow. When requesting especially to an AAA server, these messages are recommended by the WiMAX Forum to follow the DIAMETER protocol. The DIAMETER protocol is also used for Gx interface standardization by 3GPP. It contains the same QoS parameters as in DSA.

6.3.3 QoS-Related Network Elements

Figure 6.3 depicts the architecture of the Mobile WiMAX network and the related interfaces. Note that the convergence sublayer is assumed to be located at ASN-GW and the *call session control function* (CSCF) server gets involved in authenticating the QoS messages. As mentioned earlier, the QoS-related functions include QoS profile authorization, QoS admission control, PEP, PCRF, policing and monitoring, QoS parameter mapping across different QoS domains, and so on. We consider the Mobile WiMAX network elements in relation to those QoS-related functions, referring to the network architecture in this figure.

ASN-GW

ASN-GW performs various core functions, including traffic classification and service flow authorization. As to traffic classification, ASN-GW distinguishes traffic

Table 6.4 A Summary of QoS Parameters

Parameter name	Value	Description
SFID	32bit value	Service Flow ID as the primary reference of a service flow
CID	16bit value	Connection ID used for MAC PDU
Service class name	2~128 byte string	If not null, all the undefined QoS parameters follow the pre-configured (in BS by operator) set corresponding to the service class name
Global service class name	6 byte code or string	If not null, all the undefined QoS parameters follow the pre-configured (in BS by operator) set corresponding to the global service class name
QoS parameter set type	3 bit value	Each bit represents the status of a service flow, - bit 0: *provisioned* (resource is not reserved), - bit 1 *admitted* (resource has been reserved), - bit 2 *active* (resource allocation has started).
Traffic priority	0~7	Higher value means higher priority
Max sustained traffic rate	bps	Peak information rate of the service
Max traffic burst	bit	Max burst size that must be accommodated for the service
Min reserved traffic rate	bps	Min guaranteed rate for the service
Max latency	msec	Max allowable MAC to MAC delay between BS and MS
Tolerated jitter	msec	Max delay variation
Unsolicited grant interval	msec	Nominal interval btw. successive data grant opportunities
Unsolicited polling interval	msec	Nominal interval btw. successive data polling grant opportunities
Type of data delivery services	UGS, rtPS, nrtPS, BE, ertPS	Type of data delivery service as defined in Section 6.3.20 [1]
Time base	msec	Time base for rate measurement
Request /transmission policy		Capability to specify certain attributes for the associated service flow: bit 0 for broadcast polling/BW-REQ; bit 1 for multicast polling/ BW-REQ; bit 2 for piggyback BW-REQ; bit 3 for fragmentation; bit 4 for header suppression; bit 5 for packing; bit 6 for CRC; bit 7 is reserved.

based on the classifiers. If two applications (e.g. FTP, HTTP) are associated with one classifier, then ASN-GW cannot distinguish applications since it is transparent to ASN-GW. As to service flow authorization, ASN-GW interfaces with the PCRF server, from which it receives the classification rules and the QoS parameters of each service flow.

When bearer traffic arrives at the ASN-GW from the CSN (downlink), the ASN-GW identifies user traffic based on the IP address of the user. The user traffic is further classified into the traffic of service flows by the classifier based on the 6-tuple of that traffic, namely, source IP, destination IP, source port, destination port, protocol, and type of service. Bearer traffic is GRE-encapsulated with SFID

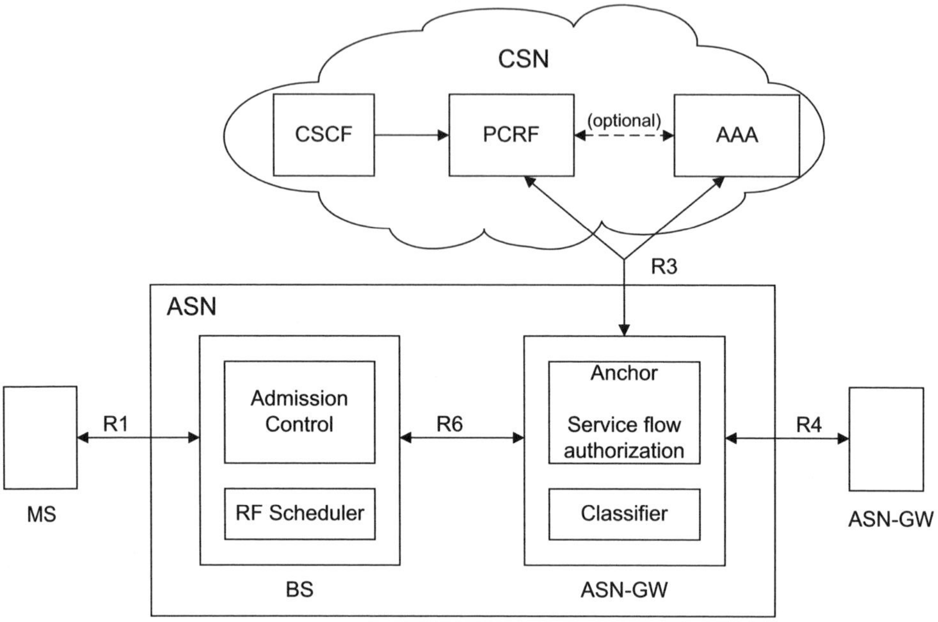

Figure 6.3 Mobile WiMAX network architecture with respect to QoS-related network elements.

information. The IP header of the GRE packet is marked with the appropriate *differentiated service code point* (DSCP) code (included in the service flow parameters obtained from the ASN-GW) so that *access network* (AN) QoS is maintained over the R6 interface. If L2 backhaul is used on the backhaul, then the ASN-GW marks the L2 header with the corresponding *class of service* (CoS) (i.e., VLAN tagging).

BS

The BS gets involved in admission control and RF scheduling functions. The BS obtains the QoS parameters from ASN-GW and passes it on to the QoS scheduler in the BS. Also the BS schedules air interface resources based on the QoS parameters of service flow. In the uplink, the BS marks the user traffic toward the ASN-GW to maintain AN QoS. As to admission control, the BS performs bandwidth-based call admission control.

PCRF

PCRF performs the decision making related to bandwidth allocation. The PCRF server may be an independent network entity or the PCRF function may be installed inside the AAA server. PCRF is an *IP multimedia subsystem* (IMS)/*multimedia domain* (MMD)–based network element for applying business rules that determine which customers and/or applications receive bandwidth priority, and when. PCRF maintains the classification rules to distinguish different service flows (or applications) and the associated QoS parameters.

ASN-GW interrogates the PCRF at the time of service flow request. PCRF may identify the user information either by searching its own database or by asking to the AAA server and then deciding the QoS parameters of service flows.

MS

MS performs various functions, including traffic classification and request for service flow authorization, as a symmetry to the ASN-GW functions. As to traffic classification, the MS MAC layer distinguishes traffic based on the classifiers. If two applications (e.g. FTP, HTTP) are associated with one classifier, then the MS cannot distinguish applications since it is transparent to the MAC layer. In relation to the request for service flow authorization, the MS may issue a service flow creation with all necessary classification rules and QoS parameters, especially for the MS-initiated mode.

When bearer traffic arrives at the MS from the application (uplink), the MS identifies user traffic based on the IP address of the user. The user traffic is further classified into the traffic of service flows by the classifier based on the aforementioned 6-tuples of that user.

6.3.4 Service Flow Setup/Release Procedures

QoS profile for a service flow is informed to ASN-GW during initial network entry or at the time of DSA procedure for each service flow. For each subscriber, the QoS profile includes the permissible number and schedule type of WiMAX service flows and the permissible range of values for the associated QoS parameters [2]. The service flow setup and release procedures differ depending on whether it is initial network entry, MS-initiated setup/release, or network-initiated setup/release, as will be described next.

Initial Network Entry QoS Setup

Figure 6.4 describes the QoS setup procedure during the initial network entry. In the figure, the dashed block represents a set of messages needed for the network entry

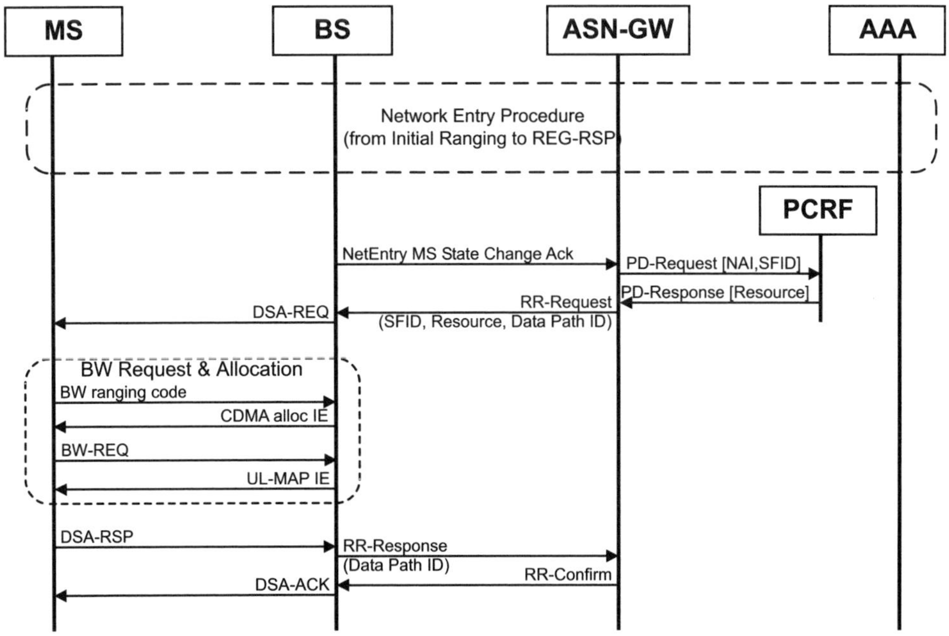

Figure 6.4 An example of QoS setup procedure during initial network entry.

procedure. A QoS profile is informed from the PCRF at the time of registration. This procedure is called *preprovisioned QoS* because the QoS service flows are prepared before they are actually used. When service flows are preprovisioned, it is hard to change the QoS profile dynamically. In this sense, it is also called *static QoS*. In contrast, *dynamic QoS* means that QoS service flows are created when a user actually requests a QoS service and that the QoS profile may be changed by request.

Specifically, the QoS setup procedure at the initial network entry is as follows: After normal network entry steps such as authentication and registration (see Figure 3.1 for detail) are finished, the ASN-GW sends PD-request to the PCRF to get QoS information (see Figure 6.4). The interface protocol between the ASN-GW and PCRF is to be determined.

MS-Initiated DSA Procedure

After an MS enters the network, it may request a new QoS service flow. In this case the MS may follow the MS-initiated DSA procedure shown in Figure 6.5, in which the MS initiates a service flow setup via the *session initiation protocol* (SIP)

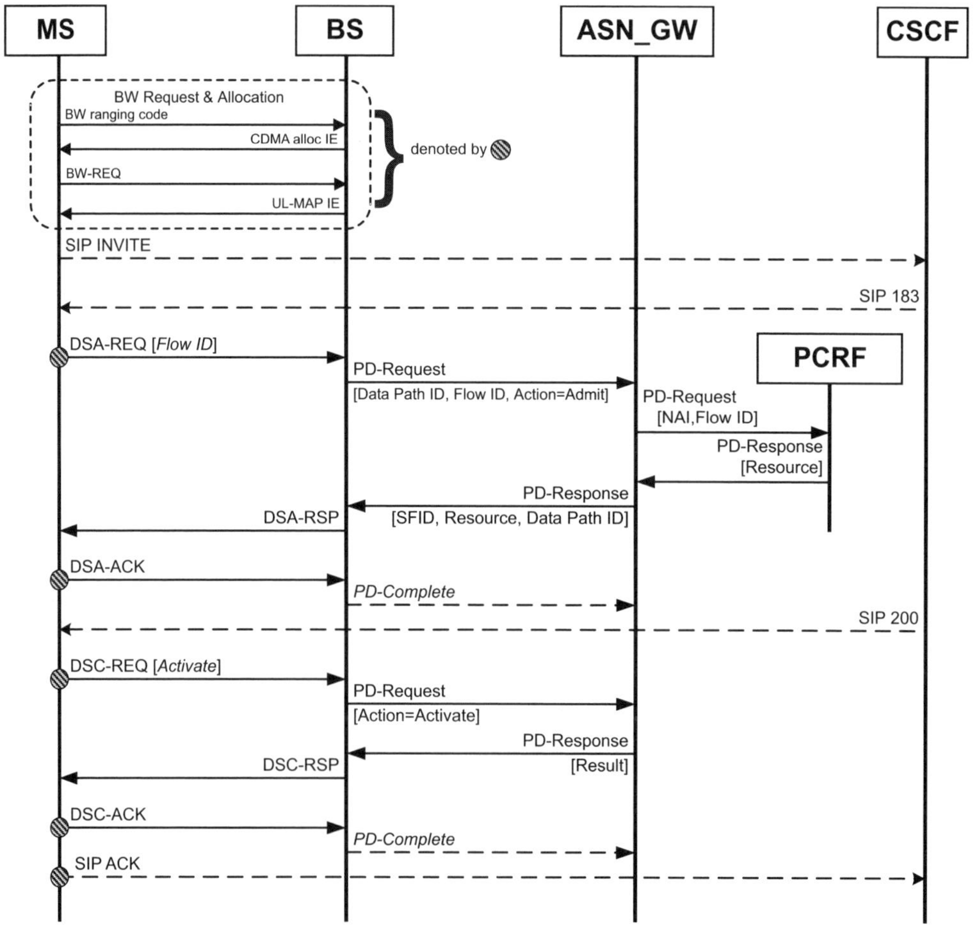

Figure 6.5 An example of MS-initiated DSA procedure.

signaling.[5] There may be DSC steps to activate the service flow between SIP 200 and SIP ACK, as activation by DSC is necessary when the service flow has been created with admitted status. The dashed arrows in the figure represent WiMAX-independent signals, which may be treated as normal data packets by BS and ASN-GW. Note that, for security reasons, the MS-initiated DSA needs to be authenticated again, although it might have already been authenticated by CSCF.

To be more specific, the MS-initiated setup procedure is as follows: An application in the MS sends a call request message, SIP INVITE in Figure 6.5, to the CSCF. In the figure, message arrows originated from the dashed block denote that the message transmission follows the BW request and allocation procedure. After the call request is accepted by the CSCF, the MS sends DSA-REQ with QoS requirements to establish a MAC connection. Then, the ASN-GW asks the PCRF to authenticate the MAC connection. Since usually the called party has not answered at this moment, the MAC connection is created with the *admitted* status. After the called party answers in the form of SIP 200, the MS sends DSC-REQ to change its status to *activated*. The interface protocol between the ASN-GW and PCRF is to be determined.

Network-Initiated DSA Procedure

In case the network initiates a service flow setup, it follows the procedure shown in Figure 6.6. When compared with the MS-initiated, SIP signaling is used in the same way as the SIP messages are transparent to the WiMAX network, but reauthentication is not needed because the RR-request from PCRF must have been authenticated.

To be more specific, the network-initiated setup procedure is as follows: An application in the MS sends a call request message, SIP INVITE in the illustration of Figure 6.6, to the CSCF. As before, the message arrows originated from the dashed circle denote that the message transmission follows the BW request and allocation procedure. After the call request is accepted by the CSCF, the CSCF informs the PCRF of the request for the connection establishment and the PCRF orders the ASN-GW to create a MAC connection with appropriate QoS requirements. As a sequel, the BS sends DSA-REQ to the MS. Since usually the called party has not answered at this moment, the MAC connection is created with the *admitted* status. After the called party answers in the form of SIP 200, the MS sends DSC-REQ to change its status to *activated*. The interface protocol between the ASN-GW and PCRF is to be determined.

Actions

The actions of each network element in relation to these procedures are as follows:

1. *Actions of ASN-GW*: In the case of the preprovisioned QoS, ASN-GW triggers the creation of a service flow based on the service flow parameters preconfigured by the operator. In the case of the dynamic QoS, the ASN-GW interrogates the PCRF to obtain the flow-based QoS parameters. ASN-GW sends BS an RR-request message with the QoS parameters of the service flow, such as minimum reserved rate and uplink scheduling type.

5. Whereas the procedures of preprovisioned QoS are well defined by the WiMAX Forum, the procedures of dynamic QoS are yet to be defined.

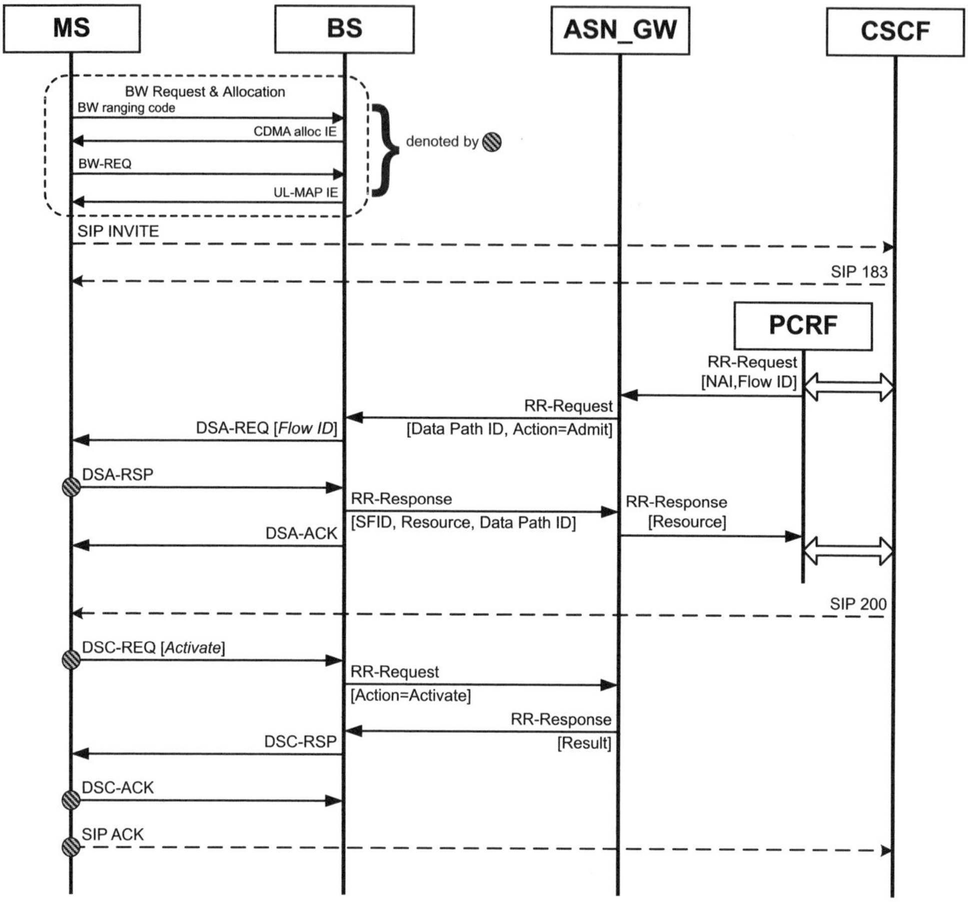

Figure 6.6 An example of network-initiated DSA procedure.

2. *Actions of BS*: When BS receives an RR-request message, it stores the QoS parameters received from ASN-GW in the serving MS context. The CAC will check for the availability of the current BS resources and then determine if the service flow can be created with the requested QoS parameters (see Section 6.3.5). If not, the BS sends an RR-reject to ASN-GW. If the flow can be created with the requested QoS parameters, then the BS sends a DSA_REQ message to the MS. After the MS accepts the service flow creation, the BS is ready to schedule data. Depending on the QoS parameters received for the service flow, the QoS scheduler will prepare the DL and UL schedulers to send data using one of the five scheduling services discussed in Section 6.1.
3. *Actions of PCRF*: The PCRF maintains the classification rules to distinguish different service flows (for the given applications) and the associated QoS parameters. ASN-GW interrogates the PCRF at the time of service flow request. PCRF may identify the user information, either by searching its own database or by asking AAA, and decides the QoS parameters of service flows. PCRF verifies whether or not a subscriber is authorized to use the type of service requested.

6.3.5 Scheduling, CAC, and Policing

Three basic elements for QoS have been dealt in previous sections—language (QoS messages), talkers (network elements), and dialogue (QoS control procedures). Whereas those three elements are standardized for interoperability, QoS enforcement functions such as scheduling, CAC, and policing are implementation issues. These three functions concern how to maximize the degree of QoS satisfaction for the admitted connections, how to minimize the blocking of connection requests and the QoS violation due to too many admitted connections, and how to protect the QoS of the contract-conforming connections against malicious connections, respectively.

Scheduling

Radio link is likely to be a typical bottleneck among all the network facilities, so the MAC scheduler plays the most important role to satisfy the given QoS requirements. The MAC scheduler must efficiently allocate the available resources according to the dynamic bandwidth request for both real-time and nonreal-time application services under the varying channel condition.

Opportunistic packet scheduling algorithms are widely considered for supporting a class of the *nonreal-time* (NRT) services. Since a tradeoff relation exists between throughput and fairness, no scheduling algorithm may be said to be absolutely best. One particular example that deals with the tradeoff between throughput and fairness is the *proportional fairness* (PF) packet scheduling algorithm. In the PF algorithm, each user evaluates its own priority score in terms of its instantaneous data rate and long-term average data rate, which reflect the current channel condition and relative sharewise fairness, respectively, into scheduling. Furthermore, it is another issue to design a PF algorithm for the multichannel system, as in the Mobile WiMAX system where multiple subchannels in each frame must be scheduled at the same time [3, 4].

As *real-time* (RT) services must also be served in the Mobile WiMAX systems, packet-scheduling algorithms for RT services were independently developed. Among many available packet-scheduling algorithms for the mobile broadband wireless access systems, the modified *largest weighted delay first* (LWDF) is one particular example of latency-aware algorithms designed for taking delay requirements into account for RT services [5]. In order to prioritize the RT service users, for example, the priority score can be exponentially weighted as the delay requirements tend to be violated in the average sense or absolute sense.

Meanwhile, in order to schedule both RT and NRT services in the Mobile WiMAX network, two different types of scheduling principles can be considered: *opportunistic scheduling* and *priority queuing*. Particular examples of opportunistic scheduling for an integrated service include the adaptive exponential (EXP)/PF algorithm [6] and *urgency and efficiency-based packet scheduling* (UEPS) algorithm [7], in which two different priority metrics are employed, with each one specified for an individual service class. The priority metric for each class is defined by considering their relative urgency, an instantaneous data rate, and fairness among RT and NRT service class users under the varying channel conditions. In this case, however, the hard QoS constraint of the RT service (e.g., in terms of maximum allowable delay) may not be warranted as the corresponding priority score is opportunistically determined for the varying conditions (e.g., the instantaneous data rate and waiting time). On the other hand, priority queuing is one particular scheduling scheme that can guarantee a hard QoS constraint as always serving the RT service class users ahead of the NRT service class users. As the delay requirement can be deterministically met for the RT service class, it is commonly adopted in the practical system. In spite of its QoS guarantee feature and simplicity, a serious disadvantage of the priority queuing scheme is that the multiuser diversity advantage of NRT users cannot be leveraged. To overcome such a disadvantage of the priority queuing scheme, an opportunistic priority queuing scheme, referred to as the *delay threshold-based priority queuing* (DTPQ) scheme, has been proposed for improving the overall system capacity subject to individual QoS requirements [8]. This scheme takes the relative urgency of the RT service into account only when its *head-of-line* (HOL) packet delays exceed a given delay threshold.

Call Admission Control

No matter how efficient the scheduler may be, QoS violation is unavoidable if the required amount of resources is greater than the available resources. Hence, CAC is also an important part of QoS. A basic function of CAC is to compare the amount of the required resources with the amount of available resources. For an efficient comparison, it is necessary to devise methods to better describe the required resources. There are various kinds of resources such as bandwidth, buffer, and CPU capacity. Among them we focus on bandwidth because it is likely to be a bottleneck resource most frequently.

IEEE 802.16e defines the QoS parameters described in Section 6.3.2. It can be inferred that the minimum reserved rate represents the bandwidth demand for nonreal-time traffic. It is more complicated for real-time traffic. Definitely real-time traffic demands bandwidth up to the maximum sustained rate. So the CAC should not be based on the minimum reserved rate. On the other hand, the CAC based on the maximum sustained rate may result in unnecessary blocking of call requests. There have been proposals for the CAC in 802.16 networks [9–12].

Another challenging problem is how to estimate the amount of available bandwidth. The available bandwidth cannot be simply represented by the amount of radio resource, or the number of available OFDM symbols in WiMAX, due to the nature of time-varying MAP overhead (mainly depending on the number of active MSs or CIDs subject to bandwidth allocation) and radio channel conditions. The bandwidth capacity of a radio link depends on many factors such as channel quality, transmission power, and the variable-size MAP overhead, especially in WiMAX.

Policing

Networks should be prepared for the occasion that service flows violate the contracted parameters (i.e., the characteristics of the corresponding traffic do not follow the traffic descriptors). In such cases the network service may be disrupted, thereby dissatisfying other contract-conforming service flows. Therefore, appropriate policing schemes are needed in conjunction with the traffic descriptors such that the QoS of the contract-conforming flows be protected without disruption. A typical example of policing is a token bucket–based rate control [13, 14].

References

[1] IEEE Std 802.16e-2005 and IEEE Std 802.16-2004/Cor1-2005, Part 16: Air Interface for Fixed and Mobile Broadband Wireless Access Systems, February 2006.

[2] WiMAX Forum, Network Architecture—Stage 3: Detailed Protocols and Procedures, Release 1.1.0, July 2007. For the latest release, refer to http://www.wimaxforum.org.

[3] Kim, H., and Y. Han, "A Proportional Fair Scheduling for Multicarrier Transmission Systems," *IEEE Communication Letters*, Vol. 9, No. 3, March 2005, pp. 210–212.

[4] Min, T., and C. G. Kang, "Multi-phase Predictive Proportional Fairness Scheduling for a Multi-Channel Wireless Packet System," *Proc. of IEEE Personal, Indoor and Mobile Radio Communications*, 2007.

[5] Andrews, M., et al., "Providing Quality of Service over a Shared Wireless Link," *IEEE Communications Magazine*, Vol. 39, No. 2, February 2001, pp. 150–154.

[6] Rhee, J.-H., J. Holtzman, and D. K. Kim "Performance Analysis of the Adaptive EXP/PF Channel Scheduler in an AMC/TDM System," *IEEE Communications Letters*, Vol. 8, No. 8, August 2004, pp. 497–499.

[7] Ryu, S., et al., "Urgency and Efficiency Based Packet Scheduling Algorithm for OFDMA Wireless System," *Proc. of IEEE International Conference on Communications*, 2005, pp. 2779–2785.

[8] Ku, J. M., et al., "Adaptive Delay Threshold-Based Priority Queueing Scheme for Packet Scheduling in Mobile Broadband Wireless Access System," *Proc. of IEEE Wireless Communications and Networking Conference*, 2006, pp. 1142–1147.

[9] Wang, L., et al., "Admission Control for Non-Preprovisioned Service Flow in Wireless Metropolitan Area Networks," *Proc. of 4th European Conference on Universal Multiservice Networks*, 2007, pp. 243–249.

[10] Chandra, S., and A. Sahoo, "An Efficient Call Admission Control for IEEE802.16 Networks," *Proc. of IEEE International Workshop on Local and Metropolitan Area Networks*, 2007.

[11] Tsai, T.-C., C.-H. Jiang, and C.-Y. Wang, "CAC and Packet Scheduling Using Token Bucket for IEEE 802.16 Networks," *Journal of Communications*, Vol. 1, No. 2, May 2006, pp. 30–37.

[12] Niyato, D., and E. Hossain, "QoS-Aware Bandwidth Allocation and Admission Control in IEEE 802.16 Broadband Wireless Access Networks: A Noncooperative Game Theoretic Approach," *Elsevier Computer Networks*, Vol. 51, No. 11, August 2007, pp. 3305–3321.

[13] Lee, T.-H., "Correlated Token Bucket Shapers for Multiple Traffic Classes," *Proc. of IEEE Vehicular Technology Conference*, 2004, pp. 4672–4676.

[14] Wu, S.-L., and W.-S. E. Chen, "The Token-Bank Leaky Bucket Mechanism for Group Connections in ATM Networks," *Proc. of International Conference on Network Protocols*, 1996, pp. 226–233.

Selected Bibliography

Ahson, S., and M. Ilyas, *WiMAX: Technologies, Performance Analysis, and QoS*, Boca Raton, FL: CRC Press, 2007.

Ali, S. H., K.-D. Lee, and V. C. M. Leung, "Dynamic Resource Allocation in OFDMA Wireless Metropolitan Area Networks," *IEEE Wireless Communications*, Vol. 14, No. 1, February 2007, pp. 6–13.

Badia, L., et al., "On the Impact of Physical Layer Awareness on Scheduling and Resource Allocation in Broadband Multicellular IEEE 802.16 Systems," *IEEE Wireless Communications*, Vol. 14, No. 1, February 2007, pp. 36–43.

Choi, Y., and S. Bahk, "Upper-Level Scheduling Integrating Multimedia Traffic in Cellular Data Networks," *Elsevier Computer Networks*, Vol. 51, No. 3, February 2007, pp. 621–631.

Cicconetti, C., et al., "Performance Evaluation of the IEEE 802.16 MAC for QoS Support," *IEEE Transactions on Mobile Computing*, Vol. 6, No. 1, January 2007, pp. 26–38.

Cicconetti, C., et al., "Quality of Service Support in IEEE 802.16 Networks," *IEEE Network*, Vol. 20, No. 2, March–April 2006, pp. 50–55.

Huang, C. Y., et al., "Radio Resource Management of Heterogeneous Services in Mobile WiMAX Systems," *IEEE Wireless Communications*, Vol. 14, No. 1, February 2007, pp. 20–26.

Iera, A., et al., "Channel-Aware Scheduling for QoS and Fairness Provisioning in IEEE 802.16/WiMAX Broadband Wireless Access Systems," *IEEE Network*, Vol. 21, No. 5, September–October 2007, pp. 34–41.

Jang, J., and K. B. Lee, "Transmit Power Adaptation for Multiuser OFDM Systems," *IEEE Journal on Selected Areas in Communications*, Vol. 21, No. 2, February 2003, pp. 171–178.

Lee, J., et al., "Realistic Cell-Oriented Adaptive Admission Control for QoS Support in Wireless Multimedia Networks," *IEEE Trans. on Vehicular Technology*, Vol. 52, No. 3, May 2003, pp. 512–524.

Niyato, D., and E. Hossain, "Queue-Aware Uplink Bandwidth Allocation and Rate Control for Polling Service in IEEE 802.16 Broadband Wireless Networks," *IEEE Trans. on Mobile Computing*, Vol. 5, No. 6, June 2006, pp. 668–679.

Rhee, W., and J. M. Cioffi, "Increase in Capacity of Multiuser OFDM System Using Dynamic Subchannel Allocation," *Proc. of IEEE Vehicular Technology Conference*, 2000, pp. 1085–1089.

Rong, B., Y. Qian, and H.-H. Chen, "Adaptive Power Allocation and Call Admission Control in Multiservice WiMAX Access Networks," *IEEE Wireless Communications*, Vol. 14, No. 1, February 2007, pp. 14–19.

Seo, H., and B. G. Lee, "Proportional-Fair Power Allocation with CDF-Based Scheduling for Fair and Efficient Multiuser OFDM Systems," *IEEE Trans. on Wireless Communications*, Vol. 5, No. 5, May 2006, pp. 978–983.

Song, G., and Y. Li, "Cross-Layer Optimization for OFDM Wireless Networks—Part I: Theoretical Framework," *IEEE Trans. on Wireless Communications*, Vol. 4, No. 2, March 2005, pp. 614–624.

Song, G., and Y. Li, "Cross-Layer Optimization for OFDM Wireless Networks—Part II: Algorithm Development," *IEEE Trans. on Wireless Communications*, Vol. 4, No. 2, March 2005, pp. 625–634.

Wong, C. Y., et al., "Multiuser OFDM with Adaptive Subcarrier, Bit, and Power Allocation," *IEEE Journal on Selected Areas in Communications*, Vol. 17, No. 10, October 1999, pp. 1747–1758.

CHAPTER 7

Mobility Support

As the name suggests, Mobile WiMAX supports mobility, which is the distinctive feature of the IEEE 802.16e system, which its predecessors (i.e., IEEE 802.16a and 802.16d) did not support. The mobility issue arises whenever the geographical region is divided into multiple cells and the frequency spectrum is used in a repetitive manner over different cells. The key element in mobility support is handover operation (HO). *Handover* (or *handoff*) refers to the operation of converting the wireless link connecting an MS to the BS in service to another wireless link connecting the MS to another BS in such a way that the communication connection is continuously maintained without degrading the QoS while the MS moves from a cell to another. The handover in the Mobile WiMAX system is founded on hard handover, which is optimized for IP data traffic. However, soft handover, which is best suited for voice traffic and so was employed in CDMA networks, is also supported in the Mobile WiMAX.

Since the mobile devices are likely to be compact and portable, the power saving is also crucial to the Mobile WiMAX operation. In support of power savings, Mobile WiMAX supports sleep mode and idle mode of operations. The sleep mode operation helps to save power by allowing the MS to be absent from the serving BS air interface while not in use, and the idle mode operation helps to save power by allowing the MS to be mostly idle and listen to the broadcast messages only periodically. In fact, the IEEE 802.16e specification amends the existing fixed broadband wireless access system to deal with all these features, including handover.

In this chapter, we will first briefly review a cellular concept, since the mobility issue arises due to the cell-based network operation. We will then discuss the handover procedure and power-saving issues specific to the Mobile WiMAX network.

7.1 Cellular Concept

Scarceness of the wireless resources essentially raises the issue of service coverage in wireless communications: If a single BS were to serve all the MSs in a certain service area, it would become hardly possible to meet all the service requirements simultaneously. In addition, it would require too much transmission power to the MSs located far away from the BS due to the rapid attenuation of the signal power. Moreover, the single BS would become the bottleneck link of communication services since the given bandwidth must be shared among all the users. This demonstrates that the service coverage of a single BS should be limited to a relatively small portion of the entire service area, and therefore the concept of cellular system arises.[1]

The concept of the cellular system was developed for the goal of solving the service coverage problem: It intended to provide a continuous service over the entire service area by dividing the service area into multiple clusters, called *cells*, and deploying one BS in each cluster. This cellular concept was able to render a simple solution to the service coverage problem, since the distance from the BS to each MS within a cell as well as the number of the MSs to service by a cell has reduced such that the desired service quality can be achieved at comparatively low transmission power and small bandwidth. However, the cellular concept has caused several new problems, including the *intercell interference* (ICI) issue, which happens due to the signals generated at the neighboring cells, and the mobility issue, which occurs when the MSs move out of the boundary of the associated cells. The cellular concept becomes practicable only when it can present effective solutions to these problems.

In the case of the Mobile WiMAX system, the cellular concept has another mission to accomplish, which is to meet the requirement of *frequency reuse factor* (FRF) of one. It is the goal initially set by the WiBro system (see Section 1.3.1 and Chapter 10 for more on WiBro) for the purpose of attaining high spectral efficiency. Due to the ICI that culminates at the cell boundary, the goal of FRF=1 is a very challenging task. However, the *adaptive modulation and coding* (AMC) technique accomplishes the goal by adjusting the modulation and coding to ICI-resilient low-data rate modes, thereby protecting the data transmission at the cell boundary. On the other side, the AMC technique yields very high data rates at the inner part of the cell by adopting high-efficiency modulation and coding schemes, and thus contributes to accomplishing the goal of achieving high spectral efficiency (see Section 2.1.3 for more discussion on AMC).

7.1.1 Intercell Interference Management

If all the BSs in a cellular system use the entire bandwidth simultaneously, an MS near the boundary of its associated BS would suffer from high interference from the adjacent cells. This ICI corrupts the user signals and degrades the QoS of the MS located at the cell boundary. One way to mitigate the ICI is to divide the entire bandwidth into several disjoint sets and *reuse* each set only at the BSs spaced far enough apart. By arranging this way, an MS near to the cell boundary is relieved from severe interference, since its adjacent cells use a different set of bandwidth. The ICI can be suppressed by this frequency reuse, but this reuse decreases the bandwidth available for use in each cell, as only a part of the entire spectrum can be utilized.

The motivation of the frequency reuse approach is that the power of transmitted signal decreases as it propagates through space. So, even though multiple BSs use the same spectrum simultaneously, the interference suffered by them can be maintained at a tolerable level if they are separated far enough apart. The set of cells that use the same channel set is called the *cochannel set* and the minimum distance between the members of cochannel set is called the *reuse distance*. The number of cochannel sets is called the *reuse factor* (or FRF). There exists a one-to-one relation between the reuse distance and the reuse factor (i.e., a reuse distance is automatically defined if a

1. The cellular concept in this section refers in large part to the same topic in [1].

system determines the reuse factor). The resource manager has to define the reuse distance considering the QoS requirement of the system because a tradeoff relationship exists between the level of ICI and the bandwidth used for each BS.

Note that a system with FRF = 1 is the most efficient in the sense that the available bandwidth in the system is fully reused in every cell. One particular example of the system that employs FRF = 1 is the cellular CDMA systems, including 3GPP-2 cdma2000 and 3GPP W-CDMA. The processing gain obtained by spreading the data signal with a higher rate pseudo-random sequence allows for the CDMA system to be robust again the ICI from neighbor cells. Meanwhile, the Mobile WiMAX system is also designed for operating with FRF=1 to maximize its cellular bandwidth efficiency. In support of this, the AMC technique renders useful means of combating the ICI by stepping down the data rate in a dynamic manner as the level of *carrier-to-interference and noise ratio* (CINR) increases, as discussed earlier. However, a robust operation can be warranted with a more loose frequency reuse in the Mobile WiMAX network as in the other typical cellular systems, as long as bandwidth efficiency is not a critical matter.

Cell Planning

In arranging the reuse of the bandwidth, the resource manager has to determine two factors the reuse distance to use in the system and the channel allocation to each cell. This task is called the *cell planning*. Figure 7.1 illustrates the cell planning for the cases of FRF = 1, 3, and 4.

The determination of reuse distance depends on the QoS of the cell boundary users, as they are likely to suffer from interference more than the users in the vicinity of BSs do. So the resource manager determines the reuse distance such that the target QoS or other performance objectives of the boundary users are satisfied.

For example, we consider a simple downlink system that has 100 channels in total and requires a minimum SIR of 6 dB to each user. We assume that the system objective is to maximize the bandwidth allocated to each BS. If we assume, in addition, for simplicity of analysis, that the effects of shadowing and small-scale fading are negligible and that each cell uses the same transmit power, then the interference caused by other cells can be determined only by the distances among the users and other cells. Noting that the interference decays fast with a path loss exponent of 3 to 4, we may consider the first tier cells within the cochannel set as the only interferers. The arrows in Figure 7.1 illustrate how to calculate the distance between a user at the cell boundary and its interfering cells in the first tier. If we set the path loss exponent to 3, the SIR of the case of FRF = 1 is determined to be

$$SIR_1 = \frac{R^{-3}}{2R^{-3} + 2(2R)^{-3} + 2\left(\sqrt{7}R\right)^{-3}} = 0.424(= -3.73\text{ dB}) \tag{7.1}$$

where R denotes the cell radius. Clearly this SIR value does not meet the minimum SIR requirement of 6 dB, so the system manager has to check with other reuse factors. Repeating the same calculation for the cases of reuse factors 3 and 4, respectively, one can get the SIR values $SIR_3 = 3.42$(=5.36 dB) and $SIR_4 = 5.72$(=7.57 dB) for the cases of FRF = 3 and FRF = 4. Based on these results the system manager will

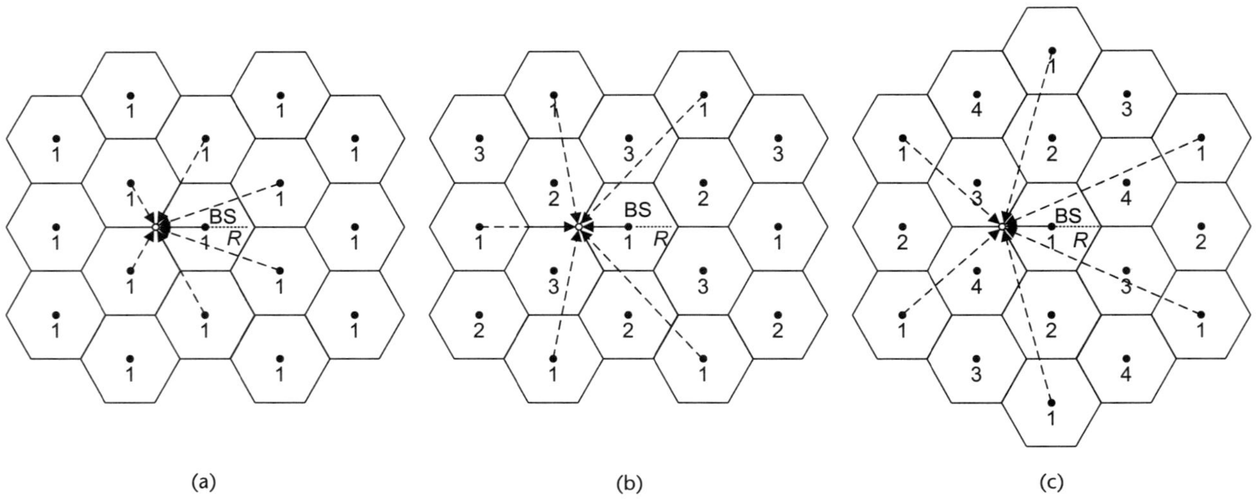

Figure 7.1 Cell planning with (a) reuse factor 1, (b) reuse factor 3, and (c) reuse factor 4.

take FRF = 4 to maximize the system objective while satisfying the QoS requirement of the boundary users.

After determining the FRF and the number of channels to assign to each cell, the system manager decides which channels to assign to each cell. This can be done in arbitrary manner, as each cell will not have any preference on the channels. Once the channels are assigned to each cell, each cell will have the same number of channels and can use them exclusively. As long as the traffic distribution is uniform among different cells, this type of equal channel assignment will be optimal in handling the traffic. However, the traffic distribution in general is not uniform and may vary with time and location. If the traffic pattern is predictable in advance, before network deployment, it would be possible to arrange channel assignment in such a way that a heavily loaded cell can borrow some number of channels from some lightly loaded cells. This borrowing arrangement can be done during the cell-planning stage, and the borrowed channel may be used permanently by the heavily loaded cell [2, 3].

Reuse Partitioning

In addition to this cell-planning arrangement, it is also possible to arrange such that different channels can adopt different reuse factors. In this case, the total bandwidth is first split into several overlaid cell plans with different reuse distances. Specifically, each cell is divided into multiple concentric subcells, and each subcell is associated with each cell plan in such a way that an inner subcell uses the cell plan with a smaller reuse distance while an outer subcell uses the cell plan with a larger reuse distance. Beyond that, the processing is similar to that for fixed channel assignment with homogeneous reuse distance. In contrast to the homogeneous arrangement for the cell planning, this heterogeneous arrangement is called *reuse partitioning*, as it involves multiple cell plans with different reuse distances. Figure 7.2 illustrates the reuse partitioning applied with each cell divided into two concentric subcells.

The principle behind the reuse partitioning is that the power level required to achieve the desired SIR in an inner subcell is in general much lower than that in an outer subcell. It is because each inner subcell is located closer to the BS at the center

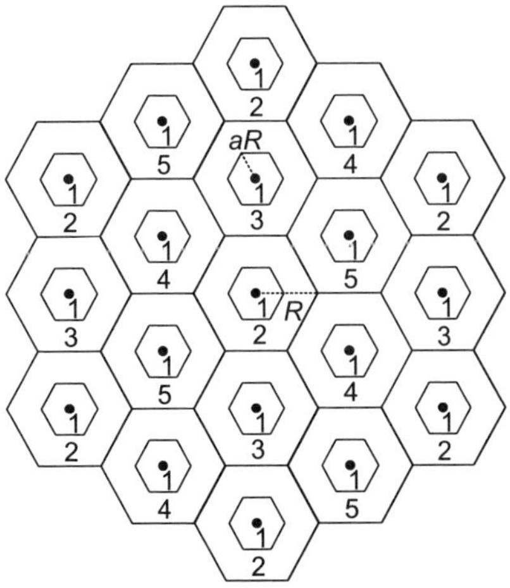

Figure 7.2 Cell structure with reuse partitioning.

of the cell than its outer cells, so for the users in the inner subcell the QoS requirements may be satisfied with a small reuse factor. Due to the tradeoff relationship that exists between the reuse factor and the bandwidth used by each BS, the reuse partitioning yields a bandwidth increase.

In order to adopt the reuse partitioning, the system manager has to determine how to partition each cell into subcells and how many channels to assign to each subcell, in addition to the tasks required for the homogeneous case.

For example, we consider the case of two-subcell-based reuse partitioning in Figure 7.2 with FRF=1 and FRF=4 applied, respectively, to the inner and the outer subcells. We first determine the radius of the inner subcell, aR, $0 \le a \le 1$, assuming the same channel conditions as in the previous example. The SIR experienced by a user at the inner subcell boundary is determined to be

$$SIR_1 \approx \frac{(aR)^{-3}}{2R^{-3} + 2(2R)^{-3} + 2\left(\sqrt{7}R\right)^{-3}} = 0.424a^{-3}\left(= -30\log_{10} a - 3.73 \text{ dB}\right) \tag{7.2}$$

If we choose a such that the SIR requirement of 6 dB is met, we get $a = 0.474$. Now we determine the number of channels to assign to each subcell. If we assume that users are distributed uniformly and each user has the same call request probability, then the traffic intensity of each subcell is proportional to the area of the subcell. In the equation, we get $c_1\!: c_2 = 0.474^2\!:\!(1-0.474^2)$ for the number of channels assigned to the inner subcell, c_1, and that assigned to the outer subcell, c_2. Considering that the reuse factor is 1 and 4, we get another equation $c_1 + 4c_2 = 100$ (channels). Solving the equations, we can determine $c_1 = 8$ and $c_2 = 23$ as the most suitable choices. Therefore the reuse partitioning enables each BS to use 31 channels exclusively, which is larger than the number 25 obtained by using homogeneous cell planning with reuse factor 4.

A notion of reuse partitioning can be also applied to the cellular OFDMA system, as long as all subcarriers or subchannels are divided into the different groups so that they can be allocated to the different regions in a way to mitigate the ICI. In this context, it is sometimes referred to as a *fractional frequency reuse* (FFR) scheme. In fact, there are many different types of FFR schemes with the varying levels of reuse efficiency.

It is noteworthy that the approach of Mobile WiMAX of fixing the FRF to one and adopting AMC in conjunction with it may be viewed as a "soft realization" of the FFR scheme. In contrast to the "hard realization" of the reuse partitioning scheme that divides the cell into concentric subregions by calculating the physical radius and allocates a fixed number of subcarriers/subchannels to each subregion, this "soft realization" allocates different numbers of subcarriers or subchannels via AMC to the "invisible subregions" determined by the channel state estimation, with the allocated subcarrier/subchannel number flexibly varying according to the channel state variation.

7.1.2 Handover Management

Another important issue to address in relation to the cellular concept is the *mobility management*. Since a cellular system separates the coverage of each BS geographically, an MS crossing the cell boundary needs to get its connection redirected to a new BS so that it can get continuous services. This requires that each wireless cellular system should be equipped with handover capability. Handover capability enables an MS to transfer the channel of an ongoing connection associated with a BS to another channel associated with another BS, as illustrated in Figure 7.3. As the MS in service moves away from the coverage of the serving BS, the quality of the connection will degrade, so the MS establishes a new connection with a new BS that can help the MS maintain the channel quality above some sustainable level. As such, handover may be interpreted as a means of maintaining the QoS of an ongoing connection in reaction to the user mobility across the cell boundary in wireless cellular systems. (Refer to Section 3.6.2 for more discussion on handover.)

Handover Criteria

Handover procedure is initiated based on the MS's measurement of the signal strengths from multiple BSs. One simple criterion of handover is that the MS redirects its connection to the BS with the highest signal strength. According to the illustration in Figure 7.4, the MS may initiate handover procedure at position A at which the signal strength of a new BS becomes stronger than that of the serving BS. This method will surely guarantee that the MS is always associated with the BS having the strongest channel, but, at the same time, it can make a handover attempt even when the current connection has acceptable quality and thereby induce unnecessary handovers.

Another criterion for handover initiation is to redirect the connection to a new BS only when the signal strength of the new BS is higher than the current signal strength by a predetermined margin. According to the illustration in Figure 7.4, this criterion makes the MS initiate the handover at point B, where the new signal strength is higher than the existing one by the margin M. This method can prevent the ping-pong effect, or the repeated handovers between two BSs, which may possibly occur at the cell boundary. In this case a larger value of margin may be taken to shift the handover point closer to the new BS (e.g., to point C), thereby reducing the handover attempts further, but an excessive margin could delay the handover initialization and consequently deteriorate the connection quality.

A third handover criterion may be to initiate handover only when the current signal strength falls below a given threshold and a new BS sends signal with a higher

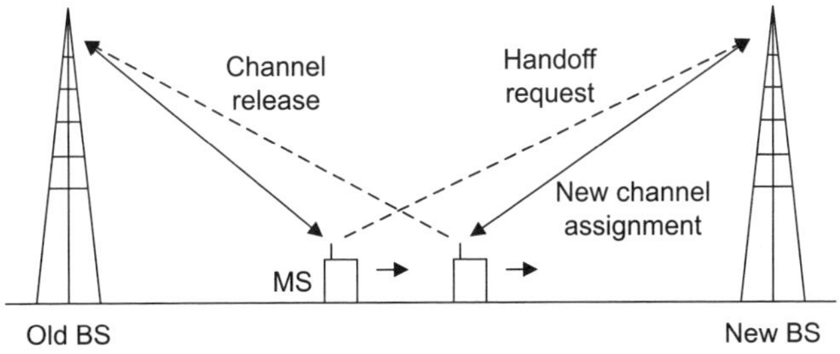

Figure 7.3 Illustration of handover operation.

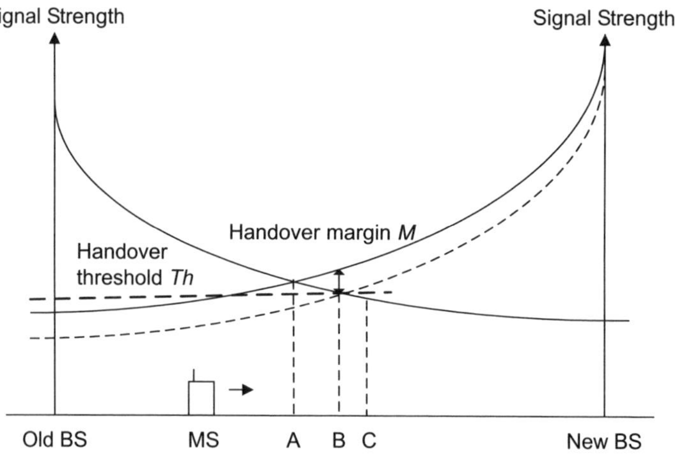

Figure 7.4 Illustration of handover initialization criteria.

strength. According to the illustration in Figure 7.4, this criterion makes handover initiated at point C at which the current signal strength drops below the threshold *Th*. In this criterion, the decision of the threshold level governs the handover initiation: If the threshold is too high, the signal strength of the current BS would always be below the threshold at the cell boundary, and, if it is too low, handover may not take place even when the current connection falls into poor state. Therefore the threshold should be set adequately by considering these three possible criteria to prevent unnecessary handovers while maintaining connection quality above an acceptable level.

Guard Channel Policy

In view of resource management, handover raises the issue of resource reservation and admission control. As noted earlier, handover is a means of maintaining the QoS of the ongoing connection at a prespecified level. This task requires reserving a certain amount of resources, as otherwise handover may fail due to lack of available channels. Resource reservation for handover is closely related to the admission control of newly originating connections, as the total resource has to be shared by the handover connections and the newly generated connections. In this sharing, it is reasonable to set a higher priority on the handover connections, since a forced termination of the ongoing connections makes much more severe impact on the connection QoS, as well as customer satisfaction, than the denial of access (i.e., blocking) of newly originating connections does.

From the resource manager's point of view, this prioritization policy may be interpreted as an assignment of different admission priorities to two different types of traffic arrivals—handover arrivals and newly originating arrivals—with a higher priority set to the former. This admission priority is directly related to the issue of reserving the resource to the two different types of traffic arrivals. Therefore the essence of handover management is how to allocate the resource to the two different traffic arrivals such that it can satisfy the QoS requirement of the handover connections and, at the same time, provide a reasonable level of performance to the new connections.

This prioritization policy may be implemented in such a way that a fixed number of channels, called the *guard channels* (GCs), are reserved for the handover arrivals [4]. This GC policy may be operated by setting a channel threshold T as follows: Any handover connection is admitted as long as a channel is available, but a new connection is admitted only when the number of ongoing connections is fewer than the threshold T. This implies that, out of the total resource of C channels, the GC policy always reserves $(C–T)$ channels for admitting the handover connections.

The GC policy focuses on two key performance metrics: one is the *dropping probability* P_D that a handover attempt is not successful due to the lack of resources, and the other is the *blocking probability* P_B that the communication system rejects a newly originating connection. The dropping probability represents the average fraction of handover attempts that are unsuccessful, indicating how well the QoS is maintained on the readily admitted connections, whereas the blocking probability reflects how the GC policy affects the QoS of the newly originating connections.

Apparently, a tradeoff relationship exists between the dropping probability and the blocking probability. In order to balance the two performance metrics, we may consider formulating a linear objective function as a weighted sum of the two probabilities (i.e., $F = \alpha_B P_B + \alpha_D P_D$) for some positive constants α_B and α_D to be determined concurrently with the GC policy and then minimizing the objective function for the given number of channels, C. In view of this optimization problem, the GC policy is optimal, as it minimizes the given linear objective function [5], and, therefore, the GC policy renders a simple but efficient operation while optimally exploiting the tradeoff relationship between the blocking and dropping probabilities.

There are several variants of the GC policy, including the *fractional guard channel* (FGC) policy [6, 7] and the dynamic reserved channel adjustment method [8–11]. In the broadband mobile systems, such as the Mobile WiMAX network, the GC policy becomes more complicated simply because multiple traffic priorities must be handled for the multimedia service classes with the varying level of QoS requirements. For example, [12] deals with the GC policy with more than two traffic priorities.

7.2 Handover Procedure

The Mobile WiMAX system basically supports the *hard handover* (also known as *break-before-make*) scheme, which disconnects the existing link to the in-service BS first and then makes a new connection to another BS, but it is also possible to implement a *soft handover* scheme that can support a reliable handover by maintaining connection with two or more BSs in the transition period.

Handover is performed in two main processes: one is the network topology acquisition process, and the other is handover execution process. Network topology acquisition refers to periodically updating the parameter values needed for making handover decisions between the MS and the BS. Handover execution refers to practically executing the handover through neighbor scanning, handover capability negotiation, MS release, and network reentry processes. Each of these processes is detailed in the following subsections and all handover-related MAC management messages herein are included in Table 5.3.

7.2.1 Network Topology Acquisition

In preparation for handover, MS and BS periodically acquire and update the parameter values needed for making handover decisions. Such network topology acquisition takes place through network topology advertisement, neighbor BSs scanning, and association processes.

Persiodically, the MS receives the channel information of the neighboring BSs (e.g., the number of neighboring BSs, DCD/UCD, and available radio resources of individual neighbor BSs) through the advertisement (MOB_NBR-ADV) message via the BSs in service. The MS sends to the BSs in service a scanning request (MOB_SCN-REQ) message to request parameters needed to perform the scanning process for measuring the quality of the signal received from the neighboring BSs. The BS sends back a scanning response (MOB_SCN-RSP) to the MS to explain when and how long the MS can measure the signal quality of the neighboring BSs. Then during the given scanning period, the MS first acquires synchronization with each neighboring BS, then measures the *carrier-to-interference and noise ratio* (CINR) and other parameters, and finally determines whether or not each neighboring BS is adequate as the handover target BS. Note that no data traffic can be transmitted either by the BS or by the MS during scanning period, since the MS is solely dedicated to synchronization process. As CINR may continuously vary, the scanning interval must be periodically incurring.

Next, the MS may perform an association process to store the ranging information obtained according to the scanning type included in the MOB_SCN-RSP message. In this stage the MS performs a contention-based initial ranging or a noncontention-based ranging with the neighboring BSs. The ranging information obtained through the association process helps to select an appropriate handover target BS, and acquire and record the information needed for executing handover to the target BS, thereby enabling fast ranging with the target BS. (Refer to Sections 3.2.1 and 3.7.2 for a more detailed operation of ranging process.)

Figure 7.5 illustrates the network topology acquisition process for a serving BS with two neighbor BSs. The number of two neighbor cells is indicated by an advertisement message (e.g., N_Neighbors = 2 for the current example). According to information in MOB_SCN-REQ and MOB_SCN-RSP messages, MS starts scanning in *M*th frames, which lasts for *N* frames. Once synchronized with each BS, MS measures the CINR for the corresponding BS during the scanning period. MS performs an association process with individual neighbor BS, which eventually ensures the service availability in the neighbor BS by referring to the value of service level prediction in RNG-RSP.

7.2.2 Handover Execution

Handover is executed through handover decision and initiation, synchronization to target BS downlink, ranging, and termination of MS context processes. In addition, MS may perform cell reselection prior to handover decision and/or perform handover cancellation in the middle of the handover process.

1. *Cell reselection*: The MS may collect and examine the information needed to hand over to a potential target BS. In support of this, the MS may acquire the

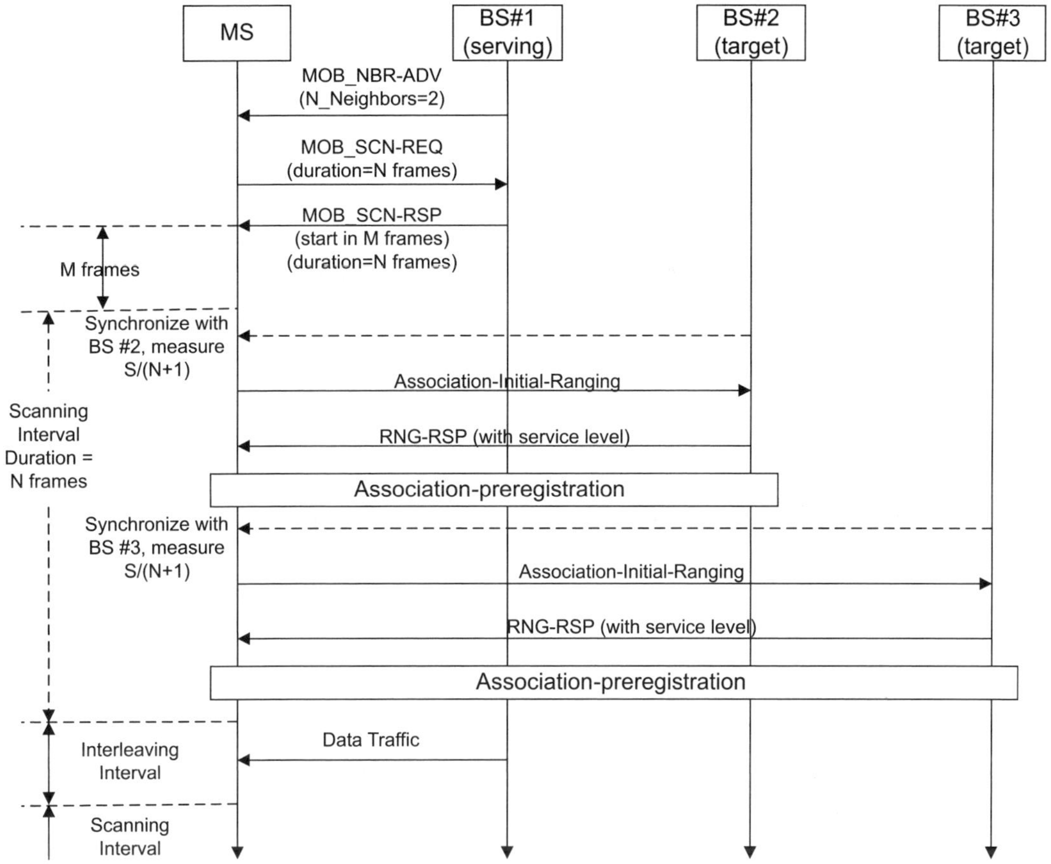

Figure 7.5 Illustration of network topology acquisition process.

information from a decoded MOB_NBR-ADV message or may request the information from the neighbor BS. This cell reselection process does not necessarily occur in conjunction with any specific handover decision.

2. *HO decision and initiation:* Handover begins with the decision by the MS to hand over from a serving BS to a target BS. The decision may originate either at the MS (i.e., *MS-initiated handover*) or at the serving BS (i.e., *network-initiated handover*). The *handover* decision is notified to the BS through MOB_MSHO-REQ message in the case of the MS-initiated handover and to the MS through the MOB_BSHO-REQ message in the case of the network-initiated handover
3. *Synchronization to target BS downlink*: MS synchronizes to the downlink transmissions of the target BS to obtain DL and UL transmission parameters. This process may be shortened if MS had previously received a MOB_NBR-ADV message including the target BSID, physical frequency, DCD, and UCD.
4. *Ranging*: MS and the target BS conduct initial ranging or handover ranging. If the MS RNG-REQ message includes the serving BSID, then the target BS may make a request to the serving BS for information on the MS over the backbone network and the serving BS may respond. Regardless of having

received the MS information from the serving BS, the target BS may request the MS information from the backbone network. If the target BS had previously received HO notification from serving BS over the backbone, then the target BS may allocate a noncontention-based initial ranging opportunity for fast ranging, which can significantly reduce the delay associated with initialization with the target BS.

5. *Termination of MS context*: The serving BS terminates the context of all connections belonging to the MS and the context associated with them (i.e., the information in queues, ARQ state-machine, counters, timers, and header suppression information).
6. *HO cancellation*: The MS may cancel the handover at any time prior to the expiration of the resource retain time interval after sending the MOB_HO-IND message.

Depending whether the handover decision originates from the MS or BS, two different cases can be considered. In the sequel, each of these cases is detailed.

MS-Initiated Handover

Specifically, the MS-initiated handover is divided into two different phases—handover preparation and action—see Figure 7.6. During the handover preparation phase, handover capability negotiation is made through backbone messages for prehandover notification. If the MS judges the CINR values of the neighboring BSs to be good enough to conduct handover, it requests to start handover to the serving BS by sending the handover request (MOB_MSHO-REQ) message and the CINR values of the neighboring BSs. On receiving the handover request, the serving BS notifies the neighboring BSs of the MAC address, the requested resources, and the expected QoS level of the MS. Then it receives the QoS value that the neighboring BSs can support. Next, the serving BS selects an appropriate target BS that can provide the higher QoS value, and then notifies it to the MS through the handover response (MOB_BSHO-RSP) message. Now, handover capability has been negotiated by selecting a target BS that can provide the best QoS level for the handover traffic.

Subsequently, during the handover action phase, handover execution is commanded by the MS, and then network reentry to the target BS is attempted through initial ranging. The more detailed steps are as follows: On receiving the MOB_BSHO-RSP message, the MS notifies the serving BS of its final decision on disconnecting the link with the serving BS or canceling/rejecting the handover by sending the MOB_HO-IND message that commences the handover action. Once handover is confirmed and acknowledged by the newly selected target BS, the newly selected target BS can offer a noncontention-based fast-ranging opportunity to the MS using the MAC address received from the backbone network, so that the MS can join the new BS as fast as possible. In other words, the target BS can reserve a bandwidth for the incoming MS that can be used for initial ranging without resorting to the contention-based ranging process, thus minimizing the handover break time. Therefore, a network reentry process can start immediately in L frames after the MOB_HO-IND message is transmitted by MS.

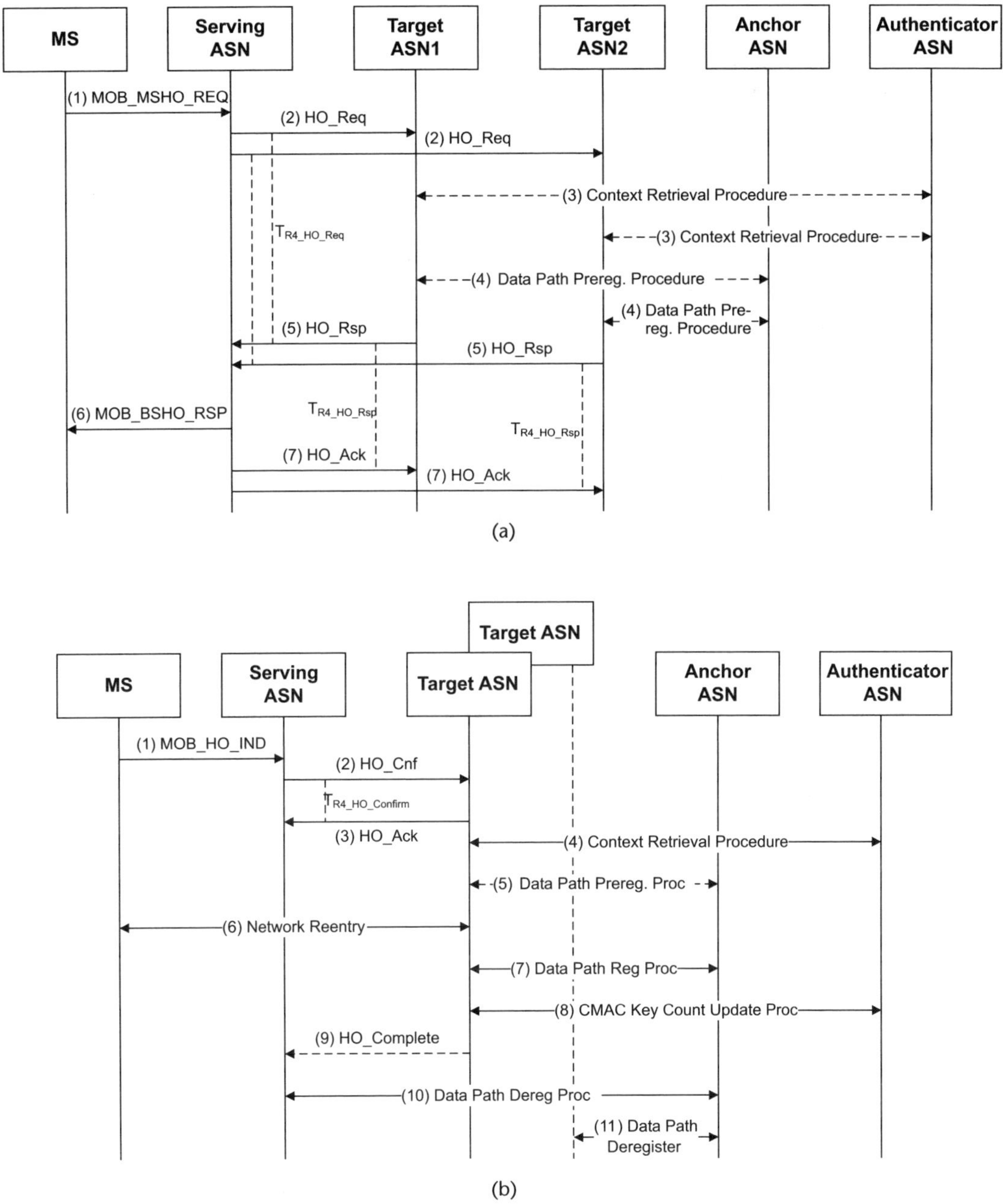

Figure 7.6 Illustration of MS-initiated handover execution process: (a) preparation phase; and (b) action phase. (*After:* [13], Figures 4-41 and 4-48, redrawn.)

Note that IEEE 802.16e specification only defines the handover procedure and the related MAC management messages (e.g., MOB_HO-RES/RSP and MOB_HO-IND) that are specific to the air link between MS and BS. The backbone messages for prehandover notification and handover confirmation/acknowledgment are defined by [13]. For example, Figure 7.6 illustrates one scenario of the end-to-end message sequences for handover preparation and action phases in association with MAC-layer handover [13].

Network-Initiated Handover

The network-initiated handover is executed in a way similar to the MS-initiated case. Only difference is that HO capability negotiation is made beforehand through backbone messages between candidate target BSs. In this case, the serving BS initiates the handover process utilizing the scanning results periodically reported by the MS. The BS transmits a MOB_BSHO-REQ message when it wants to initiate a handover. Other than using the MOB_BSHOS-REQ message, the procedure of the network-initiated handover execution is the same as that of the MS-initiated handover. Figure 7.7 illustrates the preparation phase of the handover execution process initiated by the network. The figure shows that the overall process is identical to the case of the MS-initiated preparation stage in Figure 7.6(a) except that the network now makes the initiation.

7.2.3 Soft Handover

In order to increase the cell coverage and improve the QoS performance at the cell boundary, we may optionally use soft handover techniques, namely, the *macro diversity handover* (MDHO) technique, which takes advantage of the diversity gain, and the *fast BS switching* (FBSS) technique, which relies on the anchor BS. MDHO or FBSS handover is enabled under several conditions, including that the involved BSs are synchronized based on a common time source, that the frames sent by the BSs arrive at the MS within the cyclic prefix interval, and that the BSs have synchronized frame structure. Enabling or disabling these optional soft handover techniques is determined through the exchange of the REG-REQ/RSP message.

MDHO

In the case of the MDHO, the MS communicates simultaneously with the *diversity set* of the BSs that allocate wireless resource to the MS. The diversity set refers to a set of BSs that are active to the MS and are managed by the serving BS and the MS.

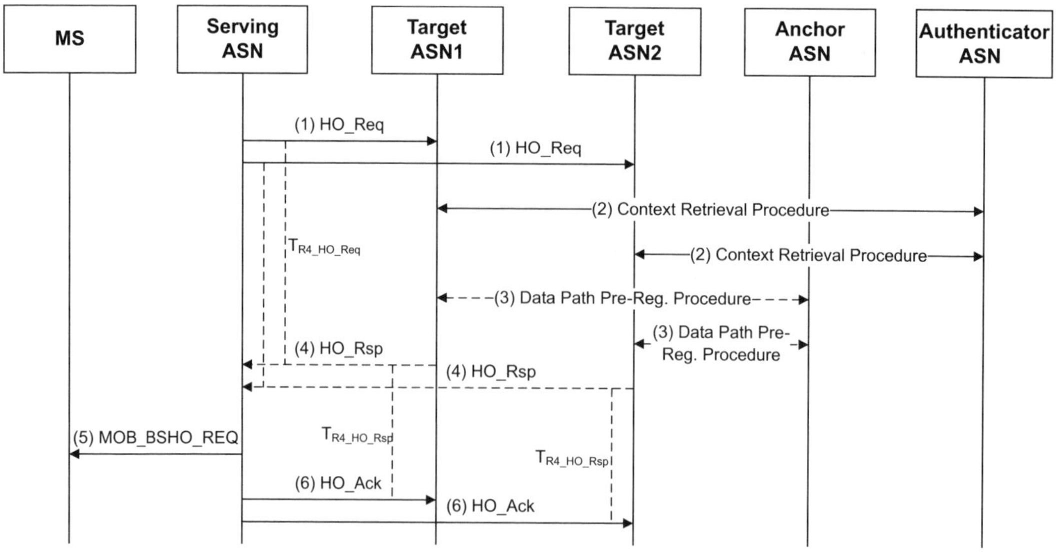

Figure 7.7 Illustration of network-initiated handover execution process (preparation phase). (*After:* [13], Figure 4-45, redrawn.)

The MS scans the neighbor BSs and selects the BSs that are suitable to be included in the diversity set. The selected BSs are reported by the MS, and the diversity set update procedure is performed by the BS and the MS.

An MDHO begins when the MS decides to transmit/receive unicast messages and traffic from multiple BSs at the same time interval. For DL MDHO, two or more BSs provide synchronized transmission of the MS downlink data such that the MS can perform a diversity combining on the data. For UL MDHO, multiple BSs receive the transmission from the MS and perform a selection diversity of the received information.

In the MDHO scheme, the MS receives the MAP message notifying the MS's burst information from multiple BSs in the diversity set, which requires that the MS must be equipped with two or more receivers. In reality, however, it is more practical to receive the MAP message from one BS only (i.e., the anchor BS). According to the current specifications, it is possible to make the anchor BS send the burst allocation information of other BSs to the MS through some information element of the MAP message, thus creating the effect of receiving the same information from multiple BSs in the diversity set to the MS. Then the MS can apply a combining technique to the received RF signals or a soft-combining technique by combining the repeatedly received signals to obtain the handover-related information. However, it is challenging to do scheduling for a fast information exchange among the constituent BSs and to apply the identical permutation patterns for diversity.

FBSS

The FBSS handover relies on the anchor BS, which corresponds to the serving BS with which the MS communicates the data in the given frame. Specifically, the anchor BS refers to the BS that the MS initially registered with and synchronized to, which the MS performs the ranging process with and receives the downlink control information from. The MS communicates only with the *anchor BS* among all BSs in the diversity set for downlink and uplink messages, including management and traffic connections. The MS monitors the broadcast messages such as MAP and FCH and performs fast switching depending on the channel state.

The MS continuously monitors the signal strength of the BSs that are included in the diversity set, selects one BS from the current diversity set to be the anchor BS, and then reports the selected anchor BS on CQICH or MOB_MSHO-REQ message. An FBSS handover begins with the decision by the MS to receive/transmit data from/to the anchor BS, which may change within the diversity set. The transition of the anchor BS, which is called *BS switching* is performed without invocating the HO procedure described previously. The FBSS handover is triggered by either MOB_MSHO-REQ or MOB_BSHO-REQ messages.

When the MS has more than one BS in the diversity set that is updated by fast feedback, the MS can transmit *fast anchor BS selection* information to the current anchor BS using the fast-feedback channel. For the transmission of the anchor BS selection information, the MS transmits the codeword corresponding to the selected anchor BS via its fast-feedback channel.

Figure 7.8 illustrates the fast anchor BS selection mechanism for the case $L = 2$. The time axis is slotted by an *anchor switch reporting* (ASR) slot, which is M frames long. For the current frame number, N, the ASR slot number is determined to be the

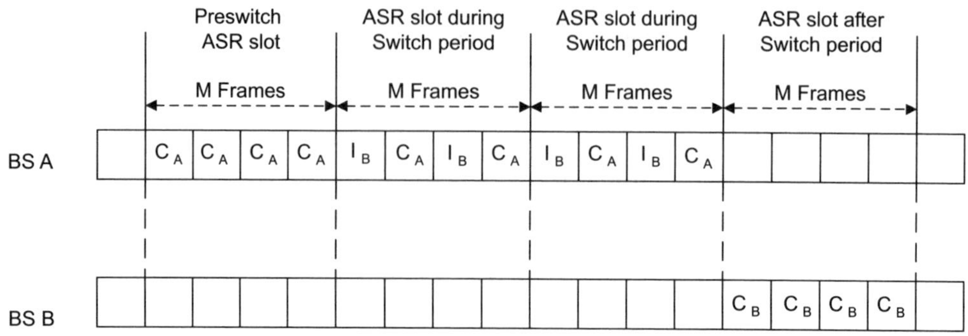

Figure 7.8 Illustration of fast anchor selection mechanism ($L = 2$). (*After:* [14].)

integer quotient of N divided by M. The ASR slot starts at the frame where frame number modulus M equals zero. The switching period whose duration is equal to L ASR slots is introduced. The MS receives from the anchor BS the length of the ASR, M (in number of frames), and the length of the switch period, L (in ASR slots), through the DCD.

The operation is as follows: The MS decides to switch the anchor BS from the current one e.g., BS A) to another one (e.g., BS B) in the diversity set that has a better signal level. It then transmits the CQI report for BS A during the ASR slot duration through the CQI channel allocated by BS A and also transmits the anchor BS switch indicator of BS B, in an alternative way. On receiving this message, BS A sends an Anchor_Switch_IE that decides the switching admission/cancelation and CQI channel allocation to the MS within the switch period. Then the MS changes the anchor BS to BS B according to the received information after the corresponding switch period and receives a MAP message from BS B.

7.3 Power Saving

When the mobility is considered on the user terminals, power consumption of the terminal becomes one of the most important design issues. In order to maintain an efficient terminal state, IEEE 802.16e standards specify *sleep mode* and *idle mode*-based operations in such a way that the terminal can operate in those power-saving modes if not in use but can return to the normal operation mode whenever needed. Sleep mode saves power by allowing MS to be absent from the serving BS air interface for a predetermined periods time, and idle mode saves power by allowing MS to become periodically available for DL broadcast traffic messaging without registering any specific BS.

7.3.1 Sleep Mode

Sleep mode, which refers to the state in which the MS makes itself absent from the serving BS air interface for prenegotiated periods of time, is characterized by the unavailability of the MS to DL or UL traffic. Sleep mode helps to decrease MS power usage and decrease the usage of the serving BS air interface resources as well.

In sleep mode, the time intervals are divided into unavailability and availability. *Unavailability interval* does not overlap with any listening window of any active power saving class, whereas *availability interval* does not overlap with any unavailability interval. During an unavailability interval, the BS does not transmit to the MS, so the MS may power down one or more physical operation components or perform other activities that do not require communication with the BS. In contrast, during an availability interval, the MS is expected to receive all DL transmissions in the same way as in the normal operation state, and in addition, the MS maintains synchronization with the BS.

The procedure of transitioning to and from sleep mode is depicted in Figure 7.9(a, b), respectively, for the MS-initiated and the BS-initiated operations. Note that MOB_SLP-REQ, MOB_SLP-RSP, and MOB_TRF-IND are the MAC management messages listed in Table 5.3. In the case of the MS-initiated operation, the MS sends the MOB_SLP-REQ message and the BS replies with the MOB_SLP-RSP message to request and respond for the transition to the sleep mode, respectively. The messages include the time to start the transition to sleep mode, the minimum and the maximum length (in frames) of the duration of sleep mode, and the time period to listen to the signals from the BS after waking up from sleep mode. Note that mutual agreement with when and how long the MS will be in sleep mode between the MS and BS is important because otherwise the BS may transmit to a MS that is already in sleep mode. Furthermore, the MS must wake up periodically to make sure that there is any traffic awaiting and the BS must know when the MS wakes up. All these actions can be supported by the parameters in MOB_SLP-REQ and MOB_SLP-RSP messages.

The sleep period consists of one or more variable-length, consecutive sleep windows, with interleaved listening windows, through one or more sleep-window itera-

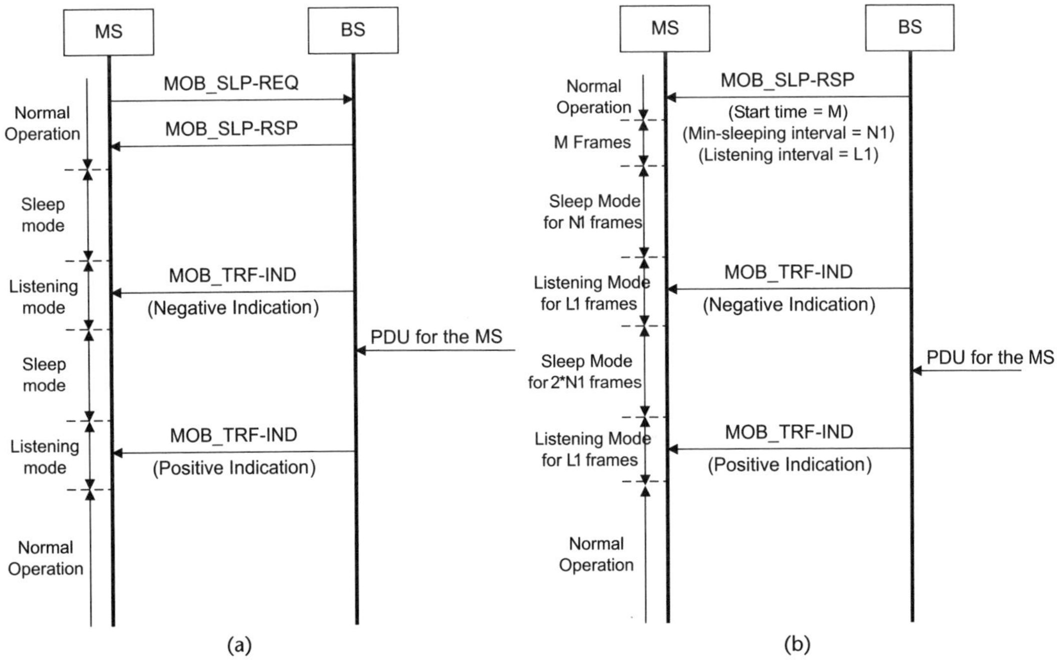

Figure 7.9 Sequence diagram for sleep mode: (a) MS initiated; and (b) BS initiated. (*After:* [14].)

tions. During a listening interval, an MS synchronizes with the serving BS downlink and listens for an appropriate traffic indication message MOB_TRF-IND. The MS decides whether to stay awake or go back to sleep based on the MOB_TRF-IND (with negative indication) message from the serving BS. During the consecutive sleep windows and listening windows that comprise a single sleep interval, the sleep window is updated using an exponentially increasing algorithm. In the case of the BS-initiated operation, the procedure is similar to the MS-initiated case except that the transition to the sleep mode is initiated by the BS.

For each MS in sleep mode, the BS may allocate, during the listening window of the MS, an UL transmission opportunity for periodic ranging. Alternatively, the BS may return the MS to normal operation by deactivating at least one power-saving class (see the following) to keep it in an active state until a UL transmission opportunity is assigned for periodic ranging. The BS may also let the MS know when the periodic ranging opportunity should occur using the next periodic ranging TLV in the last successful RNG-RSP. When the periodic ranging operation processes successfully between the MS and the BS, the BS may inform the MS of the frame number to start the next periodic ranging operation.

Figure 7.10 presents one particular example of sleep mode operation for the following scenario: First, the MS requests to transition to sleep mode, and the BS sends a response to accept the request. The MS wakes up after 2 frames of sleep, then receives the BS signal for 3 frames. Recognizing that there is no traffic through the MOB_TRF-IND message with negative indication, the MS decides to double the sleep time duration. On receiving the MOB_TRF-IND message with positive indication next, the MS will wake up and move to normal operation.

For each MS in sleep mode, the BS keeps one or more contexts, with each one related to a certain power-saving class. *Power-saving class* refers to a group of connections that have common demand properties. Different scheduling service classes may be categorized into different power-saving classes, which then support different QoS requirements via different service classes more effectively. There are three types

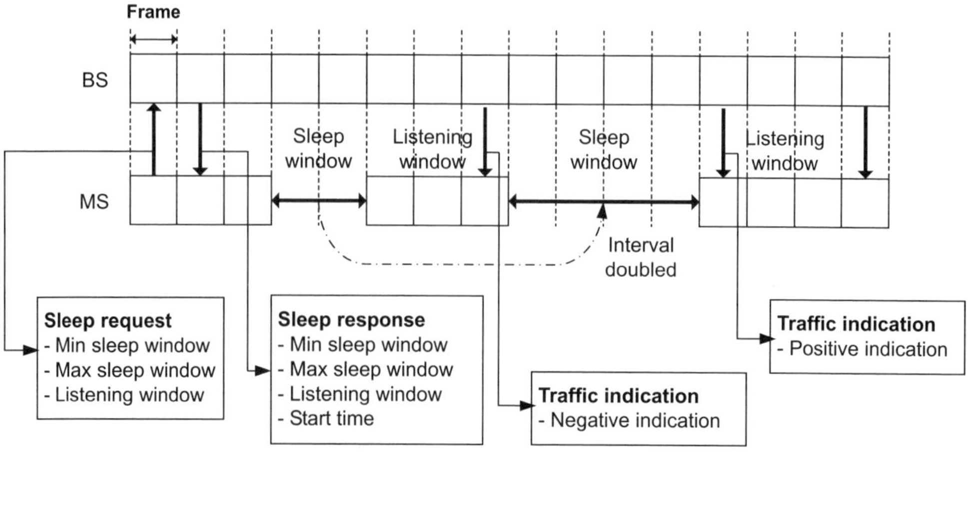

Figure 7.10 Illustration of sleep mode operation.

of power saving classes—types I, II, and III—each differing by parameter sets, procedures of activation/deactivation, and the policies of MS availability for data transmission. If a power-saving class is activated, then the sleep/listening windows sequence associated with this particular class starts. The sleep mode pattern may be applied differently to different types of power-saving classes.

Power-Saving Class Type I

Power-saving class type I is recommended for connections of BE and NRT-VR type traffic. For definition and/or activation of one or several power-saving classes of type I, a MS sends an MOB_SLP-REQ or *bandwidth request* (BR) and uplink sleep control header (for activation only) to a BS. Then, the BS responds with an MOB_SLP-RSP message or DL sleep control extended subheader. Alternatively, power-saving class may be defined/activated/deactivated by the power-saving class parameter TLVs transmitted in the RNG-REQ/RSP message. The relevant parameters are initial sleep window, final sleep window base, listening window, final sleep window exponent, the start frame number for the first sleep window, and traffic triggered wakening flag.

The power-saving class becomes active at the frame specified as "the start frame number for the first sleep window." The sleep window size grows double each time until it exceeds the specified final size. The sleep window is interleaved with a listening window of fixed duration. The BS terminates the active state of the power-saving class by sending an MOB_TRF-IND message over the broadcast CID or sleep mode multicast polling CID during the listening window. When the MS receives a UL allocation after receiving a positive MOB_TRF-IND message indication, the MS transmits at least a BR message with the BR field of the BR PDU set to 0. During the active state of a power-saving class of type I, the MS does not send or receive any MAC SDUs or their fragments, but, in contrast, during the listening windows, the MS is expected to receive all DL transmissions in the same way as in the normal operation state. The power-saving class is deactivated if one of the following events happens: (1) the BS transmits during the availability window a MAC SDU or its fragmentation over the connection belonging to the power saving class, (2) the MS transmits a BR with respect to the connection belonging to the power saving class, or (3) the MS receives an MOB_TRF-IND message indicating the presence of buffered traffic addressed to the MS. Figure 7.11 illustrates the sleep mode operation for a power-saving class of type I.

Power-Saving Class Type II

Power-saving class type II, is recommended for connections of UGS and RT-VR type traffic. The connections of this power saving class become active at the frame specified as "start frame number for first sleep window." Once started, the active state continues until it is explicitly terminated by the MOB_SLP-REQ/MOB_SLP-RSP messages or BR and UL sleep control header/DL sleep control extended subheader. Alternatively, this power-saving class may be defined and/or activated/deactivated by the TLVs transmitted in the RNG-REQ/RSP messages. The relevant parameters are initial sleep window, listening window, and the start frame number for the first sleep window.

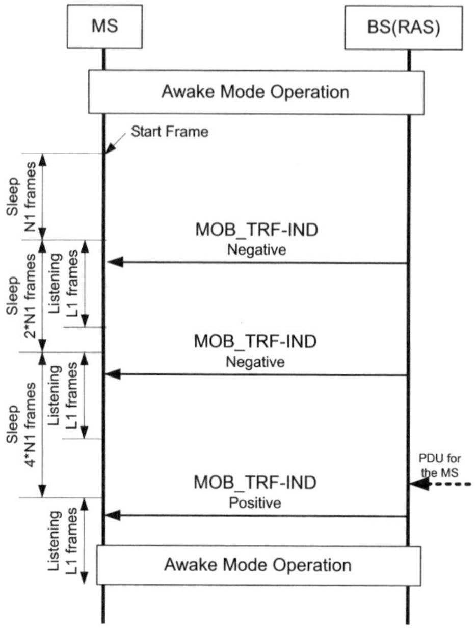

Figure 7.11 Illustration of sleep mode operation for power-saving class of type I.

Different from the case of the power-saving class type I, all sleep window sizes of the power saving class type II remain the same as the first sleep window, and the sleep windows are interleaved with the listening windows of fixed duration. During the listening windows, unlike the power-saving class type I, the MS in the power-saving class type II may send or receive any MAC SDUs or their fragments at the connections that comprise the power saving class as well as the acknowledgments to them. During the sleep windows, MS does not receive or transmit MAC SDUs, as was the case for the power-saving class type I. Figure 7.12 illustrates the sleep mode operation for power-saving class of type II.

Power-Saving Class Type III

Power-saving class type III is recommended for multicast connections as well as for management operations including periodic ranging, DSx operations, and MOB_NBR-ADV. This power-saving class type is defined/activated by MOB_SLP-REQ/RSP or BR and UL sleep control header/DL sleep control extended subheader transactions. Once started, the active state continues until it is explicitly terminated by MOB_SLP-REQ/RSP messages or BR and UL sleep control header/DL sleep control extended subheader. Alternatively, this power-saving class may be activated/deactivated by the TLVs transmitted in RNG-RSP messages. The relevant parameters are final sleep window base, final sleep window exponent, and the start frame number for the sleep window.

Power-saving class type III becomes active at the frame specified as "start frame number for the first sleep window." The duration of the sleep window is specified as base/exponent. Once the sleep window expires, the power-saving class automatically becomes inactive. For multicast services, as the BS may guess when the next portion of data will appear, the BS allocates sleep windows for all the time when the multicast traffic is not expected to arrive. After the sleep window expires, any avail-

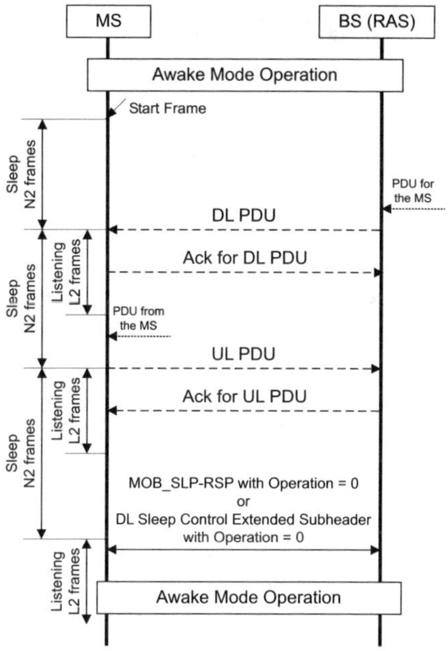

Figure 7.12 Illustration of sleep mode operation for power-saving class of type II.

able multicast data may be transmitted to the relevant MSs, and then the BS may decide to reactivate power-saving class.

Figure 7.13 illustrates the operation of multiple power-saving class sleep mode for the case of two different power-saving class types. Type I contains connections of BE and NRT-VR type, and type II contains a single connection of UGS type. For power-saving class type I, the BS allocates a sequence of listening windows of constant size and sleep windows of doubling sizes. For power-saving class type II, the BS allocates a sequence of listening windows of constant size and sleep windows of constant size too. The MS is considered unavailable (and may power down) within the unavailability windows, which are the intersections of the sleep windows of type I and type II.

7.3.2 Idle Mode

Idle mode is designed to allow the MS to be idle most of the time but become periodically available for DL broadcast traffic messaging without registering at a specific BS as the MS moves over a large geographical area populated with BSs. Idle mode restricts MS activity to scanning only at discrete intervals, so it benefits MSs by lifting all the burdens required for handover and for normal operations. Consequently, idle mode allows the MS to conserve power and operational resources. On the other hand, idle mode provides a simple and timely method for alerting the MS to the pending DL traffic directed toward the MS, so it benefits the network and BS by eliminating air interface and the network handover traffic from the essentially inactive MS.

Specifically, idle mode operates in such a way that the MS does not register to a particular BS while moving from one cell to other but receives the downlink broadcast traffic only periodically. Different from sleep mode, idle mode does not per-

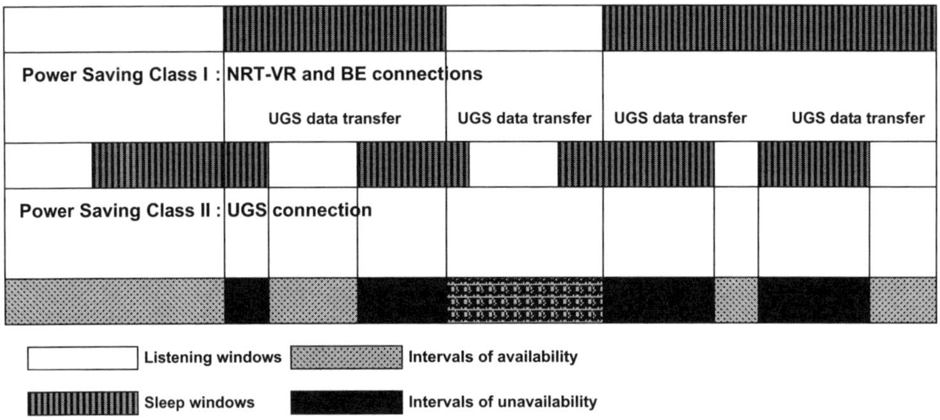

Figure 7.13 Illustration of two power-saving classes with different sleep times. (*After:* [14].)

form any functions required for activation and operation of mobile communications, such as handover, but does the scanning operation only for some discrete time period. This limited operation helps to save terminal power and operation resources. The MS and the BS exchange DREG-REQ/CMD messages (see Table 5.3) to perform such operation. The terminal and the BSs in the paging group support the renewal of the MS position by dividing time into *idle time* and *listen time* just like the sleep mode case in which time is divided into sleep time and listen time. When the BS receives packets to forward to the MS in the idle state, it broadcasts a paging message to access the terminal.

Idle Mode Initiation

If the MS does not transmit/receive data traffic for certain period of time, its status is changed from awake mode to idle mode through a deregistration procedure. The deregistration procedure may be initiated by either the MS (i.e., MS-initiated idle mode) or the BS (i.e., BS-initiated idle mode).

- *MS-initiated idle mode*: Refer to Figure 7.14 to describe the procedure of this MS-initiated deregistration to idle mode. When the MS shifts into an idle state, it creates the DREG-REQ message and transmits it to the BS (with deregistration request code set to 0×01). Then the BS transmits the DREG-CMD message including the received information to the MS (with action code 0×05). If the MS does not receive the DREG-CMD message within the specified time (i.e., T45 timer shown in Figure 7.14, whose default value is 250 ms) after sending DREG-REQ message, then it retransmits the DREG-REQ message up to the DREG request retry count. Also, the BS starts a management resource holding timer to maintain connection information with the MS immediately after it sends out the DREG-CMD message. If this management resource holding timer has expired, the BS releases connection information with the MS.
- *BS-initiated idle mode*: Refer to Figure 7.15 to describe the procedure of the BS-initiated deregistration to idle mode. The serving BS may signal for an MS to begin idle mode by sending a DREG-CMD (with action code 0×05) in an unsolicited manner. In this case, the serving BS starts the T46 timer in

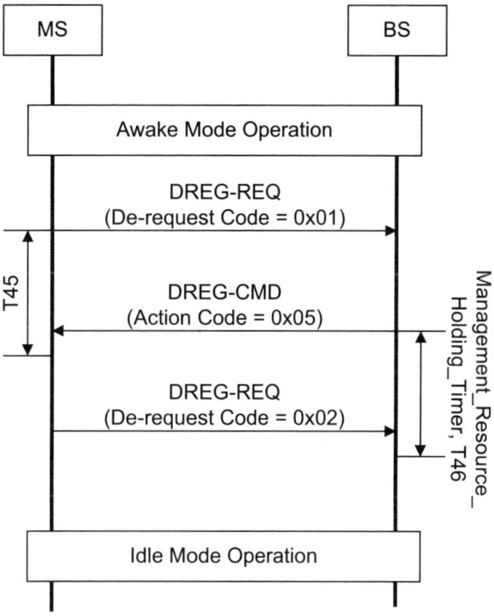

Figure 7.14 MS-initiated deregistration to idle mode.

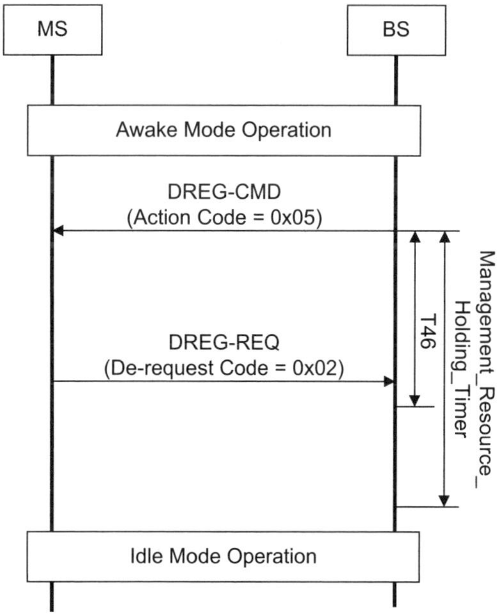

Figure 7.15 BS-initiated deregistration to idle mode.

Figure 7.15 and the management resource holding timer at the same time. If the BS does not receive the DREG-REQ message (with deregistration request code 0×02) from the MS in response to this unsolicited DREG-CMD message within T46 timer expiry, the BS retransmits the DREG-CMD message in an unsolicited manner up to the DREG command retry count. The MS enters idle mode after it sends the DREG-REQ message in response to the unsolicited DREG-CMD.

As another case of BS-initiated idle mode, the serving BS may also include a REQ-duration TLV (with action code 0×05) in the DREG-CMD, signaling to the MS to initiate an idle mode request through a DREG-REQ (with action code 0×01) before the REQ-duration expiration. This request is for the MS to deregister from the serving BS and to initiate idle mode. In this case, the BS does not start the T46 timer. The MS may include *idle mode retain information TLV* within the DREG-REQ message (with action code 0x01) transmitted at the REQ-duration expiration. Then BS transmits another DREG-CMD message (with action code 0×05) including idle mode retain information TLV. Figure 7.16 describes this BS-initiated deregistration with retain information negotiation.

Paging Operation

In support of the idle mode operation, the BSs are divided into logical groups called *paging groups*. This grouping is to offer a contiguous coverage region in which the MS does not need to transmit in the uplink but can be paged with any traffic destined to it in the downlink. The paging groups are arranged to be large enough to contain most MSs within the same paging group at most times and small enough to maintain the paging overhead at a reasonably low level. (Refer to Section 3.5 for more discussion on paging.)

Figure 7.17 illustrates four paging groups defined over multiple BSs arranged in hexagonal grids. Note that a BS may be a member of one or more paging groups.

A paging message is transmitted during the MS paging listening interval if there is any MS that needs paging. A BS broadcast paging message is an MS notification

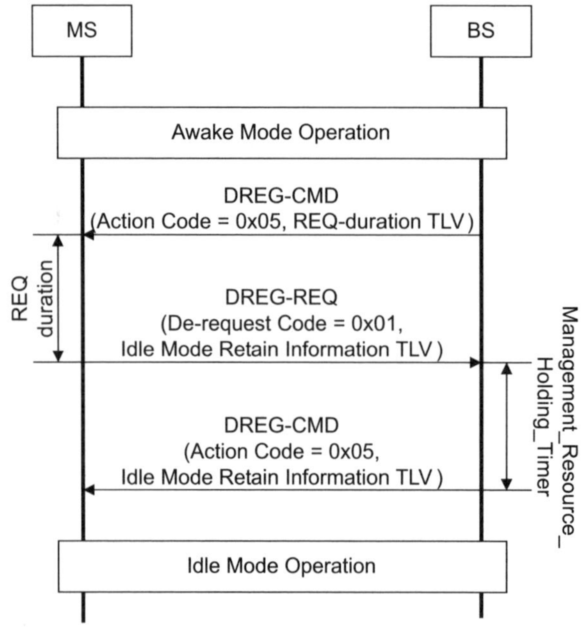

Figure 7.16 BS-initiated deregistration with retain information negotiation.

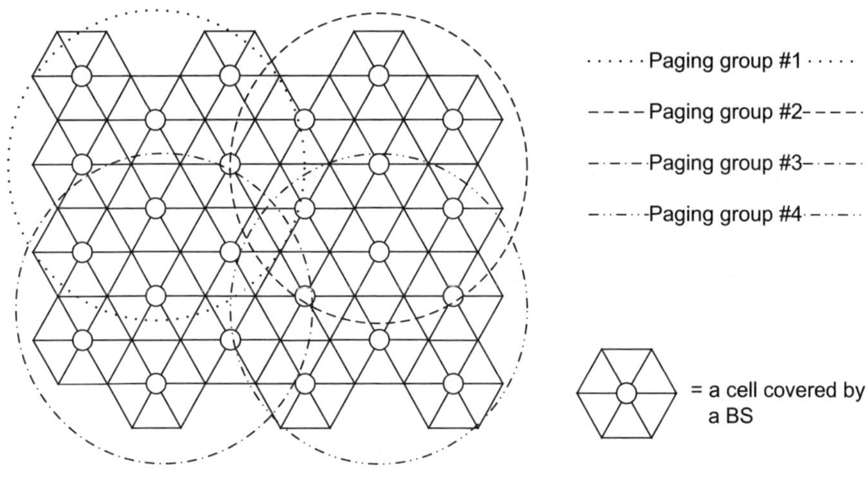

Figure 7.17 Illustration of paging groups. (*After:* [14].)

message that indicates the presence of DL traffic pending, through the BS or some network entity, for the specified MS. The BS broadcast paging message is also used as an MS notification message that polls the MS and requests a location update without requiring a full network entry.

Paging may be classified into two types as follows:

1. The BS broadcasts the MOB_PAG-ADV message periodically to notify the MS of the corresponding paging group. Then the MS in idle mode checks the MOB_PAG-ADV message periodically to check if the paging group of the corresponding MS has been changed.
2. If the BS has incoming traffic destined to a particular MS in idle mode, the BS triggers the MOB_PAG-ADV message to change the corresponding MS into awake mode. Figure 7.18 describes the paging operation that changes the MS in idle mode to awake mode, so the incoming traffic can be delivered.

Location Update

An MS in idle mode performs a location update process if any of the following four location update conditions are met—paging group update, timer update, power down update, and MAC hash skip threshold update. In addition, the MS may also perform a location update process at its own decision.

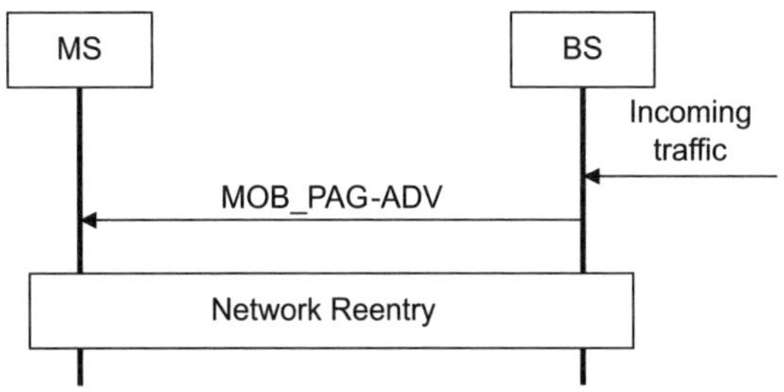

Figure 7.18 Paging operation for MS network reentry.

Specifically, the four location update conditions are as follows: First, the MS performs a location update process when the MS detects any change in the paging group. Second, the MS periodically performs a location update process before the idle mode timer expires. Third, the MS attempts to complete a location update once as a part of its orderly power-down procedure. Fourth, the MS performs a location update process when the MS MAC hash skip counter exceeds its threshold successively. In this case, the BS and MS reinitialize their respective MAC hash skip counters after successful location update.

Figure 7.19 illustrates the inter-BS location update process. When an MS moves to another paging group, the location update process takes place in the following way: We consider the scenario that the idle mode MS in paging group 1 moves to paging group 2. First, the MS receives the MOB_PAG-ADV message from the BS and recognizes that its location has been changed. Then the MS transmits the RNG-REQ message to a new BS (BS 2), including the MAC address, the ranging purpose indication TLV (with bit #1 set to 1), the location update request, and the paging controller ID. On receiving the request message, the BS 2 transmits the RNG-RSP message, including the location update response to the MS, and the location update procedure gets completed.

Idle Mode Termination

When an MS in idle mode has incoming or outgoing traffic, it has to terminate idle mode by responding to the paging. In both cases, the MS performs the network reentry procedure to change the status from idle mode to awake mode. The network reentry procedure becomes simplified if the network can reuse the authentication information from the initial entry. This simplified procedure is achieved by using the RNG-RSP message, including the HO optimization flag.

Figure 7.20 describes this network reentry procedure from idle mode. If the idle mode MS is changed into the awake mode, the MS creates the RNG-REQ message, including the MAC address and the paging controller ID value, and transmits the message to the BS. Then the ranging purpose indication field is set to network reentry (with bit #0 set to 1). Next, the BS responds by using the RNG-RSP message,

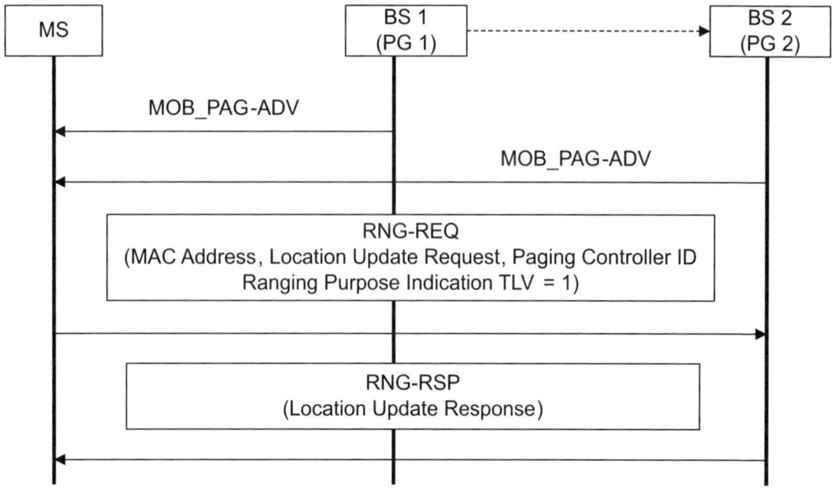

Figure 7.19 Inter-BS location update procedure.

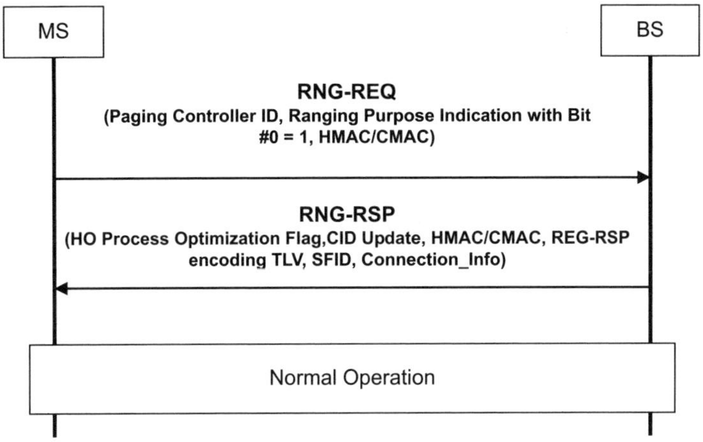

Figure 7.20 Network reentry procedure from idle mode.

including the HO optimization flag and the related CID update. On receiving this RNG-RSP message, the MS may reestablish the connection with the BS and resume normal operation at the serving BS.

References

[1] Lee, B. G., D. Park, and H. Seo, *Wireless Communications Resource Management*, New York: Wiley-IEEE, 2008.

[2] Anderson, L., "A Simulation Study of Some Dynamic Channel Assignment Algorithms in a High Capacity Mobile Telecommunications System," *IEEE Trans. on Communications*, Vol. 21, No. 11, November 1973, pp. 1294–1301.

[3] Engel, J. S., and M. Peritsky, "Statistically Optimum Dynamic Server Assignment in Systems with Interfering Servers," *IEEE Trans. on Vehicular Technology*, Vol. 22, No. 4, November 1973, pp. 203–209.

[4] Hong, D., and S. S. Rappaport, "Traffic Model and Performance Analysis for Cellular Mobile Radio Telephone Systems with Prioritized and Nonprioritized Handoff Procedures," *IEEE Trans. on Vehicular Technology*, Vol. 35, No. 3, August 1986, pp. 77–92. See also the comments on this paper in *IEEE Trans. on Vehicular Technology*, Vol. 49, No. 5, September 2000, pp. 2037–2039.

[5] Ramjee, R., D. Towsley, and R. Nagarajan, "On Optimal Call Admission Control in Cellular Networks," *Wireless Networks*, Vol. 3, No.1, March 1997, pp. 29–41.

[6] Cruz-Perez, F. A., D. Lara-Rodriquez, and M. Lara, "Fractional Channel Reservation in Mobile Communication Systems," *IEE Electronics Letters*, Vol. 35, No. 23, November 1999, pp. 2000–2002.

[7] Markoulidakis, J. G., et al., "Optimal System Capacity in Handover Prioritised Schemes in Cellular Mobile Telecommunication Systems," *Computer Communications*, Vol. 23, No. 5, March 2000, pp. 462–475.

[8] Naghshineh, M., and M. Schwartz, "Distributed Call Admission Control in Mobile/Wireless Networks," *IEEE Journal on Selected Areas in Communications*, Vol. 14, No. 4, May 1996, pp. 711-717.

[9] Levine, D. A., I. F. Akyildiz, and M. Naghshineh, "A Resource Estimation and Call Admission Algorithm for Wireless Multimedia Networks Using the Shadow Cluster Concept," *IEEE/ACM Trans. on Networking*, Vol. 5, No. 1, February 1997, pp. 1–12.

[10] Choi, S., and K. G. Shin, "Adaptive Bandwidth Reservation and Admission Control in QoS-Sensitive Cellular Networks," *IEEE Trans. on Parallel and Distributed Systems*, Vol. 13, No. 9, September 2002, pp. 882–897.

[11] Chiu, M.-H., and M. A. Bassiouni, "Predictive Schemes for Handoff Prioritization in Cellular Networks Based on Mobile Positioning," *IEEE Journal on Selected Areas in Communications*, Vol. 18, No. 3, March 2000, pp. 510–522.

[12] Li, B., et al., "Call Admission Control for Voice/Data Integrated Cellular Networks: Performance Analysis and Comparative Study," *IEEE Journal on Selected Areas in Communications*, Vol. 22, No. 4, May 2004, pp. 706–718.

[13] WiMAX Forum, Network Architecture—Stage 3: Detailed Protocols and Procedures, Release 1.1.0, July 2007. For the latest release, refer to http://www.wimaxforum.org.

[14] IEEE Std 802.16e-2005 and IEEE Std 802.16-2004/Cor1-2005, Part 16: Air Interface for Fixed and Mobile Broadband Wireless Access Systems, February 2006.

Selected Bibliography

Ahson, S., and M. Ilyas, *WiMAX: Standards and Security*, Boca Raton, FL: CRC Press, 2007.

Akyildiz, I. F., et al., "Mobility Management in Next-Generation Wireless Systems," *Proceedings of the IEEE*, Vol. 87, No. 8, August 1999, pp. 1347–1384.

Akyildiz, I. F., J. Xie, and S. Mohanty, "A Survey of Mobility Management in Next-Generation All-IP-Based Wireless Systems," *IEEE Wireless Communications*, Vol. 11, No. 4, August 2004, pp. 16–28.

Chiussi, F. M., D. A. Khotimsky, and S. Krishnan, "Mobility Management in Third-Generation All-IP Networks," *IEEE Communications Magazine*, Vol. 40, No. 9, September 2002, pp. 124–135.

Choi, Y., K. Lee, and S. Bahk, "All-IP 4G Network Archtecture for Efficient Mobility and Resource Management," *IEEE Wireless Communications Magazine*, Vol. 14, No. 2, April 2007, pp. 42–46.

Lee, J.-R., and D.-H. Cho, "Performance Evaluation of Energy-Saving Mechanism Based on Probabilistic Sleep Interval Decision Algorithm in IEEE 802.16e," *IEEE Trans. on Vehicular Technology*, Vol. 56, No. 4, July 2007, pp. 1773–1780.

Sarikaya, B., and S. Gurivireddy, "Evaluation of CDMA2000 Support for IP Micromobility Handover and Paging Protocols," *IEEE Communications Magazine*, Vol. 40, No. 5, May 2002, pp. 146–149.

Wu, W., et al., "SIP-Based Vertical Handoff Between WWANs and WLANs," *IEEE Wireless Communications Magazine*, Vol. 12, No. 3, June 2005, pp. 66–72.

Xiao, Y., "Energy Saving Mechanism in the IEEE 802.16e Wireless MAN," *IEEE Communications Letters*, Vol. 9, No. 7, July 2005, pp. 595–597.

Yabusaki, M., T. Okagawa, and K. Imai, "Mobility Management in All-IP Mobile Network: End-to-End Intelligence or Network Intelligence?" *IEEE Communications Magazine*, Vol. 43, No. 12, December 2005, pp. S16–S24.

Zhang, Q., et al., "Efficient Mobility Management for Vertical Handoff Between WWAN and WLAN," *IEEE Communications Magazine*, Vol. 41, No. 11, November 2003, pp. 102–108.

Zhang, Y., and M. Fujise, "Energy Management in the IEEE 802.16e MAC," *IEEE Communications Letters*, Vol. 10, No. 4, April 2006, pp. 311–313.

CHAPTER 8

Security Control

The Mobile WiMAX system distinguishes itself from other existing mobile systems in that it offers a strong security function by installing a dedicated security sublayer between the MAC layer and physical layer. In the case of the existing TDMA- or CDMA-based cellular systems, user channels were protected by the circuit mode operation of the wireless communications, similar to the case of the wireline telephone network. On the other side, in the case of the WiFi systems, security was not taken very seriously in the early stages, even if packet-mode (i.e., IP-based) communications were used, because it was originally designed for use in the closed domain of wireless local area network, similar to wireline LAN.[1] Different from those two cases, the Mobile WiMAX system is an IP-based *wide area network* (WAN), so a special care was needed from the beginning on the security of the network and the protection of user information.

The security sublayer in the Mobile WiMAX system provides users with privacy, authentication, and confidentiality across the fixed and mobile broadband wireless network. The security function also provides operators with strong protection from unauthorized access to the data transport services by securing the associated service flows across the network. Further, the security sublayer employs an authenticated client/server key management protocol in which the BS, the server, controls the distribution of keying materials to client MS. The basic security mechanisms are strengthened by adding the digital certificate–based MS device-authentication capability to the key management protocol [1].

This chapter discusses the architecture and operation of the Mobile WiMAX security system. It first outlines the fundamentals of cryptography that provide theoretical background of the secured communications and the cryptographic tools. Then it provides an overall sketch of the Mobile WiMAX security functions. Next, it describes the architecture of the Mobile WiMAX security system in terms of encryption, authentication, and key management. Finally, it discusses, in more detail, the key management protocols and their practical operations using the state machines and the state transition matrices.

8.1 Fundamentals of Cryptography and Information Security

As a preliminary study for the security sublayer functions of the Mobile WiMAX, this section introduces the fundamental concepts of cryptography and information security.[2] Specifically, it deals with the objective of cryptography in communica-

1. Originally, the baseline MAC of the 802.11 included security mechanisms for confidentiality and authentication, but they were too weak to protect the security of the WiFi users. Later, IEEE 802.11i enhanced the security features by defining the robust security network. Refer to Chapter 15 for details.
2. For more detailed discussions of fundamental cryptography, refer to introductory cryptographic references (e.g., [2]).

tions, the mathematical model of encrypted communications, the symmetric-key and public-key ciphers, the practical systems that use those cryptographic techniques, and other auxiliary security functions. In the last section, we introduce the overall procedure of security message flows, citing where the cryptographic elements are practically used in the Mobile WiMAX system.

8.1.1 Cryptography

Cryptography is a study of mathematical techniques that are related to the aspects of information security. The objectives of information security are many, including confidentiality and privacy, data integrity, entity authentication, nonrepudiation, availability, message authentication, digital signature, and access control. *Confidentiality* and *privacy* are to keep information secret from all but those who are authorized to see it. *Data integrity* is to ensure that information has not been altered by unauthorized or unknown means. *Entity authentication* (or *identification*) is to corroborate the identity of an entity (i.e., person, processor, device, and so on). *Nonrepudiation* is to prevent an entity from denying previous commitments or actions. *Availability* is to ensure that the system responsible for storing, processing, and delivering the information is accessible when needed. *Message authentication* (or *data origin authentication*) is to corroborate the source of the information. *Signature* is a means to bind information to an entity. *Authorization* is to convey to another entity an official sanction to do or be something. *Access control* is to restrict access to resources to the privileged entities. In addition, there are other objectives such as validation, certification, time-stamping, witnessing, and so forth.

Cryptographic tools used to provide information security are called *primitives*. Examples of primitives include the encryption schemes, hash functions, and digital signature schemes. Depending on the use of keys, primitives may be divided into three categories: (1) the *unkeyed primitives* such as hash functions, one-way permutations, and random sequences; (2) *symmetric-key primitives* such as symmetric-key ciphers, hash functions, pseudorandom sequences, and identification primitives; and (3) *public-key primitives* such as public-key ciphers, signatures, and identification primitives.

A *cryptographic protocol* is a distributed algorithm defined by a sequence of steps with cryptographic primitives that precisely specify the actions required for two or more entities to achieve a specific security objective. The evaluation criteria of cryptographic protocols are several, including the level of security, functionality, methods of operation, performance, and ease of implementation. The *level of security* means the upper bound of *work factor* (or the amount of work required to break a cryptographic system) to defeat the intended security object. *Functionality* means the capability to collaborate with others to meet various information security objectives. *Method of operation* means how different characteristics the protocol exhibits when applied in various ways and with various inputs. *Performance* means the efficiency of the particular method of operation. *Ease of implementation* means the level of easiness in implementing the protocol in the given software and hardware environment.

8.1.2 Encrypted Communication

The *encrypted communication* refers to the secured communications with security technology incorporated as described in Figure 8.1. Given the plaintext, the sender generates and sends the ciphertext over the channel to the receiver, which reconstructs the original plaintext and sends it to the destination. Here, *channel* is a means of conveying the information from the sender to the receiver. The channel is called an *unsecured channel* if it is subject to reading, deleting, inserting, reordering, and any other intrusive operation on the conveyed information, and the *adversary* is the entity or source that conducts such intrusive operations. The channel is called a *secured channel* if it is secured by applying physical or cryptographical means.

For a more rigorous description of the encryption and decryption process, we define the following terminology: $\boldsymbol{A}$ denotes a finite set called *alphabet of definition*, such as binary alphabet $\boldsymbol{A}$={0,1}, English alphabet, and octet alphabet. $\boldsymbol{M}$ denotes a set called *message space*. An element of $\mathbf{M}$, m, is called a *plaintext message*, *cleartext*, or *message*. $\mathbf{C}$ denotes a set called *ciphertext space*. $\mathbf{C}$ may be different from $\mathbf{M}$. An element of $\mathbf{C}$, c, is called a *ciphertext*.

In addition, $\boldsymbol{K}$ denotes a set called *key space*. An element of $\boldsymbol{K}$, k, is called a *key*. E_{k_E} is an *encryption function* such that each element $k_E \in \boldsymbol{K}$ uniquely determines a bijection (or one-to-one mapping) from M to C (i.e., $E_{k_E} : \boldsymbol{M} \rightarrow \mathrm{C}$). D_{k_D} is an *decryption function* such that each element $k_D \in \boldsymbol{K}$ uniquely determines a bijection (or one-to-one mapping) from $\mathbf{C}$ to $\boldsymbol{M}$ (i.e., $D_{k_D} : \boldsymbol{C} \rightarrow \boldsymbol{M}$).

Then from Figure 8.1, the operations of the sender and the receiver devices take the expressions $c = E_{k_E}(m)$ and $m = D_{k_D}(c)$, respectively. Therefore, if encryption and decryption are properly done, the relation $D_{k_D} = E_{k_E}^{-1}$ holds, and the encrypted communication yields the operation $D_{k_D}(E_{k_E}(m)) = m$ for all $m \in \mathbf{M}$.

8.1.3 Ciphers and Hash Functions

Among the multitude of cryptographical primitives discussed in Section 8.1.1, we consider the symmetric-key ciphers, public-key ciphers, and hash functions in this section, as they are practically applied in the Mobile WiMAX system for secured transmission of keys and data traffic.

Symmetric-Key Cipher

A *cipher* is defined by an encryption set $\{E_{k_E} : k_E \in \boldsymbol{K}\}$ and a corresponding decryption set $\{D_{k_D} : k_D \in \boldsymbol{K}\}$ with the property that for each encryption key $k_E \in \boldsymbol{K}$ there is a

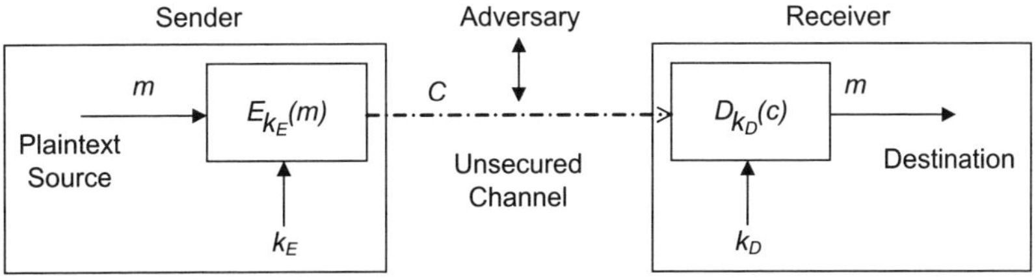

Figure 8.1 Schematics of encrypted communications.

unique decryption key $k_D \in \mathbf{K}$ such that $D_{k_D} = E_{k_E}^{-1}$ (i.e., $D_{k_D}(E_{k_E}(m)) = m$) for all $m \in \mathbf{M}$.

Depending on the data-processing structure, a cipher is divided into block cipher and stream cipher. A *block cipher* is a cipher that breaks up the plaintexts into blocks of a fixed-length block over an alphabet $\mathbf{A}$ and enciphers one block at a time. A *stream cipher* is a special case of the block cipher with the block length 1.

A cipher is called a *symmetric-key cipher* (or single-key cipher or one-key cipher) if the determination of k_E from k_D, or vice versa, is computationally easy. In the symmetric-key cipher, the two keys are normally put equal, or $k_E = k_D$. Figure 8.2 shows the diagram of a symmetric-key cipher communication system. As indicated in the figure, we assume that there exists a secure means for sending the key from the source to the destination.

In practical systems, in general, multiple ciphers are used in combined form, as illustrated in Figure 8.3. In the figure, the *product cipher* in Figure 8.3(a) is composed of multiple ciphers in cascade structure, and the *DES cipher* in Figure 8.3(b) is composed of multiple ciphers in the so-called Feistel structure.

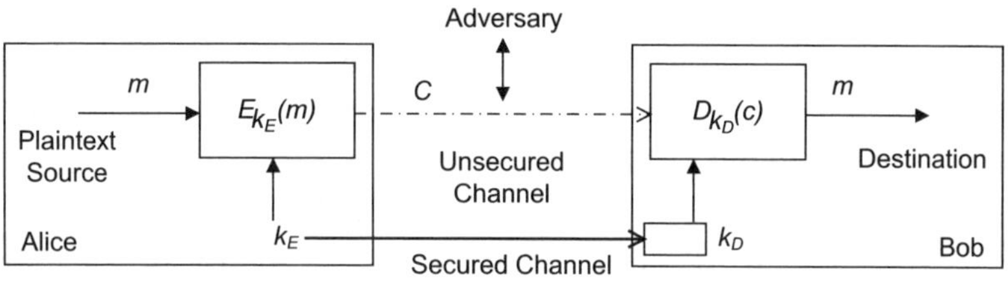

Figure 8.2 Diagram of symmetric-key cipher communication system [2].

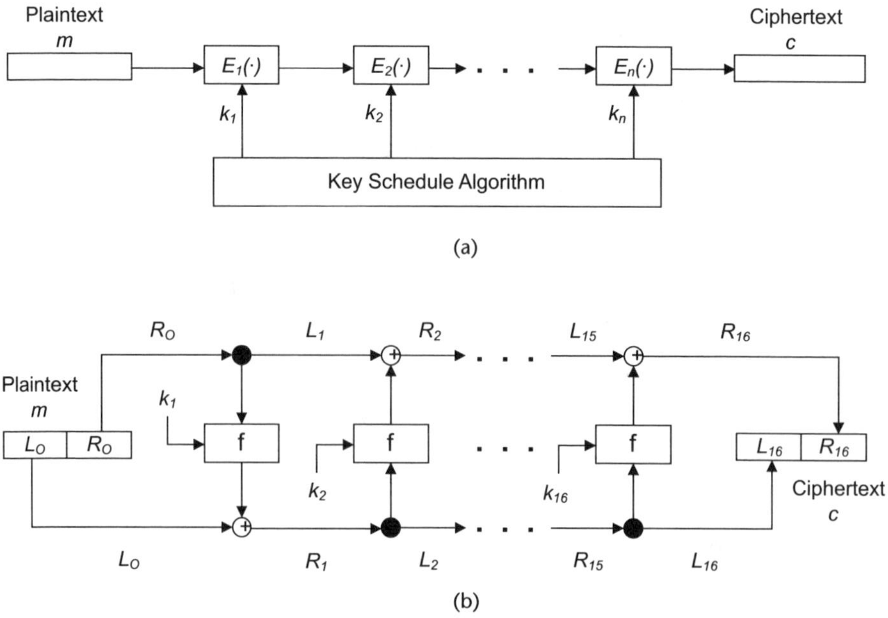

Figure 8.3 Ciphers in combined form: (a) product cipher; and (b) DES cipher (in 16-rounded Feistel structure) [2].

Public-Key Cipher

Whereas symmetric-key cipher is a cipher in which the determination of k_D from k_E is computationally easy, *asymmetric-key cipher* (or *two-key cipher*, or *public-key cipher*) is a cipher in which the determination of k_D from k_E is computationally infeasible. In this case, k_E is called the *public encryption key* and k_D the *private decryption key*. If public-key cipher is used, the sender can send a message to the receiver using the receiver's public encryption key k_E, which is confidential. Figure 8.4 shows the diagram of public-key cipher communication system. It differs from the symmetric-key cipher in Figure 8.2 in the aspect of key provision.

A function $f: X \rightarrow Y$ is called a *one-way function* if $f(x)$ is easy to compute for all $x \in X$, but for most elements $y \in Y$ it is computationally infeasible to find any $x \in X$ such that $f(x) = y$. For example, given a set $X = \{1, 2, ..., 251\}$, it is easy to compute $f(x) = x^7 \bmod 251$, but it is very hard to compute the inverse of it. As a practical example, we consider the RSA problem: Given a number n, which is a product of two distinct odd prime numbers p and q (i.e., $n = pq$), let the *great common divisor* (gcd) of a third number $e > 0$ and the Euler number of n, $\varphi(n)$ be 1 (i.e., gcd $(e, \varphi(n) = 1)$. Then, for a given c, determine a number $m > 0$ such that $m^e = c \bmod n$. For sure, this problem is computationally infeasible.

A one-way function $f: X \rightarrow Y$ is called a *trapdoor one-way function* if it has the additional property that if some extra information (called the *trapdoor information*) is given it becomes feasible to find an $x \in X$ such that $f(x) = y$ for any given $y \in Y$. This concept is very useful in designing the public-key cipher, as is incorporated in the RSA-based cipher. In the previous example, if we compute a number d in $1 < d < \varphi$ such that $ed \bmod \varphi = 1$, then the solution becomes feasible. The RSA public-key system takes (n, e) as the public key and d as the private key (see Section 8.1.4). In view of Figure 8.4, (n, e) and d correspond to k_{EB} and k_{DB}, respectively. The destination Bob generates both of them and sends (n, e) to the plaintext source Alice, keeping d itself.

The public-key cipher, however, is subject to the attack of an active adversary in the unsecured channel. Specifically, if the identity of the owner of a public key cannot be verified, then a *man-in-the-middle attack* can succeed. Figure 8.5 illustrates this problem: If the active adversary C intercepts the public key, k_{EB}, of the destination Bob and generates its own public key, k_{EC}, and passes it to the plaintext source Alice, then the adversary C can decipher the ciphertext $c = E_{k_{EC}}(m)$ generated by the source Alice using the key $E_{k_{DC}}$. Then it can change the original plaintext to m', generate a different ciphertext $c' = E_{k_{EB}}(m')$, and send it to the destination Bob. In order to protect the public key, the public-key cipher adopts public key certificate system.

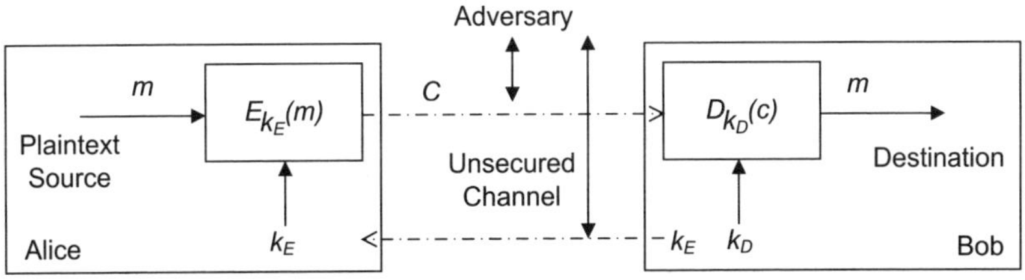

Figure 8.4 Diagram of public-key cipher communication system [2].

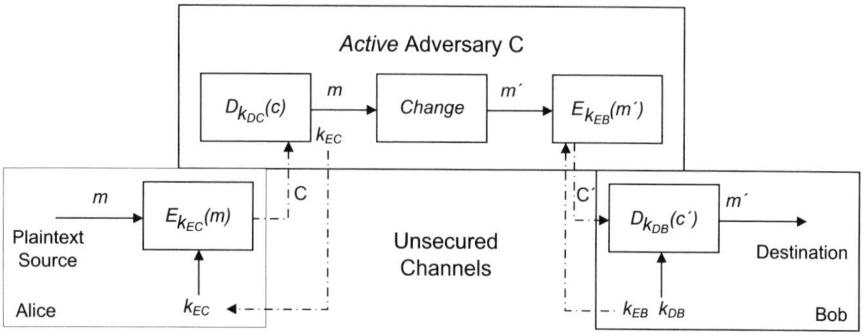

Figure 8.5 Illustration of the man-in-the-middle attack [2].

Symmetric-Key Versus Public-Key

Symmetric-key and public-key cryptographies are comparable in various aspects: Symmetric-key cryptography is advantageous in that it has higher rates of data throughput and shorter key size, so is employed as the primitives for pseudorandom number generators, hash functions, and so on. However, it is disadvantageous in that: (1) the keys must remain secret at both ends, that a large number of keys are to be managed in large networks so an unconditionally trusted *trusted third party* (TTP) is necessary for key management, (2) key life is shorter than for the case of public-key cryptography, and (3) active TTP is required for digital signature.

On the other hand, public-key cryptography is advantageous in that: (1) only one key needs to be kept secret, (2) it requires only a functionally trusted TTP, (3) a public/private key pair may remain unchanged for considerable period of time, (4) it yields a relatively efficient digital signature system, and (5) the number of keys needed in large network is considerably smaller than the symmetric-key case. However, it is disadvantageous in that: (1) the public-key system is much slower than the symmetric-key system, (2) key size is much larger than the symmetric-key case, and (3) it has a much shorter history and has not proved to be secure, so it may be subject to unknown attack.

In practical communication systems, data encryption is the longest-running part of the encryption process, whereas the key establishment is a small fraction of the total encryption process. As a consequence, it will be most desirable if public-key encryption schemes are used to establish keys for the symmetric-key schemes to conduct the main data encryption processing for the secured communications. In this arrangement, efficiency comes from the long-life nature of the public/private keys of the public-key schemes and the high-throughput nature of the symmetric-key schemes. Therefore, in practice, symmetric-key cryptography is used for encryption and data integrity purposes, and public-key cryptography is used for efficient signatures and key-management purposes.

Hash Functions

Hash function is a computationally efficient function mapping binary strings of arbitrary length to binary strings of some shorter fixed length, which are called *hash values*. For a hash function that outputs n-bit hash values (e.g., $n = 128$ or 160) to be useful cryptographically, it should have the one-way function and collision-resistant function properties. The *one-way* property is that given a specific hash-value z, it is computationally infeasible to find an input x such that $h(x)=z$. The *collision-resistant* property is that it is computationally infeasible to find two distinct inputs x and y that yield a common value after hashing (i.e., $h(x) = h(y)$).

Figure 8.6 shows an example of a hash function, which is a general model for iterative hash functions. First, the input data x is padded and formatted into t blocks of length L_x (typically 512 or 1,024), $x_1, x_2, .., x_t$. Given an *initial vector* (IV) H_0, H_i is determined by the iterative relation $H_i = f(H_{i-1}, x_i)$, $i = 1, 2, ..., t$, for a compression function f. After the completion of the iteration, the hash value is obtained by $h(x) = g(H_t)$.

The most common applications of hash functions are digital signatures and data integrity checks. For digital signatures, a long message is hashed to a shorter length hash value and the hash value is encrypted before transmission. The receiving party first decrypts the digital signature and repeats hashing on the received message and then compares whether the calculated hash value is identical to the received hash value. In this application, it is crucial that no two messages yield the same hash value, as otherwise the signer who signs one may claim later to have signed another.

For data integrity, hash functions may be used as follows: for a given data, the hash function is computed and protected in some way at one time, and the computation is repeated at another time; then the two resulting hash values can be compared for integrity check. In case the data integrity over a communication channel is to be checked, the first hash value the sender computed should be sent to the receiver in some secure method.

As such, hash functions are publicly known, usually involve no secret keys, and are commonly used for signatures and data integrity check. When used to check the data integrity, detecting whether or not the data has been altered, the relevant unkeyed hash functions are called *modification detection codes* (MDCs).

8.1.4 Practical Cryptographic Systems

The symmetric-key cipher and public-key cipher are used in practical cryptographic systems. The most commonly used among them are the DES and the RSA systems,

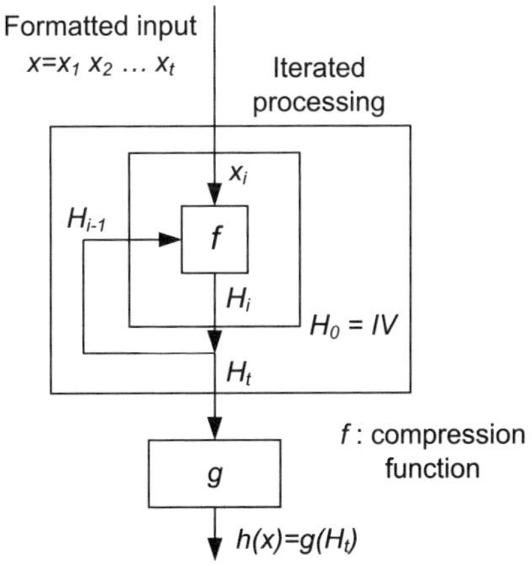

Figure 8.6 A model for illustrating iterative hash function. (*After:* [2].)

which are based on the symmetric-key and public-key-based ciphers, respectively. For practical use of the public-key systems, public-key certificates are needed in order to protect the encrypted system from the man-in-the-middle attack, for which X.509 system is most widely used.

DES—A Symmetric-Key System

Data encryption standard (DES) is the first commercial-grade modern algorithm that openly and fully specifies the implementation details. It is the most popular and extensively used cipher, especially by the banking community. It is a very well-designed cipher and has withstood almost all attacks. Specifically, DES is a symmetric-key block cipher having a 16-rounded (i.e., $r = 16$) Feistel structure shown in Figure 8.3(b). The length of the plaintext and ciphertext blocks are both 64 (i.e., $n = 64$). The input key is 64 bits long (i.e., $k = 64$), which is composed of 56 bits of key and 8 bits of parity.

Even if the DES cipher performs superbly in every aspect, there is one concern that its key length is comparatively short. An exhaustive search can attack on the DES cipher in 2^{55} trials for this 56-bit key, which is comparatively short. So, three DES ciphers are cascaded together to form the *triple DES* (3DES or TDES) cipher. In this case, the overall encryption function takes the form $E(m) = E_{\text{k3}}^{(3)}(E_{\text{k2}}^{(2)}(E_{\text{k1}}^{(1)}(m)))$. It is common to use a two-key triple-encryption type of 3DES in which two of the three keys are put identical, or k1 = k3. In the case of 3DES cipher, the block size, the key size, and the round number take the values $n = 64$, $k = 168$ (for 3 keys) or 112 (for 2 keys), and $r = 16 \times 3$, respectively.

ECB, CBC—Block Cipher Systems

In the case of the block cipher, there are several modes of operations—*electronic codebook* (ECB) mode, *cipher-block chaining* (CBC) mode, *cipher feedback* (CFB) mode, *output feedback* (OFB) mode, and *counter* (CTR) mode. In the case of the ECB mode, encryption and decryption functions are given by $c_j = E_k(x_j)$, $x'_j = D_k(c_j)$, respectively. In the case of the CBC mode, the functions changes to $c_j = E_k(c_{j-1} \oplus x_j)$, $x'_j = c_{j-1} \oplus D_k(c_j)$, respectively. The operations of the two modes are depicted in block diagram in Figure 8.7(a, b), respectively. Note that k-bit keys and n-bit plaintext blocks are used in both modes.

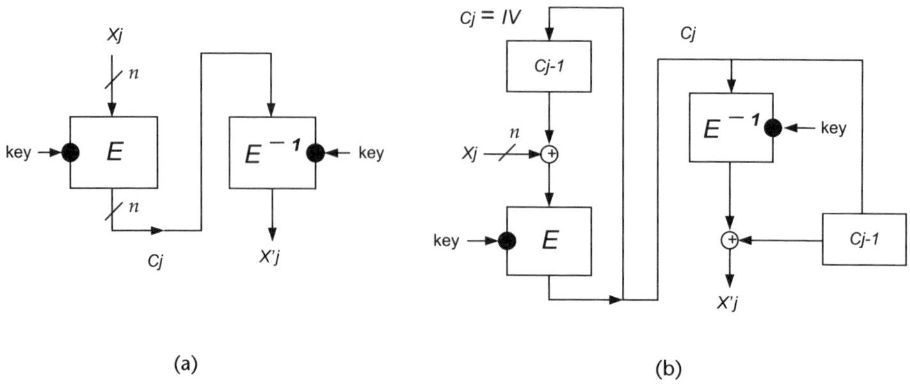

Figure 8.7 Examples of block ciphers: (a) EBC; and (b) CBC [2].

The main difference between the two modes is that the CBC mode has feedback structure in it, which indicates that a ciphertext c_j depends on x_j and all the previous x_j's. Due to this feedback structure, a single bit error in a ciphertext block causes a long string of errors in the CBC case. The *initial vector* (IV) may or may not be kept secret, but, if it is kept secret, the level of security increases. The CFB and OFB modes also contain feedback structures of different forms. In the case of CTR, counter is employed. The counter value is encrypted and then added modulo-2 to the input data.

HMAC, CMAC—Message Authentication Codes

Message authentication code (MAC) is a code intended to provide assurance of the source of data as well as the integrity of data. It can be constructed out of hash functions or block ciphers. In the former case, the *hash-based MAC* (HMAC) is constructed by the relation $\text{HMAC}(x) = h(k\|p_1\|(h(k\|p_2\|x)))$, where x is input data block, $h(x)$ is a hash function, k is a key, p_1 and p_2 are padded bits, and the symbol $\|$ represents the concatenation operation. Note that the hash functions used for constructing HMACs are keyed hash functions, in contrast to the case of MDC where unkeyed hash functions were used. In the latter case, the CBC block cipher is commonly used in building MACs and the resulting *CBC-based MAC* (CMAC) takes the structure of the CBC in Figure 8.7(b), with the encoded bit c_j becoming the MAC value H_j, $j = 2, 3, \ldots, t$.

AES—An Advanced Encryption System

Advanced encryption standard (AES) is an advanced form of ISO standard block ciphers selected as the post-DES standard in 2000. In the case of AES, the block size n is 128; the key size k is 128, 192, or 256; and the round number r is 10, 12, or 14.

AES-CCM is a block cipher that operates in CTR mode, employing CMAC for data authentication and AES for internal encryption. The abbreviation CCM is out of the initials of CTR, CBC, and MAC. In the AES-CCM system, input data of length L (bytes) is first augmented to $L + 8$ bytes by appending the first 8 bytes of the CMAC value of the original L byte data, then added by modulo-2 to the encrypted counter value, and finally prefixed with a 4-byte packet number. AES-CCM is usually used for privacy and data authentication.

RSA—A Public-Key System

The RSA (acronym for the three inventors, *Rivest, Shamir, and Adelman*) problem is used in generating a public key as follows: Given a number $n = pq$ for two strong prime numbers p and q of nearly the same size, let $e > 0$ be a number in $1 < e < \varphi$ for $\varphi = \varphi(n) = (p - 1)(q - 1)$, such that $\gcd(e, \varphi) = 1$. Then we compute a number d in $1 < d < \varphi$ such that $ed \bmod \varphi = 1$. In this setting, we take (n, e) as the public key and d as the private key. In this RSA public-key cipher, the encryption process computes $c = m^e \bmod n$ for $m \in [0, n - 1]$ and the decryption process computes $m' = c^d \bmod n = m^{ed} \bmod n$. Once the trapdoor information d is known, decryption is straightforward, and, therefore, the RSA cipher is a trapdoor one-way function.

In practical RSA public-key cipher systems, the recommended size of modulus is 1,024 bits for commercial use and 2,048 bits for government use. In commonly used public encryption systems, the exponent e is used as a public parameter, not as a key. The only public key is n. For the encryption exponent e, a small number is used for efficiency.

8.1.5 Additional Security Components

The basic functions of information security are to support confidentiality/privacy and data integrity of the information. The cryptographic theory and the derived cryptographic systems that have been discussed so far are mainly targeted at conducting those basic functions. Additional components are needed for information security, such as public-key certificate, digital signature, and entity authentication.

X.509 Public-Key Certificate

In order to protect the public-key cipher system from the man-in-the-middle attack, a *public-key certificate* is needed. It certifies that the public key belongs to the legitimate owner. As the public-key certificate, the *X.509 certificate* is most commonly used, which is the RFC 3280 standard. This standard certificate identifies the communicating parties. It defines the X.509 certificate profile requiring the following fields: *X.509 certificate format* version 3; *certificate serial number*; *certificate issuer's signature algorithm*; *certificate subject* (i.e., the certificate holder's identity, which, if the subject is the MS, includes the MS's MAC address); *subject's public key*, which provides the certificate holder's public key, identifies how the public key is used, and is restricted to RSA encryption; *certificate validity period*; *certificate issuer*; *signature algorithm*, which is identical to the certificate issuer's signature algorithm; and *issuer's digital signature*.

Digital Signature

Digital signature is a means to bind information to an entity (i.e., to confirm that an electronic document is issued by the true issuer). There are several requirements on the digital signature: It should be designed such that everyone can easily recognize that the issuer signed the document by intention; a third party cannot counterfeit it; the contents of the signed document cannot be modified; the signature cannot be copied on another document; and the issuer cannot dissent about the signature. In effect, the digital signature scheme becomes valid under the condition that every issuer does not lend its signature to a third party.

In general, the digital signature scheme is implemented on a public-key cipher system, and an arbitrator is needed in case a symmetric-key cipher system is to be used. Specifically, a digital signature scheme may be implemented on a reversible *public-key cipher* (PKC) as follows: Note that a PKC is called *reversible* if $\mathbf{M} = \mathbf{C}$ and $E_{k_E}(D_{k_D}(m)) = D_{k_D}(E_{k_E}(m)) = m$ for all m in the message space $\mathbf{M}$, as is the case with the RSA PKC. We set $S_{k_S} = D_{k_D}$, $s = S_{k_S}(h(m)) = D_{k_D}(h(m))$ for a hash function h. Then (m, s) is the signed message. The verification $V_{k_V}(m, s)$ turns out true if $E_{k_E}(s) = h(m)$, and false otherwise. It is also possible to implement digital signature scheme with message recovery as follows: Let $\mathbf{M}'$ be an extremely small portion of $\mathbf{M}$, and each element m' in $\mathbf{M}'$ has a well-defined special structure. Then $s = D_{k_D} m'$ is the signed message. The verification $V_{k_V}(s)$ turns out true if $E_{k_E}(s) = m'$ in $\mathbf{M}'$, and false otherwise. Consequently, if the signature s is transmitted, then the receiver can recover the message m' by $m' = E_{k_E}(s)$.

Entity Authentication

Entity authentication, or *identification*, is to corroborate the identity of an entity (i.e., a person, a processor, a device, or others). An identification technique assures that two parties (i.e., an entity A and a verifier B) were both involved and that the second party (i.e., the entity) was active at the time the evidence was created or acquired. The identification protocols are designed to meet the following properties: A is able to successfully authenticate itself to B; the probability that a third party C (playing the role of A) causes B to complete verification and accept it as A's identity is negligibly low; and B cannot reuse an identification exchange information with A so as to successfully impersonate A to a third party C. Those three points should remain true even if a large number of previous authentications exist; even if C has participated in previous protocol executions; and even if multiple instances of the protocol by C were executed.

Entity authentication contrasts to message authentication (or data origin authentication), which corroborates the source of the information, in the following aspects: whereas entity authentication does real-time execution, message authentication is not time critical. Whereas entity authentication deals with fixed-length, nonmeaningful messages, message authentication deals with variable-length, meaningful messages.

8.1.6 Mobile WiMAX Security Overview

The Mobile WiMAX security system is designed to provide secured communications of the data traffic by encrypting the data traffic using the *traffic encryption key* (TEK) and, therefore, puts the major thrust on generating and distributing the TEK between BS and MS in a secured manner. The overall operation for the generation of the TEK and the flow of messages for the distribution of the TEKs may be summarized as illustrated in Figure 8.8. Note that in the figure the asterisk (*) denotes the encryption process and its inverse denotes the decryption process.

To begin with, the MS sends an authentication information to the BS so that the BS can authenticate the MS. Then the BS performs authentication (i.e., entity identification) on the MS using the received authentication information. For the authentication, RSA-based or EAP-based methods are used (see Section 8.2.3). Soon after sending the authentication information, the MS sends an authorization request message to the BS to request an *authorization key* (AK). On receiving the request, the BS generates an AK and sends it to the MS.

Once a common AK is shared between the BS and MS, each station derives, independently, two additional keys—*key encryption key* (KEK) and HMAC key—out of the AK. KEK is used in encrypting TEK, and an HMAC key is used for protecting the TEK request and reply messages.

Now the MS transmits a TEK request message to the BS to request a TEK. The HMAC value calculated using the HMAC key is appended to the TEK request message before transmission to protect the data integrity of the message. Receiving the message, the BS verifies the HMAC value using the HMAC key. Next, the BS gener-

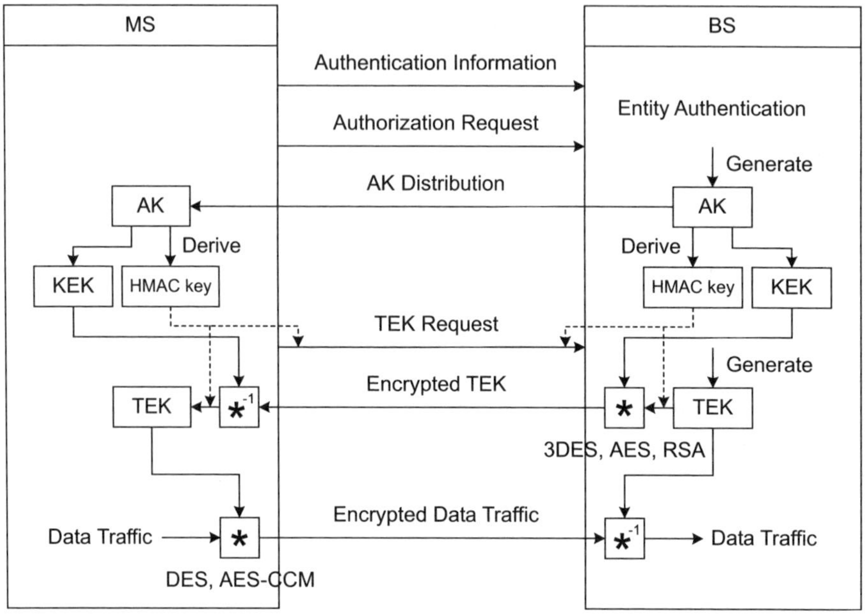

Figure 8.8 Operation of Mobile WiMAX security system.

ates a TEK, encrypts it using a symmetric-key cipher (such as 3DES, AES), and then distributes it to the MS. On receiving the encrypted TEK, the MS decrypts it using the same symmetric-key cipher and keeps the decrypted TEK for use in data traffic encryption.

From then on, the MS uses the TEK as the key for encrypting data traffic using a symmetric-key cipher (such as DES, 3DES, AES-CCM).

8.2 Security System Architecture

The Mobile WiMAX security system has two component protocols as follows: The first is *encapsulation protocol* for securing packet data across the fixed or Mobile WiMAX network. This protocol defines a set of supported cryptographic suites (i.e., a pair of data encryption and authentication algorithms) and the rules for applying those algorithms to a MAC PDU payload. The second is the *key management protocol*, which provides a secure distribution of keying data from BS to MS. The key management protocol enables the MS to synchronize the keying data with the BS and, in addition, enables the BS to enforce conditional access to network services.

Figure 8.9 shows the protocol stack for the security sublayer. Authentication is either RSA-based or *extensible authentication protocol* (EAP)–based [3]. The dashed block on the top indicates that the EAP-based authentication is done by the EAP method on the AAA server and the MS, and the relevant authentication messages are encapsulated in the MAC security sublayer. For each type of authentication method, an authorization process is needed, which generates an authorization key to authorize the MS. After authorization, there follows the *privacy key management* (PKM) control process, which eventually generates the *traffic encryption key* (TEK) and conveys it to the MS. Once the TEK is shared by the BS and the MS suc-

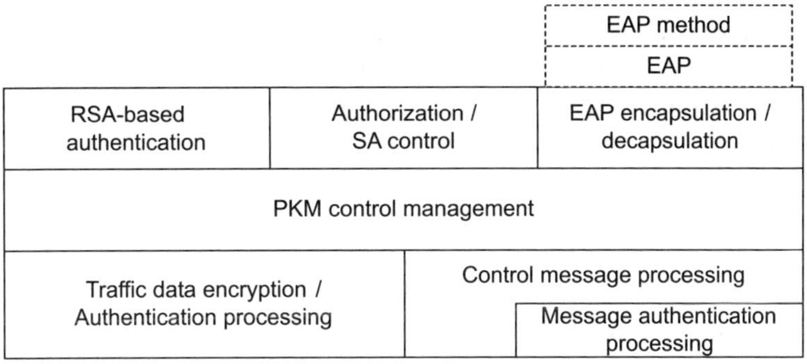

Figure 8.9 Protocol stack for the security sublayer. (*After:* [1].)

cessfully, the traffic data encryption and authentication process follow in the data plane. On the other hand, control message processing and message authentication process take place in the control plane.

8.2.1 Security Association

Security association (SA) refers to a set of security information that a BS shares with its client MSs to support secure communications. SA is identified by *security association identifiers* (SAIDs). There are three types of SAs—primary, static, and dynamic. *Primary SAs* are established during the MS initialization process. *Static SAs* are provisioned within the BS. *Dynamic SAs* are established and eliminated dynamically in response to the initiation and termination of specific service flows. Each MS establishes an exclusive primary SA with its BS. Multiple MSs may share both static and dynamic SAs. For each MS, the SAID of its primary SA is equal to the basic CID of that MS.

The security information that constitutes an SA includes the employed cryptographic suite such as TEKs, the *initialization vectors* of the relevant symmetric-key ciphers, and the TEK lifetime. The content of the SA varies depending on the cryptographic suites.

Each MS can access only to the SAs that it is authorized to access. The MS requests to its BS the keying material (e.g., DES key and CBC initialization vector) of the authorized SA. Then the BS provides to the MS the keying material and the remaining lifetime of the material, as the keying material has a limited lifetime. The MS has to request new keying material to the BS before the current keying material expires. If the MS fails to do so, it has to restart the network entry process.

8.2.2 Encapsulation

Each MAC PDU is transmitted over a connection, which is mapped to an SA. When transmitting each MAC PDU, the sender performs encryption and data authentication on the MAC PDU payload as specified by that SA. When receiving the MAC PDU, the receiver performs decryption and data authentication of the MAC PDU payload, as specified by that particular SA.

MAC PDU payload part is encrypted as required by the selected ciphersuite but the *generic MAC header* (GMH) part is not encrypted. All MAC management messages in the GMH are sent in the clear without encryption so as to facilitate registration, ranging, and normal operation of the MAC. If a MAC PDU received on a connection mapped to an SA that requires encryption turns out not encrypted, it is discarded.

The MAC PDU has the format shown in Figure 5.5, and the GMH contains the encryption-related information as shown in Figure 5.6. As indicated in the figure, the contained encryption information is *encryption control* (EC), *encryption key sequence* (EKS), and *connection identifier* (CID). These are needed to decrypt the encrypted payload at the receiver. The EC field indicates whether or not the payload carried by the GMH is encrypted: EC = 1 indicates that the payload is encrypted, and EC = 0 not encrypted. The EKS field becomes meaningful when EC = 1, in which case it carries the index of the TEK. The 2 bits in the EKS field carry the key sequence number.

Since the keying material associated with an SA has a limited lifetime, the BS refreshes the SA's keying material periodically. The BS manages the 2-bit key sequence number independently for each SA and distributes this key sequence number along with the SA's keying material to the client MS. The sequence number is to identify the specific generation of the keying material that is used to encrypt the attached MAC PDU payload. Comparing the key sequence number of a received MAC PDU with the "current" key sequence number, the MS or the BS can easily recognize whether or not the key is synchronized. The BS increments the key sequence number whenever it generates new keying materials, and the number wraps around to 0 when it reaches 3.

In order to maintain uninterrupted service during key transition in each SA, it is necessary to keep on hand the two most recent key generations. Thus, the MS maintains the two most recent generations of keying material for each SA.

8.2.3 Authentication

The PKM protocol allows for both unilateral authentication (i.e., BS authenticates MS, but not vice versa) and mutual authentication (i.e., both BS and MS authenticate each other). The former is specified under *PKM version 1* (PKMv1) and the latter under *PKM version 2* (PKMv2). The PKM also supports periodic reauthentication/reauthorization and key refresh. The PKM protocol supports two distinct authentication protocol mechanisms—the RSA-based and the EAP-based protocols.

An authentication credential is used as a means for identifying each MS. The MS presents its credential to the BS when requesting authorization, and then the BS authenticates the MS using the credential during the initial authorization exchange. The credential is a unique X.509 digital certificate issued by the manufacturer in the case of RSA authentication and an operator-specified credential in the case of EAP-based authentication.

RSA-Based Authentication

The RSA authentication protocol uses X.509 digital certificates together with the RSA public-key encryption algorithm. The X.509 certificate contains the RSA encryption public key and the MAC address of the MS.

BS authenticates a client MS during the initial authorization exchange. When an MS requests an AK to the BS, the MS presents its digital certificate together with its cryptographic capability and its CID. The BS verifies the digital certificate, then uses the verified public key to encrypt an AK, and finally sends the AK back to the requesting MS.

All the MSs that use RSA authentication must have manufacturer-issued RSA private/public key pairs in addition to manufacturer-issued X.509 digital certificates, or retain an internal algorithm to generate such key pairs dynamically. In the latter case of using internal algorithm, the MS must generate the key pair prior to its first AK exchange and support a mechanism for installing a manufacturer-issued X.509 certificate following key generation.

EAP-Based Authentication

Authentication uses EAP in conjunction with an operator-selected EAP method, which uses a particular type of credential—such as the X.509 digital certificate in the case of EAP-TLS (*transport layer security*) or a *subscriber identity module* (SIM) in the case of EAP-SIM. The particular credentials and EAP methods that are to be used must fulfill the "mandatory criteria" listed in RFC 4017 [4]. At initial authorization, the EAP transfer messages are protected with *EAP integrity key* (EIK), and during reauthorization they are protected with an HMAC/CMAC-tuple. So, any unprotected EAP transfer messages or any EAP transfer messages with invalid EIK or HMAC/CMAC digests are discarded by the BS and MS. When the BS receives valid EAP transfer messages, it sends them to the AAA server via the DIAMETER protocol. Figure 8.10 shows the relevant protocol stack.

8.2.4 Key Management

The key management protocol is to securely distribute the authorization and encryption keying materials to the MS. As discussed earlier, two PKM protocols are

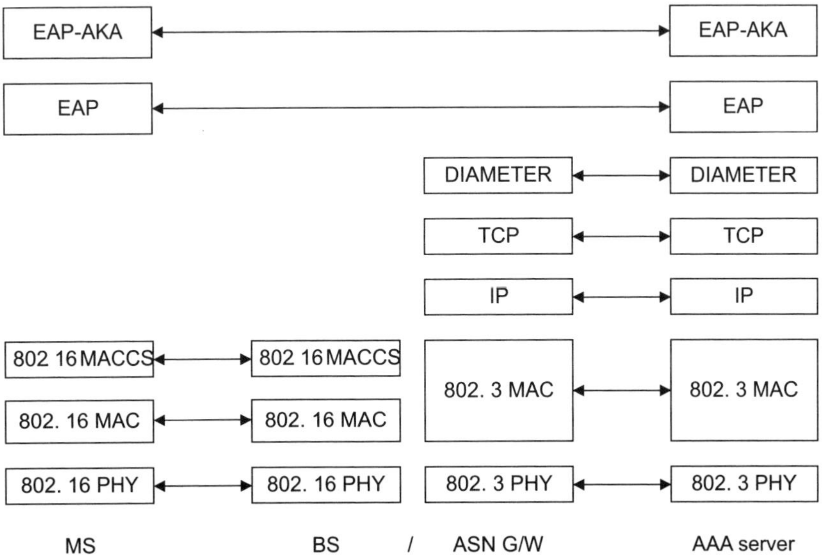

Figure 8.10 Protocol stack for EAP-based authentication.

defined in the IEEE 802.16e standards—PKMv1 and PKMv2, respectively, for unilateral authentication and mutual authentication: PKMv1 is to authenticate the MSs and PKMv2 is to authenticate both MSs and BS, mutually.

The PKM's authentication protocol establishes a shared secret key between the MS and BS, called the *authorization key* (AK), which is used to ensure the secure PKM exchanges of TEKs. To be specific, AK exchange enables the BS to determine the authenticated identity of a client MS and the specific TEKs that the MS is authorized to access. The procedure of deriving AK differs between the RSA-based and the EAP-based authentication.

In the case of the RSA-based authentication, the AK may be generated directly by the BS and delivered to the MS (in the case of PKMv1), or a *preprivacy authorization key* (pre-PAK) may be generated and delivered to the MS, and the BS and MS can both derive the identical AK from the pre-PAK (in the case of PKMv2).

In the case of the EAP-based authentication (which belongs to PKMv2), key exchange begins with the *master session key* (MSK), which was obtained at both the AAA server and the MS through the authentication process. The AAA server sends the MSK to the authenticator in the midway to the BS. Then both the authenticator and the MS derive, independently, a *pairwise master key* (PMK) out of the MSK, and then they derive the AK from the PMK. Finally, the authenticator sends the AK to the BS.

Once AK is secured at both MS and BS, both stations derive, independently, two keys, KEK and HMAC key (HMAC/CMAC in the case of PKMv2), out of the AK. BS uses KEK later in encrypting TEK, and MS uses KEK in decrypting the encrypted TEK. HMAC/CMAC is used for authenticating the TEK request and reply messages.

The exchange of the AK (or pre-PAK) messages adheres to a client/server model, where the MS (i.e., a PKM "client") requests keying material, and the BS (i.e., a PKM "server") responds to those requests, ensuring that each individual client MS receives only the keying material for which it is authorized. In this procedure, the MAC management messages such as PKM-REQ and PKM-RSP are used (see the type numbers 9 and 10 in Table 5.3).

Figure 8.11 shows the structure of AK generation for the PKMv2 case. It illustrates that the RSA-based and the EAP-based methods take different procedures until generating AK but take a common path beyond that.

8.3 Key Management

The two privacy key management protocols, PKMv1 and PKMv2, differ in various aspects. As authentication methods, PKMv1 supports RSA-based authentication, whereas PKMv2 supports both RSA-based and EAP-based authentications. For key generation and distribution, in the case of PKMv1, BS generates AK, encrypts it using the public key of the MS, and distributes the encrypted AK to MSs; in PKMv2, it differs depending on the RSA case and the EAP case. In the former case, the BS generates and encrypts the pre-PAK using the public key of the MS, and then both the BS and the MS derive AK independently. In the latter case, the AAA server exchanges MSK with the MS and authenticator, then the MS and authenticator first

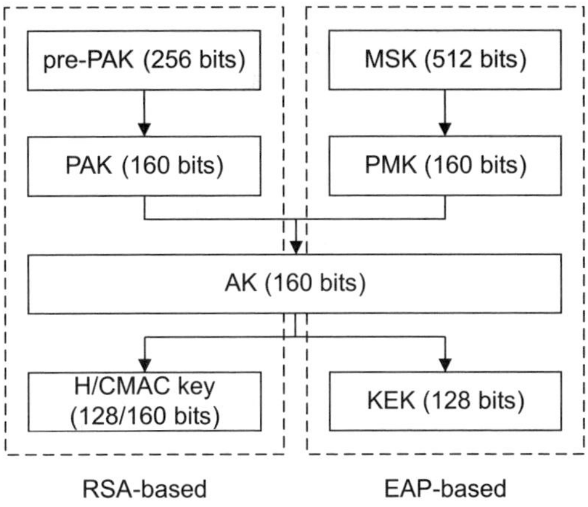

Figure 8.11 Structure of authorization key generation.

derive PMK and then derive AK, and finally the authenticator sends the AK to the BS. For TEK encryption, PKMv1 uses 3DES, AES, and RSA; PKMv2 uses 3DES, AES, RSA, and AES-ECB/KEY-WRAP schemes. For data encryption PKMv1 uses DES, AES-CCM, or no encryption; PKMv2 uses DES, 3DES, AES-CCM/CBC/CTR, or no encryption. For data integrity, PKMv1 uses HMAC or no MAC; PKMv2 uses HMAC/CMAC or no MAC. Table 8.1 lists a summary of the comparison between PKMv1 and PKMv2.[3]

8.3.1 PKMv1

The initial authentication and key management process, under PKMv1, takes place in the following procedure (see Figure 8.12):

1. The MS sends an authentication information message to the BS. The authentication information message contains the X.509 certificate of the MS manufacturer issued by the manufacturer itself.
2. The MS sends an authorization request message to the BS to request for an AK and the SAIDs. The SAIDs identify the static SAs that the MS is authorized to participate in. The RSA request message includes the manufacturer-issued X.509 certificate, a description of the cryptographic algorithms that the requesting MS supports, the basic CID of the MS. The basic CID refers to the first static CID that was assigned by the BS during the initial ranging stage.
3. The BS takes the following actions in reply to the MS's authorization request: it validates the identity of the requesting MS, determines the encryption algorithm and protocol support, activates an AK for the MS, encrypts it with the public key of the MS, and then sends it back to the MS in an authorization reply message. The authorization reply message includes the following information: the AK encrypted with the public key of the MS, a 4-bit key sequence number to use to distinguish between successive

3. As to the abbreviated terminology in Table 8.1, refer to Section 8.1 and Chapter 7 of [1].

Table 8.1 Comparison of PKMv1 and PKMv2

	PKMv1	PKMv2
Authentication direction	Unilateral	Bilateral
Authentication method	RSA based	RSA based EAP based
Authentication object	MS	MS, BS
Keys involved in authorization	AK	RSA based: Pre-PAK, AK EAP based: MSK, PMK, AK
TEK encryption	3DES, AES, RSA	3DES, AES, REA, AES-ECB/KEY-WRAP
Data encryption	No encryption, DES, AES-CCM	No encryption, DES, 3DES, AES-CCM/CBC/CTR
Data integrity	No MAC, HMAC	No MAC, HMAC/CMAC

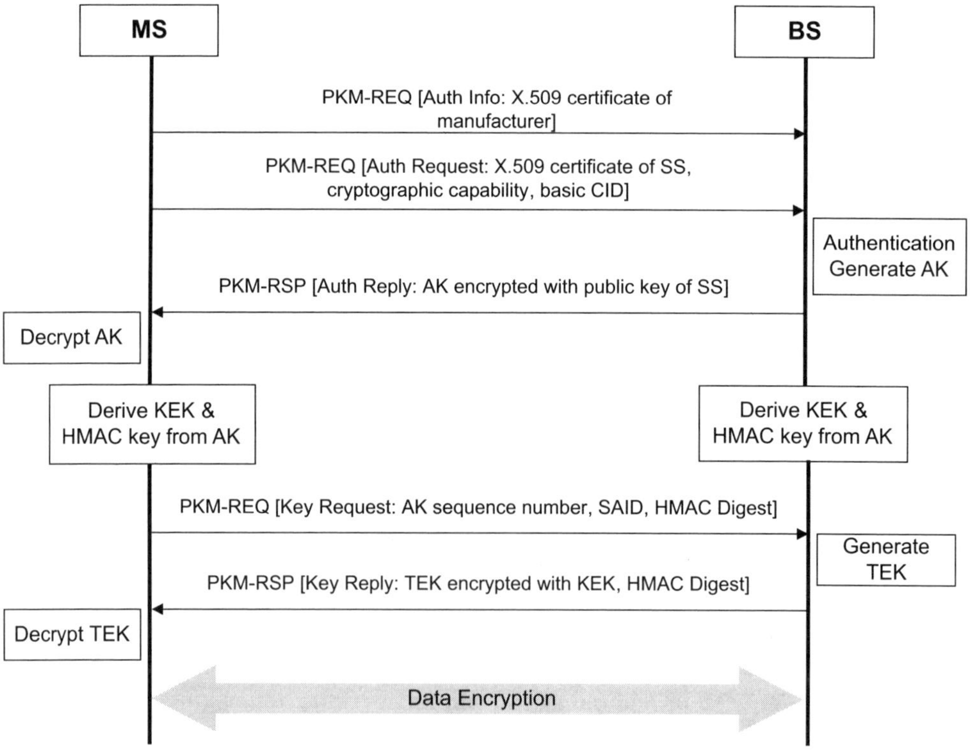

Figure 8.12 PKMv1 authentication and key exchange.

generations of AKs, the key lifetime, the identities (i.e., SAIDs), and the properties of the authorized SAs.

4. The BS and MS derive, independently, a *key encryption key* (KEK) and message authentication key (e.g., HMAC).
5. The MS sends a key request message to request the BS to generate a TEK and send it back.
6. The BS generates the TEK, encrypts it using a KEK, and sends the encrypted TEK to the MS in a key reply response message.
7. The MS decrypts the TEK and uses it in encrypting/decrypting the forthcoming data traffic.

While in service, the MS periodically refreshes the AK by reissuing an authorization request to the BS. The procedure for reauthorization is identical to that for initial authorization except that MS does not send authentication information messages. To avoid service interruptions during reauthorization, successive AKs are generated with overlapping lifetimes. Both the MS and BS support up to two simultaneously active AKs during the transition periods.

Though the PKMv1 may perform right in normal situation, it has some weak points that can cause disrupted communications or information leakage if attacked by malicious users.

First, since PKMv1 uses unilateral authentication that BS authenticates MS but not vice versa, it is subject to the attack of malicious BSs who masquerade as a normal BS. In this case, if a normal MS requests authorization, then the attacker can intercept it and send its own generated fake authorization key to the MS, thereby having the MS access the network through the attacker. As a consequence, the attacker can eavesdrop, counterfeit, and fabricate all the traffic of the MS from then on.

Second, since the main contents of the authentication request and the authentication reply messages in PKMv1 remain the same at every authorization request and reply, it is subject to replay attack (i.e., an attacker can intercept the messages and use them to replay the same messages).

Third, in PKMv1, since BS transmits the AK directly to MS, and KEK and message authentication keys are generated out of the AK, once the AK is exposed to an attacker, it can damage the whole encrypted communications.

8.3.2 PKMv2

In correction to the weak points of PKMv1, PKMv2 strengthened its security capability in three different ways. First, it adopted bilateral authentication. Specifically, the BS includes its own BS certificate in the authorization reply message. Second, it added a 64-bit random number in the authorization request message and an additional 64-bit random number in authorization reply message. Third, it avoided sending the AK directly and arranged to send pre-PAK or MSK, instead, from which the AK can be derived at the BS and MS independently.

The initial authentication and authorization process, under PKMv2, takes place in different ways for the RSA-based and the EAP-based methods, as illustrated in Figure 8.10. So we consider the two cases separately next.

RSA-Based PKMv2

The initial authentication and key management process is done in the following procedure (see Figure 8.13):

1. The MS sends an authentication information message to the BS. The authentication information message contains the X.509 certificate of the MS manufacturer.
2. The MS sends an RSA request message to the BS to request for a pre-PAK and the SAIDs. The SAIDs identify the static SAs that the MS is authorized to participate in. The RSA request message includes the manufacturer-issued X.509 certificate, a description of the cryptographic algorithms that the requesting MS supports, the basic CID of the MS, and a 64-bit random number.
3. The BS takes the following actions in reply to the MS's RSA request: it validates the identity of the requesting MS, determines the encryption algorithm and protocol support, activates a pre-PAK for the MS, encrypts it

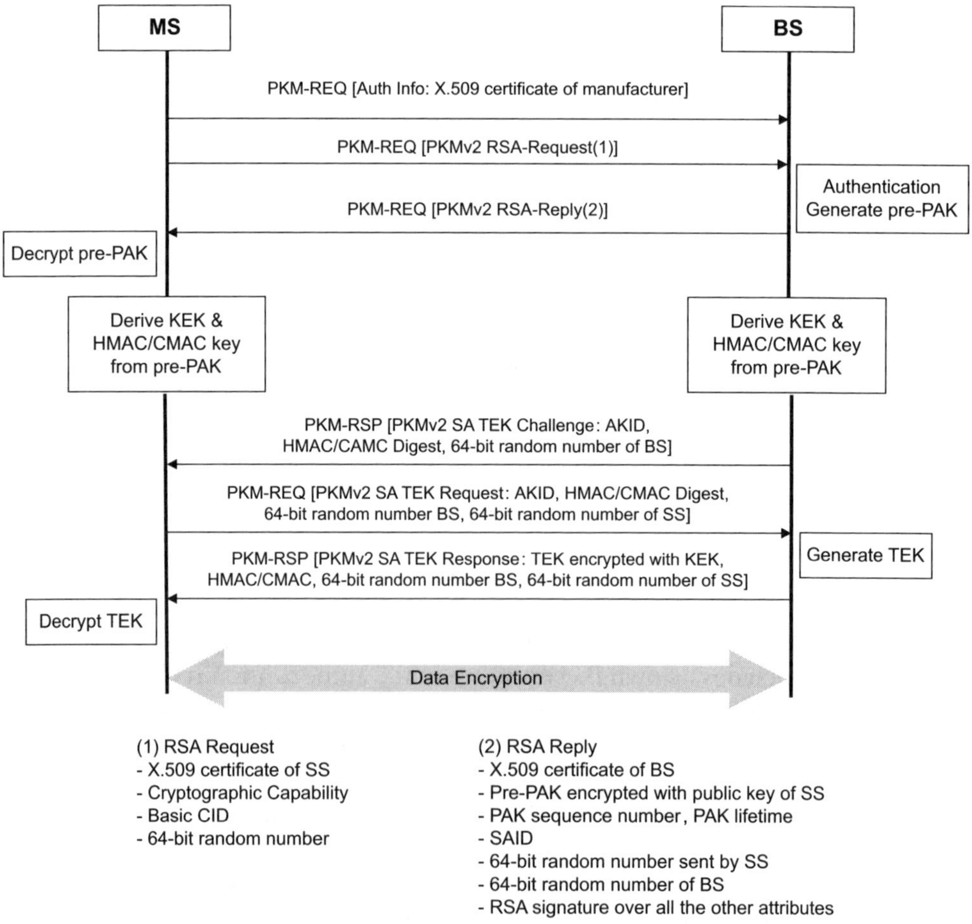

Figure 8.13 PKMv2 RSA-based authentication and key exchange.

with the public key of the MS, and then sends it back to the MS in an RSA reply message. The RSA reply message includes the following information: the BS certificate, the pre-PAK encrypted with the public key of the MS, a 4-bit PAK sequence number, PAK lifetime, SAIDs, the 64-bit random number that the MS sent, the 64-bit random number of the BS, and the RSA signature over all other attributes in the message.

4. The BS and MS derive, independently, the PAK from the received pre-PAK, the AK from the derived PAK, and finally the KEK and HMAC/CMAC key from the derived AK.
5. The BS sends a SA-TEK challenge message to the MS to check if the MS possesses a valid AK.
6. The MS sends SA-TEK request message to request the BS to generate a TEK and send it back.
7. The BS generates the TEK, encrypts it using a KEK, and sends the encrypted TEK to the MS in a SA-TEK response message.
8. The MS decrypts the TEK and uses it in encrypting/decrypting the forthcoming data traffic.

While in service, the MS periodically refreshes the pre-PAK by reissuing an RSA request to the BS. The procedure for reauthorization is identical to that for initial authorization except that MS does not send authentication information messages. To avoid service interruptions during reauthorization, successive pre-PAKs are generated with overlapping lifetimes. Both the MS and BS support up to two simultaneously active pre-PAKs during the transition periods.

EAP-Based PKMv2

In dealing with the EAP-based PKMv2, we consider the case of the *double EAP mode* (or "authenticated EAP after EAP" mode), in which the authenticated EAP messages carry the second EAP messages. It cryptographically binds the previous EAP authentication and the following EAP authentication sessions while protecting the second EAP messages, thereby enhancing the protection capability.

The initial authentication and key management process in this double EAP mode is done in the following procedure (see Figure 8.14):

1. In order to initiate the *first-round EAP* of the double EAP, the MS sends a (PKMv2) EAP start message without attributes.
2. The MS and BS perform the first-round *EAP conversation* with the EAP transfer message without the HMAC/CMAC digest. The first-round EAP conversation includes the bilateral authentication between the MS and AAA server and the distribution of the MSK from AAA server to the MS and BS. The BS then generates an *EAP integrity key* (EIK) and a PMK from the MSK (see Figure 8.15).
3. During the first EAP conversation, if the BS has to send an EAP-success message, the BS sends the EAP payload to MS with an EAP complete message encrypted by the EIK. The BS resends the the EAP complete message by second EAP timeout. The total number of resending EAP complete messages is limited to EAP complete resend. After receiving the

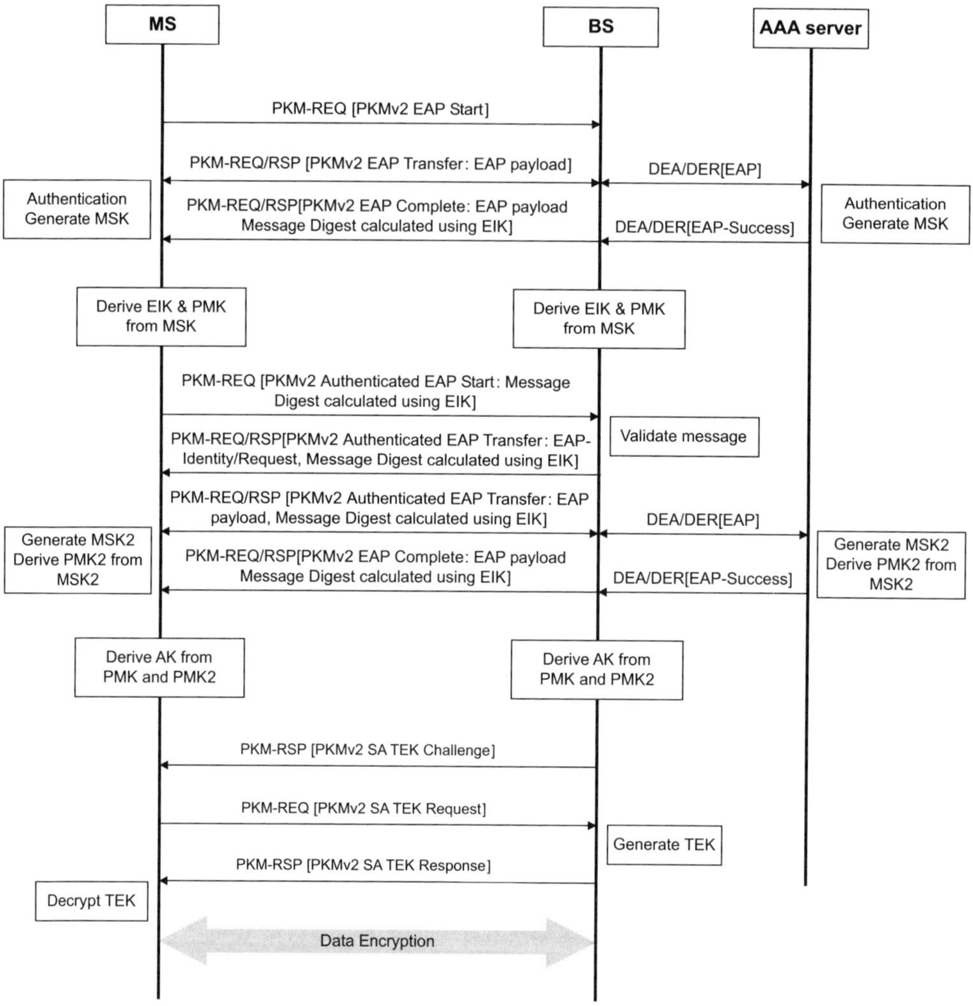

Figure 8.14 PKMv2 EAP-based authentication and key exchange.

EAP complete message, which includes the EAP success payload, the MS can possess the MSK so can derive the EIK and PMK from the MSK. In this case, the MS can validate the EAP complete message using the EIK. If the MS receives EAP failure or cannot validate the message, the MS fails authentication. After the BS transfers the EAP complete message to the MS, the BS activates the second EAP timeout in order to wait for an authenticated EAP start message. When the timer expires, the BS regards the authentication as failure.

4. After the successful first-round EAP, the MS sends an authenticated EAP start message encrypted by EIK to initiate the *second-round EAP* conversation. If BS validates the EAP start message by EIK, the BS initiates the second-round EAP by sending an authenticated EAP message, including

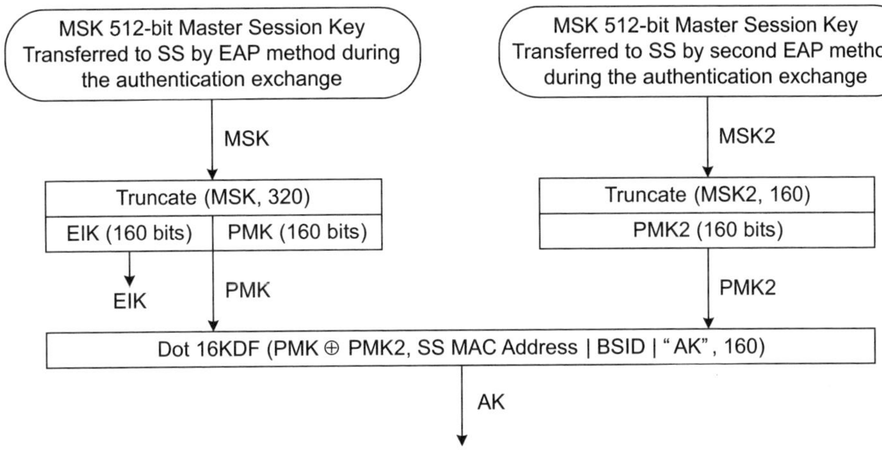

Figure 8.15 AK derivation in double EAP mode. (*After:* [1].)

an EAP-identity/request to MS. If the BS cannot validate the authenticated EAP start message, the BS regards the authentication as failure.

5. The MS and BS perform the second EAP conversation with an authenticated EAP transfer message encrypted by EIK. During the second EAP conversation period, the BS receives MSK2 from the AAA server and generates PMK2 from the MSK2.
6. If the second-round EAP succeeds, both the MS and authenticator generate the AK from PMK and PMK2.
7. Once the AK is generated, the BS and MS perform SA-TEK three-way handshake in the same way as in steps 5 to 7 of the RSA-based PMKv2.

After the successful initial authentication, the MS and BS perform reauthentication by PMK/PMKv2 lifetime. If double EAP was used in the initial authentication, the same is used in the reauthentication. Otherwise, the MS and BS can perform EAP once. The reauthentication procedure is similar to the initial authentication, except that the AK is readily available so it can be used from the beginning: Consequently the initial authentication procedure changes to the following in the reauthentication procedure:

1. In the first round, the EAP start message is encrypted by HMAC/CMAC_KEY_U derived from the AK.
2. The EAP transfer message for the first-round EAP includes HMAC/CMAC digest.
3. The EAP complete message is encrypted by the AK.
4. In the second round, the EAP start message is encrypted by HMAC/CMAC_KEY_U.
5. The EAP transfer message is encrypted by the AK.

TEK Management

As indicated in Figure 8.8, the BS encrypts the TEK before transmitting it to the MS in the key reply message. For SAs using a ciphersuite employing 64-bit DES-CBC, the TEK in the key reply is triple DES (3DES) encrypted; using a two-key, 3DES KEK derived from the AK. For SAs using a ciphersuite employing 128-bit keys, such as AES-CCM mode, the TEK is AES encrypted using a 128-bit key derived from the AK.

At all times BS maintains two diversity sets of TEK per SAID. The lifetimes of the two TEKs overlap such that each TEK generation becomes active halfway through the life of its predecessor TEK and expires halfway through the life of its successor TEK. The BS includes in its key reply messages both of the SAID's active generations of keying material.

For the SAs using a ciphersuite employing DES-CBC mode encryption, the key reply message provides the requesting MS with the TEKs, the CBC IV, and the remaining lifetime of the two sets of TEKs. For the SAs using a ciphersuite employing AES-CCM mode encryption, the key reply message provides the requesting MS with the TEKs and the remaining lifetime of the two sets of TEKs. The MS uses the remaining lifetimes in estimating when the BS will invalidate a particular TEK and, therefore, when to schedule future key request messages such that the MS requests and receives new TEKs before the currently used TEKs expire. For the AES-CCM mode, when more than half the available PN numbers in the 31-bit PN number space are exhausted, the MS schedules a future key request in the same fashion as if the key lifetime were approaching expiration. As such, the MS becomes capable of continually exchanging encrypted traffic with the BS due to the operation of the key request scheduling algorithm of the TEK state machine, combined with the control of the BS for updating and using an SAID's keying materials.

Figure 8.16 illustrates the TEK management in the BS and MS: the BS and MS maintain two TEKs at all time by exchanging key request and key reply messages. The MS sends a key request message to the BS to request TEKs of the SAID soon after the scheduled expiration time of the older of its two TEKs and before the expiration of its newer TEK, called *TEK grace time*. With the receipt of the key reply, the BS carrying two active sets of TEKs for the SAID sends back a key reply message, which includes the TEKs of the SAID, encrypted with a KEK derived from the AK. The MS always updates its records with the TEK parameters from both TEKs contained in the key reply message.[4]

8.3.3 State Machines for Key Exchange

The request, generation, and distribution of the encryption keys is a complicated process that is involved with many states, messages, events, parameters, and actions. State machines, or state transition diagrams, render a simplified and systematic means of describing the process. Many different state machines are needed to fully describe the generation and exchange of the encryption keys, such as AK and TEK, from the MS and BS aspects. Among them we consider the AK and the TEK state machines on the MS side, which describe the operation of the MS with respect to the request/wait of the relevant keys. In addition, among the two versions of key man-

4. According to [1] (Table 343), TEK lifetime is 12 hours by default and its minimum and maximum values are 30 minutes and 7 days, respectively. TEK grace period is 1 hour by default with its minimum and maximum values being 5 minutes and 3.5 days, respectively. For comparison, AK lifetime is 7 days by default and its minimum and maximum values are 1 day and 70 days, respectively.

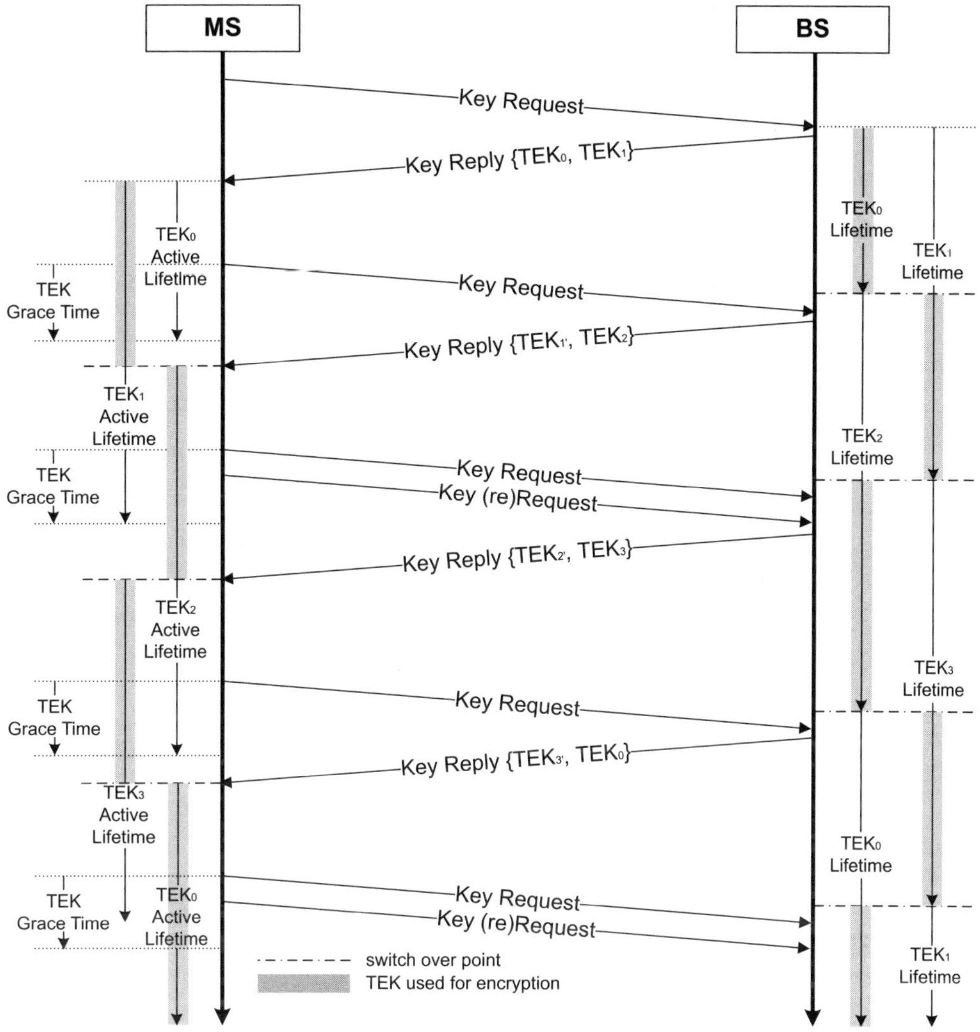

Figure 8.16 TEK management in the BS and MS. (*After:* [1].)

agement, PKMv1 and PKMv2, we consider the case of PKMv1, as it gives a simpler view of the overall operation. (Refer to Section 7.2 of [1] for a more complete description of the state machines.)

AK State Machine

Figure 8.17 depicts the authorization state machine flow diagram, which describes the operation of the MS in relation to requesting an AK to BS. The state flow diagram depicts the protocol messages transmitted and the internal events generated for each state transition, but it does not indicate additional internal actions, such as the clearing or starting of timers. The authorization state machine consists of six states and eight distinct events that trigger state transitions. The states are put in the ovals and the events are given in italics. In addition, the messages are given in normal font and state transitions are labeled in the form "cause/event or message."

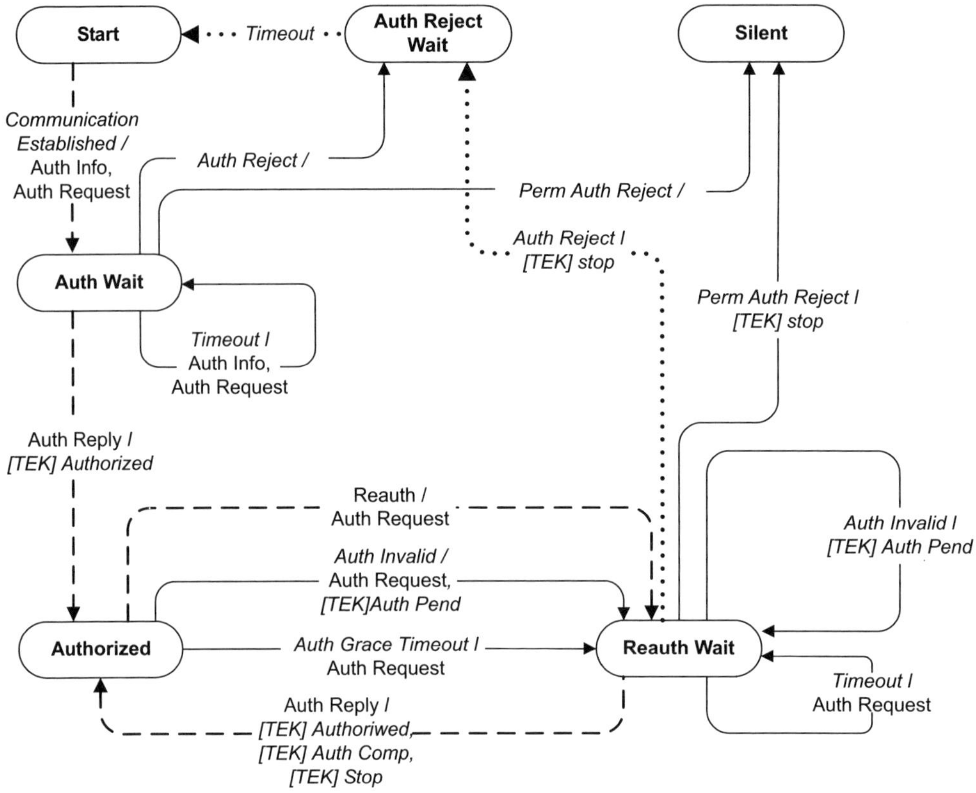

Figure 8.17 Authorization state machine flow diagram. (*After:* [1].)

The six states in the authorization state machine are Start, Auth Wait, Authorized, Reauth Wait, Auth Reject Wait, and Silent states, with "Auth" and "Reauth" representing "authorization" and "reauthorization," respectively.

1. *Start* is the initial state, at which all timers are in the off state and no processing is scheduled.
2. *Auth Wait* is the state that the MS is waiting for authorization after sending an authentication information and an authorization request message to the BS.
3. *Authorized* is the state that the MS has received Auth Reply message, so has valid AK and SAID list. At this state, the MS creates a TEK state machine.
4. *Reauth Wait* is similar to the Auth Request state in that the MS is waiting for authorization after sending an authorization request message but Reauth Wait state is caused by either near expiration or invalidity of the current authorization.
5. *Auth Reject Wait* is the state that the MS received an authorization reject message, so is waiting until the timer expires. It means that the relevant error is not of permanent nature.
6. *Silent* is the state that the MS enters into after receiving an authorization reject message in response to its last authorization request. It means that the relevant error is of a permanent nature, so the MS is not permitted to pass subscriber traffic.

We illustrate the operation of the AK state machine using the dashed and dotted lines in Figure 8.17. The dashed line going from Start state to Authorized state exhibits the path of getting authorization (i.e., the MS is receiving an AK from the BS) without experiencing rejection or timeout. The other dashed line, going from Authorized state to Reauth Wait state and then coming back, indicates the case of getting reauthorization at the first trial. The dotted line, going from Reauth Wait state to Start state, represents the case when the MS experiences reauthorization reject, so waits until the timer expires and then starts authorization request anew.

Table 8.2 is the state transition matrix that is a tabular form representation of the AK state machine flow graph in Figure 8.17. The six states are listed in the top row, the eight "causes" are listed on the leftmost column, and the resulting "events" are listed the cells inside the matrix. Each cell represents a specific combination of state and event, with the next state displayed within the cell. Each blank cell indicates that the relevant event cannot or should not occur within the relevant state, so if the event does occur, the state machine ignores it. The two dashed paths and a dotted path in the table correspond to the three illustrations in Figure 8.17.

TEK State Machine

Upon achieving authorization, the MS starts a separate TEK state machine for each of the SAIDs identified in the authorization reply message (or PKMv2 SA-TEK-REP message), if data traffic encryption is provisioned for one or more service flows. Each TEK state machine operating within the MS is responsible for managing the keying material associated with its respective SAID. The TEK state machine periodically sends a key request message to the BS, requesting refresh of the keying material for its SAID. The TEK state machine remains active as long as the MS is authorized to operate in the security domain of the BS (i.e., it has a valid AK) and the MS is authorized to participate in that particular SA (i.e., the BS continues to provide fresh keying material during the rekey cycles).

Figure 8.18 depicts the TEK state machine flow diagram, which describes the operation of the MS in relation to requesting a TEK to the BS. The TEK state machine consists of six states and nine distinct events that trigger state transitions.

Table 8.2 Authorization State Transition Matrix

Event, or Received message	State					
	(A) Start	(B) Auth Wait	(c) Authorized	(D) Reauth Wait	(E) Auth Rej Wait	(F) Silent
(1) Communication Established	Auth Wait					
(2) Auth Reject		Auth Reject Wait		Auth Reject Wait		
(3) Perm Auth Reject		Silent		Silent		
(4) Auth Reply		Authorized		Authorized		
(5) Timeout		Auth Wait		Reauth Wait		
(6) Auth Grace Timeout			Reauth Wait		Start	
(7) Auth Invalid			Reauth Wait	Reauth Wait		
(8) Reauth			Reauth Wait			

Source: [1].

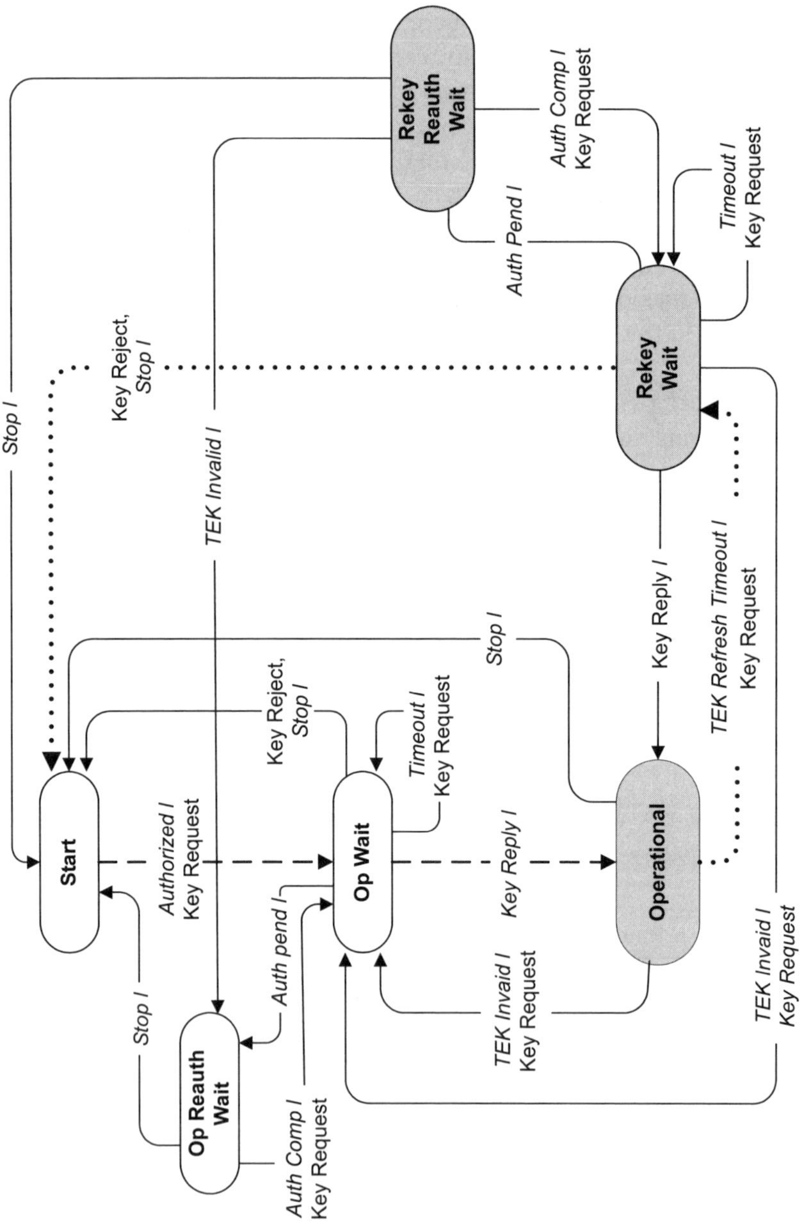

Figure 8.18 TEK state machine flow diagram. (*After:* [1].)

The six states in the TEK state machine are Start, Op Wait, Op Reauth Wait, Operational, Rekey Wait, and Rekey Reauth Wait states, with "Op" representing "Operational."

1. *Start* is the initial state, at which all timers are in off state and no processing is scheduled.
2. *Op Wait* is the state that the MS is waiting for reply from the BS after sending an initial key request for its SAID's keying material (e.g., TEK and CBC IV).
3. *Op Reauth Wait* is the state that the MS does not have a valid TEK while the authorization state machine is in the middle of reauthorization cycle.
4. *Operational* is the state that the MS has valid keying material for the associated SAID.
5. *Rekey Wait* is the state that the TEK refresh timer has expired and the MS has requested a key update for the relevant SAID. Note that the newer TEK is still valid and is used for encrypting and decrypting data traffic.
6. *Rekey Reauth Wait* is the state that the MS has a valid TEK, the MS has an outstanding request for the latest keying material, and the authorization state machine is in reauthorization cycle.

The authorization state machine and the TEK state machine share a parent-child relation. If the parent authorization state machine stops, all its child TEK state machines stop, which happens when the MS receives from the BS an authorization reject message during a reauthorization cycle. Individual TEK state machines may be started or stopped during a reauthorization cycle if the static SAID authorizations of an MS change between successive reauthorizations. The authorization state machine communicates with the TEK state machines by passing the events (e.g., stop, authorized, authorization pending, and authorization complete events)

Table 8.3 TEK State Transition Matrix

Event, or Received message	State					
	(A) Start	(B) Op Wait	(c) Op Reauth Wait	(D) Op	(E) Rekey Wait	(F) Rekey Reauth Wait
(1) Stop		Start	Start	Start	Start	Start
(2) Authorized	Op Wait					
(3) Auth Pend		Op Reauth Wait			Rekey Reauth Wait	
(4) Auth Comp			Op Wait			Rekey Wait
(5) TEK Invalid				Op Wait	Op Wait	Op Reauth Wait
(6) Timeout		Op Wait			Rekey Wait	
(7) TEK Refresh Timeout				Rekey Wait		
(8) Key Reply		Operational			Operational	
(9) Key Reject		Start			Start	

Source: [1].

and protocol messaging. However, the TEK state machines do not have any events targeted at the parent authorization state machine, so they affect it indirectly through the messages that the BS sends in response to the requests of the MS (e.g., authorization invalid message).

We illustrate the operation of the TEK state machine using the dashed and dotted lines in Figure 8.18. The dashed line going from Start state to Operational state exhibits the path of getting into operational mode without experiencing rejection or timeout (i.e., the MS requests the TEK to the BS and receives it from BS at its first trial). The dotted line, going from Operational state to Start state, represents the case when the MS experiences TEK refresh timeout, so requests new TEK but the request is rejected.

Table 8.3 is the state transition matrix for the TEK state machine flow graph in Figure 8.18. The dashed and dotted paths in the table correspond to the two previous illustrations.

References

[1] IEEE Std 802.16e-2005 and IEEE Std 802.16-2004/Cor1-2005, Part 16: Air Interface for Fixed and Mobile Broadband Wireless Access Systems, February 2006.

[2] Menezes, A. J., P. C. van Oorschot, and S. A. Vanstone, *Handbook of Applied Cryptography,* Boca Raton, FL: CRC Press, 1996.

[3] IETF RFC 3748, Extensible Authentication Protocol (EAP), June 2004.

[4] IETF RFC 4017, Extensible Authentication Protocol (EAP) Method Requirements for Wireless LANs, March 2005.

Selected Bibliography

Beker, H., and F. Piper, *Cipher Systems: The Protection of Communications*, New York: John Wiley & Sons, 1982.

Burr, W. E., "Selecting the Advanced Encryption Standard," *IEEE Security and Privacy*, Vol. 1, No. 2, March 2003, pp. 43–52.

Davies, D. W., and W. L. Price, *Security for Computer Networks*, 2nd ed., New York: John Wiley & Sons, 1989.

Diffie, W., and M. E. Hellman, "Privacy and Authentication: An Introduction to Cryptography," *Proc. of the IEEE*, Vol. 67, No. 3, March 1979, pp. 397–427.

FIPS PUB 46-3, Data Encryption Standard (DES), National Institute of Standards and Technology, October 1999, http://csrc.nist.gov/publications/fips/fips46-3/fips46-3.pdf.

FIPS PUB 197, Advanced Encryption Standard (AES), National Institute of Standards and Technology, Nov. 2001, http://csrc.nist.gov/publications/fips/fips197/fips-197.pdf.

IETF RFC 3394, Advanced Encryption Standard (AES) Key Wrap Algorithm, September 2002.

IETF RFC 3447, Public-Key Cryptography Standards (PKCS) #1: RSA Cryptography Specifications Version 2.1, February 2003.

IETF RFC 3610, Counter with CBC-MAC (CCM), September 2003.

Johnston, D., and J. Walker, "Overview of IEEE 802.16 Security," *IEEE Security & Privacy*, Vol. 2, No. 3, May–June 2004, pp. 40–48.

Kartalopoulos, S. V., "A Primer on Cryptography in Communications," *IEEE Communications Magazine*, Vol. 44, No. 4, April 2006, pp. 146–151.

Korea Information Security Agency, *WiBro Security Technology*, August 2006.

Rivest, R., A. Shamir, and L. Adleman, "A Method for Obtaining Digital Signatures and Public-Key Cryptosystems," *Communications of the ACM*, Vol. 21, No. 2, February 1978, pp. 120–126.

Salomaa, A., *Public-Key Cryptography*, New York: Springer-Verlag, 1990.

Schneier, B., *Applied Cryptography: Protocol, Algorithms and Source Code in C*, 2nd ed., New York: John Wiley & Sons, 1996.

Simmons, G. J., *Contemporary Cryptology: The Science of Information Integrity*, New York: IEEE Press, 1992.

Stallings, W., *Cryptography and Network Security: Principles and Practice*, 2nd ed., Englewood Cliffs, NJ: Prentice-Hall, 1999.

Stinson, D. R., *Cryptography: Theory and Practice*, Boca Raton, FL: CRC Press, 1995.

Yang, H., et al., "Securing a Wireless World," *Proc. of the IEEE*, Vol. 94, No. 2, February 2006, pp. 442–454.

CHAPTER 9

Multiple Antenna Technology

One of the most distinctive features of Mobile WiMAX is that it adopts multiple antenna technology. Specifically, the Mobile WiMAX system adopts *multi-input multi-output* (MIMO) and *beamforming* (BF) technologies to improve system throughput. A MIMO system with 2 × 2 transmit-receive antennas, for example, exhibits doubled downlink and uplink peak data rates of a *single-input multi-output* (SIMO) system.

In this chapter, we discuss the theory, design, and implementation issues of multiple antenna technology in relation to Mobile WiMAX. As it is rather unique to the Mobile WiMAX and nonexistent in the conventional circuit-mode cellular wireless systems, we will make the discussion rather comprehensive, by including the related information-theoretic background, open-loop and closed-loop designs of the multiantenna systems, and the algorithms of multiantenna receivers.

9.1 Fundamentals of Multiple Antenna Technology

As discussed in Section 1.1, the mobile wireless channel is subject to channel fading and cochannel interference, which are the major performance degradation factors in mobile wireless communications. The channel fading problem has been handled in various ways by the conventional mobile communication systems (e.g., a user scheduling technique that exploits multiuser diversity was used in packet-mode wireless systems [1], and a channel interleaving technique that exploits time diversity was used in circuit-mode wireless systems [2]). The cochannel interference problem has been handled by adopting power control [3] and interference cancellation techniques, such as *single antenna interference cancellation* (SAIC) and *successive interference cancellation* (SIC) [4].

The channel fading and cochannel interference problems can be better handled by employing a multiple antenna technology. If multiple antennas are used in mobile wireless systems, it is possible to achieve more reliable and/or higher data rate transmissions. The reliability originates from the spatial diversity effect and the high data rate originates from the spatial multiplexing effect of the multiple antenna system. The effects of spatial diversity and spatial multiplexing can be achieved by arranging the multiantenna system in two different operation modes, namely, open-loop and closed-loop operations. Closed loop is the case when the *channel state information* (CSI) is available at the transmitter through feedback, whereas open loop is the case the CSI is not available.

9.1.1 Multiple Antenna Techniques

The benefit and objective of multiple antenna technology can be summarized into three categories—space diversity, spatial multiplexing, and beamforming with interference nulling. Figure 9.1 illustrates the three concepts. The concepts may be realized individually, or the three concepts may be implemented simultaneously.

Space Diversity

Space diversity is intended to combine multiple signals that were transmitted from the same source but have passed through statistically independent channels. It can be implemented by either sending the same signal through an array of transmit antennas or combining multiple signals obtained through an array of receive antennas. In the former case, the transmitter may be operated in both open-loop and closed-loop manners. In space diversity systems, the SNR required for the given link error probability renders a main performance metric. In addition, diversity gain, which is the ratio of the link error probability change with respect to the SNR change in space diversity system to that in single antenna system, renders another useful performance metric. Both performance metrics, in general, depend on the channel correlation among antennas as well as the transmit/receive operation schemes.

Spatial Multiplexing

Spatial multiplexing is intended to transmit multiple independent signals over the same frequency at the same time by employing multiple transmit and multiple receive antennas. If N_t transmit antennas and N_r receive antennas are equipped in the spatial multiplexing system, then the maximum number of independent signals that can be transmitted reliably is min $\{N_t, N_r\}$. Spatial multiplexing systems may be arranged in different forms: In downlink, the N_r receive antennas may belong to the same user or may be used by N_r different users. Likewise, in uplink, the N_t transmit antennas may belong to the same user or may be used by N_t different users. CSI, if available, can help to choose the best subset of transmit antennas in case all N_t transmit antennas are used by a single user and can help to choose the best subset of users in case the transmit antennas are used by N_t different users.

Beamforming with Interference Nulling

Beamforming is intended to maximize the power of the desired signals while minimizing (or nulling) the power of the interfering signals by using multiple transmit/receive antennas. Different forms of beams can be shaped by controlling the relative magnitudes and phases of the signals transmitted/received by the antenna

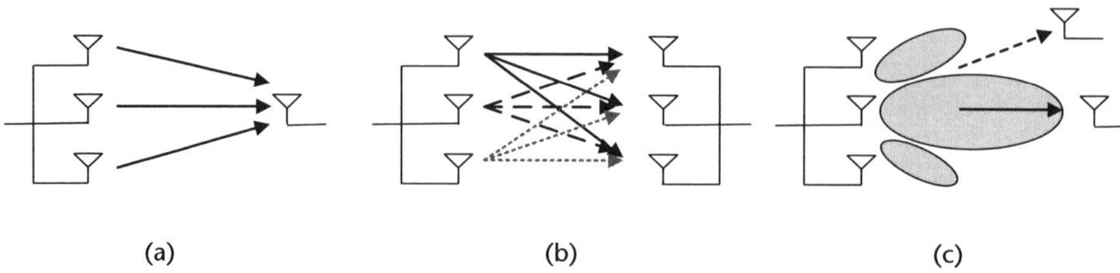

Figure 9.1 Illustration of (a) space diversity, (b) spatial multiplexing, and (c) beamforming.

elements. So, in the transmit direction, CSI feedback is necessary to aid transmit beamforming, but, in the receive direction, the estimated channel information retrieved from the user pilots in the received signals may be used.

9.1.2 Capacity of MIMO Channels

It is of fundamental importance to examine the capacity of MIMO channels, as the objective of adopting multiantenna technology is essentially increasing the capacity. In the following, we introduce the information-theoretical results of MIMO channel capacity with respect to the narrowband static and fading MIMO channels.

Narrowband Time-Invariant MIMO Channels

Channel capacity refers to the maximum data rate at which one can communicate through the channel while making the error probability arbitrarily small. It can be closely achieved by using the recently developed high-performance error correction codes, such as turbo code or *low-density parity check* (LDPC) code. Channel capacity provides us with a measure to assess communication systems and a good insight for their improvements.

The narrowband time-invariant MIMO channel is modeled by

$$\mathbf{y} = \mathbf{Hx} + \mathbf{n} \tag{9.1}$$

where $\mathbf{y}$, $\mathbf{H}$, $\mathbf{x}$, $\mathbf{n}$ denote the received signal, MIMO channel matrix, transmitted signal, and noise vector, respectively. $\mathbf{H}$ is an $N_r \times N_t$ matrix for the numbers of transmit and receive antennas, N_t and N_r, respectively. Under this modeling, the channel capacity takes the expression [5, 6]

$$C = \max_{\mathrm{Tr}(\mathbf{K}_x) \le P} \log_2 \left| \mathbf{I} + \frac{1}{N_0} \mathbf{H}\mathbf{K}_x\mathbf{H}^H \right| \text{(bits/s/Hz)} \tag{9.2}$$

where $\mathbf{I}$ denotes an identity matrix, N_0 denotes the variance of Gaussian noise $\mathbf{n}$, $\mathbf{K}_x$ denotes the covariance matrix of $\mathbf{A}$, and $|\mathbf{A}|$ denotes the determinant of matrix $\mathbf{A}$. The capacity can be maximized by adjusting $\mathbf{K}_x$ with the sum-power constraint, $\mathrm{tr}(\mathbf{K}_x) \le P$, where $\mathrm{tr}(\cdot)$ is the trace operator.

In case full CSI is available at the transmitter, it is possible to determine the optimal $\mathbf{K}_x$ that maximizes the capacity by using the *singular value decomposition* (SVD) method. Singular value decomposition operation decomposes $\mathbf{H}$ to

$$\mathbf{H} = \mathbf{U}\Lambda\mathbf{V}^H \tag{9.3}$$

for the diagonal matrix of singular values $\Lambda = \mathrm{diag}\{\lambda_1, \lambda_2, \ldots\}$ Then, the MIMO channel modeling in (9.1) can be converted to the parallel Gaussian channel modeling

$$\tilde{\mathbf{y}} = \Lambda\tilde{\mathbf{x}} + \tilde{\mathbf{n}} \tag{9.4}$$

where $\tilde{\mathbf{y}} = \mathbf{U}^H\mathbf{y}$, $\tilde{\mathbf{x}} = \mathbf{V}^H\mathbf{x}$, and $\tilde{\mathbf{n}} = \mathbf{U}^H\mathbf{n}$. Thus, the capacity of MIMO channels with full CSI at the transmitter is given by

$$C^{\mathrm{w/CSI}} = \max_{P_1+\ldots+P_{N_{\min}}\le P} \sum_{i=1}^{N_{\min}} \log_2\left(1+\frac{P_i\lambda_i^2}{N_0}\right) \text{(bits/s/Hz)} \tag{9.5}$$

where P_i denotes the power allocated to the ith parallel channel under the sum-power constraint, $N_{\min} \equiv \min\{Nr, Nt\}$ represents the number of nonzero singular values or the number of parallel channels, and λ_i^2 denotes the effective channel gain of the ith parallel channel. The problem is how to allocate the power P_i optimally to the ith parallel channel under the sum-power constraint. This is a convex optimization problem and its optimal solution can be obtained using *Karush-Kuhn-Tucker* (KKT) conditions [7]. As well known, water-filling power allocation provides the solution, or

$$P_i^{\mathrm{WF}} = \begin{cases} \left(\mu - \dfrac{N_0}{\lambda_i^2}\right), & \mu \ge \dfrac{N_0}{\lambda_i^2} \\ 0, & \text{otherwise} \end{cases} \quad \text{for } 1 \le i \le N_{\min} \tag{9.6}$$

for the water level μ chosen to meet the sum-power constraint with equality (i.e., $\sum_{i=1}^{N_{\min}} P_i^{\mathrm{WF}} = P$).

When the CSI is not available at the transmitter, we may simply allocate the equal amount of power across the transmit antennas (i.e., $\mathbf{K}_{\mathbf{x}} = \frac{P}{N_t}\mathbf{I}$). Then, by (9.2) and the relation $\mathbf{HH}^H = \mathbf{U}\Lambda\Lambda^H\mathbf{U}^H$, the capacity of MIMO channel without CSI takes the expression

$$C^{\mathrm{w/oCSIT}} = \sum_{i=1}^{N_{\min}} \log_2\left(1+\frac{P}{N_t N_0}\lambda_i^2\right) \text{(bits/sec/Hz)} \tag{9.7}$$

Fading MIMO Channels

As in the SISO case, two different types of channel capacity are defined—ergodic capacity and outage capacity. The *ergodic capacity* is the expected value of the instantaneous channel capacity over the variation of the channel, and the *outage capacity* is the largest data rate that a channel can convey for the required outage probability.

In view of (9.2), the ergodic capacity takes the expression

$$C_{Ergodic} = E_{\mathbf{H}}\left\{\log_2\left|\mathbf{I}+\frac{1}{N_0}\mathbf{HK}_{\mathbf{x}}\mathbf{H}^H\right|\right\} \text{(bits/sec/Hz)} \tag{9.8}$$

Theoretically, the ergodic capacity can be achieved by using good error correction codes whose block length is long enough to accommodate much channel fluctuations.

In general, ε-outage capacity refers to the largest target data rate R for a given outage probability, ε. Outage probability means the probability that the instantaneous channel capacity is lower than the target data rate R, or

$$P_{Outage}(R) = \Pr\left\{\log_2 \left| \mathbf{I} + \frac{P}{N_t N_0} \mathbf{H}\mathbf{H}^H \right| < R \right\} \tag{9.9}$$

Thus the ε-outage capacity may be expressed as

$$C_{Outage} = \max\left\{R \middle| P_{Outage}(R) \le \varepsilon\right\}$$

For real-time applications, the outage capacity gives more realistic performance bound than ergodic capacity does, because the relevant delay constraint limits the manageable size of the block length of error correction codes.

9.1.3 System Models

Figure 9.2 depicts the functional block diagram of the OFDM system that employs MIMO techniques. At the transmitter, the input packets of size N_{ep} each, including CRC bits, are encoded at the FEC encoder block, and the encoded data are modulated in the modulation block. The encoded and modulated signal forms a *layer*. Unlike SISO systems, MIMO systems can accommodate multiple, or L, layers. To process multiple layer signals, MIMO systems employ multiple (i.e., L) sets of FEC encoders and symbol mappers, as shown in the figure.

Space-time coding (STC) encoder matrix is an $M_t \times L$ matrix that characterizes the employed multiantenna technique among *transmit diversity* (TD), *spatial multiplexing* (SM), and a hybrid scheme of both. The number M_t may be chosen to be the same as N_t, the number of transmit antennas, or less than that. In the case of the TD scheme like Alamouti's *space-time block coding* (STBC) [9], for example, the STC encoder matrix has an orthogonal structure, which is designed to exploit the diversity gain efficiently by using multiple receivers with low complexity. On the other hand, the SM scheme transmits different symbols through different streams, thereby taking advantage of multiplexing gain. The SM scheme divides into two different types, depending on how the error-correction code is applied to the data streams. In the case of *vertical encoding* (VE), one coded and modulated layer is mapped to multiple streams ($L = 1 < M_t$), whereas L independent and separately encoded layers are mapped to multiple streams ($L = M_t$) in the case of *horizontal encoding* (HE).

In Figure 9.2, the precoding matrix $\mathbf{W}$ is included in the case of closed-loop systems to incorporate the feedback information for the improvement of the link performance. $\mathbf{W}$ is an $N_t \times M_t$ matrix, making the M_t-dimensional streams transformed into N_t-dimensional transmit signals. In the case of open-loop systems, in which case $M_t = N_t$, the precoding matrix is an $N_t \times N_t$ identity matrix. In either system, each output of the precoder is separately mapped to the subcarriers for different transmit antennas and then converted to OFDM symbols through the IFFT processing.

In the receiver, the N_r streams of received signals are individually converted to the frequency domain signals through the FFT processing and then demapped at the subcarrier demapping blocks. The demapped signals, in group of N_r streams, are decoded at the STC decoder block into L layers of FEC-coded data streams, which are individually put into the FEC decoding and CRC checking processes. In case

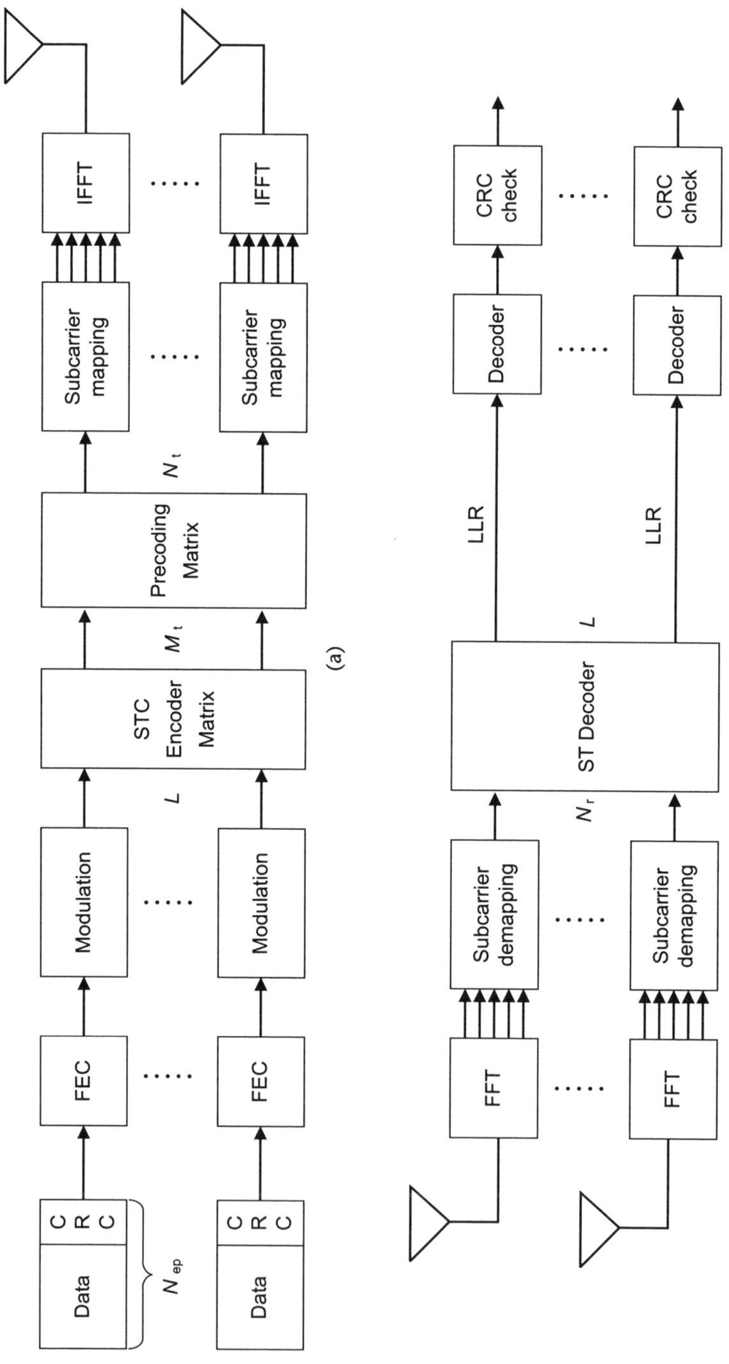

Figure 9.2 MIMO/BF-OFDM system: (a) transmitter; and (b) receiver. (*After:* [8].)

turbo coding is used at the FEC encoding, *log-likelihood ratio* (LLR) information is extracted, in addition, and delivered to the FEC decoder together with the FEC-coded data stream.

In the following, we illustrate the MIMO-OFDM technology by introducing the MIMO schemes specified in the IEEE 802.16e standards. In the standards, the STC encoding matrix is defined for the cases of two, three, and four antennas in the downlink and for the case of two antennas in the uplink. In the Mobile WiMAX system profiles, up to 2 × 2 MIMO technology is adopted, with the user terminal receiving two streams and transmitting one stream.

Open-Loop Systems

In the IEEE 802.16e standards, on which Mobile WiMAX systems are founded, the STC encoder matrices for the downlink are specified as follows: for $N_t = 2$,

$$\mathbf{A} = \begin{bmatrix} s_1 & -(s_2)^* \\ s_2 & (s_1) \end{bmatrix}, \mathbf{B} = \begin{bmatrix} s_1 \\ s_2 \end{bmatrix} \tag{9.10}$$

and for $N_t = 4$,

$$\mathbf{A} = \begin{bmatrix} s_1 & -(s_2)^* & 0 & 0 \\ s_2 & (s_1)^* & 0 & 0 \\ 0 & 0 & s_3 & -(s_4)^* \\ 0 & 0 & s_4 & (s_3) \end{bmatrix}, \mathbf{B} = \begin{bmatrix} s_1 & -(s_2)^* & s_5 & -(s_7)^* \\ s_2 & (s_1) & s_6 & -(s_8)^* \\ s_3 & -(s_4)^* & s_7 & (s_5)^* \\ s_4 & (s_3)^* & s_8 & (s_6)^* \end{bmatrix}, \mathbf{C} = \begin{bmatrix} s_1 \\ s_2 \\ s_3 \\ s_4 \end{bmatrix} \tag{9.11}$$

In the case $N_t = 2$, matrix **A**, the Alamouti's STBC matrix, represents that the STC encoder is used for TD, and matrix **B** represents that it is used for SM. Matrix **A** has a space-time coding rate of 1 since two different symbols are transmitted through two different time slots or subcarriers. On the other hand, the space-time coding rate is 2 for matrix **B**, since it transmits different symbols through the same air link.

In the case $N_t = 4$, matrix **A** represents that the STC encoder is used for TD, and matrix **C** represents that it is used for SM. In contrast, matrix **B** contains two Alamouti's STBC blocks, thus exploiting both TD and SM. Such a scheme is called a hybrid STC scheme. The space-time coding rates of matrices **A**, **B**, and **C** are 1, 2, and 4, respectively. The details on the principles and operations of open-loop MIMO systems are discussed in Section 9.2.

Closed-Loop Systems

While open-loop systems require only CQI feedback for use in AMC operation, closed-loop systems require additional feedback information for precoding. The feedback information is delivered from receiver to transmitter over the fast feedback channels (i.e., CQICH) or the uplink sounding channels. Once the appropriate precoding matrix **W** is determined, the output of the STC encoder is weighted by **W**, as indicated in Figure 9.2.

The process of determining and signaling of the precoding matrix can be replaced with a small amount of feedback information, as only a small amount of feedback information often improves the link performance significantly. In the case of the TD (i.e., matrix A) or hybrid type of STC encoder (i.e., matrix B), antenna grouping strategy is useful, whereas antenna selection strategy is useful in the case of the SM type (i.e., matrix C). Table 9.1 shows what MIMO/BF scheme to use for different numbers of layers and streams. It also indicates the applicability of the antenna grouping technique for various MIMO/BF schemes in the IEEE 802.16e standards. The details of the closed-loop MIMO/BF schemes are discussed in Section 9.3.

Frame Structure of MIMO/BF Systems

Figure 9.3 shows the possible frame structure of a MIMO/BF Mobile WiMAX system. It supports MIMO/BF zones using the designated time intervals as indicated in Figure 9.3. In the downlink, the TD and SM schemes are supported only by the MIMO PUSC zone, while the BF scheme is supported by both PUSC and AMC zones. In support of the DL BF zone, a *sounding* (SND) zone is allocated in the uplink. A *collaborative spatial multiplexing* (CSM) scheme is defined in the UL PUSC zone so that two users with a single transmit antenna each can share the same frequency.[1]

9.2 Open-Loop Technology

Due to fading in wireless communication channels, channel coefficients vary slowly or rapidly in time, frequency, and space, and thus the transmitter often operates without having channel state information. As a consequence, the full capacity of a given channel is hardly achieved unless the exact channel information can be fed back fast enough to combat against the channel variation. Moreover, if the channel is in deep fade, the system is likely to fall into outage and the error probability at the receiver cannot be made arbitrarily small.

Diversity is a powerful communication technique that deals with such fading phenomena by supplying the receiver with the transmitted signal that has passed over independently faded multiple channels. Transmit diversity techniques are devised to mitigate the outage probability by combining multiple replicas of the sig-

Table 9.1 MIMO/BF Schemes for Different Numbers of Layers and Streams

Layers (L)		1				2			3		4
Streams (M_t)		1	2	3	4	2	3	4	3	4	4
STC encoder matrix	A	Beam forming	TD	TD*	TD*	-	-	-	-	-	-
	B		-	VE*	VE*	-	HE*	HE*	-	-	-
	C		VE	VE	VE	HE	-	-	HE	-	HE

* Antenna grouping is applicable.

Applicability of antenna selection for SM is not illustrated in the table, as antenna selection changes the number of streams, M_t.

1. Shown in the figure is the case of perfect overlap CSM, where two single-antenna users start and finish transmission at the same time.

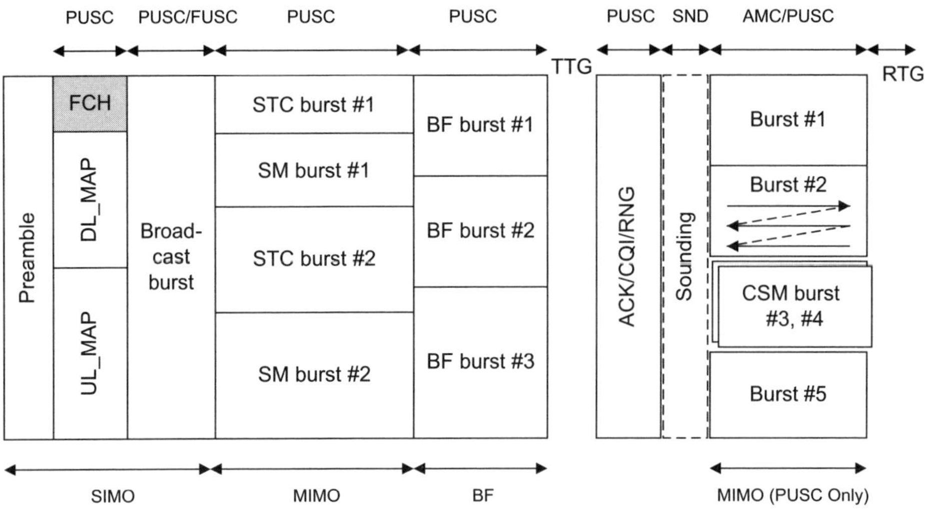

Figure 9.3 Possible frame structure of a MIMO/BF Mobile WiMAX system.

nal transmitted through independent channels formed by multiple antennas. Although the diversity gain is useful for the reduction of outage probability, the degree of performance improvement decreases as the diversity gain increases, so the transmit diversity schemes are more useful for the channel that is neither frequency-selective nor time-selective.

If the channel has a sufficient diversity gain in frequency or time, the channel capacity is limited by the degree-of-freedom gain. Spatial multiplexing exploits the independent fluctuations of fading MIMO channels to increase the degree-of-freedom gain. SM transmits different symbols through multiple antennas and, therefore, achieves higher data rate. At the receiver, the spatially multiplexed symbols can be detected by the well-known *maximum-likelihood* (ML) receiver, *zero-forcing* (ZF) linear receiver, *minimum mean-squared error* (MMSE) linear receiver, *successive interference cancellation* (SIC) receiver, and so on. MIMO receivers are discussed in more detail in Section 9.4.

9.2.1 Transmit Diversity

There are two major transmit diversity schemes—*cyclic delay diversity* (CDD) and *space-time transmit diversity* (STTD). The CDD is for downlink control channels and does not require additional antenna pilot, whereas the STTD is for traffic channels and two orthogonal pilots are needed for estimating the channels in the case of two antenna systems.

Cyclic Delay Diversity

If the same OFDM symbol with different cyclic delays, with each delay smaller than the length of cyclic prefix, is transmitted through different antennas, as shown in Figure 9.4 for the 4-antenna case, the symbol, in effect, experiences a longer multipath spread channel. It is because the channel fluctuation becomes larger across the subcarriers so the channel becomes more frequency selective. As a result,

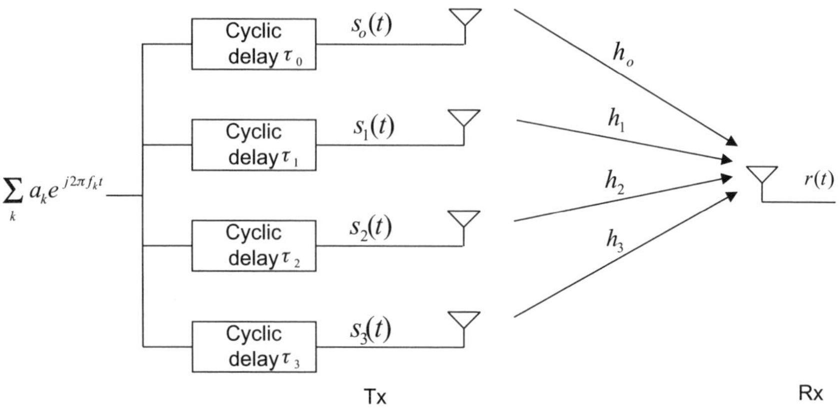

Figure 9.4 Illustration of CDD for a multiple transmit antenna system.

spatial diversity is transformed into frequency diversity. The effective diversity gain depends on the correlation of the transmit antennas.

For each antenna, the delayed signal takes the expression

$$s_m(t) = \sum_k a_k e^{j2\pi f_k(t-\tau_m)},\ m = 0, 1, 2, 3 \tag{9.12}$$

where f_k is the kth subcarrier frequency, and τ_m is the intended delay for the mth transmit antenna. Then the received signal is expressed by

$$r(t) = \sum_{m=0}^{3} s_m h_m \tag{9.13}$$

The CDD is equivalent to *delay diversity* (DD) in that they both increase the delay spread by transmitting delayed signals of the original data signal but differs from DD in that it uses cyclic delay (or phase-shift delay) whereas DD uses time-shift delay. CDD is advantageous over DD as it can completely avoid intersymbol interference. Figure 9.5 illustrates a CDD signal in comparison with a reference signal and a DD signal. In the figure, $\tau_{\max}$ is the length of cyclic prefix, which is the maximum allowable delay spread of the DD signal, and δ_{cd} is the cyclic delay of the CDD signal. The CDD technique does not require any special processing on the receiver, so no additional pilot overhead or control signal is necessary. Therefore, the CDD renders a very useful means for multiple transmit antenna systems.

Space-Time Transmit Diversity

Diversity gain can be obtained by applying coding techniques across space and time. Such space-time diversity codes exploit the space-time transmit diversity. Among the space-time diversity codes, the *space-time trellis code* (STTC) is an extension of the conventional trellis codes to multiantenna systems, whereas the *space-time block code* (STBC) is an extension of the conventional block codes to multiantenna systems. The STTC, in its original form [11], is a two-dimensional trellis code, so its decoding complexity increases exponentially with the number of transmit antennas and the constellation order. Alamouti's code [9] is an orthogonal STBC designed for

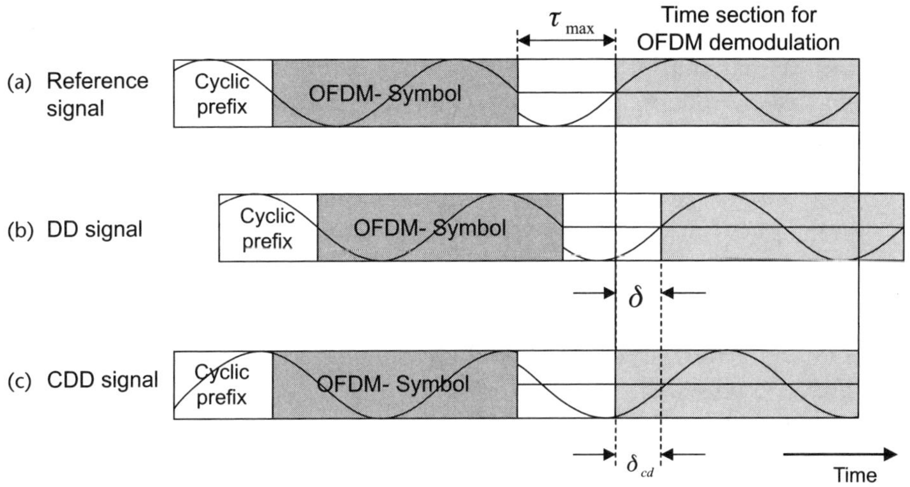

Figure 9.5 Comparison of (a) reference; (b) DD; and (c) CDD signals [10].

two transmit antennas. The Alamouti's encoder implements an optimal decoder only by applying linear processing to attain a full diversity gain and a space-time coding rate of one. For the STBCs with more than two transmit antennas, orthogonal code design with coding rate one does not exist, so some modified schemes such as a quasi-orthogonal design [12] and a coordinated interleaving STBC [13] are pursued.

In the following, we discuss the Alamouti's code in more detail. We assume a quasi-stationary channel so that the channel coefficients do not change during two symbol periods. The STC encoder matrix of the STBC is given by

$$\mathbf{A} = \begin{bmatrix} s_1 & -(s_2)^* \\ s_2 & (s_1)^* \end{bmatrix} \tag{9.14}$$

as described in (9.10). Each column denotes the vector signals transmitted at the same time through two transmit antennas. For simplicity, we deal with the case of single receive antenna. In this case, the signals received for two symbol periods are represented by

$$\begin{cases} r_1 = h_1 s_1 + h_2 s_2 + n_1 \\ r_2 = -h_1 s_2^* + h_2 s_1^* + n_2 \end{cases} \tag{9.15}$$

where h_i, $i = 1, 2$, is the channel coefficient from the ith transmit antenna to the receive antenna, and n_t, $t = 1, 2$, is *additive white Gaussian noise* (AWGN) signal at the tth symbol period. By taking the complex conjugate on the second equation, we obtain

$$\begin{pmatrix} r_1 \\ r_2^* \end{pmatrix} = \begin{pmatrix} h_1 & h_2 \\ h_2^* & -h_1^* \end{pmatrix} \begin{pmatrix} s_1 \\ s_2 \end{pmatrix} + \begin{pmatrix} n_1 \\ n_2^* \end{pmatrix} \tag{9.16}$$

This equation shows that the transmission during two symbol periods changes into spatial multiplexing with the degree-of-freedom of 2. As the resulting channel matrix is orthogonal, the optimal (i.e., ML) detection can be done by linear processing to yield

$$\begin{pmatrix} \hat{s}_1 \\ \hat{s}_2 \end{pmatrix} = \begin{pmatrix} s_1 \\ s_2 \end{pmatrix} + \frac{1}{|h_1|^2 + |h_2|^2} \begin{pmatrix} h_1^* & h_2 \\ h_2^* & -h_1 \end{pmatrix} \begin{pmatrix} n_1 \\ n_2^* \end{pmatrix} \tag{9.17}$$

Since the multiplication by an orthogonal matrix does not change the probabilistic characteristics of an AWGN vector, except for the noise power, the noise term in (9.15) is another AWGN vector.

In the case of multiple receive antennas, we may apply the *maximal ratio combining* (MRC) technique after the detection process in (9.17).

In the case of the OFDM systems, we may use the adjacent subcarrier instead of the adjacent time (or symbol period), since the two adjacent subcarriers are orthogonal to each other and their channel coefficients are highly correlated. The resulting block code may be called *space-frequency block code* (SFBC), in contrast to the name STBC.

9.2.2 Spatial Multiplexing

Spatial multiplexing can increase spectral efficiency by transmitting multiple data streams simultaneously using the same frequency channel [6]. Multiple antennas for spatial multiplexing may be used by a single user or may be shared among multiple users. In the single-user case, a single coded symbol sequence may be demultiplexed into multiple antennas or multiple coded symbol sequences may be mapped into multiple antennas. In the multiple-user case, each user transmits and receives its data sequence using a single antenna.

Single-User Spatial Multiplexing

We consider, for example, the single-user spatial multiplexing with $N_t = 2$, $N_r = 2$, and the single data stream $\{s_1, s_2, s_3, s_4, \ldots\}$. At the first symbol period, s_1 and s_2 are transmitted through the first and the second transmit antennas, respectively, and, similarly, s_3 and s_4 are transmitted through the two transmit antennas at the second symbol period. The degree-of-freedom gain depends on the smaller side of the transmitter and the receiver antennas, so the degree-of-freedom in this case is 2.

The received signal vector takes the expression

$$\begin{pmatrix} r_1 \\ r_2 \end{pmatrix} = \begin{pmatrix} h_{11} & h_{12} \\ h_{21} & h_{22} \end{pmatrix} \begin{pmatrix} s_1 \\ s_2 \end{pmatrix} + \begin{pmatrix} n_1 \\ n_2 \end{pmatrix} \tag{9.18}$$

where h_{ki} is the channel coefficient from the ith transmit antenna to the kth receive antenna, and r_k and n_k are the received signal and the noise, respectively, at the kth receive antenna. The equation states that, differently from the STBC case, which transmitted two symbols during two symbol periods (i.e., STBC has a coding rate of one), SM transmits two symbols during one symbol period (i.e., SM has a coding rate of two).

Suppose, in this example, that the four symbols s_1, s_2, s_3, s_4 form a coding block. Then some of the four symbols that pass through poor-quality channels may be corrupted by noise and interference while traveling to the receiver, and some other symbols that pass through good-quality channels may be delivered to the receiver in reliable state. The reliably received symbols help to decode the original coding block, and, in this sense, the error correction code has the effect of diversity gain among the channels of different quality.

There are two different types of SM techniques, which differ in the composing method of the channel coding blocks—*vertical encoding SM* (VESM) and *horizontal encoding SM* (HESM), as illustrated in Figure 9.6. In the case of VESM, there exists one coding block, which is *demultiplexed* (DMUX) into multiple streams of symbols (i.e., one layer and multiple streams). As discussed earlier, the multiple streams that originally belonged to the same encoding block generate the effect of SINR balancing while traveling through different channels, thereby reducing the outage probability. In the case of HESM, there are multiple encoding blocks, each of which generates one stream. Since each stream would experience different SINR, a different MCS level is chosen for each stream. For enhanced performance, an interference cancellation technique like SIC may be additionally applied after error correction.

Collaborative Spatial Multiplexing

For a MIMO channel with fixed average gains, the capacity of MIMO channels is maximized when the channel gains of all the constituent antennas are mutually independent. The channels formed by the multiple antennas at the same MS are hardly expected to be independent or weakly correlated. Such highly correlated antennas are likely to yield an ill-conditioned channel matrix and, thus, result in

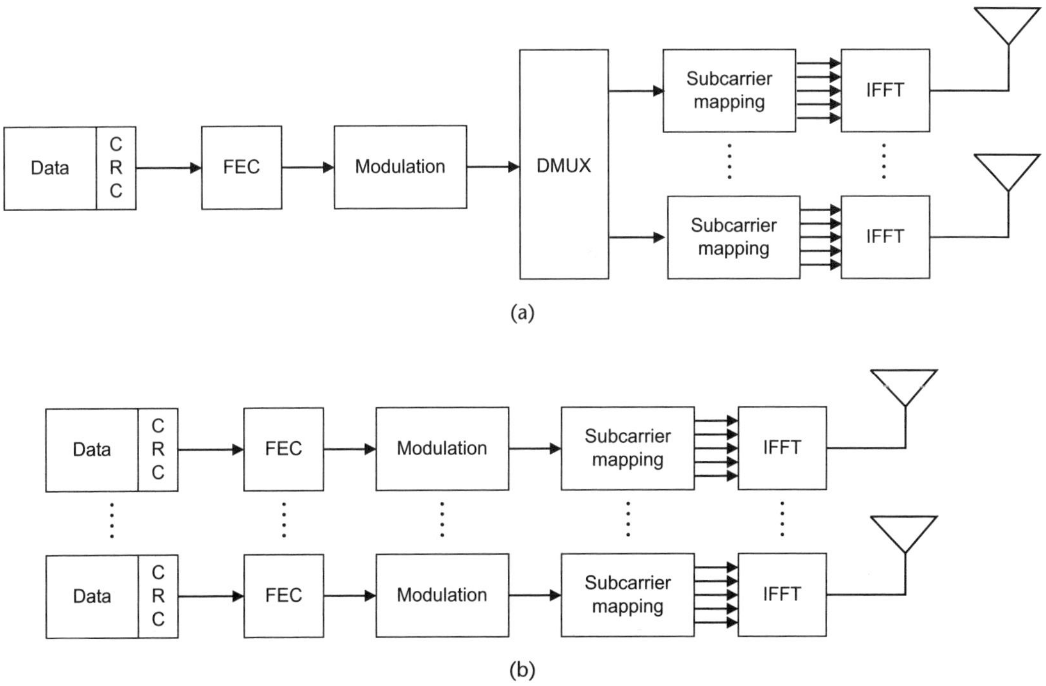

Figure 9.6 Transmitters of (a) VESM, and (b) HESM [8].

capacity loss. Therefore, it is better off in this case to arrange such that multiple MSs of spatially separated users to share the same frequency.

As discussed earlier, the HESM for one user launches independent and separately encoded layers to multiple transmit antennas. This HESM can be applied to the uplink of multiuser systems by allocating each layer to each user. In this arrangement, each user uses the common air resource and uploads its data stream as if it occupies the channel by itself. This technique is called *collaborative spatial multiplexing* (CSM). The BS receives the multiple streams in mixed form and detects the encoding block of each individual user. The detection in CSM is totally equivalent to the detection in single-user SM.

The CSM is advantageous over the single-user SM in that it does not require multiple antennas at MSs and that it has multiuser diversity. Specifically, in case the number of users is larger than that of antennas of the BS, only a part of the users can transmit through the common channel at the same time and, in selecting those users to transmit, multiuser diversity gets involved.

The CSM may operate in two different modes, depending on the bandwidth allocation strategy: perfect overlap operation and partial overlap operation.

When SIMO and CSM users are intermixed, the BS decides which of the two transmission modes to allocate for each subchannel or frequency band. For the subchannels allocated to SIMO users, a single user is selected in the same way as for the conventional noncollaborative systems. For the subchannels allocated to CSM users, multiple users are selected among the least-correlated users considering the required bandwidths. Figure 9.7(a) illustrates this strategy for the case $N_r = 2$, which is called the *perfect overlap operation*. As the perfect overlap operation allows the BS to choose the best user set for each CSM subchannel, multiuser diversity gain is maximally attained. The uplink pilot tones are allocated to a single user or two users, depending on the transmission mode.

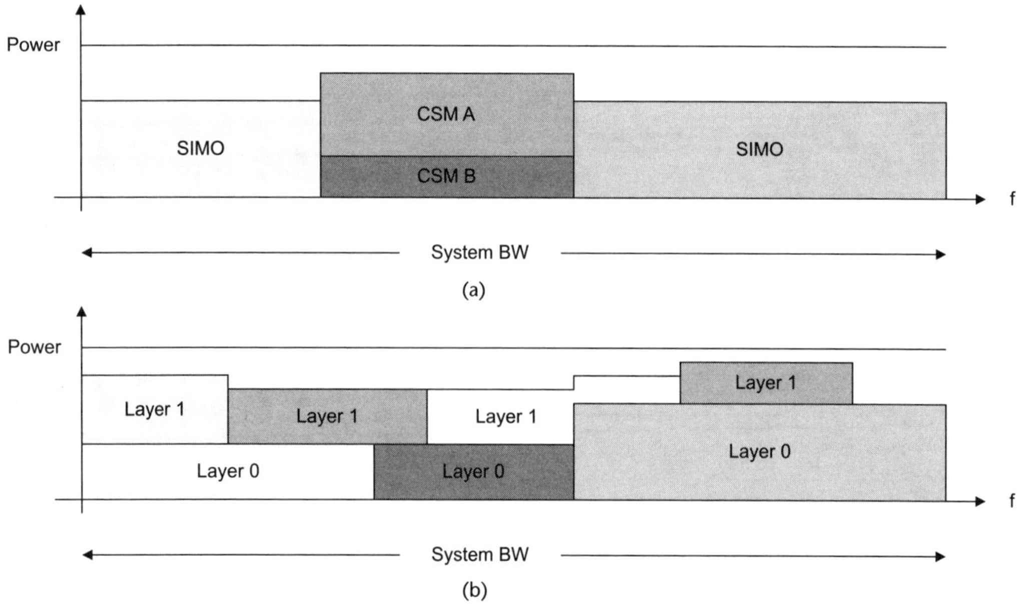

Figure 9.7 Illustration of (a) perfect, and (b) partial overlap operation ($N_r = 2$).

The perfect overlap operation has limitations in accommodating bursts of different lengths. For example, a communication system that employs HARQ with Chase combining would request a retransmission of the erred bursts and, in this case, the identical-lengthed accommodation of the perfect overlap operation does not properly work for the CSM mode. Therefore, an alternative CSM bandwidth allocation method, called *partial overlap operation*, is needed where each layer may be allocated with an arbitrary subchannel as long as the total number of layers does not exceed the number of the receiver antennas at any frequency. Figure 9.7(b) illustrates the partial overlap operation for $N_r = 2$. As the partial overlap operation has no restriction on the length of bursts, it can be adopted for the systems employing HARQ or other error-control methods.

9.2.3 Mobile WiMAX Examples

The Mobile WiMAX system employs a multiple antenna technology, as in the case of several other recent communication systems. For the illustration of open-loop MIMO technology applied to the Mobile WiMAX system, we consider the examples of two-antenna STC schemes in the downlink and the single-antenna collaborative SM scheme in the uplink.

Two-Antenna Downlink STC Transmission

For the downlink of Mobile WiMAX, there are several types of subchannelization schemes such as PUSC, FUSC, and AMC. Among them we consider the case of PUSC, as it can be easily extended to other cases. In the case of the SISO system, the cluster structure for DL PUSC is composed of two consecutive OFDM symbols as depicted in Figure 4.37. It consists of pilot and data subcarriers. The SISO cluster structure is expanded two-fold as depicted in Figure 9.8 in the case of the MIMO system with two transmit antennas, which we deal with in the following.

The pilot subcarriers for each transmit antenna are distinctly allocated in a period of four symbols, as indicated in Figure 9.8. The pilot subcarriers are allocated distinctly for each transmit antenna so that the receiver can get the channel state information from each transmit antenna separately. As the number of pilot

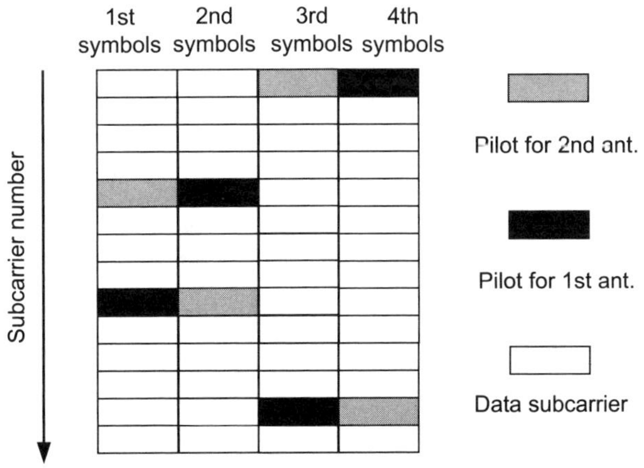

Figure 9.8 Cluster structure for two transmit antenna MIMO systems. (*After:* [14].)

subcarriers reduces to a half when compared with the single transmit antenna case, the channel estimation performance may be degraded. So the pilot subcarriers in the two transmit antenna system are boosted, in power, by 5.5 dB above data subcarriers.[2]

In mapping input data to data subcarriers for each transmit antenna, the mapping pattern differs depending on the STC scheme employed. Among the two typical STC schemes, STBC is employed for diversity and VESM for multiplexing. In the case of the STBC, data are mapped in groups of 48 symbols per subchannel, and, in the case of VESM, data are mapped in groups of 96 symbols. Figure 9.9 illustrates how the data mappings are done for the STBC and VESM cases.

Specifically, the mapping of the STBC data is done as follows: Assuming the 48 input data are given by $\{x_0, x_1, x_2, x_3, \ldots, x_{47}\}$, STBC is applied with coding rate 1 in such a way that each set of two consecutive data symbols form a pair (i.e., $(s_1, s_2) = \{x_{2n}, x_{2n+1}\}$, $n = 0, \ldots, 23$) and the pair is encoded by the STBC encoder defined by (9.14). The encoded streams are transmitted on the same subcarrier in two consecutive OFDM symbols using two transmit antennas, as described in Table 9.2.

In the case of VESM, the data mapping is done as follows: Assuming the 96 input data are given by $\{x_0, x_1, x_2, x_3, \ldots, x_{95}\}$, the VESM is applied with coding rate 2 in such a way that each set of two consecutive data symbols form a pair (i.e., $(s_1, s_2) = \{x_{2n}, x_{2n+1}\}$, $n = 0, 1, .., 47$) and the pair is transmitted on one subcarrier using two transmit antennas, as described in Table 9.3.

In case multiple subchannels are needed, subchannel allocation is done in the order of increasing subchannel and symbol indices. Within each subchannel, the subcarriers are allocated in the order of increasing subcarrier index.

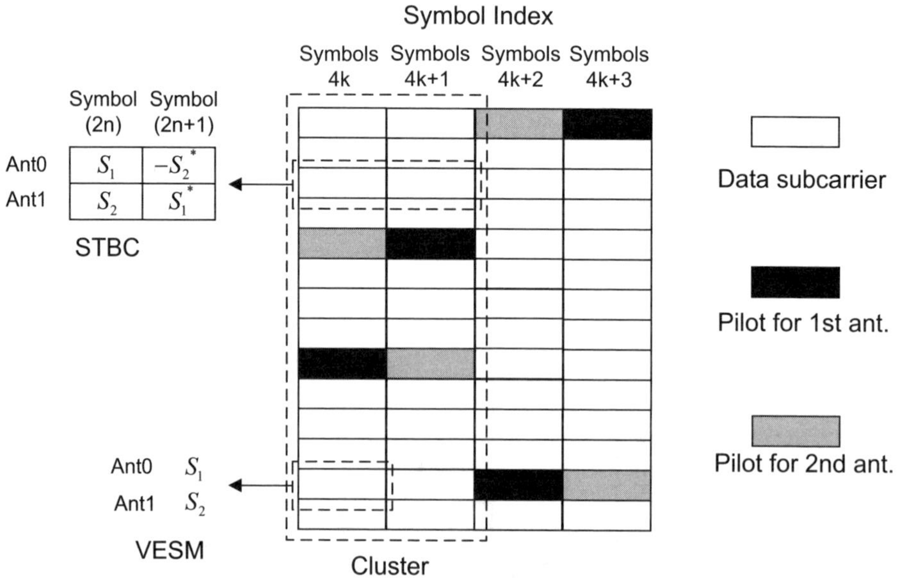

Figure 9.9 Mapping of STBC and VESM data in a two-transmit antenna system.

2. The pilot subcarrier power is boosted by 2.5 dB over the data subcarrier power in the case of SIMO and is boosted by additional 3.0 dB in the case of two transmit antenna MIMO to compensate for the halved pilot subcarrier density.

Table 9.2 Mapping of STBC Data in DL PUSC

Antennas	Antenna #0		Antenna #1	
Symbols	Even symbol	Odd symbol	Even symbol	Odd symbol
Subcarrier #0	x_0	$-x_1^*$	x_1	x_0^*
Subcarrier #1	x_2	$-x_3^*$	x_3	x_2^*
⋮	⋮	⋮	⋮	⋮
Subcarrier #22	x_{44}	$-x_{45}^*$	x_{45}	x_{44}^*
Subcarrier #23	x_{46}	$-x_{47}^*$	x_{47}	x_{46}^*

Table 9.3 Mapping of VESM Data in DL PUSC

Antennas	Antenna #0		Antenna #1	
Symbols	Even symbol	Odd symbol	Even symbol	Odd symbol
Subcarrier #0	x_0	x_{48}	x_1	x_{49}
Subcarrier #1	x_2	x_{50}	x_3	x_{51}
⋮	⋮	⋮	⋮	⋮
Subcarrier #22	x_{44}	x_{92}	x_{45}	x_{93}
Subcarrier #23	x_{46}	x_{94}	x_{47}	x_{95}

Now we consider how to do mode selection between the STBC and VESM. According to the Mobile WiMAX profile, the BS may initiate a mode selection between the STBC and VESM. If requested by the BS, the MS calculates the average CINR of each mode and selects the preferred mode. The actual algorithm is implementation-specific and dependent on the receiver algorithm. One simple algorithm is as follows:

$$Mode = \underset{\{VESM,STBC\}}{\arg\max} \left\{2Avg\_CINR_{VESM}[\text{dB}], Avg\_CINR_{STBC}[\text{dB}]\right\} \tag{9.19}$$

where the average CINR is doubled in the case of the VESM because the spatial multiplexing order is two. In the case of the VESM, the average CINRs for both ML and linear MMSE demodulators are defined in the IEEE 802.16e standard based on the narrowband signal model in (9.1) [15] (i.e., $\mathbf{y} = \mathbf{Hx} + \mathbf{n}$) as

$$Avg\_CINR = 2^{C(\mathbf{x},\mathbf{y}|\mathbf{H})} - 1 \tag{9.20}$$

The mutual information is determined by

$$C(\mathbf{x},\mathbf{y}|\mathbf{H}) = \frac{1}{2}\log_2\left|\mathbf{I}_2 + \mathbf{H}^H\mathbf{R}^{-1}\mathbf{H}\right| \tag{9.21}$$

for the ML receiver, where **R** denotes the covariance matrix of interference plus noise; and by

$$C(\mathbf{x},\mathbf{y}|\mathbf{H}) = \frac{1}{2}\sum_{n=1}^{2}\log_2(1+CINR_n) \tag{9.22}$$

for the linear MMSE receiver, where $CINR_n$ denotes the postprocessing $CINR$ for layer $n = 1, 2$. In the case of STBC, the average CINR may be computed based on the same equation in (9.21) but the mutual information is determined by

$$C(\mathbf{x},\mathbf{y}|\mathbf{H}) = \log_2\left(1+\left\|\mathbf{R}^{-1/2}\mathbf{H}\right\|_F^2\right) \tag{9.23}$$

where $\|\ \|_F$ denotes the Frobenious norm of matrix.

In the case of wideband channels, CINR can be calculated after taking the average of the mutual information over multiple (i.e., K) subcarriers as follows:

$$C_{avg} = \frac{1}{K}\sum_{k=1}^{K-1} C_k(\mathbf{x}_k,\mathbf{y}_k|\mathbf{H}_k) \tag{9.24}$$

$$Avg\_CINR[\text{dB}] = 10\log_{10}\left(2^{C_{avg}}-1\right) \tag{9.25}$$

Collaborative SM for UL PUSC

Among the subchannelization schemes for the uplink of Mobile WiMAX, we consider the PUSC, in line with the downlink case.

In the case of UL PUSC, each tile contains 12 subcarriers, including four pilot tones located at the four corner points, as depicted in Figure 4.40. The four pilot tones are fully used by one MS in the case of the single-transmit antenna scheme. However, if two transmit antennas are available at the MS, the four pilot tones are divided into two pairs, as shown in Figure 9.10, and each pair is allocated to one transmit antenna. In compensation for the decreased number of pilot tones, the pilot tones are boosted, in power, by 3.0 dB for in the two-transmit antenna case.

In the case of CSM, two MSs, each with a single transmit antenna, transmit their own data stream through the same subchannels. Since the receiver needs the channel state information of both MSs to reliably decode the received signal, the resources of the pilot tones are shared in the same way as in the STBC and VESM schemes. That is, one MS uses the pilot pattern A and the other MS uses the pilot pattern B shown in Figure 9.10, so the pilot signals of two MSs do not collide.

From the viewpoint of each MS, the transmission scheme for CSM is the same as that for SISO, except that pilot tones are used in half. In UL PUSC, six tiles form a subchannel, so each subchannel contains 48 data subcarriers. If, for example, the data sequences for two users are given by $\{x_0^1, x_1^1, x_2^1, x_3^1, \ldots, x_{47}^1\}$ and $\{x_0^2, x_1^2, x_2^2, x_3^2, \ldots, x_{47}^2\}$, where $\{x_t^k\}$ denotes the tth data symbol of the kth user, then the two sets of the 48 data symbols are mapped to the 48 data subcarriers in the subchannels of MS#1 and MS#2, respectively, as shown in Table 9.4.

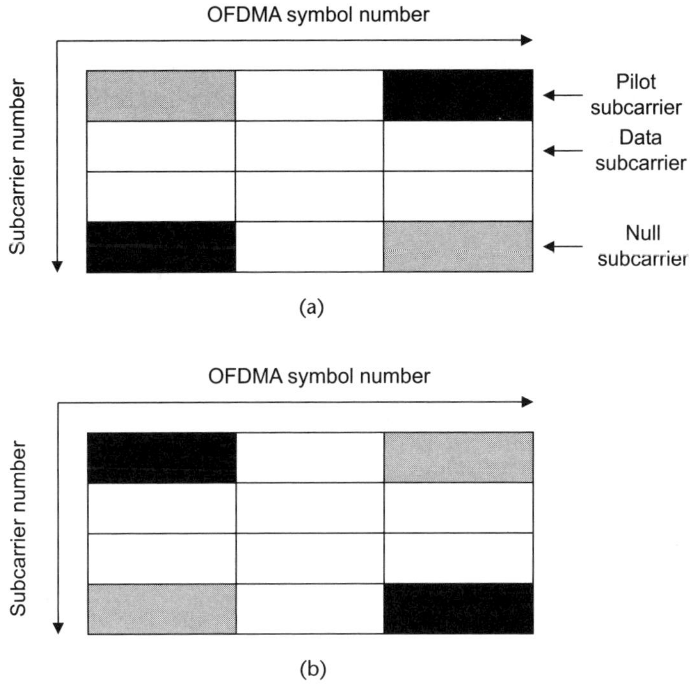

Figure 9.10 Pilot patterns in UL PUSC tile: (a) pattern A; and (b) pattern B.

Table 9.4 Mapping of CSM Data in UL PUSC

MSs	MS #1	MS #2
Subcarrier #0	x^1_0	x^2_0
Subcarrier #1	x^1_1	x^2_1
⋮	⋮	⋮
Subcarrier #46	x^1_{46}	x^2_{46}
Subcarrier #47	x^1_{47}	x^2_{47}

9.3 Closed-Loop Technology

Whereas the open-loop technology processes transmit data without *channel state information* (CSI), the closed-loop technology utilizes the CSI to process transmit data such that the spectral efficiency can be maximized. The CSI is estimated by the receiver by analyzing the received reference signals and sent back to the transmitter over a feedback channel. Specifically, in the case of the Mobile WiMAX system, MS estimates the CSI by analyzing the downlink pilot and preamble signals, and sends the estimated CSI back to the BS over the uplink feedback channel.

The channel information is processed such that the amount of information to feed back does not exceed the allowed capacity limit of the feedback channel. The processing scheme, as well as the processed feedback information, is chosen differently depending on the transmission scheme employed by the BS. For example, if the antenna selection or grouping schemes are employed by the BS, antenna index or group index may be fed back; if beamforming scheme is employed, quantized CSI may be fed back.

In the case of the TDD-based Mobile WiMAX systems, it is usually assumed that the downlink and uplink channels have a reciprocal relation (i.e., the transpose of the downlink channel matrix is the uplink channel matrix), provided that the RF chains of BS and MS are perfectly calibrated. If the MS sends reference signals in the uplink interval, called *sounding signals*, the BS can obtain the downlink channel information by analyzing the sounding signals. Note that such closed-loop technology is effective when the MS moves slowly, as otherwise the time-varying channel characteristic would cause performance degradation due to the outdatedness of the CSI.

9.3.1 Precoding

Precoding refers to the process of prearranging the transmit signals in multiple transmit antennas in consideration of the channel state in such a way that the receiver can combine the multiple antenna signals to detect the transmitted signal reliably. The channel state is estimated by analyzing the DL and UL reference signals: The DL reference signals are per-antenna common pilot signals and the UL reference signals are user-specific sounding signals. In general, the MS estimates and abstracts the CSI out of the DL pilot signals and then sends it to the BS over the UL feedback channels, but, in the case of the TDD system, the BS can estimate the CSI directly from the UL sounding signals. We consider both the pilot and the sounding signal–based precoding schemes in the following.

Precoding with DL Pilot Signals

In case the CSI is estimated by MS using the DL pilot signals, the CSI should be condensed to the level that the feedback channel can carry within its capacity. For example, it will be more desirable to transmit only the information essential for the transmitter than to transmit the channel matrix itself. In practice, there are two efficient ways introduced to date: one is the antenna selection or antenna grouping method that feeds back the index of the transmit antenna or the index of the transmit antenna group, and the other is the codebook-based method that provides the index of the precoding matrix listed in the predefined codebook.

Apparently, the equal-power allocation among multiple antennas is a simple method to take without knowing the channel state, but, if the channel information is available, it is more efficient to allocate the power only to the antennas that have good channel gains. So in the case of the precoding scheme based on antenna selection, the MS examines the channel quality for each antenna and reports to BS the set of antennas that would yield the best performance. When reporting, the MS sends the index of the antenna combination that would maximize the received SNR or the capacity.

An antenna selection method may be applied to both diversity and spatial multiplexing schemes [16, 17]. In the case of the diversity scheme, a maximum SNR criterion may be employed. Specifically, if we denote by N_s the number of antennas selected and by A_s the set of all possible N_s-antenna combinations, then the best antenna combination a is determined by

$$a = \arg\max_{a \in A_s} \left\| \mathbf{H}_a \right\|_F^2 \tag{9.26}$$

where $\mathbf{H}_a$ denotes the channel matrix corresponding to the selected antenna combination. In the case of the spatial multiplexing scheme, maximum capacity criterion may be employed for the multiplexing gain as follows:

$$a = \arg\max_{a \in A_s} \log_2 \left| \mathbf{I} + \frac{P}{N_s N_0} \mathbf{H}_a^H \mathbf{H}_a \right| \tag{9.27}$$

where P denotes the total power and N_0 the noise power.

Whereas antenna selection is intended to select and utilize only the well-performing antennas, antenna grouping is intended to use all the antennas by dividing the antennas into multiple groups. For example, in the case of the *double STTD* (DSTTD), four transmit antennas are divided into two 2-antenna groups, and then each antenna group transmits data encoded by Alamouti's STBC. Then each stream attains a diversity order of 2 from the Alamouti's STBC. When grouping the four antennas, we arrange such that the least-correlated pairs belong to the same group or the mean square error gets minimized.

The IEEE 802.16e standard addresses the antenna grouping technique for the STBC scheme with 3 and 4 transmit antennas. It is reported that such an antenna grouping technique achieves about 2.0 dB SNR gain over the open-loop transmit diversity scheme for 3 transmit antenna cases. This SNR gain is decreased to about 1.0 dB for 4 transmit antenna cases [18].

As discussed in Section 9.1.2, the closed-loop MIMO scheme based on SVD combined with the waterfilling power allocation strategy yields an optimal way to achieve the channel capacity. To implement this method, transmitter has to know the unitary matrix $\mathbf{V}$ in (9.3) whose columns are the right singular vectors of the instantaneous channel matrix $\mathbf{H}$. In support of this, the receiver must feed back the channel matrix $\mathbf{H}$ or its unitary matrix $\mathbf{V}$ to the transmitter. However, it requires a high computing overhead and a large information feedback. So it is more practical to adopt a codebook–based precoding scheme that feeds back the index of the particular matrix in the predesigned codebook that best matches to the estimated channel matrix.

Specifically, the codebook-based precoding scheme operates in the following manner: The MS first estimates the downlink channel matrix from the pilot signals, determines the best-matching unitary matrix (i.e., the precode matrix) out of the predesigned codebook, and sends the index of the selected unitary matrix to the BS. Then the BS uses the selected precoding matrix to predistort the transmission signal in the next transmission opportunity. Such an operation has low computational complexity and requires a minimal information feedback.

In the codebook-based precoding scheme, codebook design is an important issue. The codebook should be designed such that the SNR is maximized or the error is minimized and such that the codebook size is reasonably large. There are various codebook design methods reported in the literature [19–21].

Precoding with UL Sounding Signals

Unlike the FDD systems, which allocate different frequency bands to downlink and uplink channels, TDD systems use the same frequency band for the downlink and uplink transmissins. This enables TDD systems to take advantage of the channel reciprocity between the downlink and uplink channels, provided that the BS and MS hardwares are perfectly calibrated. So the BS in a TDD system can estimate the downlink channel matrix, which is assumed to be the same as the uplink channel matrix, by analyzing the uplink reference signals, or the sounding signals, during the uplink interval. This arrangement makes the operation simple since the MS does not need to send any feedback information to aid the channel matrix determination. To make it better, the channel matrix estimated directly by the BS is much more accurate than the one reconstructed from the limited feedback information sent by MS. In fact, the precoding technique based on channel sounding turns out to be the best choice, especially for the multiple transmit antenna systems, in terms of the complexity in MSs and the performance in downlink beamforming.

Sounding signals are the reference signals that the MS sends up to the BS to aid the channel matrix estimation. Sounding signals are conveyed on the sounding zone in the uplink interval (see Figure 9.3), which consists of one or more OFDMA symbol intervals in the UL frame. The MS transmits sounding signals to enable the BS to rapidly determine the channel response between the BS and the MS. The BS enables the uplink sounding by transmitting UIUC=13 in UL-MAP to indicate the allocation of a UL sounding zone within the frame. The BS transmits a UL-MAP message UL-Sounding_Command_IE() to provide the detailed sounding instructions to the MS.

The sounding channel is divided into two types, depending on the allocation method of the subcarriers in the sounding zone. In type A, the subcarriers within the sounding zone are partitioned into no-overlapping sounding frequency bands, where each sounding frequency band contains 18 consecutive subcarriers. For example, in the case of 1,024 FFT size with 864 used subcarriers, the sounding zone contains a maximum of 48 sounding frequency bands. In type B, the frequency band is allocated according to the specified downlink subcarrier permutation.

As an illustration of the sounding signal–based precoding scheme, we consider the *transmit antenna array* (TxAA) beamforming system and examine how to determine the beamforming weight [22]. If d is the modulated data symbol to transmit and P is the transmission power, then the $N_t \times 1$ precoded symbol vector $\mathbf{x}$ is represented by

$$\mathbf{x} = \sqrt{P} d \mathbf{v} \tag{9.28}$$

where $\mathbf{v}$ is the transmit beamforming weight vector to determine. The signal received at the MS with N_r receive antennas is as given in (9.1) (i.e., $\mathbf{y} = \mathbf{H}\mathbf{x} + \mathbf{n}$). Assuming a *maximal ratio combining* (MRC) receiver is used, an estimate for d is obtained by multiplying the receive beamforming weight vector $\mathbf{w}$ to the received signal (i.e.,

$\hat{d} = \mathbf{wy}$). Since $\mathbf{w}$ is the Hermitian vector of the effective channel [i.e., $\mathbf{w} = (\mathbf{Hv})^H$], it yields

$$\hat{d} = \left(\mathbf{v}^H \mathbf{H}^H \mathbf{H} \mathbf{v}\right)\left(\sqrt{P} d\right) + \mathbf{v}^H \mathbf{H}^H n \tag{9.29}$$

and the resulting SNR is represented by

$$\gamma = \frac{P\mathrm{E}\left[|d|^2\right]}{N_0}\left(\mathbf{v}^H \mathbf{H}^H \mathbf{H} \mathbf{v}\right) \tag{9.30}$$

It is well known that the optimal transmit beamforming weight vector $\mathbf{v}$ that maximizes γ is the eigenvector corresponding to the maximum eigenvalue of $\mathbf{H}^H\mathbf{H}$. In the special case of single antenna receiver, the channel matrix $\mathbf{H}$ turns into a channel vector $\mathbf{h}$, and the optimal transmit beamforming weight vector is determined to be

$$\mathbf{v}_{\text{opt}} = \frac{\mathbf{h}^H}{\|\mathbf{h}\|} \tag{9.31}$$

The transmission scheme that utilizes such transmit beamforming weight is called *maximal ratio transmission* (MRT) [23, 24], named after the symmetric receiver structure MRC. In this case, the corresponding received SNR is given by

$$\gamma_{\text{opt}}^{(N_r=1)} = \frac{P\mathrm{E}\left[|d|^2\right]}{N_0}\|\mathbf{h}\|^2 \tag{9.32}$$

Note that the beamforming scheme requires dedicated pilot signals, rather than the common pilot signals, since the pilot signals also should be weighted and the weighting vector depends on each user channel. In the case of PUSC, since each major group uses common pilot signals, each major group should be allocated to one user to make all the pilot signals dedicated to one user. The user in beamforming mode estimates the effective channel using only the dedicated pilot signals in the allocated major group.

9.3.2 Multiuser MIMO

The capacity of MIMO channel grows linearly with the smaller side of the number of antennas in the BS and MS. Usually the dimension of the MS is much smaller than that of the BS, making it difficult to put in the MS as many antennas as in the BS, so the capacity is limited by the number of MS antennas. In the environment where multiple MSs communicate with one BS, the antennas in different MSs can aid to overcome this limitation. It is because multiple user data are transmitted through multiple antennas simultaneously, with each antenna transmitting a combination of multiple user data. As a consequence, the multiuser MIMO technique brings forth a new method of increasing the capacity.

For the uplink, CSM is a good example of multiuser MIMO technique.[3] In this case, different data for multiple users are transmitted over the same air resource provided by multiple receive antennas at the BS and multiple MSs, each with a single transmit antenna. At the BS, different data from multiple users are detected by the same MIMO detection algorithm that is used for detecting the multiple streams from a single user. Therefore the total data rate increases over the case of a single-user MIMO system.

For the downlink, the multiple user data streams transmitted from the BS cannot be separated out at each MS that has a less number of antennas, unless some prearrangement is made on the user data streams before transmission. An example of such prearrangement is to subtract the *interuser interferences* (IUIs) at the BS using the CSI fed back from the MSs. In particular, *dirty-paper coding* (DPC) approaches were introduced as means for securing the sum-capacity of multiuser MIMO downlink channels [25–27], but they require a high computing burden and perfect CSI at the BS.

As a practical example, we consider the simplified multiuser MIMO downlink scheme called *per-user unitary rate control* (PU^2RC). This scheme intends to suppress the IUIs at the BS, as otherwise the MSs cannot detect the destined data streams out of the received signals. If perfect CSI were available at the BS, it would be possible to eliminate all the interferences by premultiplying the pseudo-inverse matrix of the channel matrix before transmission. In practice, however, it is not easy to know the perfect instantaneous channel information and, moreover, the channel matrix becomes ill-conditioned when some channels are highly correlated. PU^2RC is a practical multiuser MIMO downlink scheme that is capable of supporting multiple users simultaneously using only limited feedback information.

Figure 9.11 depicts the operational structure of the PU^2RC-MIMO transmitter, which has the set of precoder matrices $\mathbf{E} = \{\mathbf{E}^{(0)}, \ldots, \mathbf{E}^{(G-1)}\}$, where $\mathbf{E}^{(g)} = [\mathbf{e}_0^{(g)}, \ldots, \mathbf{e}_{M-1}^{(g)}]$, $g = 0, 1, .., G-1$, is the gth precoding matrix, and $\mathbf{e}_m^{(g)}$, $m = 0, 1, \ldots, M-1$, is the mth precoding vector in the set. G denotes the number of user groups to which K users are mapped into. Each MS will calculate a CQI value for every vector $\mathbf{e}_m^{(g)}$ in every matrix $\mathbf{E}^{(g)}$ in the set E. The amount of feedback overhead can be traded off with the scheduling flexibility at the BS, by choosing an appropriate number of precoding matrices, G, and deciding the amount of information that the MS feeds back to the BS. Here, we assume that each MS feeds back the M (or the number of transmit antennas, N_t) CQI values of the channels corresponding to its preferred/best group.

The overall operation of the PU^2RC-MIMO transmitter with partial feedback takes the following procedure [28, 29]: The transmitter first gathers the feedback information that indicates "a preferred precoding matrix" and the CQI values for all the precoding vectors in the matrix, then groups the users who declare the same preferred precoding matrix, and then selects the group with the highest group priority. How to define the group priority depends on the scheduling policy. After that, the transmitter selects the code words of multiple users with the highest priority in the selected group. How to define the codeword priority (or user priority) depends on

3. As discussed in Section 9.2.2, CSM is an open-loop technology, but we deal with it in this section because it renders a good example of multiuser MIMO.

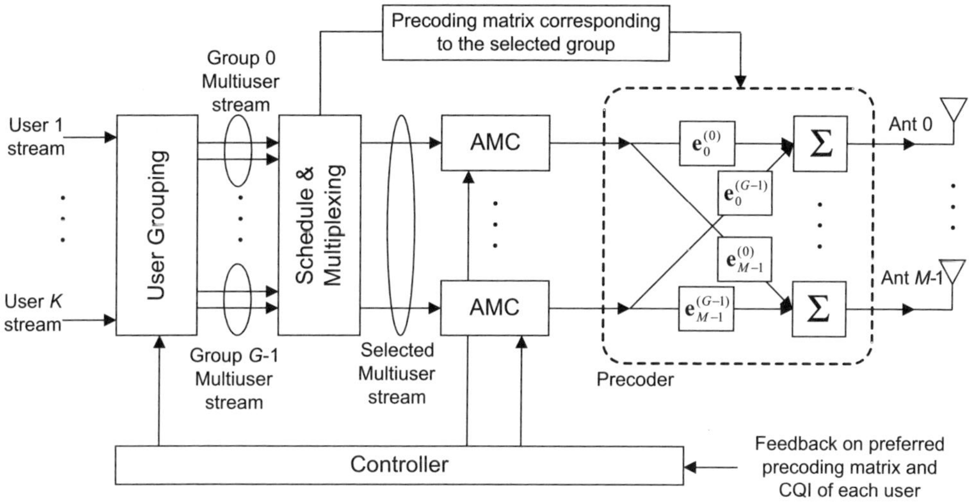

Figure 9.11 Operational structure of PU^{2}RC transmitter [28].

the scheduling policy. Finally, the transmitter applies appropriate AMC schemes to the selected code words and applies a precoding scheme corresponding to the selected group.

9.4 MIMO Receiver Algorithms

It is important to choose an adequate MIMO receiver algorithm among many available algorithms for multiple antenna systems. Without using the optimal receiver, one cannot achieve the full diversity gain that space-time codes can support. The challenge in implementing the receivers for actual MIMO systems is the computational complexity. For example, ML detection is often impractical even for a modest number of antennas due to the computational complexity that increases significantly as the number of antennas and the modulation order increase. So we investigate in this section various MIMO receiver algorithms for spatial multiplexing schemes, including ML detection, linear detection such as ZF and MMSE, SIC detection, and some near-optimal detections such as sphere decoding, modified ML detection, and QRM-MLD.

9.4.1 Maximum Likelihood Detection

We consider the MIMO system model in Figure 9.12. From the model we get the relation

$$\mathbf{y} = \mathbf{Hx} + \mathbf{n} = \mathbf{h}_0 x_0 + \mathbf{h}_1 x_1 + \ldots + \mathbf{h}_{N_t-1} x_{N_t-1} + \mathbf{n} \tag{9.33}$$

where N_t is the number of transmit antennas; N_r is the number of receive antennas. Note that x_i, $i = 0, 1, \ldots, N_t - 1$ are M-QAM modulated symbols $x_i \in C$ where $C = \{c_0, c_1, \ldots, c_{M-1})$ is the set of M-QAM complex signal constellation.

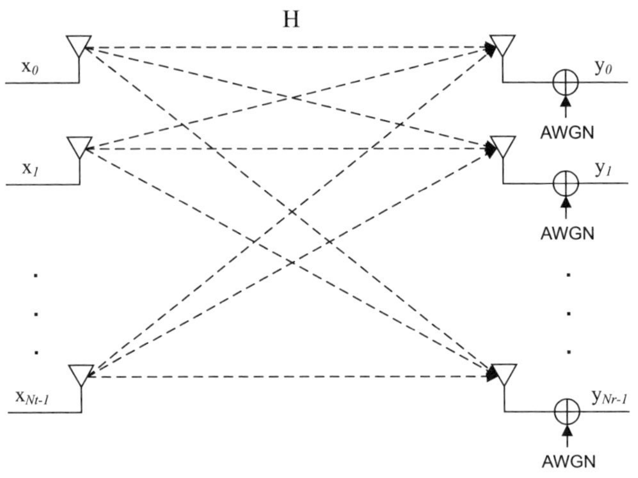

Figure 9.12 MIMO system model.

As is well known, the optimal detector for the MIMO system is the ML detector, which detects the transmitted signal vector as

$$\hat{\mathbf{x}}_{ML} = \arg\min_{\mathbf{x}} \left\| \mathbf{y} - \mathbf{H}\mathbf{x} \right\|^2 \tag{9.34}$$

The ML detection is performed through an exhaustive search over all candidate vector symbols. Obviously, the ML detector has to compute M^{Nt} distance metrics, which is very complex for large signal constellations and large transmit antennas. For example, in four transmit antenna system with 16-QAM, the number of searches that the ML receiver performs is $16^4 = 65{,}536$. Table 9.5 demonstrates the computational complexity of various conventional modulation schemes for two different antenna sizes.

9.4.2 Linear Detection

The linear receivers such as *zero-forcing* (ZF) and *minimum mean square error* (MMSE) detector separate the transmit streams and decode them individually. Decoding complexity is not high even for a large number of transmit antennas and/or a high modulation order, but they usually do not achieve full diversity gain.

Table 9.5 Computational Complexity of ML

Modulation	N_t	
	2	4
QPSK	16	256
16-QAM	256	65,536
64-QAM	4,096	16,777,216
256-QAM	65,536	4,294,967,296

As a consequence, the detection error rate is high, especially in high SNR region, when compared with that of the ML receiver. In contrast, in the case of the *successive interference cancellation* (SIC) detector, which employs the ZF or MMSE detector for detecting the transmit signals in a sequential manner, the detection error is reduced at the cost of increased complexity.

ZF Detection

In the case of the ZF detector, we suppress the interference among the transmit streams by multiplying the received signal vector $\mathbf{y}$ with the Moore-Penrose pseudo-inverse of the channel matrix, $\mathbf{H}^{+} = (\mathbf{H}^{H}\mathbf{H})^{-1}\mathbf{H}^{H}$, to get

$$\tilde{\mathbf{x}}_{ZF} = \mathbf{H}^{+}\mathbf{y} = \mathbf{x} + \left(\mathbf{H}^{H}\mathbf{H}\right)^{-1}\mathbf{H}^{H}\mathbf{n} \tag{9.35}$$

Once $\tilde{\mathbf{x}}_{ZF}$ is calculated, the demodulation is done by determining the N_t-vector $\hat{\mathbf{x}}$, whose ith element is the constellation point closest to the ith element of $\tilde{\mathbf{x}}_{ZF}$.

MMSE Detection

In the case of the MMSE detector, we minimize the mean square error by multiplying the received signal with $(\mathbf{H}^{H}\mathbf{H} + \sigma_n^2\mathbf{I}_{N_t})^{-1}\mathbf{H}^{H}$, which corresponds to a modified version of $\mathbf{H}^{+} = (\mathbf{H}^{H}\mathbf{H})^{-1}\mathbf{H}^{H}$, with the noise power σ_n^2 incorporated, to get

$$\tilde{\mathbf{x}}_{MMSE} = \left(\mathbf{H}^{H}\mathbf{H} + \sigma_n^2\mathbf{I}_{N_t}\right)^{-1}\mathbf{H}^{H}\mathbf{y} \tag{9.36}$$

The demodulation process is the same as that of the ZF detection.

In general, the MMSE detection is perceived as a better detection scheme than the ZF detection, as it reflects the noise effect in the matrix inversion process. Note that the ZF detection becomes identical to ML detection, in case the channel matrix is orthogonal. The interference suppression process of the ZF detection scheme leads to noise amplification, which affects the detection performance significantly when the SNR is low. As the SNR increases, however, the ZF detection approaches, in the detection algorithm as well as in performance, to the MMSE detection. The maximum diversity order that the ZF detection can achieve is $(N_r - N_t + 1)$.

SIC Detection

The SIC detection is a scheme that detects the transmit signal vector sequentially by detecting one signal at each iteration and eliminating the relevant interference in the received signal before continuing the next iteration. In the case of the *ordered SIC* (OSIC) detection, the strongest signal is detected, at each iteration, among the remaining set of the transmit signals. By ordering in that way, it can minimize the accumulation of detection errors iteration to iteration.

The conventional OSIC algorithm based on the MMSE receiver may be described as follows:

$$\text{Initialization: } i \leftarrow 1,\ \mathbf{G}^{1} = \left(\left(\mathbf{H}^{H}\mathbf{H} + \sigma_n^2\mathbf{I}\right)^{-1}\right)\mathbf{H}^{H}$$

$$k = \arg\max_{j} SINR_j^{i}$$

Recurrence: $i = 1{:}N_t$

$$\mathbf{w}_k = \mathbf{g}_k^i,\ \tilde{\mathbf{x}}_k = \mathbf{w}_k \mathbf{y}^i,\ \hat{\mathbf{x}}_k = Q(\tilde{\mathbf{x}}_k)$$

$$y^{i+1} = y^i - h_k x_k$$

$$\mathbf{H}^{i+1} = \left[\mathbf{h}_1^i \ldots \mathbf{h}_{k-1}^i \mathbf{h}_{k+1}^i \ldots \mathbf{h}_{N_r-i+1}^i\right]$$

$$\mathbf{G}^{i+1} = \left(\left(\mathbf{H}^{i+1}\right)^H \mathbf{H}^{i+1} + \sigma_n^2 I\right)^{-1} \left(\mathbf{H}^{i+1}\right)^H$$

$$k = \arg\max_j SINR_j^{i+1}$$

$$i \leftarrow i+1$$

In the algorithm, $SINR_j^i$ indicates the SINR of the jth transmit signal at the ith iteration, $\mathbf{g}_k^i$ is the kth row vector of $\mathbf{G}^i$, and Q is the demodulation function as described earlier.

9.4.3 Near-Optimal Algorithms

The ML detector exhibits optimal performance but requires the cost of a complicated exhaustive search. There are near-optimal detection algorithms that can reduce the search complexity to a substantially low level. In essence, they reduce the set of candidate constellation points to those near the received symbols. There are several methods of reducing the candidate sets, including tree searching, candidate selection, and symbol replica selection methods. Among them we consider in the following the three candidate selection methods—sphere decoding, QRM-MLD, and modified ML decoding.

Sphere Decoding

Sphere decoding is intended to search only over the lattice points inside a hyper-sphere of radius r around the received signal vector $\mathbf{y}$, thereby reducing the involved computations. Apparently, the constellation point closest to $\mathbf{y}$ inside the hyper-sphere will be the closest constellation point in the whole lattice. The sphere-decoding algorithm starts from a big hyper-sphere centered at the received vector $\mathbf{y}$ and checks if the hyper-sphere contains at least one symbol candidate inside. If there is no such a candidate within the hyper-sphere, the radius of the sphere is increased. Otherwise, the radius of the sphere is updated to the Euclidean distance between the received vector and the symbol candidate located inside the hyper-sphere. Then a candidate is searched within the newly updated hyper-sphere again, and this process continues until no candidate exists within the last updated hyper-sphere. Figure 9.13 illustrates this process. The complexity of the sphere-decoding algorithm does not depend on the lattice constellation size, which renders a very useful decoding means for high data rate transmission.

Sphere decoding is a kind of *integer least-square problem*, which is ML based. Some common heuristics include ZF detection, SIC, and OSIC. The three heuristics are known to have the same complexity $O(N_t \cdot N_r^2)$ [30].

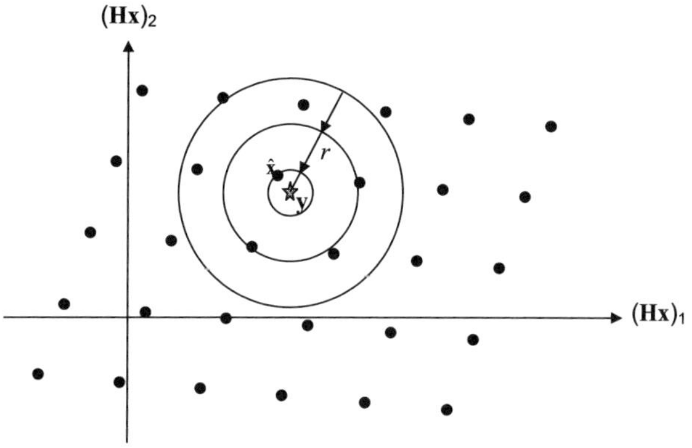

Figure 9.13 Illustration of sphere decoding.

QRM-MLD

The *maximum likelihood detection with QR decomposition and M-algorithm* (QRM-MLD) is intended to first determine the symbol replica candidates of the transmitted signals in a successive manner by applying the QR decomposition on the channel matrix **H** and then perform ML detection on the reduced set of symbol replica candidates. Specifically, we apply QR decomposition on the channel matrix to get

$$\mathbf{H} = \mathbf{QR} \tag{9.37}$$

for an $N_t \times N_t$ upper triangular matrix **R** and an $N_t \times N_t$ unitary matrix with orthonormal columns **Q**. If we multiply the received signal vector **y** with $\mathbf{Q}^H$, the resulting effective channel reduces to an upper triangular matrix **R** to yield the relation

$$\mathbf{z} = \mathbf{Q}^H \mathbf{y} = \mathbf{Rx} + \mathbf{Q}^H \mathbf{n} \tag{9.38}$$

Noting that R is an upper triangular matrix, we first determine the symbol candidate of $\hat{x}_{N_t-1}$ at the bottom. Then feeding back this value, we determine the symbol candidate replica of $\hat{x}_{N_t-2}$. By continuing this successive calculation process, we can determine all the symbol candidate replicas $\hat{x}_1, \hat{x}_2, \ldots, \hat{x}_{N_t-1}$. Note that it is also possible to choose multiple symbol replica candidates at each stage for each set of symbol replica candidates determined at the previous stage. Then we finally perform the ML detection on the sets of the symbol replica candidates to estimate the original transmit symbols. Figure 9.14 illustrates this QRM-MLD process. The QRD-MLD process can reduce the search space to a minimal level by performing the preprocessings of QR decomposition and successive calculation of symbol replica candidates.

Modified ML Algorithm

The *modified ML* (MML) algorithm is intended to reduce the computational complexity of ML detection of N_t symbols to that of $N_t - 1$ symbols by precalculating the possible symbol replica candidate points (or the constellation points) of one

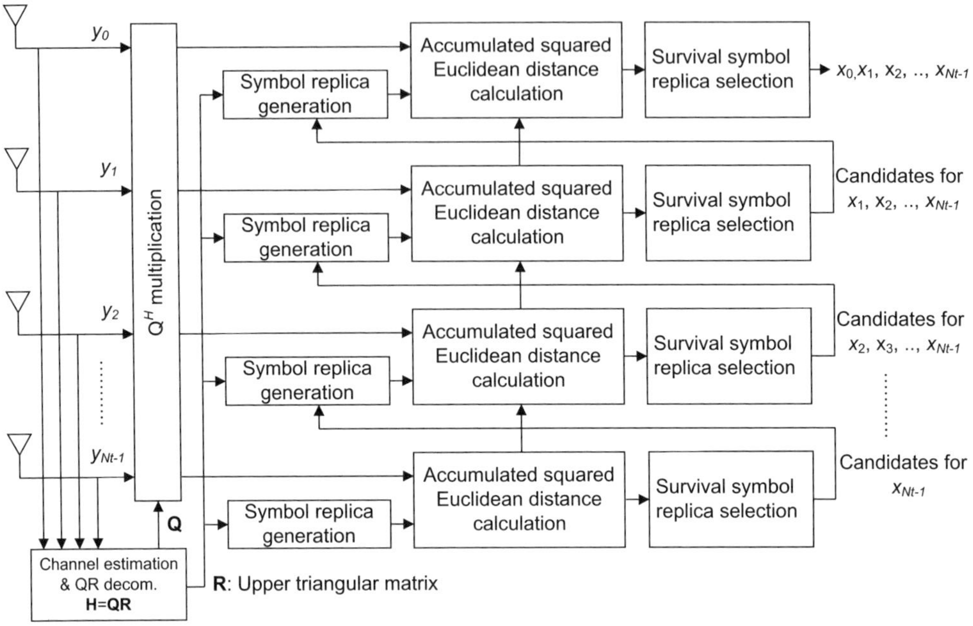

Figure 9.14 Block diagram of QRM-MLD process [31].

symbol with respect to all possible combinations of the other $N_t - 1$ symbols [32]. Specifically, we precalculate one symbol (e.g., x_0) with respect to the other $N_t - 1$ symbols (i.e., $x_i, i \in \{1, 2, \ldots, N_t - 1\}$) as follows:

$$x_0\left(x_1, x_2, \ldots, x_{N_t-1}\right) = Q\left\{\frac{\mathbf{h}_0^H}{\left\|\mathbf{h}_0\right\|^2}\left(\mathbf{y} - \sum_{i\in\{1,2,\ldots,N_t-1\}} \mathbf{h}_i x_i\right)\right\} \tag{9.39}$$

where $Q(\cdot)$ is the slicing (or demodulation) function and $\mathbf{h}_j$ is the jth column vector of the channel matrix $\mathbf{H}$. The equation indicates that x_0 is determined for each symbol set $(x_1, x_2, \ldots, x_{N_t-1})$ which takes M^{N_t-1} combinations in the case of the M-QAM modulation. We represent the kth N_t-vector $(x_0, x_1, x_2, \ldots, x_{N_t-1})$ by $\mathbf{x}_k$ and apply the ML detection for the $N_t - 1$ symbols $x_1, x_2, \ldots, x_{N_t-1}$. Then we get the MML expression

$$\mathbf{x}_{MML} = \arg\min\left\|\mathbf{y} - \mathbf{H}\mathbf{x}_k\right\|, \; k = 0, 1, \ldots, M^{N_t-1} - 1 \tag{9.40}$$

Therefore, the MML detector computes M^{N_t-1} distance metrics, which is $1/M$ of that of the ML detector. Figure 9.15 illustrates the MML detection process.

The MML process may be applied successively to further reduce the symbols one by one. Selection of the reference symbol (e.g., x_0 in the previous example) may follow the recently proposed *sorted MML* (S-MML) detection scheme [33]. It is reported that the S-MML detection outperforms the QRM-MLD at a half computational complexity [33].

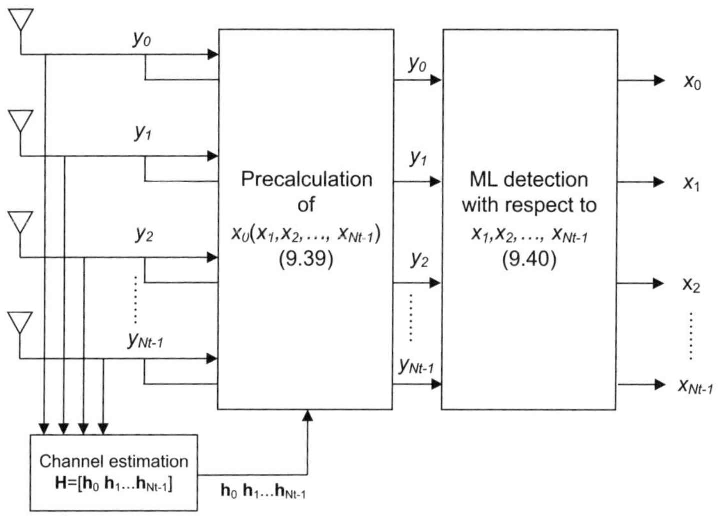

Figure 9.15 Block diagram of MML detection process.

References

[1] Viswanath, P., D. N. C. Tse, and R. Laroia, "Opportunistic Beamforming Using Dumb Antennas," *IEEE Trans. on Information Theory*, Vol. 48, No. 6, June 2002, pp. 1277–1294.

[2] Wicker, S. B., *Error Control Systems for Digital Communication and Storage*, Englewood Cliffs, NJ: Prentice-Hall, 1995.

[3] Viterbi, A. J., *CDMA: Principles of Spread Spectrum Communications*, Reading, MA: Addison-Wesley, 1995.

[4] Andrews, J. G., "Interference Cancellation for Cellular Systems: A Contemporary Overview," *IEEE Wireless Communications*, Vol. 12, No. 2, April 2005, pp. 19–29.

[5] Telatar, E., "Capacity of Multi-Antenna Gaussian Channels," *European Transactions on Telecommunications*, Vol. 10, No. 6, November 1999, pp. 585–596.

[6] Foschini, G. J., "Layered Space-Time Architecture for Wireless Communication in a Fading Environments When Using Multi-Element Antennas," *Bell Laboratories Technical Journal*, Vol. 1, No. 2, Autumn 1996, pp. 41–59.

[7] Boyd, S., and L. Vandenberghe, *Convex Optimization*, Cambridge, U.K.: Cambridge University Press, 2004.

[8] Roh, W., et al., "Framework for Enabling Closed-Loop MIMO for OFDMA," IEEE C802.16e-04/552r7, January 2005.

[9] Alamouti, S. M., "A Simple Transmit Diversity Technique for Wireless Communications," *IEEE Journal on Selected Areas in Communications*, Vol. 16, No. 8, October 1998, pp. 1451–1458.

[10] Dannann, A., S. Kaiser, "Transmit/Receive-Antenna Diversity Techniques for OFDM Systems," *European Transactions on Telecommunications*, Vol. 13, No. 5, September–October 2002, pp. 531–538.

[11] Tarokh, V., N. Seshadri, and A. R. Calderbank, "Space-Time Codes for High Data Rate Wireless Communications: Performance Criterion and Code Construction," *IEEE Trans. on Information Theory*, Vol. 44, No. 2, March 1998, pp. 744–765.

[12] Yuen, C., Y. L. Guan, and T. T. Tjhung, "Quasi-Orthogonal STBC with Minimum Decoding Complexity," *IEEE Trans. on Wireless Communications*, Vol. 4, No. 5, September 2005, pp. 2089–2094.

[13] Khan, M. Z. A., and B. S. Rajan, "Space-Time Block Codes from Co-Ordinate Interleaved Orthogonal Designs," *Proc. of IEEE International Symposium on Information Theory*, 2002, p. 275.

[14] IEEE Std 802.16e-2005 and IEEE Std 802.16-2004/Cor1-2005, Part 16: Air Interface for Fixed and Mobile Broadband Wireless Access Systems, February 2006.

[15] Zhang, J., et al., "UL CQICH SNR Feedback for MIMO Systems," IEEE C802.16e-05/118, March 2005.

[16] Heath, R. W., S. Sandhu, and A. J. Paulraj, "Antenna Selection for Spatial Multiplexing Systems with Linear Receivers," *IEEE Communications Letters*, Vol. 5, No. 4, April 2001, pp. 142–144.

[17] Gore, D. A., and A. J. Paulraj, "MIMO Antenna Subset Selection with Space-Time Coding," *IEEE Trans. on Signal Processing*, Vol. 50, No. 10, October 2002, pp. 2580–2588.

[18] Chae, C.-B., et al., "Enhancement of STC with Antenna Grouping," IEEE C802.16e-04/554r4, January 2004.

[19] Love, D. J., and R. W. Heath, Jr., "Grassmannian Beamforming for Multiple-Input Multiple-Output Wireless Systems," *IEEE Trans. on Information Theory*, Vol. 49, No. 10, October 2003, pp. 2735–2747.

[20] Mukkavilli, K. K., et al., "On Beamforming with Finite Rate Feedback in Multiple-Antenna Systems," *IEEE Trans. on Information Theory*, Vol. 49, No. 10, October 2003, pp. 2562–2579.

[21] Xia, P., and G. B. Giannakis, "Design and Analysis of Transmit-Beamforming based on Limited-Rate Feedback," *IEEE Trans. on Signal Processing*, Vol. 54, No. 5, May 2006, pp. 1853–1863.

[22] Siemens, "Description of the Eigenbeamformer Concept (Update) and Performance Evaluation," 3GPP R1-01-0203, February–March 2001.

[23] Lo, T. K. Y., "Maximum Ratio Transmission," *IEEE Trans. on Communications*, Vol. 47, No. 10, October 1999, pp. 1458–1461.

[24] Cavers, J. K., "Single-User and Multiuser Adaptive Maximal Ratio Transmission for Rayleigh Channels," *IEEE Trans. on Vehicular Technology*, Vol. 49, No. 6, November 2000, pp. 2043–2050.

[25] Zamir, R., S. Shamai, and U. Erez, "Nested Linear/Lattice Codes for Structured Multiterminal Binning," *IEEE Trans. on Information Theory*, Vol. 48, No. 6, June 2002, pp. 1250–1276.

[26] Hochwald, B. M., C. B. Peel, and A. L. Swindlehurst, "A Vector-Perturbation Technique for Near Capacity Multiantenna Multiuser Communication—Part II: Perturbation," *IEEE Trans. on Communications*, Vol. 53, No. 3, March 2005, pp. 537–544.

[27] Windpassinger, C., R. F. H. Fischer, and J. B. Huber, "Lattice-Reduction-Aided Broadcast Precoding," *IEEE Trans. on Communications*, Vol. 52, No. 12, December 2004, pp. 2057–2060.

[28] Samsung, "Downlink MIMO for EUTRA," 3GPP R1-060335, February 2006.

[29] Kim, S. J., et al., "On the Performance of Multiuser MIMO Systems in WCDMA/HSDPA: Beamforming, Feedback and User Diversity," *IEICE Trans. on Communications*, Vol. E98-B, No. 8, August 2006, pp. 2161–2169.

[30] Hassibi, B., and H. Vikalo, "On the Sphere-Decoding Algorithm I. Expected Complexity," *IEEE Trans. on Signal Processing*, Vol. 53, No. 8, August 2005, pp. 2806–2818.

[31] Higuchi, K., et al., "Adaptive Selection of Surviving Symbol Replica Candidates Based on Maximum Reliability in QRM-MLD for OFCDM MIMO Multiplexing," *Proceedings of IEEE Global Communications Conference*, 2004, pp. 2480–2486.

[32] Kim, J., and S. H. Nam, *Spatial Demultiplexing in 4x4 SM MIMO Systems: Modified ML (MML) and Recursive MML*, Technical Report, Samsung Advanced Institute of Technology, May 2005.

[33] Hwang, K. C., et al., "Iterative Joint Detection and Decoding for MIMO-OFDM Wireless Communications," *Proceedings of IEEE 40th Asilomar Conference on Signals, Systems and Computers,* 2006, pp. 1752–1756.

Selected Bibliography

Al-Dhanir, N., "Single-Carrier Frequency Domain Equalization for Space-Time Block Coded Transmissions over Frequency Selective Fading Channels," *IEEE Communications Letters*, Vol. 5, No. 7, July 2001, pp. 304–306.

Baro, S., G. Bauch, and A. Hansmann, "Improved Codes for Space-Time Trellis-Coded Modulation," *IEEE Communications Letters*, Vol. 4, No. 1, January 2000, pp. 20–22.

Bellofiroe, S., et al., "Smart-Antenna Systems for Mobile Communication Networks Part I: Overview and Antenna Design," *IEEE Antennas and Propagation Magazine*, Vol. 44, No. 3, June 2002, pp. 145–154.

Bellofiroe, S., et al., "Smart-Antenna Systems for Mobile Communication Networks Part II: Beamforming and Network Throughput," *IEEE Antennas and Propagation Magazine*, Vol. 44, No. 4, August 2002, pp. 106–113.

Biglieri, E., G. Taricco, and A. Tulino, "Decoding Space-Time Codes with BLAST Architectures," *IEEE Trans. on Signal Processing*, Vol. 50, No. 10, October 2002, pp. 2547–2552.

Blum, R. S., "MIMO Capacity with Interference," *IEEE Journal on Selected Areas in Communications*, Vol. 21, No. 5, June 2003, pp. 793–801.

Bolcskei, H., D. Gebert, and A. J. Paulraj, "On the Capacity of OFDM-Based Spatial Multiplexing Systems," *IEEE Trans. on Communications*, Vol. 50, No. 2, February 2002, pp. 225–234.

Byun, M.-K., D. Park, and B. G. Lee, "On the Performance Analysis of Space-Time Codes in Quasi-Static Rayleigh-Fading Channels," *IEEE Trans. on Information Theory*, Vol. 50, No. 11, November 2004, pp. 2865–2873.

Byun, M.-K., and B. G. Lee, "New Bounds of Pairwise Error Probability for Space-Time Codes in Rayleigh Fading Channels," *IEEE Trans. on Communications*, Vol. 55, No. 8, August 2007, pp. 1484–1493.

Costa, M., "Writing on Dirty Paper," *IEEE Trans. on Information Theory*, Vol. 29, No. 3, May 1993, pp. 439–441.

Foschini, G. J., and M. Gans, "On Limits of Wireless Communications in a Fading Environment When Using Multiple Antennas," *Wireless Personal Communications,* Vol. 6, No. 3, March 1998, pp. 311–315.

Gesbert, D., et al., "From Theory to Practice: An Overview of MIMO Space-Time Coded Wireless Systems," *IEEE Journal on Selected Areas in Communications*, Vol. 21, No. 3, April 2003, pp. 281–302.

Golden, G. D., et al., "Detection Algorithm and Initial Laboratory Results Using V-BLAST Space-Time Communication Architecture," *Electronics Letters*, Vol. 35, No. 1, January 1999, pp. 14–16.

Goldsmith, A., et al., "Capacity Limits of MIMO Channels," *IEEE Journal on Selected Areas in Communications*, Vol. 21, No. 5, June 2003, pp. 684–702.

Hammons, A. R., and H. El Gamal, "On the Theory of Space-Time Codes for PSK Modulation," *IEEE Trans. on Information Theory*, Vol. 46, No. 2, March 2000, pp. 524–542.

Hassibi, B., and H. Vikalo, "On the Expected Complexity of Sphere Decoding," *Proc. of Asilomar Conference on Signals, Systems and Computers*, 2001, pp. 1051–1055.

Heath, R., and A. J. Paulraj, "Switching Between Diversity and Multiplexing in MIMO Systems," *IEEE Trans. on Communications*, Vol. 53, No. 6, June 2005, pp. 962–968.

Jafarkhani, H., "A Quasi Orthogonal Space Time Block Code," *IEEE Trans. on Communications*, Vol. 49, No. 1, January 2001, pp. 1–4.

Jongren, G., M. Skoglund, and B. Ottersten, "Combining Beamforming and Orthogonal Space-Time Block Coding," *IEEE Trans. on Information Theory*, Vol. 48, No. 3, March 2002, pp. 612–627.

Lu, B., W. Wang, and K. Narayanan, "LDPC-Based Space-Time Coded OFDM Systems over Correlated Fading Channels: Performance Analysis and Receiver Design," *IEEE Trans. on Communications*, Vol. 50, No. 1, January 2002, pp. 74–88.

Narula, A., et al., "Efficient Use of Side Information in Multiple-Antenna Data Transmission over Fading Channels," *IEEE Journal on Selected Areas in Communications*, Vol. 16, No. 8, October 1998, pp. 1423–1436.

Paulraj, A., R. Nabar, and D. Gore, *Introduction to Space-Time Wireless Communications*, Cambridge, U.K.: Cambridge University Press, 2003.

Paulraj, A. J., et al., "An Overview of MIMO Communications—A Key to Gigabit Wireless," *Proceedings of the IEEE*, Vol. 92, No. 2, February 2004, pp. 198–218.

Rayleigh, G., and J. M. Cioffi, "Spatio-Temporal Coding for Wireless Communication," *IEEE Trans. on Communications*, Vol. 46, No. 3, March 1998, pp. 357–366.

Scaglione, A., et al., "Optimal Designs for Space Time Linear Precoders and Decoders," *IEEE Trans. on Signal Processing*, Vol. 50, No. 5, May 2002, pp. 1051–1064.

Schubert, M., and H. Boche, "Solution of the Multiuser Downlink Beamforming Problem with Individual SINR Constraints," *IEEE Trans. on Vehicular Technology*, Vol. 53, No. 1, January 2004, pp. 18–28.

Sharif, M., and B. Hassibi, "On the Capacity of MIMO Broadcast Channels with Partial Side Information," *IEEE Trans. on Information Theory*, Vol. 51, No. 2, February 2005, pp. 506–522.

Shiu, D., et al., "Fading Correlation and Its Effect on the Capacity of Multielement Antenna Systems," *IEEE Trans. on Communications*, Vol. 48, No. 3, March 2000, pp. 502–513.

Simon, S., and A. Moustakas, "Optimizing MIMO Antenna Systems with Channel Covariance Feedback," *IEEE Journal on Selected Areas in Communications*, Vol. 21, No. 3, April 2003, pp. 406–417.

Spencer, Q. H., et al., "An Introduction to the Multi-User MIMO Downlink," *IEEE Communications Magazine*, Vol. 42, No. 10, October 2004, pp. 60–67.

Tarokh, V., H. Jafarkhani, and A. R. Calderbank, "Space-Time Block Codes from Orthogonal Designs," *IEEE Trans. on Information Theory*, Vol. 45, No. 5, July 1999, pp. 1456–1467.

Vishwanath, S., N. Jindal, and A. J. Goldsmith, "Duality, Achievable Rates, and Sum-Rate Capacity of Gaussian MIMO Broadcast Channels," *IEEE Trans. on Information Theory*, Vol. 49, No. 10, October 2003, pp. 2658–2668.

Wolniansky, P. W., et al., "V-BLAST: An Architecture for Realizing Very High Data Rates over the Rich-Scattering Wireless Channel," *Proc. of URSI International Symposium Signals, Systems, and Electronics*, 1998, pp. 295–300.

Yu, W., et al., "Iterative Water-Filling for Gaussian Vector Multiple Access Channels," *IEEE Trans. on Information Theory*, Vol. 50, No. 1, January 2004, pp. 145–152.

Zheng, L., and D. N. C. Tse, "Diversity and Multiplexing: A Fundamental Tradeoff in Multiple-Antenna Channels," *IEEE Trans. on Information Theory*, Vol. 49, No. 5, May 2003, pp. 1073–1096.

CHAPTER 10

WiBro: The First Mobile WiMAX System[1]

WiBro, an abbreviation of *wireless broadband*, refers to the 2.3-GHz frequency-based Mobile WiMAX system developed and deployed in Korea. The WiBro system has selected a collection of features out of the mandatory and many optional specifications adopted in the IEEE 802.16e standards to form its unique profile.

In 2002, the Ministry of Information and Communications (MIC) in Korea reallocated the 100-MHz frequency band at 2.3-GHz spectrum for portable Internet services (or WiBro services), instead of fixed *wireless local loop* (WLL) services. The 100-MHz band was divided into three triplets of 9-MHz band each and some guard bands, with each triplet consisting of three 9-MHz bands. Figure 10.1(a) shows the resulting frequency allocation for the WiBro services in relation to the frequency allocation of overall wireless communication services in Figure 10.1(b). In 2004, MIC and the Telecommunication Technology Association (TTA), the Korean domestic standardization body, issued the basic requirements on WiBro. In 2005, the MIC issued WiBro licenses to three operators including KT (formerly, Korea Telecom) and SK Telecom.

In 2005, Samsung Electronics developed the world's first commercial Mobile WiMAX system based on the 2.3-GHz WiBro profile and KT deployed the WiBro network based on the 27-MHz band in the middle (i.e., B-band).[2] In June 2006, KT made a large-scale trial of the 2.3-GHz WiBro network deployed in the Seoul metropolitan area, and, in April 2007, KT started full commercial WiBro services in the Seoul metropolitan area and its vicinity for the first time in the world.

In this chapter, we deal with the technical aspects of this WiBro system, focusing on the system design, network deployment, and services. Specifically, we discuss various issues on the requirements and configuration of the WiBro system; on the design of the *radio access station* (RAS), or *base station* (BS), and the *access control router* (ACR), or *access service network gateway* (ASN-GW); and on the radio network planning, implementation, and optimization of the WiBro network.[3] In addition, we introduce various application services that make WiBro unique and differentiate it from other types of existing mobile services. In all the WiBro-related discussions, we will rely on the practical systems, networks, and services that Samsung Electronics and KT have developed and deployed.

1. The content of this chapter is generated out of the cowork of H. Kim at KT, J. Lee at Samsung Electronics, and B. G. Lee, who also coauthored a similar chapter in [1]. Publisher Wiley kindly agreed to put the content in this book.
2. ETRI developed a prototype WiBro system in December 2004, confirming the technical feasibility of the WiBro system.
3. The WiBro system uses the terminology ACR for ASN-GW and the terminology RAS for BS. While either terminology may be used, we choose to use the terms ACR and RAS in this chapter, as they give a realistic flavor of the WiBro system.

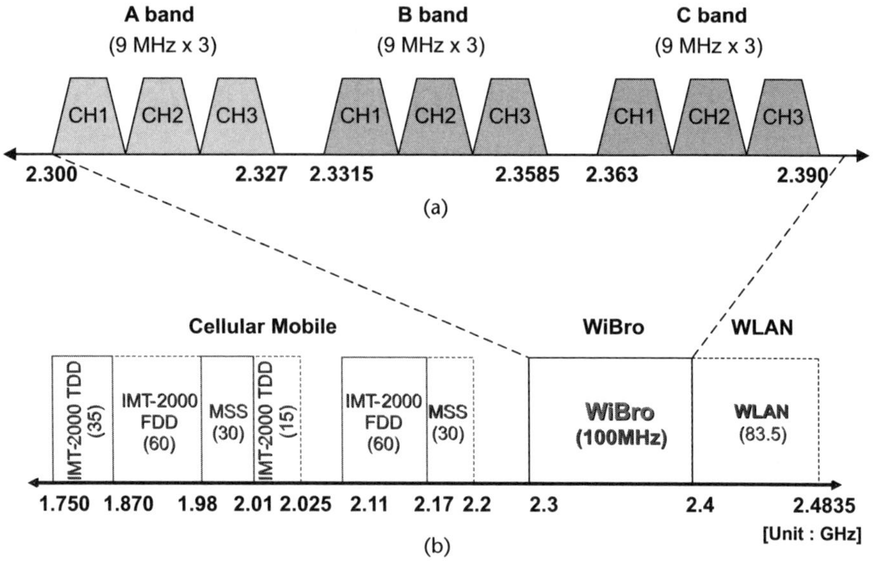

Figure 10.1 Frequency allocation for (a) WiBro and (b) overall wireless communication services in Korea.

10.1 WiBro Network Configuration

As is the case of Mobile WiMAX, WiBro is designed to be an IP-mode mobile network, in contrast to the existing cellular mobile networks such as WCDMA/HSDPA or cdma2000/1x-EVDO, which are circuit-mode-based with packet-mode hybridization. Since WiMAX adopts an all-IP network structure tailored for Internet service provision, the network structure is simple, inexpensive to construct, and adequate for providing a diverse set of services.

Let's revisit earlier chapters, where Figure 2.11 illustrates the configuration of the existing circuit-mode cellular mobile communication network and Figure 1.6 illustrates the configuration of the Mobile WiMAX network. Comparing the two figures, we observe that the WiBro network includes neither the *base station controller* (BSC) and MSC of the IS-95/EV-DO cellular mobile family nor the RNC, SGSN, and GGSN of the GSM/WCDMA cellular family. It includes only the ASN-GW instead.

The IP packets generated by user terminals can be delivered to the Internet via the BS to ASN-GW path. This demonstrates how simple it is to provide Internet services over the WiBro network. Consequently, a diverse set of services including the *voice over IP* (VoIP) services can be provided over the WiBro network at low cost.

10.1.1 WiBro Network Architecture

Figure 10.2 shows the architecture of the WiBro network. The WiBro network is composed of MSs, an *access service network* (ASN), and a *connectivity service network* (CSN).

The ASN contains RASs (or BSs) and ACR (or ASN-G/W), which are managed by Mobile WiMAX *system manager* (WSM). Located in the user side of the ASN are

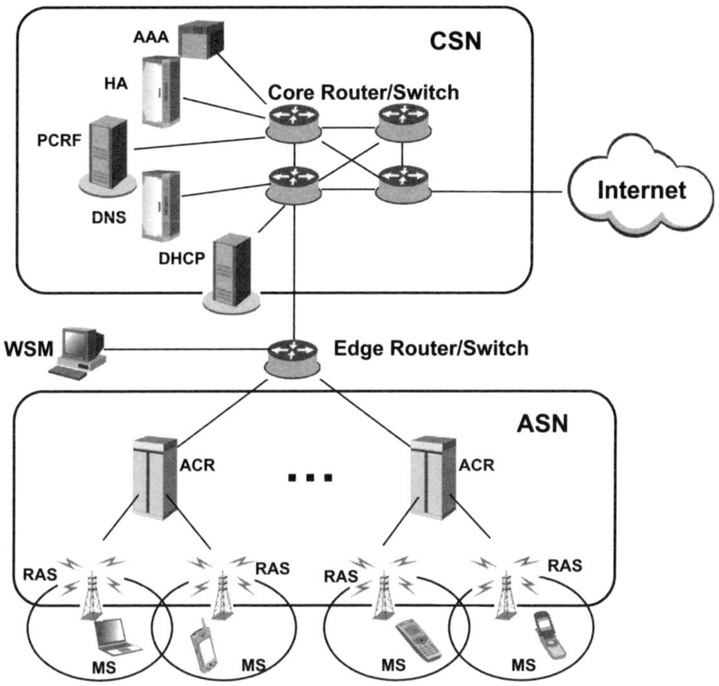

Figure 10.2 WiBro network architecture.

the MSs. As such, the WiBro network architecture is much simpler than that of existing mobile communication networks.

CSN is composed of various servers, namely, an *authentication, authorization, and accounting* (AAA) server, a *home agent* (HA), a *dynamic host configuration protocol* (DHCP) server [2], a *domain name service* (DNS) server, and a *policy and charging rules function* (PCRF) server. ASN is connected with CSN via a router or a switch.

10.1.2 ASN-GW

ACR is the central system of the WiBro network, which connects the CSN and RAS. It enables multiple RASs to interwork with CSN and IP networks and sends and receives traffic between the external network and MS. Basically, it performs the routing function for transferring data between the RAS and the Internet, and the control function for controlling the WiBro users, services, and mobility. It corresponds to the combined functions of the circuit-mode-based BSC and the *packet data serving node* (PDSN) in the existing cellular mobile network.

The functions of the ACR include handover control, IP routing and mobility management, user service profile information provision, billing service provision to billing server, and security function. The ACR performs the packet classification and *packet header suppression* (PHS) functions in support of the *convergence sublayer* (CS). It performs the header compression function, and supports *robust header compression* (ROHC) [3]. In addition, it performs the paging and location register functions for the MSs in idle mode. For authentication, the ACR performs the authentication and key distribution functions by interworking with the AAA server, with the RAS performing the key receiver function to receive the security key from the key distributor. ACR interworks with the AAA server of the CSN for charging services too. It also interworks with the HA of CSN for *mobile IP* (MIP) services, and supports both *proxy MIP* (PMIP) and *client MIP* (CMIP).

Mobile WiMAX System Manager (WSM)

WSM provides the management environment for network operators to operate and maintain the ACR and RAS. The WSM performs ACR and RAS architecture management, fault management, statistics management, configuration management, command line interface, and software version management. Operators can inquire the status of the system, sectors, *frequency assignments* (FAs), and repeaters to the WSM. The WSM classifies and selects the commands that can be executed in the ACR and RAS.

10.1.3 RAS (or BS)

RAS is an entry system of the WiBro network, which connects ACR and MS. It receives subscriber data via wireless path, passes the data to the ACR in the upstream, and distributes the data received from the ACR to user terminals in the downstream. The RAS performs various functions, including transmission and reception of physical layer signals with MSs, modulation/demodulation, coding, packet scheduling for QoS assurance, allocation of radio resources, service flow management, ARQ processing, and ranging function. In addition, the RAS controls the connection for packet calls and handover.

RAS performs *service flow management* (SFM) function to create/change/release connections for each *service flow* (SF). An admission control function is required while creating/changing the connections. In support of the SFM function of the RAS, the ACR performs an *SF authentication* (SFA) function to obtain QoS information from the *policy function* (PF) and applies it when creating the SF, and also performs an *SF identification* (SFID) management function to create/change/release SFID and map the SF according to the packet classification.

In relation to handover, the RAS performs a handover control function to determine the initiation of the corresponding handover procedure. It checks the neighbor RAS list and relays the handover signaling message to the right target RAS system. At that time, the ACR and RAS conduct the context function to exchange the context information between the target RAS system and the serving RAS system.

For radio resource management, the RAS performs *radio resource control* (RRC) and *radio resource agent* (RRA) functions to collect/manage the radio resource information from MSs and the RAS itself.

Mobile Station (MS)

MS is the end-user device of the WiBro network that performs the input/output function and various other functions to process the information, to access IP-based WiBro networks, and to perform the various functions required for the terminating network element. The functions of MS include wireless access to the ASN, IP-based call services, the support of IP mobility, the authentication and security of MS and subscribers, and the reception of multicast services.

10.1.4 CSN Servers

CSN servers include the basic network servers for service provision and various application servers for providing application services. The network servers include the HA for management of home address, the AAA server for security and accounting functions, the DNS server for conversion of IP addresses and system names, the DHCP server for dynamic allocation of IP, the PCRF server for managing the service policy and for sending QoS setting and accounting rule information. The application servers include the servers for *push-to-talk* (PTT), *instant messaging* (IM), *multimedia messaging system* (MMS), *location-based service* (LBS), games, and other services.

The HA accesses other networks and enables MIP users to access the Internet. The HA interworks with the ACR that performs a *foreign agent* (FA) function in mobile IPv4 environment and interworks with MS to exchange data in mobile IPv6 environment.

The AAA server interfaces with the ACR and carries out subscriber authentication, authorization, and accounting functions. It interfaces with the ACR via the DIAMETER protocol and provides *extensible authentication protocol* (EAP) certification [4].

The DNS server manages the domain names. It interprets the domain or host names to the IP addresses in the form of binary digits.

The DHCP server manages the setup and the IP addresses of MSs. It performs the management and allocation of the IP addresses and other setup information for MSs. When external DHCP server does not exist, since the ACR includes the DHCP server and relay agent functions, the ACR performs the DHCP server function.

The PCRF server manages the service policy and sends both QoS setting information for each user session and accounting rule information to the ACR.

10.2 WiBro System Requirements

The WiBro system specifications require various different types of parameters and functions in radio access, network, and services. For example, the WiBro system is basically required to use TDD and OFDMA technologies among all the options available within the IEEE 802.16e standards. The basic requirements encompass the channel bandwidth, frequency reuse factor, spectral efficiency, handover, and others. In addition, there are various functional requirements in the network and services levels as well.

10.2.1 Requirements on Radio Access

The requirements on radio access are addressed in terms of three basic system parameters and six additional requirements. The basic system parameters include duplexing, channel bandwidth, and multiple access and the additional requirements include *frequency reuse factor* (FRF), spectral efficiency, per-subscriber transmission rate, handover, mobility, and service coverage. Besides, there are other items to consider as well, including the support of QoS parameters (namely, jitter, delay, packet error rate, and transmission rate), interworking with other networks, the number of concurrently serviceable subscribers, power-saving functionality, the support of AMC function, authentication and encryption, robustness to multipath environment (or delay spread), and round-trip delay.

The three basic parameters initially specified in the WiBro system are: (1) a frequency band of 2.3 GHz, (2) a channel spacing of 9 MHz (effective bandwidth: 8.75 MHz each), and (3) *time-division duplexing* (TDD). In addition to three basic parameters, the WiBro profile has additional parameters: It uses the OFDMA as a multiple access technology in conjunction with an FFT size of 1,024 and a TDD frame length of 5 ms. It uses QPSK, 16-QAM, and 64-QAM for modulation, convolutional turbo code for channel coding, and hybrid ARQ for data retransmission. When compared with the main parameters of the Mobile WiMAX given in Table 4.2, WiBro profile takes a subset of them, as listed in Table 10.1(a).

10.2.2 Requirements on Networks and Services

The requirements of WiBro network and services are specified to achieve the fundamental goal of providing high-data-rate services to the users in macrocell, microcell, and picocell environments. The WiBro network is required to enable MSs to receive services while moving at vehicle speed. It is also required to support L2 handover capability so that MSs crossing over a cell boundary can receive IP-based services continuously without interruption. In addition, it is required to support security functions to protect the subscriber information, equipment, and the network from the abuse or attack of unauthorized third parties. Further, it is required to support differentiated QoS services by providing real-time and nonreal-time services as well as best-effort services.

There are additional functionalities to be supported: WiBro network should provide versatile forms of billing-related data that can support various different types of billing systems. It should be capable of interworking with the diverse set of existing wireless data networks such as wireless LANs and mobile data networks. Besides, it should support multicast and broadcast services. In particular, it should

Table 10.1 WiBro Profile: (a) Main Parameters and (b) Main Requirements

Parameters	Values
Frequency band	2.3 GHz
Effective bandwidth	8.75 MHz
Duplexing	TDD
Multiple access	OFDMA
TDD frame length	5 ms
FFT size (N_{FFT})	1,024
Modulation	QPSK,16-QAM,64-QAM
Channel coding	Convolutional turbo code
ARQ	Hybrid ARQ

(a)

Parameters	Values
Frequency reuse factor	1
Peak throughput	19 Mbps, DL 5 Mbps, UL
Spectral efficiency	6/2 bps/sector DL/UL, max 2/1 bps/sector DL/UL, avg
Mobility	120 km/h
Handoff	150 ms
Service coverage	Picocell: 100 m Microcell: 400 m Macrocell: 1 km

(b)

support the change of the network configuration and network operation of the uplink/downlink portion of the TDD frame in accordance with the asymmetrical characteristics of the Internet traffic.

Specifically, the main requirements and the performance goals of the WiBro system are as follows: The target peak throughput is 19 Mbps downstream and 5 Mbps upstream. The target FRF among cells is 1. The target spectral efficiency is 6 bps/Hz/cell at maximum and 2 bps/Hz/cell on average in the downlink, and 2 bps/Hz/cell at maximum and 1 bps/Hz/cell on average in the uplink. The maximum moving speed of the user who can receive the service is 120 km/h and the maximum allowed time of service interrupt due to handover is 150 ms. Cell service coverage is 100m in picocells, 400m in microcells, and 1 km in macrocells, all in radius. Table 10.1(b) lists a summary of the main requirements specified in the WiBro profile.

10.2.3 Requirements on ACR and CSN

Security is an important issue in WiBro network. For authentication and encryption key exchange, the ACR and CSN are required to support the *extensible authentication protocol* (EAP)–based authentication and security protocol and, as necessary, to include the *public key infrastructure* (PKI)–based functions as well. For the exchange of authentication information, they need to support the DIAMETER protocols. In addition, they need to support various types of subscriber and terminal authentication functions.

WiBro network is required to interwork with other networks in support of the handover and roaming services. Interworking is important in order to maintain the IP-based services when the MSs in service move into other networks. By interworking with other networks, WiBro network can exchange the authentication and billing services with the other networks, thus enabling the practical operation of mobile IP-based IP mobility. In addition, WiBro network is required to support L2- and L3-based mobility functions so that WiBro services can be maintained continuously even when an MS in service moves into another cell under different ACR.

For network management, WiBro network is required to support *simple network management protocol* (SNMP)–based network management functions (e.g., failure management, configuration management, performance monitoring).

In order to support various billing management functions, WiBro system is required to support a set of basic data depending on the per-subscriber service characteristics, such as the start and the finish times of the service, the number of the served data packets, the identification numbers of the RAS and MS, the service class and QoS level, and the reason of error or failure if it occurred.

The WiBro network is required to provide various control services for MSs to access the WiBro network and receive services, namely, the network access control, traffic connection and control, and network release control functions. The network access control function includes authentication, registration and address assignment, and billing start functions; the traffic connection control function includes the setup, change, and termination of the traffic connection; and the network release control function includes deregistration, address withdrawal, and billing stop.

In order to support the QoS attributes defined at the RAS, the core network is required to support providing differentiated QoS depending on the service attributes.

10.2.4 Requirements on RAS

From the RAS point of view, it is important to meet the high data rate requirement of the WiBro system. So the RAS is to be designed such that it can ensure high spectral efficiency. To achieve the goal, it adopts an FRF of one and, in addition, the AMC technology that varies the modulation and coding scheme depending on the MS location (e.g., urban area or rural area; picocell, microcell, or macrocell; center or boundary of a cell) and the channel condition (e.g., the load of the neighboring cells, the current channel state of the user). User transmission rate is related to the transmission capacity that a network operator can provide in the service aspect, which may be different from the minimum transmission rate that the physical layer can provide.

The WiBro system specifies various parameters for QoS, such as jitter, delay, frame loss rate, and transmission rate. Specifically, *jitter* refers to the variation of the arrival times of the consecutive frames transmitted in the wireless access link; *delay*, the time duration that takes to deliver frames to the destination in the wireless access link; *frame loss rate*, the ratio of the unsuccessfully received frame to the total transmitted frames in the wireless access link; and *transmission rate*, the data rate required to meet the service quality when traffic is generated. The RAS should be designed and operated to meet these specifications, whose values may be determined differently depending on the service types.

In addition, the WiBro system is required to offer various types of handover functions in order to maintain IP services continuously, even when MSs in service move to different sectors within the same cell, move into different cells, or switch to different frequency bands. The RAS is required to be designed to meet the relevant handover requirement properly.

Requirements on MS

There are several functions and capability required for the MS to provide services properly: It is required to support power-saving techniques to minimize power consumption, and to support L2- and L3-based mobility in order to maintain IP-based service to the MSs in service even after moving into another RAS. It is required to be capable of receiving the multicast and broadcast information that is transmitted from the network. It is required to provide a proper means for access control when interworking with other networks is in progress. It is required to support the EAP-based authentication and security protocols and, if necessary, to expand them to PKI-based ones. Also, it is required to support various subscriber and terminal authentication functions and various encryption functions.

10.3 RAS System Design

The RAS performs air-interface processing based on the Mobile WiMAX profiles of IEEE 802.16e specification. Table 10.2 lists a design specification of the WiBro sys-

Table 10.2 WiBro System Design Specification

Parameters	Value or technology
Multiple Access	OFDMA
Duplexig	TDD
FA separation	9 MHz
Effective bandwidth	8.75 MHz
MIMO technology	2x2
Peak data rate	45 Mbps DL, 12 Mbps UL @10MHz
Mobility	120 km/h
Service coverage	1~15 km
Cyclic prefix	12.8 us (1/8)

tem that meets the Mobile WiMAX profile. Note that the data rates of 45 Mbps *downlink* (DL) and 12 Mbps *uplink* (UL) can be obtained in the WiMAX Wave 2 system using 10-MHz bandwidth (based on 64-QAM with 5/6 code rate in DL and 16-QAM with 3/4 code rate in UL, utilizing 26 symbols in DL and 12 symbols in UL). The WiBro profile is a subset of the Mobile WiMAX profile. The WiBro profile uses an effective 8.75-MHz bandwidth. The maximum DL data rate is 34.56 Mbps (where 24 symbols are used for data burst, 2 symbols are used for DL-MAP) and the maximum UL date rate is 8.64 Mbps (when using UL CSM at highest UL MCS). Note that there is a 10-MHz profile in mobile WiMAX (where possible DL/UL symbol ratios are 30:18, 27:21, and so on). The system supports the 2×2 MIMO technology, 120 km/h mobility,[4] and 1–15 km of coverage. It adopts the *time-division duplexing* (TDD) technology for duplexing and the *orthogonal frequency division multiple access* (OFDMA) technology for multiple access, taking the *cyclic prefix* (CP) of 1/8 size. Multiple RASs are controlled by a single ACR, and they process functions to link to the ACR and their own MSs. The functions include modulation/demodulation, radio resources management, packet scheduling to guarantee QoS, and interworking with the ACR for handover.

10.3.1 RAS Architecture

An RAS, in general, is composed of five functional units, namely, a *global positioning system* (GPS) receiver and clock unit, an RF system unit, a baseband unit, a network processor unit, and a network interface unit, as illustrated in Figure 10.3.

4. The WiBro speed requirement was originally set to 60 km/h in consideration of the existing cellular service providers. However, the WiBro system is designed to support 120 km/h and even higher speeds at the cost of some performance loss. Further, WiBro wave 2 targets at the speed of 350 km/h.

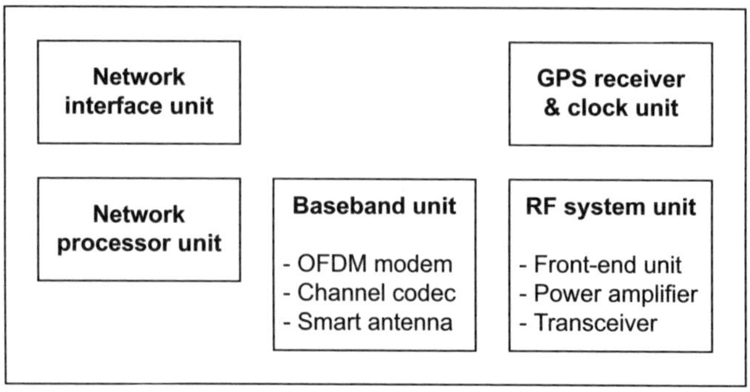

Figure 10.3 Functional architecture of RAS.

GPS Receiver and Clock Unit

The GPS receiver and clock unit receives the GPS signal and generates timing reference to maintain system synchronization.

RF System Unit

The main components of the *radio frequency* (RF) system include a front-end unit, power amplifier/low-noise amplifier, and transceiver.

The front-end unit sends out RF transmit signals to the antenna and bandpass-filter, and amplifies the received signals. It uses a switch for TDD. In addition, it may support the diagnosis function on the RF transmit/receive paths.

Specifically, the key functions of the RF system unit include the following: transmission of RF signals through antenna; suppression of spurious out-of-band signals that are emitted from the received RF signal; low-noise amplification of the received pass-band RF signals; distribution of the down-converted RF signals to several channel cards; and TDD switching function for the RF. These functions are performed in the *front-end board* (FEB) containing power amplifier, *low-noise amplifer* (LNA), and transceiver.

Baseband Unit

The baseband unit includes an OFDM modem, channel codec, and smart antenna/ *space time coding* (STC) unit.

The OFDM *modulator/demodulator* (modem) performs the following functions: modulation and demodulation of OFDM signals, synchronization for packet traffic burst, link control (such as power control, frequency offset control, and timing offset control), and intercarrier interference cancellation.

The channel *coder/decoder* (codec) is responsible for coding/decoding the duo-binary *convolutional turbo code* (CTC) and memory management for HARQ support.

The smart antenna performs TDD RF calibration and beamforming for supporting the *space division multiple access* (SDMA) scheduler. The STC and frequency hopping diversity coding are supported for two and four transmit antennas.

Network Processor Unit

The network processor unit of RAS is responsible for such functions as scheduling multiuser traffic with QoS support, radio resource management, and handover support in cooperation with ACR.

Network Interface Unit

The network interface unit supports the interface between the ACR and RAS.

10.3.2 RAS Functions

The main function of RAS is to perform the air interface processing based on the IEEE 802.16e specifications. Multiple RASs are controlled by a single ACR. The RAS processes functions to link the ACR with MSs, which include modulation/demodulation, call processing, radio resources management, packet scheduling to support QoS, interworking with ACR for handover, operation and maintenance, and other additional functions. Table 10.3 lists a summary of the WiBro RAS functions.

Call Processing Function

The RAS performs the call processing function to provide an initial access to MSs, allocates the *connection identifier* (CID) to MSs, and supports handover. In addition, the RAS supports the location update and registration between MSs and ACR, and transmits subscriber data between MSs and ACR.

Handover Optimization Function

The RAS may support the handover between sectors (i.e., *intersector HO*), the handover between RASs (i.e., inter-RAS HO), the handover between ACRs (i.e., *inter-ACR HO*), and the handover between FAs (i.e., *inter-FA HO*). To maintain the call quality, it is necessary to optimize packet loss rate and handover delay time.

The RAS may use a flexible frequency management scheme to obtain higher SINR for MSs in the area. Also, it can raise the link performance during handover by supporting soft handover, *macro diversity handover* (MDHO), between sectors in the upper link.

Table 10.3 A Summary of WiBro RAS Functions

RAS Functions	Description	
Call processing function	Call processing for initial access	
Handover optimization function	Maintaining call quality during handover , Minimizing delay time	
Resource management function	Radio resource management, Overload control	Scheduling,
Operation and maintenance function	Configuration management, Measuring and statistics	Status management,
Additional functions	System test, Failure diagnosis and processing, Auto-switching,	Operation test, Alarm, Remote access

To minimize the handover delay, the target RAS may reuse the session data setup between the serving RAS and the MSs, and minimize the time for re-entry of the MSs to the target RAS in the mobile area. It can be achieved by reducing the IEEE 802.16e MAC messages between the RAS and MS, such as *subscriber-station basic capability* (SBC), *registration* (REG), and *dynamic service addition* (DSA).

Resource Management Function

The RAS may detect the service status of the MS for each FA/sector and may not assign additional calls to the FA/sector where the service is not available, but assigns calls to the service available FA/sector. The RAS manages the MS awake/sleep mode and the MS connection status. It also controls the traffic volume received from the ACR by limiting the number of assigned calls for a certain time period based on the specified overload grades.

The RAS may manage the failure status, MS connection status, and the awake/sleep mode status of MSs for each FA/sector.

To support QoS, the RAS performs the QoS scheduling function, where the scheduler manages QoS parameters for each subscriber. The QoS parameters can be set up separately.

If the system gets overloaded, the RAS may limit new calls not to exceed the threshold value defined on the overload level for a certain period of time, based on the specified overload grades.

Operation and Maintenance Function

In the events of failure, cancellation, board status change, link status change, board switching, or link switching, the corresponding data are reported to the network management system (i.e., WSM) in real time. Through the WSM, a network operator can inquire the configuration data of RAS and support the expansion/reduction of the network.

While operating the system, network operators can change or delete the system operating parameters, and also change the configuration data.

Upper processors of RAS may manage the status (such as operation status, duplex status) of lower processors. Operators can inquire about the status of the system, sector, FA, and repeater services. The network management system can classify and select the commands that can be executed in the ACR and RAS.

The RAS counts the number of events in the system to create the statistics data by checking the status, maintenance, and the performance of the system, and sends the data to the network management system so that operators can check the statistics data as necessary.

Additional Functions

The RAS may provide the diagnosis and failure detection functions for the *interprocessor communication* (IPC) and RF path of the boards, which are related to system services except for the power supply module.

The RAS may support the diagnostic function for the call path in the link and the system. The diagnostic results and the quality measurements are obtained, through the network management system, in the format that can be analyzed.

The RAS may generate alarms in the case of a system failure or an operation failure and report the status to the network management system in real time according to the severity criterion that operators had preset. If there occurs any mechanical failure that may affect the service, it switches the boards and the links, and executes the overload control function for the corresponding device to stop the service. In addition, if there is any failure in the duplex boards or links, it switches the failed board or link, and reports the changeover status to the network management system. In the case of a software failure, it restarts the failed block or recovers it by restarting or auto-loading.

Alarms can be classified into multiple levels depending on the level of critical effects on the system (e.g., critical alarm, major alarm, minor alarm, and warning alarm). Alarm is released if the system is recovered back to normal operation.

Duplex boards and devices, except for the channel cards, are designed to switch to each other automatically in case a failure occurs, without affecting the services that the system is providing. In channel card switching, the service is dropped since the standby channel card does not backup the data of the present active channel card. Nevertheless, it does not affect new services.

Operators may perform maintenance and debugging remotely using the remote access function of the RAS.

10.4 ACR System Design

The ACR processes the control signal of the air interface and the bearer user traffic. For interworking with CSN, ACR provides the authentication, accounting, IP QoS, and other functions.

10.4.1 ACR Architecture

The ACR interfaces with the RAS, other ACRs, WSM, and CSN servers such as HA, DNS, AAA, DHCP, and PCRF servers. In principle, an ACR can control up to 1,000 RASs using their own RAS IDs. The maximum number of acceptable RASs varies depending on the call model, RAS configuration, and RAS throughput. Figure 10.4 depicts how the ACR interfaces with the RAS, other ACR, WSM, and CSN servers.

The interface between an ACR and an RAS in the same ASN has a reference point R6, as defined in Mobile WiMAX NWG. The R6 interface consists of signaling plane (IP/UDP/R6) and bearer plane (IP/*generic routing encapsulation* (GRE)) [5]. The physical access of the interface can be implemented over *gigabit Ethernet* (GE)/*fast Ethernet* (FE).

The interface between an ACR and another ACR in different ASN is specified as R4 interface, as defined in Mobile WiMAX NWG. The R4 interface consists of signaling plane (IP/UDP/R4) and bearer plane (IP/GRE). Its physical access method is GE/FE.

The interface between an ACR and a WSM complies with the SNMPv2c/SNMPv3 (which are the IETF standards), *secure file transfer protocol* (sFTP), or a manufacturer's proprietary standard. Its physical access method is GE/FE.

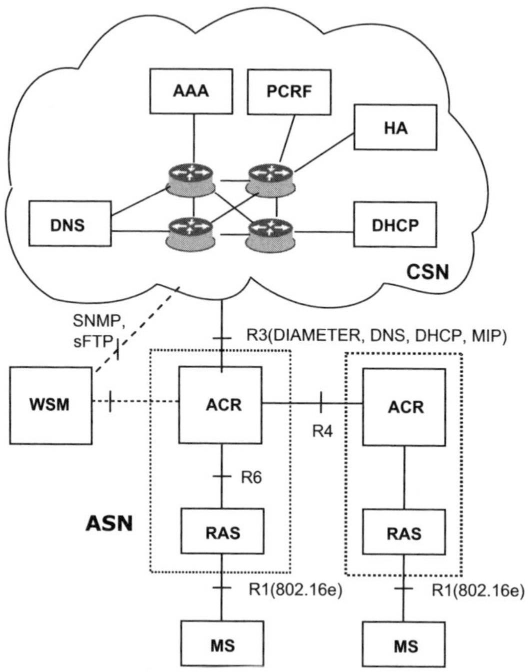

Figure 10.4 Interfaces of ACR.

The interfaces between an ACR and CSN servers are as follows: The interface with a HA complies with the IETF standard MIP specification and its physical access method is GE/FE. The interface with an AAA server complies with the IETF standard DIAMETER specification and its physical access method is GE/FE. The interface with a PCRF server complies with the IETF standard DIAMETER specification and the 3GPP standard *policy and charging control* (PCC) specification and its physical access method is GE/FE. The interface with a DNS server complies with the IETF standard DNS specification and its physical access method is GE/FE. The interface with a DHCP server complies with the IETF standard DHCP specification and its physical access method is GE/FE.

10.4.2 ACR Functions

The ACR performs various ASN gateway functions including mobility support, call processing, bearer processing, and interworking functions. For mobility support, it performs ranging control, MIP, context, handover, and other functions; for call processing, it does authentication, security key management, service flow authentication, accounting, and other functions; for bearer processing, it does IP packet forwarding and routing, header compression, QoS, and other functions; and for interworking, it does HA interworking, AAA server interworking, PCRF server interworking, IP address allocation, and other functions. Table 10.4 lists a summary of the ACR functions.

Table 10.4 A Summary of WiBro ACR Functions

ACR Functions	Description
Mobility support functions	Optimized hard handover, MIP function, Handover processing under L2 layer, Handover processing including L3 layer, Context transmission, Paging controller, Fast BS switching
Call processing functions	Service flow authorization, MS authentication, Subscriber authentication, Security key management, Accounting information collection & report
Bearer processing functions	Packet classification, Packet header suppression, Robust header compression, Data path function, IP QoS function, End-to-end QoS structure, Simultaneous IPv4/IPv6 support, IP routing function, Ethernet/VLAN interface
Interworking functions	HA interworking, AAA server interworking, PCRF server interworking, IP address allocation

Mobility Support Function

The ASN consists of several ACRs and each ACR group's several MSs, and manages each group as one IP subnet. Thus the CSN communicating with an MS considers the MS as an end host connected to an IP subnet.

The ACR provides the mobility support functions listed in Table 10.4 to maintain the connection between the MS and the network wherever the MS moves in such an IP network structure at any status.

The ACR performs an L2 handover without reauthentication, keeping the anchor function of MIP FA. So the ACR can carry out the a hard handover optimized up to the highest level defined in the IEEE 802.16e standards by minimizing the break time caused by the handover.

The ACR supports simple IP and MIP, and provides both PMIP and CMIP as the FA of MIP. In the PMIP, when an MS does not support the MIP stack, the ACR performs the FA function of MIP and the MIP client function replaces the MS. Thus, the ACR enables continuous services even when the MS supports only simple IP. In the CMIP, an MS supports the MIP stack and the ACR acts only as the FA of the MIP.

When an MS in either awake mode or sleep mode moves, the ACR makes a handover processed only under L2 layer to perform the handover quickly. Although the MS moves to another IP subnet, the L3 layer between the ACR, which is the current session anchor, and the MS is not changed, and only the L2 layer is extended to transmit traffic. In these handover methods, the R3 relocation procedure is performed to change the L3 layer from a serving ACR to a target ACR when the MS status is changed into idle mode after handover. In general, if an MS supporting MIP moves in a new subnet area, MIP handover is initiated and a new FA is accessed. However, the ACR extends L2 path and deals with handover quickly when the MS in awake mode/sleep mode moves in a new subnet area.

When an idle mode MS not being served moves to another ACR area or an MS in awake mode or sleep mode moves to a new ASN area consisting of another network operator's devices, the handover function including the L3 layer takes place. The handover function including the L3 layer relocates the anchor point of R3, which interfaces with the CSN, to a target ASN when an MS moves to another ASN area. In the handover including L3 layer, the R3 relocation procedure is performed after the handover under L2 layer is performed to minimize the break time.

The ACR stores and updates various types of context information to manage all the status information of MSs, such as awake mode, sleep mode, and idle mode. The context information includes the MS ID information, service flow information, security information, paging information, and other MS information. The ACR transmits the context information of an MS to a target RAS or a target ACR when the MS performs a handover, a location update, and a *quick connection setup* (QCS), and this enables the MS to access the target RAS or the target ACR quickly.

The ACR as a paging controller performs paging and location management functions. The ACR transmits a paging message to the designated paging group area to enable the idle-mode MS to enter a new network and changes the status of the MS from idle mode to awake mode. The ACR can perform paging functions by composing the paging groups diversely.

In addition, the ACR manages the location of idle-mode MS in a paging group. If an MS in idle mode receives a paging message from the paging controller, then the MS acquires the current location information, performs a location update procedure, if necessary, and notifies a new paging group of its location update.

The ACR manages the active set defined in the IEEE 802.16e standards and can quickly provide anchor BS switching between RASs by supporting *fast BS switching* (FBSS) effectively. The FBSS deals with signaling for handover in advance and sends an indication message via the dedicated channel, *channel quality indicator channel* (CQICH), so that it can increase the handover success rate and shorten the break time.

Call Processing Function

The ACR provides the various call-processing functions listed in Table 10.4, which are related to authorization, authentication, accounting, and key management functions.

When an MS accesses the Mobile WiMAX network initially or requests the creation/change/deletion of service flow during the access, ACR can create, change, or delete the connection corresponding to the service flow via the process of DSA/DSC/DSD.

The ACR performs an MS authentication function using EAP by interworking with the AAA server to determine whether or not the MS is valid. In the MS authentication using EAP, the ACR performs the authenticator function and transfers the EAP payload between the MS and the AAA server by acting as a "pass-through" agent independent of the EAP method on the upper EAP. The MS sends the EAP payload to RAS via the *privacy key management* (PKM) message. The EAP payload sent to the RAS is delivered to the AAA server through R6 interface (between RAS and ACR) and R3/DIAMETER interface (between ACR and AAA server). Since the ACR maintains the anchor authenticator function, the reauthentication procedure of an MS may be omitted unless the designated ACR has been changed during a handover or in reentry of an MS in idle mode. At this time, the EAP-*transport layer security* (TLS) method based on X.509 certificate is supported for MS authentication. The detailed EAP method may vary depending on the network operator's policy.

The ACR performs a subscriber authentication function by using EAP after interworking with an AAA server. At this time, the EAP authentication and key agreement method or EAP tunnel TLS method is supported for subscriber authentication. The detailed EAP method may vary depending on the network operator's policy.

If MS certification or subscriber authentication is successfully completed, ACR receives the upper security key, the *master session key* (MSK), from an AAA server and then it creates and manages a security key for the MAC management message authentication and traffic encryption. The ACR manages such security information as *security association* (SA) and shares the SA information between MS and ACR via the authentication process.

The ACR collects *call detail record* (CDR) in order to charge the Mobile WiMAX service to subscribers. The ACR collects the accounting information about the session time, the number of data packets, service level, QoS, and so on. In addition, the ACR sends the collected accounting information to the AAA server via the DIAMETER protocol. Since the accounting information is collected for each service flow, some differentiated accounting policy can be provided for each service flow.

Bearer Processing Function

The ACR provides various bearer processing functions listed in Table 10.4 to process and deliver user data packets end to end while satisfying the required QoS.

Since the IEEE 802.16e standard adopts a connection-oriented method, all uplink/downlink packets are mapped to a specific connection for the packet exchange. The ACR performs a packet classification function, in which packets are classified and mapped into the MAC connections depending on each service flow. The IEEE 802.16e standard defines the packet classification rule including ATM, IP, Ethernet/*virtual local area network* (VLAN), and *robust header compression* (ROHC). Among them, the Mobile WiMAX specifies only IP and ROHC as mandatory requirements, and thus the ACR provides packet classification for IP and ROHC.

The IEEE 802.16e standard defines *packet header suppression* (PHS) and enables the suppression of the repeated part of a packet header after the packets are classified, for efficient use of radio resources. The MS and ACR set PHS parameters and determine the part to be deleted from the packet header, and this is called the PHS rule. The ACR exchanges the PHS rule with the MS via the DSA procedure while setting a connection. The ACR and MS suppress or restore the packet header according to the PHS rule during the packet exchange.

The ACR provides the ROHC function. ROHC is to compress packet headers including the IP header and the algorithm defined in IETF RFC3095. Whereas PHS is defined in the IEEE 802.16e standard and applies to Mobile WiMAX, ROHC is applied to various other technologies as well, including WCDMA, and has high compatibility. Whereas PHS simply suppresses and restores a part of packet, ROHC can manage the packet status dynamically and yield high efficiency in header compression. Further, the feedback path of ROHC is managed to enhance the robustness of the protocol. However, the ROHC algorithm is more complicated than the PHS algorithm.

The ACR interfaces with multiple RASs or ACRs and the interface function on these bearer planes are called *data path function* (DPF). The NWG standard classifies the type of DPFs into Type 1 and Type 2. For Type 1, IP packets are exchanged, but, for Type 2, MAC *service data units* (SDUs) are exchanged. ACR supports DPF Type 1. The WiMAX NWG R6 interface is used to communicate with RASs, and the R4 interface is used to communicate with other ACRs (see Figure 10.4).

The ACR provides *differentiated service* (DiffServ)–based QoS. DiffServ is to apply differentiated scheduling by varying the *differentiated services code point* (DSCP) value according to different QoS levels. The ACR provides IP QoS by varying the DSCP value according to the QoS level of service flows. In such a way, the DSCP value is used for providing QoS in ASN.

For MSs to feel actual QoS, it is important to ensure the end-to-end QoS, not the QoS in a particular link. The section between MS and RAS is an air link, and a service is provided to the section according to the QoS defined in the IEEE 802.16e standard. The section between ACR and RAS is the MAC section, and the QoS to be applied to each service flow is set after classifying the service flow. The QoS set in the MAC section is mapped with the QoS in ASN. The information on QoS classes and QoS parameters of the service flow for the end-to-end QoS is stored in the AAA server or the PCRF server. The ACR receives the QoS information from the AAA server or the PCRF server when the relevant service flow is created. Based on this QoS information, the ACR sends the QoS information corresponding to the service class of the air link to RAS and sets the DSCP value for the QoS of the MAC section.

The ACR supports the dual stack of IP (i.e., IPv4 and IPv6) simultaneously. The dual stack function of the ACR is implemented for both MS access and interworking between *network elements* (NEs). When an MS accesses IPv6 network, the normal service can be provided even if the ACR and RAS are connected via an IPv4 network. Since the ACR is connected with RAS via a tunnel, the IP protocol version for the tunnel can be independently selected from the protocol version used in the MS.

Since the ACR provides several Ethernet interfaces, it stores the information on the Ethernet interface to route IP packets according to the routing table. The network operator organizes the routing table for the ACR operation. The approach to organize and set the routing table is similar to the standard setting of the router. The routing table of the ACR is configured depending on the operator's setting and configuration. The ACR can set the routing table to support the static and the dynamic routing protocols, such as *open shortest path first* (OSPF), *intermediate system to intermediate system* (IS-IS), and *border gateway protocol* (BGP). In addition, the ACR supports the IP packet routing function to transmit the packets that are handled inside the system via the interface specified by the system routing table. The ACR also supports the function to forward the IP packets received from external networks according to the routing information of the routing table.

The ACR provides the Ethernet interface and supports the link grouping function, VLAN function, and Ethernet *class of service* (CoS) function under IEEE 802.3ad for the Ethernet interface. The MAC bridge function defined in IEEE 802.1d is excluded. The ACR enables several VLAN IDs to be set in one Ethernet interface and maps the DSCP value of IP header with the CoS value of the Ethernet header in the transmitted packet to support the Ethernet CoS.

Interworking Function

The ACR provides the interworking functions listed in Table 10.4, including the internetworking with various servers such as HA, AAA, and PCRF servers.

For support of MIP service, the ACR supports CMIP, which is the IETF standard MIP, and PMIP, which is defined in WiMAX NWG standard, for the interface between the ACR and HA. As an FA, the ACR allocates a *care-of-address* (CoA) to MS and the MS sends the CoA to HA. The ACR exchanges the MS traffic via the tunneling interaction with HA. In addition, the ACR performs the PMIP client function for the MS not providing the MIP stack. The ACR can interwork with multiple HAs and specify the HA interworking for each MS. The information on the HA that interworks with each MS is informed from the AAA server to ACR in the initial MS authentication stage.

The ACR interworks with the AAA server under the IETF standard DIAMETER specification and performs the MS authentication and subscriber authentication functions according to the EAP method by interworking with the AAA server. The EAP method is implemented in the MS and the AAA server. The ACR relays the EAP payload to the AAA server. In addition, the ACR collects the accounting information on the subscriber access, the session time, the number of data packets, the service level, and QoS, and then transmits the information to the AAA server via the DIAMETER protocol.

The PCRF server determines the accounting rule on the basis of the QoS policy and the service flow and sends it to the ACR. According to the policy received from the PCRF server, ACR can create/change/release the service flow dynamically, and it controls the QoS complying with the service flow. In addition, the ACR creates accounting data by using the accounting rule received from the PCRF server. The interface between the ACR and the PCRF server is based on the Gx interface of 3GPP Rel 7, and the Gx interface is determined by using the DIAMETER protocol. The Gx interface is for provisioning the service data flow-based charging rules between the *traffic plane function* (TPF) and the *charging rules function* (CRF), also known as the service data flow-based charging rules function.

The ACR allocates an IP address to an MS by using the DHCP or MIP method. How the ACR allocates an IP address to an MS depends on the service type provided to the MS. When simple IP service is provided to an MS, the ACR allocates an IP address to the MS by using DHCP. At this time, the ACR acts as a DHCP server or a DHCP relay agent. When PMIP service is provided to MS, the ACR allocates an IP address to the MS by using DHCP and uses MIP for HA. When CMIP service is provided to MS, the ACR delivers a home IP address to the MS by using MIP.

Figure 10.5 shows the RAS (or BS) and ACR (or ASN-GW) systems that are developed by Samsung Electronics to perform the various RAS and ACR functions described so far.

10.5 Access Network Deployment

Deployment of the ASN is a very important process for operators, as it is the dominant portion of the total investment for providing services and, in addition, it is highly correlated with the service quality. In order to achieve cost-effective network

(a) (b)

Figure 10.5 Example of WiBro systems developed by Samsung Electronics (a) RAS, and (b) ACR.

deployment, it is necessary to determine the optimal location and height, the type of the WiBro systems (such as indoor/outdoor, BS, repeaters, FA, sector), and the sector configuration. It is also necessary to conduct medium-scale tuning process by determining the position, type, azimuth, and down-tilt of the antenna and then doing fine-scale tuning by adjusting the engineering parameters such as transmit power allocation and hand-over parameters. Such an optimal cell planning enables us to ensure good service quality and maximum coverage with minimal investment.

10.5.1 Access Network Planning

In planning the access radio network, the following two factors should be considered: First, it is important to understand the system and the radio environment. In support of this, we need to analyze the propagation characteristics of the 2.3-GHz radio signal and determine the service quality based on the *received signal strength indicator* (RSSI) and the CINR values. Second, it is important to do efficient cell planning of the ground. In support of this, we need to choose the potential RAS sites by considering the centers of high data traffic areas, the relatively high buildings to ensure the *line-of-sight* (LOS) site is as wide as possible, and the good locations for indoor services of tall buildings adjacent to the main roads.

Figure 10.6 shows the overall procedure of *radio network planning* (RNP). The RNP process, in general, is composed of three stages: dimensioning, preliminary planning, and final planning. If there is a network readily existing, all the steps may not be required. In addition, the steps marked by dotted lines in the figure may be omitted depending on the specific project.

Dimensioning

Dimensioning process is the first stage of the RNP. Its goal is to determine the number of RASs and the network configuration based on the analysis of the coverage

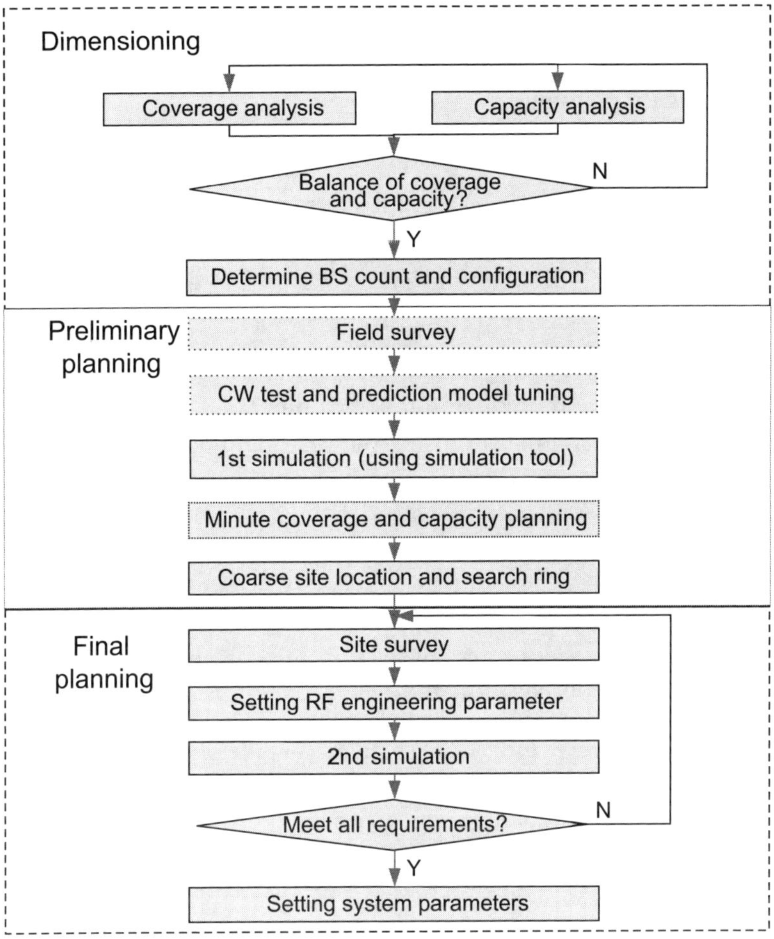

Figure 10.6 Overall procedure of radio network planning.

and capacity. The dimensioning process is composed of two key components: coverage planning and capacity planning.

Coverage planning is done based on the link budget. It is composed of the following four steps:

1. Analyze the link budget parameters such as slow fading margin, handover gain, interference margin, receiver sensitivity, and noise figure.
2. Calculate the *maximum allowable path loss* (MAPL).
3. Analyze the prediction model.
4. Determine the RAS counts and configuration.

Capacity planning is done based on the user traffic. It is composed of the following four steps:

1. Generate the user traffic model.
2. Calculate the system capacity.
3. Determine the traffic loading.

4. Determine the RAS counts and configuration. The RAS counts and configuration may be determined by examining the balance of the coverage and capacity planning.

Preliminary Planning

Preliminary planning is the second stage of the RNP. It is the preparation process before doing onsite cell planning based on the RAS counts and the configuration obtained in the dimensioning stage. The following five steps are performed in this stage:

1. Field survey;
2. CW test and prediction model tuning;
3. The first simulation (using the cell planning tools);
4. Fine coverage and capacity planning;
5. Coarse site location and search ring.

Final Planning

Final planning is performed based on the results of the preliminary planning. In this stage, site survey is done and all the system parameters are determined. Specifically, this stage is composed of the following four steps:

1. Site survey: First, site location is determined, and equipment type is determined as well, by considering all the aspects including the coverage, capacity, and interference. Throughout the overall RNP procedures and network optimization process, the site location and height are the most important factors that affect the network performance. Second, antenna position, as well as antenna type and configuration (azimuth and down-tilt) values, are determined to ensure the desired coverage.
2. Setting all the RF engineering parameters.
3. The second simulation (using the detailed RF engineering parameters).
4. Setting system parameters.

10.5.2 RNP Case Studies

As the case studies of RNP, we briefly introduce two real RNP results obtained during the Mobile WiMAX network deployment in Seoul by KT: one is the congested Yeoksam 5 area located in downtown Seoul and the other is the suburban Pangyo IC area to the south of Seoul.

Case 1: Yeoksam 5 Area

Yeoksam 5 area covers the crossroads near Gangnam Station and Gangnam Street in downtown, Seoul. The Mobile WiMAX systems installed are the wall-mounted type and the environment-friendly type. Since this area was one of the most thriving and congested areas in metropolitan Seoul, with tall buildings, it was important to decentralize the traffic. Antennas were installed with a tilt over 20 degrees, with 15 dB gain, and with a wide horizontal beam angle. The α sector was designed to be partially covered by in-building repeaters.

The problems observed while installing the antennas were as follows: Due to the building structure in Kangnam area, it was not possible to install antennas at the edge of the building. In addition, there existed a weak signal area under the γ antenna. Therefore, coverage adjustment or additional repeater installment was needed during the network optimization process.

Figure 10.7 shows the RF design map for the case of Yeoksam 5 area. For RF design forms for the installation of RAS and antennas, refer to [1].

Case 2: Pangyo IC Area

Pangyo IC area is located to the south of Seoul and includes Gyeongbu Expressway, having about 10 lanes and an *interchange* (IC) that connects to the nearby city Bundang. Antennas were installed on a 45-m-high pole. Since the area is very noisy due to high-speed automobiles, high-gain antennas were installed to avoid the *pseudo-random noise* (PN) generated on the expressway and to achieve wide coverage. Antennas with narrow horizontal beam angles were installed to reduce the interference among sectors. While installing the antennas, a slope road was observed in the β1 direction. So an antenna with a wide vertical beam angle would have to be adopted if there were weak signal areas on that road.

Figure 10.8 shows the RF design map for the case of Pangyo IC area. For RF design forms for the installation of RAS and antennas, refer to [1].

10.5.3 Access Network Implementation and Optimization

In order to achieve the target performance of the network, network optimization process is necessary. In network optimization, there are two different types: RF optimization and system optimization. The *RF optimization* is the overall process of improving the RF parameters (such as CINR and RSSI). Specifically, the process includes the adjustment of antenna such as gain, tilt (mechanical and electrical), and directivity; relocation of equipment; and implementation of additional antennas.

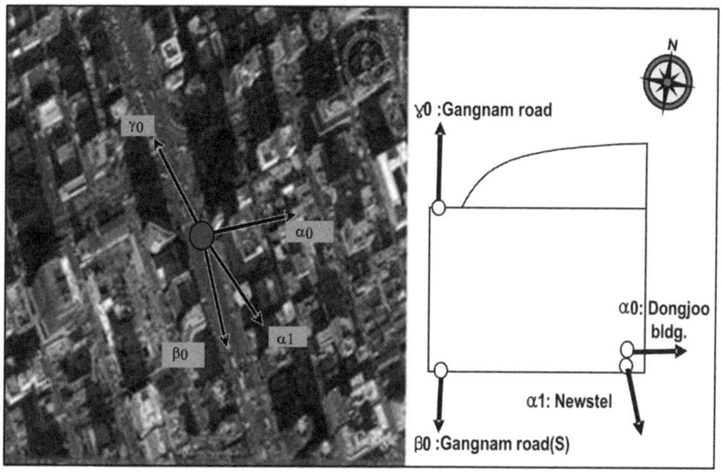

Figure 10.7 RF design map of Yeoksam 5 area.

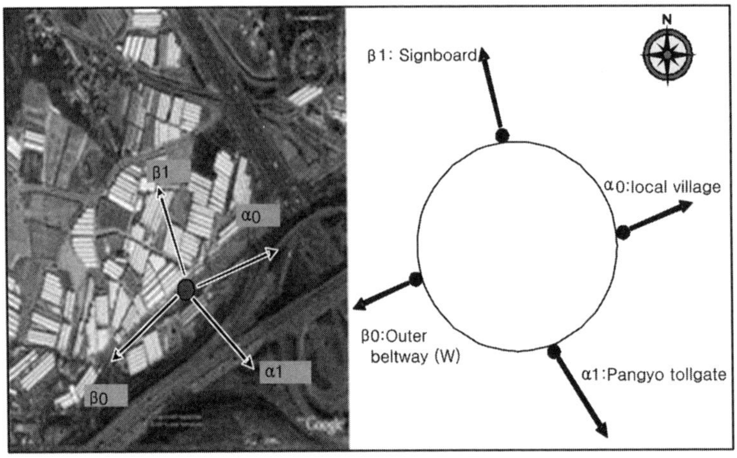

Figure 10.8 RF design map of Pangyo IC area.

The *system optimization* is the process of upgrading the software package in the system. It adjusts the system parameters such as output and timer.

KT installed RASs, different types of repeaters (optical and RF), and feeder lines to cover the entire metropolitan Seoul area and its vicinities. While performing a large-scale trial in the whole coverage area, KT set the target performance of some major items as listed in Table 10.5. The target performance for actual commercial services was set to be much higher than that given on the table. In order to achieve the target performance, network optimization process was conducted iteratively, and the target performance was finally achieved for the commercial services.

10.6 Other Network Elements Deployment

In addition to the RAS in the ASN, it is necessary to install other network elements in the CSN, such as ACR, an *elementary management system* (EMS), AAA, HA, a DNS server, a DHCP server, NMS, an aggregation switch, and a router. Besides, it is

Table 10.5 Target Performance of Some Selected Items

	Items		Target	Remarks
Data	Success rate of network connection		>98%	
	Transmission completion rate		>97%	FTP file transfer
	Throughput per user (minimum)	Downlink	512 kbps	FTP file transfer at cell edge
		Uplink	128 kbps	
	Throughput per user (average)	Downlink	3 Mbps	FTP file transfer at random place
		Uplink	1 Mbps	
	Handover latency		<150ms	
Service	Streaming service	Completion rate	>95%	

needed to deploy the transmission lines connecting those network elements and the connection to backbone network for commercial services.[5]

10.6.1 Core Network Planning

In order to deploy an efficient core network and to provide quality services, we need to establish suitable design criteria. The design criteria adopted for the core network deployment were targeted at the following goals: accommodation of high capacity of data traffic according to traffic prediction, efficient mobility support with minimized handover traffic between ACR, cost-effective implementation of the network, reliable network for ensuring high-quality services, and flexible and scalable network architecture for easy expansion, removal, and substitution.

In general, the core network design process is divided into three stages, namely, network design, equipment planning, and implementation planning, as will be detailed next.

Core Network Design

The core network design stage is composed of the following four steps:

1. Analyze the market forecast data.
2. Predict the data traffic. As a means for the data prediction, first, classify terminal type; second, calculate annual data traffic of each terminal type; and third, calculate the average monthly data traffic, the hourly data traffic at the busiest hour, and the peak data traffic.
3. Establish a design standard for each network element.
4. Determine the network topology and the routing policy.

In order to decentralize the traffic and avoid service discontinuity caused by network failure, each node is protected by dualization. The main and local nodes in the existing network are dualized, and the center nodes are located in two different places. It is possible to easily expand the network by taking advantage of the layered network architecture readily deployed nationwide. Note that the existing network has 217 branches, 31 local nodes of IP premium network, which can be used as the backbone for WiBro services.

The connection between the WiBro network and the Internet backbone is done in the following two types:

1. *Direct connection type*: The WiBro network is directly connected to the Internet backbone. So it is adequate to handle high data traffic and to support broad coverage. The connection has the following route: RAS > aggregation switch > ACR > PE router (local node).
2. *Interworking type*: RAS is connected to the local nodes of IP premium network. So it is adequate to save network implementation cost and support small coverage at remote sites. The connection has the following route: RAS > aggregation switch > PE router (local node) > ACR.

5. KT implemented AAA, billing system, server farm, back-end platform and other network elements when deploying the CSN. Refer to [8].

Equipment Planning

The equipment planning stage is the process of making basic equipment planning (i.e., network element planning) and is composed of the following three steps:

1. Determination of the target coverage;
2. Collection of the information to plan, such as the predicted traffic data and the on-site data at the installation place;
3. Determination of the specification and the quantity needed at each location, including the general requirements (or equipment specifications), installation plan, the number of lines required, and the detailed list of the required equipment quantity.

Implementation Planning

The implementation planning is the final stage of the core network planning to prepare for the detailed layout of the equipment and the construction plan. It is composed of the following three steps:

1. On-site investigation;
2. Equipment layout drawing;
3. Construction planning.

10.6.2 Servers and Other Elements

Once core network planning is done, various servers, aggregation switches, and transmission lines are installed as discussed next.

AAA and Other Servers

The AAA and supplementary servers are needed to manage subscribers' access to the network and services and to generate the billing data. The WiBro service provided by KT adopts the *single sign-on* (SSO) concept by interworking network connection and service authentication for specific value-added application services. In addition, the system is designed such that it generates different billing rates depending on the contents type and the service class. Such functionality is made possible by getting the support of several different servers as follows:

1. Authentication server for authentication of the user;
2. Session server for session DB;
3. *Operation and maintenance platform* (OMP) server for operation and management of AAA;
4. *Authentication center* (AuC) server for management of authentication key;
5. Statistics server for authentication and account statistics;
6. Billing server for the generation of packet data record.

Aggregation Switches

Taking into account the installation cost of *wavelength division multiplexing* (WDM), aggregation switch (L2 switch) is deployed to accommodate multiple

RASs. In the case of one RAS, it is directly connected to WDM without aggregation switch.

Transmission Lines

The transmission line between ACR and the aggregation switch is deployed on WDM to ensure network reliability. Both primary and backup lines are installed. The transmission line between the aggregation switch and RAS is deployed on metro-Ethernet, because its per-line cost is much cheaper than that of the synchronous optical transmission system and it is more flexible to accommodate RAS traffic over 10 Mbps.

10.7 WiBro Services

From the perspective of network performance, the Mobile WiMAX network outperforms any other existing mobile networks, as described in several documents [9, 10]. Such superiority may be characterized by the keywords *mobility*, *broadband*, *all-IP*, *always-on*, *low-cost*, and so forth.

WiBro services pursue mobile *triple play service* (TPS) (i.e., the convergence of communication, Internet, and broadcasting services by keeping up with the market demand and by utilizing the advantage of the Mobile WiMAX network). To provide differentiated services to the users, the following three distinctive capabilities of WiBro network and services may be exploited:

- First is the capability of supporting open networks and services based on all-IP network, which is effective in supporting the managed PDA, interactive e-learning, mobile commerce, charge per sale, and so on.
- Second is the capability of offering Web 2.0–based service in a mobile environment, which is effective in supporting personal mobile media, location-based community service, customized Web contents, and so on.
- Third is the capability of offering larger upload throughput, which is effective in supporting multiparty videoconferencing, integrated communicators, MIP channels, online games, and so on.

In order to provide differentiated WiBro services, efficient service platform and software architecture are needed as well, as addressed next.

10.7.1 Service Platform

In order to provide a diverse set of services to users, client software is needed at user devices and an application service platform is needed at the access network. Figure 10.9 shows the overall architecture of the WiBro service platform. The presence server and the call control manager belong to the core part of the service platform. Servers related to messaging functions and other applications are also implemented in the service platform. The WiBro application services include user interface, information service, and entertainment service.

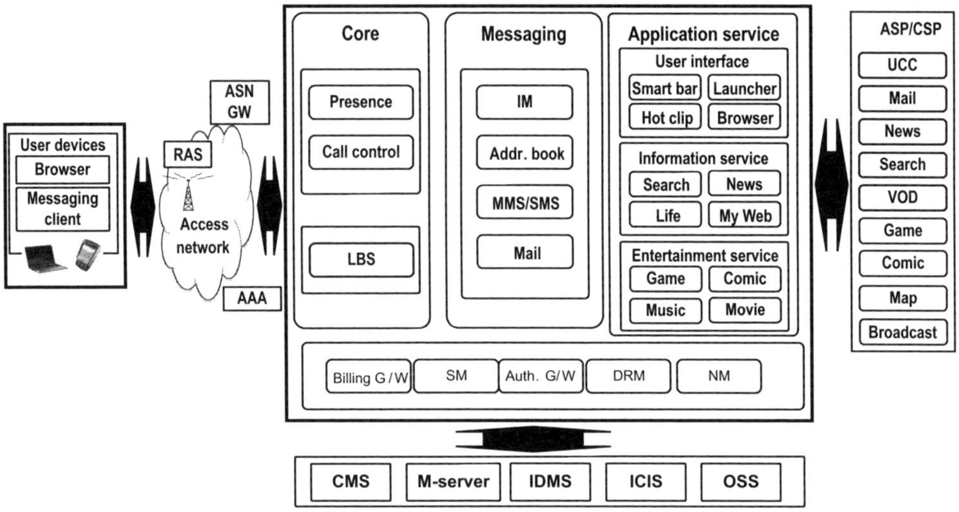

Figure 10.9 Service platform architecture.

Software Architecture

User devices for WiBro services are divided into two groups: the *communication module-only* type such as PCMCIA card and USB dongle, and the *user devices with built-in communication module* such as smart phone, PDA, PMP, and embedded laptop computer. To support various application services in the second group, an efficient software architecture including OS, middleware, and applications is needed. For example, a dual-mode smart phone (WiBro + CDMA) was introduced to the market, which can find versatile applications in WiBro and CDMA services.

Connection Manager

The connection manger controls network entry based on user's configuration (automatic or manual network entry). Besides, the connection manager can display and manage various useful information on the network status, including signal strength, connection time, and transmission rate.

Launcher

The launcher is an integrated *user interface* (UI) platform to accommodate flexible requirements of operators, application service providers, and users. Major application services can be quickly started on the launcher screen. Users can use the search function by entering the keyword on the launcher without opening a Web browser.

10.7.2 Core Application Services

WiBro services are categorized based on the mobile TPS concept, which enables the offering of a variety of application services in addition to the basic Internet connection. The WiBro service categories are core, differentiated, and competitive service groups, as shown in Figure 10.10. The *core service group* contains the most fundamental services including Web Mail, Multi-Board, My Web, PC Control, and Mobile UCC. The *differentiated service group* contains the WiBro differentiated services, including entertainment contents, online game, e-learning, and *location-based*

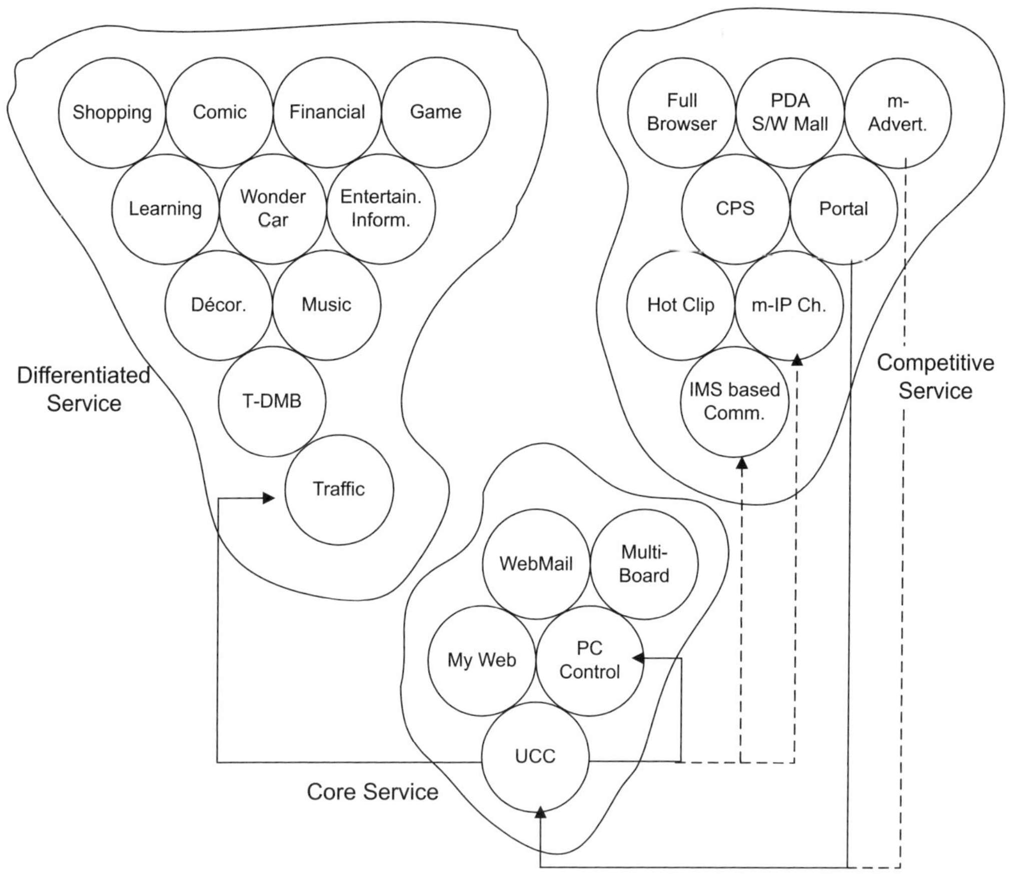

Figure 10.10 Grouping of WiBro application services.

service (LBS). The *competitive service group* utilizes the high-performance feature of the Mobile WiMAX network, including full Web browsing and m-IP channel services.

Among the three service categories, the core service was offered in the initial stage by KT, with the differentiated and competitive services being planned for introduction in later stages. In the following, we briefly discuss the five services belonging to the core service group.

My Web

My Web is a *really simple syndication* (RSS) service providing user's favorite information without visiting each website unnecessarily. Through My Web service, user can obtain the desired information, check the blogs of friends, and even obtain the entire posts without logging on and worrying about costly packets.

In the My Web service, users can define the preferred group and the channel that belongs to each group, where channel refers to the web sites providing the posts, such as blogs, news feeds, or podcasts. To use this service, user needs to register specific web sites by entering the address of the blog, the *xml* address, or the RSS feeding address in the URL. Once registration is done, the most recent posts are provided to the subscriber. When a new page of the channel is loaded, a number appears that indicates the amount of new information.

There is a search function in the My Web services. There are two types of searching: online search (or Internet search) and offline search (in the user device). In general, the information provided by the RSS may not load the entire post. However, when My Web searches the load data, the entire post is loaded without accessing or visiting the original post. Consequently, there is no waste of packets. The service is now being expanded to show the entire post. An auto-search function enables reserved search. If a user subscribes to a keyword, then the post with the keyword is saved automatically.

Web Mail

Web Mail is an integrated e-mail managing service on the smart phone or PDA that manages multiple POP3 e-mails or Web mails that users usually use, such as Yahoo mail, Gmail, or Hotmail. To use Web Mail service, the user needs to register his/her e-mail accounts on the registration page of the Web Mail service. Multiple e-mail accounts from different mail services can be registered and managed by single client software of the Web Mail service.

There are Inbox, Outbox, Draft, Temp, Storage, and Recycled directories in the mailbox. In the Inbox directory, retrieved mail list is shown, including e-mail service name, sender's name, the subject of the e-mail, the received date, and the attached file icon. By clicking the desired e-mail to read, user can read the whole e-mail contents on the pop-up window after the contents are received. Basically, the contents are shown in text form, but optionally the user can see the original web page in *html* form also. The file attached to e-mail can also be read in the Web Mail service. After the attached file is downloaded, the proper application of the file is executed. When sending, the default e-mail account that the user has chosen when registering the e-mail account is automatically selected as the sender. The receiver can be chosen from the *personal information management system* (PIMS) contact list of the device.

Multi-Board

Multi-Board is the next generation rich media communication platform that enables users to access anytime and anywhere. Multi-Board provides business people a virtual space where people can collaborate with others beyond the geographical restriction. Multi-Board provides multiway conference environments from anywhere. Even if users are away, they still can join the conference through the Multi-Board service. People at remote sites can get real-time training and education effectively through the Multi-Board service. The trainer can use various materials such as office documents, pictures, and even DVDs for the trainees away from the classroom as if both parties were having a class sitting in the same place.

Multi-Board service supports multiparty video conferencing for up to 12 people. During videoconferencing, a clear sound and video is provided, and more than 500 audiences can join a single session. Multi-Board supports diverse screen layout modes. Depending on the conference and meeting needs, screen layout may be chosen. The rich media communication features of the Multi-Board are especially useful for business collaboration. Meeting participants can share the applications on their desktop computer and coedit spreadsheet and word documents. It can remotely access a third party's PC by getting authorization with application sharing and can easily manage and control the workspace. The participants can share streaming media, DVD, and multimedia contents without any buffering. One of the most distinguished features that Multi-Board provides is the triple service in a single session. That is, people can communicate while talking and looking at each other and can also share movies and data as well.

Mobile UCC

User created content (UCC), or *user generated content* (UGC), refers to the media content produced by the end users. One of the major trends in current Internet industry is highly related to UCC. By using Mobile UCC, developed by KT, users can experience video recording, live broadcasting, Hot UCC, My Album, and other services, and access other popular UCCs providing Web sites.

After recording video or taking photos, users can easily upload the contents to the associated UCC portals, including SeeU or MBox. UCC live broadcasting function is also available in the Mobile UCC service. If the user sets video to start broadcasting after completing the creation of the title, content, category setting, and the number of viewers, then the viewers will be able to see the user's station on the on-air list of the associated UCC portals. User may check the number of viewers and end the broadcasting. User can enjoy diverse contents via Hot UCC menu and can easily check the contents in the My Album menu. User can easily upload the contents from My Album to the associated UCC portals as well.

PC Control

PC Control service provides managing functionalities of the various contents in remote terminal. When PC Control is running, local WiBro mobile device, home PC, and office PC are linked together. By opening mobile window explorer at a local mobile device, a user can see the files stored in the office PC. A user can copy or move specific files between the linked devices by the PC Control service. By operating a local mobile device, the user can edit office files or play multimedia files stored at remote devices. It is also possible to share multimedia files with friends via the PC Control function over the WiBro network.

10.7.3 Other Major Services

Even with a diverse set of new services newly developed for Mobile WiMAX, communication service remains as a fundamental service, to which new features can be added. In addition, the *multicast-broadcast service* (MBS) that takes advantage of the unique broadcast/multicast capability of Mobile WiMAX is expected to become another major service.

Communicator Service

Voice call is still a fundamental communication service even today, and VoIP is now popular at the IP data network. For an efficient VoIP service, Mobile WiMAX network supports different service classes, including *unsolicited grant service* (UGS) and *extended real-time polling service* (ertPS) (refer to Section 6.1).

In addition to VoIP service, different types of services, including voice, message, and e-mail, are integrated into single communicator application in the WiBro services. They include the call-type services such as mobile VoIP, *push-to-talk* (PTT), *push-to-view* (PTV); the messenger-type services such as chatting, file/folder transfer; the message-type services such as *short message service* (SMS), *multimedia message service* (MMS), and e-mail; and the value-added services such as file and application sharing.

m-IP Channel Service

Broadcast or multicast information that needs to be delivered to multiple users in a single or multiple cells can share the radio resources dedicated to this service in the Mobile WiMAX network. This is the MBS, which is an efficient way of utilizing the radio resources without allocating radio resources to each user.

MBS is viewed as an attractive service that can differentiate Mobile WiMAX network from all other cellular mobile networks by providing real-time, high-quality, and interactive multimedia contents to the users. Push-type broadcasting services based on zone are one of the examples of MBS. The m-IP channel services developed by KT support multiple channels of broadcasting contents (i.e., 30 fps, 320x240, H.264/AAC+ video) at a rate of 512 Kbps/channel.

References

[1] Kim, H., J. Lee, and B .G. Lee, “WiBro—A 2.3 GHz Mobile WiMAX: System Design, Network Deployment, and Services,” in *Mobile WiMAX*, New York: Wiley-IEEE Press, AUTHOR: UPDATE PUBLICATION INFORMATION.

[2] IETF RFC 2131, Dynamic Host Configuration Protocol (DHCP), March 1997.

[3] IETF RFC 3095, Robust Header Compression (ROHC): Framework and Four Profiles: RTP, UDP, ESP, and Uncompressed, July 2001.

[4] IETF RFC 3748, Extensible Authentication Protocol (EAP), June 2004.

[5] IETF RFC 1701, Generic Rounting Encapsulation (GRE), October 1994.

[6] Kim, S., “WCDMA RNP Special Topic Guidance Engineering Parameter Analysis,” *Huawei Tech.*, October 2004.

[7] KT Corporation, “WiBro Engineering Project Report,” June 2007.

[8] KT Corporation, “Mobile Internet Business Project Report,” May 2006.

[9] WiMAX Forum, “Mobile WiMAX—Part I: A Technical Overview and Performance Evaluation,” white paper, August 2006.

[10] WiMAX Forum, “A Compartive Analysis of Mobile WiMAX Deployment Alternatives in the Access Network,” white paper, May 2007.

Selected Bibliography

TTA TTAR-0016, Evaluation Criteria of Radio Access Technology for 2.3GHz Portable Internet, August 2004.

TTA TTAR-0020, 2.3GHz Portable Internet Service Requirements and Network Reference Model, August 2004.

TTA TTAS.KO-06.0082/R1, Specifications for 2.3GHz Band Portable Internet Service, Phase II-revision version, December 2005.

WiMAX Forum, Network Architecture—Stage 2: Architecture Tenets, Reference Model and Reference Points [Part 0–3], Release 1.1.0, July 2007.

WiMAX Forum, Network Architecture—Stage 3: Detailed Protocols and Procedures, Release 1.1.0, July 2007.

PART II

WiFi: Wireless Local Area Networks

IEEE 802.11 *wireless local area networks* (WLANs) or *WiFi* have been extensively deployed in the recent years in many different environments for enterprise, home, and public networking. The 802.11 is probably the most widely accepted broadband wireless networking technology, providing the highest transmission rate among standard-based wireless networking technologies. The original first generation 802.11 devices, introduced in the late 1990s, provided an Ethernet-like best-effort service with the transmission rate up to 2 Mbps. Today's state-of-the-art 802.11 devices can provide multimedia applications including VoIP and video streaming with strong security and transmission rates up to 54 Mbps. In fact, this technology is evolving today for higher transmission rates, seamless mobility, and so on.

The chapters in this part present various aspects of the 802.11. The core ideas and underlying philosophies are emphasized, while the detailed bits and bytes are also touched whenever desired. The chapters are organized into topical features of the protocols. That is, the PHY and the baseline MAC, based on IEEE 802.11-1999, are first presented. Then, some additional features, namely, *quality-of-service* (QoS) provisioning, security mechanisms, mobility support, and spectrum and power management, are individually presented in separate chapters. Finally, the last chapter is dedicated to a discussion on some ongoing standardization efforts.

First, Chapter 11 introduces the WiFi and IEEE 802.11 networks. Two different network architectures, namely, the infrastructure and ad hoc modes, are presented. After explaining the reference model, the layer interaction issues are discussed. Those include the interactions between: (1) PHY and MAC, (2) MAC and IEEE 802.2 LLC, and (3) MAC and IEEE 802.1D bridge. Finally, some key technologies, including multiple access, duplexing, multiple rate support, power saving, mobility, confidentiality, spectrum and power management, and QoS support, are briefly discussed.

Chapter 12 presents various PHY protocols, including the 802.11a/b/g. We first describe the general operations of the 802.11 PHY involving frame transmission and reception as well as the channel sensing mechanism, called *clear channel assessment* (CCA). Then, each of 802.11a, 802.11b, and 802.11g are individually presented. Along with their original and mandatory schemes (e.g., *orthogonal frequency division multiplexing* of the 802.11a/g and *complementary code keying*), some optional schemes are also briefly discussed. Those include the reduced-clock operations of the 802.11a and *packet binary convolutional code* (PBCC) of the 802.11b/g. The coexistence of the 802.11b and 802.11g is also briefly discussed.

Chapter 13 presents the baseline MAC protocols. We first start with the definitions of the MAC frame format for three different types (i.e., data, management, and control frames). Then, the mandatory *distributed coordination function* (DCF) based on *carrier-sense multiple access with collision avoidance* (CSMA/CA) and the optional *point coordination function* (PCF) based on a poll-and-response mechanism are presented. The performance of the DCF is also briefly presented. After discussing the multiple transmission rate support, the MAC management functions, including time synchronization, power management, and (re)association are discussed.

Chapter 14 presents the QoS extension of the MAC (i.e., the 802.11e). We first discuss the limitation of the baseline MAC to support QoS. Then, some key concepts of the 802.11e, including prioritized versus parameterized QoS, *traffic stream* (TS), and *transmission opportunity* (TXOP), are introduced. We then present the *hybrid coordination function* (HCF) composed of *enhanced distributed channel access* (EDCA) and *HCF controlled channel access* (HCCA). After presenting the admission control and scheduling issues, some optional features of the 802.11e, developed to make the MAC more efficient, are also presented. Those include *direct link setup* (DLS), block Ack, and *automatic power save delivery* (APSD).

Chapter 15 presents the security features found in both the baseline MAC and the enhancement via the 802.11i. We first introduce the authentication methods and encryption scheme, called *wired equivalent privacy* (WEP), defined in the baseline MAC. After presenting the limitations of these legacy schemes, the new mechanisms defined in the 802.11i are presented. We first present IEEE 802.1X-based authentication, and then various keys and key management via four-way handshake are presented. Finally, two newly defined ciphers, namely, the optional *temporal key integrity protocol* (TKIP) and the mandatory *countermode with CBC-MAC protocol* (CCMP), are presented.

Chapter 16 presents the mobility support of the WLAN. First, the handoff procedures, including scanning, 802.11 authentication, reassociation, 802.11i authentication and key management, and finally 802.11e TS setup, are briefly discussed. Then, IEEE 802.11F *interaccess point protocol* (IAPP), a recommended practice for the communication among the APs, is presented. Then, finally two emerging standards, namely, IEEE 802.11k allowing fast scanning and IEEE 802.11r for fast roaming, to enable fast handoff in the 802.11 WLAN are introduced based on their draft specifications.

Chapter 17 presents the spectrum and power management for 5-GHz band as specified in the 802.11h. We first discuss the regulatory requirements for *dynamic frequency selection* (DFS) and *transmit power control* (TPC) for the 5-GHz operations in the United States and Europe based on the FCC and ETSI documents. We then present the DFS and TPC protocols defined in the 802.11h, which is defined to meet the regulatory requirements. Some example algorithms for these two functions are also briefly presented.

Chapter 18 presents some ongoing evolutions of the WLAN via standardization, including the 802.11n for higher throughput support, the 802.11s for mesh networking, and finally the 802.11k for *radio resource measurement* (RRM). For the 802.11n, new MAC schemes, including the frame aggregation, and PHY schemes, including multi-input multi-output (MIMO), are introduced. For the

802.11s, new frame formats and routing protocols in the mesh network are briefly presented. Finally, for the 802.11k, a number of newly defined measurement schemes are briefly introduced.

CHAPTER 11

Introduction to WiFi Networks

IEEE 802.11 WLAN, or WiFi, is probably the most widely accepted broadband wireless networking technology, providing the highest transmission rate among standard-based wireless networking technologies. Today's WiFi devices, based on IEEE 802.11a [1] and 802.11g [2], provide transmission rates up to 54 Mbps and, further, a new standard IEEE 802.11n [3], which supports up to 600 Mbps, is being standardized. The transmission range of a typical WiFi device is up to 100m, where its exact range varies depending on the transmission power, the surrounding environments, and others. The 802.11 devices operate in unlicensed bands at 2.4 and 5 GHz, where the exact available bands depend on each county.

Most of today's laptop computers as well as many PDAs and smart phones are shipped with embedded WLAN interfaces. Moreover, many electronic devices, including VoIP phones, personal gaming devices, MP3 players, digital cameras, and camcorders are being equipped with WLAN interfaces as well. The most typical applications of the 802.11 WLAN include Internet access of portable devices in various networking environments, including campus, enterprise, home, and hot-spot environments, where one or more *access points* (APs) are deployed to provide Internet service in a given area. The 802.11 could be used for a peer-to-peer communication among devices where APs are not deployed. For examples, laptops and PDAs in proximity can use the 802.11 to share their local files. Also, people in proximity can do networked gaming using their gaming devices with the 802.11 interface. It is primarily being used for the indoor purpose. However, it can be also used in outdoor environments, and some level of mobility (e.g., walking speed) can be also supported.

IEEE 802.11 *working group* (WG) has generated a family of standards for WLAN. The IEEE 802.11 standard specifies the protocols for both the *medium access control* (MAC) sublayer and the *physical* (PHY) layer. As illustrated in Figure 11.1, existing higher-layer protocols, which were originally developed for wireline networking such as TCP, UDP, IP, and IEEE 802.2 *logical link control* (LLC), can work on top of the 802.11 MAC since the 802.11 was developed basically to provide the services in a similar way that IEEE 802.3 Ethernet does.

IEEE 802.11 WG [4] started its standardization activities in 1991. It published the first standard specification in 1997, and then a revision in 1999 [5]. The protocols described in this book are based on IEEE 802.11-2007, which was published in 2007 [6]. IEEE 802.11-2007 revision describes the IEEE 802.11 standard for WLANs with all the amendments that have been published until June 2007. We refer to the original protocols found at IEEE 802.11-1999 as the *baseline protocols* (i.e., *baseline MAC* and *baseline PHY*, respectively) in the remaining chapters of Part II.

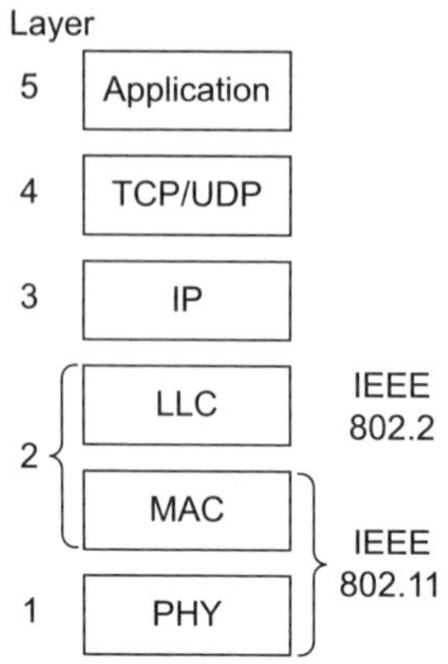

Figure 11.1 The relationship between IEEE 802.11 MAC/PHY and other higher layer protocols.

The amendments, which were standardized after the publication of the baseline protocols, and were then rolled into IEEE 802.11-2007, include:

- IEEE Std 802.11a-1999 (Amendment 1) for "High-Speed Physical Layer in the 5 GHz Band";
- IEEE Std 802.11b-1999 (Amendment 2) for "Higher-Speed Physical Layer Extension in the 2.4 GHz Band";
- IEEE Std 802.11b-1999/Corrigendum 1-2001 for "Higher-speed Physical Layer (PHY) extension in the 2.4 GHz band—Corrigendum1";
- IEEE Std 802.11d-2001 (Amendment 3) for "Specification for Operation in Additional Regulatory Domains";
- IEEE Std 802.11g-2003 (Amendment 4) for "Further Higher Data Rate Extension in the 2.4 GHz Band";
- IEEE Std 802.11h-2003 (Amendment 5) for "Spectrum and Transmit Power Management Extensions in the 5GHz Band in Europe";
- IEEE Std 802.11i-2004 (Amendment 6) for "Medium Access Control (MAC) Security Enhancements;
- IEEE Std 802.11j-2004 (Amendment 7) for "4.9 GHz–5 GHz Operation in Japan";
- IEEE Std 802.11e-2005 (Amendment 8) for "Medium Access Control (MAC) Quality of Service Enhancements."

For amendment called IEEE 802.11x-yyyy, x represents an alphabet assigned to the particular project, which generated the amendment, and yyyy represents the year when the amendment was published (i.e., when the project for the amendment was completed). The alphabet has been assigned beginning with "a," where a new project is assigned a letter next to the preceding one's. We learn from the list that some projects took more time than others. For example, IEEE 802.11e took more than five years from its inception to the completion, and it became the eighth amendment after the 802.11g to 802.11j [7], of which the projects started before that of the 802.11e.

In this book, we cover the baseline protocols as well as many of the amendments included in IEEE 802.11-2007. Those include 802.11a, 802.11b, 802.11e, 802.11g, 802.11h, and 802.11i [1, 2, 8–11]. We also cover a recommended practice, called 802.11F, for *interaccess point protocol* (IAPP) [12]. Finally, we present some of the emerging amendments that are currently being standardized, including 802.11k, 802.11n, 802.11r, and 802.11s [3, 13–15].

Figure 11.2 illustrates the relationship among the protocols defined in both the baseline 802.11-1999 and the newly published 802.11-2007 along with the chapter, where the corresponding protocol is presented. In the figure, the direction of an arrow specifies the original and amended protocols. For example, the 802.11e MAC is an amendment of the 802.11-1999 baseline MAC. An important fact is that the 802.11e, an amendment of the baseline MAC, is actually a superset of the baseline MAC, and hence, it is backward compatible with the baseline MAC. Accordingly, an 802.11e device can communicate with a legacy device (operating with the baseline MAC) using the baseline protocol. It is the same with the 802.11b and 802.11g devices. An 802.11g device can communicate with an 802.11b device using the 802.11b transmission schemes, such as *modulation and coding schemes* (MCSs), since the 802.11g is a superset of the 802.11b. Note that some specifications like 802.11h involve both MAC and PHY amendments. Also, a recommended practice 802.11F for IAPP specifies the operations above the MAC. A major reason why the 802.11F was defined as a recommended practice is that it is meant for the operations above MAC while the 802.11 standards are meant for MAC and PHY. Further details about each amendment will be described throughout the rest of Part II.

For Part II, we use the term *MAC frame* or simply *frame* in order to represent a transmission unit of data. In fact, for packet-switched networks, the term *packet* is generally used for the same purpose. The readers are recommended to understand a

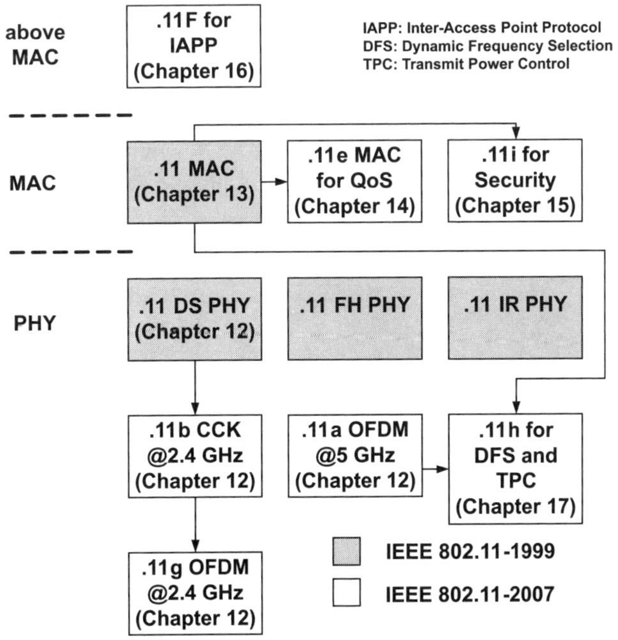

Figure 11.2 Protocols in IEEE 802.11-1999 and IEEE 802.11-2007 with the corresponding chapters.

frame as a layer-2 packet. Whenever needed, we will use more specific (and technically more accurate) terms like *protocol data unit* (PDU) and *service data unit* (SDU). A unit of data that arrives at a protocol layer from the higher layer is called an SDU, while a unit of data that is forwarded by a protocol layer to the lower layer after a data processing is called a PDU. For example, a unit that arrives at the MAC from the higher layer (e.g., IEEE 802.2 LLC layer) is called *MAC service data unit* (MSDU). The 802.11 MAC processes the MSDU and generates one or more *MAC protocol data units* (MPDUs) by appending a MAC header and a *frame check sequence* (FCS) to the MSDU or its fragment. The MPDU is forwarded to the lower layer (i.e., the PHY layer). Similarly, we can define the *PHY service data unit* (PSDU), arriving from the MAC, and the *PHY protocol data unit* (PPDU), which is transmitted to the channel. Note that PSDU is actually the same as MPDU, since the PSDU is what the PHY receives from the MAC, while the MPDU is what the MAC forwards to the PHY. Exceptionally, when a frame aggregation scheme of the emerging IEEE 802.11n, referred to as *aggregate MPDU* (A-MPDU), is used, a PSDU comprises a number of MPDUs [3]. Further details will be discussed in Section 18.1. The exact format of each PDU depends on the corresponding protocol and will be presented in the following chapters. Figure 11.3 illustrates the relationship among MSDU, MPDU, PSDU (=MPDU), and PPDU, assuming that fragmentation is not employed.

11.1 Network Architecture

The very basic form of IEEE 802.11 WLAN is called a *basic service set* (BSS). There are two types of BSS: *infrastructure BSS* and *independent BSS* (IBSS). The former is composed of an *access point* (AP), which works as the interface between the (wireline) infrastructure and the wireless link, and a number of stations associated with the AP. On the other hand, the latter is composed of a number of stations, which are communicating directly with one another. This type of BSS is also referred to as an ad hoc mode.[1] A BSS is identified by the corresponding *BSS identification* (BSSID). The BSSID is the MAC address of the AP in the case of an infrastructure BSS, while it is randomly chosen in the case of an IBSS by the station initializing the IBSS.

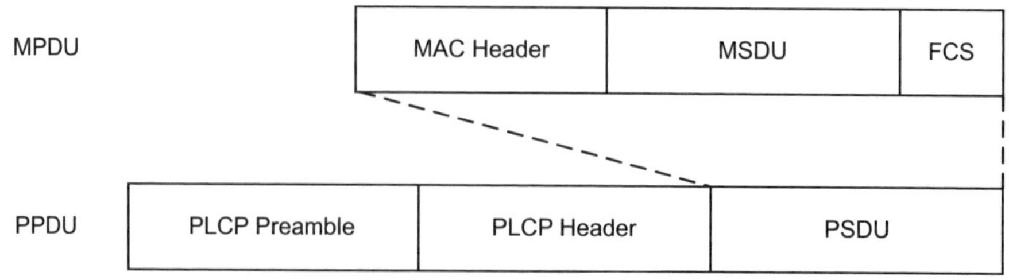

Figure 11.3 The relationship among MSDU, MPDU, PSDU, and PPDU.

1. This ad hoc mode of IEEE 802.11 should be differentiated from mobile ad hoc networking, for which the protocols are being defined by the *Internet engineering task force* (IETF) *mobile ad hoc networking* (MANET) working group [16].

The term *station* can be often replaced by other commonly used terms like wireless terminal, mobile node, and so on. However, this term should be more carefully understood in this book. Specifically, it should be noted that an AP itself is a station with extra functionalities. Therefore, exactly speaking, a station that is not an AP should be referred to as a *non-AP station*. However, when the distinction between AP and non-AP station is clear, a non-AP station might be simply referred to as a station.

All the stations in a given BSS, including the AP in the case of the infrastructure BSS, operate in the same frequency channel. Under the default MAC operation, the AP itself contends with non-AP stations for the wireless channel usage. Accordingly, the duplexing scheme for the infrastructure case can be classified as *time division duplexing* (TDD), in which the uplink (i.e., station-to-AP) and downlink (i.e., AP-to-station) transmissions times share a given frequency channel. Moreover, it should be clear that *frequency division duplexing* (FDD), in which two separate frequency channels are allocated for uplink and downlink transmissions, respectively, is not an option for the 802.11 WLANs.

11.1.1 Ad Hoc Network

As shown in Figure 11.4, a pair of stations in the 802.11 IBSS communicate directly in a peer-to-peer basis. The IBSS architecture assumes that all the stations are within their transmission ranges, and, hence, any pair of stations can directly communicate each other. In reality, two stations might not be within their communication ranges, and they have to rely on a relay station, which can forward their frames. The 802.11 architecture according to the baseline MAC does not support such multihop transmissions. Accordingly, transmissions other than single hop transmissions are out of the scope of the standard. However, if one implements layer-3 routing protocols (e.g., those defined in IETF MANET WG [16]) into the 802.11 devices in an IBSS, the stations can support multihop transmissions using such a layer-3 routing.

By definition, an IBSS does not involve an infrastructure, which could connect the WLAN to an outside network (e.g., the Internet). However, in reality, connect-

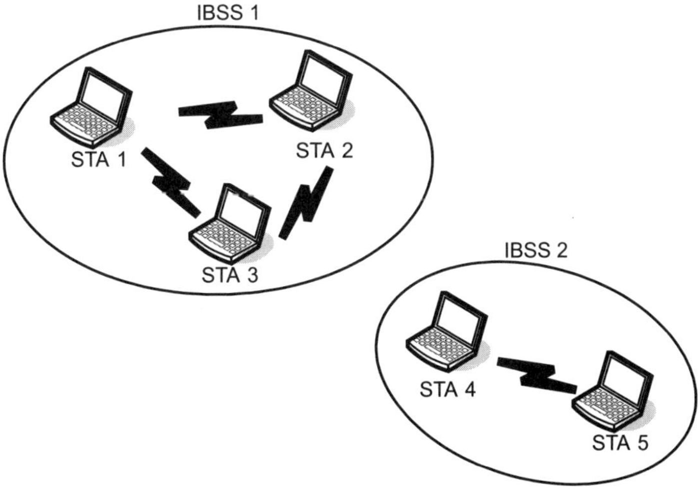

Figure 11.4 Illustration of IEEE 802.11 IBSS.

ing an IBSS with the Internet can be easily achieved. For example, if one station in an IBSS is connected with the Internet, and this station runs an Internet connection sharing functionality, which employs a routing protocol at layer 3, the station can work as a gateway between the 802.11 IBSS and the Internet without violating the architectural concept of the 802.11 IBSS. This is because the scope of the 802.11 protocols is limited only up to the layer-2 MAC.

11.1.2 Infrastructure Network

As shown in Figure 11.5, an infrastructure BSS is composed of a station working as an AP (e.g., STA 3 in BSS 2) and a number of non-AP stations that are associated with the AP (e.g., STA 4 and STA 5 in BSS 2). Note that this figure particularly shows that an AP is also a station with more functionalities than non-AP stations. A major extra functionality of the AP is the bridging function (i.e., the frame routing). A station with an outgoing frame in a BSS just sends this frame to its AP, and then the AP forwards the received frame to a relevant link. For typical APs with a single WLAN interface and a single backhaul link, the forwarding should be either back to its BSS or to the backhaul link.

The area containing the members of a BSS is called *basic service area* (BSA). It can be also understood as the transmission and reception range of the AP in a BSS. Note that the BSA of an infrastructure BSS is conceptually the same as a "cell" in a cellular network. In order to transmit and receive data frames in an infrastructure BSS, a station has to first associate with the corresponding AP by exchanging a few management frames, including authentication and association request/response frames. A station can be associated with a single AP at a given time, implying a hard handoff support in multi-AP network.

Under the baseline MAC, stations in a BSS communicate only via their AP. That is, in Figure 11.5, STA 4 and STA 5 are not allowed to transmit and receive frames directly even if their transmission and reception ranges overlap, but do only through STA 3, which is their AP. Since direct transmission is not allowed per the baseline MAC, a station does not need to worry about where the destination station is

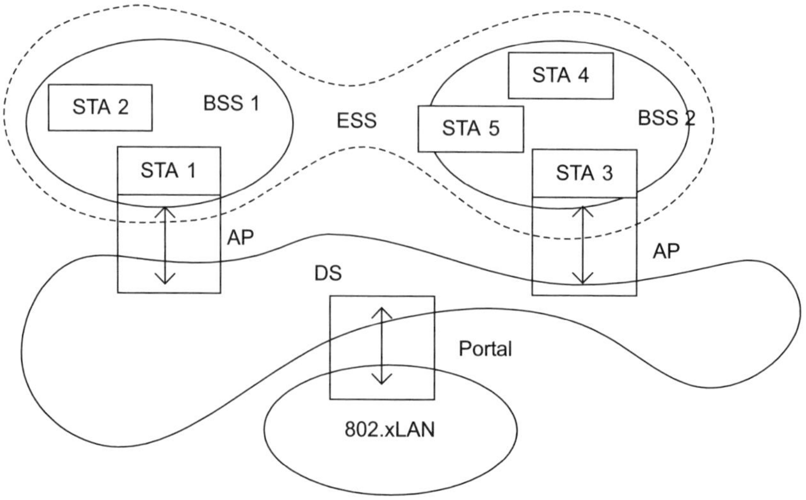

Figure 11.5 Illustration of infrastructure BSS and the formation of an ESS via DS.

located or, more specifically, whether or not the destination is within its transmission range. It simply needs to send any frame to its AP, which in turn forwards the frame properly according to the destination.

Another major reason why such a direct transmission is not allowed per the baseline MAC is related to the power-save support. As described in Section 13.5.2, a non-AP station is allowed to go to the doze state, in which the station is not involved with transmission/reception of frames in order to save its energy, and the AP needs to support such power-saving stations by buffering frames destined to these stations and then transmitting buffered frames when they are awake. Since a direct transmission is not allowed, a station can transmit a frame to another station in the BSS without worrying about whether or not the destination station is in doze state, since the frame will be first transmitted to the AP, which in turn forwards the frame to the destination station in power-saving mode when the station is awake.

The disallowance of direct transmissions is no longer valid per IEEE 802.11e, which optionally allows two neighboring stations to set up a direct link when desired, as presented in Section 14.5.1. The 802.11e simply resolves this issue by disallowing the destination station to go to the doze state once a direct link is set up. In fact, there is also an ongoing effort for the standardization of IEEE 802.11z for the enhancement of the 802.11e *direct link setup* (DLS). The enhancement will allow an operation with non-DLS capable APs and allow stations with an active DLS session to enter the doze state.

11.1.3 Distribution System (DS) and Extended Service Set (ESS)

While a single AP's BSA may be large enough to cover a small geographical area including a home or a small hot spot, multiple APs are usually deployed to cover large hot spots or enterprise networks. Figure 11.5 shows a WLAN with two APs, which are connected via *distribution system* (DS). The DS in a WLAN represents a conceptual system used to interconnect a set of BSSs and integrated LANs to create an *extended service set* (ESS). One can interpret a DS as a backhaul, which is typically constructed using a wireline networking technology (e.g., IEEE 802.3 Ethernet), but it can even be implemented using wireless networking technologies (e.g., IEEE 802.11 itself). The emerging 802.11s for ESS mesh networking will allow the 802.11 APs (called mesh access points in the 802.11s terms) to communicate directly in a peer-to-peer basis over the 802.11 links, and also enable multihop transmissions among the APs using a layer-2 routing [15]. The 802.11s basically provides a means to establish a DS wirelessly using the 802.11 technology. Accordingly, the 802.11s is not meant for non-AP stations. Further details will be presented in Section 18.2.

An ESS is identified by a *service set identification* (SSID), which is a character set of up to 32 octets. SSID is often referred to as the *network name* in commercial 802.11 WLAN devices. The APs periodically broadcast a management frame, called *beacon* frame, which includes a field indicating the corresponding SSID so that the stations can identify the ESS of the APs. The beacon includes many other information fields crucial for the 802.11 WLAN operations, such as time synchronization, power save support, and handoff. All the devices within an ESS, including the APs and non-AP stations, belong to the same subnet. Accordingly, a frame transmission from a station within an ESS to another station in the same ESS does not involve a layer-3 routing operation. The 802.11 additionally defines a logical entity called *portal* to conceptually represent a function providing the delivery of an MSDU between the DS and a non-IEEE 802.11 LAN as shown in Figure 11.5. While the DS itself is implemented using non-IEEE 802.11 technologies in most deployments, the concept of portal could be useful in order to separate the 802.11 and non-802.11 networks conceptually.

A frequency channel planning might be an issue for multiple AP deployment in a given geographical area. To reduce cochannel interference, it is desired to use nonoverlapping frequency channels for immediately neighboring APs. However, there may not be enough nonoverlapping channels available (e.g., there are only three for the 802.11b/g). Although one can avoid cochannel interference by allocating three nonoverlapping channels to neighboring APs in a two-dimensional structure, it becomes problematic in three-dimensional environments (e.g., multistory buildings). Therefore, it could be practically impossible to avoid cochannel interference, at least in the 802.11b/g WLANs. Fortunately, the 802.11 *distributed coordination function* (DCF) MAC, which is based on *carrier sense multiple access with collision avoidance* (CSMA/CA) protocol, allows APs (or BSSs) in the same frequency channel to share the channel rather smoothly thanks to its carrier sense-based access mechanism.

11.2 Reference Model

The reference model of IEEE 802.11 is shown in Figure 11.6. As described earlier, the 802.11 defines both MAC and PHY, and IEEE 802.2 LLC sits on top of the 802.11 MAC. Sublayers and entities in the reference model communicate via an interface, called a *service access point* (SAP). A set of primitives are defined for each SAP for the internal (i.e., within a station) communication between two sublayers (or entities).

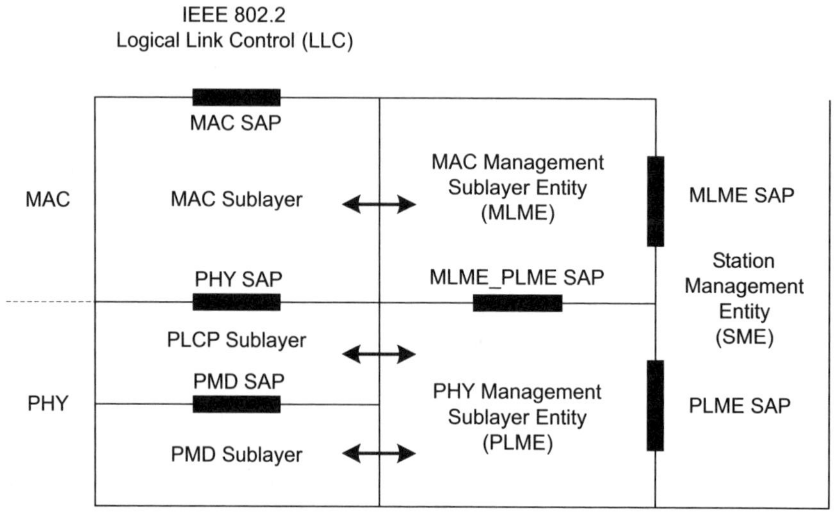

Figure 11.6 IEEE 802.11 reference model. (*After:* [5].)

The MAC is divided into two sublayers—MAC sublayer and *MAC management sublayer entity* (MLME). The MAC sublayer deals with frame construction, frame transmission, frame reception, and error recovery, involving when and how to transmit and receive frames. On the other hand, the MLME deals with management-related functions including time synchronization, power saving, association, handoff, and finally the support of *management information bases* (MIBs). A MIB is a type of database used to manage the devices in a communications network and is accessed and modified using *simple network management protocol* (SNMP) [17]. Normally, the MAC sublayer operations are more time critical than the MLME operations.

The PHY is divided into three sublayers—*physical layer convergence procedure* (PLCP) sublayer, *physical medium dependent* (PMD) sublayer, and *physical layer management entity* (PLME). The PMD defines the characteristics of, and the method of transmitting and receiving data through, a wireless channel between two or more stations. That is, the PMD basically defines the MCSs along with the relevant *radio frequency* (RF) characteristics, which are used for the transmission and reception of frames. On the other hand, the PLCP adapts the capabilities of the PMD system to the PHY service by defining a method of mapping the MPDUs into PPDUs. Each PHY specification defines its own PMD and PLCP. Finally, the PLME provides the MIB service by supporting a number of MIBs.

As shown at the right side of Figure 11.6, the reference model includes a conceptual entity called *station management entity* (SME). The SME is a cross-layer entity as shown in the reference model in the sense that it can internally communicate with multiple layers. For example, the SME has SAPs with both MLME and PLME. In fact, the SME can have an interface with the application layer as well. One can understand the SME as a conceptual entity, which monitors and controls the operations of an 802.11 device. When a user uses an 802.11 device, the user can control the operation of the device in various ways (e.g., by specifying desired SSID, desired BSSID, desired channel number, and desired security key). Moreover, the user can see the current status of the device, such as the current channel number, *received signal strength* (RSS), and so on. All these are specified and obtained by the application layer program (e.g., an 802.11 connection manager) through the SME, which in turn communicates with MLME or PLME in order to achieve what was commanded by the application layer program. For example, when a user specifies a list of desired BSSIDs, the SME communicates with the MLME in order to obtain the list of available APs (or BSSID) in the neighborhood, and then requests the MLME to associate with one of these APs. A number of implementation-dependent algorithms, which control and optimize the performance of an 802.11 device, are implemented as part of the SME.

11.3 Layer Interactions

In this section, we discuss how the layers interact within an 802.11 device for data communication. The MAC interacts with either IEEE 802.2 LLC or IEEE 802.1D MAC bridge as its higher layer and with the underlying PHY as its lower layer. The PHY interacts with the MAC as its upper layer.

11.3.1 MAC Message Types

Before delving into layer interactions, we first briefly discuss various MAC messages defined by the MAC layer. The 802.11 MAC layer defines three types of MAC messages, namely, data, management, and control types. First, the data messages are those arriving from the higher layer (i.e., either LLC or MAC bridge) at the transmitter side. That is, the MAC message arrives at the MAC in the form of MSDU. Second, the management messages are used to support the 802.11 services and are locally generated by the MAC according to the control by the MLME and SME. Finally, the control messages are used to support the delivery of data and management messages, and are locally generated by the MAC. Further details of these three types of messages are presented in Section 13.1.

Note that only the data messages are related with the interaction of the MAC with its higher layers. On the other hand, all three types of messages are related with the interaction between the MAC and the PHY, since a MAC message, which is one of three types, is forwarded to the underlying PHY, as it is encapsulated by an MPDU.

11.3.2 Interaction Between MAC and PHY

The MAC generates an MPDU out of a MAC message (i.e., one of data, management, and control messages). A data message is actually an MSDU, which arrived from the higher layer through the MAC SAP. A management message, which is also referred to as a *MAC management protocol data unit* (MMPDU), is locally generated according to the control of the MLME sublayer. Finally, a control message is generated by the MAC sublayer to assist the transmissions of data and management messages. In fact, an MPDU can be generated using a fragment of a MAC message in the case of data and management messages. We further discuss the possibility of fragmenting a message into multiple fragments in Section 13.2.6.

As shown in Figure 11.7, an MPDU is generated by encapsulating a MAC message or its fragment by appending both a MAC header and an FCS. Note that this figure is slightly different from Figure 11.3 in which "MSDU" is replaced by "MAC message or its fragment." The format of a MAC header depends on the message types and conveys various kinds of information needed by the receiver stations, including the addresses of both transmitter and receiver. The FCS is generated using CRC-32 and is used by the receiver to check whether the frame is received without any error. A generated MPDU is forwarded to the PHY for the transmission via the PHY SAP along with the transmission rate and transmit power used for the frame transmission. That is, according to the 802.11 standard, the algorithms for both the

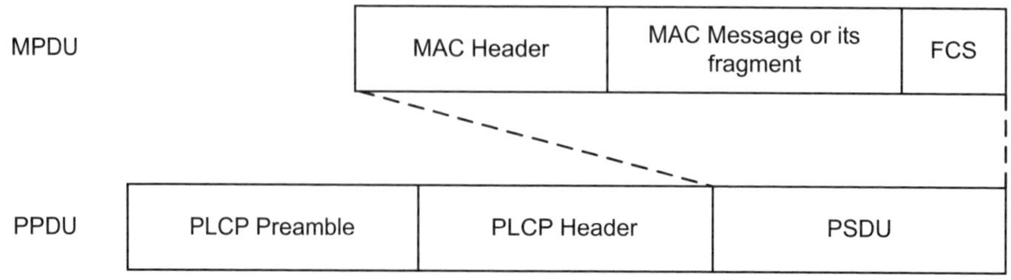

Figure 11.7 Relationship between MPDU and PPDU.

link adaptation (determining the transmission rate of a frame) and the transmit power control (determining the transmit power used for a frame transmission) are implemented in the MAC.

As the forwarded MPDU is arriving at the PHY as a PSDU, the PHY generates a PPDU using the PSDU by appending a PLCP preamble and a PLCP header. The preamble is a set of known symbols, which are used by a receiver to detect the incoming frame. Under the basic operation of the 802.11 MAC, a receiver cannot know in advance when a frame is arriving, and, hence, detecting an incoming frame is very important. The PLCP header actually conveys some information, including the transmission rate and the frame length. A key fact is that the PPDU is generated while a PSDU is arriving from the MAC. In fact, the PHY should start transmitting the first symbol of the PLCP preamble as soon as the MAC commands the transmission of an MPDU because of the tight timing requirement. Then, the following bits from the PSDU can be transmitted after the PLCP header transmission with short delays. This operation should be achievable as long as the interface speed between the MAC and the PHY is faster than the link speed (i.e., the transmission rate over the channel).

11.3.3 Interaction Between MAC and IEEE 802.2 LLC

IEEE 802.2-1998 LLC [18] sits on top of the 802.x MAC, and works as the interface between the MAC and a layer-3 protocol (e.g., IP layer). As shown in Figure 11.8, the LLC generates an *LLC protocol data unit* (LPDU) out of an *LLC service data unit* (LSDU) arriving from a higher layer. In the figure, as an LSDU arrives from the IP layer, the LSDU is an IP datagram. The main service of the LLC is the multiplexing of multiple layer-3 protocols. That is, multiple layer-3 protocols can sit on top of the 802.11 MAC inside a device. Most typical layer-3 protocols include IP and *address resolution protocol* (ARP). The LLC header includes a field indicating which protocol generated the LSDU encapsulated within the corresponding LPDU, and, hence, the LLC at the receiver can determine to which higher-layer protocol the received LSDU should be forwarded.

The 802.11 LLC provides three modes of operation: (1) type 1 unacknowledged connectionless mode, (2) type 2 connection mode, and (3) type 3 acknowledged connectionless mode. For type 2 connection mode, a data link connection is estab-

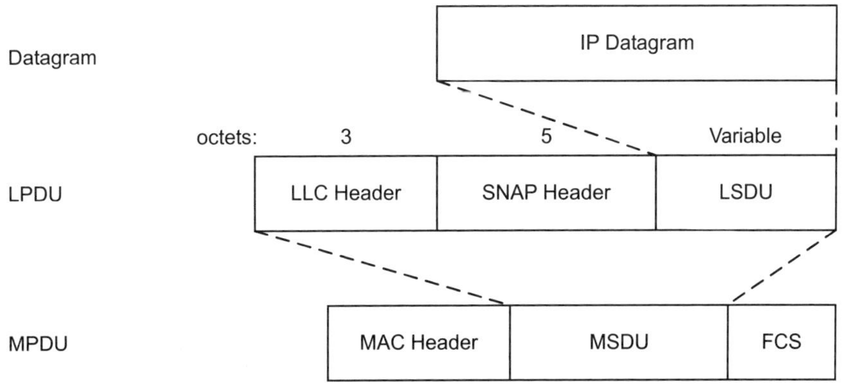

Figure 11.8 Relationship between LPDU and MPDU.

lished between LLCs before exchanging information LPDUs. On the other hand, types 1 and 3 do not establish a connection, where the LLC at the receiver under type 3 acknowledge connectionless mode acknowledges the reception of an LPDU for error recovery. Type 1 unacknowledged connectionless mode is meant for a best-effort service, and it is typically used for the LLC sitting on top of the 802.11 MAC.

An LPDU format is found at Figure 11.9, where DSAP and SSAP represent destination and source service access points, respectively. These SAPs basically specify the higher-layer protocols sitting on top of the LLC at both source and destination stations for the multiplexing service. An LLC header may be followed by a *subnet access protocol* (SNAP) header, as shown in Figure 11.10. When a SNAP header is used, both DSAP and SSAP fields are set to 0xaa, and the control field occupies one octet. Then, the LLC header is followed by a SNAP header of 5 octets, which comprises an IEEE *organizationally unique identifier* (OUI) field and an *Ethernet type* (EtherType) field.

The EtherType field is again used for the multiplexing purpose. For example, 0×0800 is used to specify that the LSDU is actually an IP datagram. Since the EtherType field is 2 octets, while the SAP field is one octet, more protocols can be identified using the EtherType. In fact, the EtherType field is rooted in the popular Ethernet; the same type field and its encoding are used for the Ethernet frames. For the 802.11 devices, the LLC/SNAP header format is typically used for the LPDU carrying an IP datagram. The OUI is 3 octets long, and represents a vendor-specific code, which is uniquely assigned to a vendor. In fact, the first three octets of a 6-octet 802.11 MAC address also represent the OUI of the vendor, which produced the corresponding 802.11 device. For the LLC/SNAP, the OUI is typically set to all zero. If it is set to a specific nonzero OUI, the EtherType field can be encoded using vendor-specific private protocol identifiers.

Figure 11.11 illustrates the relationship among LPDU, MPDU, and PPDU by considering the delivery of an IP datagram over IEEE 802.11b WLAN without being fragmented. Further details of the PPDU and MPDU formats will be presented in Chapters 12 and 13, respectively. Note that the LLC generates an LPDU out of an LSDU (i.e., an IP datagram), and forwards it to the MAC. Then, the MAC receives it

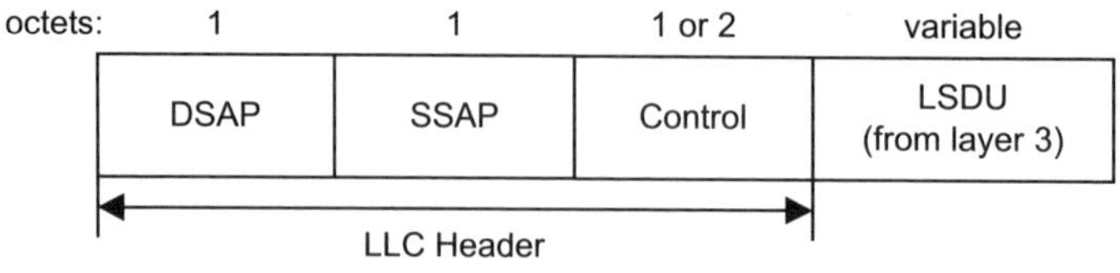

Figure 11.9 LPDU format. (*After:* [18].)

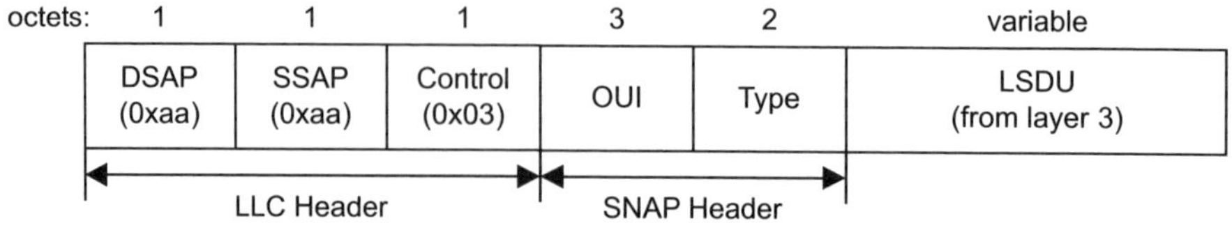

Figure 11.10 LPDU with SNAP header.

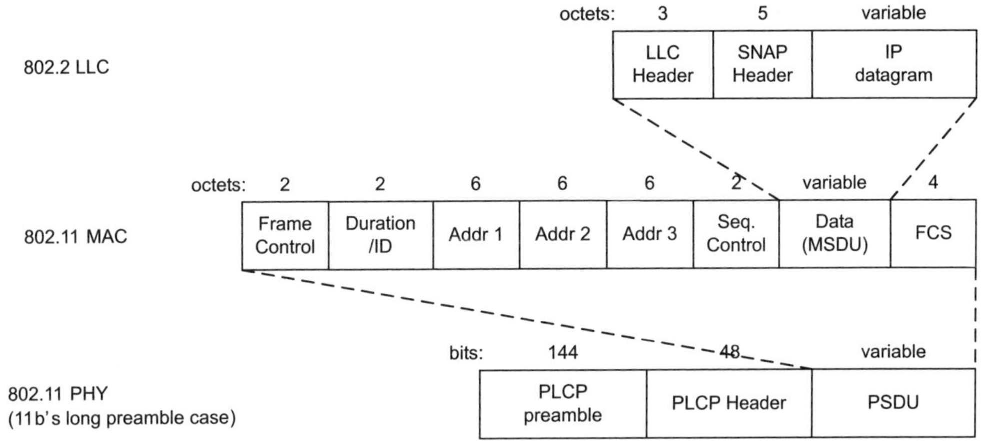

Figure 11.11 Relationship among LPDU, MPDU, and PPDU assuming IEEE 802.11b PHY.

as its MSDU and then generates an MPDU out of the MSDU. The MPDU is forwarded to the PHY, which receives it as its PSDU. The PHY then generates a PPDU out of the PSDU and transmits over the channel.

11.3.4 Interaction Between MAC and IEEE 802.1D MAC Bridge

IEEE 802.1D-2004 MAC bridge [19] allows communication between end stations attached to separate LANs, which could be of different kinds (e.g., IEEE 802.11 and IEEE 802.3). A MAC bridge is transparent to LLC and network layer protocols, just as if the stations were attached to the same LAN. Figure 11.12 illustrates the relationship among the LLC, MAC, and bridge. A bridge in the middle includes two network interfaces, where these two interfaces can be of different kinds. Note that this bridge is connecting two separate LANs. In fact, a bridge can include more than two interfaces. A typical example of a bridge is an Ethernet switch with four or more Ethernet ports. Another good example of interest is the 802.11 AP, which typically includes one or more Ethernet interfaces and one or more 802.11 WLAN interfaces. In the figure, there are two LLC entities at both ends. From the LLC's perspective, two LANs, which are connected via the bridge in the middle, look like a single LAN. Note that a set of LANs connected via a bridge form a single IP subnet.

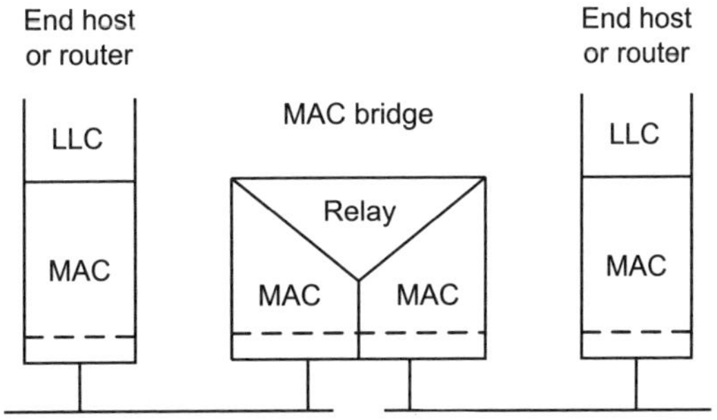

Figure 11.12 MAC versus LLC and bridge. (*After:* [19].)

As a bridge connects multiple LANs, it should provide a routing function. Different from layer-3 routing employing sophisticated routing protocols, layer-2 routing employed by the 802.1D bridge is based on *self-learning*. Let us consider a bridge with four interfaces. When a frame arrives at one interface, the bridge has to determine which interface it has to forward the received frame to. If the destination interface is known, the frame is forwarded to the particular interface. Otherwise, the bridge forwards the frame to the other three interfaces. Note that upon receiving a frame at an interface, the bridge happens to learn that the source station of the frame can be reached via the incoming interface, and, hence, when it receives a frame destined to the station, it can select a correct interface to forward a frame to. As time goes, the bridge is likely to learn more about the location of stations in terms of the corresponding interfaces, thus making more correct routing decisions. This is the reason it is called self-learning routing.

Figure 11.13 illustrates the relationship among router, bridge, and AP, where a layer-3 router connects two subsets. Each subset is composed of a single bridge; two APs, where the APs are connected by a bridge; and stations associated with APs. Figure 11.14 shows the protocol stacks involved with the end-to-end path from station 1 to station 2 in Figure 11.13. We observe that the end-to-end path is composed of two layer-3 hops and six layer-2 hops.

11.4 Key Technologies

In this section, we briefly discuss the key technologies employed by IEEE 802.11 WLAN.

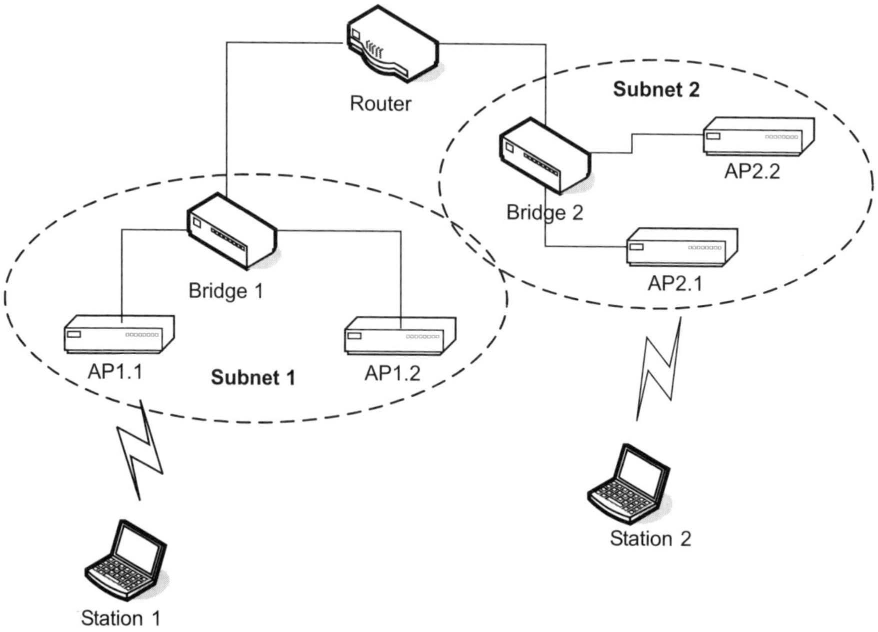

Figure 11.13 Router versus bridge.

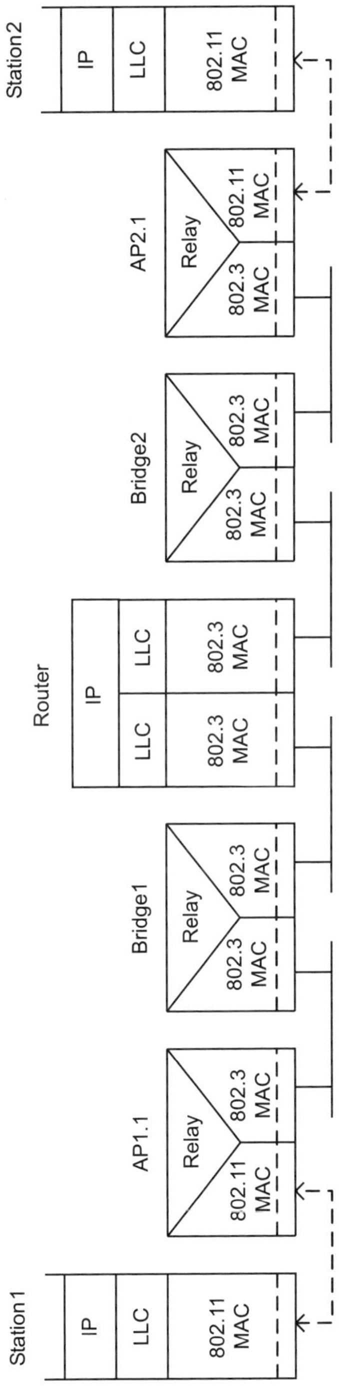

Figure 11.14 End-to-end path from station 1 to station 2.

11.4.1 Multiple Access, Duplexing, and MAC

The 802.11 basically operates with a *time division duplexing* (TDD) scheme for the sharing between uplink and downlink transmissions. That is, a single frequency channel is used for all the transmissions in a BSS. *Frequency division duplexing* (FDD) is not an option for the 802.11.

Moreover, the multiple access of the 802.11 is based on time division; it is a kind of *time division multiple access* (TDMA) scheme even if its MAC is very different from conventional TDMA MAC schemes. The 802.11 baseline standard defines connectionless MAC for the best-effort service. The baseline MAC is composed of two coordination functions, namely, the mandatory contention-based *distributed coordination function* (DCF) and the optional contention-free *point coordination function* (PCF). The DCF is based on *carrier sense multiple access with collision avoidance* (CSMA/CA) and the PCF is a poll-and-response MAC. In fact, the PCF has been rarely implemented in real products due to its complexity, the lack of demands, the lack of desirable operational functions, and so on. Under the DCF, which has been employed by most, if not all, 802.11 devices, a station transmits only when it determines that the channel is not occupied by other transmissions, and this makes the 802.11 DCF a perfect fit to the operation at unlicensed bands, where various types of devices should co-exist with some etiquette.

11.4.2 Multiple Transmission Rate Support

The 802.11 PHYs support multiple transmission rates by using different combinations of MCSs. Both the 802.11a [1] and 802.11g [2] support up to 54 Mbps, which make the 802.11 the fastest standard-based wireless technology right now. In fact, as will be discussed in Section 18.1, the emerging 802.11n PHY will support up to 600 Mbps by utilizing multiple antenna technologies (i.e., MIMO schemes) and channel bonding (i.e., using 40 MHz bandwidth instead of 20 MHz) [3]. Table 1.15 summarizes various PHYs of the 802.11 along with their transmission schemes, frequency bands, and supported transmission rates.

As 802.11 PHYs support multiple transmission rates, selecting a rate for a given frame transmission is a very important issue for the performance optimization of the network. In general, the higher the transmission rate, the shorter the transmission range, since high-order modulation schemes require higher *signal-to-interference-and-noise ratio* (SINR) for successful transmissions. The rate selection should be made in an adaptive manner along with the time-varying channel condition from the transmitter to the receiver. This problem is known as a *link adaptation*, where it is an implementation-dependent algorithmic issue.

For an intelligent link adaptation, a protocol support is typically needed (i.e., closed-loop link adaptation). That is, as the receiver is ideally the best one to determine the optimal rate, a feedback mechanism is needed as part of the protocol. However, the 802.11 does not support such a feedback mechanism. Accordingly, the 802.11 link adaptation has been typically an open-loop approach so that the transmitter determines the rate to use based on the history of the frame transmission successes/failures. A close-loop link adaptation mechanism is being defined as part of IEEE 802.11n today [3].

11.4.3 Power-Saving Schemes

Power saving is one of the major concerns for battery-powered portable mobile communication devices. The 802.11 MAC defines *power-saving mode* (PSM) operation, in which a station switches back and forth between the active and the doze states. The station consumes minimal energy in the doze state, since it can neither transmit nor receive frames while staying in that state. It is well known that an 802.11 interface consumes a considerable amount of energy even during the channel sensing process, while a station in the active state continues to sense the channel unless it transmits or receives a frame. Accordingly, the best strategy for the power saving is putting the 802.11 station into the doze state as frequently as possible without sacrificing the delay performance too much. A station under the PSM basically wakes up periodically in order to receive a beacon frame transmitted by its AP, since the AP buffers frames destined to such PSM stations and announces the buffered frame information via beacons. The PSM station goes back to the doze state if there is no buffered frame. Otherwise, it stays awake, and requests the AP to transmit the buffered frames to itself.

The PSM operation is designed for the power saving of stations with no traffic or traffic with a periodic pattern. The 802.11e further enhances the power-saving scheme, thus defining a scheme called *automatic power save delivery* (APSD), which allows a station to save some power even during a QoS stream operation, such as *voice over WLAN* (VoWLAN) operation [9]. That is, since many QoS applications generate traffic with some periodic patterns, the 802.11e APSD allows a station with ongoing QoS applications to save energy while there is no traffic to transmit/receive even during the runtime of such QoS applications. Power-saving mechanisms involve again implementation-dependent algorithms (e.g., specifically to determine when to sleep and when to wake up while maintaining an acceptable delay performance).

11.4.4 Mobility Support

Mobility support has not been a major concern of the 802.11 WLAN, since people rarely use their laptops or PDAs to access the Internet via WLAN while they are moving around. However, some level of mobility is supported by the 802.11. For example, the walking speed mobility is surely supported. That is, an 802.11 station can switch from one AP to another in an ESS while it moves by reassociating with a new AP. Note that mobility is not supported in the case of IBSS. The 802.11 allows a station to be associated with a single AP at a given time. That is, a hard handoff is supported.

Today, along with the emergence of the VoWLAN applications, supporting seamless and smooth handoffs in the 802.11 WLAN is becoming a hot topic. For a handoff to occur, a station has to first detect neighboring APs via a scanning process. Then, it has to determine which AP to reassociate with. Once this is determined, a reassociation process is conducted along with an authentication with the new AP. For most of today's 802.11 devices, the scanning process takes the most time, and there are ongoing efforts (e.g., in IEEE 802.11k [13]) to reduce the scanning time. Moreover, IEEE 802.11r, which is also being standardized, attempts to reduce the handoff time while maintaining QoS and security [14].

11.4.5 Access Control and Confidentiality Support

The baseline MAC of the 802.11 has security mechanisms for confidentiality by encrypting frame payload using a cipher called RC4 and authentication by basically checking if both communication parties have the same security key. However, these schemes were found to be too weak to protect the security of the WiFi users. The problems include the cryptographic weakness of the RC4 and the lack of key management. For example, under the legacy security mechanism, the same security key is basically used for every station in the network, and the key is rarely updated over time. Such a security hole of the 802.11 was a big hurdle for the wide acceptance of WiFi at one point. For enterprise networking in particular, a strong security support is a must.

Then, IEEE 802.11i [11] was developed with stronger security features by defining the *robust security network* (RSN), composed of a stronger encryption scheme based on *advanced encryption scheme* (AES), enhanced mutual authentication based on IEEE 802.1X [20], per-frame authentication, per-station key management, and so on.

11.4.6 Spectrum and Transmit Power Management

IEEE 802.11h defines mechanisms for spectrum and transmit power managements including *dynamic frequency selection* (DFS) and *transmit power control* (TPC) [10]. While the 5-GHz bands, where the 802.11a operates, are unlicensed bands, there are in fact primary users that also use these bands. Those primary users are satellite and radar systems. Today, many regulatory bodies require a WLAN device operating at 5-GHz bands to have both DFS and TPC functions to minimize the interference of the WLAN to these primary users. That is, when a radar system is detected, the WLAN devices should leave the current channel switching to another channel, and when a satellite system is detected, the WLAN devices are required to reduce their transmission power.

11.4.7 Traffic Differentiation and QoS Support

The baseline MAC is enhanced by the 802.11e to support *quality of service* (QoS) for multimedia applications such VoWLAN and video streaming [9]. The 802.11e MAC is called *hybrid coordination function* (HCF); it comprises the contention-based *enhanced distributed channel access* (EDCA) and the poll-and-response *HCF controlled channel access* (HCCA). EDCA and HCCA enhance DCF and PCF, respectively. According to the 802.11e, both EDCA and HCCA are mandatory, while in the baseline MAC only the DCF is mandatory.

The EDCA provides prioritized channel access to frames with different priorities. Basically, eight different priority values are supported (following the convention of IEEE 802.1D [19]), while only four different levels of channel access prioritization are provided according to the EDCA. On the other hand, the HCCA relies on the polling and downlink frame scheduling of the AP to meet the QoS described by a set of parameters. While both the PCF of the baseline MAC and the HCCA of the 802.11e are polling MACs, there are a number of differences. The

802.11e also defines various features needed for QoS provisioning, including the means for admission control of QoS streams.

Another key feature of the 802.11e MAC is a concept called a *transmission opportunity* (TXOP), under which a station is allowed to transmit multiple frames back to back with a given time limit, called a TXOP limit, once it successfully transmit a frame into the channel. This feature makes the stations' transmission times more controlled and predictable, so that QoS can be more tightly provisioned. Moreover, it can enhance the protocol efficiency, thanks to the reduced protocol overhead.

The 802.11e also defines a number of features that enhance the performance of WLANs even if they are not directly related with the QoS. A *direct link setup* (DLS) between neighboring stations in an infrastructure BSS is defined. Moreover, a *block acknowledgment* (BACK) mechanism could be more efficient than the immediate *acknowledgment* (ACK) scheme of the baseline MAC in the sense that a smaller amount of the wireless bandwidth is wasted for the ACK feedback. Note that both direct link setup and BACK are related with the enhancement of the protocol efficiency.

References

[1] IEEE 802.11a-1999, Amendment 1 to Part 11: Wireless LAN Medium Access Control (MAC) and Physical Layer (PHY) Specifications: High-Speed Physical Layer in the 5 GHz Band, 1999.

[2] IEEE 802.11g-2003, Amendment 4 to Part 11: Wireless LAN Medium Access Control (MAC) and Physical Layer (PHY) Specifications: Further Higher Data Rate Extension in the 2.4 GHz Band, 2003.

[3] IEEE 802.11n/D3.0, Draft Supplement to Part 11: Wireless LAN Medium Access Control (MAC) and Physical Layer (PHY) Specifications: Enhancements for Higher Throughput, September 2007.

[4] IEEE Working Group (WG), http://www.ieee802.org/11.

[5] IEEE 802.11-1999, IEEE Standard for Local and Metropolitan Area Networks—Part 11: Wireless LAN Medium Access Control (MAC) and Physical Layer (PHY) Specifications, Reference Number ISO/IEC 8802-11:1999(E), IEEE Std 802.11, 1999 edition, 1999.

[6] IEEE 802.11-2007, IEEE Standard for Local and Metropolitan Area Networks—Part 11: Wireless LAN Medium Access Control (MAC) and Physical Layer (PHY) Specifications, IEEE Std 802.11-2007, (Revision of IEEE Std 802.11-1999), June 12, 2007.

[7] IEEE 802.11j-2004, Amendment 7 to Part 11: Wireless LAN Medium Access Control (MAC) and Physical Layer (PHY) Specifications: 4.9 GHz–5 GHz Operation in Japan, 2004.

[8] IEEE 802.11b-1999, Supplement to Part 11: Wireless LAN Medium Access Control (MAC) and Physical Layer (PHY) Specifications: Higher-Speed Physical Layer Extension in the 2.4 GHz Band, IEEE Std 802.11b-1999, 1999.

[9] IEEE 802.11e-2005, Amendment 8 to Part 11: Wireless LAN Medium Access Control (MAC) and Physical Layer (PHY) Specifications: Medium Access Control (MAC) Enhancements for Quality of Service (QoS), 2005.

[10] IEEE 802.11h-2003, Amendment 5 to Part 11: Wireless LAN Medium Access Control (MAC) and Physical Layer (PHY) Specifications: Spectrum and Transmit Power Management Extensions in the 5 GHz Band in Europe, 2003.

[11] IEEE 802.11i-2004, Amendment 6 to Part 11: Wireless LAN Medium Access Control (MAC) and Physical Layer (PHY) Specifications: Medium Access Control (MAC) Security Enhancements, 2004.

[12] IEEE 802.11F-2003, IEEE Recommended Practice for Multi-Vendor Access Point Interoperability Via an Inter-Access Point Protocol Across Distribution Systems Supporting IEEE 802.11™ Operation, 2003.

[13] IEEE 802.11k/D9.0, Draft Supplement to Part 11: Wireless LAN Medium Access Control (MAC) and Physical Layer (PHY) Specifications: Specification for Radio Resource Measurement, September 2007.

[14] IEEE 802.11r/D8.0 Draft Supplement to Part 11: Wireless LAN Medium Access Control (MAC) and Physical Layer (PHY) Specifications: Fast ESS Transition, September 2007.

[15] IEEE 802.11s/D1.07, Draft Supplement to Part 11: Wireless LAN Medium Access Control (MAC) and Physical Layer (PHY) Specifications: Mesh Networking, September 2007.

[16] IETF MANET WG, http://www.ietf.org/html.charters/manet-charter.html.

[17] IETF RFC 4789, Simple Network Management Protocol over IEEE 802 Networks, 2006.

[18] IEEE 802.2-1998, IEEE Standard for Local and Metropolitan Area Networks—Part 2: Logical Link Control, 1998.

[19] IEEE 802.1D-2004, IEEE Standard for Local and Metropolitan Area Networks—Media Access Control (MAC) Bridges (Incorporates IEEE 802.1t-2001 and IEEE 802.1w), 2004.

[20] IEEE 802.1X-2004, IEEE Standard for Local and Metropolitan Area Networks—Port-Based Network Access Control, 2004.

Selected Bibliography

Geier, J., *Wireless LANs: Implementing High Performance IEEE 802.11 Networks*, 2nd ed., Indianapolis, IN: SAMS, 2002.

O'Hara, B., and A. Patrick, *IEEE 802.11 Handbook, A Designer's Companion*, 2nd ed., New York: IEEE Press, 2005.

Prasad, A. R., and N. R. Prasad, *802.11 WLANs and IP Networking: Security, QoS, and Mobility*, Norwood, MA: Artech House, 2005.

Walke, B. H., S. Mangold, and L. Berlemann, *IEEE 802 Wireless Systems: Protocols, Multi-Hop Mesh/Relaying, Performance and Spectrum Coexistence*, New York: Wiley, 2006.

CHAPTER 12

PHY Protocols

IEEE 802.11 PHYs have been evolving dramatically. IEEE 802.11-1999 defines three PHY protocols, namely, *direct-sequence spread spectrum* (DSSS), *frequency-hopping spread spectrum* (FHSS), and *infrared* (IR), where all three PHYs support only the transmission rates of 1 and 2 Mbps [1]. The extensions of the 802.11 PHY include the 802.11a-1999 supporting up to 54 Mbps based on the *orthogonal frequency division multiplexing* (OFDM), the 802.11b-1999 supporting up to 11 Mbps based on the *complementary code keying* (CCK), and the 802.11g-2003 again based on OFDM to support up to 54 Mbps transmission rates [1–4]. Both 802.11b and 802.11g also support optional transmission modes based on *packet binary convolutional code* (PBCC) modulations [3, 4].

The 802.11 PHYs operate in unlicensed bands at 2.4 GHz and 5 GHz. While most other PHYs, including DSSS, FHSS, 802.11b, and 802.11g, operate at the 2.4-GHz bands, the 802.11a operates at the 5-GHz bands. The 802.11g, in fact, includes the 802.11b, while the 802.11b includes the baseline DSSS PHY. Accordingly, the 802.11g is backward compatible with the 802.11b, while the 802.11b is backward compatible with the baseline DSSS PHY. Today, the most popular 802.11 PHY is the 802.11g, thanks to its fast transmission rate as well as low-cost chipset availability, even if the 2.4-GHz bands, in which the 802.11g operate, are much more crowded than the 5-GHz bands of the 802.11a. The supported transmission rates as well as the frequency bands for each PHY are summarized in Table 12.1.

In this chapter, we present the 802.11a, 802.11b, and 802.11g PHYs. The DSSS PHY is presented as part of the 802.11b, while the other two baseline PHYs (i.e., FHSS and IR) are not presented since these two PHYs are obsolete and not likely to be a basis for any future PHYs of the 802.11. The emerging IEEE 802.11n supporting up to 600 Mbps will be presented in Section 18.1.

12.1 IEEE 802.11 PHY Operations

Before presenting individual PHYs, we first explain the general functions of the 802.11 PHY related with the frame transmission and reception.

12.1.1 Frame Transmission

When a PSDU arrives at the PHY from the MAC, it arrives along with a set of parameters, which are collectively known as TXVECTOR. These parameters specify the options to be used for the transmission of the PPDU, which contains the

Table 12.1 Various PHYs of IEEE 802.11

PHY	Transmission schemes	Frequency bands	Transmission rates (Mbps) supported
Baseline	DSSS, FHSS and IR	DSSS, FHSS – 2.4 GHz IR – 850~950 nm	1, 2
802.11a	OFDM	5 GHz	6, 9, 12, 18, 24, 36, 48, 54
802.11b	CCK	2.4 GHz	5.5, 11 + DSSS rates
802.11g	OFDM	2.4 GHz	6,9, 12, 18, 24, 36, 48, 54 + 802.11b rates
802.11n	OFDM, MIMO	2.4 GHz, 5 GHz	Up to 600

PSDU. Whereas the exact set depends on the individual PHY, the parameters in the TXVECTOR might include:

- *Rate:* indicates the transmission rate to be used for the PPDU transmission.
- *Length:* indicates the length of the PSDU.
- *Preamble type:* indicates the type of the PLCP preamble to be used for the PPDU transmission. The preamble is provided so the receiver can perform the necessary frame detection and synchronization operations. For the same transmission rate, various preambles might be available.
- *Modulation:* indicates the modulation scheme to be used for the PPDU transmission. Different modulation schemes might be defined for the same transmission rate.
- *Transmit power level:* indicates the transmit power level to be used for the PPDU transmission.

Note that all these parameters including the rate and the transmit power level are determined by the MAC. The selection of these parameters for a given environment affects the network performance. The rate adaptation is one of the most important implementation-dependent algorithmic issues in the 802.11 WLAN, and this will be further discussed in Section 13.4.2. Typically, the 802.11 stations use a fixed transmit power, while the power level can be adapted. The issues related with the transmit power control are discussed in Section 17.3.

12.1.2 Frame Reception

Now, when a PPDU arrives at a PHY from the channel, the PHY extracts the PSDU from the PPDU, and forwards it to the MAC along with a set of parameters, which are collectively known as RXVECTOR. The parameters include rate, length, preamble type, modulation, service, and *received signal strength indicator* (RSSI). While all others are the same as those defined for the TXVECTOR, the RSSI indicates the RF energy level measured during the reception of the PPDU. Normally, 8 bits are used to indicate 256 levels, where the mapping from an energy level to the RSSI value is not standardized, but implementation dependent.

At the end of the PPDU reception, whether or not the PPDU was received correctly is also indicated to the MAC. There are three cases when the reception could be in error. Note that an erroneous frame reception triggers the use of *extended interframe space* (EIFS) instead of *distributed interframe space* (DIFS) by the DCF, as will be explained in Section 13.2.2.

- *Format violation:* when a received PPDU is not in the correct format (e.g., due to an error in the PLCP header as detailed later);
- *Carrier lost:* when the carrier is lost in the middle of the PPDU reception (e.g., due to a deep fade of the channel);
- *Unsupported rate:* when the received PPDU is transmitted at unsupported rate; note that there exist optional transmission rates, which do not need to be implemented.

As explained earlier, the PLCP preamble is provided so the receiver can perform the necessary frame detection and synchronization operations. The preamble is composed of a set of known symbols so that a receiver PHY can detect the start of an incoming frame. As will be detailed in Chapter 13, an 802.11 station basically does not know when a frame will be arriving at its PHY, and, hence, it continues to sense the channel in order to receive any incoming frame. The receiver PHY should be able to detect the incoming preamble within a *clear channel assessment* (CCA) time as explained next.

12.1.3 CCA Operations

A major function of the PHY (or PLCP more exactly) is to indicate the channel status (i.e., busy or idle) to the MAC, and this function is referred to as the CCA. The PHY continues to sense the channel, irrespective of whether or not the station has any frame to transmit, and, hence, the CCA indication is a flag to the MAC informing whether the channel is busy or idle. The CCA busy indicates that the channel is not available due to another frame transmission on the channel or unknown signal with the energy level above a threshold.

The CCA operation depends on each individual PHY, but is basically based on two possible methods—*energy detection* (ED) and *carrier sensing* (CS).

- ED-based CCA busy status is triggered by the detection of any signal with the RF energy above an ED threshold.
- CS-based CCA busy status is triggered by the detection of a PHY-specific signal (e.g., PLCP preamble).

When either of the events occurs, the CCA busy status should be indicated to the MAC within a CCA time, where the value of the CCA time is dependent on the individual PHY, as shown in Table 12.2. The other related parameters include SlotTime, which is used for the backoff in the DCF, as explained in Section 13.2.1, and RxTxTurnaroundTime, which is the delay for the PHY to switch from the reception mode to the transmission mode, where both are again dependent on the PHY. Within a SlotTime, the station should be able to determine whether or not the

Table 12.2 CCA Parameters (in μs)

Parameter	802.11a	802.11b	802.11g	
SlotTime	9	20	20	9
RxTxTurnaroundTime	< 2	≤ 5	5	5
CCA time	< 4	≤ 15	15	4

channel is busy, and then, to transmit a frame at the end of the SlotTime. Accordingly, the sum of the CCA time and the RxTxTurnaroundTime should be less than or equal to the SlotTime as shown in Table 12.2. Note that for the 802.11g, two different SlotTime values are supported, where the short SlotTime (i.e., 9 μs) can be used in a pure 802.11g BSS, in which no 802.11b stations coexist.

For the CCA operation, the PLCP header, which follows immediately after the PLCP header in the PPDU, makes a role as follows. The PLCP header includes a field, called LENGTH, which indicates either the length of the PSDU, conveyed in the PPDU, or the transmission time, as defined by the individual PHY. If the PLCP header is correctly received, the receiver PLCP can determine the duration of the incoming PPDU based on the LENGTH information. Note that the channel status is set to busy at the beginning of a PPDU reception. However, in the middle of a PPDU reception, the carrier of the incoming PPDU might be lost. In this case, even if the carrier is lost, the receiving PLCP assumes that the channel is busy, and, hence, holds the CCA busy indication to the MAC until the period indicated by the LENGTH field has expired. At the end of the frame transmission period, the PHY indicates two facts to the MAC: (1) the CCA idle status, and (2) the end of a PPDU reception with an error (i.e., carrier lost).

12.2 IEEE 802.11a OFDM PHY in 5 GHz

IEEE 802.11a-1999 employs OFDM to provide the transmission rates of 6, 9, 12, 18, 24, 36, 48, and 54 Mbps at the 5-GHz bands. The 6-, 12-, and 24-Mbps rates are mandatory for both transmission and reception. The 802.11a uses 52 subcarriers (i.e., 48 data subcarriers and 4 pilots) that are modulated using one of *binary phase shift keying* (BPSK), *quadrature phase shift keying* (QPSK), *16-* and *64-quadrature amplitude modulation* (16-QAM and 64-QAM). The convolutional coding with a code rate of 1/2, 2/3, or 3/4 is also used for error correction.

12.2.1 Modulation and Coding Schemes

The 802.11a supports 8 different transmission rates from 6 to 54 Mbps by using different combinations of modulation and coding schemes. Table 12.3 summarizes the supported rates, along with the corresponding modulation and code rates. Four different modulation schemes, namely, BPSK, QPSK, 16-QAM, and 64-QAM, are used, and three different codes rates, namely, 1/2, 2/3, and 3/4, are used. The number N_{DBPS} of data bits per OFDM symbol is determined by the combination of the modulation and coding schemes as follows:

Table 12.3 IEEE 802.11a PHY Transmission Rates

Transmission Rate (Mbps)	Modulation	Code Rate (R)	Coded bits per subcarrier (N_{BPSC})	Coded bits per OFDM symbol (N_{CBPS})	Data bits per OFDM symbol (N_{DBPS})
6	BPSK	1/2	1	48	24
9	BPSK	3/4	1	48	36
12	QPSK	1/2	2	96	48
18	QPSK	3/4	2	96	72
24	16-QAM	1/2	4	192	96
36	16-QAM	3/4	4	192	144
48	64-QAM	2/3	6	288	192
54	64-QAM	3/4	6	288	216

Source: [1].

$$N_{DBPS} = N_{CBPS} \times R,$$
$$N_{CBPS} = N_{BPSC} \times N_{SD}$$

where R is the convolutional code rate, N_{CBPS} is the number of coded bits per OFDM symbol, N_{BPSC} is the number of coded bits per subcarrier, and N_{SD} is the number of data subcarriers in the OFDM signal. For an M-ary modulation, $N_{BPSC} = \log_2 M$. For example, for 64-QAM, $N_{BPSC} = 6$. For IEEE 802.11a PHY, N_{SD} is fixed at 48 as detailed in Section 12.2.2. Finally, the transmission rate is determined as N_{DBPS} divided by the OFDM symbol duration (including the guard interval), which is 4 μs.

Figure 12.1 shows the error performance in terms of *bit error rate* (BER) as the *signal-to-noise ratio* (SNR) increases over the *additive white Gaussian noise* (AWGN) channel. Naturally, the higher the transmission rate, the less reliable. For a given environment, the best transmission rate maximizing the performance exists. We further discuss the rate adaptation issue in Section 13.4.2.

12.2.2 OFDM PLCP Sublayer

PPDU Format

The PPDU format of IEEE 802.11a PHY is illustrated in Figure 12.2. The PPDU starts with the PLCP preamble field. The preamble is composed of 10 short symbols (each of 0.8 μs) and two long symbols (each of 4 μs).

The PLCP header appearing immediately after the PLCP preamble is composed of the SIGNAL and SERVICE fields. The SIGNAL field comprising a single OFDM symbol is modulated with BPSK and rate-1/2 convolutional code (i.e., 6 Mbps rate) and, hence, includes 24 bits (see Table 12.3). The SIGNAL field includes two subfields, namely, RATE and LENGTH fields, where the RATE field indicates the transmission rate used for the remaining part of the PPDU, and the LENGTH field indicates the length of the PSDU contained in the PPDU.

The RATE field should indicate one of the eight transmission rates. Note that only 3 bits are actually needed to represent eight rates, while there are 4 bits in the

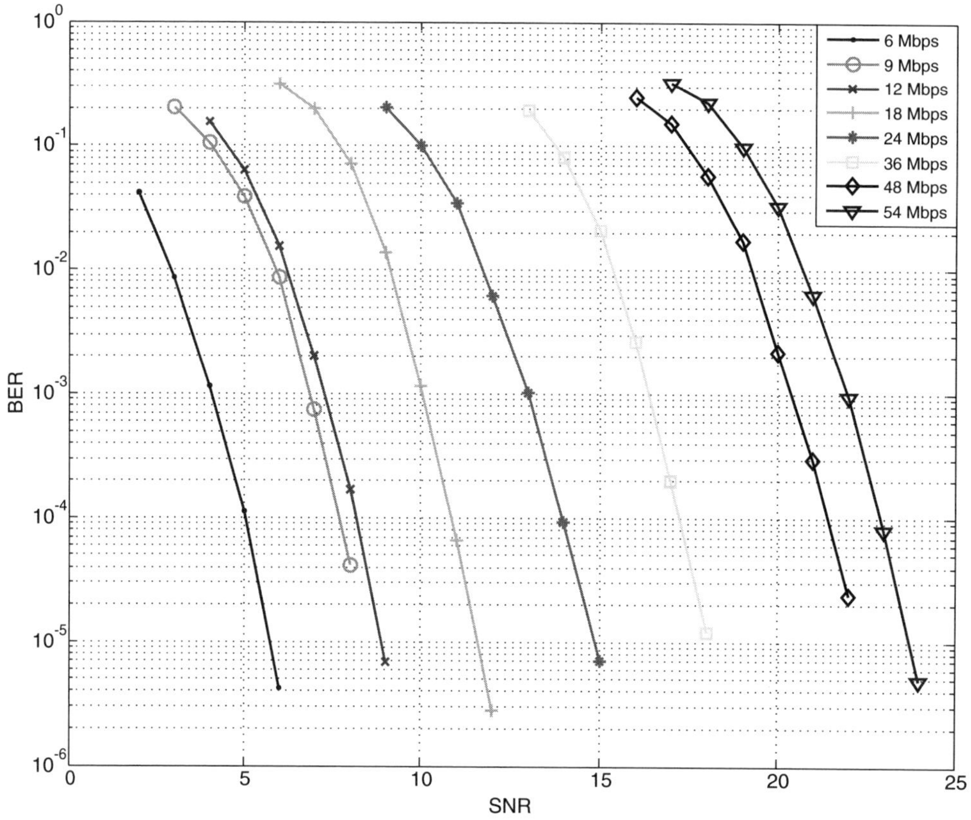

Figure 12.1 BER versus SNR of IEEE 802.11a PHY over AWGN channel.

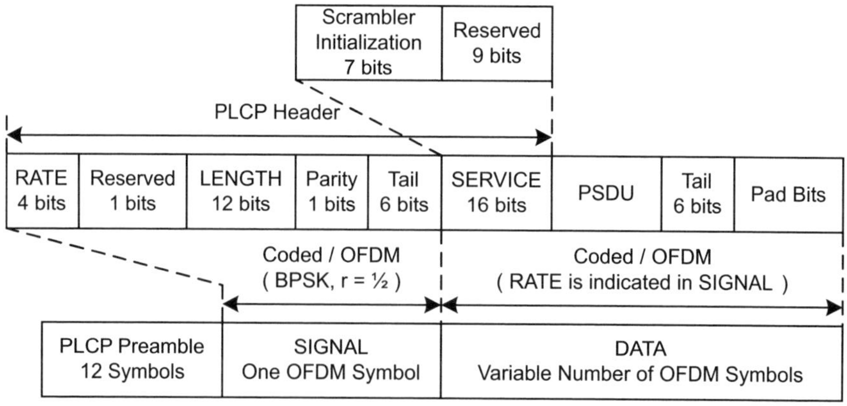

Figure 12.2 IEEE 802.11a PPDU format. (*After:* [1].)

field. The LENGTH field is 12 bits long, and, hence, can indicate the length up to 4,095 octets. According to IEEE 802.11-2007, the maximum PSDU length is 2,360, as shown in Figure 13.2, and, hence, it can be accommodated by this LENGTH field. The parity bit is used to check the error in the SIGNAL field; if an error is detected by the parity bit, the PLCP declares an erroneous reception of the incoming frame. Note that this parity bit–based error detection is not a very strong one, since it can detect only odd numbers of errors. The tail bits return the convolutional encoder to the zero state at the receiver PHY.

The SERVICE field includes the scrambler initialization bits, which are used for the scrambler. The pad bits are used to make the number of bits in the DATA field (i.e., the combination of SERVICE, PSDU, tail, and pad bits) be an integer multiple of N_{DBPS}, which is determined by the transmission rate indicated by the RATE field, as presented in Table 12.3. Accordingly, the length of the pad bits ranges from 0 to $N_{DBPS} - 1$ (bits).

OFDM Modulation

The OFDM is known to have advantages for high-speed wireless communications, thanks to the signal transmission via multiple orthogonal subcarriers. Depending on the number of subcarriers, the symbol duration is increased proportionally, and, then, the multipath fading effect can be easily eliminated by introducing a small guard interval. The OFDM signals are practically generated and detected using the *fast Fourier transform* (FFT) algorithm.

Figure 12.3 illustrates the subcarrier allocation for the 802.11a OFDM signal. In total, 52 subcarriers (–26 to 26) are employed, where 48 subcarriers for data, and 4 subcarriers (–21, –7, 7, 21) are for pilot, respectively. Accordingly, 64-FFT is used for the 802.11a OFDM. The pilot signals are used to make the coherent detection robust against frequency offsets and phase noise. The pilots are BPSK modulated by a pseudo-binary sequence to prevent the generation of spectral lines. Note that all the data subcarriers employ the same modulation scheme, determined by the RATE field in the PLCP header.

The gap between two consecutive subcarriers is 312.5 kHz, which makes the OFDM symbol duration—(excluding the *guard interval* (GI)—3.2 μs. Including the GI of 0.8 μs, the OFDM symbol duration becomes 4 μs.

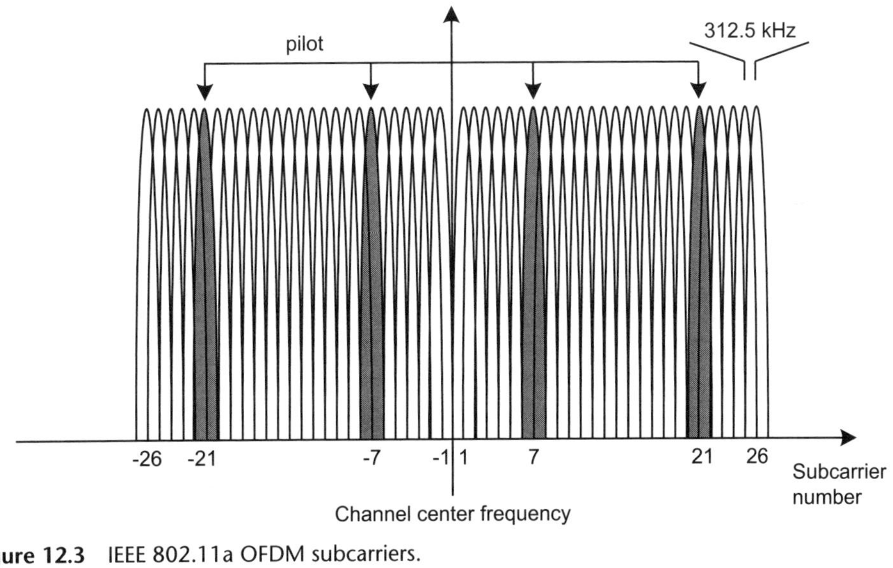

Figure 12.3 IEEE 802.11a OFDM subcarriers.

PLCP Preamble

The PLCP preamble field is used for the signal detection and synchronization. As shown in Figure 12.4, it comprises 10 short training symbols and 2 long training symbols. Short symbols are used for *automatic gain control* (AGC) convergence, diversity selection, timing acquisition, and coarse frequency acquisition at the receiver, while long symbols are used for channel estimation and fine frequency acquisition at the receiver. Each short OFDM training symbol of 0.8 μs uses 12 subcarriers, while each long OFDM training symbol of 3.2 μs uses 53 subcarriers (including a zero value at DC). A GI of 1.6 μs precedes the long symbols. Accordingly, the total duration of the PLCP preamble becomes 16 μs. Table 12.4 summarizes various parameters related with IEEE 802.11a PHY.

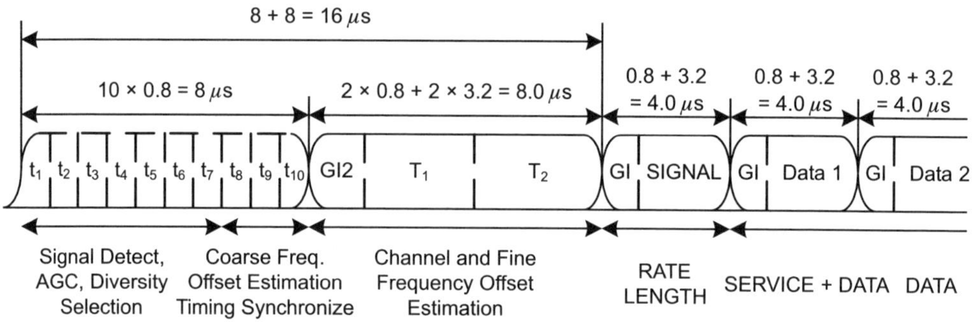

Figure 12.4 OFDM training sequence. (*After:* [2].)

Table 12.4 Various Parameters for IEEE 802.11a PHY

Parameter	Value
N_{SD}: Number of data subcarriers	48
N_{SP}: Number of pilot cubcarriers	4
N_{ST}: Number of subcarriers, total	52 (N_{SD} + N_{SP})
Δ_F: Subcarrier frequency spacing	0.3125 MHz (=20 MHz/64)
T_{FFT}: IFFT/FFT period	3.2μs ($1/\Delta_F$)
$T_{PREAMBLE}$: PLCP preamble duration	16μs ($T_{SHORT}+T_{LONG}$)
T_{SIGNAL}: Duration of the SIGNAL BPSK-OFDM symbol	4.0μs ($T_{GI}+T_{FFT}$)
T_{GI}: GI duration	0.8μs ($T_{FFT}/4$)
T_{GI2}: Training symbol GI duration	1.6μs ($T_{FFT}/2$)
T_{SYM}: Symbol interval	4μs ($T_{GI} + T_{FFT}$)
T_{SHORT}: Short training sequence duration	8μs ($10 \times T_{FFT}/4$)
T_{LONG}: Long training sequence duration	8μs ($T_{GI} + 2 \times T_{FFT}$)

Source: [2].

PLCP Data Scrambler

The DATA field (including SERVICE, PSDU, tail, and pad bits) is scrambled with a length-127 frame-synchronous scrambler shown in Figure 12.5. The generator polynomial of the scrambler is $S(x) = x^7 + x^4 + 1$. When transmitting, the initial state of the scrambler will be set to a pseudo-random nonzero state. The scrambler initialization field of the SERVICE field is set to all zero before scrambling. Then, after the scrambling, this field will indicate the initial state of the scrambler so that the receiver can obtain the initial state for the descrambling.

Convolutional Coding

The encoding scheme for the rate-1/2 convolutional code is illustrated in Figure 12.6. The convolutional encoder uses the generator polynomials, $g_0 = 133_8$ and $g_1 = 171_8$, of rate $R = 1/2$ with the constraint length = 7. The rates of 2/3 and 3/4 are achieved by puncturing the rate-1/2 convolutional mother code. First, the code rate of 2/3 is obtained by erasing the fourth bit out of every 4 encoded bits generated by the mother encoder. Second, the code rate of 3/4 is obtained by erasing the fourth and fifth bits out of every 6 encoded bits generated by the mother.

Data Interleaving

All encoded data bits are interleaved by a block interleaver, which is employed to prevent long sequences of adjacent noisy bits from entering the convolutional decoder. The interleaver block size is equal to the number of bits in a single OFDM

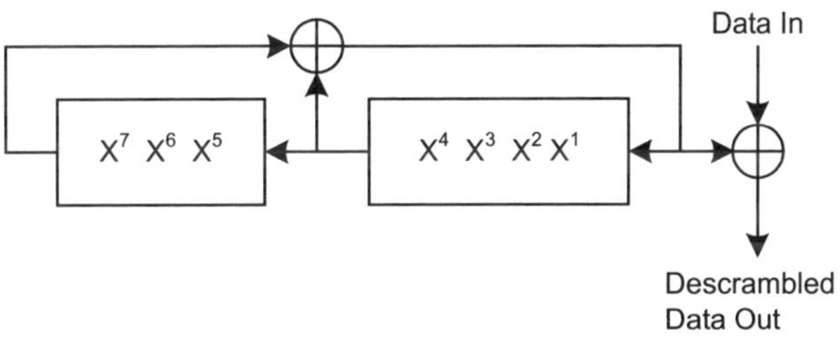

Figure 12.5 PLCP data (de)scrambler. (*After:* [1].)

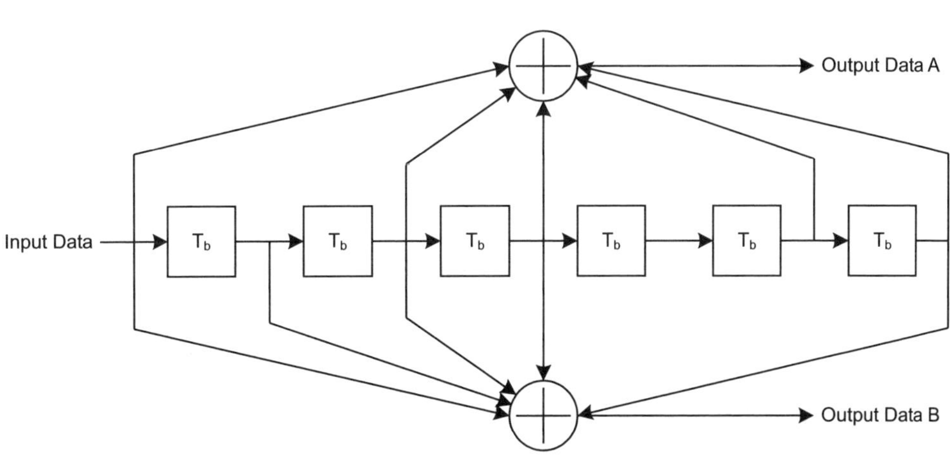

Figure 12.6 Convolutional coding. (*After:* [1].)

symbol (i.e., N_{CBPS}). The interleaver is defined by a two-step permutation. The first permutation ensures that adjacent coded bits are mapped onto nonadjacent subcarriers. The second permutation ensures that adjacent coded bits are mapped alternately onto less and more significant bits of the signal constellation, and, hence, long runs of low reliability bits are avoided.

The permutations work as follows, where the index of the coded bit before the first permutation is denoted by k, the index after the first and before the second permutation is denoted by i, and, finally, the index after the second permutation is denoted by j. The first permutation is defined by

$$i = (N_{CBPS}/16)(k \bmod 16) + \text{floor}(k/16), \text{ where } k = 0, 1, \ldots, N_{CBPS}$$

The function floor(.) denotes the largest integer less than or equal to the parameter. The second permutation is defined by

$$j = s \times \text{floor}(i/s) + \left(i + N_{CBPS} - \text{floor}(16 \times i/N_{CBPS})\right) \bmod s$$

where $i = 0, 1, \ldots, N_{CBPS} - 1$.

The value of s is determined by $s = \max(N_{BPSC}/2, 1)$, where N_{BPSC} is the number of coded bits per subcarrier.

Subcarrier Mapping and OFDM Modulation

The OFDM subcarriers are modulated by using BPSK, QPSK, 16-QAM, or 64-QAM. The encoded and interleaved binary serial input data is divided into groups of N_{BPSC} (1, 2, 4, or 6) bits and converted into complex numbers representing BPSK, QPSK, 16-QAM, or 64-QAM constellation points. The conversion is performed according to Gray-coded constellation mappings as shown in Figure 12.7.

The output values, d, are obtained via the resulting $(I + jQ)$ value multiplied by a normalization factor K_{MOD} [i.e., $d = (I + jQ) \times K_{MOD}$]. The normalization factor, K_{MOD}, depends on the employed modulation as summarized in Table 12.5. Note that the modulation scheme can change during the course of a PPDU transmission when the signal changes from SIGNAL to DATA, as shown in Figure 12.2. The purpose of the normalization factor is to achieve the same average power for all mappings.

Then, the stream of the complex numbers (i.e., d's) is divided into groups of N_{SD} (= 48) complex numbers. Each complex number is mapped into a data subcarrier, as illustrated in Figure 12.3. Therefore, all the 48 complex numbers are transmitted in a single OFDM symbol. Note that every forty-eighth complex number is mapped into the same subcarrier.

12.2.3 Physical Medium–Dependent (PMD) Operations

The general block diagram of the transmitter and receiver for the 802.11a OFDM PHY is shown in Figure 12.8. The 64-IFFT block converts a block of constellation points from 52 subcarriers to a time domain block (i.e., an OFDM symbol, excluding GI, of 3.2 μs). While an 802.11a signal is said to occupy a 20-MHz spectrum, its actual occupied bandwidth is 16.6 MHz, which is determined by 312.5 kHz (i.e., the

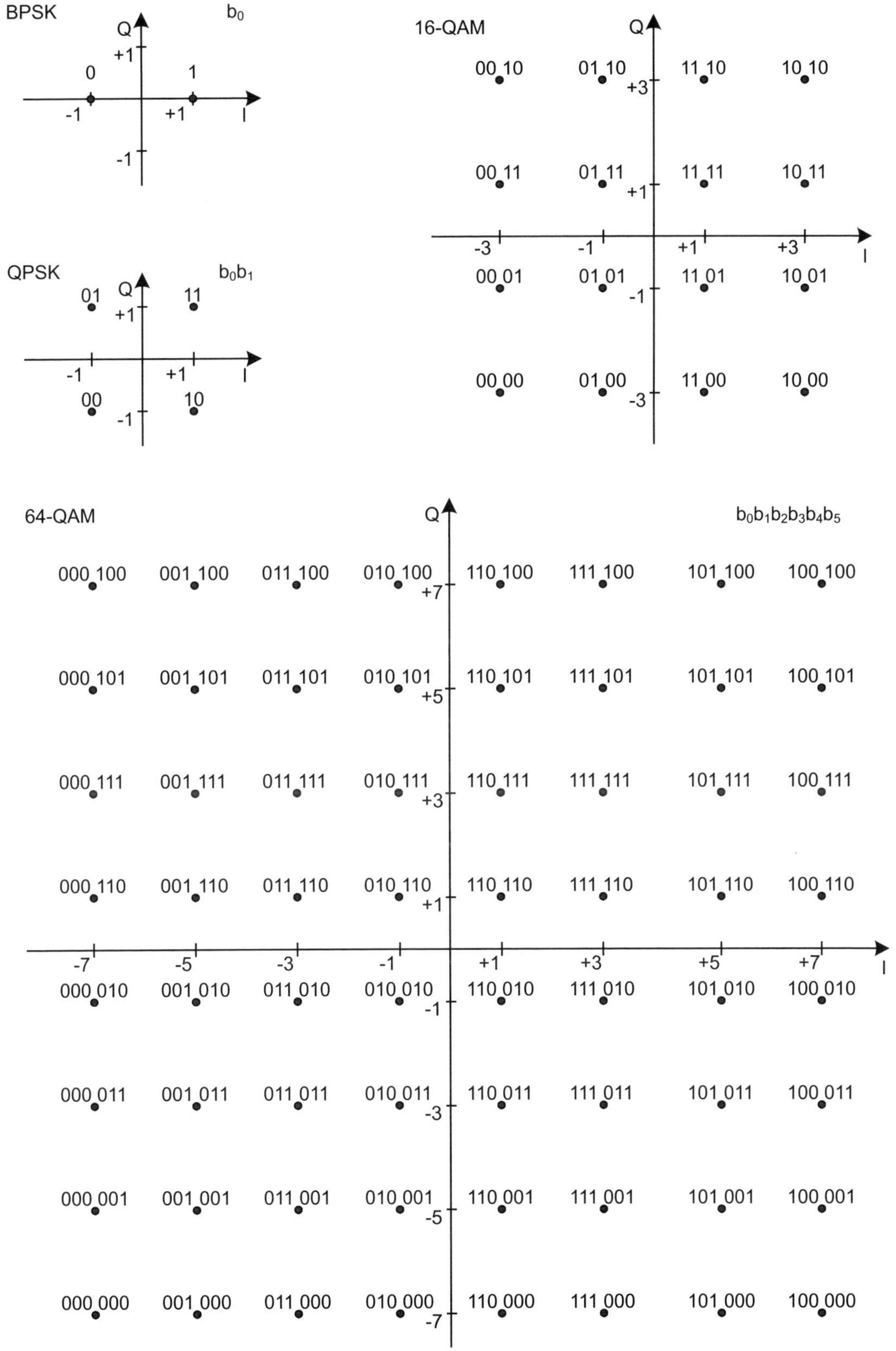

Figure 12.7 BPSK, QPSK, 16-QAM, and 64-QAM constellation bit encoding. (*After:* [1].)

Table 12.5 Modulation-Dependent Normalization Factor

Modulation	K_{MOD}
BPSK	1
QPSK	$1/\sqrt{2}$
16-QAM	$1/\sqrt{10}$
64-QAM	$1/\sqrt{42}$

Source: [1].

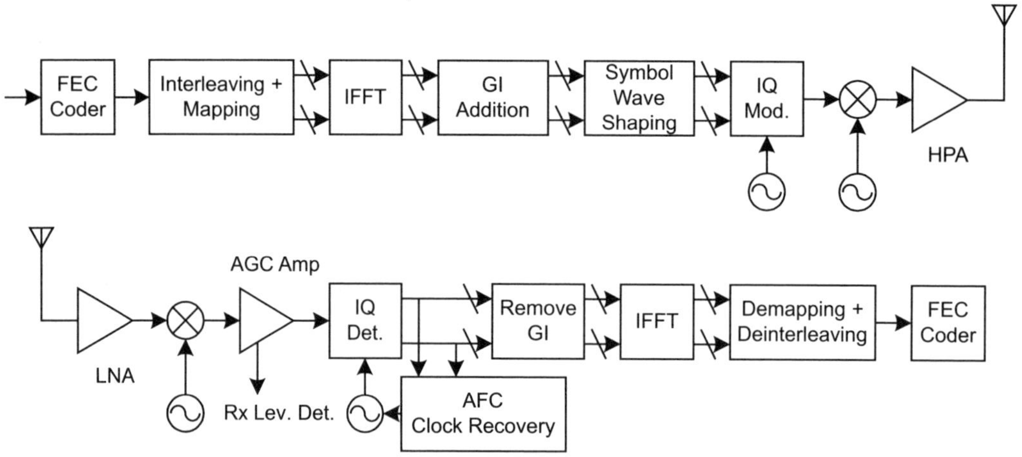

Figure 12.8 Transmission and reception block diagram. (*After:* [1].)

gap between two adjacent subcarriers), multiplied by 53 (i.e., the number of subcarriers including the DC).

Operating Frequency Channels

IEEE 802.11a operates in 5-GHz bands by occupying a 20-MHz spectrum. The center frequency of each channel is determined by

$$\text{Center frequency (MHz)} = 5{,}000 + 5 \times N_{ch}$$

where N_{ch} represents the channel number.

Table 12.6 summarizes the frequency channels, which are allowed in various regulatory domains, along with their center frequencies.[1] The regulatory requirements for each subband in each regulatory domain in terms of the maximum transmit power and other requirements—*transmit power control* (TPC) and *dynamic frequency selection* (DFS)—are described in Section 1.2.2. Moreover, the regulations related with TPC and DFS are further discussed in Section 17.1.

Figure 12.9 illustrates the channelization for IEEE 802.11a with FCC frequency allocation at the United States as an example. Note that according to Table 12.6, the

1. The information about the allowed channels for the United States, Europe, and Japan is from [1]. Therefore, Korea is from [4]. Finally, information for Canada, China, and Japan is from [6].

Table 12.6 Frequency Channels for IEEE 802.11a

Frequency sub-Band (GHz)	Channel number	Center frequency (GHz)	Regulatory domain
5.15~5.25	36 40 44 48	5.180 5.200 5.220 5.240	Canada, China, Europe, Japan, Korea, US
5.25~5.35	52 56 60 64	5.260 5.280 5.300 5.320	Canada, China, Europe, Korea, US
5.47~5.725	100 104 108 112 116 120 124	5.500 5.520 5.540 5.560 5.580 5.600 5.620	Canada, China, Europe, Korea, US
	128 132 136 140	5.640 5.660 5.680 5.700	Canada, China, Europe, US
5.725~5.850	149 153 157 161	5.745 5.765 5.785 5.805	Korea, US
	165	5.825	US

United States has allocated the largest spectrum for IEEE 802.11a (i.e., 24 channels are allocated in total across 580-MHz spectrum).

Transmit Spectrum Mask

The transmitted spectrum will have a 0 dBr (decibel relative to the maximum spectral density of the signal) up to the bandwidth of 18 MHz, –20 dBr at 11-MHz frequency offset, –28 dBr at 20-MHz frequency offset, and –40 dBr at 30-MHz frequency offset and above. The transmitted spectral density of the transmitted signal should fall within the spectral mask, as shown in Figure 12.10. It should be noted that the signals transmitted at two adjacent channels with a 20-MHz gap could interfere with each other according to the spectrum mask.

Receiver Minimum Input Sensitivity

The *frame error rate* (FER) for a PSDU length of 1,000 octets should be under 10 percent with the *minimum sensitivity* summarized in Table 12.7. The minimum sensitivities are measured at the antenna connector. Apparently, the lower the transmission rate, the lower the minimum sensitivity. That is, the lower transmission rate scheme should work in worse channel environments.

CCA Operations

The start of a valid OFDM transmission at a reception level equal to or greater than the minimum modulation and coding rate sensitivity (i.e., –82 dBm) should make the CCA to indicate the busy status with a probability > 90 percent within the CCA delay of 4 μs. If the preamble portion was missed, the receiver should indicate the

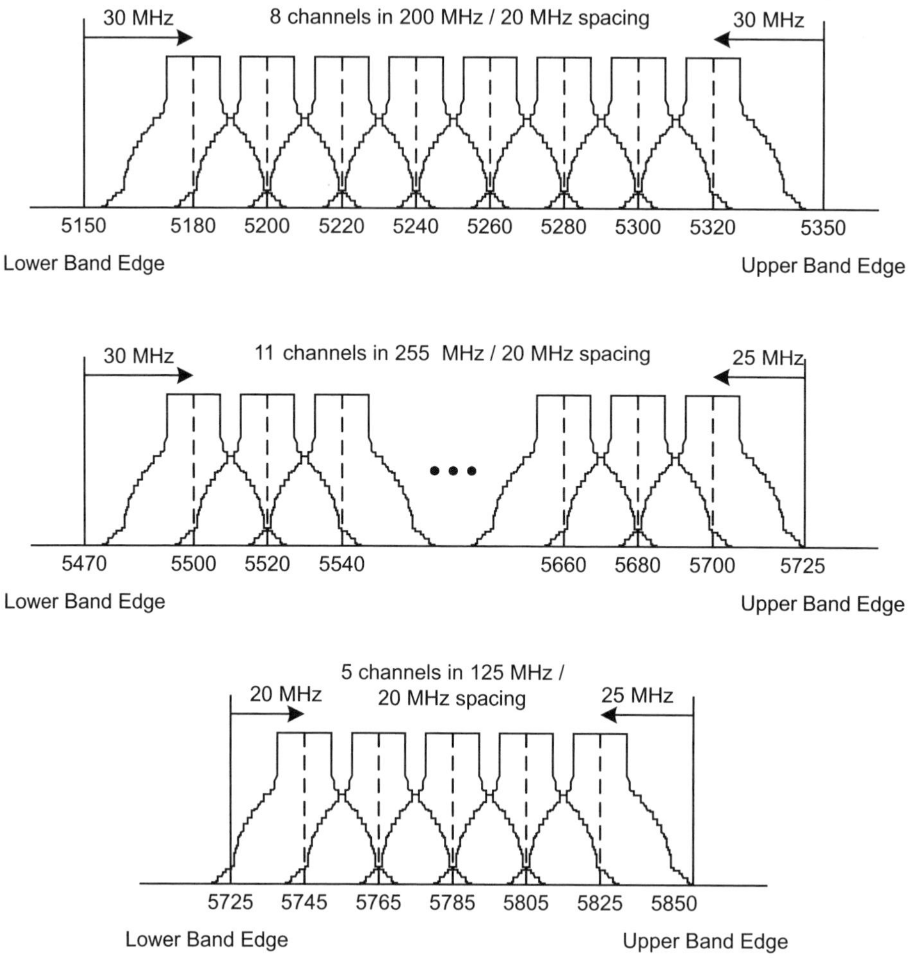

Figure 12.9 The 5-GHz frequency channels for operation. (*After:* [1].)

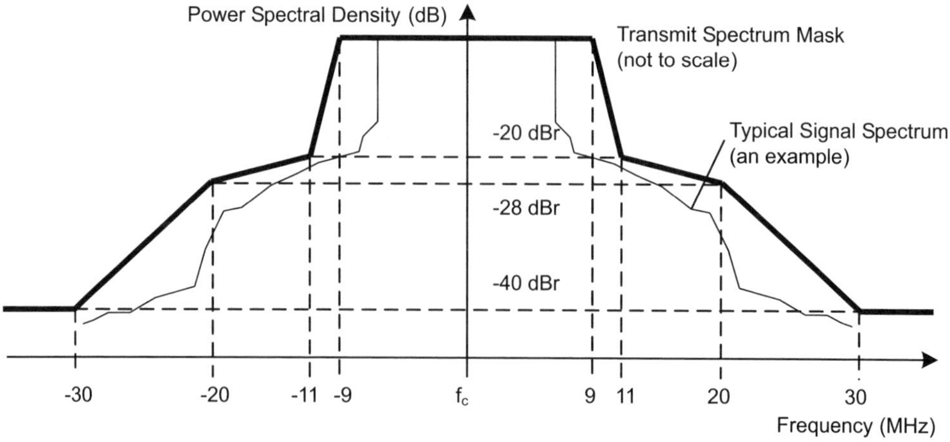

Figure 12.10 Transmit spectrum mask. (*After:* [1].)

Table 12.7 Receiver Minimum Sensitivity

Rate (Mbps)	Modulation	Code Rate (*R*)	Minimum sensitivity (dBm)
6	BPSK	1/2	−82
9	BPSK	3/4	−81
12	QPSK	1/2	−79
18	QPSK	3/4	−77
24	16-QAM	1/2	−74
36	16-QAM	3/4	−70
48	64-QAM	2/3	−66
54	64-QAM	3/4	−65

Source: [1].

CCA busy status for any signal 20 dB above the minimum modulation and coding rate sensitivity (i.e., –62 dBm).

12.2.4 Reduced-Clock Operations

Per IEEE 802.11-2007 [1], the OFDM PHY based on IEEE 802.11a is also able to support two reduced-clock operations as follows:

The OFDM system provides a *half-clocked* operation using a 10-MHz spectrum with the transmission rates of 3, 4.5, 6, 9, 12, 18, 24, and 27 Mbps, where 3, 6, and 12 Mbps are mandatory. The half-clocked operation doubles OFDM symbol times and CCA times. Both SlotTime and SIFSTime, as defined in Table 13.8, are also doubled.

The OFDM system also provides a *quarter-clocked* operation using a 5 MHz spectrum with the transmission rates of 1.5, 2.25, 3, 4.5, 6, 9, 12, and 13.5 Mbps, where 1.5, 3, and 6 Mbps are mandatory. The quarter-clocked operation quadruples OFDM symbol times and CCA times. Both SlotTime and SIFSTime, as defined in Table 13.8, are also quadrupled.

These reduced-clock operations can be used for the operations in the U.S. public safety band at 4.94–4.99 GHz, defined in FCC CFR47 [7], Section 90.1209. The available channels as well as the transmit power limits are summarized in Tables 12.8 and 12.9. Note that for each subband, two transmit power limit options are available. The reduced-clock operations are also allowed at 4.9-GHz and 5-GHz bands in Japan [1, 8].

12.3 IEEE 802.11b HR/DSSS PHY in 2.4 GHz

IEEE 802.11b-1999, which is referred to as *high-rate DSSS* (HR/DSSS) PHY, extends the DSSS PHY by providing 5.5- and 11-Mbps data transmission rates in addition to the 1- and 2-Mbps rates. To provide the higher rates, 8-chip CCK is employed as the modulation scheme. The chipping rate is 11 Mchips/s, which is the same as the DSSS PHY, thus providing the same occupied channel bandwidth. The basic transmission mode uses the same PLCP preamble and header as the DSSS PHY, and, hence, both PHYs can coexist in the same BSS.

12.3.1 PLCP Sublayer

For IEEE 802.11b PHY, two different preambles and headers are defined: (1) the mandatory long preamble and header, which are interoperable with the baseline DSSS PHY, and (2) the optional short preamble and header. The PPDU formats with two preamble and header options are illustrated in Figures 12.11 and 12.12.

A PPDU includes a PLCP preamble, a PLCP header, and a PSDU. The PLCP preamble contains *synchronization* (SYNC) and *start frame delimiter* (SFD). The PLCP header contains *signaling* (SIGNAL), *service* (SERVICE), *length* (LENGTH), and CRC-16.

The long PLCP preamble and header use the 1 Mbps Barker code spreading with DBPSK modulation, and the PSDU is transmitted at 1, 2, 5.5, or 11 Mbps. The short PLCP preamble and header is defined as optional for the 802.11b. The short preamble and header may be used to minimize overhead, and, hence, maximize the net-

Table 12.8 The U.S. Public Safety Transmit Power Limits

Frequency band (GHz)	U.S. public safety (mW)		
	20 MHz channel	10 MHz channel	5 MHz channel
4.94-4.99 low power	100	50	25
4.94-4.99 high power	2000	1000	500

Source: [1].

Table 12.9 The 4.9-GHz U.S. Public Safety Bands in the United States

Channel starting frequency (GHz)	Channel Spacing (MHz)	Channel set	Transmit Power limit (mW)
4.9375	5	1, 2, 3, 4, 5, 6, 7, 8, 9, 10	25
4.9375	5	1, 2, 3, 4, 5, 6, 7, 8, 9, 10	500
4.89	10	11, 13, 15, 17, 19	50
4.89	10	11, 13, 15, 17, 19	1000
4.85	20	21, 25	100
4.85	20	21, 25	2000

Source: [1].

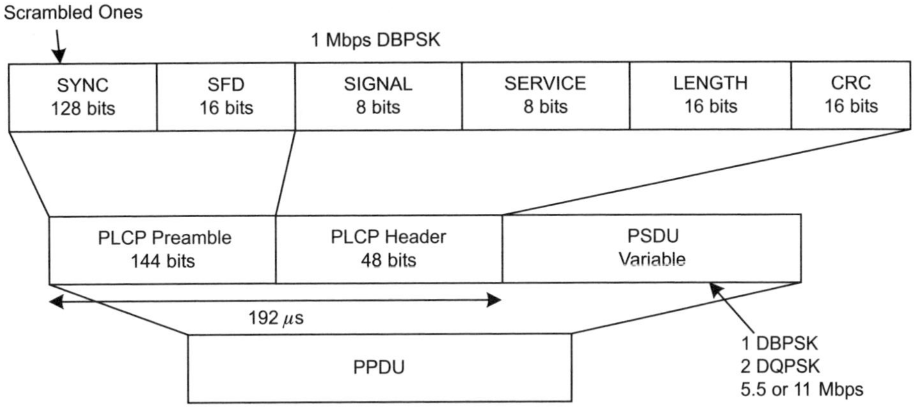

Figure 12.11 IEEE 802.11b long PPDU format. (*After:* [1].)

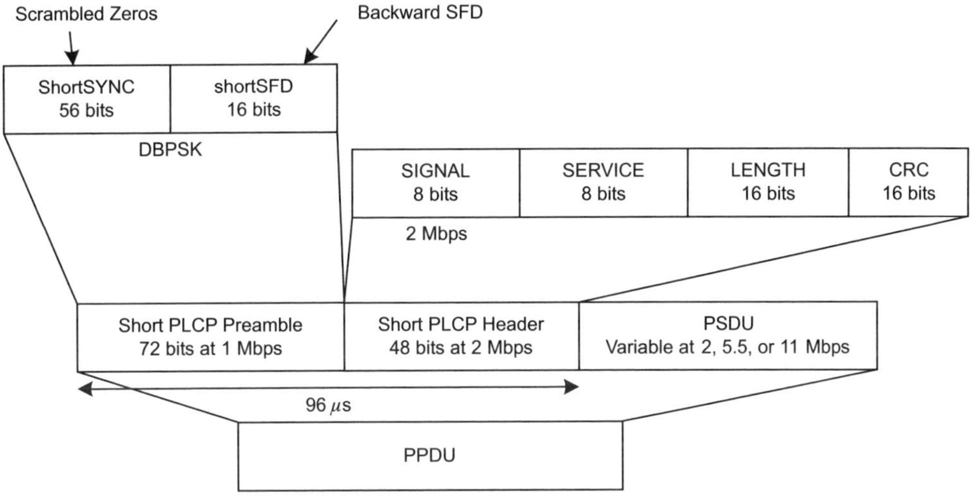

Figure 12.12 IEEE 802.11b short PPDU format. (*After:* [1].)

work throughput. To interoperate with a receiver that is not capable of receiving a short preamble and header, the transmitter should use the long PLCP preamble and header. The short PLCP preamble uses the 1 Mbps Barker code spreading with DBPSK modulation. The short PLCP header uses the 2 Mbps Barker code spreading with DQPSK modulation, and the PSDU is transmitted at 2, 5.5, or 11 Mbps.

For the long PLCP preamble, the SYNC field consists of 128 bits of scrambled "1" bits. The initial state of the scrambler (seed) is [1101100]. The SFD is provided to indicate the start of the PLCP header. The SFD is a 16-bit field, [1111 0011 1010 0000]. For the short PLCP preamble, the shortSYNC field consists of 56 bits of scrambled "0" bits. The initial state of the scrambler (seed) is [001 1011]. The shortSFD is a 16-bit field and is the time reverse of the field of the SFD in the long PLCP preamble. That is, the field is the bit pattern [0000 0101 1100 1111]. The polynomial $G(z) = z^{-7} + z^{-4} + 1$ is used to scramble all bits transmitted, and it can be implemented as shown in Figure 12.13.

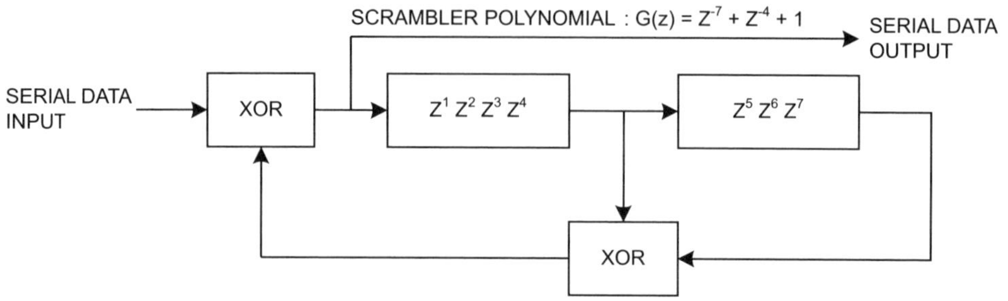

Figure 12.13 IEEE 802.11b data scrambler. (*After:* [1].)

The 8-bit SIGNAL field indicates the modulation that is used for the PSDU. As shown in Table 12.10, the SERVICE field includes the following three bits. Bit 7 is used to supplement the LENGTH field as described next. Bit 3 is used to indicate whether the modulation method is CCK or PBCC. Bit 2 is used to indicate that the transmit frequency and symbol clocks are derived from the same oscillator. This locked clock bit is set based on its implementation configuration.

The LENGTH field indicates the number of microseconds required to transmit the PSDU. Because there is an ambiguity in the number of octets that is described by a length in integer microseconds for any data rate over 8 Mbps, a length extension bit in the SERVICE field indicates when the smaller potential number of octets is correct.

- 5.5 Mbps CCK Length = number of octets × 8/5.5, rounded up to the next integer.
- 11 Mbps CCK Length = number of octets × 8/11, rounded up to the next integer; the length extension bit is set to 0 if the rounding took less than 8/11 or to 1 if the rounding took more than or equal to 8/11.

At the receiver, the number of octets in the MPDU is calculated as follows:

- 5.5 Mbps CCK Number of octets = Length × 5.5/8, rounded down to the next integer.
- 11 Mbps CCK Number of octets = Length × 11/8, rounded down to the next integer, minus 1 if the service field (b7) bit is a 1.

The calculation works slightly differently for PBCC. The SIGNAL, SERVICE, and LENGTH fields are protected by the CRC-16 FCS, which is the ones complement of the remainder generated by the modulo 2 division of the protected fields by the polynomial $x^{16} + x^{12} + x^5 + 1$.

Table 12.10 IEEE 802.11b SERVICE Field Definitions

b0	b1	b2	b3	b4	b5	b6	b7
Reserved	Reserved	Locked clock bit 0 = not 1 = locked	Mod. selection bit 0 =CCK 1 =PBCC	Reserved	Reserved	Reserved	Length extension bit

Source: [1].

12.3.2 Modulation Schemes

Four modulation schemes and data rates are specified for IEEE 802.11b PHY. The 1-Mbps rate is based on DBPSK modulation, the 2-Mbps rate is based on DQPSK modulation, and finally the CCK modulation is used for both the 5.5- and 11-Mbps rates. An optional PBCC mode is also provided for potentially enhanced performance.

DSSS Modulation

The following 11-chip Barker sequence is used as the *pseudo-noise* (PN) code sequence for the 1 Mbps and 2 Mbps modulations:

$$+1, -1, +1, +1, -1, +1, +1, +1, -1, -1, -1$$

The symbol duration is exactly 11 chips long. For 1- and 2-Mbps transmission rates, the DBPSK and DQPSK are used, respectively, where the encoding is specified in Tables 12.11 and 12.12, respectively. As *differential phase shift keying* (DPSK) is employed, the phase values specified in the tables represent the phase change of the current symbol relative to the preceding symbol, not the absolute phase. Each DPSK symbol is multiplied by the 11-chip Barker sequence for spreading.

CCK Modulation

The CCK is a kind of *M*-ary orthogonal modulation scheme. A set of orthogonal symbols are established, where each symbol corresponds to either 4 data bits (for 5.5 Mbps) or 8 data bits (for 11 Mbps). Each orthogonal symbol is represented by a CCK code word with 8 chips.

$$\begin{aligned} c &= \{c_0, \ldots, c_7\} \\ &= \left\{ e^{j(\varphi_1+\varphi_2+\varphi_3+\varphi_4)}, e^{j(\varphi_1+\varphi_3+\varphi_4)}, e^{j(\varphi_1+\varphi_2+\varphi_4)}, \right. \\ &\left. -e^{j(\varphi_1+\varphi_4)}, e^{j(\varphi_1+\varphi_2+\varphi_3)}, e^{j(\varphi_1+\varphi_3)}, -e^{j(\varphi_1+\varphi_2)}, e^{j\varphi_1} \right\} \end{aligned} \tag{12.1}$$

Table 12.11 The 1 Mbps DBPSK Encoding

Bit input	Phase change($+j\omega$)
0	0
1	π

Source: [1].

Table 12.12 The 2-Mbps DQPSK Encoding

Dibit pattern(d0,d1) D0 is first in time	Phase change($+j\omega$)
00	0
01	π
11	$\pi/2$
10	$3\pi/2(-\pi/2)$

Source: [1].

The terms φ_1, φ_2, φ_3, and φ_4 are determined according to the data bits as explained next. The codes are known as *complementary codes*. Each chip takes one of four phases (i.e., QPSK). Accordingly, the CCK uses either 2^4 or 2^8 codes out of 4^8 possible codes. The chosen codes are with good autocorrelation and cross-correlation properties. Note that this code is based on a generalized Hadamard transform encoding. The code length is 8, and the chipping rate is 11 Mchips/s. The symbol duration is exactly 8 complex chips long, and, hence, the symbol rate is 1.375 Msymbols/s. Accordingly, the data transmission rate becomes either 5.5 or 11 Mbps depending on the number of data bits per symbol.

Being in every chip, the term φ_1 modifies the phase of all code chips of the sequence and rotates the whole symbol by the appropriate amount relative to the phase of the preceding symbol. The fourth and seventh chips are rotated 180 degrees by a cover sequence to optimize the sequence correlation properties and minimize DC offsets in the codes. This can be seen by the minus sign on the fourth and seventh terms in (12.1).

For the 5.5 Mbps transmission rate, the PSDU is divided into 4-bit nibbles, where each of 4 bits (d0 to d3; d0 first in time) is transmitted in a symbol. The data bits d0 and d1 encode φ_1 based on DQPSK using the encoding specified in Table 12.13. The phase change for φ_1 is relative to the phase φ_1 of the preceding symbol. All odd-numbered symbols from the PSDU octets are given an extra 180-degree (π) rotation, in addition to the standard DQPSK modulation, as shown in Table 12.13, where the first symbol is assigned number 0. The data bits d2 and d3 determine φ_2, φ_3, and φ_4 as follows: $\varphi_2 = (\text{d2} \times \pi) + \pi/2$, $\varphi_3 = 0$, and $\varphi_4 = \text{d3} \times \pi$. The resulting encoding table for h_0 to h_7 is shown in Table 12.14, where the actual code c is determined by $c = \{c_0, ..., c_7\} = e^{j\varphi_1} \cdot \{h_0, ..., h_7\}$.

Table 12.13 DQPSK Encoding for CCK Modulation

Dibit pattern (d0,d1) (d0 is first in time)	Even symbols Phase change (+jω)	Odd symbols Phase change (+jω)
00	0	π
01	π/2	3π/2(−π/2)
11	π	0
10	3π/2(−π/2)	π/2

Source: [1].

Table 12.14 CCK Encoding Table for 5.5 Mbps

d2, d3	h_0	h_1	h_2	h_3	h_4	h_5	h_6	h_7
00	1j	1	1j	−1	1j	1	−1j	1
01	−1j	−1	−1j	1	1j	1	−1j	1
10	−1j	1	−1j	−1	−1j	1	1j	1
11	1j	−1	1j	1	−1j	1	1j	1

Source: [1].

For the 11 Mbps transmission rate, the PSDU is divided into bytes, where each of 8 bits (d0 to d7; d0 first in time) is transmitted in a symbol. The first dibit (d0, d1) encodes φ_1 based on DQPSK using the encoding specified in Table 12.13. The phase change for φ_1 is relative to the phase φ_1 of the preceding symbol. All odd-numbered symbols from the PSDU octets are given an extra 180-degree (π) rotation, in addition to the standard DQPSK modulation, as shown in Table 12.13, where the first symbol is assigned number 0. The data dibits (d2, d3), (d4, d5), and (d6, d7) encode φ_2, φ_3, and φ_4, respectively, based on QPSK as specified in Table 12.15.

Optional PBCC Modulation

The 802.11b optionally defines PBCC modulations by employing a rate-1/2 convolutional coding with the constraint length = 7 for 5.5 and 11 Mbps transmission rates. Accordingly, the PSDU part in the PPDU can be optionally transmitted using the PBCC. It basically encodes the PSDU with the convolutional encoder and then modulates with either BPSK (for 5.5 Mbps) or QPSK (11 Mbps). The mapping from the coded bits to PSK constellation points is determined by a 256-bit pseudo-random cover sequence. The convolutional encoder uses the generator polynomials, $g_0 = 155_8$ and $g_1 = 137_8$, which are different from those used for IEEE 802.11a PHY.

12.3.3 PMD Operations

Operating Frequency Channels

IEEE 802.11b operates in 2.4-GHz bands by occupying a 22-MHz spectrum. For most countries, the frequency channels are in the ISM band ranging from 2.4 GHz to 2.4835 GHz. The center frequency of each channel is determined by

$$\text{Center frequency (MHz)} = 2{,}407 + 5 \times N_{ch} \tag{12.2}$$

where N_{ch} (ranging from 1 to 13) represents the channel ID. Table 12.16 summarizes the frequency channels, which are allowed in various regulatory domains, along with their center frequencies. Note that channel 14 does not follow (12.2) for the center frequency mapping.

Two adjacent channels are only 5 MHz apart, while the channel spectrum occupies 22 MHz. In fact, any of available channels can be used for an 802.11b WLAN. However, when neighboring WLANs operate in the channels that are overlapping, the performance might be severely compromised. Note that there are only three

Table 12.15 QPSK Encoding for CCK Modulation

Dibit pattern [di,d(i+1)] (di is first in time)	Phase
00	0
01	π
11	$\pi/2$
10	$3\pi/2(-\pi/2)$

Source: [1].

Table 12.16 The 2.4-GHz Bands for IEEE 802.11b PHY; "X" Indicates That the Channel Is Available

CHNL_ID	Frequency (MHz)	Regulatory domains							
		FCC	IC	ETSI	Spain	France	Japan	China	Korea
1	2412	X	X	X	—	—	X	X	X
2	2417	X	X	X	—	—	X	X	X
3	2422	X	X	X	—	—	X	X	X
4	2427	X	X	X	—	—	X	X	X
5	2432	X	X	X	—	—	X	X	X
6	2437	X	X	X	—	—	X	X	X
7	2442	X	X	X	—	—	X	X	X
8	2447	X	X	X	—	—	X	X	X
9	2452	X	X	X	—	—	X	X	X
10	2457	X	X	X	X	X	X	X	X
11	2462	X	X	X	X	X	X	X	X
12	2467	—	—	X	—	X	X	X	X
13	2472	—	—	X	—	X	X	X	X
14	2484	—	—	—	—	—	X	—	—

Source: [1].

nonoverlapping channels where 11 or 13 channels are available. As shown in Figure 12.14, channels 1, 6, and 11 are nonoverlapping, and, hence, can be used for the channel planning in multi-AP WLANs in the country where 11 channels are available. Where 13 channels are available, a set of channels 1, 7, and 13 might be a better choice. Another possible option where 13 channels are available could be using 4 partially overlapping channels (i.e., channels 1, 5, 9, and 13).

Transmit Spectrum Mask

As shown in Figure 12.15, the transmitted spectrum should be under –30 dBr for 1 MHz $< |f - f_c| <$ 22 MHz, and should be under –50 dBr for $|f - f_c| >$ 22 MHz, where f_c is the channel center frequency.

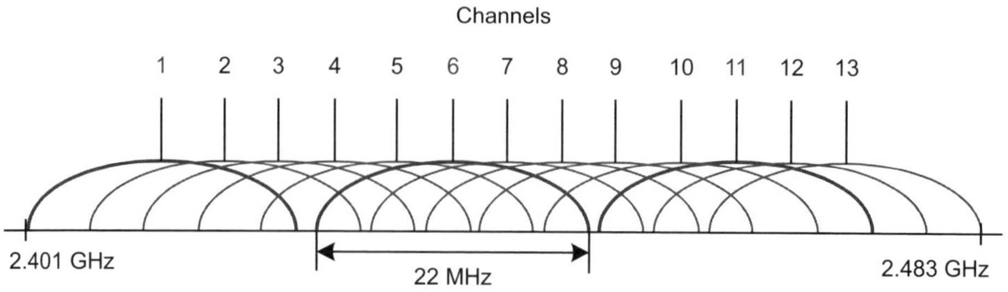

Figure 12.14 IEEE 802.11b channels.

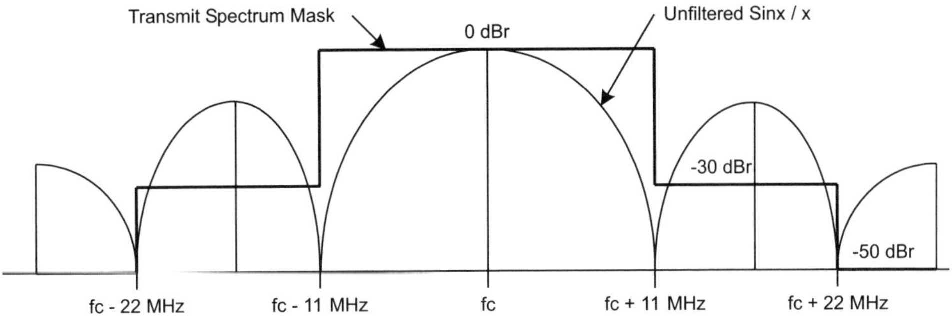

Figure 12.15 Transmit spectrum mask of IEEE 802.11b PHY. (*After:* [1].)

Receiver Minimum Input Sensitivity

Unlike the 802.11a, the 802.11b does not specify the receiver minimum input sensitivity per transmission rate. The only requirement is that the FER should be less than 8×10^{-2} for a PSDU length of 1,024 octets transmitted at 11 Mbps CCK modulation for an input level of –76 dBm measured at the antenna connector.

CCA Operations

The 802.11b PHY provides the CCA according to at least one of the following three methods:

- ED CCA mode (energy above threshold): CCA reports a busy medium upon detecting any energy above the ED threshold.
- CS CCA mode (CS with timer): CCA starts a timer whose duration is 3.65 msec and reports a busy medium only upon the detection of an 802.11b signal. CCA should report an idle channel after the timer expires and no 802.11b signal is detected. The 3.65-ms timeout is the duration of the longest possible 5.5 Mbps PSDU. Upon reception of a correct PLCP header, the timer will be overridden by the PPDU transmission time found in the LENGTH field.
- CS/ED CCA mode (a combination of CS and energy above threshold): CCA reports busy at least while an 802.11b PPDU with energy above the ED threshold is being received at the antenna.

The ED threshold affects the sensitivity to the incoming signals. However, the 802.11b does not specify the ED threshold value. Instead, for the CS/ED CCA mode, the following constraints are specified according to the transmit power of the station. That is, if a valid 802.11b signal is detected during its preamble within the CCA time, the ED threshold should be set to less than or equal to –76 dBm for transmit power > 100 mW; –73 dBm for 50 mW < transmit power = 100 mW; and –70 dBm for transmit power = 50 mW.

When the ED CCA mode is not employed, the CCA will indicate an idle channel while a non-802.11b signal is on the channel. Such non-802.11b signals include those from Bluetooth and IEEE 802.11g WLAN. Accordingly, the coexistence of the 802.11b stations with the 802.11g stations becomes an issue. We further discuss this in Section 12.4.2.

12.4 IEEE 802.11g ER PHY in 2.4 GHz

IEEE 802.11g-2003 PHY is referred to as the *extended rate PHY* (ERP). The 802.11g combines the 802.11b PHY and the 802.11a PHY modified to work at the 2.4 GHz. Accordingly, the 802.11g supports 12 different transmission rates, namely, 1, 2, 5.5, 6, 9, 11, 12, 18, 24, 36, 48, and 54 Mbps, where all the 802.11b rates (i.e., 1, 2, 5.5, and 11 Mbps) and the mandatory rates of the 802.11a (i.e., 6, 12, and 24) are mandatory rates of the 802.11g. The 802.11g PHY is backward compatible with the 802.11b PHY so that an 802.11g station can communicate with an 802.11b station using an 802.11b transmission rate.

12.4.1 Mandatory and Optional Modes

IEEE 802.11g defines a set of mandatory transmission modes as well as optional transmission modes, where the mandatory modes are rooted in IEEE 802.11a.

Mandatory ERP-OFDM

The OFDM PHY from the 802.11a is referred to as *ERP-OFDM* in the 802.11g. The ERP-OFDM employs the transmission schemes and the PPDU format exactly the same as that of the 802.11a, which are presented in Section 12.2. As in the case with the 802.11a, the support of 6, 12, and 24 Mbps is mandatory.

The 802.11g should support the 1 and 2 Mbps rates of the DSSS modulation as well as 5.5- and 11-Mbps rates of the CCK modulation, as defined in Section 12.3.2. Moreover, the short PPDU format, which is optional per the 802.11b, is mandatory for the 802.11g as well. An 802.11g station should be ready to detect both the 802.11b preamble and the ERP-OFDM preamble because either of them might arrive at any time.

Optional Modes

The 802.11g additionally defines two optional modes, namely, *ERP-PBCC* and *DSSS-OFDM*. We briefly introduce these schemes here.

The ERP-PBCC is a single carrier modulation scheme, which encodes the payload using a packet binary convolutional code with the constraint length of 9. These are extensions to the optional PBCC modulations of the 802.11b, as defined in Section 12.3.2. The ERP-PBCC modes support 22 Mbps and 33 Mbps.

The DSSS-OFDM is a hybrid modulation combining a DSSS preamble and header with an OFDM payload transmission. The PPDU format is illustrated in Figure 12.16. The DSSS-OFDM modes support data transmission rates of 6, 9, 12, 18, 24, 36, 48, and 54 Mbps.

12.4.2 Coexistence with IEEE 802.11b

The 2.4-GHz ISM band is a shared medium, and coexistence with other devices including the 802.11b stations is an important issue for maintaining the high performance of the 802.11g WLAN. The optional ERP-PBCC and DSSS-OFDM mode frames start with a PLCP preamble, which is compatible with the 802.11b stations. However, the PLCP preamble of the ERP-OFDM is not decodable by the 802.11b

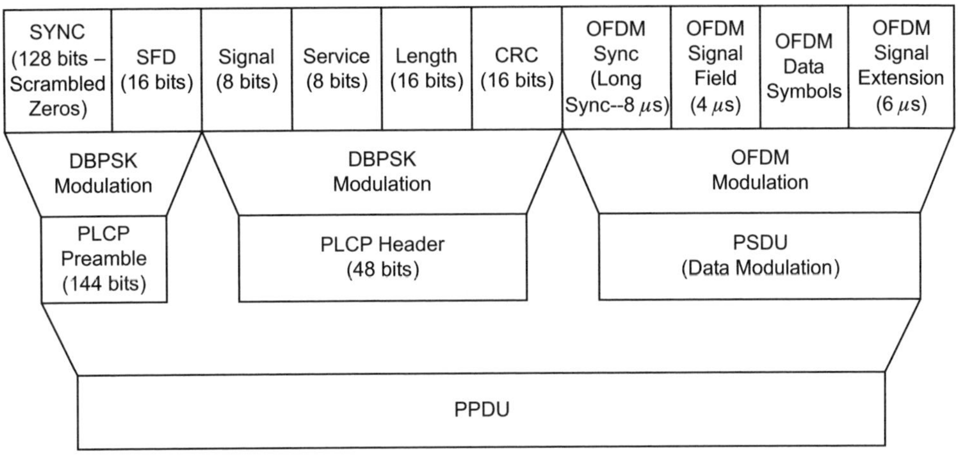

Figure 12.16 Long preamble PPDU format for DSSS-OFDM. (*After:* [1].)

stations, and, hence, depending on the CCA mode of the 802.11b station (i.e., if the ED CCA mode is not employed), the ERP-OFDM signal will not be detected by the 802.11b stations. It is known that many implementations of the 802.11b PHY did not employ the ED CCA. In such a case, the 802.11b station will not assess the channel to be busy while there is an ongoing ERP-OFDM signal on the channel. This can cause severe performance degradation, since the 802.11g stations are essentially hidden from such 802.11b stations so that the 802.11 MAC based on carrier sensing will not work properly.

Accordingly, the 802.11g stations transmitting the ERP-OFDM frames should protect themselves from the coexisting 802.11b stations. Such schemes are referred to as *self-protection mechanisms*. These are rooted in the virtual carrier sensing mechanism as presented in Sections 13.2.3 and 13.2.5. Basically, an 802.11g station transmits an RTS frame transmitted at one of the 802.11b rates, and then the receiver station in turn responds with a CTS frame again transmitted at the same rates so that the neighboring 802.11b stations can be made silent when the ERP-OFDM modulated frames are being transmitted.

Moreover, transmitting a CTS frame without receiving an RTS is also allowed. This CTS frame is specifically referred to as the *CTS-to-self*, which has Address 1 (or RA) set to its own address. A CTS-to-self frame transmitted at an 802.11b rate can be used in order to protect subsequent frame exchanges from neighboring 802.11b stations. While the CTS-to-self is not as robust as the RTS/CTS exchange since it does not protect the transmissions from the neighbors of the receiver station, it might be a less costly scheme. Other mechanisms for the coexistence have been also proposed in the literature (e.g., [9]).

References

[1] IEEE 802.11-2007, IEEE Standard for Local and Metropolitan Area Networks—Part 11: Wireless LAN Medium Access Control (MAC) and Physical Layer (PHY) Specifications, IEEE Std 802.11-2007 (Revision of IEEE Std 802.11-1999), June 12, 2007.

[2] IEEE 802.11a-1999, Amendment 1 to Part 11: Wireless LAN Medium Access Control (MAC) and Physical Layer (PHY) Specifications: High-Speed Physical Layer in the 5 GHz Band, 1999.

[3] IEEE 802.11b-1999, Amendment 2 to Part 11: Wireless LAN Medium Access Control (MAC) and Physical Layer (PHY) Specifications: Higher-Speed Physical Layer Extension in the 2.4 GHz Band, 1999.

[4] IEEE 802.11g-2003, Amendment 4 to Part 11: Wireless LAN Medium Access Control (MAC) and Physical Layer (PHY) Specifications: Further Higher Data Rate Extension in the 2.4 GHz Band, 2003.

[5] MIC Notice No. 2007-63, Ordinances and Technical Standards for Radio Waves; Part 13: Technical Standards for Radio Equipments, Radio Research Laboratory, Korea Ministry of Information and Communication (MIC), September 10, 2007.

[6] O'Hara, B., and A. Patrick, *IEEE 802.11 Handbook, A Designer's Companion*, 2nd ed., New York: IEEE Press, 2005.

[7] FCC CFR47, Title 47 of the Code of Federal Regulations, Part 15: Radio Frequency Devices, Federal Communication Commission, September 20, 2007.

[8] IEEE 802.11j-2004, Amendment 7 to Part 11: Wireless LAN Medium Access Control (MAC) and Physical Layer (PHY) Specifications: 4.9 GHz–5 GHz Operation in Japan, 2004.

[9] Choi, S., and J. del Prado, "802.11g CP: A Solution for IEEE 802.11g and 802.11b Inter-Working," *Proc. IEEE VTC'03-Spring*, Jeju, Korea, April 2003.

CHAPTER 13

Baseline MAC Protocol

In this chapter, we present the baseline protocols of IEEE 802.11 MAC defined in IEEE 802.11-1999. In general, a MAC protocol provides a number of functions, where core functions include: (1) determining when to transmit and when to receive frames, (2) providing error control mechanisms, and (3) providing the frame formats. Other additional functions could include: (1) security support, (2) *quality-of-service* (QoS) provisioning, and (3) mobility support. The 802.11 MAC supports all of these functionalities.

The IEEE 802.11 MAC is based on the logical functions, called the *coordination functions*, which determine when a station operating within a BSS is permitted to transmit and may be able to receive frames via the wireless channel. According to the baseline standard, two coordination functions are defined, namely, the mandatory *distributed coordination function* (DCF), for a distributed, contention-based channel access, based on *carrier-sense multiple access with collision avoidance* (CSMA/CA), and the optional *point coordination function* (PCF), for a centralized, contention-free channel access, based on a poll-and-response mechanism. The DCF has been the most dominant form of the 802.11 MAC, while the PCF was rarely implemented in reality.

Figure 13.1 shows the conceptual relationship between PCF and DCF. As shown in the figure, the PCF sits on top of the DCF, which means that the PCF operation relies on that of the mandatory DCF. The 802.11 MAC operates with time-division-based frame-by-frame transmissions. Every station in a BSS, which is the basic unit of the network like a cell in a cellular network, uses the same frequency channel for all the frame transmissions. Unlike many other wireless systems, the 802.11 does not employ any of transmission slots, control channels, and pilot channels. Moreover, under the DCF, the AP accesses the channel in the exactly same manner as non-AP stations do. It is known that the AP's downlink transmissions might be the bottleneck of the entire network performance due to this fact.

13.1 MAC Frame Formats

In this section, we first present the formats of the MAC frames (i.e., MPDUs). An MPDU is used to convey one of three types of MAC messages—data, management, and control. First, the data messages are those arriving from the higher layer (i.e., either LLC or MAC bridge) at the transmitter side. That is, an MAC message arrives at the MAC in the form of MSDU. Second, the management messages, called *MAC management protocol data units* (MMPDUs), are used to support the 802.11 services, and are locally generated by the MAC according to the control by the MLME

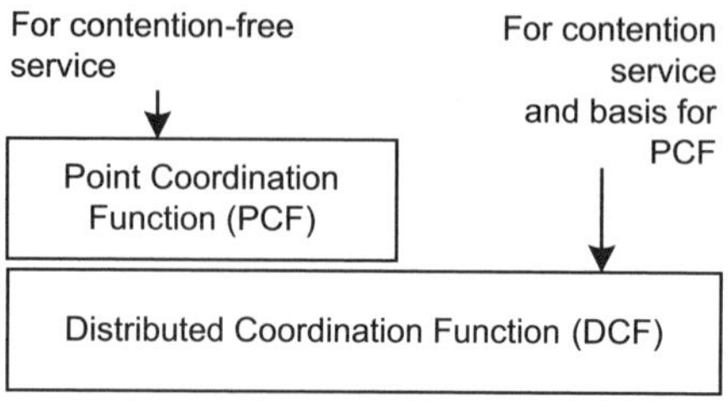

Figure 13.1 Relationship between PCF and DCF. (*After:* [1].)

and SME. Finally, the control messages are used to support the delivery of data and management messages, and are locally generated by the MAC.

13.1.1 General Frame Format

Figure 13.2 shows the general format of IEEE 802.11 MAC frame (or MPDU more specifically). As shown in the figure, an MPDU is composed of a *MAC header*, a *frame body*, and finally a *frame check sequence* (FCS). The frame body contains an MSDU (in the case of a data type frame) or an MMPDU (in the case of a management type frame) or their fragment (if fragmentation is used, as presented in Section 13.2.6). Note that for each individual frame type, some fields may not be present, and specific frame formats are discussed in the Sections 13.1.2 to 13.1.4. In general, most fields in Figure 13.2 are present for both data and management type frames while many fields are not present in the case of control type frames, where the specific format depends on each individual control frame.

MAC Header—Frame Control

The very first subfield in a MAC header is the *frame control* field, which consists of the following subfields: *protocol version, type, subtype, to DS, from DS, more fragments, retry, power management, more data, protected frame,* and *order*. Figure 13.3 shows the detailed format of the frame control field.

- *Protocol version* field: for the current standard, the value of the protocol version is fixed to 0.
- *Type and subtype* fields: the type field indicates the type of the frame (i.e., data, management, and control). The subtype field indicates the subtype of the frame, where various subtypes are defined for different types of frames. Valid combinations of types and subtypes are summarized in Tables 13.1 and 13.2. We further discuss individual subtypes of frames in Sections 13.1.2 through 13.1.4.
- *To DS* and *from DS* fields: to DS = 1 and from DS = 0 when the frame is destined to the DS while to DS = 0 and from DS = 1 when the frame is arriving from the DS. For the data type frames, which are transmitted between stations within an IBSS, between stations via a direct link in an 802.11e WLAN, as well as management and control type frames, both fields are set to 0. Both fields might be set to one in the frame transmitted in a *wireless distribution system* (WDS). This is further discussed as part of the Address 4 explanation.

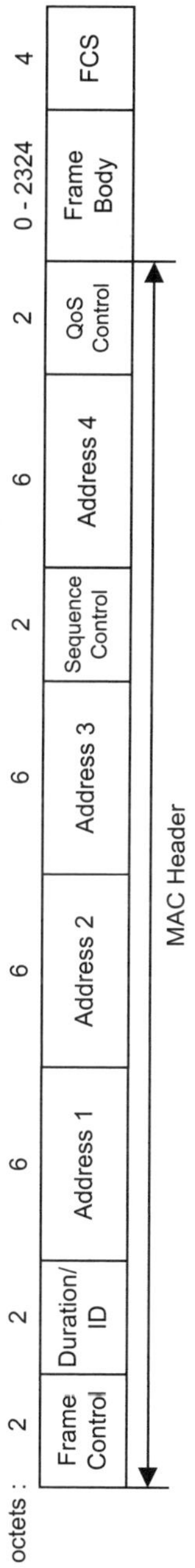

Figure 13.2 General format of IEEE 802.11 MAC frame. (*After:* [2].)

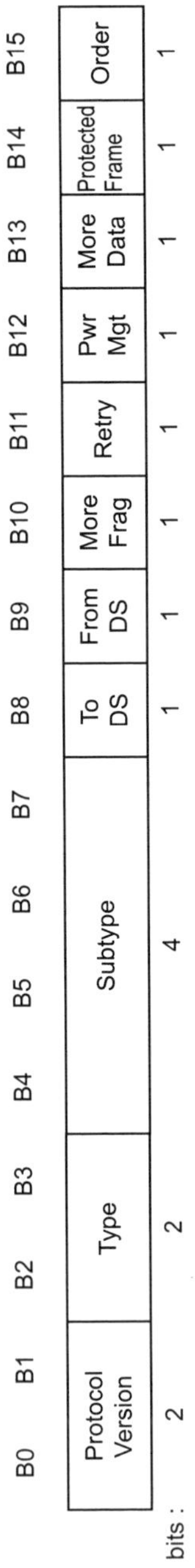

Figure 13.3 Frame control field. (*After:* [2].)

Table 13.1 Type and Subtype Combinations for Management and Control Frames

Type value B3 B2	Type description	Subtype value B7 B6 B5 B4	Subtype description
00	Management	0000	Association request
00	Management	0001	Association response
00	Management	0010	Reassociation request
00	Management	0011	Reassociation response
00	Management	0100	Probe request
00	Management	0101	Probe response
00	Management	0110-0111	Reserved
00	Management	1000	Beacon
00	Management	1001	ATIM
00	Management	1010	Disassociation
00	Management	1011	Authentication
00	Management	1100	Deauthentication
00	Management	1101	Action
00	Management	1110-1111	Reserved
01	Control	0000-0111	Reserved
01	Control	1000	Block Ack Request (BlockAckReq)
01	Control	1001	Block Ack (BlockAck)
01	Control	1010	PS-Poll
01	Control	1011	RTS
01	Control	1100	CTS
01	Control	1101	ACK
01	Control	1110	CF-End
01	Control	1111	CF-End + CF-Ack

- *More fragments* field: this field is set to 1 in all data or management type frames that have another fragment following the current MSDU or the current MMPDU.

Table 13.2 Type and Subtype Combinations for Data Frames

Type value B3 B2	Type description	Subtype value B7 B6 B5 B4	Subtype description
10	Data	0000	Data
10	Data	0001	Data + CF-Ack
10	Data	0010	Data + CF-Poll
10	Data	0011	Data + CF-Ack + CF-Poll
10	Data	0100	Null (no data)
10	Data	0101	CF-Ack (no data)
10	Data	0110	CF-Poll (no data)
10	Data	0111	CF-Ack + CF-Poll (no data)
10	Data	1000	QoS Data
10	Data	1001	QoS Data + CF-Ack
10	Data	1010	QoS Data + CF-Poll
10	Data	1011	QoS Data + CF-Ack + CF-Poll
10	Data	1100	QoS Null (no data)
10	Data	1101	Reserved
10	Data	1110	QoS CF-Poll (no data)
10	Data	1111	QoS CF-Ack + CF-Poll (no data)
11	Reserved	0000-1111	Reserved

Source: [2].

- *Retry* field: this field is set to 1 in any data or management type frame that is a retransmission of an earlier frame.
- *Power management* field: this field is used to indicate the power management mode—*power save mode* (PSM) or *active mode* (AM)—of the transmitter. The value indicates the mode in which the station will be after the successful completion of the frame exchange sequence, as discussed in Section 13.5.2.
- *More data* field: this field is used to indicate to a station in PSM that more frames are buffered for that station at the AP. The more data field may be also set to 1 in frames transmitted by a station to the AP in response to a CF-Poll in a *contention free* period (CFP) to indicate that the station has more buffered frames to transmit, which will be discussed more in detail in Section 13.3.

- *Protected frame* field: this field is set to 1 if the frame body field contains information that has been processed by a security encapsulation algorithm as presented in Chapter 15.
- *Order* field: this field is set to 1 in any frame, which is strictly ordered. Frames might be intentionally reordered at the transmitter in order to support the power management and QoS.

MAC Header—Address Fields

Each address field contains a 48-bit (6-octet) address as defined in IEEE 802-2001 [3]. The first 3 octets of an address represent the *organizationally unique identifier* (OUI), which is uniquely assigned to a manufacturer. Two stations manufactured by a manufacturer might have a common value at the first 3 octets of the addresses, while the remaining 3 octets should be different. Although the OUIs are 3 octets long, their true address space is 22 bits because two bits—least significant bit (LSB) and the next of octet 0—are reserved to indicate whether the address is (1) *individual/group* (I/G) address and (2) *universally* or *locally administered* (U/L) address, respectively.

The I/G address bit (LSB of octet 0) is used to indicate the destination address as an individual address or a group address. If the I/G address bit is 0, it indicates that the address field contains an individual address of a station. If this bit is 1, the address field contains a group address that identifies one or more (or all) stations connected in the network. A special predefined group address of all 1s is the broadcast address for all the stations in the network. The U/L address bit is the bit of octet 0 adjacent to the I/G address bit. This bit indicates whether the address has been assigned by a local or universal administrator. Universally administered and locally administered addresses have this bit set to 0 and 1, respectively.

There are five address fields in the MAC frame format. These fields are used to indicate the *basic service set identification* (BSSID), *source address* (SA), *destination address* (DA), *transmitter address* (TA), and *receiver address* (RA).

- BSSID represents the address of the AP in the case of an infrastructure BSS, while it is a locally administered individual address in the case of an IBSS, where it is randomly selected by the station initializing the IBSS.
- SA represents an individual address that identifies the MAC entity from which the transfer of the MSDU (or fragment thereof) contained in the frame body field was initiated.
- DA represents an individual or group address that identifies the MAC entity or entities intended as the final receiver(s) of the MSDU (or its fragment) contained in the frame body field.
- TA represents an individual address that identifies the station that has transmitted the MPDU onto the wireless channel.
- RA represents an individual or group address that identifies the intended immediate receiver(s) of the MPDU over the wireless channel. A frame with an individual address as its RA is referred to as a *directed* or unicast frame.

Address 1 always specifies the RA while Address 2 always specifies the TA. The content of Address 3 depends on an individual frame, and Address 4 is used only for data type frames transmitted over a WDS, which is a *distribution system* (DS), connecting multiple APs, implemented using the 802.11 WLAN. The contents of the four address fields are summarized in Table 13.3.

To DS = 0 & From DS = 0, such as a transmission within an IBSS: Address 1 = RA = DA and Address 2 = TA = SA, while Address 3 = BSSID, identifying the BSS. This combination is used for transmissions in an IBSS as well as for management and control type frames in an infrastructure BSS. It is also used for a direct link in an infrastructure BSS running IEEE 802.11e (to be discussed in Section 14.5.1).

To DS = 0 & From DS = 1 (i.e., a downlink transmission): Address 1 = RA = DA and Address 2 = TA = BSSID, while Address 3 = SA is the MAC address of the source in the subnet, which could be either a router or another non-AP station.

To DS = 1 & From DS = 0 (i.e., an uplink transmission): Address 1 = RA = BSSID and Address 2 = TA = SA, while Address 3 = DA is the MAC address of the destination in the subnet, which could be either a router or another non-AP station.

To DS = 1 & From DS = 1 (i.e., a transmission within a WDS): Address 1 = RA and Address 2 = TA, while Address 3 = DA and Address 4 = SA. Note that for a transmission within a WDS (e.g., a wireless system connecting multiple APs), the destination and the source should be different from the receiver and the transmitter, respectively. Note that in WDS, both transmitter and receiver should be basically APs.

MAC Header—Duration and Sequence Control

The *duration/ID* field is 16 bits long. The contents of this field vary with frame type and subtype, with whether the frame is transmitted during the CFP, and with the QoS capabilities of the sending station. In control frames of subtype PS-Poll, this field carries the *association identifier* (AID) of the station, transmitting the PS-Poll. The AID ranges from 1 to 2,007, and is uniquely assigned to a station by the AP upon its association with the AP. Otherwise, this field is used to indicate the duration (in μsec), which the frame exchange including the frame in consideration is expected to last. The indicated duration is used by the transmitting station to reserve the wireless channel usage. The usage of this field is further detailed in Section 13.2.3 and Section 14.3.

The *sequence control* field is 16 bits long, and is composed of two subfields—the *sequence number* and the *fragment number*. Sequence control field is not present in control type frames. The sequence number field is a 12-bit field indicating the sequence number of an MSDU or MMPDU. Each MSDU or MMPDU transmitted

Table 13.3 Contents of Five Address Fields Depending on the Values in "To DS" and "From DS" Fields

To DS	From DS	Address1	Address2	Address3	Address4	Direction
0	0	DA	SA	BSSID	N/A	Direct Link
0	1	DA	BSSID	SA	N/A	Downlink
1	0	BSSID	SA	DA	N/A	Uplink
1	1	RA	TA	DA	SA	WDS

by a station is assigned a sequence number. The fragment number field is a 4-bit field indicating the number of each fragment of an MSDU or MMPDU. The fragment number is set to 0 in the first or only fragment (i.e., for no fragmentation) of an MSDU or MMPDU and is incremented by one for each successive fragment of that MSDU or MMPDU. The fragment number remains constant in all retransmissions of the fragment. The fact that the fragment number field is 4 bits long implies that the maximum number of fragments is 16. In fact, in the case of the baseline MAC, the maximum number is determined to 11, as discussed further in Section 13.2.6.

The *QoS control* field exists in QoS data type frames, and it contains various kinds of information related with QoS provisioning. This field will be discussed in Section 14.2.4.

Frame Body Field

The length of the frame body field is variable. The maximum length is 2,324 octets, which is determined by the maximum MSDU size (i.e., 2,304 octets) plus any overhead due to security encapsulation. Accordingly, when a security option is not employed, the maximum length of the frame body is 2,304 octets. The maximum overhead due to the security encapsulation (i.e., 20 octets) occurs when the TKIP encapsulation is employed. See Figure 15.17 for further details. Even if the maximum MSDU size is 2,304, practically the maximum length is 1,508 octets. This is because the 802.11 AP is typically attached to an Ethernet, which has the maximum transfer unit (MTU) of 1,500 octets. Note that the MTU represents the maximum IP datagram size supported by a specific link technology. Including the LLC header of 8 octets with the SNAP option as shown in Figure 11.10, the maximum MSDU size in a practical 802.11 WLAN becomes 1,508 octets.

The format of the frame body field is dependent on each frame subtype. For a data type frame, the frame body field contains an MSDU or its fragment. If an encryption scheme is employed, the content of the frame body is encrypted. For a management type frame, the frame body field contains a number of information elements according to the definition of each individual management type frame, as presented in Section 13.1.4. Management type frames are not encrypted, but IEEE 802.11w, currently being standardized, will enable encrypting of some management type frames.

Frame Check Sequence (FCS) Field

The FCS field is a 32-bit field containing a 32-bit CRC (or CRC-32). CRC-r is known to be able to detect all burst errors less than r+1 bits. The FCS is calculated over all the fields of the MAC header and the frame body field, which are referred to as the calculation fields. The FCS is calculated using the following standard generator polynomial of degree 32:

$$G(x) = x^{32} + x^{26} + x^{23} + x^{22} + x^{16} + x^{12} + x^{11} + x^{10} + x^{8} + x^{7} + x^{5} + x^{4} + x^{2} + x + 1$$

The FCS is the ones complement of the sum (modulo 2) of the following:

- The remainder of $x^k \times (x^{31} + x^{30} + x^{29} + \dots + x^2 + x + 1)$ divided (modulo 2) by $G(x)$, where k is the number of bits in the calculation fields;
- The remainder after multiplication of the contents (treated as a polynomial) of the calculation fields by x^{32} and then division by $G(x)$.

At the receiver, the initial remainder is preset to all ones and the serial incoming bits of the calculation fields and FCS, when divided by $G(x)$, results in the following unique nonzero remainder value when there is no transmission error:

$$x^{31} + x^{30} + x^{26} + x^{25} + x^{24} + x^{18} + x^{15} + x^{14} + x^{12} + x^{11} + x^{10} + x^{8} + x^{6} + x^{5} + x^{4} + x^{3} + x + 1$$

13.1.2 Data Frames

According to Figure 13.2, the length of the MAC header is 32 octets. However, the QoS control field is used only for the data type frames of IEEE 802.11e MAC, which will be presented in Chapter 14. Moreover, the address 4 field is present only in data type frames transmitted within a WDS. Accordingly, for the baseline MAC, the MAC header length of a data type frame is typically 24 octets.

In Table 13.1, for data type frames, four bits in the subtype field represent specific subtypes. That is, bits b7, b6, b5, and b4 represent QoS, Null (no data), CF-Poll, and CF-Ack, respectively. QoS means that the data type frame is transmitted by an 802.11e MAC, and the frame includes the QoS control field. Null (no data) means that the frame body of the data type frame is not present. That is, even if it is a data type frame, the frame does not carry any data. Null data frames are actually more like control frames. For example, CF-Ack is a data type frame, but its function is basically the same as that of the ACK control frame. Depending on which bits are set to 1, the exact subtype is determined. For example, a data type frame with subtype fields (b7, b6, b5, b4) = (0, 0, 0, 1) is Data + CF-Ack. This is a data frame conveying an MSDU or its fragment with the piggybacked control information (i.e., CF-Ack).

For data type frames, found in Table 13.2, some new names, which are not found in the table, are often used in order to refer to a set of frames in an aggregated manner. Parentheses enclosing portions of names or acronyms are used to designate a set of related names that vary based on the inclusion of the parenthesized portion. For example,

- QoS +CF-Poll frame refers to the three QoS data subtypes that include +CF-Poll, namely, QoS Data+CF-Poll frame (subtype 1010), QoS Data+CF–Ack+CF-Poll frame (subtype 1011), and QoS CF–Ack+CF-Poll (no data) frame (subtype 1111).
- QoS CF-Poll frame refers specifically to the QoS CF-Poll frame (subtype 1110).
- QoS (+)CF-Poll frame refers to all four QoS data subtypes with CF-Poll, namely, QoS CF-Poll frame (subtype 1110), QoS CF-Ack+CF-Poll (no data)

frame (subtype 1111), QoS Data+CF-Poll frame (subtype 1010), and QoS Data+CF-Ack+CF-Poll frame (subtype 1011).

- QoS (+)Null frame refers to all three QoS data subtypes with no data, namely, QoS Null (no data) frame (subtype 1100), QoS CF-Poll (no data) frame (subtype 1110), and QoS CF-Ack+CF-Poll (no data) frame (subtype 1111).
- QoS +CF-Ack frame refers to the three QoS data subtypes that include +CF-Ack, namely, QoS Data+CF-Ack frame (subtype 1001), QoS Data+CF-Ack+CF-Poll frame (subtype 1011), and QoS CF-Ack+CF-Poll (no data) frame (subtype 1111).
- (QoS) CF-Poll frame refers to both QoS CF-Poll (no data) frame (subtype 1110) and CF-Poll (no data) frame (subtype 0110).

Under the mandatory baseline MAC (i.e., DCF), data subtypes other than Data—those with subtype fields (b7, b6, b5, b4) = (0, 0, 0, 0)—are not used at all. All other non-QoS subtypes are used as part of the optional baseline MAC (i.e., PCF). The remaining subtypes (i.e., all QoS subtypes) are defined according to IEEE 802.11e. Each individual subtype frames as well as its usage will be presented in Sections 13.3 and Chapter 14.

13.1.3 Control Frames

The values in the frame control field within control frames are illustrated in Figure 13.4.

There are eight subtypes of control frames as follows. As shown in Figures 13.5 through 13.8, the control frames basically have no frame body field, whereas the MAC header is significantly reduced by omitting many redundant fields. From the viewpoint of the data transfer, control frames are necessary overheads, and, hence, minimizing the control frame sizes is a natural approach.

- *Acknowledgment* (ACK): this frame is used to acknowledge a successful reception of a directed data or management frame. The frame format is shown in Figure 13.6.
- *Request-to-send* (RTS): this frame is transmitted to reserve the wireless channel for a subsequent frame exchange. The frame format is shown in Figure 13.5.
- *Clear-to-send* (CTS): this frame is transmitted in response to an RTS by the receiver of the RTS in order to acknowledge the successful reception of the RTS. This frame also reserves the wireless channel for a subsequent frame exchange. The frame format is shown in Figure 13.6.
- *Contention-free end* (CF-End): this frame is transmitted by the AP in order to indicate the end of a *contention-free period* (CFP). The frame format is shown in Figure 13.8.
- CF-End + CF-Ack: this frame is transmitted by the AP in order to simultaneously indicate both the end of a CFP and the acknowledgment of a successful reception of the preceding frame from a station. The frame format is shown in Figure 13.8.

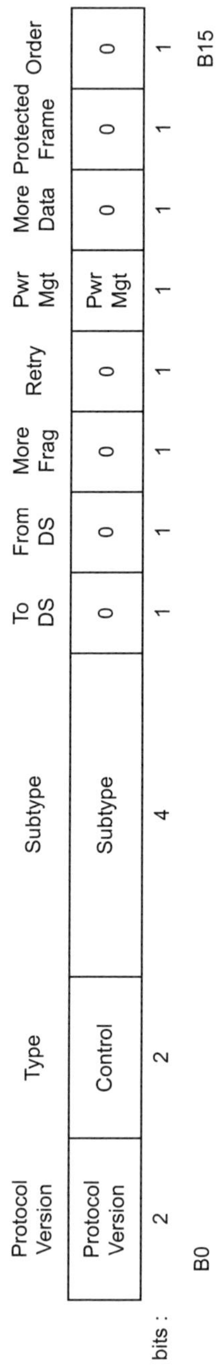

Figure 13.4 Values in the frame control field within control frames. (*After:* [2].)

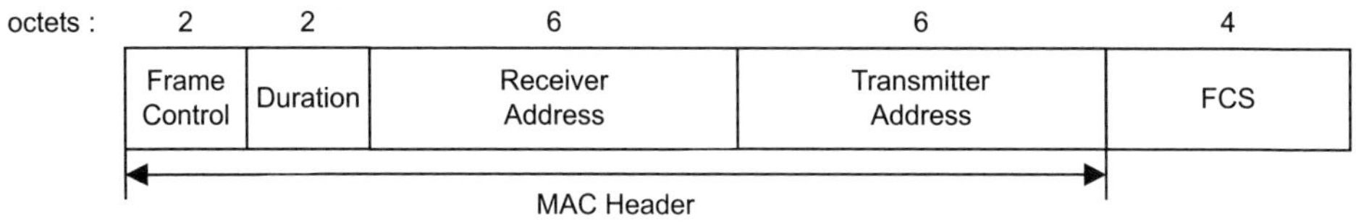

Figure 13.5 Format of RTS frame. (*After:* [2].)

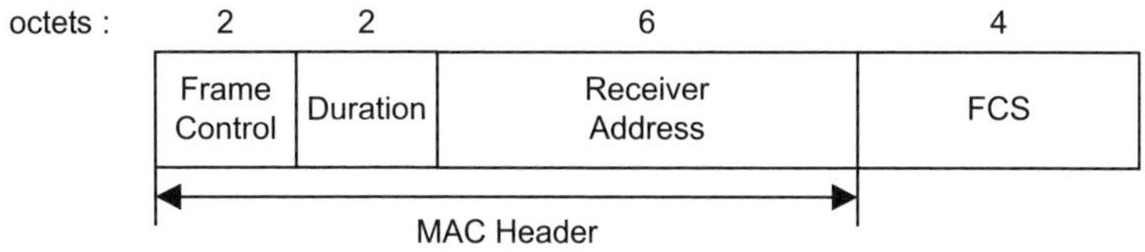

Figure 13.6 Format of CTS and ACK frames. (*After:* [2].)

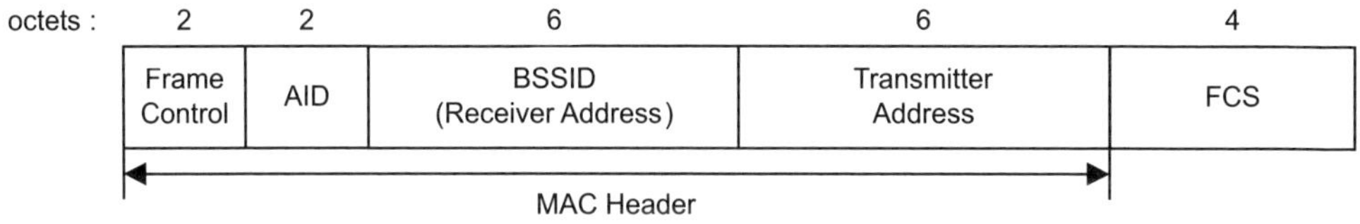

Figure 13.7 Format of PS-Poll frame. (*After:* [2].)

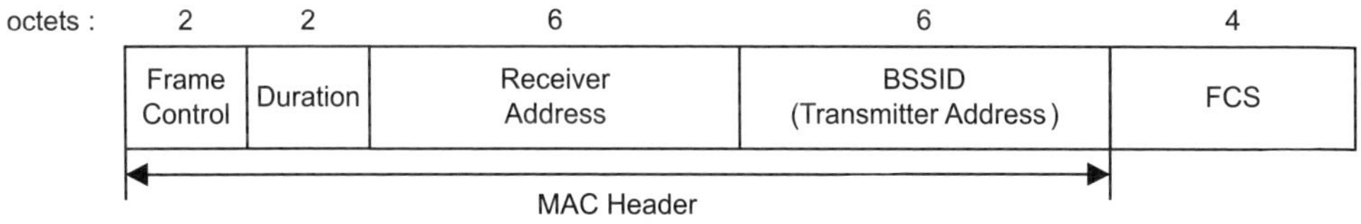

Figure 13.8 Format of CF-End and CF-End+CF-Ack frames. (*After:* [2].)

- *Power save poll* (PS-Poll): this frame is transmitted by a station in PSM in order to request its AP to transmit the buffered frames destined to the station. The frame format is shown in Figure 13.7.
- *Block acknowledgment* (BlockAck or BA): being defined in IEEE 802.11e, this frame is used instead of ACK in order to acknowledge successful frame receptions. Different from the normal ACK, BlockAck acknowledges multiple frame receptions. This frame will be further discussed in Section 14.5.2.
- *Block acknowledgment request* (BlockAckReq or BAR): this frame is used by the transmitter of data frames in order to request the recipient of the data frames to transmit a BlockAck. This frame will be further discussed in Section 14.5.2.

13.1.4 Management Frames

Management type frames do not include the QoS control and address 4 fields in Figure 13.2, as shown in Figure 13.9. Management type frames cannot have address 4 since these frames are supposed only for the internal BSS purpose. The frame body of a management type frame contains various kinds of information depending on each individual frame.

There are 12 subtypes of management frames as follows:

- *Beacon*: this frame conveys various types of information related to the operation of the BSS. For the baseline MAC, those include: (1) the capability supported/needed in the BSS, (2) timestamp for time synchronization, (3) beacon interval, (4) SSID, (5) transmission rates supported in the BSS, (6) CF information, (7) IBSS information (only for IBSS), and (8) *traffic indication map* (TIM) for power management support. In an infrastructure BSS, the AP basically transmits beacon frames periodically, while beacons are transmitted by stations in a contentious manner in an IBSS.
- *Probe request*: when a station searches neighboring BSSs, it broadcasts probe request frames. This frame contains the information related to the station's capability and the type of BSSs that the station is looking for.
- *Probe response*: upon the reception of a probe request, an AP transmits a probe response frame to indicate the information of its BSS. The format of the probe response is almost identical to that of the beacon frame.
- *Authentication*: this frame is used for a station to be authenticated with an AP. According to IEEE 802.11i, the usage of this frame became basically obsolete.
- *Deauthentication*: this frame is transmitted by a station or an AP in order to terminate the authentication status of a station.
- *Association request*: after getting authenticated, a station can get associated with an AP. This frame is transmitted by a station to an AP in order to request an association.
- *Association response*: this frame is transmitted by an AP in response to an association request frame from a station by indicating the success or failure of the association request.
- *Reassociation request*: this frame is transmitted from a station, which would like to hand off from an AP to another AP. The receiver of the frame is the new AP. The format of the reassociation frame is almost identical with that of the association frame, except that the reassociation frame contains the MAC address of the current AP.
- *Reassociation response*: this frame is transmitted by an AP in response to a reassociation request frame. The format is identical to that of the association response frame.
- *Disassociation*: this frame is transmitted by a station or an AP in order to terminate the association status of a station.
- *Announcement traffic indication message* (ATIM): this frame is used in order to support power management in an IBSS.

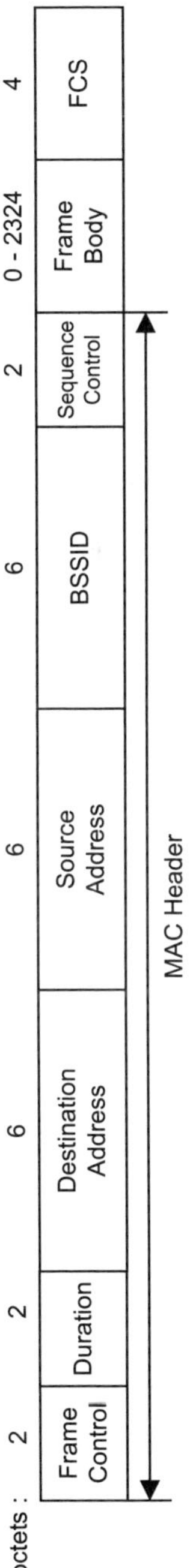

Figure 13.9 Management frame format. (*After:* [2].)

- *Action*: this frame is a kind of wrapper, which can be very flexibly extended. Since the subtype field in the MAC header is 4 octets, up to 16 subtypes can be defined. However, as many MAC extensions, including 802.11e, 802.11h, and so on, were being defined, more and more management type frames were also defined. In order to resolve the limited management subtype space, the action frame was developed. An action frame includes the *action* field (as illustrated in Figure 13.10) at the very first of the frame body field. The action field is composed of a one-octet category subfield and the action details of a variable size. The available codes for the field are summarized in Table 13.4. Code 0 is for IEEE 802.11h, while codes 1 to 3 are for IEEE 802.11e. Action frames of a given category are referred to as *<category name> action frames*. For example, frames in the QoS category are called *QoS action frames*. The format of the action details is dependent on a specific protocol, and, hence, an unlimited number of management frames can be defined using this action frame format. For each category, a number of action frames are defined, and they are identified using different action values as shown in Table 13.5. Specific action frames will be discussed in Chapters 14 and 17.

Table 13.6 summarizes the frame body contents of all the management frames except ATIM and action frames. Note that the frame body of the ATIM frame is null. A field in the frame body can be classified into either a fixed field or an information element, where a fixed field has a fixed length, and an information element can have a variable length. The second column in the table specifies whether the cor-

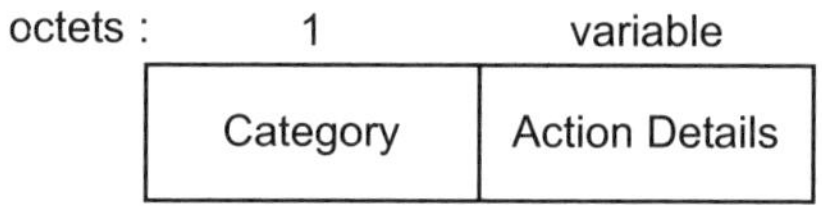

Figure 13.10 Action field in action frame. (*After:* [2].)

Table 13.4 Category Values of the Action Field in Action Frame

Code	Reference	Meaning
0	802.11h	Spectrum management
1	802.11e	QoS
2	802.11e	Direct-Link Setup (DLS)
3	802.11e	BlockAck
4-126		Reserved
127		Vendor-specific
128-255		Error

Source: [2].

Table 13.5 Various Types of Action Frames

Action Value	Action Frame Category			
	Spectrum Management (Category = 0)	QoS (Category = 1)	DLS (Category = 2)	BlockAck (Category = 3)
0	Measurement Request	ADDTS (Add TS) Request	DLS Request	ADDBA (Add Block Ack) Request
1	Measurement Report	ADDTS (Add TS) Response	DLS Response	ADDBA (Add Block Ack) Response
2	TPC Request	DELTS (Delete TS)	DLS Teardown	DELBA (Delete Block Ack)
3	TPC Report	Schedule	—	—
4	Channel Switch Announcement	—	—	—

responding field is a fixed field (F) or an information element (IE). The third column in the table then specifies the standard or the amendment, in which the corresponding field is defined. Specific fixed fields and information elements are explained in the corresponding chapters.

13.2 Distributed Coordination Function (DCF)

The DCF is the mandatory part of the 802.11 baseline MAC protocol. It is designed to provide a best-effort service. Its carrier sensing–based polite transmissions make it a perfect choice for the operations at unlicensed bands. In this section, we explain how the DCF works by considering the transmission of data type frames for the simplicity of the explanation. However, the same access mechanism is used for the transmission of management type frames including beacon frames as well. The performances of the 802.11 DCF in terms of throughput and delay have been mathematically analyzed in the literature [4–7].

13.2.1 CSMA/CA Basic Access Procedure

The 802.11 DCF works with a single *first-in-first-out* (FIFO) transmission queue. The CSMA/CA constitutes a distributed MAC based on a local assessment of the channel status (i.e., whether the channel is busy, such as somebody transmitting a frame or idle, such as no transmission over the channel). Basically, the CSMA/CA of the DCF works as follows: when a frame arrives at the head of the transmission queue, if the channel is busy, the MAC waits until the channel becomes idle and then defers for an extra time interval, called the *DCF interframe space* (DIFS). If the channel stays idle during the DIFS deference, the MAC then starts a backoff procedure by selecting a random backoff number for its backoff counter. For each idle slot time interval, during which the channel stays idle, the backoff counter is decre-

Table 13.6 Frame Body Contents of Management Frames

Frame body contents	F/IE	Reference	Beacon	Probe Request	Probe Response	Authentication	Deauthentication	Association Request	Association Response	Reassociation Request	Reassociation Response	Disassociation
Authentication Algorithm Number	F	Baseline				O						
Authentication Transaction Sequence Number	F	Baseline				O						
Beacon Interval	F	Baseline	O		O							
Capability	F	Baseline	O		O			O	O	O	O	
Current AP Address	F	Baseline								O		
Listen Interval	F	Baseline						O		O		
Reason Code	F	Baseline					O					O
AID	F	Baseline							O		O	
Status Code	F	Baseline				O			O		O	
Timestamp	F	Baseline	O		O							
SSID	IE	Baseline	O	O	O			O		O		
Supported Rates	IE	Baseline	O	O	O			O	O	O	O	
Frequency Hopping Parameter Set	IE	Baseline	O		O							
Direct Sequence Parameter Set	IE	Baseline	O		O							
Contention Free Parameter Set	IE	Baseline	O		O							
TIM	IE	Baseline	O									
IBSS Parameter Set	IE	Baseline	O		O							
Challenge Text	IE	Baseline				O						
Country	IE	802.11d	O		O							
Frequency Hopping Parameters	IE	802.11d	O		O							
Frequency Hopping Pattern Table	IE	802.11d	O		O							
Request Information	IE	802.11d		O	O							
ERP Information	IE	802.11g	O		O							
Extended Supported Rates	IE	802.11g	O	O	O			O	O	O	O	
Power Constraint	IE	802.11h	O		O							
Power Capability	IE	802.11h						O		O		
Transmit Power Control Report	IE	802.11h	O		O							
Supported Channels	IE	802.11h						O		O		
Channel Switch Announcement	IE	802.11h	O		O							
Quite	IE	802.11h	O		O							
IBSS DFS	IE	802.11h	O		O							
RSN	IE	802.11i	O		O			O		O		
BSS Load	IE	802.11e	O		O							
EDCA Parameter Set	IE	802.11e	O		O				O		O	
QoS Capability	IE	802.11e	O					O		O		

mented. When the counter reaches zero, the frame is transmitted. On the other hand, when a frame arrives at the head of the queue, if the MAC is in either the DIFS deference or the random backoff procedure (this situation is possible due to the "post" backoff requirement as described later), the processes described earlier are applied again. That is, the frame is transmitted only when the random backoff has finished successfully. When a frame arrives at an empty queue and the channel has been idle longer than the DIFS time interval, the frame is transmitted immediately. The timing of DCF channel access is illustrated in Figure 13.11.

The fact that carrier is sensed before transmission makes the DCF a carrier sense multiple access (CSMA) protocol, while the fact that a frame is transmitted after a random amount of time via a backoff procedure makes the DCF a collision avoidance (CA) protocol. However, it should be noted that this backoff procedure cannot completely eliminate collisions. Note that frames from different stations collide if they finish backoff procedures at the same backoff slot boundary. The collision probability depends on the CW value as well as the number of actively contending stations in the network.

Each station maintains a contention window (CW), which is used to select the random backoff count. The backoff count is determined as a pseudo-random integer drawn from a uniform distribution over the interval [0,CW]. How to determine the CW value is further detailed later. If the channel becomes busy during a backoff procedure, the backoff is suspended. When the channel becomes idle again, and stays idle for an extra DIFS time interval, the backoff procedure resumes with the latest backoff counter value.

For each successful reception of a directed (i.e., unicast) frame, the receiving station immediately acknowledges the frame reception by sending an *acknowledgment* (ACK) frame, as shown in Figure 13.12. The ACK frame is transmitted after a *short IFS* (SIFS), which is shorter than the DIFS. Other stations resume the backoff procedure after the DIFS idle time. Thanks to the SIFS interval between the data and ACK frames, the ACK frame transmission is protected from other stations' contention. If an ACK frame is not received after the data transmission, the frame is retransmitted after another random backoff.

The backoff of the DCF is often referred to as *binary exponential backoff*. The CW size is initially assigned CWmin and increases when a transmission fails (i.e., the transmitted data frame has not been acknowledged). After any unsuccessful transmission attempt, another backoff is performed using a new CW value updated

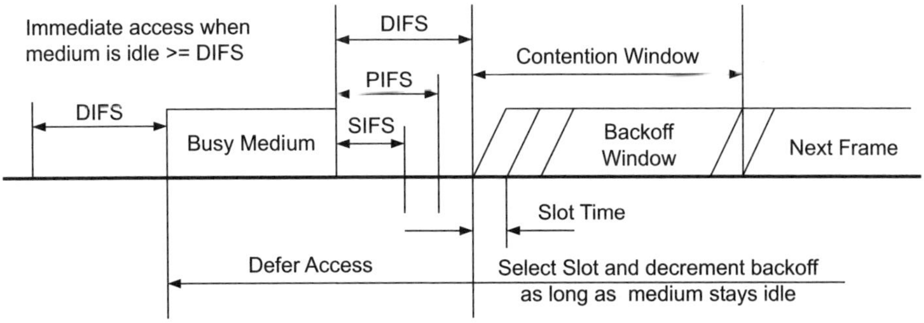

Figure 13.11 IEEE 802.11 DCF channel access. (*After:* [2].)

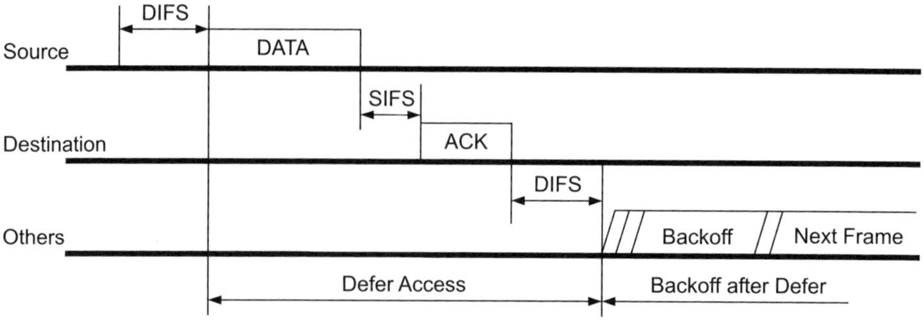

Figure 13.12 ACK transmission after a successful directed frame reception. (*After:* [2].)

by CW: = 2 × (CW + 1) − 1, with the upper bound of CWmax. The evolution of the CW value is illustrated in Figure 13.13 assuming that CWmin = 15 and CWmax = 255. The actual values of CWmin and CWmax are dependent on the underlying PHY. This exponential increase of the CW value reduces the probability of consecutive collisions in case there are multiple stations attempting to access the channel. Note that an ACK reception failure could occur due to either collision or channel error of either the directed data frame or the corresponding ACK frame. In fact, if the ACK reception failure was due to a channel error, the CW value increment is actually not desirable. However, in the 802.11, a transmitter cannot differentiate the failures due to the collision and channel error, and, hence, the CW value is increased irrespective of the cause of a transmission failure.

After each successful transmission, the CW value is reset to CWmin, and the transmission-completing station performs the DIFS deference and a random backoff even if there is no other pending frame in the queue. This is often referred to as a "post" backoff, as this backoff is done after, not before, a transmission. This post backoff ensures there exists at least one backoff interval between two consecutive frames (or more exactly speaking, two consecutive MSDUs, due to possible fragmentations as discussed later) transmissions. When a new frame arrives at the head of the queue while the post backoff is ongoing, the frame can be transmitted after the post-backoff is completed. On the other hand, if a new frame arrives at the head of the queue after a post backoff is completed, the frame might be transmitted immediately if the channel has been idle over DIFS at the moment of the frame arrival or

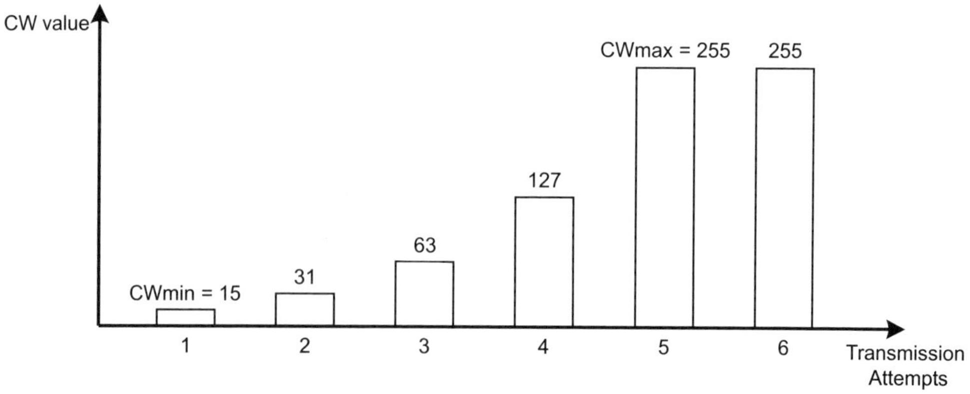

Figure 13.13 Evolution of the CW value assuming CWmin = 15 and CWmin = 255.

after another backoff procedure if the channel is busy then. Accordingly, there can be one or two backoff counter value selections between two consecutive frame transmission attempts.

Figure 13.14 shows the state diagram of the DCF for the frame transmission operations. The symbols representing operations and conditions are summarized in Table 13.7. There are the following three states:

- Tx Idle: in this state, the MAC waits for a frame arriving from the higher layer;
- Tx & Wait Ack: in this state, the MAC transmits a pending frame and waits for a corresponding ACK from the receiver MAC;
- Backoff: in this state, the MAC performs a backoff value countdown.

Note that there are multiple intended receivers in the case of group addressed (i.e., broadcast and multicast) frames. Accordingly, a group-addressed frame cannot be acknowledged. A station after transmitting a broadcast or a multicast frame assumes that the frame transmission was successful, and, hence, the frame is never retransmitted. Obviously, CWmin is used for a subsequent backoff procedure.

The values of CWmin, CWmax, *slot time* (SlotTime), and *SIFS time* (SIFSTime) are dependent on the underlying PHY, as summarized in Table 13.8. The value of CWmin of an 802.11g station is basically set to 15, while 31 should be used when it operates in an 802.11b BSS. SlotTime is the unit of time for the backoff procedure. For the case of the 802.11g, SlotTime is set to 20 μs by default, and when all the stations in the BSS are the 802.11g stations, 9 μs can be optionally used.

The DCF is known to provide a long-term fairness for the channel access among the stations in an error-free channel environment, so that every station transmits the same number of frames in the long term assuming that every station has always frames to transmit. This long-term channel access opportunity fairness makes long-term throughput fairness if every station transmits frames with the same aver-

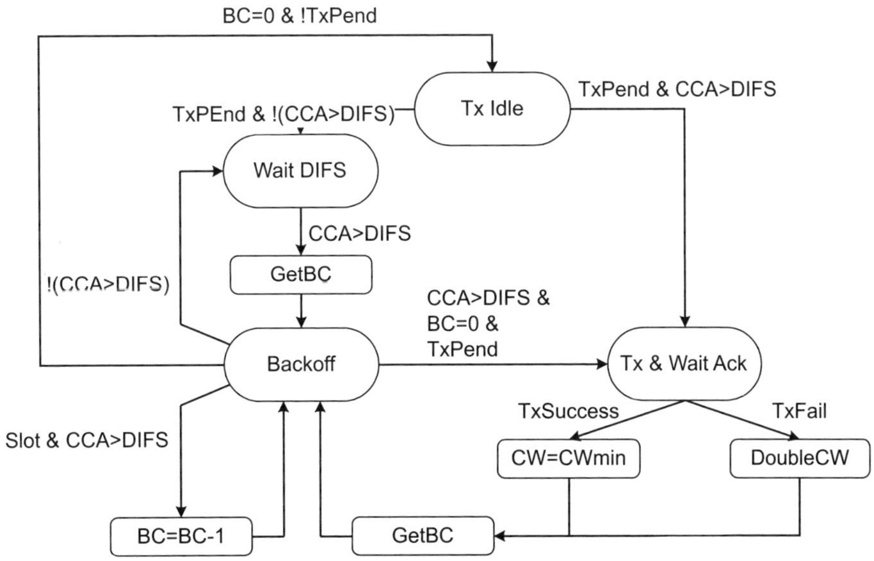

Figure 13.14 DCF state diagram for frame transmission operations.

Table 13.7 Symbols Used in the DCF State Diagram

Operations	Meaning
GetBC	Newly select a backoff counter if there is no suspended backoff
BC=BC-1	Decrease the backoff counter by one
CW=CWmin	Reset the CW to CWmin
DoubleCW	Increase CW to 2x(CW+1)-1
Conditions	**Meaning**
TxPend	A frame pending at the transmission queue
CCA>DIFS	The channel has been idle over a DIFS interval
BC=0	The backoff counter reaches zero
Slot	A slot time has passed
TxSuccess	A successful transmission
TxFail	An unsuccessful transmission

Table 13.8 MAC Parameters of Various PHYs

	IR	FHSS	DSSS/802.11b	802.11g	802.11a
SlotTime (μsec)	8	50	20	20 or 9	9
SIFSTime (μsec)	10	28	10	10	16
CWmin	63	15	31	31 or 15	15
CWmax	1023	1023	1023	1023	1023

age length. This fairness property holds independent of the transmission rates employed by different stations. Accordingly, the long-term throughputs of contending stations end up being the same irrespective of their transmission rates. This throughput fairness property of the 802.11 DCF was introduced as a *performance anomaly* in the literature [8]. We will further discuss this property in comparison with the temporal fairness supported by IEEE 802.11e in Section 14.3.

13.2.2 Interframe Spaces (IFSs)

As explained earlier, different interframe spaces are defined in order to give the priority to different frame transmissions. There are four types of IFSs defined with the following relationship:

- *Short IFS* (SIFS) is used between a frame and an immediate response (e.g., Data-ACK and RTS-CTS-Data-ACK). (Request-to-send/clear-to-send (RTS/CTS) exchange will be defined later.) The value of SIFS is dependent on the underlying PHY.
- *PCF IFS* (PIFS) is defined to be SIFS + SlotTime, where the SlotTime is used as the time unit for the backoff countdown, and its value is also dependent on the underlying PHY. PIFS is used before sending a beacon under the PCF, and also when there is no response after a polling frame. More details will be given in Section 13.3.
- *DCF IFS* (DIFS) is defined to be SIFS + 2 × SlotTime, and it is used before a backoff countdown after a busy channel interval.
- *Extended IFS* (EIFS) is used instead of DIFS after an erroneous frame reception. Accordingly, after a successful frame reception, the IFS is switched back to DIFS. More details on EIFS are discussed next.

There are basically two different cases of an unsuccessful frame reception: (1) the PHY has indicated the erroneous reception to the MAC (e.g., carrier sync lost during frame reception, and incoming frame modulated at an unknown data rate), and (2) the error is detected by the MAC via an incorrect FCS. The EIFS is defined to provide enough time for stations to wait for an ACK frame of an incorrectly received frame. Accordingly, the EIFS value is determined by the sum of one SIFS, one DIFS, and the time needed to transmit an ACK frame at the underlying PHY's lowest mandatory rate—EIFS = SIFS + ACK_Transmission_Time (at the lowest transmission rate of the PHY) + DIFS. Figure 13.15 illustrates that the stations receiving the data frame incorrectly defer for an EIFS period before starting a backoff procedure. Note that the ACK frame does not need to be transmitted at the lowest transmission rate of the PHY as discussed in Section 13.4.2, and when it is transmitted at a higher rate, the EIFS value might be much larger than the actual ACK transmission time.

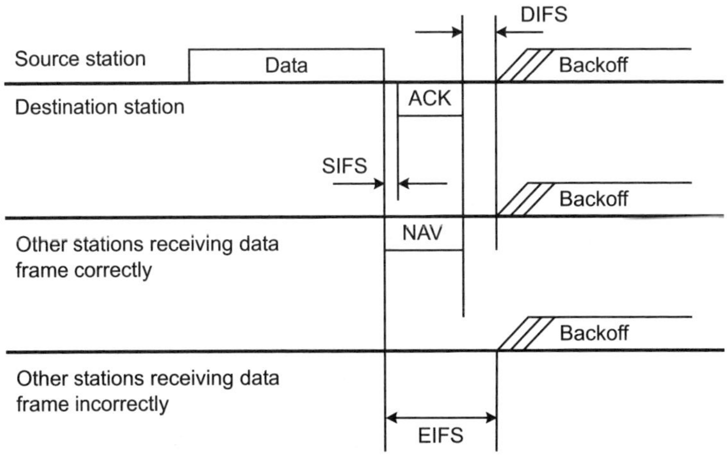

Figure 13.15 DCF access operation, where the ACK frame is transmitted at the lowest transmission rate.

The IFS values depending on the underlying PHYs are summarized in Table 13.9. Note that for the 802.11g, two different values are defined for SlotTime, namely, 9 and 20 μs, and hence, two sets of IFS values are defined accordingly.

13.2.3 Virtual Carrier Sensing

The CSMA/CA of the 802.11 DCF heavily depends on the carrier sensing functionality. There are two types of methods to determine whether the channel is idle or busy: (1) physical carrier sense, and (2) virtual carrier sense. The physical carrier sense is the *clear channel assessment* (CCA) mechanism explained in Section 12.1. Remember that the CCA mechanism depends on the underlying PHY and its implementation.

The virtual carrier sense works as follows. In the header of each MAC frame, there is a Duration/ID (or simply duration) field, which indicates the period (in µsec) of subsequent frame transmission(s). Once a station successfully receives and decodes a frame, it sets a counter, called *network allocation vector* (NAV), to the value found in the duration field, at the end of the frame reception, if its NAV counter value is smaller than the duration value. The NAV counter value decreases every μsec regardless of the channel status. The MAC considers the channel busy as long as the NAV has a nonzero value irrespective of the CCA indication from the PHY, and, hence, it is called *virtual carrier sense*. As will be discussed in Section 13.2.5, the virtual carrier sensing is meant for handling hidden stations. Only when both physical and virtual carrier sensing mechanisms declare an idle channel, is the channel determined to be idle.

13.2.4 Recovery Via ARQ

As explained earlier, the receiver of a directed (i.e., unicast) frame responds with an ACK frame after an SIFS interval from the unicast frame reception. After transmitting a directed frame, the transmitter waits for *ACK timeout* time, which is defined to be SIFS + RX_Start_Delay + SlotTime, where RX_Start_Delay represents the delay from the start of the preamble to the issuance of a frame reception start indication by the PHY. If a frame reception does not start during ACK timeout, the transmitter concludes that the previously transmitted directed frame was not delivered to the receiver correctly. Otherwise, the transmitter continues to receive an

Table 13.9 IFS Values for Various PHYs

	SlotTime	SIFS	PIFS = SIFS + SlotTime	DIFS = SIFS + 2 x SlotTime
802.11a	9 µsec	16 µsec	25 µsec	34 µsec
802.11b	20 µsec	10 µsec	35 µsec	50 µsec
802.11g	9 µsec	10 µsec	19 µsec	28 µsec
	20 µsec	10 µsec	30 µsec	50 µsec

incoming frame and checks if it is an ACK frame directed to itself. Only if a proper ACK frame is correctly received without an FCS error, does the transmitter conclude that the previous directed frame transmission was successful. Upon a failure of a directed frame transmission, a retransmission is scheduled with an increased contention window size and the *retry* bit in the MAC header of the scheduled frame set to one.

Note that an ACK frame is transmitted upon a successful reception of a directed frame irrespective of the channel status assessed by the ACK-transmitting station. Note that the channel could be busy due to the virtual carrier sensing. That is, a station might receive a directed frame correctly while its NAV value is nonzero. One might think that this is not a desirable behavior since nonzero NAV means the busy channel status for this station, and a station is not supposed to transmit a frame when the channel is busy. However, if this station does not transmit an ACK frame, even if it received a data frame successfully, the successful transmission was in vain, and the data frame has to be retransmitted. However, if an ACK frame is transmitted, the ACK frame might reach the data transmitted station successfully, and, hence, no retransmission of the data frame will be needed.

Depending on the platform, meeting the SIFS operation for an ACK transmission could be quite challenging. In typical implementations, a receiver station starts preparing an ACK after receiving a frame up to Address 1 if the address field matches with its own address. If the FCS test turns out to be successful, the ACK frame is transmitted as scheduled. Otherwise, the transmission is cancelled. This is a way to meet the time-critical SIFS operation for ACK transmissions. A detailed operation is found at [9].

Upon a failure of an ACK reception, the pending data frame should be retransmitted after another backoff with an updated CW value. The number of retransmissions is limited by *short retry limit* (ShortRetryLimit) or *long retry limit* (LongRetryLimit) depending on the length of the frame. If the frame is longer than a threshold, called *RTS threshold* (RTSThreshold), the frame is considered a long frame, and, hence, the LongRetryLimit is used while the ShortRetryLimit is used otherwise. The RTSThreshold is used to control the usage of the RTS frame, and its operation will be further detailed in Section 13.2.5. The values of ShortRetryLimit and LongRetryLimit are configurable as they are defined as *management information bases* (MIBs), while their default values are 7 and 4, respectively. If a data frame is not successfully transmitted after as many retransmissions as the corresponding retry limit times, the frame is discarded at the transmitter without further transmission attempts.

13.2.5 RTS/CTS

In the WLAN environments, there might be hidden stations. Two stations are hidden from each other when they cannot see each other (i.e., one cannot sense the other's transmission). In Figure 13.16, Stations 1 and 2 are hidden each other when the carrier sensing of a station's transmission is possible only within the circle with the station at the center. Since the DCF operates based on carrier sensing, the existence of such hidden stations might degrade the network performance severely. For example, in Figure 13.16, when both stations 1 and 2 would like to transmit frames

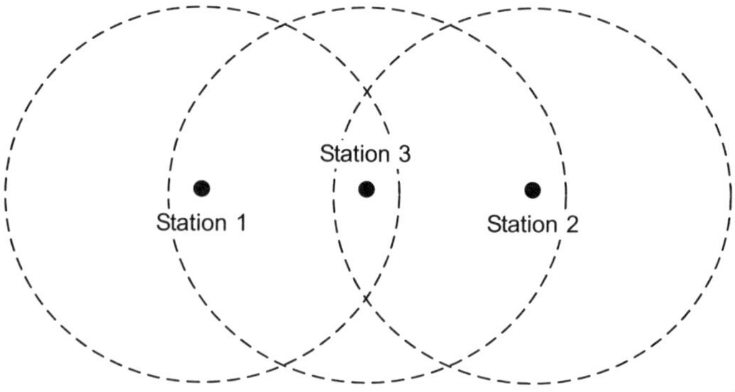

Figure 13.16 Hidden station problem.

to station 3, station 1 might initiate its transmission while station 2 is transmitting a frame to station 3 so that a collision could occur at station 3. As stations 1 and 2 are hidden from each other, such collisions might occur often, thus degrading the network performance severely.

In order to reduce the hidden station problem, the 802.11 defines a Request-to-send/Clear-to-send (RTS/CTS) mechanism. That is, if the transmitter opts to use the RTS/CTS mechanisms, before transmitting a data frame, the station transmits an RTS frame, and then it is followed by a CTS frame transmitted by the receiver. The duration/ID field of the MAC header in RTS and CTS frames specifies how long it does take to transmit the subsequent data frame and the corresponding ACK response. Therefore, other stations hearing the transmitter and hidden stations close to the receiver will not start any transmissions; their timer called *network allocation vector* (NAV) is set, and as long as the NAV value is nonzero (i.e., a busy channel due to virtual carrier sensing), a station does not contend for the channel. Between two consecutive frames in the sequence of RTS, CTS, data, and ACK frames, a SIFS is used. Figure 13.17 illustrates the timing diagram involved with an RTS/CTS frame exchange. In the figure, the duration/ID field in the RTS frame includes the total time (in μs) corresponding to SIFS + CTS_Transmission_Time + SIFS + DATA_Transmission_Time + SIFS + ACK_Transmission_Time, and the stations receiving the RTS frame set their NAV with the value in the duration/ID field at the end of the RTS reception.

Whether to transmit an RTS before a data transmission is determined by RTSThreshold, which is a configurable MIB value. If the pending MPDU length is larger than the threshold, the RTS is transmitted, and vice versa. Normally, the

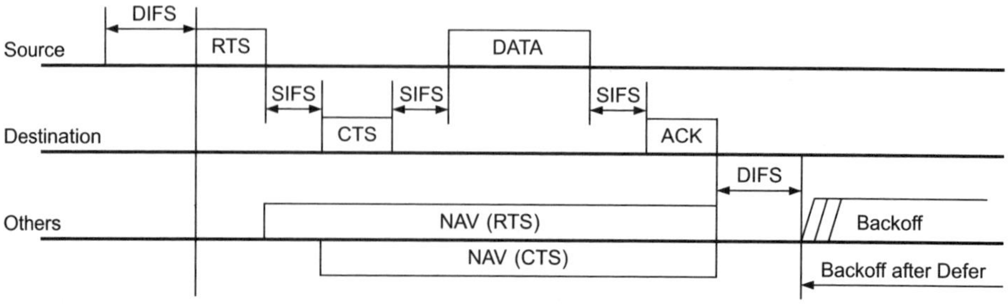

Figure 13.17 RTS/CTS frame exchange. (*After:* [2].)

threshold value is set to a value larger than the maximum MPDU size (e.g., 3,000), so that the RTS/CTS exchange is not used.

Note that a CTS frame is transmitted upon the successful reception of an RTS frame, only if the receiver has zero NAV value. This is different from the rule for the ACK transmission, which is transmitted irrespective of the NAV value. After an RTS frame is transmitted, there might not be the corresponding CTS frame transmission due to the RTS transmission failure or a nonzero NAV at the receiver. The transmitter concludes that its RTS transmission has failed if a frame reception does not start during CTS Timeout, which is defined to be SIFS + RX_Start_Delay + SlotTime. Otherwise, the transmitter continues to receive an incoming frame and checks if it is a CTS frame directed to itself. Note that the CTS frame reception procedure including the CTS timeout is basically the same as that for the ACK reception.

Upon the lack of a successful CTS reception, the transmitter of the RTS cannot initiate its data frame transmission. In this case, a problem occurs to the stations, which set their NAV values upon the reception of the preceding RTS frame since they cannot access the channel due to the virtual carrier sensing while there is no ongoing transmissions. In order to resolve this problem, a station is allowed to reset its NAV upon the failure of an RTS/CTS exchange. That is, if no frame reception starts during an interval of 2 × SIFS + CTSTxTime + RX_Start_Delay + 2 × SlotTime, where CTSTxTime represents the time needed to transmit a CTS frame, the station concludes that the RTS/CTS exchange failed, and, hence, resets its NAV value.

The reason why the usage of an RTS/CTS exchange is determined according to the size of the pending data frame is due to the fact that RTS/CTS exchange consumes the precious wireless bandwidth. For short data frames, or, more exactly speaking, for data frames with short transmission time, the collision probability and also the bandwidth waste due to collisions are relatively small so that not using RTS/CTS exchanges might be a better option. Note that the frame transmission time is actually determined by both the frame length and the employed transmission rate. Accordingly, it seems to be more reasonable to define an RTSThreshold in terms of the frame transmission time, not in terms of the frame length [10].

It should be also noted that RTS/CTS frame exchange might be useful even if there do not exist any hidden stations. Since the RTS frame is a short control frame, when two or more RTS frames collide, the waste of the bandwidth could be smaller than that due to collision of multiple data frames so that RTS/CTS frame exchanges can be used in order to enhance the throughput performance of a WLAN [4]. On the other hand, it has been also reported that in high-speed WLANs, such as IEEE 802.11a, the RTS/CTS exchange might not be very useful because of relatively short data frame transmission times [10].

13.2.6 Fragmentation

A unicast MSDU[1] might be fragmented into multiple MPDUs as illustrated in Figure 13.18. Group-addressed frames cannot be fragmented. Fragmentation creates

1. While an MMPDU can be fragmented according to the same rule described in this section, we only consider the fragmentation of MSDUs for simplicity.

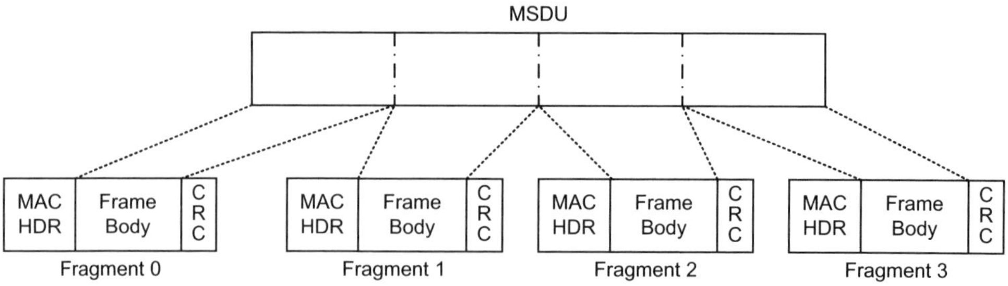

Figure 13.18 Fragmentation of an MSDU into multiple MPDUs (or fragments). (*After:* [2].)

MPDUs smaller than the original frame length in order to increase reliability. Note that for the given channel and transmission rate, the smaller the frame length, the more reliable the frame transmission could be. The process of recombining MPDUs into a single MSDU is defined as defragmentation. Both fragmentation and defragmentation are conducted at each immediate transmitter and receiver station, respectively.

If an MSDU would result in a length greater than the *fragmentation threshold* (FragmentationThreshold) when the MAC header and FCS are added, the MSDU is fragmented. The value of FragmentationThreshold can be configured, as it is a MIB value, while the default value is set to a very large value (e.g., 3,000), so that fragmentation will not occur. The length of the fragments except for the last one is the same as determined by the FragmentationThreshold, while the very last one might have a shorter length. All the fragments have virtually the same MAC header except for the duration/ID field and the fragment number in the sequence control field of the MAC header. The minimum FragmentationThreshold is 256 bytes, and the maximum MSDU size is 2,304. Accordingly, the maximum number of fragments out of a single MSDU is 11 for the baseline MAC. The maximum number becomes 12 for the IEEE 802.11-2007 MAC including the 802.11e and 802.11i due to additionally defined fields.

The fragments out of a single MSDU are transmitted back to back as shown in Figure 13.19. That is, upon the reception of an ACK corresponding to the first fragment, the second fragment is transmitted after a SIFS interval. This back-to-back transmission is referred to as a *fragmentation burst*. A fragmentation burst continues until an expected ACK is not received, upon which a retransmission of the failed fragment is attempted after a backoff with an updated CW value.

In Figure 13.19, we observe that an RTS/CTS exchange precedes the first fragment transmission. This can occur when the size of the fragment is larger than

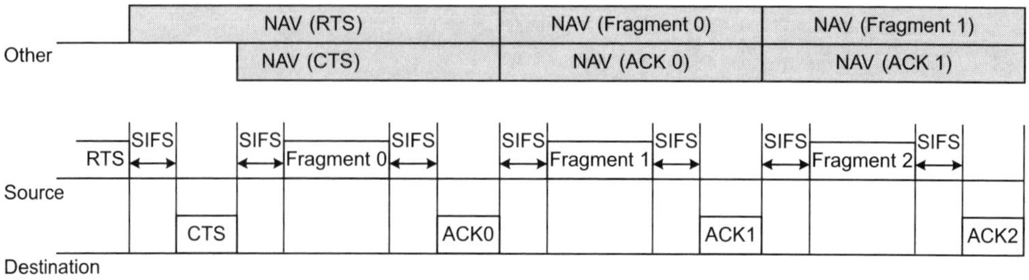

Figure 13.19 NAV setting with fragment burst. (*After:* [2].)

RTSThreshold. It should be noted that the duration/ID field in the frames within a fragmentation burst conveys the duration that can cover up to the end of the subsequent frame exchange. For example, the duration/ID field of fragment 0 in the figure protects up to the transmissions of fragment 1 and its corresponding ACK. This rule is similar to that of RTS/CTS, which protects the subsequent frame exchange (i.e., a data frame and an ACK).

13.2.7 Throughput Performance

We briefly discuss the throughput performance of the 802.11 DCF to better understand the characteristics of the DCF. We make the following assumptions:

- IEEE 802.11a BSS has the BSS basic rate set of {6, 12, 24 (Mbps)}. (The BSS basic rate set is discussed in Section 13.4.2, and it determines the transmission rate of the ACK frames.)
- Transmitter stations have an infinite number of MSDUs.
- The length of the MSDUs is fixed.
- There is no channel error.

We first evaluate the throughput performance of a BSS with a single transmitter. Since there is neither contention nor channel error, the frame transmission is always successful. We derive the throughput performance using the analysis in [11, 12]. Figure 13.20 shows the throughput as the MSDU length increases. Apparently, the higher the transmission rate is, the higher the throughput is achieved. Moreover, the longer the MSDU length is, the larger the throughput is achieved. This is due to the

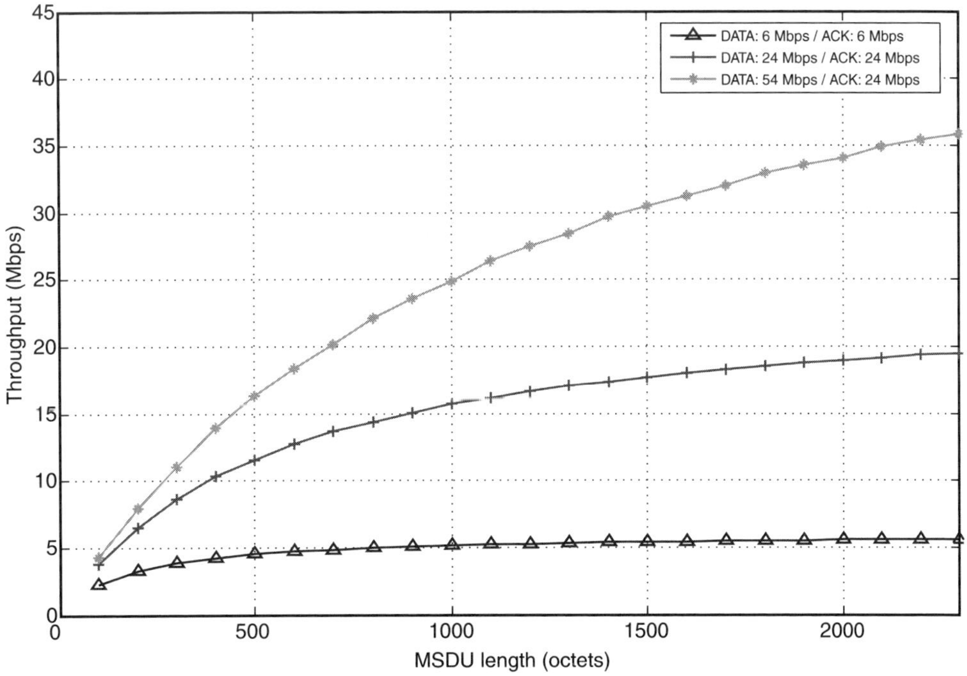

Figure 13.20 Throughput versus MSDU length.

fact that the protocol overhead (e.g., PLCP preamble/header, IFSs, backoff, and ACK frames) is basically fixed irrespective of the MSDU length, and the overhead becomes relatively smaller as the MSDU length increases. The related discussion is further made in [13].

We now consider a BSS with an increasing number of stations, where the MSDU length is fixed at 1,508 octets. As discussed in Section 13.1.1, this is practically the maximum MSDU length. Figure 13.21 shows the aggregate throughput (i.e., the sum of all individual stations' throughput values) as the number of stations increases. We derive the throughput performance using the analysis in [4, 14]. We observe that as the number of stations increases, the aggregate throughput basically decreases, and it becomes zero eventually. Note that we need about 1,000 stations, which is an unrealistic number, to make the throughput zero.

Another notable observation is the fact that the throughput for the transmission rate of 54 Mbps slightly increases when the number of stations increases from 1 to 2 before starting to decrease beginning the case with 3 stations. This is because the throughput loss due to the backoff is reduced when there are two stations compared with a single station case; unless there is collision, the average backoff duration between two consecutive frames on the channel is halved. On the other hand, the collision probability increases as the number of stations increases. The effect of the reduced backoff duration overhead is larger than that that due to increased collisions for 54 Mbps, but it is not the case for 24 and 6 Mbps for which the throughput monotonically decreases as the number of stations increases. This is due to the fact that the frame transmission times as well as the collision times are longer for these two rates, so that the positive impact to the throughput performance due to the reduced backoff duration overhead is relatively small and the loss due to increased collisions is more influential.

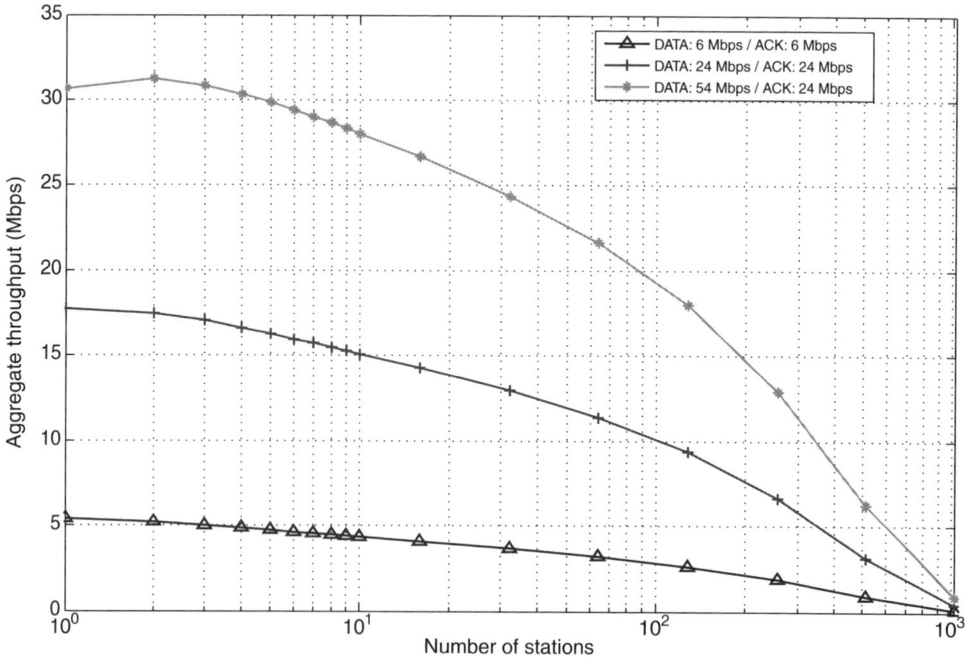

Figure 13.21 Aggregate throughput versus the number of stations.

13.3 Point Coordination Function (PCF)

The PCF, which is optionally defined, is a poll-and-response MAC for nearly isochronous service (i.e., the service for semi-real-time applications). However, as discussed in Section 14.1.1, this protocol has too many problems to be useful for the real-time applications requiring QoS. Due to such problems as well as the difficulty for the implementation, the PCF was rarely implemented, and accordingly virtually no products employing the PCF appeared in the market. In this section, we present how the PCF works briefly since the 802.11e MAC was developed by enhancing the PCF as well as the DCF.

The PCF can be used only in an infrastructure BSS, and the *point coordinator* (PC) within the AP in a BSS serves as a polling master for the stations in the BSS. The PC can be understood as a functional entity residing within the AP. For the rest of this section, the term AP should be understood as PC. As shown in Figure 13.1, the PCF sits on top of the DCF. That is, the PCF operation relies on that of the DCF, and, hence, the PCF cannot exist if the DCF does not exist. This is because: (1) the PCF channel access should be protected by the virtual carrier sensing of the DCF, and (2) a station has to be associated with an AP first in order to get a PCF service, while the association procedure relies on the DCF channel access. The support of the PCF in a BSS is signaled by the AP via beacons. A station desiring to get the PCF service requests it as part of the association procedure.

13.3.1 CFP Structure and Timing

When the PCF is used in a BSS, the time axis is divided into *superframes* or *CFP repetition intervals* (CFPRIs). Each superframe is composed of a *contention-free period* (CFP) and *contention period* (CP), as shown in Figure 13.22, where the PCF is used during a CFP and the DCF is used during a CP, respectively. A superframe is composed of a number of beacon intervals, where beacons are transmitted at every beacon interval. A superframe starts with a beacon frame. The AP generates beacon frames at regular beacon frame intervals, and, hence, every station knows when the next beacon frame is about to arrive; this instance is called *target beacon transition time* (TBTT). Note that TBTTs are periodically scheduled over time. As beacons are transmitted via contention, a beacon transmission might be delayed a bit from the scheduled time (i.e., a TBTT).

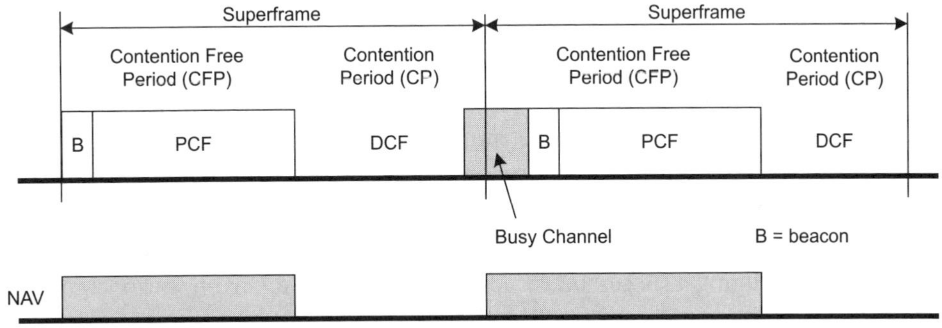

Figure 13.22 Time sharing between CFP and CP.

The PCF has higher priority than the DCF, because the CFP during which the PCF is used is protected from the DCF contention via the NAV set, as shown in Figure 13.22. For this purpose, all the stations set their NAV with the *maximum CFP duration* (CFP_Max_Duration), which is determined and announced via beacon frames by the AP, at each TBTT at which a CFP is scheduled to start. As the name stands for, the CFP_Max_Duration specifies the maximum CFP duration in a BSS. For each superframe, the actual CFP duration is dynamically determined, and hence might be shorter than CFP_Max_Duration. It is mandatory that a superframe includes a CP of the minimum length that allows at least one MSDU delivery with the maximum MPDU size with security encryption expanded, including the corresponding response frame exchange and IFSs, under the DCF at the lowest transmission rate. Accordingly, the value of the superframe length minus CFP_Max_Duration should be larger than or equal to the minimum length.

The starting time of a CFP might be delayed from a TBTT if the channel is busy at the TBTT, as shown in Figure 13.22. At a TBTT at which a CFP is scheduled to start, a beacon is transmitted after a PIFS idle time without a backoff procedure. Note that a beacon is otherwise transmitted via the DCF contention, as any other types of frames are. Since all non-AP stations set their NAV value at the TBTT, there will be no contention from other stations. Even if there somehow are contentions from other stations, the winner of the contention should be the AP, thanks to the PIFS channel access, so that the first frame transmitted after the TBTT will be the beacon from the AP. Note that the PIFS deference guarantees higher channel access priority over other DCF-based contention using the DIFS deference. When the starting time of a CFP is delayed, the maximum CFP duration is foreshortened as the maximum CFP duration is bounded by the NAV value, which is set according to the CFP_Max_Duration at the TBTT.

13.3.2 Basic Access Procedure

During a CFP, there is no contention among stations; instead, stations are polled. See Figure 13.23 for typical frame exchange sequences during a CFP. The AP polls a station asking for the transmission of a pending frame. Upon being polled, the polled station transmits a single data frame after a SIFS interval from the polling frame reception. In fact, the default interframe space used within a CFP is SIFS.

If the AP itself has pending data for this station, it uses a combined data and poll frame (i.e., Data + CF-Poll) by piggybacking the CF-Poll frame into the data frame. In such a case, the polled station acknowledges the reception of the downlink data frame and transmits its pending data frame simultaneously by transmitting a Data + CF-Ack frame. This frame exchange is illustrated in Figure 13.23 as the exchange of D1 + CF-Poll and U1 + CF-Ack, where Dx and Ux stand for the downlink (i.e., AP-to-station) and uplink (i.e., station-to-AP) data transmitted to and received from station x, respectively. As shown in the figure, a downlink data can be piggybacked with both CF-Poll and CF-Ack (i.e., Data + CF-Ack + CF-Poll frame). The frame labeled as "D2 + CF-Ack + CF-Poll" conveys data destined to station 2 while acknowledging a successful reception of U1 + CF-Ack from station 1 and polling station 2.

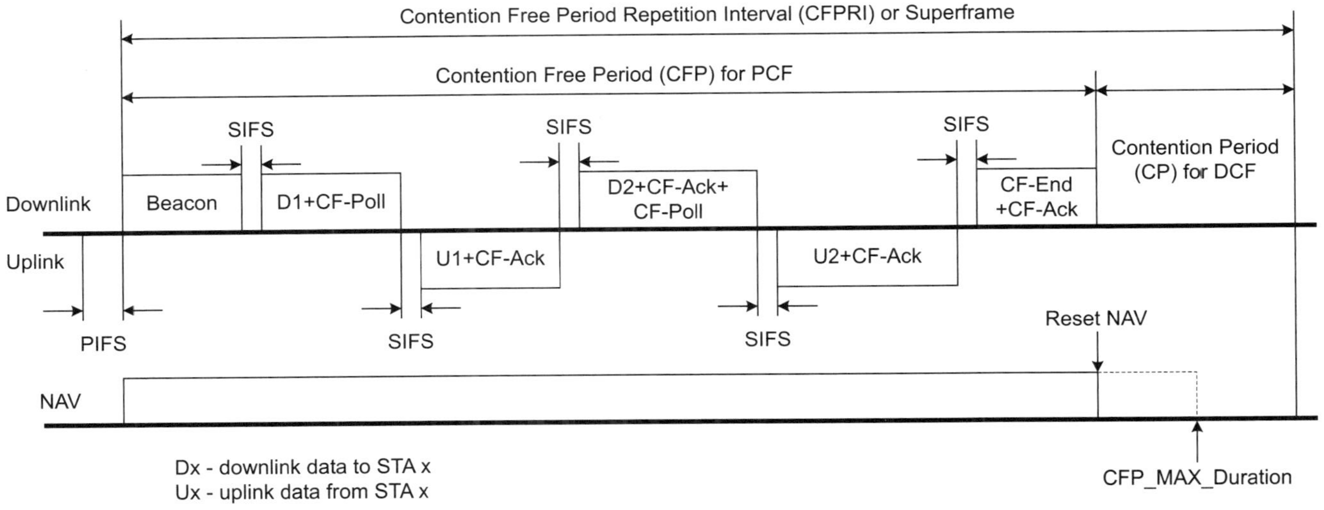

Figure 13.23 IEEE 802.11 PCF channel access during a CFP. (*After:* [2].)

When a polled station does not have any pending data, it responds with a null subtype data frame (e.g., Null or CF-Ack). On the other hand, a polling frame might be successfully received by the polled station, and in such a case, there will not be any frame transmission after the polling frame transmission. If the AP receives no response from a polled station during a PIFS interval, it polls the next station. Therefore, no idle period longer than PIFS occurs within CFP.

At a given moment, whether to transmit a downlink frame, or to poll a station, or to poll and send to a station is an implementation-dependent scheduling algorithm problem. The AP might want to end the CFP before the scheduled CFP ending time, determined by the CFP_Max_Duration value. The last frame in a CFP is either CF-End or CF-End + CF-Ack. If the AP wants to end a CFP after receiving an uplink data frame from a station, a CF-End + CF-Ack frame is transmitted instead of CF-End. Upon reception of a CF-End or a CF-End + CF-Ack, all the stations in the BSS reset their NAV values, thus resuming the DCF contention. Note that the reset of the NAV basically implies the start of a CP.

13.4 Other MAC Operations

13.4.1 Unicast Versus Multicast Versus Broadcast

A group-addressed (i.e., broadcast and multicast) frame has multiple receivers and hence cannot be acknowledged. The transmitter of a group-addressed frame always assumes that the frame transmission is successful, and hence the CW value is reset to CWmin after a group-addressed frame transmission. As group-addressed frames are never retransmitted, the broadcast/multicast transmission service of the 802.11 MAC is unreliable by definition.

In case of infrastructure BSS, non-AP stations are basically not allowed to transmit group-addressed frames. When a non-AP station likes to broadcast/multicast a frame, it first transmits the frame to its AP in a unicast manner (so with possible retransmissions). The frame from the station to the AP has the following address field values: Address 1 = BSSID, Address 2 = the station's address, and finally Address 3 = a group address. Then, the AP in turn broadcasts or multicasts the frame to both its associated stations (via downlink transmission) and the DS. Accordingly, one can assume that a broadcast/multicast frame first reaches the AP quite reliably and then is forwarded by the AP less reliably. There have been research efforts to develop reliable multicast transmissions in the 802.11 WLANs [15, 16].

The only exceptional case when a non-AP station is allowed to transmit a group-addressed frame is the probe request frame transmission. As probe request frames are transmitted by a station searching neighboring BSSs, often without being associated with an AP (as further detailed in Sections 13.5.3 and 16.1.1), these frames are broadcast over the channel.

13.4.2 Multirate Support

As presented in Chapter 12, the 802.11 PHYs support multiple transmission rates, and these rates can be used in an adaptive manner in order to maximize the network performance depending on the underlying channel condition. The algorithm for the rate adaptation is implementation-dependent, but in order to ensure coexistence and interoperability on multiple rate–capable PHYs, the protocol defines a set of rules to be followed by all stations.

BSS Basic and Operational Rate Sets

First of all, a couple of rate sets are defined, namely, *BSS basic rate set* and *operational rate set* for a BSS. Both of the rate sets are indicated in the *supported rate* field of the beacon and probe response frames. The BSS basic rate set defines a set of rates that must be supported by all the stations in a BSS. On the other hand, the operational rate set specifies a set of rates, which can be used by stations in a BSS. For example, in the case of IEEE 802.11g WLAN, the BSS basic rate set could be {1, 2, 5.5, 11}, while the operational rate set could be {1, 2, 5.5, 11, 6, 9, 12, 18, 24, 36, 48, 54}, respectively. Since the basic rate set includes only the 802.11b rates while the operational rate set includes all the 802.11g rates (including the 802.11b rates), the BSS with these two specific rate sets allows the 802.11b stations to get associated. Another set, called the *supported rate set*, is defined for each station, and this specifies a set of rates that are supported by the given station. In the previous example of an 802.11g WLAN, both 802.11b and 802.11g stations can get associated with the AP, and their supported rate sets are {1, 2, 5.5, 11} and {1, 2, 5.5, 11, 6, 9, 12, 18, 24, 36, 48, 54}, respectively. The supported rate set of a station is known the AP during the association procedure. The supported rate set of an AP should be the same as the operational rate set of the BSS.

A set of transmission rate–specific rules are defined based on these rate sets. Control frames (including ACK, RTS, and CTS) and group-addressed (i.e., broadcast and multicast) data and management frames (e.g., beacon) are transmitted with one of the rates in the BSS basic rate set, so that they will be understood by all the stations in the BSS. On the other hand, directed (i.e., unicast) data and management frames can be transmitted at any rate in the operational rate set. However, the rate should be determined by considering the supported rates of the receiver. For data frames of Data + CF-Ack, Data + CF-Poll + CF-Ack, and CF-Poll + CF-Ack, the rate chosen to transmit the frame must be supported by both the addressed receiver and the station to which the CF-Ack is intended.

The transmission rate of a unicast frame can be determined in order to maximize the network performance (e.g., the network throughput), as discussed further later. On the other hand, the transmission rate of the beacon frames can be determined in order to control the size of the *basic service area* (BSA) (i.e., the geographical coverage of the BSS). Note that all the stations in a BSS should be able to receive beacons from its AP. The higher the beacon transmission rate is, the smaller the BSA will be.

To allow the transmitter to calculate the value of the duration/ID field, the receiver transmits its control response (e.g., CTS and ACK) at the highest rate in the BSS basic rate set that is less than or equal to the rate of the immediately previous frame in the frame exchange sequence. For example, in the case of IEEE 802.11a WLAN with the BSS basic rate set of {6, 12, 24} and the operational rate set of {6, 9, 12, 18, 24, 36, 48, 54}, if a data frame is transmitted at 6, 9, 12, 18, 24, 36, 48, 54 Mbps, then the ACK for this data frame will be transmitted at 6, 6, 12, 12, 24, 24, 24, 24 Mbps, respectively. In addition, the control response frame should be transmitted using the same PHY option as the received frame. As RTS frames must be transmitted at one of the rates in the BSS basic rate set, both RTS and CTS frames are always transmitted at the same rate, and this rate must belong to the BSS basic rate set. However, for the ACK frames, the situation is different. Since the data frames can be transmitted at any operational rate, which might not belong to the BSS basic rate set, the ACK frames could be transmitted at any rate in the BSS basic rate set, not necessarily at the same rate as the previous data frame.

Rate Adaptation

Multiple transmission rates should be exploited in an adaptive manner depending on the underlying channel condition in order to maximize the system performance. The 802.11 MAC can determine and command the PHY on which transmission rate to use for a particular frame transmission. We here consider the rate adaptation problem with the rule presented earlier. The rate adaptation is often referred to as *link adaptation* in the literature.

In general, a transmitter can change its transmission rate with or without feedback from the receiver, where the feedback information could be either SINR or the desired transmission rate determined by the receiver. Depending on whether to use the feedback from the receiver, rate adaptation schemes can be classified into two categories: *closed-loop* and *open-loop* approaches. In the baseline 802.11 protocol, there is no means for the receiver to send the feedback. Accordingly, the open-loop approach has been the only option for standard-compliant rate adaptation algorithms. In fact, the emerging IEEE 802.11n is expected to have a mechanism for a receiver to send the feedback on the desired transmission rate, as discussed in Section 18.1.1.

A very simple and widely implemented open-loop rate adaptation algorithm is *automatic rate fallback* (ARF), which was originally developed for Lucent Technologies' WaveLAN-II WLAN devices [17]. We briefly explain how the ARF algorithm works. It alternates the transmission rates by keeping track of a timing function as well as missing ACK frames. If two consecutive ACKs are not received correctly by the transmitter, the second retry of the data frame and the subsequent transmissions are made at a lower transmission rate and a timer is started. When either the timer expires or the number of successfully received ACKs reaches 10, the transmission rate is raised to the next higher transmission rate and the timer is cancelled. However, if an ACK is not received for the very next data frame, the transmission rate is lowered again and the timer is restarted.

Apparently, ARF has a purely heuristic and conservative nature, and, hence, it cannot react quickly when the wireless channel condition fluctuates. In other words, the transmitter station may attempt to increase its transmission rate to probe the wireless channel condition upon consecutive successful ACK receptions and decrease its rate upon consecutive (re)transmission failures without any consideration of the actual cause of the transmission failures (i.e., channel errors or frame collisions). However, thanks to its simplicity, ARF is still widely employed in commercial 802.11 WLAN devices, and many proposed open-loop rate adaptation schemes (e.g., [18–20]) are rooted in ARF.

13.5 MAC Management

There are basically four different MAC management functions: (1) synchronization; (2) power management; (3) association and reassociation; and (4) *management information base* (MIB) support. These management functions are controlled by *MAC layer management entity* (MLME) at the MAC.

13.5.1 Time Synchronization

All the stations within a single BSS are synchronized to a common clock. Each station maintains a local timer, called the *timing synchronization function* (TSF) timer, with modulus 264 counting in increments of μsec. The accuracy of the TSF timer is supposed to be no worse than ±0.01 percent. Since the accuracies of the TSF timers are different, two TSF timers are expected to have different values at a given time, and the gap between two timer values is expected to become larger as time goes. Accordingly, the timer values should be synchronized periodically in order to maintain the gap within a desired bound. The synchronization of TSF timers in a BSS is basically achieved via beacon frames. Beacons are transmitted every beacon interval, where the unit of the beacon interval is a *time unit* (TU), which is 1,024 μs. A beacon interval value, which is used in typical WLANs, is 100 TUs or 102.4 ms.

Beacon Transmissions in Infrastructure BSS

In the infrastructure BSS, the AP, which serves as the timing master, periodically transmits beacon frames. *Target beacon transmission times* (TBTTs) appear periodically, and at every TBTT, the AP generates a beacon frame and places the beacon at the head of the MAC queue so that the beacon frame becomes the next frame to transmit. Even if the TBTT instances are periodic, the actual beacon transmission times might not be periodic, since beacons are transmitted via contention. Normally, beacons are transmitted using the normal DCF contention with an exception that the beacon transmission after a PIFS defers to TBTTs, at which a CFP is scheduled to start as discussed in Section 13.3.1. Periodic beacon transmissions are illustrated in Figure 13.24.

A beacon frame includes a timestamp, which is obtained from the AP's local TSF timer value. The timestamp value is set to the value of the station's TSF timer at the time that the data symbol containing the first bit of the timestamp appears at the wireless channel. This can be obtained by considering the delay needed to transmit a

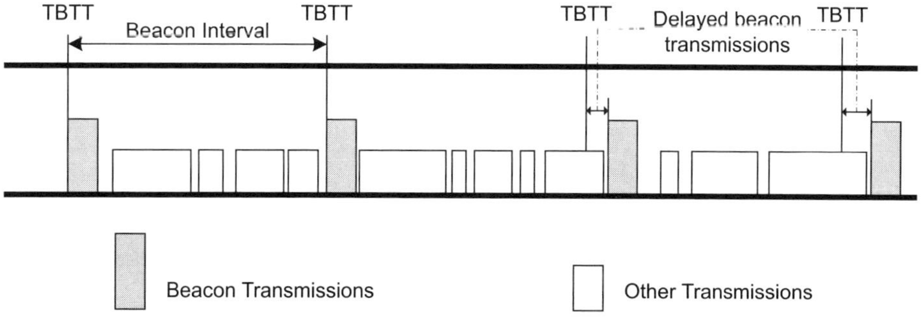

Figure 13.24 Periodic beacon transmissions by the AP in an infrastructure BSS. (*After:* [2].)

frame at the PHY. Upon the reception of a beacon, a station updates its local TSF timer value by using the timestamp in the received beacon. It first adjusts the received timestamp value by adding the receiving delay at the PHY (i.e., the time delay since the first bit of the timestamp was received from the wireless channel). Then, the receiving station set its local TSF timer to the adjusted timestamp value.

Beacon Transmissions in IBSS

In an IBSS, there is no single timing master since there is no AP. Instead, stations transmit beacon frames in a contentious manner. At each TBTT, every station generates a beacon frame and schedules the beacon frame as the next frame to transmit. Then, a beacon transmission is attempted via the DCF contention, as illustrated in Figure 13.25. The scheduled beacon transmission is cancelled if the station receives a beacon frame transmitted by another station in the BSS. After either transmitting or receiving a beacon frame, a station resumes its contention for other nonbeacon frames. Note that under this beacon transmission rule, there might be more than one beacon transmission within a beacon interval. For example, if two beacons are transmitted simultaneously (i.e., a beacon collision), none of the beacons might be correctly received by other stations, which did not transmit the previously collided beacons. In such a case, these stations will continue the channel access for their beacon transmission, and hence there will be more beacon transmissions within the same beacon interval. Even if a beacon does not collide, a station might not be receiving the beacon correctly due to the channel error, and this station might transmit another beacon in the same beacon interval.

Upon the reception of a beacon, a station updates its local TSF timer value with a certain condition. That is, only if the adjusted timestamp value (i.e., the received timestamp value plus the receiving delay at the PHY) is *later than* the station's TSF timer value does the receiving station update its local TSF timer with the adjusted timestamp value. As the timing master (i.e., beacon transmitter) might change every

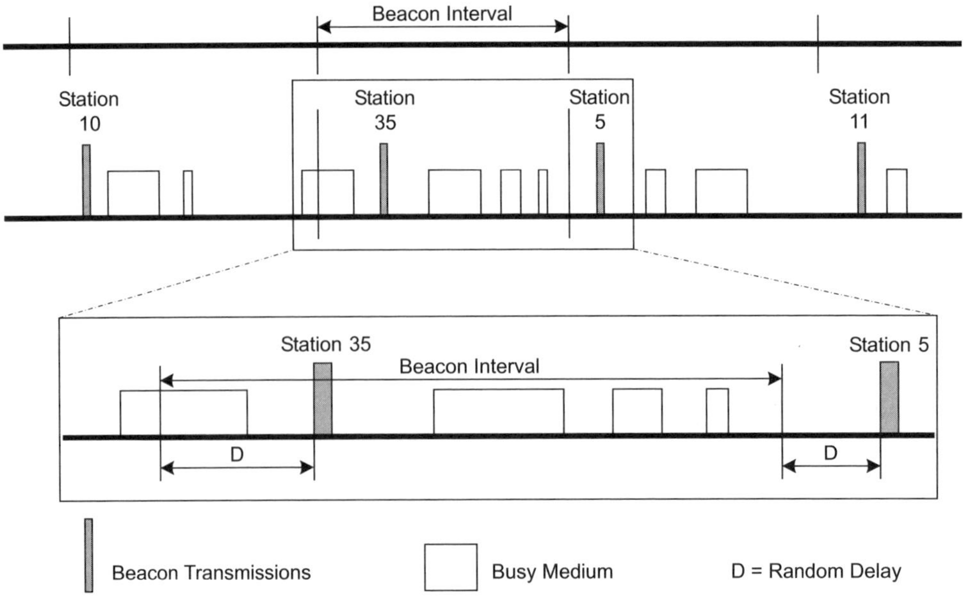

Figure 13.25 Contention-based beacon transmissions in an IBSS. (*After:* [2].)

beacon interval, and even worse, there might be more than one timing master (i.e., multiple beacon transmissions) within a given beacon interval, this condition can at least avoid the oscillation of the TSF timer speed. It basically attempts to synchronize the local TSF timers to the fastest TSF timer in the IBSS.

Needs for Synchronization

One might ask why the synchronization among stations is needed in an 802.11 WLAN at the beginning. Note that in the 802.11, the transmissions do not need to be synchronized as in typical *time-division multiple access* (TDMA) systems, where the packet transmissions should be synchronized. However, even in the 802.11, some level of synchronization is needed since there are some periodic behaviors.

First, beacons are transmitted periodically with periodic TBTTs. It is required that every station in the BSS should know when the next TBTT occurs. A superframe for the PCF operation starts at a TBTT. In fact, stations have to set their NAV at the TBTT in order to protect the upcoming CFP. Most importantly, these periodic beacon transmissions are tightly related with the power saving as discussed in Section 13.5.2. Basically, a power-saving station has to wake up periodically at TBTTs. Second, the synchronization is needed for the support of FHSS PHY, even if the FHSS PHY is not practically used any longer. This is because stations have to hop from one frequency slot to another simultaneously.

13.5.2 Power Management

Under the DCF, an 802.11 station continues to check whether or not the channel is busy irrespective of whether or not it has a frame to transmit. One often predicts that the power consumption during a frame transmission is a lot while the power consumption during a frame reception or a channel sensing is not much. In fact, the power consumption during a frame reception is quite comparable with that during a frame transmission. Even worse, the power consumption during the channel sensing is quite close to that during a frame reception. For example, practically possible values of the power consumption during transmission, reception, and channel sensing are 1.5W, 1W, and 0.9W, respectively. Many of today's 802.11 devices consume much less power though. Since an 802.11 station continues to sense the channel even if there is no pending frame transmission, the station might end up consuming significant amount energy even if there is no active traffic. Accordingly, a mechanism to let the 802.11 station sleep by turning off many of its components to minimize the energy consumption is desired.

This situation makes the power management a crucial part of the 802.11 MAC, especially, for battery-powered portable devices. The 802.11 defines two operational states of stations, namely, *doze* and *awake states*. In awake state, a station can transmit, receive, and sense the channel. It actually continues to sense the channel unless it either transmits or receives a frame. On the other hand, in doze state, a station is not able to transmit or receive or sense the channel, and, hence, consumes very little energy. The energy consumption of a station can be minimized by maximizing the time during which the station stays at the doze state. A key challenge here is that the power management should not allow a station to miss incoming frames even if it spends much time at the doze state. Note that transmission of outgoing frames is rather straightforward, since upon the arrival of a data frame from the higher layer, the station in the doze state can wake up by switching to the awake state in order to transmit the pending frame. However, in the case of incoming frames, the station in the doze state is the receiver, not the transmitter, and hence a specific mechanism is needed in order to awake the station when the transmitter would like to transmit the frames to this station.

How a station switches between these two states is determined by its power management mode—*active mode* (AM) and *power-save mode* (PSM). A station in the AM always keeps operating in the awake state, while a station in the PSM can switch back and forth between the awake and doze states depending on the traffic pattern.

Power Management in Infrastructure BSS

In the infrastructure BSS, the AP buffers all the frames addressed to a station in the PSM and announces the existence of such buffered frames via the *traffic indication map* (TIM) field in beacon frames. Stations in the PSM wake up (i.e., switch from the doze state to the awake state) periodically in order to receive beacon frames. If there is no buffered frame, the station goes back to the doze state. Otherwise, the station stays awake, and requests the delivery of its buffered frames by transmitting a special control frame, called *power save-poll* (PS-Poll). In fact, upon the reception of a PS-Poll, a single frame is transmitted to the *power-save* (PS) station. When there are multiple buffered frames at the AP, a frame from the AP indicates the existence of more buffered frames via the *more data* bit at the MAC header. Only a single frame is transmitted upon the reception of a PS-Poll frame by the AP.

The wakeup period of a PS station is configurable by the PS station. That is, the station does not need to wake up in order to receive every beacon frame. The energy consumption should be proportional to the number of beacon frame receptions. On the other hand, the frame delivery from the AP to the station could be delayed if many beacon frames are skipped. Accordingly, there is a tradeoff relationship between the energy saving and the delay performance. The longer the wakeup period is, the more energy saving could be achieved, but the longer time the frame delivery from the AP to the station could take. The wakeup period should be determined adaptively depending on the traffic pattern [21]. In fact, the maximum wakeup period should be known to the AP, since the AP needs to determine how long a frame destined to a specific PS station should be buffered. It is known to the AP by the station during the (re)association procedure using an information element, called *listen interval*, in the (re)association request frames.

The AP should keep track of the power management mode of each associated station. Note that the AP should buffer frames destined to PS stations, while it can transmit frames destined to AM stations via contention as these stations must be ready to receive any incoming frame. Accordingly, a station should inform its power management mode to the AP. The *power-management* bit at the MAC header can be used for the purpose. When a station would like to switch from a mode to another, such a plan is informed to the AP by transmitting a directed frame to the AP with the proper power management bit set. Upon the reception of the corresponding ACK frame from the AP after transmitting such a directed frame transmission, the station confirms that the AP was informed of the upcoming mode switch, and hence the actual mode switch is made. However, in some cases, a station that would like to change its power management mode might not have any pending directed frame to transmit to the AP. A null data frame (i.e., without any frame body) might be used in this case. Such a null data frame is used to inform the upcoming mode switch of the transmitting station without conveying any data.

So far, we have discussed the unicast frame transmission destined to PS stations. The support for the group-addressed frames is a bit different. That is, group-addressed frames should be buffered at the AP if at least one of the associated stations is in the PSM. Then, the buffered group-addressed frames are transmitted right after a specific beacon frame, called a *delivery traffic indication message* (DTIM) beacon, before transmitting any unicast frames. A beacon could be either a DTIM beacon or a non-DTIM beacon, where the DTIM transmission period is determined as a number of beacon intervals. For example, one out of every three beacons could be a DTIM beacon. The DTIM period should be determined in consideration of the maximum delivery delay of broadcast/multicast frames. An example of power-management operations including the DTIM beacons is illustrated in Figure 13.26.

Power Management in IBSS

In an IBSS, there is no AP, which can keep track of the power-management mode of non-AP stations. Accordingly, a different type of power-saving support mechanism is developed. In an IBSS, each station keeps track of the power-management mode of other stations. Note that in an infrastructure BSS, a non-AP station does not need to worry about the power-management mode of other stations. A frame can be transmitted to the AP, which is always awake, and then the AP forwards the

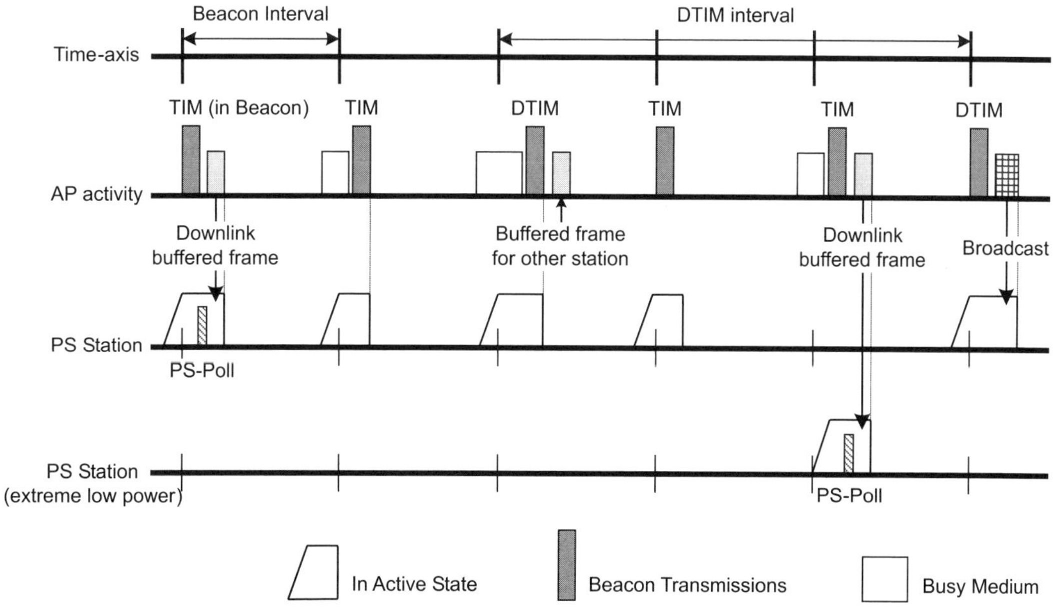

Figure 13.26 Power management operation in infrastructure BSS; trapezoids convey that the station is in the awake state. (*After:* [2].)

received frame to the destination station according to the power-management mode of the destination station.

A frame destined to non-PS stations can be transmitted via the DCF contention without being buffered. However, a frame destined to PS stations should be buffered first. A PS station in an IBSS wakes up periodically in order to transmit/receive a beacon every TBTT. Then, the station stays awake during a fixed interval, called an *announcement traffic indication message* (ATIM) window. During the ATIM window, a station that has frames destined to a PS station transmits an ATIM frame, which is a directed control frame, to the PS station. When the station has broadcast/multicast frames, a broadcast/multicast ATIM frame is transmitted during the ATIM window. Note that only beacon and ATIM frames are allowed to be transmitted during an ATIM window. Upon the successful reception of an ATIM frame during an ATIM window, the station stays awake during the rest of the beacon interval. Otherwise, the station goes back to the doze state in order to minimize the energy consumption. A station that successfully transmitted a beacon or an ATIM frame during an ATIM window also stays awake during the rest of the beacon interval in order to transmit the buffered directed frame to the PS station. The power-management operation in an IBSS is illustrated in Figure 13.27.

The ATIM window size is determined by the station initializing an IBSS, and then it is announced via beacon frames. If the ATIM window size is zero, the PSM is not supported in the IBSS. In fact, the network performance is affected by the ATIM window size. That is, if the ATIM window is too short, only few ATIM frames can be successfully transmitted during an ATIM window. Note that as ATIM frames are transmitted via the normal DCF procedures, there might be many ATIM frame collisions depending on the number of actively contending stations. On the other hand, if the ATIM window is too long, the remaining beacon interval, which can be utilized for the data transfer, could be too short to accommodate many data frame transmissions, and hence the throughput performance might be degraded.

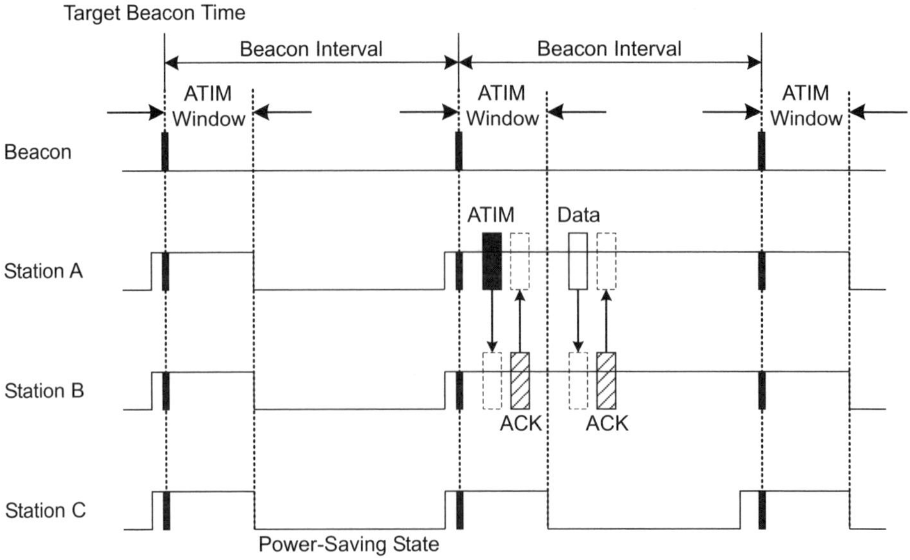

Figure 13.27 Power-management operation in IBSS. (*After:* [2].)

13.5.3 (Re)association

In an infrastructure BSS, a station has to first associate with an AP before starting any normal data transfers. Figure 13.28 illustrates the association procedure. The station first searches neighboring APs via a scanning process. There are two types of scanning—passive and active ones. Passive scanning is done by overhearing beacons transmitted by APs. On the other hand, active scanning is done by broadcasting probe request frames. Upon receiving a probe request, the AP responds with a probe response. Note that the frame format of the probe response is almost identical with that of the beacon. By receiving probe responses from APs, the station learns about these neighboring APs. Figure 13.28 assumes the active scanning.

In an IBSS, there is no AP, and hence non-AP stations have to respond to a probe request. In fact, a station that believes that it transmitted the most recent beacon frame[2] transmits a probe response upon a probe request reception. Note that such a station does not go to the doze state even after the ATIM window.

Then, the station has to get authenticated by an AP. It is allowed for a station to get authenticated by multiple APs. There are two types of authentication algorithms according to the 802.11 baseline MAC, namely, *open system* (exchanging two authentication frames) and *shared key* (exchanging four authentication frames). The open system authentication is used in Figure 13.28. Finally, a station can

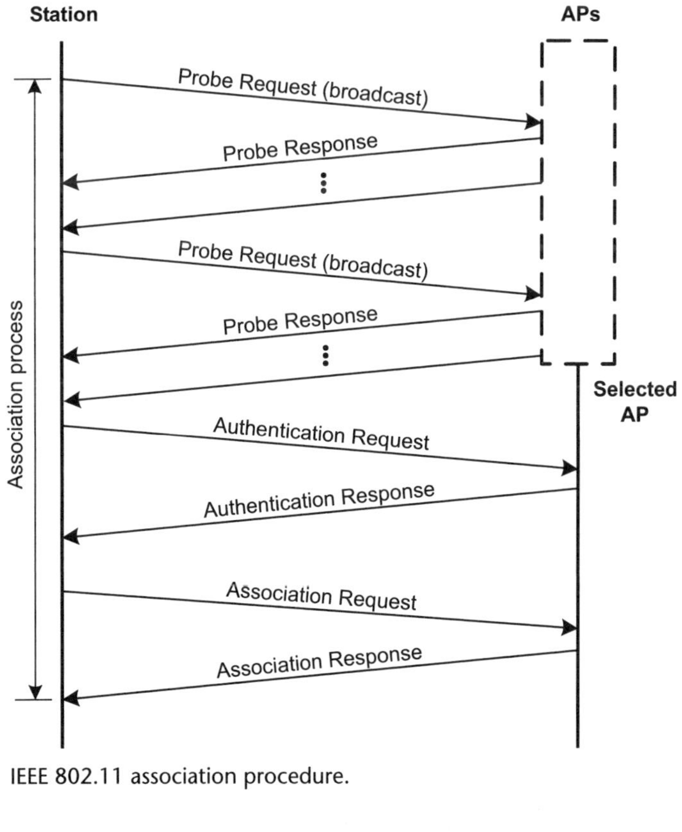

Figure 13.28 IEEE 802.11 association procedure.

2. There might be multiple such stations in a given beacon interval because of the possible beacon collisions as discussed in Section 13.5.1.

choose one of the APs with which it is authenticated in order to get associated by exchanging an association request and an association response. A station is limited to get associated with only a single AP at a given time.

When a station moves out of the coverage of its associated AP, the station performs handoff procedures by finding new AP(s) via scanning and reassociating with an AP. The detection of APs can be done via scanning processes (either passive or active scanning). The difference between the association and reassociation is basically the fact that a reassociation request frame is used instead of an association request frame in the case of the reassociation, and the reassociation request frame includes the MAC address of the current AP. The new AP can utilize the current AP's MAC address in order to communicate with the current AP for the handoff support. The details for the (re)association, mobility, and handoff support will be further discussed in Chapter 16, and the details for the authentication procedure will be presented in Chapter 15.

13.5.4 Management Information Base

The MIB comprises the managed objects, attributes, actions, and notifications required to manage a station. These MIB values can be accessed for the network management purpose by external entities, such as *simple network management protocol* (SNMP) [22]. Some MIB values are both readable and modifiable, while others are only readable. The 802.11 MAC supports various MIBs related with the station configuration and the network operational statistics. Those related with the station configuration include the current channel, supported transmission rates, and power-management mode. Those related with the MAC operational parameters include RTSThreshold, FragmentationThreshold, LongRetryLimit, and ShortRetryLimit. Various counters representing the operational statistics include FCS error count, RTS failure count, RTS success count, and transmitted frame count.

References

[1] IEEE 802.11-1999, IEEE Standard for Local and Metropolitan Area Networks—Part 11: Wireless LAN Medium Access Control (MAC) and Physical Layer (PHY) Specifications, Reference Number ISO/IEC 8802-11:1999(E), IEEE Std 802.11, 1999 edition, 1999.

[2] IEEE 802.11-2007, IEEE Standard for Local and Metropolitan Area Networks—Part 11: Wireless LAN Medium Access Control (MAC) and Physical Layer (PHY) Specifications, IEEE Std 802.11-2007, (Revision of IEEE Std 802.11-1999), June 12, 2007.

[3] IEEE 802-2001, IEEE Standard for Local and Metropolitan Area Networks: Overview and Architecture, March 2002.

[4] Bianchi, G., "Performance Analysis of the IEEE 802.11 Distributed Coordination Function," *IEEE Journal on Selected Areas Communications*, Vol. 18, No. 3, 2000, pp. 535–547.

[5] Cali, F., et al., "Dynamic Tuning of the IEEE 802.11 Protocol to Achieve a Theoretical Throughput Limit," *IEEE/ACM Trans. on Networking*, Vol. 8, No. 6, 2000, pp. 785–799.

[6] Kim, H., and J. C. Hou, "Improving Protocol Capacity with Model-Based Frame Scheduling in IEEE 802.11-Operated WLANs," *Proc. ACM 9th International Conference on Mobile Computing and Networking (MobiCom'03)*, San Diego, CA, September 14–19, 2003.

[7] Choi, S., K. Park, and C. Kim, "Performance Impact of Interlayer Dependence in Infrastructure WLANs," *IEEE Trans. on Mobile Computing*, Vol. 5, No. 7, 2006, pp. 829–845.

[8] Heusse, M., et al., "Performance Anomaly of 802.11b," *Proc. IEEE INFOCOM'03*, San Francisco, CA, March 30–April 3, 2003.

[9] Lee, I., C. E. Sundberg, and S. Choi, "A Modified Medium Access Control Algorithm for Systems with Iterative Decoding," *IEEE Trans. on Wireless Communincations*, Vol. 5, No. 2, February 2006, pp. 270–273.

[10] Tinnirello, I., S. Choi, and Y. Kim, "Revisit of RTS/CTS Exchange in High-Speed IEEE 802.11 Networks," *Proc. IEEE International Symposium on a World of Wireless, Mobile and Multimedia Networks 2005 (WoWMoM'05)*, Taormina, Italy, June 13–16, 2005.

[11] Qiao, D., S. Choi, and K. G. Shin, "Goodput Analysis and Link Adaptation for IEEE 802.11a Wireless LANs," *IEEE Trans. on Mobile Computing*, Vol. 1, No. 4, October–December 2002, pp. 278–292.

[12] Qiao, D., and S. Choi, "Goodput Enhancement of 802.11a Wireless LAN Via Link Adaptation," *Proc. IEEE International Conference on Communications (ICC'01)*, Helsinki, Finland, June 11–14, 2001.

[13] Kim, Y., et al., "Throughput Enhancement of IEEE 802.11 WLAN Via Frame Aggregation," *Proc. IEEE VTC'04-Fall*, Los Angeles, CA, September 26–29, 2004.

[14] Bianchi, G., and I. Tinnirello, "Remarks on IEEE 802.11 DCF Performance Analysis," *IEEE Comm. Letters*, Vol. 9, No., 8, August 2005, pp. 765–767.

[15] Choi, S., and K. Choi, "Reliable Multicast for Wireless LAN," in *Resource, Mobility, and Security Management in Wireless Networks and Mobile Communications*, Y. Zhang, H. Hu, and M. Fujise, (eds.), Boca Raton, FL: Auerbach Publications, 2006, pp. 113–142.

[16] Seok, Y., et al., "Leader Based Multicast," proposed to IEEE 802.11v, IEEE 802.11-07/0115r1, January 2007.

[17] Kamerman, A., and L. Monteban, "WaveLAN-II: A High-Performance Wireless LAN for the Unlicensed Band," *Bell Labs Technical Journal*, Vol. 2, No. 3, 1997, pp.118–133.

[18] Chevillat, P., et al., "A Dynamic Link Adaptation Algorithm for IEEE 802.11a Wireless LANs," *Proc. IEEE International Conference on Communications 2003 (ICC'03)*, Anchorage, AK, May 11–15, 2003.

[19] Qiao, D., and S. Choi, "Fast-Responsive Link Adaptation for IEEE 802.11 WLANs," *Proc. IEEE International Conference on Communications 2005 (ICC'05)*, Seoul, Korea, May 16–20, 2005.

[20] Kim, J., et al., "CARA: Collision-Aware Rate Adaptation for IEEE 802.11 WLANs," *Proc. IEEE INFOCOM'06*, Barcelona, Spain, April 23–29, 2006.

[21] Krashinsky, R., and H. Balakrishnan, "Minimizing Energy for Wireless Web Access with Bounded Slowdown," *Proc. ACM 8th International Conference on Mobile Computing and Networking (MobiCom'02)*, Atlanta, GA, September 23–28, 2002.

[22] IETF RFC 4789, Simple Network Management Protocol over IEEE 802 Networks, 2006.

CHAPTER 14

QoS Provisioning

IEEE 802.11-1999 was designed to allow users to experience an Ethernet-like service over wireless by virtue of supporting a best-effort service (not guaranteeing any service level to users/applications), and, hence, the 802.11 was often referred to as the *wireless Ethernet*. However, recently many multimedia applications over WLAN have emerged, including *voice over WLAN* (VoWLAN), video streaming, video conferencing, and so on, and the wireless Ethernet is not capable of supporting such multimedia applications with a proper *quality of service* (QoS) provisioning. Supporting these applications requires the differentiation of heterogeneous traffic types to meet the required QoS including delay, throughput, and so on. To satisfy such needs, IEEE 802.11e-2005 was developed by enhancing the existing 802.11 baseline MAC. The 802.11e MAC is expected to expand the 802.11 application domain by enabling such multimedia applications as voice and video services.

14.1 Introduction to IEEE 802.11e

As described in Chapter 13, the baseline MAC is composed of two coordination functions: the mandatory DCF, which is based on CSMA/CA, and the optional PCF, which is based on a poll-and-response protocol. The IEEE 802.11e MAC is defined to have a single coordination function, called *hybrid coordination function* (HCF), by combining a contention-based channel access, evolved from the DCF, and a controlled channel access, evolved from the PCF. The former is referred to as *enhanced distributed channel access* (EDCA), while the later is referred to as *HCF controlled channel access* (HCCA). Note that the EDCA is often called *enhanced DCF* (EDCF) due to its origin, and also because the name EDCF was used in earlier versions of the 802.11e draft. One might find many papers including this old name in the literature (e.g., [1, 2]). In this chapter, a station implementing IEEE 802.11e is referred to as a *QoS station*. In the same manner, a *QoS AP* represents an 802.11e-running AP.

The EDCA provides differentiated channel access to the frames with different user priorities as labeled by a higher layer in both infrastructure BSS and IBSS. With this scheme, a frame with a higher priority has a high probability to be transmitted over the air before other lower priority frames, though it is not guaranteed due to the contentious nature of the CSMA/CA. On the other hand, the HCCA can be used to provide the parameterized QoS in an infrastructure BSS. For this type of QoS support, two 802.11e QoS stations (i.e., a QoS AP and a non-AP QoS station) will set up a virtual connection, called *traffic stream* (TS), before commencing any actual QoS data transfer. As part of the traffic stream setup, the traffic characteristics and QoS requirement parameters are exchanged and negotiated. During the traffic stream runtime, the AP schedules frame transmissions by transmitting downlink frames as well as the polling frames, which grant the wireless bandwidth to non-AP QoS stations, based on the contracted QoS parameters. In the context of the HCCA, an AP is called a *hybrid coordinator* (HC). It is a similar concept as the AP being a PC when the PCF operates in the baseline MAC.

Before delving into the details of the 802.11e QoS provisioning, we first discuss the limitations of the baseline MAC per IEEE 802.11-1999 in terms of QoS support, which motivated the development of the 802.11e.

14.1.1 Limitations of Baseline MAC

There are a number of problems with the baseline MAC in terms of QoS provisioning. First of all, the 802.11 baseline MAC does not support QoS signaling and admission control. In order to get the required QoS for a specific flow, a station should be able to signal its needs to the AP. Then, the AP should determine whether or not the requested QoS can be provided. The station should be allowed to transmit or receive QoS frames in the flow only after the AP admits the requested flow. Note that if there is not enough bandwidth, QoS can never be provisioned. Moreover, each frame should carry a label, which identifies the QoS requirements of the particular frame, so that the receiver can treat the received frame accordingly. Unfortunately, none of these mechanisms is defined in the 802.11 baseline MAC.

In terms of the channel access, the 802.11 baseline MAC does not support the concept of differentiating frames with different user priorities. Basically, the DCF is supposed to provide a channel access with equal probabilities to all stations contending for the channel access in a distributed manner. However, equal access probabilities are not desirable among stations with different user priority frames. There have been efforts to provide limited QoS using the DCF by reordering frame transmissions above the MAC according to the frame priorities (e.g., [3]), but this type of approaches has a fundamental limitation coming from the DCF's characteristics. A QoS-aware MAC should be able to treat frames with different priority or QoS requirements differently.

The PCF was originally developed to support time-bounded services, which the emerging 802.11e MAC is for, but it contains many problems. We list some of them here. First, the alternating CFP and CP might introduce a lot of overhead if the superframe size becomes small. Note that in order to provide a short delay bound using the PCF, the superframe should be small. For example, in order to support voice traffic with the delay bound requirement of 10 ms using the PCF, the superframe size should be also 10 ms or so. However, it is not possible since there exists the minimum CP duration, as discussed in Section 13.3.1. Accordingly, the superframe cannot be reduced as we wish. Moreover, when the superframe size is configured small, only a very limited portion of the superframe can be used for the CFP because of the minimum CP duration. In order to handle this problem, the

802.11e HCF allows the polling-based channel access during both CPF and CP such that the superframe size can be virtually independent of the targeted delay bounds.

Second, the PCF assumes a full control over the channel during a CFP. This might be true as long as there is no neighboring AP (or PC) operating at the same channel. In reality, there may be neighboring BSSs in the same channel, and they are often referred to as *overlapping BSSs* (OBSSs). In the OBSS situation, contention-free operation during a CFP cannot be achieved properly. Because the PC assumes the full control over the channel during the CFP, especially, the PCF operation is subject to failure. The polling of 802.11e HCF during a CP is performed after a channel sensing all the time, and it can be very OBSS-friendly since it does not assume the full control over the channel.

Third, the beacon transmission or the superframe start time can vary for each superframe with the baseline MAC. At TBTT, a PC schedules the beacon as the next frame to be transmitted, and the beacon can be transmitted when the channel has been determined to be idle for at least PIFS. From the baseline 802.11 MAC, stations can start their transmissions even if the frame transmission cannot finish before the upcoming TBTT. Depending on the wireless channel at this moment of time (i.e., whether it is idle or busy around the TBTT), a delay of the beacon frame may occur. The time the beacon frame is delayed (i.e., the duration it is sent after the TBTT) delays the transmission of time-bounded frames that have to be delivered in CFP. This may severely affect the QoS, as this introduces unpredictable time delays in each CFP. An 802.11e QoS station does not transmit a frame if the frame transmission cannot be finished by the upcoming TBTT.

Finally, the channel occupancy time or the transmission time of polled stations is unpredictable with the PCF. A station that has been polled by the PC is allowed to send a single frame that may be of an arbitrary length, up to the maximum of 2,304 bytes. Depending on the underlying PHY transmission rate, the duration of a frame transmission upon being polled might be really large. For example, with the 802.11b PHY, the worst-case transmission time (i.e., for the maximum length frame transmitted at 1 Mbps) could be more than 20 ms. This may destroy any attempt to provide QoS to other stations that are polled during the rest of the CFP. As explained next, the 802.11e introduces the concept of *transmission opportunity* (TXOP), and a QoS station cannot occupy the channel longer than the corresponding TXOP limit. If a TXOP is too short to transmit even a single frame, the frame has to be fragmented (i.e., divided into multiple MPDUs).

14.2 Key Concepts

We first briefly explain some key concepts, which are introduced by the 802.11e for QoS provisioning.

14.2.1 Prioritized Versus Parameterized QoS

The 802.11e supports two different paradigms for QoS provisioning, namely, prioritized QoS and parameterized QoS. Conceptually, they are similar to *differentiated service* (DiffServ) [4] and *integrated service* (IntServ) [5], respectively, which were defined by IETF for QoS support in the IP networks.

Under the parameterized QoS paradigm, a virtual connection, called *traffic stream* (TS), is first set up between a transmitter and a receiver before commencing any QoS data frame transmissions. As part of a TS setup, a set of parameters specifying the traffic pattern of the corresponding QoS flow as well as the QoS requirements (e.g., delay and throughput) are exchanged between the AP and the station requesting a TS setup. The AP then determines whether or not the newly requested TS is acceptable to the network by considering whether there is enough residual bandwidth to support the requested QoS to this TS. Accordingly, this is basically an admission control problem. Once the AP admits a TS (i.e., if a TS is set up), the AP endeavors to support the agreed QoS to this TS. The HCCA is nicely fitted to this parameterized QoS, since the AP can schedule the polling and downlink frame transmissions in order to provide the QoS.

On the other hand, under the prioritized QoS paradigm, each frame arrives at the MAC from the higher layer along with a *user priority* (UP) value. Then, the 802.11e MAC provides differentiated channel accesses for frames with different UPs. That is, under this scheme, a higher priority frame is likely to be transmitted earlier than lower priority frames. There are eight different UPs defined, whereas only four levels of prioritized channel accesses are supported. In fact, the support of the eight UPs are rooted in IEEE 802.1D MAC bridge [6]. The EDCA is nicely fitted to the prioritized QoS paradigm, as the EDCA can support prioritized channel accesses based on UPs of frames. A TS setup is not normally needed for the EDCA for prioritized QoS support. However, a TS might need to be set up for prioritized QoS if QoS AP mandates admission control for specific priority traffic.

14.2.2 Traffic Identifier (TID)

Each MSDU arriving at the MAC from the LLC carries 1 of 16 *traffic ID* (TID) values, where the TID values ranging from 0 to 7 identify UPs and those from 8 to 15 identify *traffic stream identifiers* (TSIDs). Note that the TID is assigned to an MSDU in the layers above the MAC. Upon a TS setup, the TS is assigned a unique TSID. Having eight different TSIDs implies that there can be up to 8 TSs per QoS station per direction. That is, a single QoS station can operate up to 8 downlink TSs and 8 uplink (including direct link as discussed in Section 14.4) TSs at a given time. A term, called *traffic category* (TC), is also used to represent a label for the MSDUs of a particular UP. There is a one-to-one mapping between a TC and a UP. Each 802.11e QoS data frame then carries the TID value of the MSDU being conveyed so that the frame can be treated accordingly by the receiver.

14.2.3 Transmission Opportunity (TXOP)

A new concept, called *transmission opportunity* (TXOP), is also introduced. A TXOP is defined as an interval of time when a QoS station, which is called a TXOP holder, has the right to initiate transmissions. During a TXOP, the TXOP holder can transmit multiple frames back to back with SIFS time gaps with certain rules. If the remaining time within a TXOP is too short to transmit the pending frame, fragmentation is required to transmit the frame.

There are two ways for a station to acquire a TXOP (i.e., either via EDCA contention or via HCCA). A TXOP acquired by a successful EDCA contention is referred to as an EDCA TXOP. On the other hand, a TXOP acquired via the HCCA is called an HCCA TXOP. When the HCCA TXOP is acquired by a non-AP station via a polling from the HC (i.e., AP), it is specifically called a polled TXOP. The maximum length of a TXOP is referred to as *TXOP limit*, and the value of a TXOP limit is determined with certain rules depending on the channel access schemes. That is, the TXOP limit is determined by the AP and is announced to stations via the beacons (in case of EDCA TXOP) and the corresponding polling frame (in case of polled TXOP). The TXOP operation makes stations' transmission times more controlled and predictable from the viewpoint of the AP, since the TXOP limit values are basically determined by the AP. Moreover, the multiple consecutive frame transmissions during a TXOP can enhance the protocol efficiency. In addition, the TXOP allows the 802.11e MAC to provide the long-term (weighted) temporal fairness to the stations, as further discussed in Section 14.3.1.

Another important concept related with the TXOP is *controlled access phase* (CAP), which is defined as a time period in which the AP maintains control of the channel, after gaining the channel access by sensing the channel to be idle for a PIFS duration. It may span multiple consecutive HCCA TXOPs including polled TXOPs. In terms of the fact that the AP assumes a control over the channel, the CAP is similar to the CFP in the baseline MAC. However, different from the CFPs, CAPs do not need to appear periodically and its appearance is not bound with the beacon transmissions.

14.2.4 QoS Control Field

As shown in Figure 14.1, a QoS type data frame has a 16-bit QoS control field at the very end of the MAC header. As presented in Table 13.1 and discussed in Section 13.1.2, QoS data frames have bit b7 = 1 in the frame control field. The QoS control field identifies the TC or TS to which the frame belongs and various other QoS-related information about the frame. Each QoS control field comprises five subfields, as defined for the particular transmitter (i.e., HC or non-AP station) and frame subtype. The usage of these subfields of the QoS control field is illustrated in Table 14.1. In the table, QoS (+)CF-Poll refers to a QoS frame piggybacked with the CF-Poll as explained in Section 13.1.2; those include QoS CF-Poll (no data), QoS CF-Ack+CF-Poll (no data), QoS Data+CF-Poll, and QoS Data+CF-Ack+CF-Poll as shown in Table 13.2.

We now briefly explain each subfield in the QoS control field:

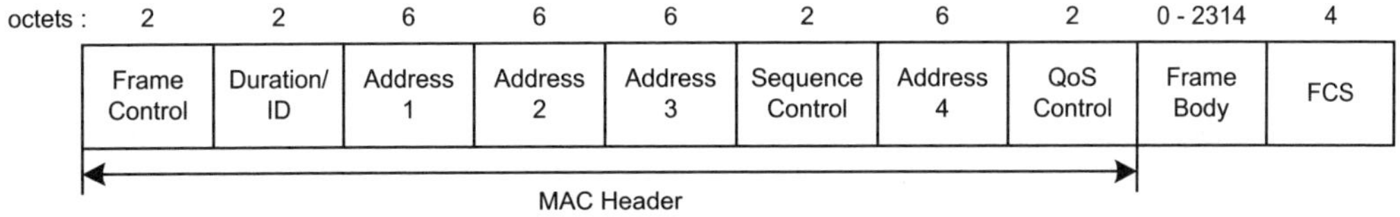

Figure 14.1 QoS data frame format. (*After:* [7].)

Table 14.1 QoS Control Field Format

Applicable Frame (sub) Types	Bits 0-3	Bit 4	Bits 5-6	Bit 7	Bits 8-15
QoS (+)CF-Poll frames sent by HC	TID	EOSP	Ack Policy	Reserved	TXOP limit
QoS Data, Qos Null, and Qos Data+CF-Ack frames sent by HC	TID	EOSP	Ack Policy	Reserved	QAP PS Buffer State
QoS data type frames sent by non-AP QoS stations	TID	0	Ack Policy	Reserved	TXOP duration requested
	TID	1	Ack Policy	Reserved	Queue size

Source: [7].

- *TID* subfield identifies the TC or TS to which the corresponding MSDU in the frame body field belongs as explained in Section 14.2.2. The TID subfield also identifies the TC or TS of traffic for which a TXOP is being requested, through the setting of the *TXOP duration requested* subfield or the *queue size* subfield. For QoS Data+CF-Poll, the TID subfield in the QoS control field indicates the TID of the data. For all QoS (+)CF-Poll frames of subtype null, the TID subfield in the QoS control field indicates the TID for which the poll is intended. However, transmitting frames of this TID value is not required, and the polled station may respond with any frame.
- *End of service period (EOSP)* subfield is used by the HC to indicate the end of the current *service period* (SP). An SP is a contiguous time during which one or more downlink unicast frames and/or one or more polled TXOPs are granted to a station. The HC sets the EOSP subfield to 1 in its transmission and retransmissions of the SP's final frame to end a scheduled/unscheduled SP and sets it to 0 otherwise. See Sections 14.4.3 and 14.5.3 for related operations.
- *Ack policy* subfield identifies the acknowledgment policy that is followed upon the delivery of the MPDU. The encoding for this subfield is found at Table 14.2. The acknowledgment policy that is for a particular frame is specified by the higher layer upon the arrival of the MSDU from the higher layer.
- *TXOP limit* subfield specifies the time limit of a polled TXOP granted by a QoS (+)CF-Poll frame from the HC. The unit for the TXOP limit is 32 μs, and the range of time values is 32 to 8,160 μs. A TXOP limit value of 0 implies that one MPDU or one QoS null frame is to be transmitted upon the reception of the QoS (+)CF-Poll frame.
- *Queue size* subfield indicates the amount of buffered traffic for a given TC or TS at the non-AP station sending this frame. The AP may use this information to determine the TXOP duration assigned to the non-AP station. The queue size value is the total size, expressed in units of 256 octets, of all MSDUs buffered at the station (excluding the MSDU of the present QoS data frame) in the delivery queue used for MSDUs with the specified TID. A queue size value of 0 indicates no buffered traffic in the queue.
- *TXOP duration requested* subfield indicates the duration, in units of 32 μs, which the sending non-AP station desires for its next TXOP for the specified TID. The range of time values is 32 to 8,160 μs. The AP may use this information to determine the TXOP duration assigned to the non-AP station. A value

Table 14.2 Ack Policy Subfield in the QoS Control Field

Bits in QoS Control field		Meaning
Bit 5	Bit 6	
0	0	Normal Ack. The addressed recipient returns an ACK or QoS +CF-Ack frame after a short interframe space (SIFS) period, according to the procedures defined in Sections 13.2.4 and 13.3.2. The Ack Policy subfield is set to this value in all directed frames in which the sender requires acknowledgement. For QoS Null (no data) frames, this is the only permissible value for the Ack Policy subfield
1	0	No Ack. The addressed recipient takes no action upon receipt of the frame. More details are provided in Section 14.5.2. The Ack Policy subfield is set to this value in all directed frames in which the sender does not require acknowledgement. This combination is also used for broadcast and multicast frames that use the QoS frame format.
0	1	No explicit acknowledgement. There may be a response frame to the frame that is received, but it is neither the ACK nor any data frame of subtype +CF-Ack. For QoS CF-Poll and QoS CF-Ack+CF-Poll data frames, this is the only permissible value for the Ack Policy subfield.
1	1	Block Ack. The addressed recipient takes no action upon the receipt of the frame except for recording the state. The recipient can expect a BlockAckReq frame in the future to which it responds using the procedure described in Section 14.5.2.

Source: [7].

of 0 in this subfield indicates that no TXOP is requested for the MSDUs for the specified TID in the current SP.

- *AP PS buffer state* subfield indicates the PS buffer state at the AP for a non-AP receiver station. The AP PS buffer state subfield is further subdivided into three subfields: *buffer state indicated* (1 bit), *highest priority buffered AC* (2 bits), and *AP buffered load* (4 bits). The buffer state indicated subfield indicates whether the AP PS buffer state is valid. The highest priority buffered AC subfield indicates the AC of the highest priority traffic that is buffered at the AP, excluding the MSDU of the present frame. The AP buffered load subfield indicates the total buffer size, expressed in units of 4,096 octets, of all MSDUs buffered at the QoS AP (excluding the MSDU of the present QoS data frame).

14.3 IEEE 802.11e Hybrid Coordination Function (HCF)

IEEE 802.11e defines a single coordination function, called the *hybrid coordination function* (HCF). The HCF combines functions from the DCF and PCF with some enhanced QoS-specific mechanisms and QoS data frames in order to allow a uniform set of frame exchange sequences to be used for QoS data transfers during both CP and CFP. Note that the 802.11e MAC is backward compatible with the baseline MAC, and, hence, it is a superset of the baseline MAC. The HCF is composed of two channel access mechanisms: (1) a contention-based channel access referred to as the *enhanced distributed channel access* (EDCA), and (2) a controlled channel access referred to as the *HCF controlled channel access* (HCCA). In fact, EDCA and HCCA are enhanced versions of DCF and PCF of the baseline MAC, respectively.

The term "hybrid" comes from various aspects of the HCF. First, it combines both contention-based and controlled channel accesses. Second, the controlled channel access (i.e., polling) under the HCCA works in both CFP and CP. While the DCF was mandatory and the PCF was optional per the baseline MAC, both EDCA and HCCA are mandatory according to the 802.11e. Figure 14.2 shows the logical relationship between the 802.11e HCF and the 802.11 DCF/PCF. As shown in the figure, the HCF sits on top of the DCF in the sense that the HCF utilizes and honors the CSMA/CA operation of the DCF.

Readers who are interested in the performance of the 802.11e WLAN are referred to [8, 9]. Even though many 802.11e papers are based on some old versions of the draft, and, hence, the exact numbers may not be true, the general tendencies are still valid. We now explain how the 802.11e HCF works.

14.3.1 Enhanced Distributed Channel Access (EDCA)

The EDCA is designed to provide differentiated, distributed channel accesses for frames with eight different *user priorities* (UPs), ranging from 0 to 7, by enhancing the DCF. Each MSDU from the higher layer arrives at the MAC along with a specific UP value. Then, this MSDU is mapped into an *access category* (AC), which is a label for the common set of EDCA parameters that are used by a QoS station to contend for the channel in order to transmit MSDUs with certain priorities.

There are four ACs defined, where AC_BK, AC_BE, AC_VI, and AC_VO are intended for background, best effort, video, and finally voice traffic services, respectively. The mapping between a UP to an AC is found at Table 14.3, along with the informative designation for each AC. As there are four ACs and eight UPs, two UPs are mapped into an AC. Support of the eight UPs are rooted in IEEE 802.1D MAC bridge [6]. It should be noted that UP 0, which is the default UP, is not the lowest priority. Actually, its priority is higher than UPs 1 and 2, and, hence, these two UPs are designated for background traffic service. For example, the best-effort service could be used for typical Internet accesses (e.g., Web browsing), while the background service could be used for other less time-critical services (e.g., background file transfer).

All management frames are assigned to AC_VO. In the case of control frames, BlockAckReq, BlockAck, and RTS frames are sent using the same AC as the corresponding QoS data or management frames. (The usage of BlockAckReq and

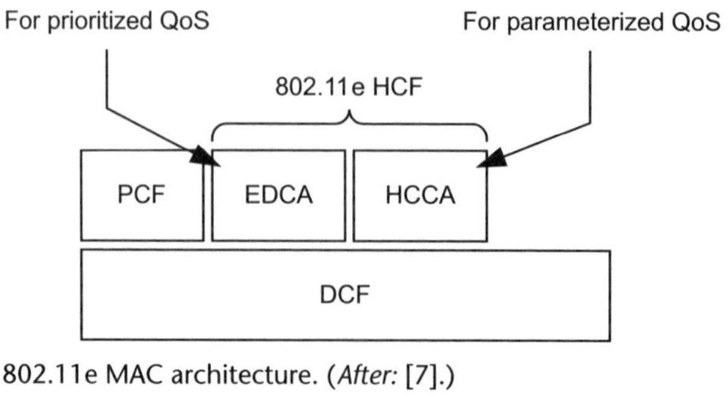

Figure 14.2 IEEE 802.11e MAC architecture. (*After:* [7].)

Table 14.3 UP to AC Mappings

Priority	User Priority (UP)	Access Category (AC)	Designation (Informative)
Lowest	1	AC_BK	Background
	2	AC_BK	Background
	0	AC_BE	Best Effort
	3	AC_BE	Best Effort
	4	AC_VI	Video
	5	AC_VI	Video
	6	AC_VO	Voice
Highest	7	AC_VO	Voice

BlockAck will be presented in Section 14.5.2.) Finally, PS-Poll frame is sent via AC_BE.

Enhanced Distributed Channel Access Functions

A QoS station implements four *enhanced distributed channel access functions* (EDACFs), where each EDCAF corresponds to an AC, as shown in Figure 14.3. An MSDU arriving from the higher layer is enqueued into a queue corresponding to an EDCAF. Each EDCAF behaves as a single enhanced DCF contending entity, where each EDCAF has its own *arbitration interframe space* (AIFS), which is used instead of the DIFS of the DCF, and maintains its own backoff counter. When there is more

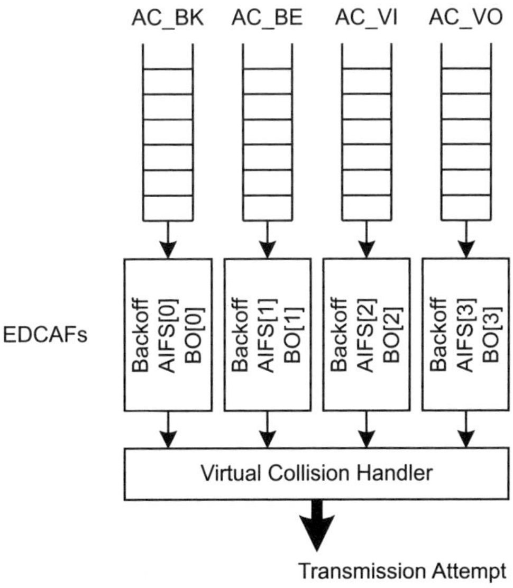

Figure 14.3 Four EDCAFs for IEEE 802.11e EDCA.

than one EDCAF finishing the backoff at the same time, the collision is handled in a virtual manner. That is, the highest priority frame among the colliding frames is chosen and transmitted, and the others perform a backoff with increased CW values. A QoS data frame carries its UP value in the QoS control field of the MAC frame header.

Basically, an EDCAF uses AIFS[AC], CWmin[AC], and CWmax[AC] instead of DIFS, CWmin, and CWmax of the DCF, respectively, for the contention to transmit a QoS frame belonging to access category AC. AIFS[AC] is determined by

$$\text{AIFS[AC]} = \text{SIFS} + \text{AIFSN[AC]} \times \text{SlotTime}$$

where AIFSN[AC] is an integer greater than one for non-AP stations and greater than zero for AP. As will be discussed further, the EDCA allows for the AP and non-AP stations to use a different set of EDCA parameters. The EIFS value is also determined per AC basis as follows:

$$\text{EIFS[AC]} = \text{EIFS} - \text{DIFS} + \text{AIFS[AC]}$$

Figure 14.4 shows the timing diagram for the channel access of an EDCAF. Compared with Figure 13.11 illustrating the DCF channel access, the only differences are the usages of AIFS[AC] and CW[AC]. However, exactly speaking, the backoff countdown rules of DCF and EDCA are slightly different as follows: the first countdown occurs at the end of the AIFS[AC] interval. Moreover, at the end of each idle slot interval, either a backoff countdown or a frame transmission occurs, but not both. Accordingly, a frame transmission occurs only a slot after the backoff counter becomes zero. Note that according to the baseline DCF, a station counts down a backoff counter, and if the counter becomes zero, it transmits a frame at the moment.

Figure 14.5 illustrates an example that shows the different backoff rules. An 802.11e EDCA station and an 802.11 DCF station are contending for the channel. For the EDCA station, only one EDCAF with AIFS = SIFS + 2 × SlotTime = DIFS is active while the other EDCAFs are inactive. Each number represents the *backoff counter* (BC) value at the given moment. We first observe that after an AIFS (= DIFS) idle interval, the EDCA station decreased the BC by one. On the other hand, the

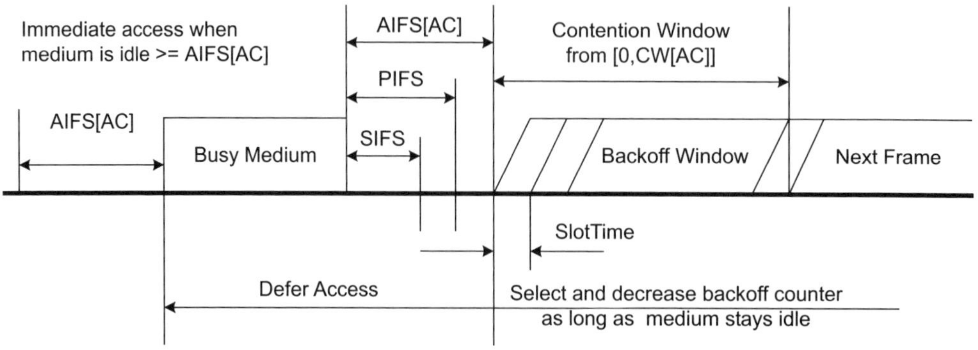

Figure 14.4 IEEE 802.11e EDCA channel access. (*After:* [7].)

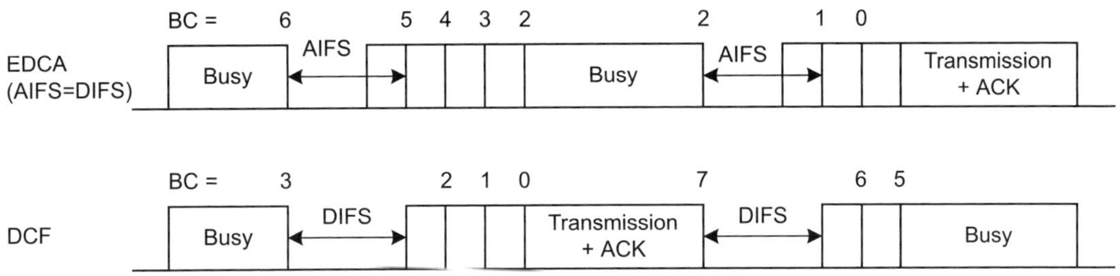

Figure 14.5 Differences between the DCF and EDCA backoff rules.

DCF station decreases the BC by one after a DIFS + SlotTime interval. The EDCA station transmits its frame one slot after the BC became zero. On the other hand, the DCF station transmits its frame as soon as the BC became zero. One might think that they operate statistically the same since with the EDCA, the BC decrement starts early by one slot, but the transmission is attempted late by one slot, compared with the DCF. However, it is not true. Note that whenever the channel becomes idle after a busy period, the EDCA advances one slot further compared with the DCF. Due to this difference, the EDCA can never be configured to operate exactly in the same manner as the DCF. This is in fact a problem since it might be desirable to make the EDCAF[AC_BE] operate in the same manner as the DCF.

EDCA TXOP

IEEE 802.11e supports a *transmission opportunity* (TXOP) as the interval of time when a particular station has the right to initiate transmissions. An EDCA TXOP is acquired when a QoS station transmits a frame into the channel via the EDCA contention and then receives the corresponding ACK. The EDCA TXOP limit is determined per AC basis by the AP. After receiving the ACK, the TXOP holder (i.e., the transmitter that checks how much residual time is remaining in the present TXOP). The TXOP holder (or more exactly speaking, the EDCAF that successfully transmitted the data frame) is allowed to transmit more QoS data frames (of the same AC) with a SIFS time interval between an ACK and the subsequent frame transmission. When the TXOP limit is zero for a particular AC, the corresponding EDCAF is allowed to transmit only a single MSDU for a given TXOP.

Figure 14.6 shows the transmission of two QoS data frames of user priority UP during an EDCA TXOP, where the whole transmission time for two data and ACK frames is less than the EDCA TXOP limit determined by the AP. If the residual time in the present TXOP is too short to transmit the pending frame, the TXOP holder can either stop transmitting or transmit a fragment of the pending frame. The

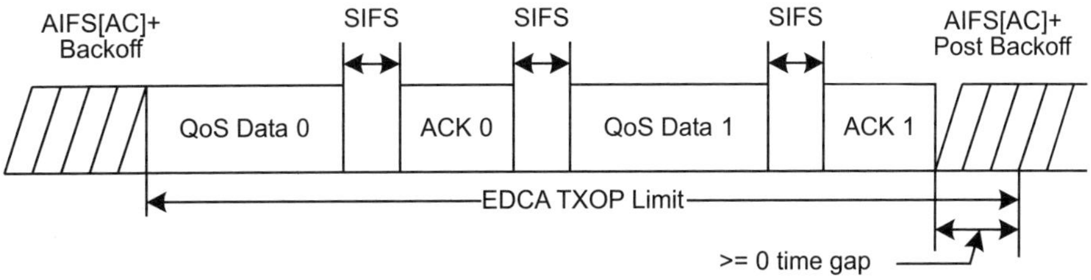

Figure 14.6 EDCA TXOP operation timing structure.

TXOP limit itself might be too short to transmit the pending frame, and then the frame has to be fragmented for a possible transmission. In Figure 14.6, the TXOP holder just stops transmitting after transmitting two frames successfully instead of transmitting an extra fragment.

A TXOP might end prematurely if a frame transmission fails (i.e., an ACK timeout occurs). As in the DCF, a retransmission of the frame occurs after another contention with a doubled CW value. An EDCA TXOP will not extend across TBTT. That is, a QoS station never transmits a frame if the corresponding frame exchange (i.e., including the ACK) is not expected to finish until the upcoming TBTT. Due to this constraint, an AP of the 802.11e is normally expected to transmit a beacon at each TBTT without any delay.

There are two different rules to set the duration/ID field in each QoS frame transmitted within a TXOP:

- The duration is set to cover until the end of the subsequent frame exchange. For example, in Figure 14.6, the duration/ID fields of QoS data 0 and ACK 0 protect up to the transmissions of QoS data 1 and its ACK 1. Note that this duration/ID field setting rule is basically the same as what is used for the fragmentation burst presented in Section 13.2.6.
- The duration is set to cover until the end of the current EDCA TXOP. This rule is simpler and possibly more robust to protect the current TXOP. However, if a TXOP ends prematurely due to a frame transmission failure, others are unnecessarily prevented from contending for the channel due to nonzero NAV value even if the TOXP holder ended its transmissions.

A transmitter can choose either of these two rules to set the duration/ID field for its transmitted frames.

EDCA Parameters

The values of AIFS[AC], CWmin[AC], CWmax[AC], and TXOP Limit[AC], which are referred to as the *EDCA parameters*, are determined and announced by the AP via the *EDCA parameter set element* in beacon frames. The AP might adapt these parameters dynamically depending on network conditions, even though frequent adaptation is not desired due to the network stability. Upon the reception of updated EDCA parameters via a beacon frame, a QoS station should update its parameters within a beacon interval. Basically, the smaller AIFS[AC] and CWmin[AC], the shorter the channel access delay for the frames with the corresponding UPs, and, hence, the more bandwidth share for a given traffic condition. These parameters can be used in order to differentiate the channel access among different user priority (or AC more accurately speaking) traffic types.

Moreover, the AP can use a set of EDCA parameters that are different from those announced by beacon frames and adopted by non-AP stations. In the baseline DCF-based WLAN, both AP and stations use the same DCF access rule, and hence the AP's transmissions were identified as a bottleneck for the whole network performance. It is because of the fact that: (1) there are more downlink frames than uplink frames in typical client-and-server scenarios (e.g., file download and Web surfing) and (2) the AP's channel opportunity is basically the same as that of each non-AP station. For example, in a BSS with an AP and N stations, when all of the AP and station are actively contending for the channel, every station including the AP is supposed to get $1/(1 + N)$ share of the total bandwidth.

The AIFS should be larger than or equal to PIFS for APs while the AIFS should be larger than or equal to DIFS for non-AP stations. This gives higher priority to the AP's transmissions. Moreover, the minimum of an EDCA TXOP is the time to transmit a 256-octet MPDU at the lowest transmission rate. While the EDCA parameters can be updated by the AP over time in an infrastructure BSS, the default values found in Table 14.4 are used in an IBSS. In the table, aCWmin and aCWmax represent the CWmin and the CWmax corresponding to the underlying PHY. For example, aCWmin = 15 for the 802.11a and aCWmin = 31 for the 802.11b according to Table 13.8.

Temporal Fairness Versus Throughput Fairness

The EDCA TXOP provides a temporal fairness to the EDCAFs of the same AC. On the other hand, a weighted temporal fairness is provided among the EDCAFs of different ACs, where the weights are determined by the TXOP limit per AC.

Let us consider a situation in which every EDCAF in each QoS station contends for the channel in an error-free channel environment. All the EDCAFs of the same AC are expected to utilize the same amount of time in the long term due to EDCA TXOP. On the other hand, the DCF is known to provide a long-term fairness for the channel access among the stations in an error-free channel environment, so that every station transmits the same number of frames in the long term, assuming that every station always has frames to transmit. This long-term channel access opportunity fairness makes long-term throughput fairness if every station transmits frames with the same average length.

This throughput fairness property of the 802.11 DCF was introduced as a *performance anomaly* in the literature [10]. Let's consider a situation in which two 802.11b stations contend for the error-free channel with infinite number of fixed-size frames to transmit. One station uses the highest 11-Mbps rate, while the other uses the lowest 1-Mbps rate. In the long term, the number of the successfully transmitted frames will be the same for both stations due to the channel access opportunity fairness. Accordingly, their long-term throughput should be the same even if their transmission rates are 11 times different. Moreover, their throughputs

Table 14.4 Default Values of EDCA Parameters

AC	CWmin	CWmax	AIFSN	TXOP Limit DC-CCK/PBCC PHY	TXOP Limit OFDM/CCK-OFDM PHY
AC_BK	aCWmin	aCWmax	7	0	0
AC_BE	aCWmin	aCWmax	3	0	0
AC_VI	(aCWmin+1)/2 - 1	aCWmin	2	6.016 msec	3.008 msec
AC_VO	(aCWmin+1)/4 - 1	(aCWmin+1)/2 - 1	2	3.008 msec	1.504 msec

Source: [7].

will be under 1 Mbps since the throughput of the 1 Mbps rate station cannot be over 1 Mbps.

This performance anomaly does not exist any longer in the 802.11e EDCA WLAN where temporal fairness is provided. As every station utilizes the same amount of time share in the long term, and the number of frames transmitted basically depends on the employed transmission rate as well as the frame length, the throughput of each station will be dependent on its employed transmission rates. Other solutions to the performance anomaly have also been proposed (e.g., [11]).

14.3.2 HCF Controlled Channel Access (HCCA)

If the EDCA is for the prioritized QoS, which supports differentiated channel accesses to eight different UP traffic types, the HCCA is for the parameterized QoS, which provides the QoS based on the contract between the AP and the corresponding QoS station(s). Before commencing the transfer of any frame requiring the parameterized QoS, a virtual connection, called *traffic stream* (TS), is first established. A traffic stream could be for one of uplink, downlink, and direct link, which are for QoS station-to-AP, AP-to-QoS station, and QoS station-to-QoS station, respectively.

In order to set up a traffic stream, a set of traffic characteristics (such as nominal MSDU size, mean data rate, and maximum burst size) and QoS requirement parameters (such as delay bound) are exchanged and negotiated between the AP and the corresponding QoS station(s), and the traffic stream should be admitted by the AP. Accordingly, the AP should implement an admission control algorithm to determine whether or not to admit a specific traffic stream into its BSS. The admission control problem is further discussed in Section 14.4.

Once a traffic stream bound with the HCCA as its access policy is set up, the HC endeavors to provide the contracted QoS by allocating the required bandwidth to the traffic stream using the HCCA. As discussed further in Section 14.4.1, a traffic stream might also be set up while the EDCA is chosen as the access policy.

HCCA Basic Access

Under the HCCA, the HC has the full control over the channel during a CFP, and during a CP it can also gain the channel access (i.e., start an HCCA TXOP) after a PIFS idle interval whenever it wants. The channel access gaining is done by initiating its downlink frame transfer or by transmitting a polling frame, such as a QoS (+)CF-Poll frame, in order to grant a polled TXOP to a QoS station. Note that an HCCA TXOP, granted to a non-AP QoS station via a QoS (+)CF-Poll, is called a polled TXOP. The beauty of the HCCA is in that: (1) the polling can be done in CP so that it is not bound with beacon transmissions, and (2) as polling during CP requires a channel sensing, it can be friendlier to overlapping BSSs.

The usage of the PIFS interval allows the HC to have the priority over non-AP stations accessing the channel based on EDCA. This is because the minimum AIFS for non-AP stations is DIFS. After gaining the channel access, the HC can maintain the control over the channel by consecutively granting HCCA TXOPs including polled TXOPs, and a time period when the HC maintains control of the channel is referred to as a *controlled access phase* (CAP). In terms of the fact that the AP assumes a control over the channel, the CAP is similar to the CFP in the baseline MAC. However, different from the CFPs, CAPs do not need to appear periodically, and their appearance is not bound with the beacon transmissions. Figure 14.7 illustrates the coexistence of CAP, CFP, and CP.

By receiving a QoS (+)CF-poll, the TXOP holder assumes the control over the channel up to the TXOP limit specified in the QoS control field of the QoS (+)CF-Poll frame (see Figure 14.1 and Table 14.1). During the TXOP, the TXOP holder transmits multiple MPDUs, in which the transmitted frames and their transmission order are determined by the TXOP holder according to their implementation-dependent scheduling algorithm. Note that this is very different from the usage rule of the EDCA TXOP, in which only frames from the same AC can be transmitted. The duration/ID field in QoS (+)CF-Poll is set to the TXOP limit plus an extra slot time. Accordingly, all other stations that receive the QoS (+)CF-Poll set the NAV accordingly such that they will not contend for the channel during that time period. The timing diagram of a polled TXOP operation is depicted in Figure 14.8. In the figure, the receiver should be the AP itself if the traffic stream is for uplink, while the receiver could be another non-AP station if the traffic stream is for a direct link.

After an HCCA TXOP, the HC might want to grant another HCCA TXOP by transmitting a downlink frame or a QoS (+)CF-Poll after a PIFS interval in order to

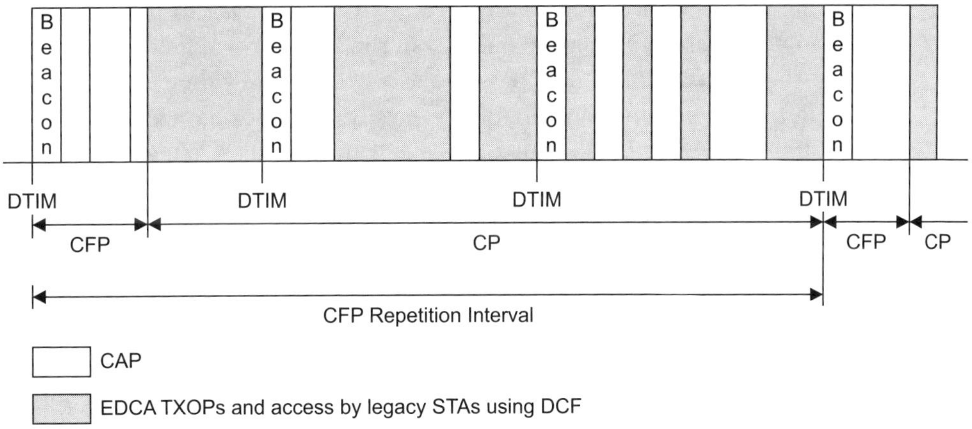

Figure 14.7 CAP/CFP/CP coexistence. (*After:* [7].)

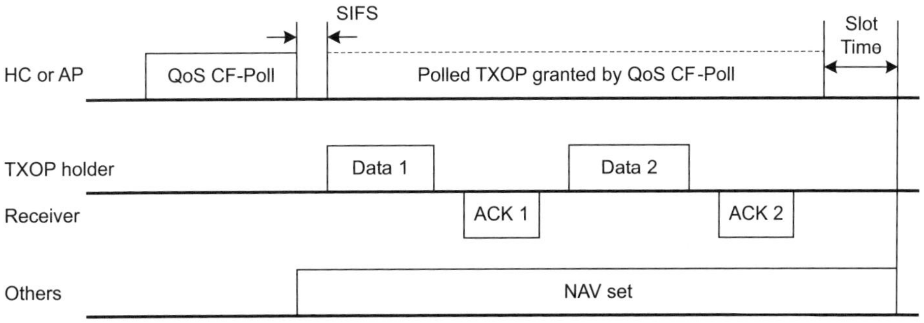

Figure 14.8 Polled TXOP timing.

continue the current CAP. A CAP ends when the HC does not reclaim the channel after a PIFS interval from the end of an HCCA TXOP. A TXOP (and hence a transmission within a TXOP) will not extend across TBTT, CFP_Max_Duration (for the TXOP during CFP), and CAPLimit, which specifies the maximum CAP length.

Error Recovery

After the HC sends a QoS (+)CF-Poll, the polled station has to respond in a SIFS interval. If the channel becomes busy within a PIFS interval after transmitting a QoS (+)CF-Poll, HC assumes that the TXOP was granted successfully. On the other hand, if the channel stays idle during the PIFS interval, the HC concludes that the previously sent QoS (+)CF-Poll was not successfully received by the polled station, and, hence tries to recover from this erroneous situation by either granting another HCCA TXOP or ending the current CAP. During a polled TXOP, every frame exchange should occur with a SIFS interval. After transmitting a directed frame, if there is no ACK response (i.e., the channel stays idle during the subsequent PIFS interval), the TXOP holder concludes that the previous frame transmission was not successful, and, hence, tries to recover from this erroneous situation by retransmitting the frame again or by transmitting another frame. Each MPDU is subject to a retry limit as well as an MSDU lifetime.

For two different cases, a non-AP QoS station might respond to a QoS (+)CF-Poll with a QoS (+)Null (i.e., without data) when a polled TXOP was granted. First, the polled station might not have any pending data to transmit, and then, a QoS (+)Null frame with zero queue size in the QoS control field is transmitted. Second, if the granted polled TXOP is too short to transmit the pending frame, a QoS (+)Null frame with a nonzero queue size value or a nonzero TXOP duration requested value in the QoS control field is transmitted. When a queue size is transmitted, the HC combines the queue size information with the rate of the received QoS (+)Null frame in order to determine the required size of the requested TXOP.

Return of Residual Polled TXOP

Within a polled TXOP, the unused portion of TXOPs should be returned back to the HC. In order to return it, the last data frame within a polled TXOP is directed to the HC, where this frame contains the zero queue size value in the QoS control field. If the TXOP holder does not have any data frame to transmit to the HC, a QoS null frame might be transmitted. The last data frame also uses the normal ACK policy so that it can be retransmitted upon the reception failure of the corresponding ACK. If the CCA busy occurs within a PIFS interval after transmitting such a last data frame, the transmission of the frame is assumed successful. Otherwise, this last frame is retransmitted after the PIFS interval.

The duration/ID field of the last data frame within a polled TXOP is set to ACK transmission time + SIFS + SlotTime, where the duration/ID field of the very last frame (i.e., the corresponding ACK) is set to a SlotTime.

NAV Operation During a TXOP

The HC sets its own NAV in order to prevent its transmission during a polled TXOP. The duration/ID field in QoS (+)CF-Poll is set to the TXOP limit plus an extra slot time. Accordingly, all other stations that receive the QoS (+)CF-Poll set the

NAV accordingly such that they will not contend for the channel during that time period. In the frames of the nonfinal frame exchange sequences within a polled TXOP, the duration/ID field is set to the remaining TXOP duration. On the other hand, in the frames of the final (or the only) frame exchange sequence within a polled TXOP, the duration/ID field is set to the actual remaining time for the frame exchange as discussed earlier.

During a polled TXOP, non-AP QoS stations save the MAC address of the TXOP holder, which is found at the corresponding QoS (+)CF-Poll frame. A QoS station then responds to an RTS frame from the TXOP holder with a CTS frame even if its NAV is nonzero. Note that under the original RTS/CTS exchange rule, a station does not transmit a CTS if it has a nonzero NAV. In the 802.11e, RTS/CTS frame exchange is allowed in the CFP as well.

There are a couple of conditions to reset NAV values at non-AP QoS stations. First, if a CF-End from its HC is received, non-AP QoS stations reset their NAV. Second, if a QoS (+)CF-Poll directed to the HC itself (i.e., Address 1 equal to the BSSID), with the duration/ID field set to zero is received, non-AP QoS stations reset their NAV. The HC can use either of two methods to reset the stations' NAV values at the end of a CAP.

14.4 Admission Control and Scheduling

For the parameterized QoS support, a TS has to be set up after a proper admission control procedure. Even for the prioritized QoS support, an admission control is desirable for controlled QoS provisioning, and in this case a TS is again set up after an admission control procedure. The underlying philosophy is that a TS should be set up only if the required QoS can be provisioned, and the HC should check whether it is feasible to support the required QoS during the admission control procedure. Note that either HCCA or EDCA or both can be associated with a TS. Once a TS, associated with HCCA, is set up, the HC has to endeavor to provide the agreed QoS by scheduling the *service periods* (SPs) properly. An SP is a contiguous time during which one or more downlink unicast frames and/or one or more polled TXOPs are granted to a station.

14.4.1 TS Operations

A TS lifetime is composed of three procedures, namely, TS setup, TS operation, and TS deletion. An example of the TS operation is illustrated in Figure 14.9. For the TS operations, four QoS action frames are utilized, namely, *ADDTS request*, *ADDTS response*, *DELTS*, and *schedule* frames, where ADDTS and DELTS imply the TS addition and TS deletion, respectively.

A TS is set up by the exchange of an *ADDTS request* and an *ADDTS response* between a non-AP QoS station and the HC. A TS setup can be initiated only by a non-AP QoS station. That is, only a non-AP QoS station can send an ADDTS request frame. Both ADDTS request and ADDTS response frames convey the *TSPEC* (traffic specification) element and optionally one or more *TCLAS* (traffic classification) elements. These two elements basically specify the traffic characteris-

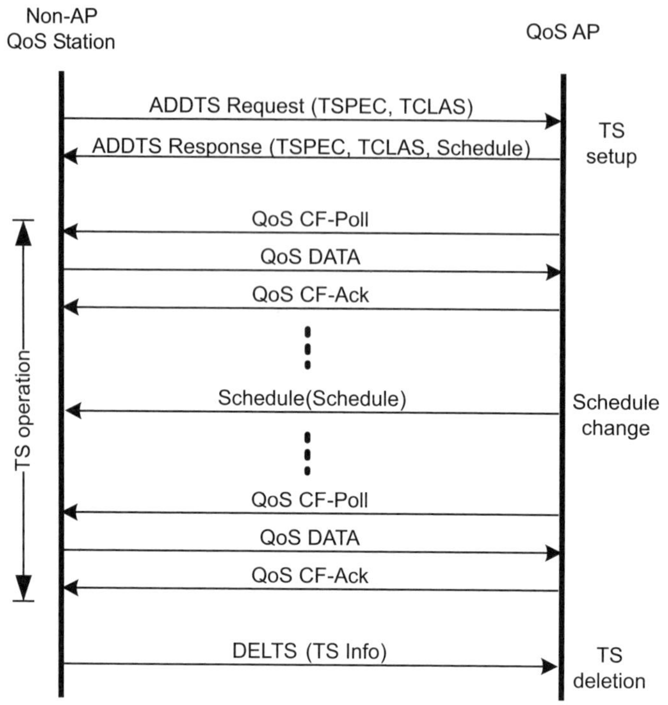

Figure 14.9 An example of the operation during a TS lifetime.

tics and QoS requirements of the requested TS. The ADDTS response frame also includes a *schedule* element, which specifies the SP schedule information. See Section 14.4.2 for further details. If the HC does not have enough resources to meet the requested QoS, the TS setup can be rejected, and it is indicated in the ADDTS response frame. Upon a TS setup failure, the non-AP QoS station might want to send an ADDTS request frame again with a revised TSPEC element.

During the TS operation, the transfer of QoS traffic occurs following the MAC protocols presented in Section 14.3. For a TS with the access policy including HCCA, the HC schedules service periods according to the TSPEC. When the schedule for service periods is changed due to various reasons (e.g., the channel condition variation, the addition or deletion of another TS, and the request by the non-AP QoS station), it is notified by the corresponding QoS station via a *schedule* frame. The schedule frame includes a schedule element reflecting the modified schedule.

A TS is deleted via a transmission of a DELTS frame by either the non-AP station or the HC.

HCCA Versus EDCA

The traffic admitted via a TSPEC could be transferred using EDCA or HCCA or *HCCA, EDCA mixed mode* (HEMM). This depends on the access policy set in the *TS info* field in the TSPEC. A TSPEC request may be set so that both HCCA and EDCA mechanisms (i.e., HEMM) are used. A TS setup via an admission control is mandatory in order to use the HCCA. On the other hand, for the EDCA, the decision whether to set up a TS depends on the policy of the AP. This is determined per

AC; the AP might mandate a TS setup for specific ACs via the *EDCA parameter set* in beacons.

While both EDCA and HCCA are mandatory per IEEE 802.11e, the HCCA is essentially optional. The reason is in that the HCCA can be used only for the data transmissions in an admitted TS. A TS setup can be initiated only by a non-AP QoS station. This implies that a QoS station can live without the implementation of the HCCA, since the AP will never request the station to use the HCCA. An AP might be requested for a TS setup. However, an AP can always reject a TS setup request by saying that the request cannot be accommodated due to some reason. Accordingly, the AP can also live without implementing the HCCA.

TS Characterized by TSPEC

A TSPEC describes the traffic characteristics and the QoS requirements of a TS. The main purpose of the TSPEC is to reserve resources within the HC and modify the HC's scheduling behavior. It also allows other parameters that are associated with the TS to be specified, such as a traffic classifier and acknowledgment policy. A TS may have one or more TCLASs associated with it. The AP uses the parameters in the TCLAS elements to filter the MSDUs belonging to this TS so that they can be delivered with the QoS parameters associated with the TS. Once a TS is set up after a successful negotiation, the TS is identified by the combination of TSID, direction, and the associated non-AP station's address.

In the case of a direct link, the non-AP station that is going to send the data requests a TS setup. In the case of flow relayed by the AP, the sending and receiving non-AP stations may both create individual TSs for the flow. A non-AP station can simultaneously support up to eight downlink TSs from the HC to itself and up to eight uplink or direct link TSs from itself to other stations, including the HC. A HC can simultaneously support up to eight downlink TSs and up to eight uplink or direct link TSs per associated non-AP station.

TS Life Cycle

Figure 14.10 illustrates a life cycle of TS. Initially TS is inactive. A non-AP QoS station cannot transmit any QoS data frames using an inactive TS. Following a successful TS setup initiated by the non-AP station, the TS becomes active, and either the non-AP station and/or the HC may transmit QoS data frames belonging to this TS depending on the direction of the TS. While the TS is active, the parameters of the TSPEC characterizing the TS can be renegotiated when the renegotiation is initiated by the non-AP station. If the negotiation is successful, the TSPEC is modified.

An active TS becomes inactive following a TS deletion process initiated by either the non-AP station or HC. It also becomes inactive if no activity is detected for a duration of an inactivity interval. When an active TS becomes inactive, all the resources allocated for the TS are released. Before a TS becomes inactive due to the inactivity, an active TS may become suspended if no activity is detected for a duration of a suspension interval. Upon detection of activity, the TS may be reinstated. Both the inactivity interval and suspension interval are specified in the TSPEC, where the suspension interval is always less than or equal to the inactivity interval.

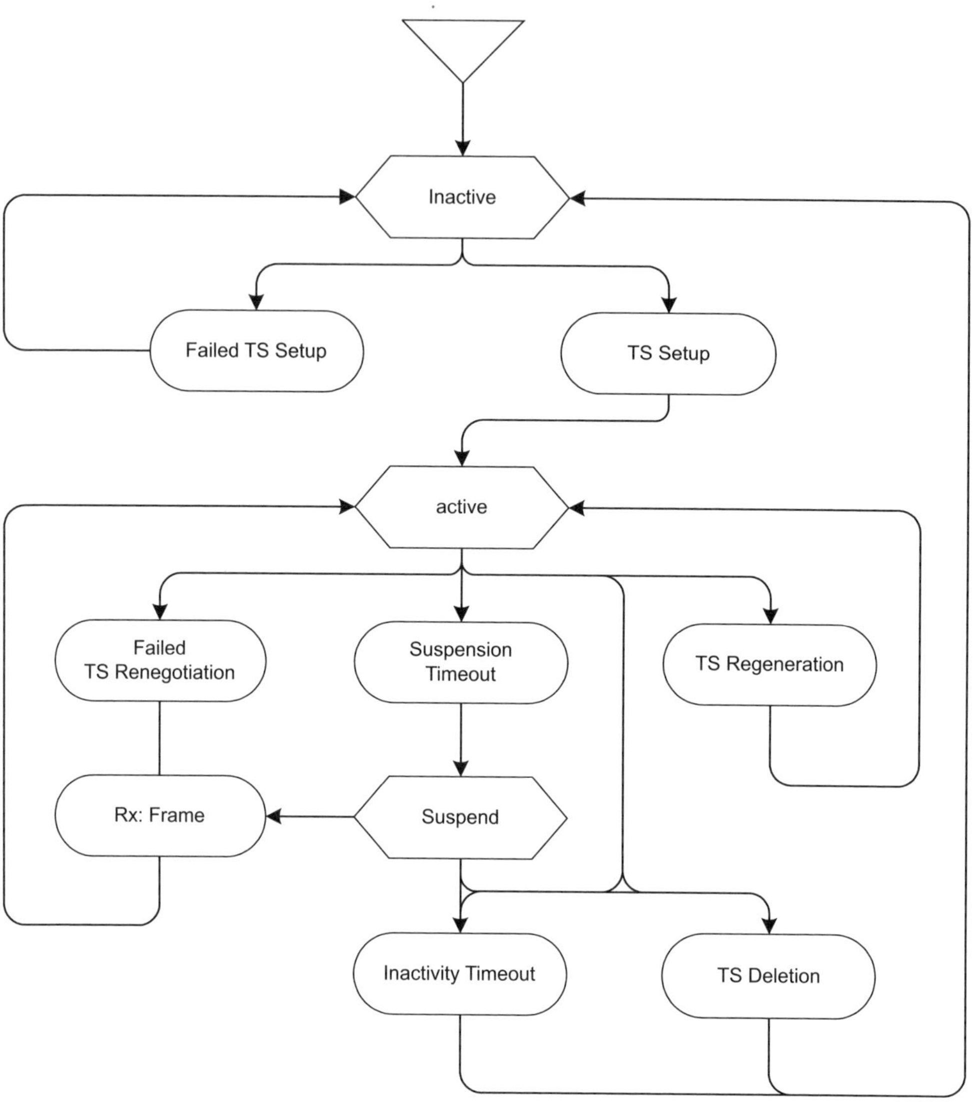

Figure 14.10 A TS life cycle. (*After:* [7].)

14.4.2 Information Elements for TS

A number of information elements are related to the TS operations. Those include traffic specification (TSPEC), traffic classification (TCLAS), and schedule information elements.

Traffic Specification

The TSPEC element is included in ADDTS request and ADDTS response frames. It contains the set of parameters that define the characteristics and QoS expectations of a traffic flow, in the context of a particular non-AP station, for use by the HC and non-AP station(s) to support QoS traffic transfer using the procedures defined Section 14.3. The element includes a number of fields and subfields as shown in Figure 14.11.

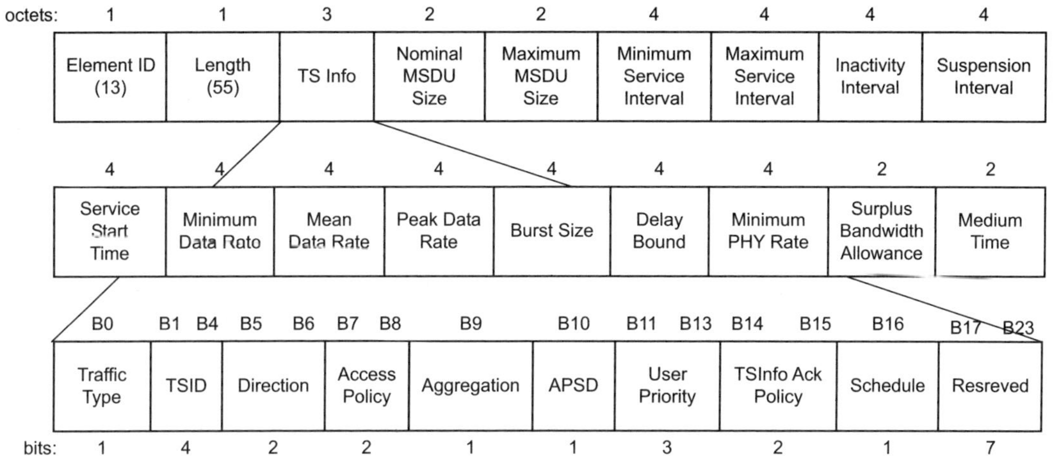

Figure 14.11 *Traffic specification* element and TS info field. (*After:* [7].)

The TSPEC allows a set of parameters more extensive than may be needed, or may be available, for a given TS of the parameterized QoS traffic. Therefore, a TSPEC might include a number of unspecified parameter values (i.e., a zero value). Non-AP stations set the value of any parameters to unspecified if they have no information for setting the parameters. The HC may change the value of parameters that have been set unspecified by the station to any value that it deems appropriate. Unspecified parameters indicate that the non-AP station does not have specific requirements for these parameters if the TSPEC was issued by that non-AP station or that the HC does not provide any specific values for these parameters if the TSPEC was issued by the HC.

The very first field of the TSPEC element is the TS info field, and specifies the basic information about the TS under this TSPEC. The TS info field includes nine subfields as follows:

- The *traffic type* subfield is set to 1 for a periodic traffic pattern (e.g., isochronous TS of MSDUs, with constant or variable sizes, that are transmitted at a fixed rate) or set to 0 for an aperiodic, or unspecified, traffic pattern (e.g., asynchronous TS of low-duty cycles).
- The *TSID* subfield contains the TSID value ranging from 7 to 15. The combination of the TSID and direction subfields identify the TS of the corresponding QoS station. For a given station, up to eight uplink TSs and up to eight downlink or direct link TSs can be set up. A bidirectional link request is equivalent to a downlink TS and an uplink TS, each with the same TSID and parameters.
- The *direction* subfield specifies the direction of data carried by the TS. There are directions, namely, uplink, downlink, direct, and bidirectional link (i.e., both uplink and downlink with the same parameters).
- The *access policy* subfield specifies the access that would be used for the TS. There are three policies, namely, EDCA, HCCA, and finally HEMM. Under the HEMM policy, both HCCA and EDCA can be used while under the other two policies, only either HCCA or EDCA is allowed.

- The *aggregation* subfield is valid only: (1) when access policy is HCCA, or (2) when the access policy is EDCA and the schedule subfield is set to 1. It is set to 1 by a non-AP station to indicate that an aggregate schedule is required. It is set to 1 by the AP if an aggregate schedule is being provided to the non-AP station.
- The *APSD* subfield indicates that *automatic PS delivery* (APSD) is to be used for the traffic associated with the TSPEC, as detailed in Section 14.5.3.
- The *UP* subfield indicates the UP value to be used for the transport of MSDUs belonging to this TS in cases where relative prioritization is required. When the TCLAS element is present in the request, the UP subfield in TS Info field of the TSPEC element is not valid.
- The *TS info Ack policy* subfield indicates whether acknowledgments are required for MPDUs belonging to this TS and the desired form of those acknowledgments. There are three Ack policies, namely, normal Ack, no Ack, and finally block Ack, as detailed in Section 14.5.2.
- The *schedule* subfield specifies the requested type of schedule. The encoding of the subfield when the access policy is EDCA is shown in Table 14.5. The schedule subfield is not valid when the access policy is not EDCA. When the schedule and APSD subfields are set to 1, the AP sets the aggregation subfield to 1, indicating that an aggregate schedule is being provided to the non-AP station. The related operations will be detailed in Section 14.5.3.

The remaining 15 fields in the TSPEC element represent the characteristics of the traffic pattern as well as the QoS requirements as follows:

- The *nominal MSDU size* field specifies the nominal size of MSDUs belonging to the TS. One bit in this field also indicates whether the MSDU size specified in the remaining 15 bits is fixed or nominal.
- The *maximum MSDU size* field specifies the maximum size of MSDUs belonging to the TS.
- The *minimum service interval* field specifies the minimum interval between the starts of two successive SPs.
- The *maximum service interval* field specifies the maximum interval between the starts of two successive SPs.

Table 14.5 Schedule Subfield Encoding

APSD	Schedule	Usage
0	0	No Schedule
1	0	Unscheduled APSD
0	1	Reserved
1	1	Scheduled APSD

Source: [7].

- The *inactivity interval* field specifies the minimum amount of time that may elapse without arrival or transfer of an MPDU belonging to the TS before this TS is deleted by the HC.
- The *suspension interval* field specifies the minimum amount of time that may elapse without arrival or transfer of an MSDU belonging to the TS before the generation of successive QoS (+)CF-Poll is stopped for this TS. The suspension interval is always less than or equal to the inactivity interval.
- The *service start time* field specifies the time when the first scheduled SP starts. The service start time indicates to AP the time when a non-AP station first expects to be ready to send frames and a power-saving non-AP station will be awake to receive frames. The field represents the four lower-order octets of the TSF timer at the start of the SP.
- The *minimum data rate* field specifies the lowest data rate at the MAC SAP for transport of MSDUs belonging to this TS.
- The *mean data rate* field specifies the average data rate at the MAC SAP for transport of MSDUs belonging to this TS.
- The *peak data rate* field specifies the maximum allowable data rate at the MAC SAP for transfer of MSDUs belonging to this TS. If p is the peak rate (in bps), then the maximum amount of data belonging to this TS arriving in any time interval $[t1,t2]$, where $t1 < t2$ and $t2 - t1 > 1$ TU, does not exceed $p \times (t2 - t1)$ bits.
- The *burst size* field specifies the maximum burst, in octets, of the MSDUs belonging to this TS that arrive at the MAC SAP at the peak data rate.
- The *delay bound* field specifies the maximum amount of time allowed to transport an MSDU belonging to the TS, measured between the arrival time of the MSDU at the MAC SAP and the completion time of the successful transmission or retransmission of the MSDU to the receiver.
- The *minimum PHY rate* field specifies the desired minimum transmission rate at the underlying PHY to use for this TS, which is required for transport of the MSDUs belonging to the TS.
- The *surplus bandwidth allowance* field specifies the excess allocation of time (and bandwidth) needed to transport an MSDU belonging to the TS. This field takes into account the retransmissions. For example, if there are 0.3 retransmissions on average to transmit an MSDU while meeting the throughput and delay bound requirements, this value becomes 1.3. Accordingly, it should be greater than 1, where a value of 1 indicates that no additional allocation of time is requested.
- The *medium time* field contains the amount of time allowed to access the medium for one second. This field is used only in the ADDTS response frame. The derivation of this field is discussed in Section 14.4.3. This field is not used for HCCA.

The mean data rate, the peak data rate, and the burst size are the parameters of the *token bucket model*, which provides the standard terminology for describing the behavior of a traffic source [12–14]. The minimum PHY rate information is intended to ensure that the TSPEC parameter values resulting from an admission control negotiation are sufficient to provide the required throughput for the TS.

The UP, minimum data rate, mean data rate, peak data rate, burst size, minimum PHY rate, and delay bound fields in a TSPEC element express the QoS expectations requested by a non-AP station if this TSPEC was issued by that non-AP station, or provided by the HC if this TSPEC was issued by the HC, when these fields are specified with nonzero values. Unspecified parameters in these fields (i.e., a zero value) indicate that the non-AP station does not have specific requirements for these parameters if the TSPEC was issued by that non-AP station or that the HC does not provide any specific values for these parameters if the TSPEC was issued by the HC.

Traffic Classification

The TCLAS element is optionally included in ADDTS request and ADDTS response frames. An MSDU arriving at the MAC should be associated with a TID value. The TCLAS element contains a set of parameters necessary to classify incoming MSDUs (from a higher layer in a station or from the DS in an AP) into a particular TS to which they belong. The operation to map between each MSDU and a particular TS is performed above the MAC SAP by utilizing the parameter values found in the TCLAS element. If required, the TCLAS element is provided in ADDTS request and ADDTS response frames only for the downlink or bidirectional links. The TCLAS element does not need to be provided for the uplink or directlink transmissions because the classification is conducted at the non-AP station, which is initiating the TS setup. The format of this element is illustrated in Figure 14.12.

The *frame classifier* field is composed of the following subfields: *classifier type*, *classifier mask*, and *classifier parameters*. The classifier type subfield specifies the type of classifier parameters in this TCLAS as defined in Table 14.6. These classifier types will be explained next. The classifier mask subfield specifies a bitmap, where bits that are set to 1 identify a subset of the classifier parameters whose values must match those of the corresponding parameters in a given MSDU in order for that MSDU to be classified to the TS of the associated TSPEC. An MSDU that fails to be classified into any active TS is classified to be a best-effort MSDU.

As shown in Figure 14.13, for classifier type 0, the following classifier parameters are contained in an Ethernet packet header: source address, destination address, and EtherType.

For classifier type 1, the frame classifier is defined separately for IPv4 and IPv6, and distinguished by the version field, as shown in Figures 14.14 and 14.15, respectively. The classifier parameters include (1) those from the TCP or UDP header,

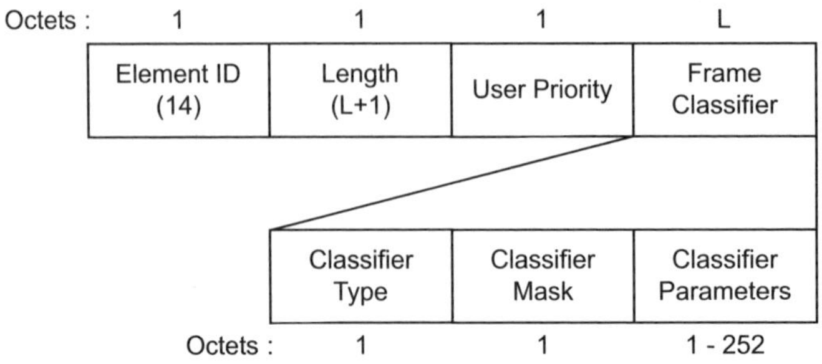

Figure 14.12 TCLAS element format. (*After:* [7].)

Table 14.6 Frame Classifier Type

Classifier type	Classifier parameters
0	Ethernet parameters
1	TCP/UDP IP parameters
2	IEEE 802.1D/Q parameters
3-255	Reserved

Source: [7].

Octets :	1	1	6	6	2
	Classifier Type (0)	Classifier Mask	Source Address	Destination Address	EtherType

Figure 14.13 Frame classifier field of classifier type 0. (*After:* [7].)

Octets :	1	1	1	4	4	2	2	1	1	1
	Classifier Type (1)	Classifier Mask	Version	Source IP Address	Destination IP Address	Source Port	Destination Port	DSCP	Protocol	Reserved

Figure 14.14 Frame classifier field of classifier type 1 for IPv4 traffic. (*After:* [7].)

Octets :	1	1	1	16	16	2	2	3
	Classifier Type (1)	Classifier Mask	Version	Source IP Address	Destination IP Address	Source Port	Destination Port	Flow Label

Figure 14.15 Frame classifier field of classifier type 1 for IPv6 traffic. (*After:* [7].)

namely, version, source address, destination address, source port, and destination port, and (2) *differentiated services code point* (DSCP) [15] and protocol fields from the IPv4 header or flow label field from the IPv6 header.

For classifier type 2, as shown in Figure 14.16, the classifier parameters contain the IEEE 802.1Q tag header [16] comprising: (1) IEEE 802.1D user priority, and (2) IEEE 802.1Q *virtual LAN* (VLAN) ID.

Schedule Element

The *schedule* element is included in ADDTS response and schedule frames. The schedule element is transmitted by the HC to a non-AP station to announce the schedule that the HC/AP will use for admitted streams originating from or destined to that non-AP station in the future. The information in this element may be used by

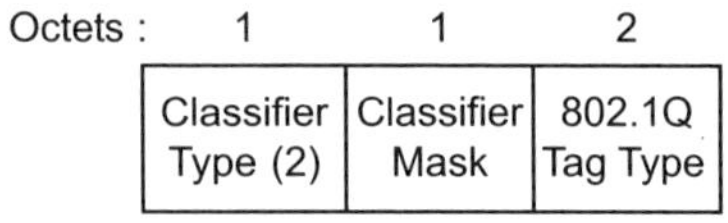

Figure 14.16 Frame classifier field of classifier type 2. (*After:* [7].)

the non-AP station for various purposes including power management and internal scheduling. The information element format is shown in Figure 14.17.

The *aggregation* subfield is to indicate if the schedule is an aggregate schedule for all TSIDs associated with the non-AP station to which the frame is directed. The *TSID* subfield indicates the TSID for which this schedule applies. The *direction* subfield specifies the direction of the TSPEC associated with the schedule. The TSID and direction subfields are valid only when the aggregation is not used.

The *service start time* field indicates the scheduled time when the service starts and represents the 4 lower-order octets of the TSF timer value at the start of the first SP. The *service interval* field indicates the time between two successive SPs and represents the measured time from the start of one SP to the start of the next SP. The HC may set the service start time field and the service interval field to 0 (unspecified) for nonpowersaving stations. We further discuss the issues related with the power saving in Section 14.5.3. The *specification interval* field specifies the time interval to verify schedule conformance, as explained in Section 14.4.3.

14.4.3 Admission Control and Scheduling Policies

Admission control is always needed when a station desires guarantee on the amount of time that it can access the channel for the QoS provisioning. As there are two channel access mechanisms, there are two distinct admission control mechanisms: one for EDCA and the other for HCCA. Admission control basically depends on vendor-specific scheduler as well as the available channel capacity.

Admission Control for EDCA

The AP advertises in the EDCA parameter set element in the beacon and (re)association response frames whether admission control is required for each of the ACs. An ADDTS request frame is transmitted by a non-AP station to the HC in order to request a TS setup in any direction (i.e., uplink, downlink, direct link, or bidirectional) employing an AC that requires admission control. The ADDTS request frame contains the UP associated with the traffic and indicates the EDCA as

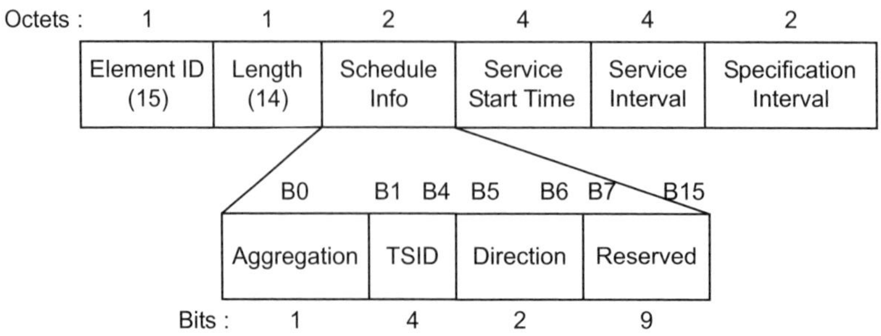

Figure 14.17 Schedule element. (*After:* [7].)

the access policy. The AP associates the received UP of the ADDTS request frame with the appropriate AC according to the UP-to-AC mappings shown in Table 14.3.

If the AP decides to admit a TS with the access policy of EDCA, the AP also derives the *medium time* from the information conveyed in the TSPEC element in the ADDTS request frame. How to determine the medium time is an implementation-dependent issue, but a possible simple calculation, based on nominal MSDU size, mean data rate, minimum PHY rate, and surplus bandwidth allowance, all found in the TSPEC, is as follows:

$$\text{Medium Time} = \text{Surplus Bandwidth Allowance} \times \text{MPDUExchangeTime} \times \text{Mean Data Rate/Nominal MSDU Size}$$

where MPDUExchangeTime is the time required to transmit an MPDU conveying an MSDU of the nominal size at the minimum PHY rate including ACK and SIFS.

The medium time is informed to the non-AP station via the ADDTS response frame. Each EDCAF maintains two variables: *admitted_time* and *used_time*. The admitted_time and used_time are reset to 0 at the time of (re)association. The parameter admitted_time is the medium time allowed by the AP during each interval of EDCA_Averaging_Period. The parameter used_time specifies the amount of time used by the non-AP station in an interval of EDCA_Averaging_Period. The value of EDCA_Averaging_Period is configurable, where the default value is 5 seconds, where its unit is a second.

Upon the reception of a TSPEC element in an ADDTS response frame indicating that a TS setup has been accepted, the non-AP station (re)computes the admitted_time for the specified EDCAF as follows:

$$\text{admitted\_time} = \text{admitted\_time} + \text{EDCA\_Averaging\_Period} \times \text{medium time}$$

For each TS setup, a new medium time is allocated, thus increasing the admitted_time. Then, the non-AP station updates the value of used_time as follows:

At the beginning of each EDCA_Averaging_Period interval,
used_time = max((used_time—admitted_time), 0)

After each MPDU (re)transmission attempt,
used_time = used_time + MPDUExchangeTime

If the used_time value reaches the admitted_time value, the corresponding EDCAF is not allowed to transmit using the EDCA parameters for that AC. However, a non-AP station may choose to temporarily replace the EDCA parameters for that EDCAF with those specified for an AC of lower priority if the admission control is not required for those ACs.

Admission Control for HCCA

Under the HCCA, the HC is responsible for granting or denying polling service to a TS based on the parameters in the associated TSPEC. If the TS is admitted, the HC is responsible for scheduling channel access to this TS based on the negotiated TSPEC parameters. The polling service based on admitted TS provides a guaranteed channel access from the scheduler in order to provision its QoS requirements. The nature of wireless communications may make it impossible to absolutely guarantee the QoS requirements. However, in a controlled environment (e.g., no interference), the behavior of the scheduler can be observed and verified to be compliant to meet the service schedule.

The scheduler is implemented so that, under controlled operating conditions, all stations with admitted TSs are offered TXOPs that satisfy the service schedule found in the schedule element of either ADDTS response or schedule frames. Specifically, if a TS is admitted by the HC, then the scheduler should serve the non-AP station during an SP. An SP starts at fixed intervals of time specified in the service interval field of the schedule element. The first SP starts when the 4 lower order octets of the TSF timer equals the value specified in the service start time field of the schedule element. Additionally, the minimum TXOP duration should be at least the time to transmit one maximum MSDU size successfully at the minimum PHY rate specified in the TSPEC.

When the HC aggregates the admitted TSs, it sets the aggregation field in the associated TSPEC to 1. An HC schedules the transmissions in HCCA TXOPs and informs the service schedule to the non-AP station. The HC should provide an aggregate service schedule if the non-AP station sets the aggregation field of the TSPEC in its ADDTS request. If the HC establishes an aggregate service schedule for a non-AP station, it aggregates all TSs for the station. The service schedule is informed to the non-AP station in a schedule element of an ADDTS response or schedule frame. The service interval field value in the schedule element should be greater than the minimum SI. The service schedule could be subsequently updated by an AP as long as it meets TSPEC requirements. The HC may update the service schedule at any time by sending a schedule element in a schedule frame.

During any time interval [$t1$, $t2$] including the interval, which is greater than the specification interval, as specified in the schedule element, the cumulative TXOP duration should be greater than the time required to transmit all MSDUs (of nominal MSDU size) arriving at the mean data rate for the stream, over the period [$t1$, $t2 - D$]. The parameter D is set to the maximum SI specified in the TSPEC. If the maximum SI is not specified, then D is set to the delay bound in the TSPEC.

The HC uses the minimum PHY rate in calculating TXOPs if the minimum PHY rate is present in the TSPEC field in the ADDTS response. Otherwise, the HC may use an observed PHY rate in calculating TXOPs.

At least, a minimum set of TSPEC parameters should be specified during the TSPEC negotiation. The specification of a minimum set of parameters is required so that the scheduler can determine a schedule for the to-be-admitted TS. These parameters are mean data rate, nominal MSDU size, minimum PHY rate, surplus bandwidth allowance, and at least one of maximum service interval and delay bound in the ADDTS request frame. In the ADDTS response frame, the minimum set includes mean data rate, nominal MSDU size, minimum PHY rate, surplus bandwidth allowance, and maximum service interval. If both maximum SI and delay bound are specified in the ADDTS request, the HC may use only the maximum SI. If a station specifies a nonzero minimum SI and if the TS is admitted, the HC should generate a schedule that conforms to the specified minimum SI.

HCCA Scheduling Issues

In order to meet the negotiated QoS requirements, the HC needs to schedule its downlink frame transmissions as well as the QoS (+)CF-Poll frame transmissions properly. Since the wireless channel involves the time-varying and location-dependent channel conditions, developing a good scheduling algorithm is a challenging problem. Note that an intelligent scheduling algorithm can result in better system performance (e.g., not violating the negotiated QoS), while admitting more TSs. A non-AP QoS station operating under the HCCA also needs a scheduling algorithm to determine which frames to transmit during a polled TXOP. Note that different from the EDCA TXOP, in which the allowed frames are restricted to those of the same AC, during a polled TXOP, the TXOP holder can transmit whatever frames it chooses.

We here introduce a very simple scheduling algorithm for the HC, which is specified in the IEEE 802.11e specification as an informative example. The basic idea is to schedule fixed batches of TXOPs at constant time intervals for non-AP QoS stations with TSs as shown in Figure 14.18.

First, the SI is determined as follows:

$$SI = \min(\min(MSI_i), \text{ a submultiple of the beacon interval})$$

where MSI_i represents the *maximum service interval* (MSI) of the *i*th TS. Second, the *TXOP duration* (TD) for each allocation is determined as follows.

For the calculation of the TXOP duration for an admitted TS, the scheduler uses the following parameters: mean data rate (ρ) and nominal MSDU size (L) from the negotiated TSPEC; the scheduled SI calculated earlier; PHY transmission rate (R) (i.e., the minimum PHY rate from the TSPEC); maximum allowable size of MSDU, or 2,304 octets (M); and overheads in time units (O). Then, the scheduler calculates the number N_i of MSDUs that arrive at station i at the mean data rate during the SI as follows.

$$N_i = \left\lceil \frac{SI \times \rho_i}{L_i} \right\rceil$$

Then, the scheduler calculates the TXOP duration TD_i as the maximum of the time to transmit N_i frames at R_i and the time to transmit one maximum size MSDU at R_i (plus overheads):

$$TD_i = \max\left(\frac{N_i \times L_i}{R_i} + O, \frac{M}{R_i} + O \right)$$

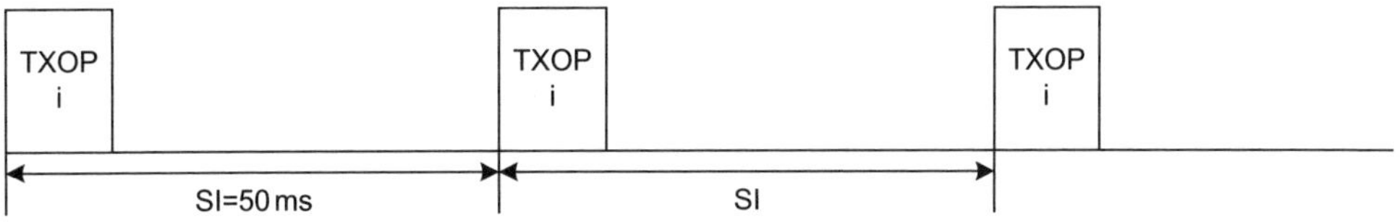

Figure 14.18 Schedule for TSs from station *i*. (*After:* [7].)

An example is shown in Figure 14.18. Stream from station *i* is admitted. The beacon interval is 100 ms and the maximum SI for the stream is 60 ms. The scheduler calculates a scheduled SI equal to 50 ms using the steps explained. As more stations establish TSs, there will be more TXOPs in each SI as shown in Figure 14.19. If a newly established TS has a maximum SI smaller than the current minimum, the SI will change.

More sophisticated scheduling algorithms have been proposed in the literature including *estimated transmission times earliest due date* (SETT-EDD) scheduling [17].

14.5 Other Optional Features

There are three more features defined as part of the 802.11e MAC, namely, *direct link setup* (DLS), *block Ack* (BlockAck), and *automatic power save delivery* (APSD) mechanisms. They are not directly related to the QoS provisioning but can increase the efficiency of the 802.11 WLAN. Per IEEE 802.11e, the support of these mechanisms is optional. In fact, all three mechanisms require a priori agreement or setup between communicating parties.

14.5.1 Direct Link Setup (DLS)

The first is the *direct link setup* (DLS) mechanism. The baseline MAC does not allow stations in an infrastructure BSS to transmit frames to each other directly, and instead the AP should always relay the frames. For certain applications (e.g., the bandwidth-intensive video streaming within a home), this limitation results in using the precious wireless bandwidth twice, and, hence, the 802.11e defines the mechanism to support the direct QoS station-to-QoS station transfer.

Basically, before commencing any direct frame transfer, a direct link is set up between two QoS stations via the DLS procedure, which involves the exchange of management frames between two QoS stations through the AP. Note that the DLS is not applicable in IBSSs, where frames are always transmitted directly between stations.

Figure 14.20 illustrates the direct link setup procedure. In the figure, there is an 802.11e AP and two QoS stations.

- Step 1a: QoS station 1 sends a *DLS request* frame to the AP.
- Step 1b: The AP forwards a DLS request frame to the recipient (i.e., QoS station 2).

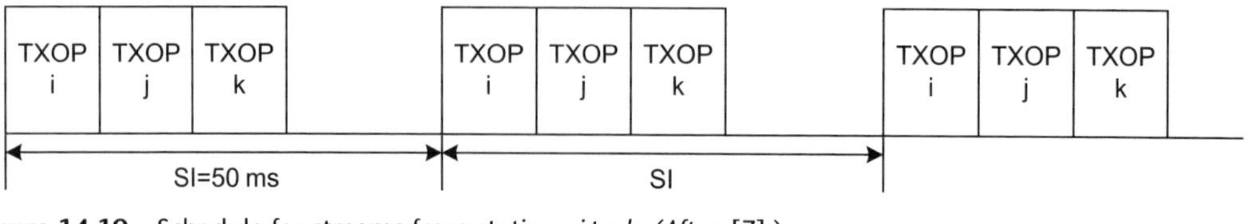

Figure 14.19 Schedule for streams from stations *i* to *k*. (*After:* [7].)

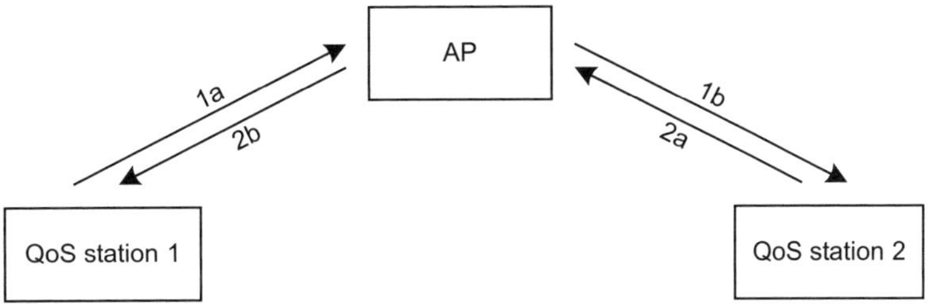

Figure 14.20 Direct link setup procedure. (*After:* [7].)

- Step 2a: QoS station 2 sends a *DLS response* frame to the AP.
- Step 2b: The AP forwards the DLS response frame to QoS station 1.

Among these steps, Step 1a may be skipped if the QoS AP initiates a DLS procedure. Note that both DLS request and DLS response frames are management type action frames (See Table 13.4). As part of the DLS request and response exchange, two QoS stations exchange their operational rates and other required information. In fact, a DLS request can be rejected by either QoS AP or recipient QoS station. Once a direct link is set up, both stations cannot go to the PSM as long as the direct link is active (i.e., there are active frame transfers). If there is not any transmission via a direct link between two stations for a given threshold, called DLS_Idle_Timeout, the direct link is declared inactive, and the direct link is torn down.

IEEE 802.11z, which is an ongoing standardization effort, is expected to enhance the DLS mechanism of the 802.11e by allowing operation with non-DLS-capable APs and also allowing stations with an active direct link session to enter PSM.

14.5.2 Block Ack

The baseline MAC defines a stop-and-wait *automatic retransmission request* (ARQ) scheme, which requires immediate ACK transmissions from the receiver after a SIFS interval from the data reception. The 802.11e defines two more acknowledgment policies, namely, *block acknowledgment* (BlockAck) and no Ack policies. The Ack policy to use for each frame is specified in the Ack policy field of the QoS control field in QoS data frames. Upon the reception of a QoS data frame with no Ack policy, the receiver does not take any action. The transmitter assumes that the frame transmission was successful as soon as the frame is transmitted. Note that the baseline MAC does not allow such a no-Ack policy for unicast frames while broadcast and multicast frames are not acknowledged. When multiple frames are transmitted with no Ack policy within a TXOP, they are transmitted back to back with SIFS intervals.

On the other hand, the block Ack mechanism allows a group of QoS data frames to be transmitted, each separated by a SIFS interval when they are transmitted within a TXOP, and then a *block Ack request* (BlockAckReq) frame is transmitted. Then, the recipient (i.e., the receiver of the data frames) transmits a single BlockAck frame in order to acknowledge the group of QoS data frames from the originator (i.e., transmitter of the data frames). The stop-and-wait ARQ of the baseline MAC involves a lot of overhead due to the immediate ACK transmissions.

However, the newly introduced block Ack allows a selective-repeat ARQ and can potentially enhance the system efficiency significantly.

Immediate and Delayed Block Ack

Two types of block Ack policies are defined, namely, immediate block Ack and delayed block Ack, depending on whether a BlockAck frame is transmitted immediately after a BlockAckReq frame reception. The former is suitable for high-bandwidth, low-latency traffic, while the latter is suitable for applications that are tolerant of moderate latency. The delayed block Ack policy requires minimal HW changes. During the setup of a block Ack between an originator and a recipient, the policy chosen to use is also agreed explicitly.

Figures 14.21 and 14.22 illustrate the frame transmissions under immediate and delayed block Ack policies, respectively. As shown in Figure 14.21, under the immediate block Ack policy, the BlockAck is transmitted after a SIFS interval from the BlockAckReq in the same TXOP. On the other hand, as shown in Figure 14.22, under the delayed block Ack policy, the BlockAckReq is responded by an Ack frame, and then the recipient transmits the BlockAck later in its own TXOP. The BlockAck is in turn acknowledged by an Ack frame. For the immediate Block Ack policy, the BlockAck should be transmitted with a SIFS interval, and it often requires specific hardware support since the SIFS is quite a short time interval.

The very first frame within a TXOP employing a block Ack policy should be followed by an immediate response. If the first frame is an RTS, then it will be followed by a CTS. On the other hand, if the first frame is a data frame, then it should be followed by an immediate Ack, as shown in Figures 14.21 and 14.22. That is, if the RTS/CTS exchange is not employed, the first data frame cannot use the block Ack policy. Note that if the first frame is not immediately acknowledged, upon the collision of the first frame, the originator cannot detect it, and, hence, the rest of the frame transmission might also involve further collisions so that the waste of the bandwidth due to the collision of the first frame can be quite enormous. The impact of using the block Ack even for the first data frame within a TXOP is evaluated in [18].

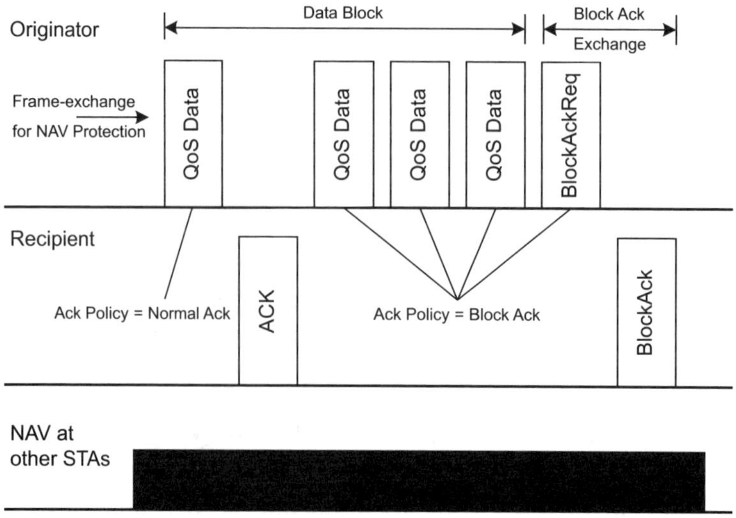

Figure 14.21 Immediate block Ack policy.

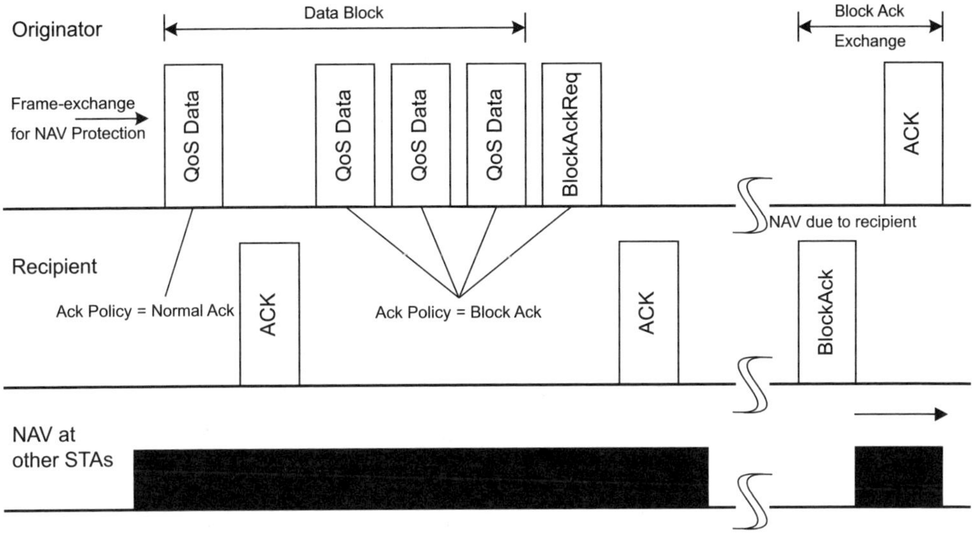

Figure 14.22 Delayed block Ack policy.

Block Ack Procedures

Figure 14.23 illustrates the frame exchanges related with block Ack mechanisms. First, the originator and the recipient exchange ADDBA request and ADDBA response frames in order to set up the block Ack mechanism for a particular TID. During this step, the originator and the recipient agree on the type of block Ack policy to use, and the recipient allocates some buffer space to support this block Ack agreement. Then, multiple data frame transfers as well as BlockAckReq and BlockAck frame exchanges occur. A BlockAck contains a bitmap, which indicates which of the previously transmitted frames were successfully received. The originator can selectively retransmit unsuccessfully transmitted frames in subsequent TXOPs.

Finally, the block Ack mechanism should be torn down explicitly by transmitting a DELBA request frame if the originator decides not to utilize the block Ack mechanism any longer. Peer failure happens if there is a timeout (i.e., no response from the other over a given threshold time interval), and then the BlockAck is torn down automatically. The ADDBA request, ADDBA response, and DELBA request frames are management type action frames (see Table 13.4).

Both BlockAckReq and BlockAck frames are control type frames. Figures 14.24 and 14.25 show the formats of BlockAckReq and BlockAck frames, respectively.

In Figure 14.24 for the BlockAckReq frame, the duration/ID field is set to the value greater than or equal to the time required to transmit one ACK or BlockAck frame, depending on whether a delayed or immediate block Ack policy is used, plus one SIFS interval. The *BAR control* field indicates the TID for which a BlockAck frame is requested. The *block Ack starting sequence control* field contains the sequence number of the first MSDU for which this BlockAckReq is sent.

If the BlockAck frame is sent under the immediate block Ack policy, the duration/ID field value is the value obtained from the duration/ID field of the immediate BlockAckReq frame, minus the time required to transmit the BlockAck frame and its SIFS interval. If the BlockAck frame is sent under the delayed block Ack policy,

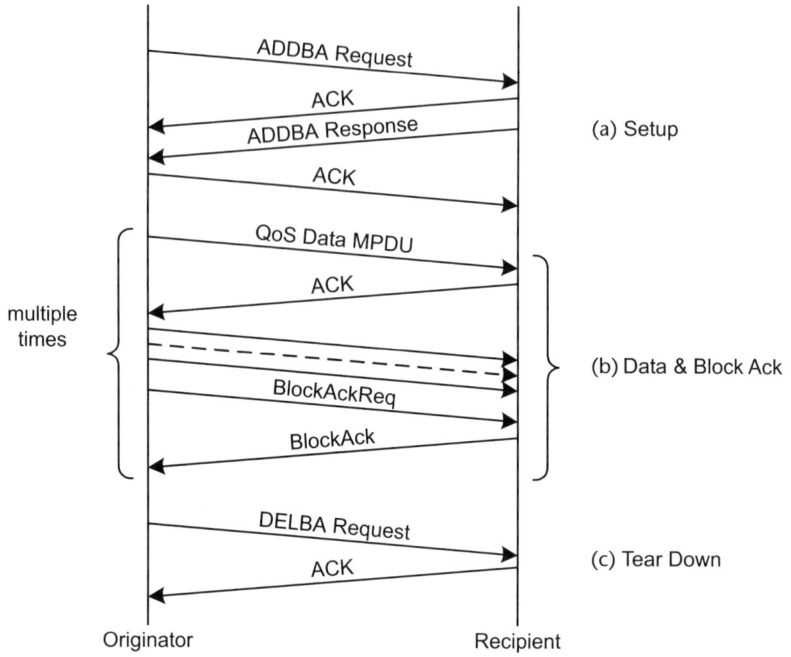

Figure 14.23 Message exchange for block Ack mechanism: (a) setup, (b) data and acknowledgment transfer, and (c) tear down. (*After:* [7].)

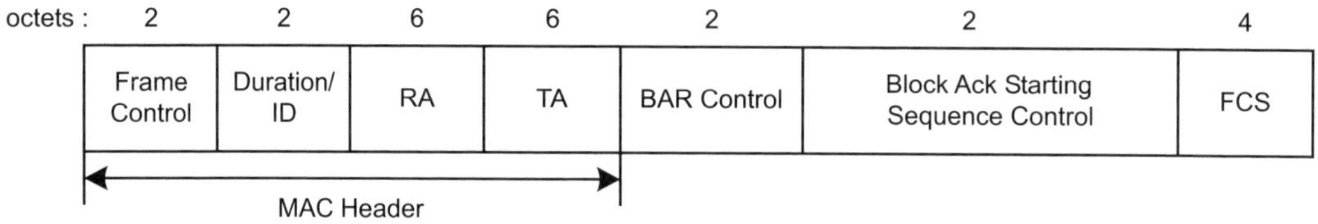

Figure 14.24 BlockAckReq frame. (*After:* [7].)

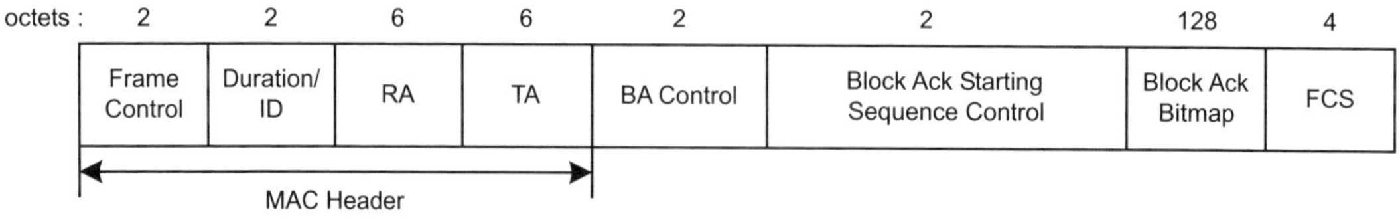

Figure 14.25 BlockAck frame. (*After:* [7].)

the duration/ID field value is set to the value greater than or equal to the time for transmission of an ACK frame plus a SIFS interval. The *BA control* field indicates the TID for which a BlockAck frame is requested.

The *block Ack starting sequence control* contains the sequence number of the first MSDU for which this BlockAck is sent, and it is actually copied from the immediately previously received BlockAckReq frame. The *block Ack bitmap* size is 128 octets long, and is used to indicate the receiving status of up to 64 MSDUs, since each of two octets are used for a single MSDU. The reason why two bytes are needed for a single MSDU is because an MSDU can be fragmented to up to 12 MPDUs. It turns out that this large block Ack bitmap size becomes a big overhead, thus making the block Ack even less efficient than the normal Ack policy unless a BlockAck frame acknowledges many data frames [18]. IEEE 802.11n, which is currently being standardized, is developing an enhancement of the block Ack mechanism, and one enhancement is the definition of a compressed bitmap with the length of only 8 octets. Further details will be presented in Section 18.1.3.

14.5.3 Automatic Power Save Delivery (APSD)

As discussed in Section 13.5.2, the PSM operation is defined in the baseline MAC for the energy-efficient operation of the 802.11 stations by staying in the doze state whenever there is no active traffic. Unfortunately, the PSM is not an agile operation in the sense that the wakeup instances are bound with TBTTs. When a station communicates with a periodic traffic pattern, where the period is less than the beacon period, it is very difficult (if not impossible) to utilize the PSM operation. The APSD is a mechanism to deliver unicast downlink frames to power-saving stations without sacrificing the QoS. This mechanism allows a station to switch back and forth between the doze and active states, where the transition timing is not bound with the beacon intervals.

In order to use the APSD, it should be set up between two stations in advance as part of the TS setup. The usage is indicated in the APSD bit in the TS info field of the TSPEC element. There are two types of APSDs, namely, *unscheduled APSD* (U-APSD) and *scheduled APSD* (S-APSD), as determined by the APSD and schedule subfields in the TS info field (see Table 14.5 for the encoding). The U-APSD is available only when the access policy associated with a TS is the EDCA, while the S-APSD is available irrespective of the access policy. Under the U-APSD, non-AP QoS stations basically transmit their frames during unscheduled SPs, while under the S-APSD, non-AP QoS stations basically transmit their frames during scheduled SPs. Irrespective of the APSD types, the non-AP QoS station basically stays in the doze state between two consecutive SPs to save the energy. The end of a SP is signaled via the EOSP bit of the QoS control field in a QoS data frame sent by the HC.

Unscheduled APSD

The U-APSD operation is based on unscheduled SPs, which are SPs triggered by the corresponding non-AP QoS station's frame transmission. That is, a non-AP QoS station wakes up in order to trigger an unscheduled SP when it expects that there are buffered downlink frames destined to itself. For the TS using the U-APSD, the EDCA is the only available access policy. We first define delivery-enabled and trigger-enabled ACs as follows:

- *Delivery-enabled AC*: an AC at QoS AP where the AP is allowed to use EDCA to deliver traffic from the AC to a non-AP QoS station in an unscheduled SP triggered by the station;

- *Trigger-enabled AC*: an AC at non-AP QoS station where QoS data and QoS null frames from the non-AP station that map to the AC trigger an unscheduled SP.

An unscheduled SP begins when the AP receives a trigger frame from a non-AP station, where a trigger frame is a QoS data or QoS null frame associated with an AC the station has configured to be trigger-enabled. An unscheduled SP ends after the AP has attempted to transmit at least one MSDU or MMPDU associated with a delivery-enabled AC and destined for the non-AP station, where the maximum number of buffered MSDUs and MMPDUs, which can be transmitted during an unscheduled SP, is determined and informed by the non-AP station.

Each of the ACs can be configured to be delivery-enabled/trigger-enabled by a non-AP QoS station and can be informed to the AP via the (re)association request frame during the (re)association procedure.

Scheduled APSD

The S-APSD operation is based on scheduled SPs, where the schedule is specified in the schedule element within the ADDTS response or schedule frames. For the TS using the S-APSD, any type of access policies can be used.

Scheduled SPs appear periodically with the period of *service interval* (SI) found in the service interval field of the schedule element. The first scheduled SP starts when the 4 lower order octets of the TSF timer equals the value specified in the service start time field. A non-AP station using scheduled SP first wakes up to receive downlink unicast frames buffered and/or polls from the HC. The station wakes up subsequently at a fixed time interval equal to the SI.

References

[1] Choi, S., et al., "IEEE 802.11e Contention-Based Channel Access (EDCF) Performance Evaluation," *Proc. IEEE International Conference on Communications 2003 (ICC'03)*, Anchorage, AK, May 2003.

[2] Mangold, S., et al., "IEEE 802.11e Wireless LAN for Quality of Service," *Proc. European Wireless (EW'02)*, Vol. 1, Florence, Italy, February 2002, pp. 32–39.

[3] Yu, J., S. Choi, and J. Lee, "Enhancement of VoIP over IEEE 802.11 WLAN Via Dual Queue Strategy," *Proc. IEEE International Conference on Communications 2004 (ICC'04)*, Paris, France, June 2004.

[4] IETF RFC 2475, An Architecture for Differentiated Services, December 1998.

[5] IETF RFC 1633, Integrated Services in the Internet Architecture: An Overview, June 1994.

[6] IEEE 802.1D-2004, IEEE Standard for Local and Metropolitan Area Networks—Media Access Control (MAC) Bridges (Incorporates IEEE 802.1t-2001 and IEEE 802.1w), 2004.

[7] IEEE 802.11-2007, IEEE Standard for Local and Metropolitan Area Networks—Part 11: Wireless LAN Medium Access Control (MAC) and Physical Layer (PHY) Specifications, IEEE Std 802.11-2007 (Revision of IEEE Std 802.11-1999), June 12, 2007.

[8] Grilo, A., and M. Nunes, "Performance Evaluation of IEEE 802.11e," *Proc. The 13th* IEEE *International Symposium on Personal, Indoor and Mobile Radio Communications (PIMRC'02)*, Lisbon, Portugal, September 2002.

[9] Ni, Q., "Performance Analysis and Enhancements for IEEE 802.11e Wireless Networks," *IEEE Network*, Vol. 19, No. 4, July/August 2005, pp. 21–27.

[10] Berger-Sabbatel, G., et al., "Performance Anomaly of 802.11b," *Proc. IEEE INFOCOM'03*, San Francisco, CA, March 2003.

[11] Yang, D., et al., "Performance Enhancement of Multi-Rate IEEE 802.11 WLANs with Geographically Scattered Stations," *IEEE Trans. on Mobile Computing*, Vol. 5, No. 7, July 2006, pp. 906–919.

[12] IETF RFC 2212, Specification of Guaranteed Quality of Service, September 1997.

[13] IETF RFC 2215, General Characterization Parameters for Integrated Service Network Elements, 1997.

[14] IETF RFC 3290, An Informal Management Model for Diffserv Routers, 2002.

[15] IETF RFC 2474, Definition of the Differentiated Services Field (DS Field) in the IPv4 and IPv6 Headers, 1998.

[16] IEEE 802.1Q-2005, IEEE Standard for Local and Metropolitan Area Networks—Virtual Bridged Local Area Networks (Incorporates IEEE 802.1Q-1998, IEEE 802.1u-2001, IEEE 802.1v-2001, and IEEE 802.1s-2002), 2006.

[17] Grilo, A., M. Macedo, and M. Nunes, "A Scheduling Algorithm for QoS Support in IEEE 802.11e Networks," *IEEE Wireless Communications Magazine*, Vol. 10, No. 3, June 2003, pp. 36–43.

[18] Tinnirello, I., and S Choi, "Efficiency Analysis of Burst Transmissions with Block ACK in Contention-Based 802.11e WLANs," *Proc. IEEE International Conference on Communications (ICC'05)*, Seoul, Korea, May 2005.

Selected Bibliography

Aad, I., and C. Castelluccia, "Differentiation Mechanisms for IEEE 802.11," *Proc. IEEE INFOCOM'01*, Anchorage, AK, April 2001.

Ansel, P., Q. Ni, and T. Turletti, "FHCF: A Simple and Efficient Scheduling Scheme for IEEE 802.11e," *Springer/Kluwer Journal on Mobile Networks and Applications (MONET)*, Vol. 11, No. 3, 2006, pp. 391–403.

Banchs, A., and L. Vollero, "Throughput Analysis and Optimal Configuration of 802.11e EDCA," *Computer Networks*, Vol. 50, No. 11, 2006, pp. 1749–1768.

Bianchi, G., I. Tinnirello, and L. Scalia, "Understanding 802.11e Contention-Based Prioritization Mechanisms and Their Coexistence with Legacy 802.11 Stations," *IEEE Network*, Vol. 19, No. 4, 2005, pp. 28–34.

Boggia, G., et al., "Feedback-Based Control for Providing Real-Time Services with the 802.11e MAC," *IEEE/ACM Trans. on Networking*, Vol. 15, No. 2, April 2007, pp. 323–333.

Chen, X., et al., "Supporting QoS in IEEE 802.11e Wireless LANs," *IEEE Trans. on Wireless Communications*, Vol. 5, No. 8, August 2006, pp. 2217–2227.

Cicconetti, C., et al., "Design and Performance Analysis of the Real-Time HCCA Scheduler for IEEE 802.11e WLANs," *Computer Networks*, Vol. 51, No. 9, 2007, pp. 2311–2325.

Chou, C. T., K. G. Shin, and S. N. Shankar, "Contention-Based Airtime Usage Control in Multirate IEEE 802.11 Wireless LANs," *IEEE/ACM on Trans. Networking*, Vol. 14, No. 6, December 2006, pp. 1179–1192.

Gao, D., J. Cai, and K. N. Ngan, "Admission Control in IEEE 802.11e Wireless LANs," *IEEE Networks*, Vol. 19, No. 4, July/August 2003, pp. 6–13.

Gao, D., and J. Cai, "Admission Control with Physical Rate Measurement for IEEE 802.11e Controlled Channel Access," *IEEE Commun. Letters*, Vol. 9, No. 8, August 2005, pp. 694–696.

Ge, Y., J. C. Hou, and S. Choi, "An Analytic Study of Tuning Systems Parameters in IEEE 802.11e Enhanced Distributed Channel Access," *Computer Networks*, Vol. 51, No. 8, June 2007, pp. 1955–1980.

Hui, J., and M. Devetsikiotis, "A Unified Model for the Performance Analysis of IEEE 802.11e EDCA," *IEEE Trans. on Communications*, Vol. 53, No. 9, 2005, pp. 1498–1510.

Hwang, G. H., and D. H. Cho, "Performance Analysis on Coexistence of EDCA and Legacy DCF Stations in IEEE 802.11 Wireless LANs," *IEEE Trans. on Wireless Communications*, Vol. 5, No. 12, December 2006, pp. 3355–3359.

Kong, Z., et al., "Performance Analysis of IEEE 802.11e Contention-Based Channel Access," *IEEE Journal on Selected Areas Communications*, Vol. 22, No. 10, December 2004, pp. 2095–2106.

Park, S., et al., "Collaborative QoS Architecture Between DiffServ and 802.11e Wireless LAN," *Proc. IEEE VTC 2003-Spring*, Jeju, Korea, April 2003.

Ramaiyan, V., A. Kumar, and E. Altman, "Fixed Point Analysis of Single Cell IEEE 802.11e WLANs: Uniqueness, Multistability and Throughput Differentiation," *Proc. ACM SIGMETRICS'05*, Banff, Alberta, Canada, June 6–10, 2005, pp. 109–120.

Skyrianoglou, D., N. Passas, and A. K. Salkintzis, "ARROW: An Efficient Traffic Scheduling Algorithm for IEEE 802.11e HCCA," *IEEE Trans. on Wireless Communications*, Vol. 5, No. 12, December 2006, pp. 3558–3567.

Tao, Z., and S. Panwar, "Throughput and Delay Analysis for the IEEE 802.11e Enhanced Distributed Channel Access," *IEEE Trans. on Communications*, Vol. 54, No. 4, 2006, pp. 596–603.

Xiao, Y., "Performance Analysis of Priority Schemes for IEEE 802.11 and IEEE 802.11e Wireless LANs," *IEEE Trans. on Wireless Communications*, Vol. 4, No. 4, July 2005, pp. 1506–1515.

Xiao, Y., H. Li, and S. Choi, "Two-Level Protection and Guarantee for Multimedia Traffic in IEEE 802.11e Distributed WLANs," *ACM Wireless Networks (WINET)*, February 2007.

Zhang, L., and S. Zeadally, "HARMONICA: Enhanced QoS Support with Admission Control for IEEE 802.11 Contention-Based Access," *Proc. Real-Time and Embedded Technology and Applications Symp.*, May 2004.

Zhu, J., and A. O. Fapojuwo, "A New Call Admission Control Method for Providing Desired Throuhput and Delay Performance in IEEE 802.11e Wireless LANs," *IEEE Trans. on Wireless Communications*, Vol. 6, No. 2, February 2007, pp. 701–709.

CHAPTER 15

Security Mechanisms

Wireless communication could be inherently insecure due to the broadcast nature of the transmissions. For example, an attacker can eavesdrop on the communication over a wireless channel easily by passively monitoring the signals from a remote place using a high-gain antenna. Accordingly, a special mechanism to protect the communication is needed. For the wide acceptance of IEEE 802.11 WLANs, a strong security mechanism is considered a key factor.

IEEE 802.11i-2004 defines the *robust security network association* (RSNA), which is established between two stations (i.e., a station and an AP in an infrastructure BSS or a pair of stations in an IBSS [1]). The RSNA relies on IEEE 802.1X [2] to transport its authentication services and to deliver key management services. Therefore, all stations and APs in an RSNA contain an 802.1X entity that handles these services, and the 802.11i defines how an RSNA utilizes the 802.1X in order to access these services.

The 802.11i RSNA enhances the security mechanisms defined in IEEE 802.11-1999. The legacy mechanisms are referred to as pre-RSNA security mechanisms, composed of data confidentiality via *wired equivalent protection* (WEP) encapsulation and authentication between two stations, and were found to have many security holes and be easy to break. The RSNA defines a number of security features on top of the WEP and IEEE 802.11 authentication including: (1) enhanced mutual authentication mechanisms for both APs and stations, (2) key management algorithms, (3) cryptographic key establishment, and (4) an enhanced data confidentiality mechanism, called *counter mode with CBC-MAC* (cipher-block chaining message authentication code) *protocol* (CCMP) and, optionally, *temporal key integrity protocol* (TKIP).

IEEE 802.11i defines the *robust security network* (RSN), where only the RSNA is allowed, as well as the *transition security network* (TSN), where both the RSNA and pre-RSNA are allowed. The TSN is supposed to support the coexistence of the 802.11i stations and the 802.11-1999 stations. We first present the pre-RNSA security mechanisms, defined in IEEE 802.11-1999, since some functions of the RSNA are built on top of the pre-RSNA.

15.1 Pre-RSNA Security

The pre-RSNA security mechanisms are composed of authentication and data confidentiality supports. Except for open system authentication, all pre-RSNA security mechanisms have been deprecated as part of IEEE 802.11i since they fail to meet their security goals. New implementations might support pre-RSNA methods only for the backward compatibility purpose.

15.1.1 Wired Equivalent Privacy

WEP was developed to protect (using a 40-bit security key) the confidentiality of data exchanged among authorized stations from eavesdropping. As the name stands for (i.e., the wired equivalent protection), it was originally believed to be capable of supporting a security level as strong as wired networking. However, since then, lots of security flaws have been identified, as discussed in Section 15.1.3. Per IEEE 802.11-1999, the implementation of WEP is optional. The same algorithms have been widely used with a 104-bit key instead of a 40-bit key in many commercial implementations, even if the usage of 104-bit key was not defined in IEEE 802.11-1999. Depending on the size of the key, WEP is referred to as WEP-40 or WEP-104. Note that WEP-104 has been a de facto standard. The WEP cryptographic encapsulation and decapsulation mechanisms are the same irrespective of the size of the key is employed. Accordingly, we present the WEP mechanism assuming the 40-bit key system as defined in the standard.

WEP Encapsulation

Figure 15.1 illustrates the WEP-encapsulated MPDU format compared with the nonencrypted one. The frame body is basically expanded by 8 octets by including the *initialization vector* (IV) field and the *integrity check value* (ICV) field. The IV field includes three subfields, including the 3-octet *initialization vector*, the 2-bit *key identifier* (ID), and 6-bit pad (set to all zero). The key ID indicates one of four possible secret key values for use in decrypting this frame body at the receiver. The WEP ICV is computed using the CRC-32, as defined in Section 13.1.1, calculated over the plaintext (i.e., original nonencrypted) MPDU data field.

The WEP scheme uses the RC4 *pseudo-random number generator* (PRNG) algorithm, based on 64-bit keys. The RC4 is a stream cipher from *RSA Data Security, Inc.* [4]. Figure 15.2 illustrates the block diagram of the WEP encapsulation. A 64-bit *seed* is generated by combining a 40-bit *secret key* (which should be known to both the transmitter and the receiver offline) and a 24-bit IV chosen by the transmitter station. It is recommended to change the IV every frame.

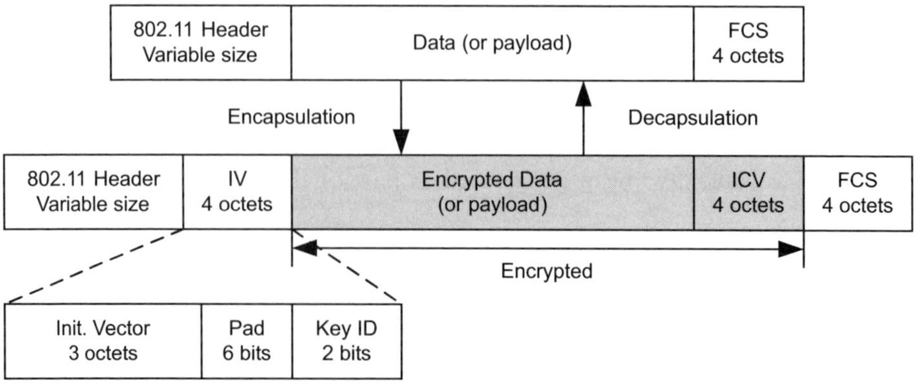

Figure 15.1 Original frame versus WEP-encapsulated MPDU. (*After:* [3].)

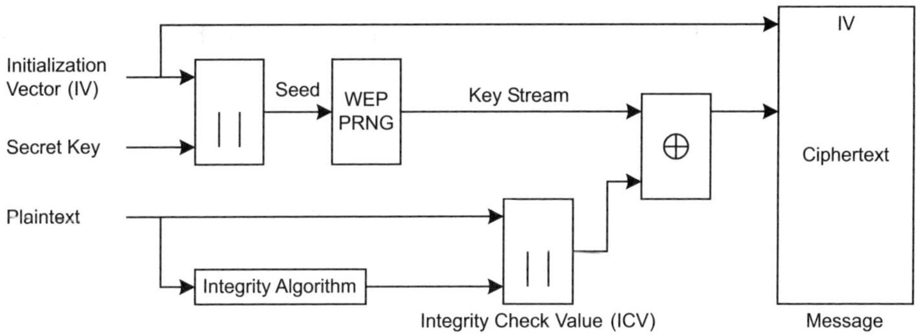

Figure 15.2 WEP encapsulation block diagram. (*After:* [3].)

There are two types of secret keys, namely, *key-mapping keys* and *default keys*. A key-mapping key can be established between a pair of a transmitter and a receiver. If a key-mapping key is not established between two stations, default keys should be used. There can be up to four default secret keys established in a BSS, and one of them is used for an MPDU encryption. The employed secret key is identified in the *key ID* subfield of the IV field within the MPDU. On the other hand, an *integrity algorithm*, based on CRC-32, is applied to the plaintext to generate an ICV. The ICV is intended for the receiver to check the integrity of the received frame. Then, the key sequence generated using the RC4 algorithm is XORed with the plaintext and the ICV to generate a ciphertext. The ciphertext along with the IV value is transmitted in the frame body of the MPDU, as shown in Figure 15.1. The fact that the MPDU was WEP-encapsulated is indicated in the *protected frame* bit within the frame control field of the MAC header, as shown in Figure 15.3.

WEP Decapsulation

The receiver station performs the reverse operation by decrypting the received frame body and checking if the decrypted frame is intact, as illustrated in Figure 15.3. Upon the reception of a WEP-encapsulated MPDU, the receiver identifies both the employed secret key and the IV from the IV field of the received MPDU. Then, by combining both the secret key and the IV, a seed is generated, which is fed

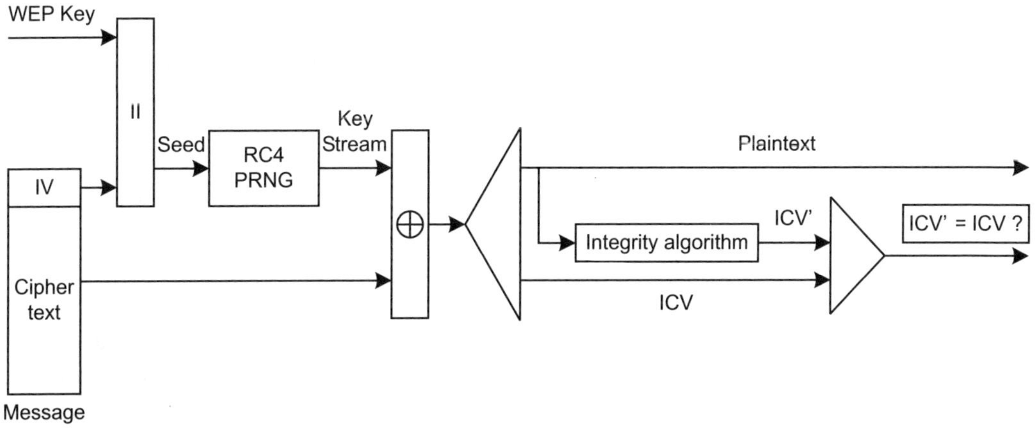

Figure 15.3 WEP decapsulation block diagram. (*After:* [3].)

into the RC4 PRNG. The key stream generated by the RC4 PRNG is XORed with the ciphertext found at the received MPDU.

Finally, using the plaintext out of the XOR operation, the ICV is calculated via the CRC-32 algorithm, and then it is bit-wise compared with the decrypted ICV from the received MPDU. If the two are bit-wise identical, the received MPDU is determined to be valid, and the plaintext excluding the ICV field is forwarded to the higher layer.

15.1.2 Pre-RSNA Authentication

The Pre-RSNA is often referred to as the *IEEE 802.11 authentication*, since a new authentication procedure under IEEE 802.11i employs IEEE 802.1X for the authentication. The 802.11 authentication is performed between two stations in either an IBSS or an infrastructure BSS. In case of the infrastructure BSS, the authentication is between a station and an AP, and only after a successful authentication can an association between the station and the AP be established. The authentication is optional in an IBSS.

There are two forms of the authentication, namely, *open system* and *shared key* authentications. The open system is virtually equivalent with no authentication since two stations simply exchange authentication frames under this type of authentication. On the other hand, with the shared key type, two stations exchange four frames to check if they have the same secret key. Unless they have the same key, the authentication process is supposed to fail.

Authentication Frame

The authentication procedure is based on the exchange of authentication frames. Unicast is the only option for the authentication frames, since the authentication is performed between a pair of stations. On the other hand, the deauthentication frames used for the deauthentication are advisory and may be sent as group-addressed frames. Shared key authentication is deprecated and should not be implemented except for backward compatibility with pre-RSNA devices.[1]

An authentication frame contains a number of fields, including: (1) *authentication algorithm number*, (2) *authentication transaction sequence number*, (3) *status code*, (4) *challenge text*, and (5) *vendor-specific information*. Table 15.1 summarizes six types of authentication frames along with the corresponding field values. The usage of each specific authentication frame will be detailed next.

Here we will use the following nicknames to represent authentication frames with different transaction sequence numbers (see Table 15.2). The station which transmits *authentication-request* and *authentication-confirm* frames are referred to as a *requester*, while the station that transmits *authentication-response* and *authentication-ack* are referred to as a *responder*.

Open System Authentication

The open system authentication is virtually equivalent with no authentication. The requester transmits an authentication-request frame (with the authentication

1. As discussed in Section 16.4, per emerging IEEE 802.11r, the shared key authentication is reinstated for the support of a fast BSS transition.

Table 15.1 Six Types of Authentication Frames

Authentication algorithm	Authentication transaction sequence no.	Status code	Challenge text
Open System	1	Reserved	Not present
Open System	2	Status	Not present
Shared Key	1	Reserved	Not present
Shared Key	2	Status	Present
Shared Key	3	Reserved	Present
Shared Key	4	Status	Not present

Source: [3].

Table 15.2 Nicknames for Various Authentication Frames

Nickname	Authentication transaction sequence number
Authentication-Request	1
Authentication-Response	2
Authentication-Confirm	3
Authentication-Ack	4

algorithm = "open system") to the responder, which in turn responds with an authentication-response with a status code. The status code basically indicates either success or failure. Even if there is no authentication algorithm involved with this open system authentication method, a rule such as the MAC address filtering could be employed. That is, the requester is authenticated only if the requester's MAC address is found in the list of reliable stations (i.e., the *access control list*) maintained by the responder. However, this simple authentication rule can be easily broken since there are methods, known as a *MAC address spoofing*, to manually change the MAC address of a station. One can easily learn valid (i.e., authenticated) MAC addresses by eavesdropping the frame transmissions over the air and then can use it by changing its own MAC address.

Shared Key Authentication

The shared key authentication involves a four-way handshake to determine whether both the requester and the responder have the same secret key, which is used for WEP encryption. In order for this authentication algorithm to work, both the requester and the responder should have communicated the secret key offline via a separate communication channel (e.g., via a manual setting by a system administrator or an end user). As illustrated in Figure 15.4, the frame exchange for the shared key authentication works as follows:

- The requester sends an authentication-request frame (with the authentication algorithm = "shared key") to the responder.

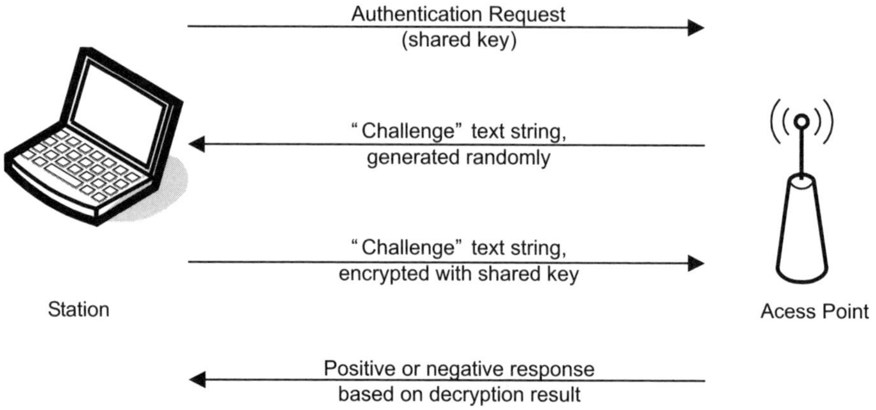

Figure 15.4 Shared key authentication.

- The responder sends an authentication-response frame, including a challenge text of up to 253 octets. The challenge text is randomly generated.
- The requester WEP-encapsulates the received challenge text and sends it to the responder via an authentication-confirm frame. The WEP encapsulation is presented in Section 15.1.1.
- The responder WEP-decapsulates the received challenge text and checks if the decrypted text matches with the original challenge text, which it sent earlier in the authentication-response via the WEP ICV check. The match implies that both the requester and the responder have the same secret key. The result is conveyed to the requester via an authentication-ack frame.

15.1.3 Limitations of Pre-RSNA

It has been known that the pre-RSNA security mechanisms are basically limited. There have been a lot of reports on the security problems of the pre-RSNA [5–8]. We briefly discuss some of them here.

Authentication

Both the authentication and the encryption should use the same secret key, which is architecturally flawed. Moreover, the authentication is only one way. That is, in an infrastructure BSS, a station is authenticated by the AP, but not the other way around. Moreover, as discussed in Section 15.1.1, the access control list-based authentication is not secure at all.

Confidentiality

The RC4 algorithm is known to be cryptographically weak so that the secret key can be easily recovered. The situation becomes worse due to the fact that the first bytes of the plaintext are known in most cases due to the LLC/SNAP header (i.e., 0xaa) as presented in Section 11.3.2. It was also known that there exist certain weak IVs. Moreover, some earlier implementation of the WEP did not even change the IV. Even if the IV is changed every frame, the IV space is too small (i.e., 2^{24}).

The lack of a key management is another big problem. The four default keys have to be set up manually, and, in many WLANs, these keys are rarely changed over time due to the administrative difficulty. When the commonly employed default keys are used, all the stations in the BSS share those keys. Accordingly, the confidentiality among the stations in the BSS is not actually achieved. Even if using key-mapping keys, established for a pair of a transmitter and a receiver, is possible, it is rarely used due to the difficulty for a manual setting of such keys.

Other Issues

The integrity is provided by using the ICV based on CRC-32, which was not originally developed for this purpose. The CRC is a linear function, and that allows the attacker to flip arbitrary bits in the ciphertext and correctly adjust the ICV so that the modified ciphertext appears valid. Finally, no replay protection is provided; any intercepted frame can be simply resent.

15.2 Robust Security Network Association (RSNA)

IEEE 802.11i-2004 defines the *robust security network* (RSN). The 802.11i defines two cryptographic algorithms, namely, mandatory CCMP and optional TKIP. The TKIP is based on the RC4 algorithm as the pre-RSNA WEP is, and the CCMP is based on a new cipher, called *advanced encryption standard* (AES). Unless stated otherwise, we consider the security in infrastructure BSSs in this section since the infrastructure BSSs are practically more popular than the IBSSs even though the 802.11i addresses the security mechanisms in IBSSs as well. We also do not consider the station-to-station direct communication in an infrastructure BSS (i.e., via a direct link established through the 802.11e DLS as presented in Section 14.5.1), even if the 802.11i addresses this situation as well.

In an RSN, two stations (i.e., a non-AP station and an AP) make a *robust security network association* (RSNA), which is established on the foundation of the following features:

- Enhanced mutual authentication mechanisms for both AP and station;
- Key management algorithms;
- Cryptographic key establishment;
- Enhanced data encapsulation mechanisms, called CCMP and TKIP.

We present the procedures to establish an RSNA in detail in the following. We first consider IEEE 802.1X port-based access control protocol, which is used for the 802.11i authentication.

15.2.1 IEEE 802.1X Port-Based Access Control

IEEE 802.11i RSNA relies on IEEE 802.1X-2004 to provide the authentication and key management services. In a network where the 802.1X is employed, there exist three entities as follows [2]:

- *Supplicant*: an entity at one end of a point-to-point LAN segment, which is being authenticated by an authenticator attached to the other end of that link;
- *Authenticator*: an entity at one end of a point-to-point LAN segment, which helps the authentication of the entity attached to the other end of that link;
- *Authentication server* (AS): an entity that provides an authentication service to an authenticator. This service determines, from the credentials provided by the supplicant, whether the supplicant is authorized to access the network services provided by the authenticator.

As illustrated in Figure 15.5, in the context of the 802.1X, a non-AP station performs the role of the supplicant, and the AP does the authenticator role. In this chapter, we will use the terms "station" and "supplicant" interchangeably, and also the terms "AP" and "authenticator" interchangeability as well. The AS is a server located in the infrastructure (or possibly the AP itself). Typically, a *remote authentication dial-in user service* (RADIUS) server can be used as an AS [9].

IEEE 802.1X controls the flow of MSDUs between the *distribution system* (DS) and supplicants by utilizing the *controlled/uncontrolled port* model. The 802.1X authentication frames are transmitted in IEEE 802.11 data frames, rather than the 802.11 management frames, and passed via the uncontrolled port of the authenticator. Until the 802.1X authentication procedures as well as the key distribution complete successfully over the uncontrolled port, the general data traffic between a supplicant and its authenticator is blocked by the controlled port.

It is assumed that a secure channel is established between the authenticator and the AS in advance, and how to establish such a secure channel is not within the scope of IEEE 802.11i. Suitable protocols for the secure channel include RADIUS [9] and DIAMETER [10].

EAP and EAPOL

The 802.1X does not define its own authentication protocol; it relies on the *extensible authentication protocol* (EAP) [11], which provides a generalized framework for different authentication methods. The RSNA depends on the use of an EAP that supports mutual authentication (e.g., *EAP-transport layer security* (EAP-TLS) [12]), between the AS and the supplicant, not just the unilateral authentication of the supplicant to an authenticator. While the 802.11i does not specify a specific EAP method, the EAP-TLS has been a de facto standard for the 802.11i EAP authentica-

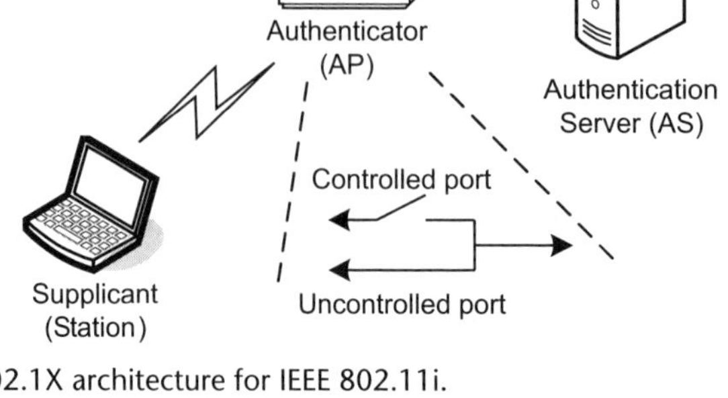

Figure 15.5 IEEE 802.1X architecture for IEEE 802.11i.

tion. Some other proprietary EAP methods, which are used in commercial products, include *EAP tunneled transport layer security* (EAP-TTLS), developed by Funk Software and Certicom [13], and *protected extensible authentication protocol* (PEAP), developed by Microsoft, Cisco, and RSA Security [14]. PEAP and EAP-TTLS make it possible to authenticate wireless LAN stations without requiring them to have digital certificates. That is, only the AS is required to have a certificate. EAP-TLS requires both the supplicant and the AS to have certificates. Both PEAP and EAP-TTLS utilize TLS to set up an end-to-end tunnel to transfer the supplicant's credentials without having to use a certificate on the supplicant. The requirements for the EAP methods to be used in IEEE 802.11 WLAN are discussed in [15].

IEEE 802.1X defines an encapsulation method to carry EAP messages between supplicant and authenticator. This encapsulation is known as *EAP over LAN* (EAPOL). For the 802.11i, three types of EAPOL frames are utilized, namely, *EAPOL-start*, *EAP-packet*, and *EAPOL-key*. There are four types of EAP-packet frames depending on the code in the frame, namely, *EAP request*, *EAP response*, *EAP success*, and *EAP failure*. EtherType 0x888e (i.e., *port access entity Ethernet type*) is used for EAPOL frames. Note that the LLC/SNAP header, discussed in Section 11.3.2, can be used for EAPOL frames. EAP-packet is a container used to transfer EAP messages. EAP request and response frames are exchanged a number of times to transfer the identification and credentials between supplicant and authenticator. The contents in the EAP request and response frames and the number of frame exchanges depend on the employed EAP method. Figure 15.6 illustrates the EAP authentication procedure. EAPOL-key frames are used to exchange keys between station and AP. The frame format of the EAPOL-key is presented in Section 15.3.2.

15.2.2 RSNA Establishment

In an ESS, a station establishes an RSNA with its AP using either IEEE 802.1X authentication and key management or using a *preshared key* (PSK). When the 802.1X is used, the station establishes an RSNA via the following procedures.

1. The station identifies an AP as RSNA-capable via scanning.

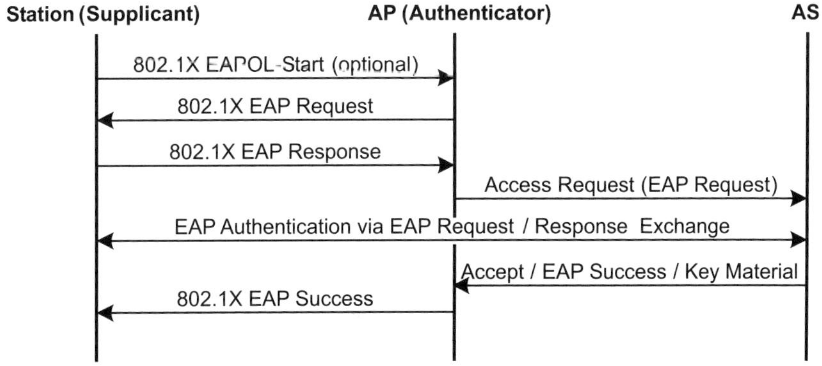

Figure 15.6 IEEE 802.1X EAP authentication. (*After:* [3].)

2. The station is authenticated by the AP via the open system authentication of IEEE 802.11 pre-RSNA.
3. The station negotiates cipher suites (e.g., either TKIP or CCMP) during the (re)association.
4. The station and the AS authenticate each other via the 802.1X authentication procedure and generate a *pairwise master key* (PMK). The PMK is then sent from the AS to the authenticator (i.e., AP).
5. Based on the PMK, a *pairwise transient key* (PTK) is derived. Then, both the PTK and *group transient key* (GTK) are established at both the station and the AP via the four-way handshake of *EAPOL-key* messages.
6. The IEEE 802.1X controlled port is unblocked. The station and the AP use the PTK and GTK for the protection of unicast and broadcast/multicast frames, respectively.

Note that the shared key authentication of the pre-RSNA is deprecated as part of the 802.11i, since the 802.11i relies on the 802.1X for authentication after the (re)association. If a PSK is used, step (4) can be skipped, and instead the PSK is used as the PMK for the next procedure.

Figure 15.7 illustrates the entire procedures for the RSNA authentication and key management. Each step will be detailed next. Steps (1) to (3) are shown in Figure 15.7(a). Then, step (4) corresponding to the 802.1X authentication is shown in

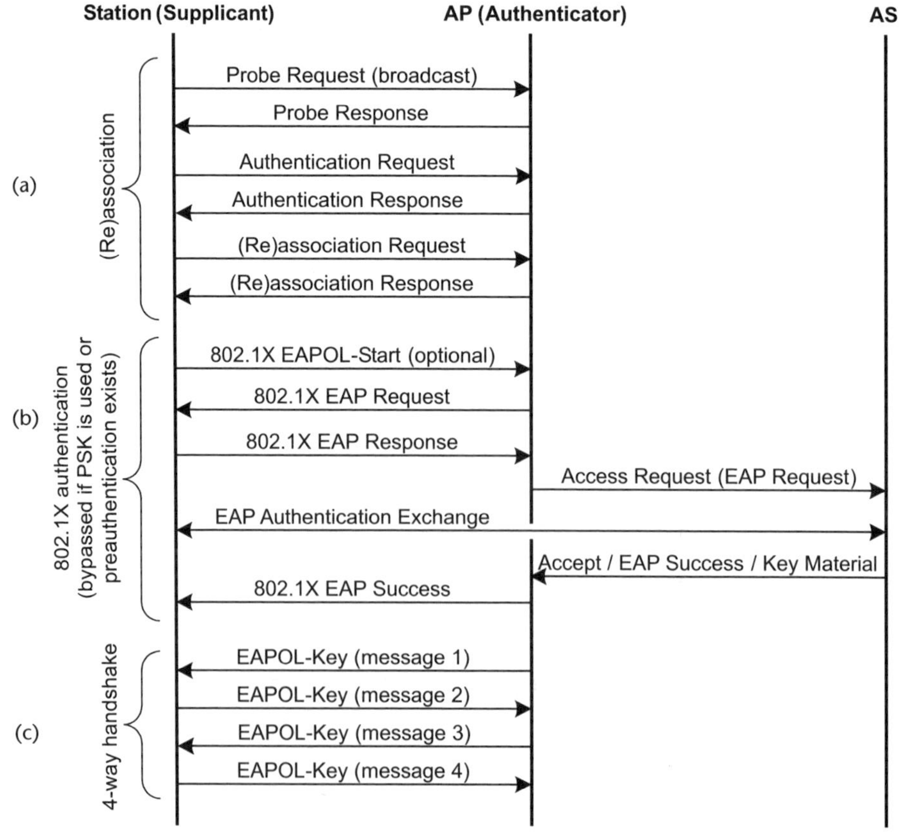

Figure 15.7 (a–c) IEEE 802.11i RSNA setup procedure.

Figure 15.7(b). Finally, step (5) corresponding to the four-way handshake is shown in Figure 15.7(c). When a GTK is newly generated by the AP, it is notified to the station via a group key handshake. Both the four-way handshake and group handshake are detailed in Section 15.3, and two supported cipher suites (i.e., TKIP and CCMP) are presented in Sections 15.4.

Security Associations

IEEE 802.11i uses the notion of a security association to describe secure operations. A security association is a context composed of a set of policy (or policies) and key (or keys) used to protect information. The security associations supported by an RSN station include the following:

- *Pairwise master key security association* (PMKSA): a result of a successful IEEE 802.lX authentication, a preshared PMK information (i.e., PSK), or a PMK cached via some other mechanism (e.g., preauthentication). A corresponding identifier, called a *pairwise master key identifier* (PMKID), is allocated for a PMKSA.
- *Pairwise transient key security association* (PTKSA): a result of a successful four-way handshake.
- *Group transient key security association* (GTKSA): a result of a successful group key handshake or successful four-way handshake.

RSN Information Element (RSNIE)

The beacon and probe response frames convey an *RSN information element* (RSNIE) in order to inform the security capability and policy provided by the BSS. In IEEE 802.11i, encryption algorithms are called *cipher suites*, where cipher suites used for unicast and broadcast/multicast frames are referred to as *pairwise cipher suites* and *group cipher suites*, respectively. In a BSS, only a single group cipher can be used, since it is for every station while multiple pairwise ciphers can be supported since any one of those pairwise ciphers can be used for each station pair. The available cipher suites are CCMP, TKIP, WEP-40, and WEP-104, where WEP mechanisms are only for group cipher suites to allow pre-RSNA stations to associate with the BSS. When WEP is allowed, the network is referred to as a TSN.

As illustrated in Figure 15.8, the RSNIE includes: (1) the group cipher suite, representing the group cipher suite used in the BSS, (2) the pairwise cipher suite list, representing pairwise cipher suites supported in the BSS, and (3) the *authentication and key management* (AKM) suite list, representing whether IEEE 802.1X or PSK or both are supported. Using the preauthentication bit in the RNS capabilities field, the AP can also indicate whether preauthentication is supported. As will be discussed further in Section 15.2.3, a station can be preauthenticated with neighboring

	Element ID	Length	Version	Group Cipher Suite	Pairwise Cipher Suite Count	Pairwise Cipher Suite List	AKM Suite Count	AKM Suite List	RSN Capabilities	PMKID Count	PMKID List
Octets :	1	1	2	4	2	4-*m*	2	4-*n*	2	2	16-*s*

Figure 15.8 RSN information element format. (*After:* [3].)

APs through the current AP and the DS in advance. As a PMK is established out of the preauthentication, the full 802.1X authentication can be skipped during the handoff (called *BSS transition* in 802.11 terms) procedure.

The RSNIE is also included in the (re)association request frames in order for a non-AP station to indicate its security capability to the AP. In this case, the PMKID list is also included in the RSNIE in order for the non-AP station to indicate the PMKIDs that the station believes to be valid for the AP. For example, this can be useful when a PMKSA is established with the AP in advance via preauthentication. Finally, the RSNIE is included in the second and third messages of the four-way handshake for the generation of PTK.

EAP Authentication

As presented earlier, a station discovers the AP's security capability and policy by receiving beacon or probe response frames via scanning. As shown in Figure 15.7(b), assuming that IEEE 802.1X authentication is used, the EAP authentication process starts when the AP sends an EAP-request or the station sends an EAPOL-start message. EAP-packet frames (i.e., EAP request and response frames) are exchanged between the station and the AS via the AP's uncontrolled port. Once the station and the AS mutually authenticate each other (e.g., using EAP-TLS, as discussed in Section 15.2.1), the controlled port at the AP is unblocked so that general data transfer can be allowed.

When the IEEE 802.1X authentication completes successfully, the station and the AS will share a secret, called a PMK. The AS transfers the PMK, within the *master session key* (MSK), to the AP. The MSK is at least 64 octets long, and the derivation of the PMK out of the MSK is discussed in Section 15.3.1. When both the station and the AP have the PMK, they have established a PMKSA and insert the PMKSA into the PMKSA cache along with the corresponding PMKID. Upon receiving the PMK from the AS, the AP also initiates a key confirmation handshake with the station as detailed in Section 15.3.3.

15.2.3 Preauthentication

The 802.11i allows a station, authenticated and associated with an AP, to preauthenticate with multiple target APs via the current AP and the DS using EAPOL frames. To initiate the preauthentication, the station transmits an EAPOL-start frame with the *destination address* (DA) (i.e., address 3 in the MAC header) being the BSSID of the target AP and the *receiver address* (RA) (i.e., address 1 in the MAC header) being the BSSID of the current AP. The current AP receives the EAPOL-start frame and forwards it to the DS. The DS delivers the EAPOL-start frame to the target AP. The authenticator at the target AP responds by sending an EAP-request destined to the station. The frame is forwarded to the DS port, and the DS delivers it to the current AP, which finally forwards it to the station. The conversation between the preauthenticating station and the target AP based on EAP-packet frames continues via the current AP and the DS, until the EAP authentication succeeds or fails. All the EAPOL frames used for the preauthentication are encapsulated in an IEEE 802 frame using the preauthentication EtherType (i.e., 0x88c7) instead of the EtherType 0x888e used for original EAPOL frames.

A successful authentication results in PMKSA, which the station uses to complete the four-way and group-key handshakes after a reassociation with the target AP. Note that an IEEE 802.1X authentication is not needed after the reassociation, since a PMKSA exists already, so that the handoff latency can be significantly reduced. A station may initiate the preauthentication with any preauthentication-enabled AP within its ESS, regardless of whether the targeted AP is within its radio range. Even if a station has preauthenticated, it is still possible that it may have to conduct a full IEEE 802.1X authentication, as the target AP might have deleted its PMKSA due to unavailability of resources, delayed association of the station, and so on.

15.3 Keys and Key Distribution

We now present the hierarchy of the keys used for RSNA and how the keys are exchanged between the station and the authenticator (i.e., the AP). The four-way handshake is defined for the distribution of PTK and GTK after the completion of the IEEE 802.1X authentication, and the group key handshake is defined for the distribution of a newly generated GTK. IEEE 802.1X EAPOL-key frames are used for the key distribution.

15.3.1 Key Hierarchy

IEEE 802.11i RSNA defines two key hierarchies: (1) *pairwise key hierarchy*, to protect unicast frames for each individual station pair; and (2) *group key hierarchy*, a hierarchy consisting of a single key to protect multicast and broadcast frames. The description of the key hierarchies uses the following two functions:

- L(Str, F, L) extracts bits F through $F + L - 1$ from Str starting from the left.
- PRF-n represents a pseudo-random function producing an output of n bits as defined next.

PRF-n is defined next. In the following, HMAC-SHA-1(·) represents the message authentication function using the SHA-1 hash function [16]; A is a unique label for each different purpose of the PRF; Y is a single octet containing 0; X is a single octet containing the parameter; and || denotes concatenation.

$$\text{H-SHA-1}(K, A, B, X) = \text{HMAC-SHA-1}(K, A \| Y \| B \| X)$$

$$\text{PRF}(K, A, B, Len)$$

$$\quad \textbf{for } i = 0 \textit{ to } (Len + 159)/160 \textbf{ do}$$

$$\quad\quad R \leftarrow R \| \text{H-SHA-1}(K, A, B, i)$$

$$\quad \textbf{return } L(R, 0, Len)$$

$$\text{PRF}-n(K, A, B) = \text{PRF}(K, A, B, n)$$

In an ESS, the IEEE 802.1X *authenticator's MAC address* (AA) and the AP's BSSID are the same, and the *supplicant's MAC address* (SPA) and the station's MAC address are the same.

Pairwise Key Hierarchy

The pairwise key hierarchy utilizes PRF-384 or PRF-512 to derive session-specific (i.e., for each individual station pair) keys from a PMK of 256 bits, as depicted in Figure 15.9. The pairwise key hierarchy takes a PMK and generates a PTK. The PTK is partitioned into *key confirmation key* (KCK), *key encryption key* (KEK), and *temporal key* (TK), which are used to protect unicast frames between the AP and station. A PTK is used between a single station and a single AP.

If a PSK is used, the PSK is directly used as the PMK. Otherwise, the PMK is derived from the MSK. The PMK is computed as the first 256 bits of the MSK—that is, PMK = L(MSK, 0, 256), where MSK consist of at least 256 bits. Then, the PTK is derived from the PMK as follows.

PTK = PRF-n (PMK, "Pairwise key expansion", Min(AA, SPA) || Max(AA, SPA) || Min(ANonce, SNonce) || Max(ANonce, SNonce))

TKIP uses n = 512 and CCMP uses n = 384. *Supplicant nonce* (SNonce) and *authenticator nonce* (ANonce) are pseudo-random values generated by the station and the AP, respectively. The AP and station normally derive a PTK only once per association.

The KCK is computed as the first 128 bits (bits 0–127) of the PTK—that is, KCK = L(PTK, 0, 128). The KCK is used to calculate the *message integrity code* (MIC) of EAPOL-key frames in the four-way handshake and group key handshake messages.

The KEK is computed as bits 128–255 of the PTK—that is, KEK = L(PTK, 128, 128). The KEK is used by the EAPOL-key frames to encapsulate GTK in the four-way handshake and group key handshake messages.

The TK is computed as bits 256–383 (for CCMP) or bits 256–511 (for TKIP) of the PTK—that is, TK = L(PTK, 256, 128) or TK = L(PTK, 256, 256). The TK is used

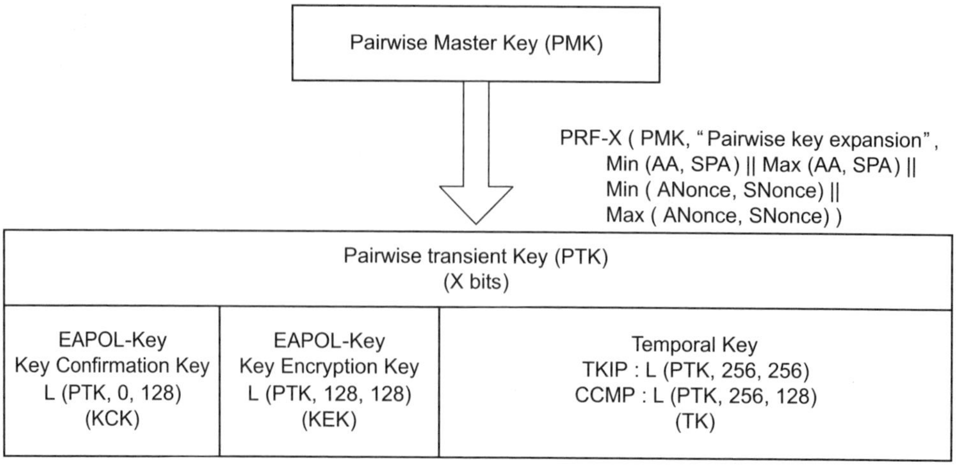

Figure 15.9 Pairwise key hierarchy. (*After:* [3].)

as the frame encryption key for protected data frames and is an input to the employed pairwise cipher suite.

A PMKID is determined as HMAC-SHA1-128(PMK, "PMK Name" || AA || SPA), where HMAC-SHA1-128 is the first 128 bits of the HMAC-SHA-1 of its argument list.

Group Key Hierarchy

The AP may want to update the GTK during the lifetime of a BSS for a number of reasons. For example, the AP may change the GTK on disassociation or deauthentication of a station.

Figure 15.10 depicts a relationship among the keys of the group key hierarchy. In this model, the group key hierarchy takes a GMK and generates a GTK. The GTK is partitioned into TKs used to protect broadcast/multicast frames. GTKs are used between a single AP and all stations authenticated to that AP. The AP derives a new GTK when desired.

Group nonce (GNonce) is a pseudo-random value generated by the AP. The GTK is derived from the GMK by

$$\text{GTK} = \text{PRF-}n(\text{GMK}, \text{“Group key expansion”} \parallel \text{AA} \parallel \text{GNouce})$$

TKIP uses $n = 256$, CCMP uses $n = 128$, and WEP use $n = 40$ or $n = 104$. The temporal key (TK) is first n bits of the GTK—that is, TK = L(GTK, 0, n), where n = 40, 104, 128, or 256.

15.3.2 EAPOL-Key Frames

The key exchange between the station and the AP is conducted using IEEE 802.1X EAPOL-key frames, which are carried in 802.11 data frames. The format of an EAPOL-key frame is illustrated in Figure 15.11, where the fields in the frame body are as follows:

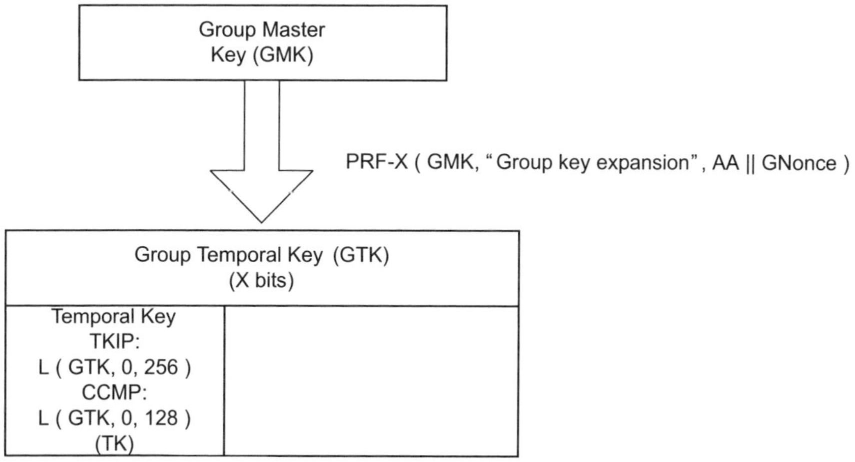

Figure 15.10 Group key hierarchy. (*After:* [3].)

Protocol Version 1 octet	Packet Type 1 octet	Packet Body Length 2 octets
Descriptor Type 1 octet		
Key Information 2 octets		Key Length 2 octets
Key Replay Counter 8 octets		
Key Nonce 32 octets		
EAPOL-Key IV 16 octets		
Key RSC 8 octets		
Reserved 8 octets		
Key MIC 16 octets		
Key Data Length 2 octets		Key Data n octets

Figure 15.11 EAPOL-key frame format. (*After:* [3].)

- The *descriptor type* field identifies the IEEE 802.11 key descriptor. Value 2 indicates IEEE 802.11i.
- The *key information* field, shown in Figure 15.12, specifies characteristics of the key.

The subfields of the key information field are as follows:

- *Key descriptor version* (bits 0–2) specifies the key descriptor version type. The value 1 is for RC4 encryption with HMAC-MD5 MIC and 2 is for AES key wrap [17] with HMAC-SHA1-128 MIC [16, 18].
- *Key type* (bit 3) specifies whether this EAPOL-key frame is part of a four-way handshake deriving a PTK—that is, 0 (group) and 1 (pairwise).
- *Install* (bit 6) indicates that IEEE 802.1X component will configure the temporal key derived from this message into its station.

B0 - B2	B3	B4 B5	B6	B7	B8	B9	B10	B11	B12	B13	B14 B15
Key Descriptor Version	Key Type	Reserved	Install	Key Ack	Key MIC	Secure	Error	Request	Encrypted Key Data	SMK Message	Reserved

Figure 15.12 Key information field format. (*After:* [3].)

- *Key ack* (bit 7) is set in messages from the AP if an EAPOL-key frame is required in response to this message.
- *Key MIC* (bit 8) indicates that a MIC is in this EAPOL-key frame.
- *Secure* (bit 9) is set once the initial key exchange is complete.
- *Error* (bit 10) is set by a station to report that a MIC failure occurred in a TKIP MSDU handshake failure.
- *Request* (bit 11) is set by a station to request that the AP initiates either a four-way handshake or group key handshake.
- *Encrypted key data* (bit 12) indicates that the key data field is encrypted.
- The *key length* field specifies the length of the PTK, as shown in Table 15.3.
- The *key replay counter* field carries a sequence number that the protocol uses to detect replayed EAPOL-key frames.
- The *key nonce* field conveys the ANonce from the AP and the SNonce from the station.
- The *EAPOL-key IV* field contains the IV used with the KEK.
- The *key RSC* field contains the *receive sequence counter* (RSC) for the GTK being installed in the BSS.
- The *key MIC* field contains the MIC of the EAPOL-key frame, generated using the KCK.
- The *key data length* field represents the length of the key data field.
- The *key data* field is used to include any additional data required for the key exchange that is not included in the fields of the EAPOL-key frame. The additional data include information element(s), such as RSNIE, and *key data cryptographic encapsulation(s)* (KDEs)—that is, GTK(s) or PMKID(s).

EAPLO-Key Frame Notation

The EAPOL-key frames are used for the four-way and group-key handshakes, and the following notation is used for the rest of this chapter, where the included parameters are summarized in Table 15.4.

EAPOL-Key(S, M, A, I, K, SM, KeyRSC, ANonce/SNonce, MIC, DataKDs)

15.3.3 The Four-Way Handshake

When an IEEE 802.1X authentication completes with an EAP-success frame transmitted by the AP, the AP initiates a four-way handshake. As shown in Figure 15.13, a four-way handshake is conducted according to the following steps:

Table 15.3 Cipher Suite Key Lengths

Cipher suite	CCMP	TKIP	WEP-40	WEP-104
Key length (octets)	16	32	5	13

Source: [3].

Table 15.4 Parameters in the EAPOL-Key Notation

Parameter	Corresponding field or meaning
S	The secure bit of the key information field
M	The key MIC bit of the key information field
A	The key Ack bit of the key information field
I	The install bit of the key information field
K	The key type bit of the key information field, i.e., P (Pairwise) or G (Group).
KeyRSC	The key RSC field
Anonce/SNonce	The key nonce field
MIC	The key MIC field
DataKDs	A sequence of zero or more information elements and KDEs, contained in the key data field, including (1) RSNIE, (2) GTK[N], representing the GTK with the key identifier field set to N, and (3) PMKID, representing the key identifier used during 4-way PTK handshake for PMK key identification

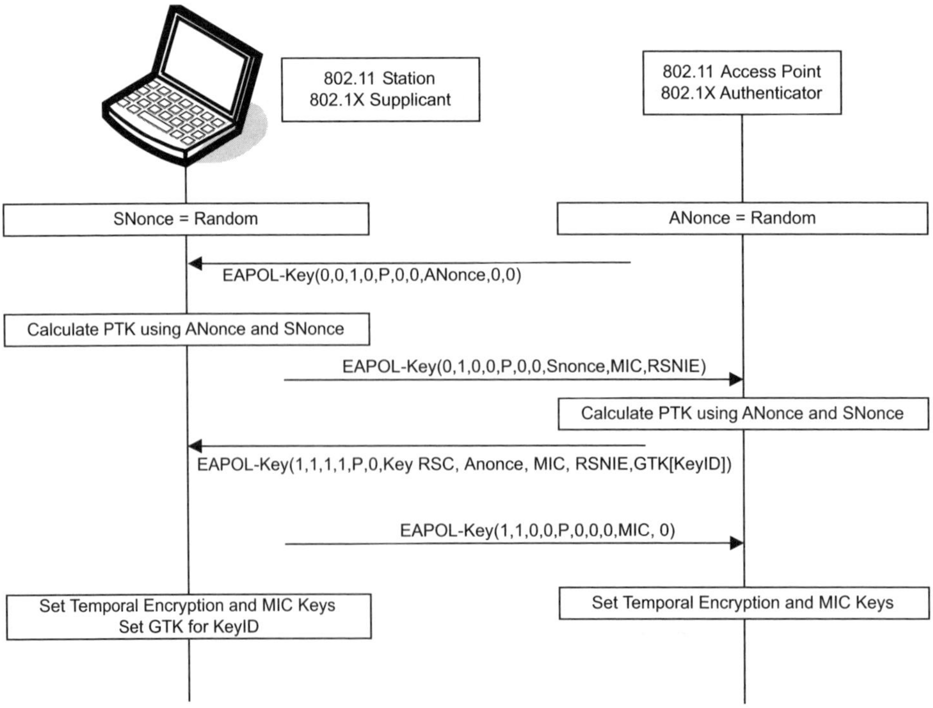

Figure 15.13 A sample four-way handshake with the corresponding EAPOL-key frames. (*After:* [3].)

- The station and the AP generate SNonce and ANonce, respectively.
- The AP sends an EAPOL-key frame (message 1) containing the ANonce.
- The station derives a PTK from ANonce and SNonce.
- The station sends an EAPOL-key frame (message 2) containing the SNonce, the RSNIE from the (re)association request frame, and a MIC.
- The AP derives PTK from ANonce and SNonce and validates the MIC in the EAPOL-key frame (message 2). If needed, a GTK is also newly derived.
- The AP sends an EAPOL-key frame (message 3) containing keyRSC, ANonce, MIC, the RSNIE from its beacon, and the GTK encapsulated using the KEK. Whether to install the temporal keys is also indicated.
- The station sends an EAPOL-key frame (message 4) to confirm that both PTK and GTK are installed.

After installing the PTK and GTK, the MAC encrypts and decrypts all subsequent MSDUs. Upon a successful completion of the four-way handshake, the AP and station have authenticated each other, and the IEEE 802.1X controlled ports are unblocked to permit general data traffic.

15.3.4 Group Key Handshake

If the AP later changes the GTK, it sends the new GTK and GTK sequence number to the station using the group key handshake to allow the stations to continue to receive broadcast/multicast messages. As shown in Figure 15.14, the group key handshake is conducted according to the following steps:

- The AP generates a new GTK from GNonce. It encapsulates the GTK using KEK and sends an EAPOL-key frame (message 1) containing the GTK along with the keyRSC (i.e., the last frame sequence number sent using the GTK).
- On receiving the EAPOL-key frame, the station validates the MIC and then decapsulates the GTK to install the GTK and the RSC.
- The station then sends an EAPOL-key frame (message 2) to the AP.
- Upon receiving the EAPOL-key frame, the AP validates the MIC. If the GTK is not already installed, after individually delivering the GTK to all associated stations, it installs the GTK.

15.4 RSNA Data Confidentiality Protocols

The 802.11i RSNA is based on two enhanced data confidentiality mechanisms, namely, the mandatory CCMP and the optional TKIP.

15.4.1 Temporal Key Integrity Protocol (TKIP)

The TKIP is a cipher suite enhancing the WEP protocol on pre-RSNA hardware, and it uses WEP with the following modifications:

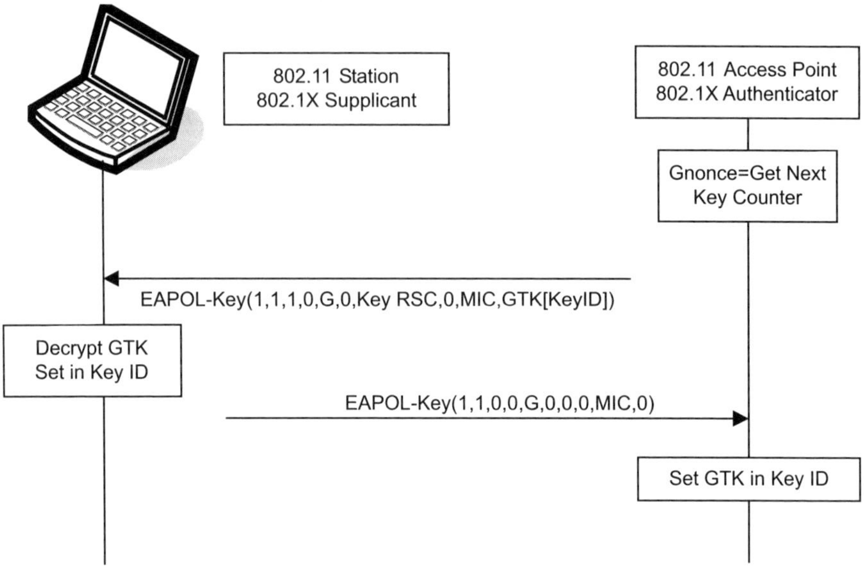

Figure 15.14 A sample group key handshake with the corresponding EAPOL-key frames. (*After:* [3].)

- A transmitter calculates a keyed cryptographic MIC, called *Michael*, over the frame source and destination addresses, the priority, and the plaintext data. Any frames with invalid MICs (i.e., possibly affected by forgery attacks) are discarded at the receiver.
- TKIP uses a packet *TKIP sequence counter* (TSC) to sequence the frames it sends, and this counter is encoded as a WEP IV and Extended IV. Any frames received out of order (i.e., possibly affected by replay attacks) are discarded at the receiver.
- TKIP uses a cryptographic mixing function to combine a TK, transmitter address, and the TSC into the WEP seed. This mixing function is designed to defeat weak-key attacks against the WEP key.

TKIP Operations

When transmitting a frame using the TKIP encapsulation, the transmitter station determines the value of the key (i.e., seed) used to initialize the RC4 stream cipher. TKIP calculates this key in two phases: both phases use a TSC of 48 bits. This TSC value is initialized to one when a new TK is established. The TSC monotonically increases for every MPDU by one. That is, each MPDU is assigned a unique TSC value. This is used for the protection against replay attacks.

As explained in Section 15.3.1, a 256-bit TK is derived from either PTK or GTK. A station uses bits 0–127 of the TK as the input to the TKIP key mixing functions. Bits 128–191 of the TK are used as the MIC key for MSDUs from the AP to the non-AP station. Bits 192–255 of the TK are used as the MIC key for MSDUs from the non-AP station to the AP.

As illustrated in Figure 15.15, the key mixing for TKIP works in two phases:

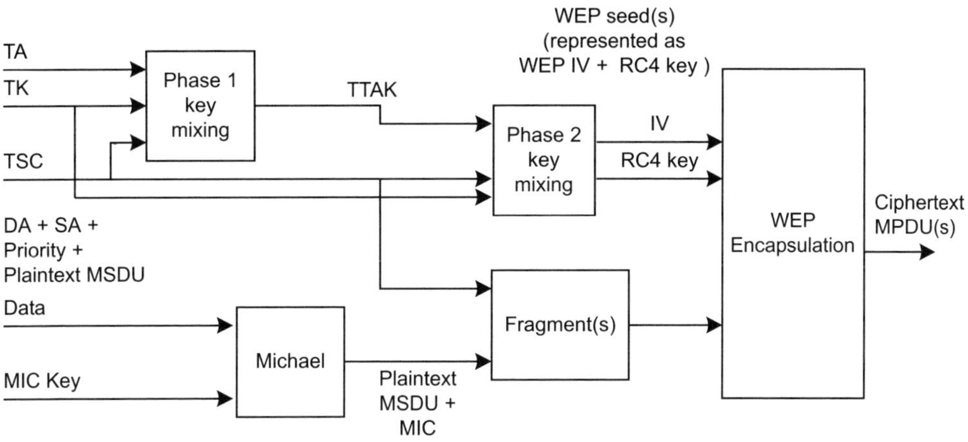

Figure 15.15 TKIP encapsulation block diagram. (*After:* [3].)

- The first phase uses the 32 MSBs of the TSC and combines it with the TA (i.e., the transmitter station's address) and TK to obtain the *TKIP-mixed transmit address and key* (TTAK) of 80 bits. The TTAK can be cached and used repeatedly because the upper portion of the TSC changes only once for every 216 frames sent from the TA.
- The second phase determines the key used for the encryption of a frame (i.e., the WEP seed), which is used to initialize the RC4 algorithm. The inputs to the second phase WEP seed calculation are the TK, TTAK, and the 16 LSBs of the TSC. The resulting WEP seed is 128 bits long.

The block diagram for the TKIP decapsulation is illustrated in Figure 15.16. The receiver station performs the reverse operations by decrypting the received MPDUs and checking whether the received MSDU (i.e., reconstructed possibly via defragmentation) is intact by checking the MIC. If the MIC is correct, the MSDU is forwarded to the higher layer.

TKIP-Encapsulated MPDU

TKIP reuses the pre-RSNA WEP MPDU format. It extends the MPDU by 4 octets to accommodate an extension to the WEP IV (i.e., the *extended IV* field) and extends the MSDU by 8 octets to accommodate the new MIC field. TKIP inserts the extended IV field between the WEP IV field and the encrypted data. TKIP appends the MIC to the data field; the MIC becomes part of the encrypted data. Once the MIC is appended to an MSDU data, the added MIC octets are considered part of the MSDU for possible fragmentation.

Figure 15.17 depicts the TKIP-encapsulated MPDU format compared with the nonencrypted one. The TSC occupies 6 octets across IV and extended IV fields. Note that the figure depicts only the case when an MSDU is encapsulated in a single MPDU (i.e., no fragmentation). Otherwise, only the last MPDU will include the MIC, or the last and the second-to-last MPDUs will include parts of the MIC.

The ExtIV bit in the Key ID octet indicates that an extended IV follows the original IV. Accordingly, the ExtIV bit will be 0 for WEP-encapsulated MPDUs. The Key ID field indicates the key used for the encapsulation of the frame. TSC5 is the

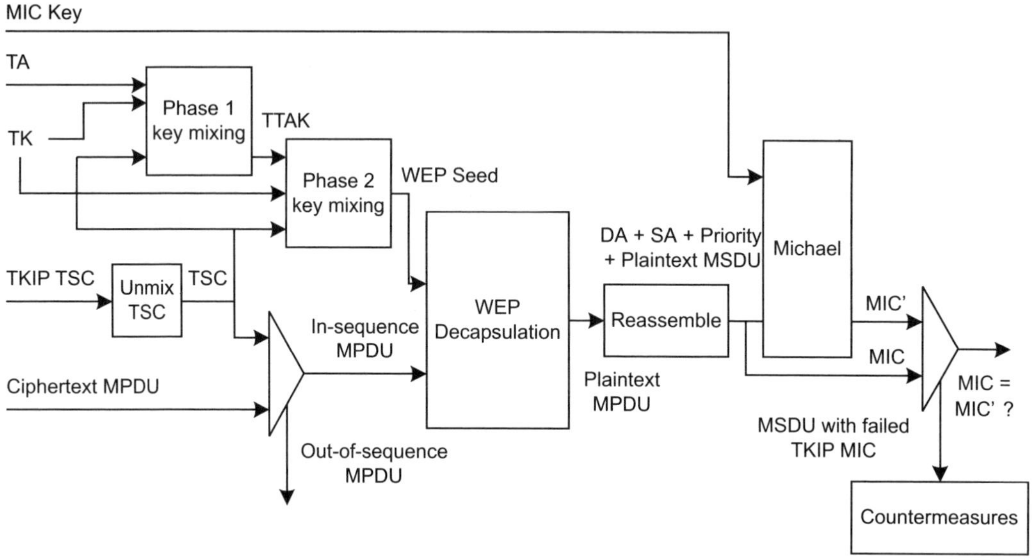

Figure 15.16 TKIP decapsulation block diagram. (*After:* [3].)

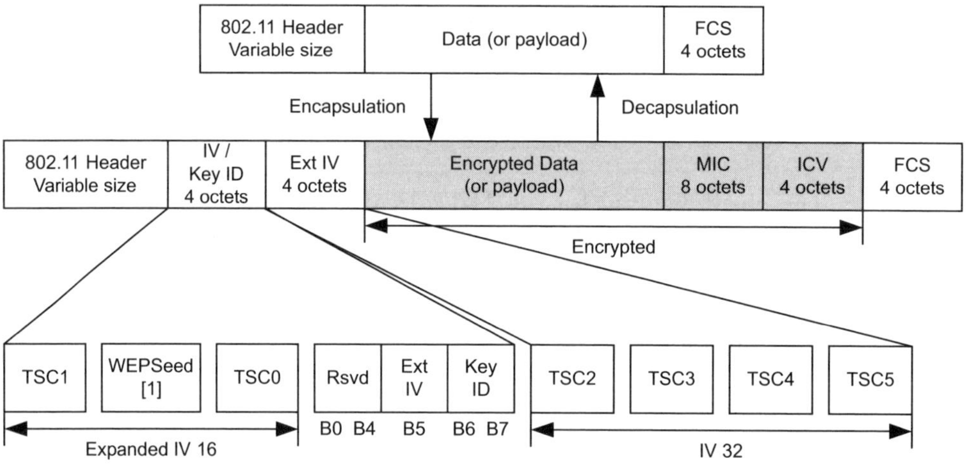

Figure 15.17 Expanded TKIP MPDU. (*After:* [3].)

most significant octet of the TSC, and TSC0 is the least significant. Octets TSC0 and TSC1 form the IV sequence number and are used with the TKIP phase 2 key mixing. Octets TSC2–TSC5, in the extended IV field, are used in the TKIP phase 1 key mixing for the generation of a TTAK. WEPSeed[1] is set to (TSC1 | 0x20) & 0x7f, where "|" represents the bitwise logical "or" function, and "&" represents the bitwise logical "and" function.

Michael MIC

To defend against active attacks, the TKIP includes a MIC, called Michael. The MIC is calculated over the MSDU DA, MSDU SA, MSDU priority, and entire unencrypted MSDU. TKIP uses different MIC keys depending on the direction of the transfer, as described earlier.

This MIC offers only weak defenses against active attacks such as message forgeries, but it provides the best that can be achieved with the legacy hardware. A failure of the MIC in a received MSDU indicates a probable active attack. If a probable active attack is detected, TKIP takes countermeasures. That is, upon detecting two MIC failure events within 60 seconds, non-AP stations and the AP disable all receptions using TKIP for 60 seconds, and then refresh both PTK and GTK. The slowdown makes it difficult for an attacker to attempt many forgery attacks in a short time.

15.4.2 Countermode with CBC-MAC Protocol (CCMP)

The mandatory CCMP employs the AES encryption algorithm [19] using the CCM mode of operation [20]. The CCM mode combines *counter mode* (CTR) for confidentiality and *cipher block chaining message authentication code* (CBC-MAC) for authentication and integrity protection. The CCM protects the integrity of both MPDU payload and selected portions of the MAC header. All AES processing used within CCMP uses AES with a 128-bit key and a 128-bit block size. Note that AES is a block cipher different from RC4, a stream cipher, used in WEP and TKIP. For the CCMP support, an additional hardware is typically required due to the high computation complexity.

CCMP Operations

CCM is a generic mode that can be used with any block-oriented encryption algorithm. CCM has two parameters, namely, M and L, and the CCMP uses $M = 8$, indicating that the MIC is 8 octets, and $L = 2$, indicating that the length field is 2 octets, which is sufficient to hold the largest possible IEEE 802.11 MPDU in octets. CCM requires a unique nonce value for each frame protected by a given temporal key, and CCMP uses a 48-bit *packet number* (PN) for this purpose.

The CCMP encapsulation is illustrated in Figure 15.18. As explained in Section 15.3.1, a TK is derived from either PTK or GTK. A station uses the 128-bit TK as the CCMP key, which is used for both confidentiality and integrity. The CCMP encrypts the payload of a plaintext MPDU and encapsulates the resulting cipher text using the following steps:

- Increment the PN to obtain a fresh PN for each MPDU so that the PN never repeats for the same TK.
- Use the fields in the MPDU header to construct the *additional authentication data* (AAD) for CCM. The CCM algorithm provides the integrity protection for the fields included in the AAD. The fields in the MAC header, which are not supposed to change for retransmissions, are included in the AAD. Those include the protected frame bit, the address fields (including address 4 if present), and the TID of the QoS control field (in case of the 802.11e data frame).
- Construct the CCM nonce block from the PN, TA (i.e., the transmitter station's address), and the priority of the MPDU.
- Place the new PN and the key identifier into the 8-octet CCMP header.

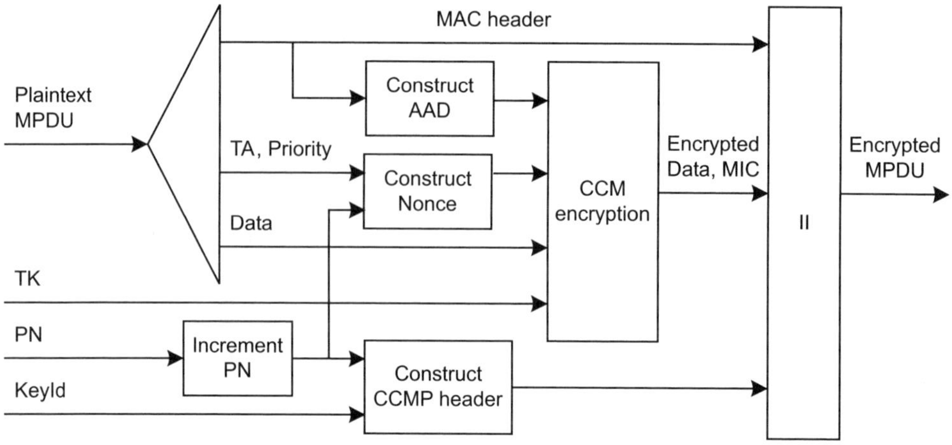

Figure 15.18 CCMP encapsulation block diagram. (*After:* [3].)

- Use the TK, AAD, nonce, and MPDU data to form the cipher text and MIC. This step is known as *CCM originator processing*, which provides authentication and integrity of the frame body and the AAD as well as data confidentiality of the frame body. The output from the CCM originator processing consists of the encrypted data and 8 additional octets of encrypted MIC.
- Form the encrypted MPDU by combining the original MAC header, the CCMP header, the encrypted data, and MIC.

As shown in Figure 15.19, the CCMP decrypts the payload of the received MPDU and decapsulates a plaintext MPDU. The decryption processing prevents the replay of MPDUs by validating that the PN in the MPDU is greater than the replay counter maintained for the session.

CCMP-Encapsulated MPDU

Figure 15.20 depicts the MPDU format using CCMP. The CCMP encapsulation expands the original MPDU size by 16 octets, where 8 octets are for the CCMP header field and 8 octets for the MIC field. The CCMP header field is constructed

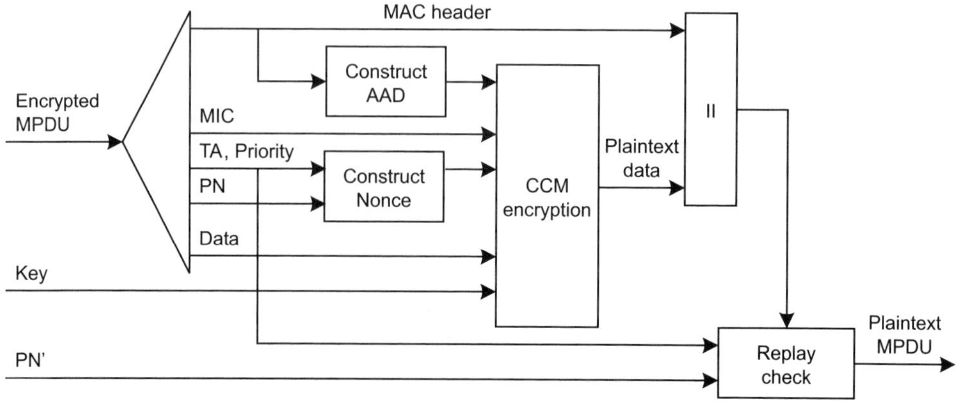

Figure 15.19 CCMP decapsulation block diagram. (*After:* [3].)

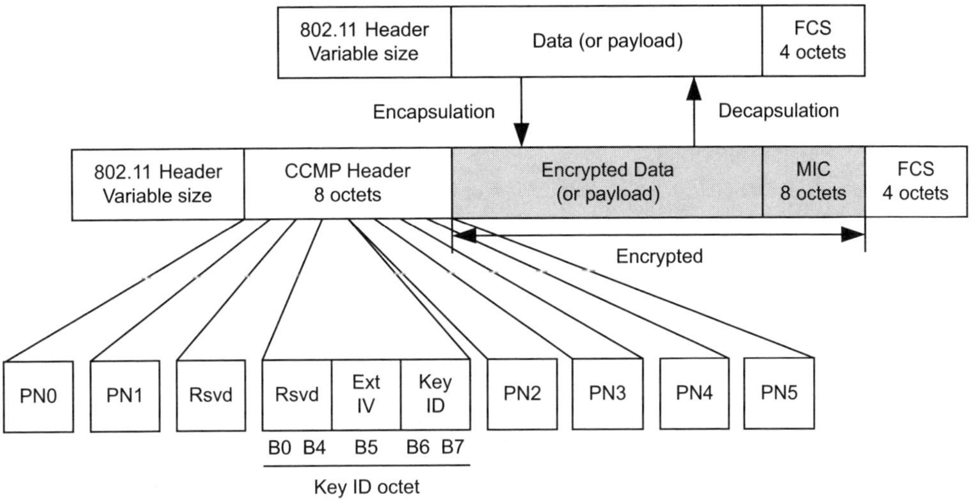

Figure 15.20 Expanded CCMP MPDU. (*After:* [3].)

from the PN, ExtIV, and Key ID subfields. The usages of ExtIV and Key ID are the same as in the TKIP-encapsulated MPDU. The 48-bit PN is represented by an array of 6 octets. PN5 is the most significant octet of the PN, and PN0 is the least significant. Note that CCMP does not use the WEP ICV. The MIC is calculated per MPDU, and, hence, the MIC field exists in every MPDU.

References

[1] IEEE 802.11i-2004, Amendment 6 to Part 11: Wireless LAN Medium Access Control (MAC) and Physical Layer (PHY) Specifications: Medium Access Control (MAC) Security Enhancements, 2004.

[2] IEEE 802.1X-2004, IEEE Standard for Local and Metropolitan Area Networks—Port-Based Network Access Control, 2004.

[3] IEEE 802.11-2007, IEEE Standard for Local and Metropolitan Area Networks—Part 11: Wireless LAN Medium Access Control (MAC) and Physical Layer (PHY) Specifications, IEEE Std 802.11-2007, (Revision of IEEE Std 802.11-1999), June 12, 2007.

[4] RSA Security, Inc., http://www.rsa.com/.

[5] Arbaugh, W. A., et al., "Your 802.11 Network Has No Clothes," *IEEE Wireless Communications*, Vol. 9, No. 6, December 2002, pp. 41–51.

[6] Edney, J., and W. A. Arbaugh, *Real 802.11 Security: Wi-Fi Protected Access and 802.11i*, Reading, MA: Addison-Wesley, 2004.

[7] Prasad, A. R., "WLANs: Protocols, Security, and Deployment," Ph.D. Thesis, Delft University Press, Delft, the Netherlands, December 2003.

[8] Borisov, N., I. Goldberg, and D. Wagner, "Intercepting Mobile Communications: The Insecurity of 802.11," *Proc. ACM 7th International Conference on Mobile Computing and Networking (MobiCom'01)*, Rome, Italy, July 16–21, 2001.

[9] IETF RFC 2865, Remote Authentication Dial In User Service (RADIUS), June 2000.

[10] IETF RFC 3588, Diameter Base Protocol, 2003.

[11] IETF RFC 3748, Extensible Authentication Protocol, 2004.

[12] IETF RFC 2716, PPP EAP TLS Authentication Protocol, 1999.

[13] Funk, P., and S. Blake-Wilson, "EAP Tunneled TLS Authentication Protocol (EAP-TTLS)," Work in Progress, August 2004.

[14] Palekar, A., et al., "Protected EAP Protocol (PEAP)," Work in Progress, July 2004.

[15] IETF RFC 4017, Extensible Authentication Protocol (EAP) Method Requirements for Wireless LANs, 2005.

[16] IETF RFC 2104, HMAC: Keyed-Hashing for Message Authentication, 1997.

[17] IETF RFC 3394, Advanced Encryption Standard (AES) Key Wrap Algorithm, 2002.

[18] FIPS PUB 180-1-1995, Secure Hash Standard, National Inst. of Technology and Standards, Federal Information Processing Standards (FIPS) Pub. 180-1, 1995.

[19] FIPS PUB 197-2001, Advanced Encryption Standard (AES), National Inst. of Technology and Standards, Federal Information Processing Standards (FIPS) Pub. 197, 2001 (http://www.nist.gov/aes).

[20] ETF RFC 3610, Counter with CBC-MAC (CCM), 2003.

CHAPTER 16

Mobility Support

IEEE 802.11 WLANs have been widely deployed and popularized as a dominant means of broadband wireless access networks in recent years. It is being deployed in many different environments, such as home, enterprise, and hot-spot areas. In the enterprise networks as well as hot spots, it is typical to have multiple APs to cover a geographical area in service. While the mobility support via handoff to support a seamless service is a key mechanism in most cellular-based wireless networking, it has not been a major concern for the 802.11 WLAN, since people rarely use their laptops or PDAs to access the Internet via WLAN while they are moving around. On the contrary, some level of mobility is supported by the 802.11. For example, a walking speed mobility is surely supported. That is, an 802.11 station can switch from one AP to another in an ESS while it moves by reassociating with a new AP.

Today, along with the emergence of the *voice over WLAN* (VoWLAN) applications, supporting seamless and smooth handoffs in the 802.11 WLAN is becoming a hot topic. For a handoff to occur, a station has to first detect neighboring APs via a scanning process. Then, it has to determine which AP to reassociate with. Once this is determined, a reassociation process is conducted along with an authentication with the new AP. For most of today's 802.11 devices, the scanning process takes the most time, and there are ongoing efforts (e.g., in IEEE 802.11k [1]) to reduce the scanning time. Moreover, IEEE 802.11r, which is also being standardized, tries to reduce the handoff time while maintaining QoS and security [2]. In this chapter, we present various mechanisms to enable smooth WLAN mobility support.

Where a WLAN is composed of multiple APs, the system that connects the multiple APs is called a *distribution system* (DS), and a set of BSSs and the DS connecting these BSSs is called *extended service set* (ESS). In today's WLANs, the DS is typically constructed with the Ethernet. One can easily imagine that this kind of WLAN structure is similar to that of the wide-area cellular systems, where multiple base stations are connected via wireline links, and each base station serves an area called a *cell*.

16.1 IEEE 802.11 Handoff Procedures

In an infrastructure BSS, a station should be associated with an AP before commencing any normal data transfer. However, before the association can be made, scanning and authentication should precede. The same procedure is made for a handoff from an AP to another, which is referred to as a *BSS-transition*, where a *reassociation* instead of an association occurs. In the following, we use the terms *handoff* and *BSS-transition* interchangeably. Figure 16.1 illustrates the (re)association procedure assuming active scanning and the open system authentication. We explain the detailed procedures needed until the reassociation in the following.

Being associated with an AP is similar to the situation in which two stations are connected via an Ethernet cable. However, depending on the employed security mechanism, a station might need to perform an IEEE 802.1X authentication [3] as well as a security key management before the AP allows this station to transmit any data frame. Moreover, if this station intends to transmit QoS data, a TS might need to be established. Accordingly, for a VoWLAN device, which runs the 802.11e for QoS provisioning [4] and the 802.11i for security support [5], the whole handoff should involve (1) reassociation, (2) the 802.11i *authentication and key management* (AKM), and finally (3) the 802.11e TS setup, as illustrated in Figure 16.2. We discuss the whole procedures in the following. For detailed operations of IEEE 802.11i and IEEE 802.11e, the readers are referred to Chapters 15 and 14, respectively.

16.1.1 Scanning

Scanning is a process for a station to search neighboring APs. There are two different types of scanning: active and passive scanning.

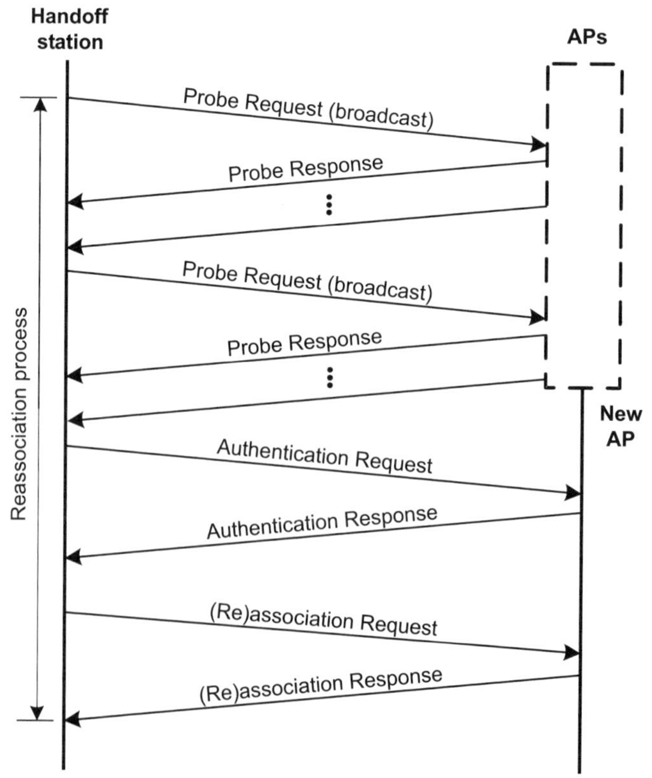

Figure 16.1 IEEE 802.11 (re)association procedure.

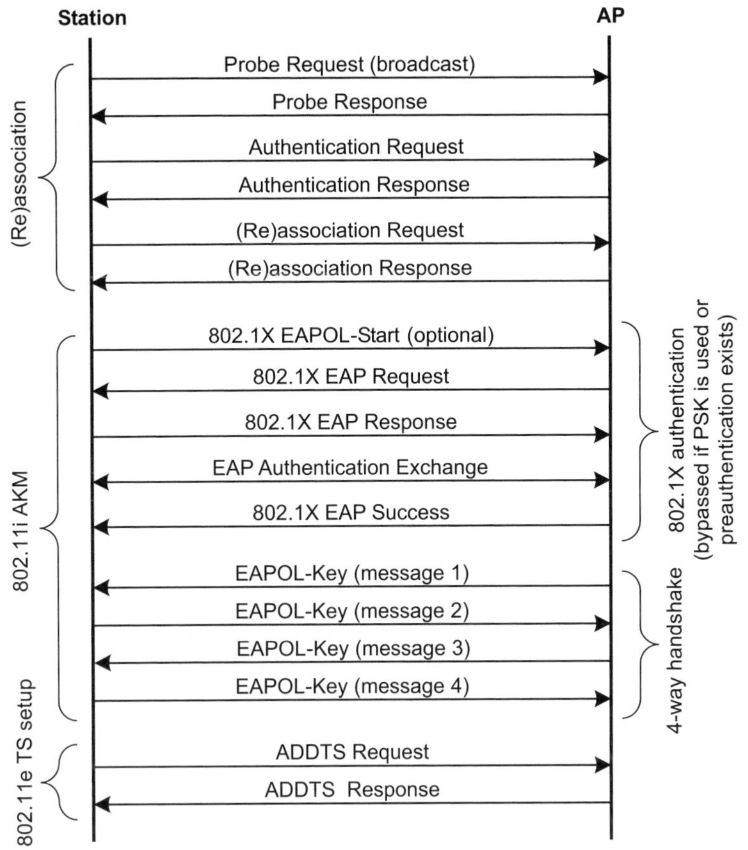

Figure 16.2 Connectivity setup procedure involving (re)association, IEEE 802.11i AKM, and IEEE 802.11e TS setup.

Active Versus Passive Scanning

Typically, the faster scanning process is the active scanning. With this type of scanning, the station broadcasts a probe request frame, and then APs receiving this probe request respond with a probe response frame. The probe response includes virtually the same information as the beacon frame except the *traffic indication map* (TIM) field used for the *power-save mode* (PSM) support. Therefore, the station receiving a probe response acquires all the necessary information to be associated with the AP. When there are multiple APs found in the neighborhood, the station chooses one of them based on its implementation-dependent decision criteria.

Active scanning is prohibited in some frequency bands and regulatory domains. With the passive scanning, the station detects the neighboring APs by receiving the beacon frames transmitted by these APs. This can be a slow and conservative scanning process since the beacon transmission interval is typically about 100 ms (or 100 TUs or 102.4 ms more exactly). Basically, the passive scanning can be used to detect neighboring APs in the current channel. Note it does not cost much since a station receives all the incoming frames anyway, while staying in the awake state.

Active Scanning Procedure

As explained in Chapter 12, a number of frequency channels are defined for each PHY of the 802.11. By default, a station might not have any information about which channels neighboring APs might be operating in. In such a case, the station has to scan all possible channels to find neighboring APs. Apparently, the total time to scan and find APs depends on the number of frequency channels as well as how long the station spends in each channel.

The active scanning time in each channel is determined by two parameters in particular, namely, *MinChannelTime* and *MaxChannelTime*. In each channel, the active scanning follows the steps described here:

1. Upon the entrance into a channel, wait until the ProbeDelay time has expired or a frame reception starts.
2. Broadcast a probe request frame after a DCF or EDCA contention.
3. Clear and start a ProbeTimer.
4. If the channel stays idle until the ProbeTimer reaches MinChannelTime, go to Step 6.
5. Wait until the ProbeTimer reaches MaxChannelTime, and then process all the received probe response frames until then.
6. Clear NAV and scan the next channel.

Step 1 is intended to make a scanning station not too aggressive. Otherwise, a scanning station's probe request frame might have a higher chance to collide with other ongoing frames over the channel, especially, when these frames are transmitted by stations that are hidden from the scanning station. Under the EDCA, both probe request and probe response frames are transmitted via AC_VO, as they are management frames. The values of the three parameters involved with the active scanning—ProbeDelay, MinChannelTime, and MaxChannelTime—are implementation-dependent. There have been reports on the optimal values for these parameters (e.g., [6]).

Figure 16.3 illustrates how the active scanning works in a given channel. Note that the probe response frames are unicast frames so that there is a corresponding ACK transmission by the scanning station.

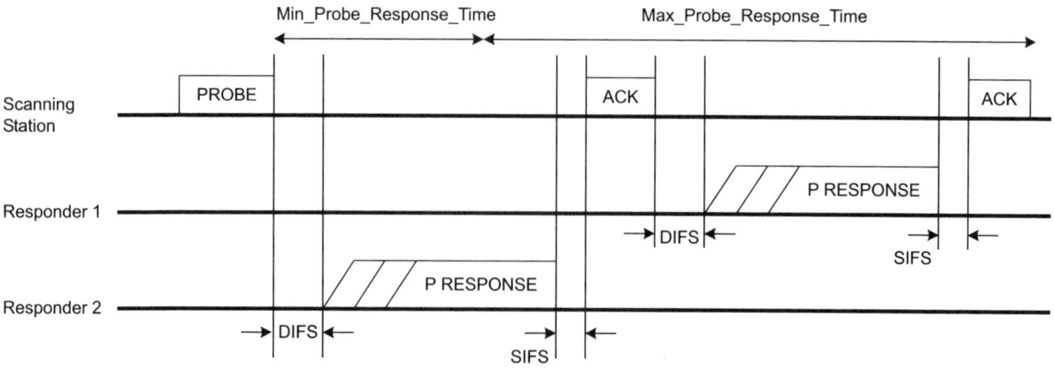

Figure 16.3 Active scanning procedure. (*After:* [7].)

Scanning Notification and Algorithm

When a station would like to scan other channels, it has to inform the fact to its AP because during the scanning process, especially, at other channels, the downlink frames destined to this station have to be buffered at the AP. Unfortunately, in the 802.11 MAC, there is no specific frame defined for this purpose. However, the PSM can be used for this purpose. That is, the AP buffers all the frames destined to a PS station by believing that the PS station is not ready to receive frames normally. A PS station can then scan other channels instead of saving its energy consumption in the doze state. This means that a station that would like to scan other channels can notify its AP that it would enter the PSM while its plan is to scan channels, not to save energy.

When to initiate a scanning process is an important implementation-dependent algorithmic issue. In order to hand off to a new AP, which can provide the best network performance, in a timely manner, a station has to keep track of the information of neighboring APs via regular scanning. However, as discussed previously, scanning APs can sacrifice the performance with the current AP. For example, when a station has infinitely many frames to transmit, it is desired to contend for the channel continuously in order to maximize the throughput performance, and scanning APs in other channels will degrade the throughput performance apparently.

The simplest policy is to trigger scannings periodically where the period can be determined by considering the tradeoff relationship. Another policy is to trigger a scanning only after a performance metric with the current AP goes below a threshold. The performance metric can be simply the received signal strength determined by *received signal strength indication* (RSSI) from the PHY. Others possible metrics include (1) the number of transmission failures within a time window, (2) the transmission rate employed by the rate adaptation algorithm, and (3) the throughput or delay performance. Note that if the employed transmission rate is the lowest rate, the channel condition with the current AP must be bad. Once triggered, scannings might be conducted periodically.

16.1.2 Authentication

Authentication is for the station to be authenticated with the AP by proving itself a valid station. While this was presented in Section 15.1.2, we briefly recap it for the completeness of this chapter. There are two types of authentication methods: *open system* and *shared key*. With the open system, the station sends an authentication-request frame, and the requested AP responds with an authentication-response frame. There is no specific security information conveyed in the authentication-request frame in this case. With the shared key method, a four-way handshake is used for the AP to check whether or not the requesting station has the same security key. As discussed in Section 15.2, IEEE 802.11i [5] requires using the open system authentication as a new stronger authentication scheme, based on IEEE 802.1X [3], as defined for the authentication purpose.

16.1.3 (Re)association

Finally, a station can be (re)associated with an AP that the station has chosen. A (re)association is made by exchanging a (re)association request frame and a (re)association response frame. Through this exchange, the station is assigned an *association identification* (AID), and the AP is informed of the information required for the proper communication (e.g., the transmission rates supported by the newly associated station). The association is for a new association with a WLAN, while the reassociation is for the handoff from one AP to another. The procedures for both association and reassociation are basically the same (i.e., the exchange of request and response frames). However, in order for the AP to differentiate between a new association and a handoff, the 802.11 defines two different frame types for the association request. The request for the handoff is called reassociation request, where the corresponding frame includes the MAC address of the old AP (in the current AP address field) so that the new AP can communicate with the old AP to support the handoff process, as discussed in Section 16.2.

AP Selection Criteria

The criteria to choose an AP is implementation-dependent. The most typical criterion could be selecting the AP with the strongest received signal strength or RSSI, where the signal strengths of various APs can be measured during the scanning process. The strongest signal strength might indicate the best channel condition between the station and the AP. However, the AP with the strongest signal strength might not be often the best choice. Note that the AP might be heavily loaded with many contending stations. Especially, in a WLAN with the baseline DCF operation, if there are stations employing low transmission rates, the maximum throughput of a station might be heavily compromised due the channel access opportunity fairness of the DCF, as discussed in Section 14.3.1.

Alternative criteria might be based on the BSS traffic load information. According to IEEE 802.11e, the beacon conveys the *BSS load* element, which indicates: (1) station count (i.e., the total number of stations associated with the BSS), (2) channel utilization (i.e., the percentage of time which the AP sensed the channel busy), and (3) available admission capacity (i.e., the remaining amount of the medium time). A station can obtain this traffic load information during the scanning process, and, hence, can choose an AP to be associated with based on the traffic load information.

The usage of a handoff metric combining both signal strength and BSS traffic load can be a good criterion. Basically, the transmission rate, which can be used between a station and an AP, depends on the received signal strength. Accordingly, the available or achievable bandwidth can be determined by considering both the signal strength and the BSS traffic load. The station can select the AP that is expected to provide the largest achievable bandwidth.

Another important issue, besides the selection of the AP, is when to hand off to this AP. One might say that it is an easy problem, since a station can hand off from the current AP to a new AP, with the best metric value, as soon as the metric value with the new AP turns out to be better than that of the current AP. However, this naive policy might result in very frequent handoffs by oscillating between two APs. One should be careful in this decision because frequent handoffs are not desirable at all since: (1) the reassociation procedure itself consumes the precious wireless bandwidth, and (2) there is always a nonzero incurring delay involved with the reassociation.

An easy solution for this is setting a *hysteresis* for the handoff decision. That is, in order to hand off to a new AP, two conditions are checked: (1) the handoff metric value of the new AP should be larger than a threshold, and (2) the handoff metric value of the new AP minus that of the current AP should be larger than another threshold. Depending on the second threshold value, the handoff frequency can be controlled. Note that a large threshold results in less frequent handoffs.

Hard Handoff and Preauthentication

It should be noted that a station can be associated with a single AP at a given time. It is natural since a simultaneous communication with multiple APs (i.e., soft handoff of *code-division multiple access* systems [8]) is not supported in the 802.11 WLAN. Therefore, the 802.11 handoff is a hard handoff.

On the other hand, a station can be authenticated with multiple APs. Note that this can reduce the handoff time. However, this requires a station to search neighboring APs and get authenticated (i.e., 802.11 authentication) in advance. This is often referred to as *preauthentication*. While the names are the same, this preauthentication should be differentiated from the 802.11i's preauthentication (discussed in Section 15.2.3). However, the basic concept of getting authenticated before an actual handoff is the same for both schemes. The utility of the preauthentication actually depends on how many preauthentication states neighboring APs allow and how long the APs keep the authenticated states. As keeping such states consumes the memory space, an AP cannot keep it forever.

16.1.4 IEEE 802.11i Authentication and Key Management

After a station is (re)associated with an AP, it might want to establish an RSNA based on IEEE 802.11i. As illustrated in Figure 16.2, the 802.11i AKMP involves IEEE 802.1X authentication and key management via four-way handshake. The 802.1X authentication might be skipped if a PSK is used. During a BSS transition, the 802.1X authentication might be skipped if a preauthentication was performed with the new AP in advance. After the RSNA is established, the station is allowed to participate in the general data transfer. The detailed operations are referred to Section 15.2.

16.1.5 IEEE 802.11e TS Setup

If the station after establishing an RSNA desires to transfer QoS traffic, depending on the policy of the AP (i.e., whether an admission control is mandated), the station might need to set up a TS. In order to set up a TS, the station has to send an ADDTS request frame, and then the AP responds with an ADDTS response frame by indicating whether the requested TS is admitted or not. After a TS is set up, a QoS traffic can be transferred with proper QoS provisioning. The detailed operations are referred to Section 14.4.

16.1.6 Layer-2 Versus Layer-3 Mobility

Figure 16.4 illustrates two different types of handoff events within a WLAN depending on whether the two APs involved in the handoff are in the same subnet or

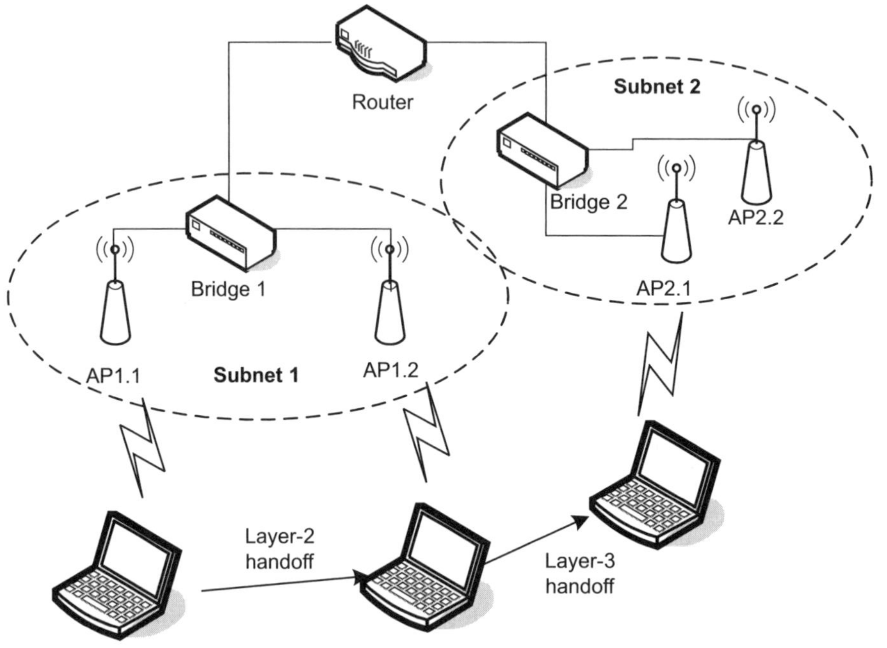

Figure 16.4 Layer-2 versus layer-3 handoffs.

not. All the APs and stations in an ESS can be in the same subnet (i.e., all the APs are connected via IEEE 802.1D MAC bridges, such as Ethernet switches), and the APs work in the bridge mode. When the two APs involved in a station handoff are in the same subnet, the handoff only requires a layer-2 mobility support without any intervention from the layer 3. (See the handoff event labeled as "layer-2 handoff" in Figure 16.4.) However, when the two APs are in different subnets, or the APs are in the router mode, a handoff involves layer-3 operations since the station cannot use the same IP address after reassociating with the new AP. This should be handled by mobile IP [9, 10] or *dynamic host configuration protocol* (DHCP) [11].

After reassociating with a new AP, a broadcast frame (originating from the handed off station) should be transmitted to the same subnet so that the layer-2 route tables in the MAC bridges including the APs can be updated. This operation basically completes the layer-2 handoff procedure, thanks to the IEEE 802.1D's self-learning process of the forwarding table [12]. This is further discussed in Section 16.2 as part of IEEE 802.11F *Inter-Access Point Protocol* (IAPP).

16.2 IEEE 802.11F for Inter-Access Point Protocol (IAPP)

As explained in the previous section, within an infrastructure BSS, a station is associated with an AP, and this station communicates with any other station through this AP. A WLAN can be composed of multiple APs. A key function in this multi-AP WLAN is the handoff or roaming (i.e., a station can switch from an AP to another as it moves). The handoff involves the communication between the APs, which relies on the DS.

While the 802.11 defines the concept of the DS, it does not define how to implement the DS. The reasons behind include the following: (1) the DS involves the protocols belonging to the above MAC, which is out of scope of the 802.11, dealing with only the MAC and PHY, and (2) it could be desirable to have the flexibility for the DS construction. Note that the DS can be constructed with any network link, even with the WLAN link, which is referred to as *wireless distribution system* (WDS). We further discuss this possibility in Section 18.2. However, the lack of the standardized DS construction caused APs from different vendors not to interoperate, especially, in the context of the handoff support. In the 802.11 WLAN (or more specifically, ESS), a station should have only a single association (i.e., the association with a single AP). However, the enforcement of this restriction is unlikely to be achieved due to the lack of the communication among the APs within the ESS.

IEEE 802.11F-2003 for IAPP is a recommended practice that specifies the information to be exchanged between APs among themselves and higher layer management entities to support the 802.11 DS functions [13]. According to the IEEE standards terms, the recommended practice is defined as a document in which procedures and positions preferred by the IEEE are presented. On the other hand, standards like IEEE 802.11-2007 are defined as documents specifying mandatory requirements (along with some optional functions). It should be noted that the 802.11F does not define anything related to the station operation for the handoff. The 802.11 MAC management (i.e., MLME) defines the AP scanning of the stations and reassociation procedures for the basic handoff support, as discussed in Section 16.1.

In fact, IEEE 802.11F was officially withdrawn by IEEE *standard association* in February 2006. The 802.11F was a trial-use recommended practice, and it had a two-year lifetime. During this lifetime, no one seemed to have implemented an IAPP as defined in IEEE 802.11F or had any interest in developing systems using the 802.11F. Moreover, the 802.11F was not updated to change the operation of AP in which a layer-2 frame is sent immediately after deciding to accept the association of a station. Under IEEE 802.11i, the layer-2 update frame should not be sent by the AP until the authenticator allows frames to pass over the IEEE 802.1X controlled port between the station and the DS. Eventually, no one requested the trial-use period to be extended. Moreover, if all the APs deployed in a WLAN are manufactured by the same vendor, they can operate with a proprietary IAPP. An example is the IAPP used for KT (Korea Telecom) WLAN [14].

Specifically, the emergence of *WLAN switch* architectures, in which the 802.11 MAC is actually divided into a time-critical lower MAC, including the frame transmission/reception, and less time critical upper MAC related with the network management, including the mobility and security support, made changes as well. Under this architecture, an AP can be implemented as a lower layer-2 device. That is, an AP might include only the lower MAC functions, and then less time-critical upper MAC functions are implemented in a so-called WLAN switch, which connects multiple APs including only lower-MAC functions. In this architecture, as the upper MAC functions of all the APs are implemented in the same device, there is no need for an external communication among APs even for the handoff support, thus eliminating the needs for IAPP.

We present IEEE 802.11F in this section for the completeness of the discussion by introducing a possible solution for the interoperability of the APs. The related issues for the interoperability of the APs are also currently being discussed in IETF *control and provisioning of wireless access points* (CAPWAP) working group with the charter of developing a CAPWAP protocol to provide interoperability among WLAN backend architectures [15].

16.2.1 Inter-AP Communication

The IAPP uses TCP/IP or UDP/IP to carry IAPP packets between APs and describes the use of *remote authentication dial in user service* (RADIUS) protocol [16], so that APs may obtain information about one another. Note that RADIUS is an *authentication, authorization, and accounting* (AAA) protocol for applications such as network access or IP mobility. As discussed in Section 15.2, this protocol is used as part of IEEE 802.11i for security support as well. A proactive caching mechanism is also defined in order to provide faster roaming by sending the station context to neighboring APs before the actual handoff event.

Figure 16.5 shows the architecture of the AP with IAPP. The *AP management entity* (APME) is a function that is external to the IAPP and is typically the main operational program of the AP, implementing an AP manufacturer's proprietary features and algorithms. As presented in Section 11.2, the 802.11-2007 defines an entity called *station management entity* (SME), which works as the brain of an 802.11 station, and the APME of the AP includes the SME functions. As shown in the figure, the APME can manage/control the IAPP, 802.11 MAC, and 802.11 PHY via the IAPP *service access point* (SAP), *MAC sublayer management entity* (MLME) SAP, and *PHY layer management entity* (PLME) SAP, respectively.

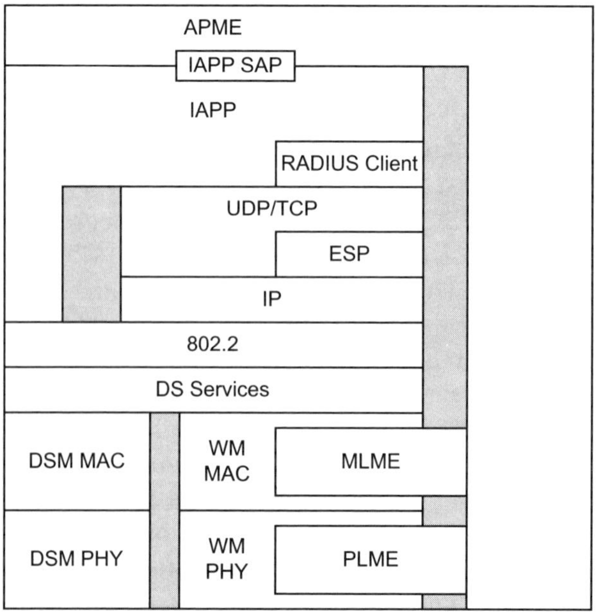

Figure 16.5 AP architecture with IAPP. (*After:* [13].)

Some functions of the IAPP rely on the RADIUS protocol for the correct and secure operation. In particular, the IAPP entity (i.e., the AP) should be able to find and use a RADIUS server to look up the IP addresses of other APs in the ESS when given the BSSIDs of those APs, and to obtain security information to protect the content of certain IAPP packets. Actually, the RADIUS server must provide extensions for IAPP-specific operations.

The IAPP supports: (1) DS services, (2) address mapping between AP's MAC and IP addresses, (3) formation of DS, (4) maintenance of DS, (5) enforcement of a single association of a station at a given time, and (6) transfer of station context information between APs.

16.2.2 IAPP Operations

There are basically three different IAPP operations: (1) *station add* (station ADD) operation, (2) *station move* (station MOVE) operation, and (3) proactive caching. These operations are explained next.

Station ADD Operation

The station ADD operation is triggered when a station is newly associated with an AP. When a station is associated, the AP transmits two packets to the DS or the wired infrastructure: a layer-2 update frame and an IAPP ADD-notify packet.

The layer-2 update frame is an 802.2 type 1 LLC *exchange identifier* (XID) update response frame. This frame is sent using a MAC source address equal to the MAC address of the station that has newly associated. Upon the reception of this frame, any layer-2 devices (e.g., bridges, switches, and other APs) update their forwarding tables with the correct port to reach the new location of the station according to the IEEE 802.1D bridge table self-learning procedure [12]. The format of an XID update frame carried over 802.3 is shown in Figure 16.6. The 802.3 MAC header is shown only as an example. A MAC protocol other than IEEE 802.3 may be used depending on the MAC protocol used for the implementation of the DS. The MAC DA is the broadcast MAC address. The MAC SA is the MAC address of the station that has newly associated. The length field is the length of the information following this field, 8 octets. The value of both the DSAP and SSAP is null. The control field and XID information field are defined in IEEE 802.2 [17].

The IAPP ADD-notify packet is sent using the IAPP over UDP/IP to notify the APs that the station identified in the packet has associated at the AP sending the packet. The IP datagram is sent at the destination IP address of the IAPP IP multicast address (i.e., 224.0.1.178). Any APs receiving this packet within the ESS removes stale association information about the associated station. If the newly associated station operates in a standard-compliant manner, there should not be any stale association information about the associated station in other APs. If another AP has

	MAC DA	MAC SA	Length	DSAP	SSAP	Control	XID Information Field
octets :	6	6	2	1	1	1	3

Figure 16.6 Layer-2 update frame format. (*After:* [13].)

such information, this AP must be the station's old AP so that the station has to reassociate, not associate, with the new AP. However, this erroneous situation can occur if the station is not implemented in a standard-compliant manner, so that the station transmits an association request frame instead of a reassociation request frame even if it hands off from an AP.

Station MOVE Operation

The station MOVE operation is triggered when a station is reassociated with an AP, which happens when this station hands off from one AP to another AP. The station, which is handing off from an AP, transmits a reassociation request frame to the new AP, where the reassociation request frame includes the MAC address of the old AP. The new AP transmits two packets in this case as well: a layer-2 update frame and an IAPP MOVE-notify packet. The IAPP MOVE-notify packet, over TCP/IP, is transmitted to the old AP, where TCP is used instead of UDP to achieve a reliable transmission. The old AP then responds by transmitting an IAPP MOVE-response packet, over TCP/IP, where TCP is again used for a reliable transmission. The response packet carries the context block for the station's association from the old AP to the new AP. The examples of the context include security and accounting information of the corresponding station.

Since the reassociation request frame from the station contains the old AP's MAC address, the new AP needs to look up the IP address of the old AP via the help of the RADIUS server, which provides the mapping information between the MAC and IP addresses of all the APs in the WLAN. The purpose of the layer-2 update frame is the same as with the station ADD operation case. One important fact is that the layer-2 update frame is broadcast only after the IAPP MOVE-response packet is received from the old AP, as the final step of the hand-off procedure.

Proactive Caching Operation

The proactive caching is triggered when a station is (re)associated with an AP or the context of the station changes. Basically, when the proactive caching is triggered by the APME of an AP, the AP (or the AP's IAPP entity, more specifically) transmits the IAPP CACHE-notify packets, over TCP/IP, to its neighboring APs, which respond with an IAPP CACHE-response packet, over TCP/IP. The notify packet includes the context of the newly (re)associated station. This proactive caching can significantly reduce the handoff delay by broadcasting the layer-2 update frame without waiting for the IAPP MOVE-response packet upon a reassociation (or handoff) of a station when the new AP has the context of the newly reassociated station, where the context was received from the old AP of this station via the IAPP CACHE-notify packet earlier [18].

One may question about how to know the neighboring APs in advance. This can be achieved via a dynamic learning. That is, an AP can learn that another AP is its neighbor when a station hands off from this AP to itself. The neighboring AP list can grow over time as more and more stations move around across the ESS. On the other hand, the network administrator can of course manually preconfigure the list of neighboring APs.

16.3 Mechanisms for Fast Scanning

As presented in Section 16.1, a number of steps involving frame exchanges are needed in order for a station to conduct a BSS-transition or handoff. Several independent empirical studies on the 802.11 handoff showed that the scanning procedure consumed the most time in the overall handoff latency [14, 19]. To search for target APs, a station scans one channel at a time, using either passive or active scanning, and waits on that channel during a preset timeout before moving on to the next channel. Scanning all the 11 (or 13 depending on the particular regulatory domain) channels of the 802.11b can take a few hundreds of milliseconds using active scanning. If passive scanning is used, or if there are more channels to scan, the scanning process might take even longer. Note that the number of channels for the 802.11a could be even more depending on the countries.

16.3.1 Need for Fast Scanning

Figure 16.7 shows experimental results measuring the time needed to hand off from one AP to another. The experiments were conducted in a commercial WLAN network environment illustrated in Figure 16.8, where a person with an 802.11 station walks through the building following a fixed route during each measurement run. The station sends periodic ping messages to the network in order to maintain and display the connectivity. Therefore, as the station moves around, it conducts handoffs whenever it leaves a BSS and enters another. For further details about the experimental methodology, refer to [14].

The delay is measured from the start of scanning to the completion of a reassociation (i.e., the procedures illustrated in Figure 16.1). As shown in Figure 16.7, the incurring delay is around 500 ms, which is quite notable. Moreover, the majority of the delay is due to the scanning process. It should be noted that the actual delay performance heavily depends on individual 802.11 devices. As we learned in Section 16.1, a number of parameters including ProbeDelay, MinChannelTime, and MaxChannelTime affect the scanning time in each channel.

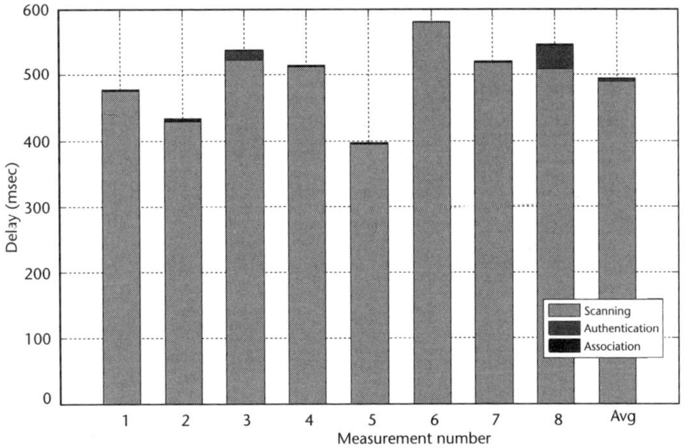

Figure 16.7 IEEE 802.11b handoff time measurement.

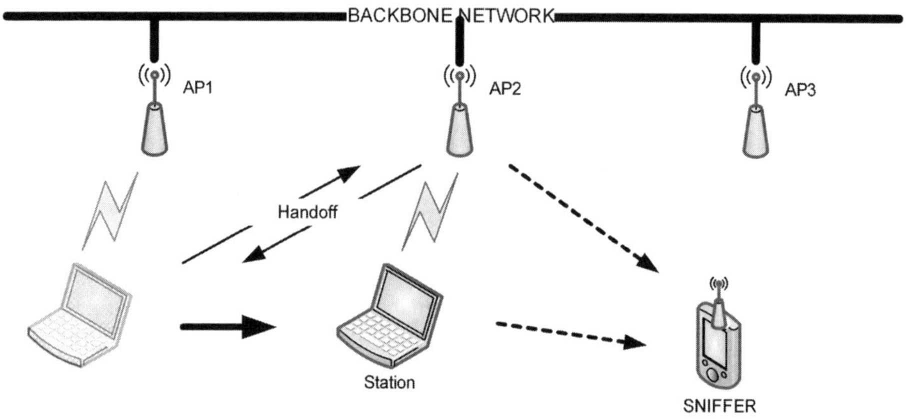

Figure 16.8 Experimental environment.

Moreover, the number of available channels, which depend on the underlying PHY as well as the particular country, determines the total scanning time. However, for most 802.11 stations, the scanning is expected to consume the most time.

16.3.2 IEEE 802.11k for Fast Scanning

IEEE 802.11k *radio resource measurement* (RRM) protocol is an emerging standard enabling stations to understand the radio environment that they operate in. The RRM enables stations to measure and gather data on radio link performance and on the radio environment. The measured data can be used for the optimization of the network performance. While there are a lot of measurements defined by the 802.11k, some of the measurement results and measurement mechanisms can facilitate fast handoff, especially, by expediting the scanning procedure, and they are presented in this section. In this book, we refer to IEEE 802.11k/D9.0 draft specification [1], and as the standardization has not been finalized yet, the detailed protocols are subject to change.

There can be basically two ways to shorten the scanning procedure: reducing the number of channels to scan and reducing the scanning time on each channel.

Limiting Scanning Channels

Under the 802.11k, the beacon frame is expanded to include the *AP channel report element*, which contains a list of channels in which a station is likely to find an AP. The same idea was also proposed in the literature [20]. That is, the channels in which the neighboring APs operate are indicated. The format of the AP channel report element is shown in Figure 16.9. The *regulatory class* field specifies the region

	Element ID	Length	Regulatory Class	Channel List
octets :	1	1	1	variable

Figure 16.9 AP channel report element. (*After:* [1].)

or country in which the channel list is valid. The channel list contains a number of channel numbers.

Upon the reception of beacons from its AP, a station can learn about the list of channels in which neighboring APs are operating, and, hence, can limit the channels to be scanned to the set of channels found in this list. When the station scans all these channels, there might be some channels in which the station fails to detect an AP, even if the information in the AP channel report is not stale. This is because the AP channel report does not give any information about the neighboring APs' locations. For example, when a neighboring AP operating in channel 1 is located near one end of a BSA, while the station is located near the other end of a BSA, the station in this BSA might not detect this AP in channel 1 even if its current AP informed the possible existence of this neighboring AP in channel 1 via its AP channel report. As a station can possibly reduce the number of channels to scan, the scanning time might be significantly reduced.

Neighbor Report Request/Response

Another mechanism for a station to obtain the information about neighboring APs is the *neighbor report request/response frame* exchange. The neighbor report request is sent to the current AP, which returns a neighbor report containing the information about known neighboring APs that could be candidates for a handoff of its associated stations. An AP may get the information of neighboring APs from the measurements received from the stations within the BSS (according to the 802.11k measurement request/response mechanisms), from management interface such as *simple network management protocol* (SNMP) [21], or from DS via IAPP.

A neighbor report response frame contains a number of the *neighbor report elements*, where the number corresponds to the number of neighboring APs. The neighbor report element contains various types of information about a single neighboring AP. As shown in Figure 16.10, each neighbor report element describes an AP, and consists of BSSID, BSSID information, channel number, regulatory class, and PHY type, and can include optional subelements. The BSSID is the BSSID of the BSS being reported. The subsequent fields in the neighbor report element pertain to this BSS.

The BSSID information field can be used to help determine neighbor BSS transition candidates. It contains a number of fields, including the following:

- *AP reachability* field, indicating whether the neighboring AP accepts preauthentication;
- *Security* bit, indicating whether the neighboring AP supports the same security provisioning as used by the station in its current association;
- *Key scope* bit, indicating whether the neighboring AP has the same authenticator as the current AP;

	Element ID	Length	BSSID	BSSID Information	Regulatory Class	Channel Number	PHY Type	Optional subelements
octets :	1	1	6	4	1	1	1	variable

Figure 16.10 Neighbor report element. (*After:* [1].)

- *Capabilities* subfield, containing selected capability information for the neighboring AP about whether spectrum management (i.e., the 802.11h), QoS (i.e., the 802.11e), APSD (from the 802.11e), radio measurement (i.e., the 802.11k), and delayed or immediate BlockAck (from the 802.11e) are supported.

The channel number field indicates the last known operating channel of the neighboring AP. The PHY type field indicates the PHY type of the neighboring AP. The optional subelements field may contain a number of subelements, and one interesting subelement is the TSF information subelement. This subelement contains both the TSF offset subfield and the beacon interval field, where the former contains the neighboring AP's TSF timer offset and the latter contains the beacon interval of the neighboring AP. The TSF timer offset is the time difference, in TU units, between the current AP and a neighboring AP. This value can be useful to determine the TBTTs of the neighboring AP so that a station can minimize the passive scanning time.

Fast Active Scanning

The probe request for active scanning is transmitted at the broadcast address, which may result in simultaneous responses from multiple APs. To avoid collision, probe responses follow the same DCF (or EDCA with AC_VO) rule to access the channel as data frames from any other station do. As a result, the probe response from an AP could be potentially delayed by other frames from any station. For this reason, the MaxChannelTime cannot be too short, as the station may otherwise miss a delayed probe response. Another reason to keep the MaxChannelTime long is that the station has no prior knowledge on the number of probe responses to expect and may want to get as many probe responses back as possible.

This analysis points out that the broadcast probe request makes it difficult to shorten the active scanning timeout. Thanks to the 802.11k neighbor report, a station will have prior knowledge of the identities (i.e., BSSIDs) of the neighboring APs. The station may, therefore, choose to send a unicast probe request destined to a particular AP instead. Two benefits arise from using unicast probe request. First, there is no potential collision from multiple simultaneous probe responses. Second, there is no need to wait for more than one response. To take advantage of these benefits, the authors of [22] proposed a fast active scanning scheme, which utilizes unicast active probe request transmissions. Nevertheless, this scheme is not included in the current draft due to the usage of unicast probe request, which is not standard compliant.

Fast Passive Scanning

Passive scanning is attractive since it causes no additional network traffic loads. Furthermore, in certain frequency bands in certain countries, active scanning is banned under certain conditions so that passive scanning may be the only choice. Despite its attractiveness, passive scanning usually takes longer than active scanning, as the scanning station needs to wait up to a beacon interval (typically 100 ms) to receive a beacon. If the TBTTs of the neighboring APs are known to a station in advance, a fast passive scanning can be achieved. That is, knowing when to expect a beacon from a particular neighboring AP on a particular channel, a station can switch to the channel right before the TBTT and switch back to the original channel right after receiving a beacon. A station can optionally obtain the TBTT information of neighboring APs from the 802.11k neighbor report, explained next.

The 802.11k also newly defines the *measurement pilot* frame. The measurement pilot frame is a short action frame transmitted pseudo-periodically by an AP at a small interval relative to a beacon interval. The measurement pilot frame provides a subset of the information provided in a beacon frame, whereas it is smaller than a beacon and is transmitted more often than a beacon. The measurement pilot frames are basically transmitted at every *target measurement pilot transmission time* (TMPTT) scheduled periodically with the period of MeasurementPilotPeriod.

The purpose of the measurement pilot frame is to assist a station with fast passive scanning. That is, a passive scanning can now be performed by receiving either a beacon or a measurement pilot frame, where the chance to receive a measurement pilot frame is higher, thanks to its relatively higher transmission frequency. Moreover, since these frames are transmitted frequently, a station can collect the channel condition measurements via passive scanning rapidly as well. Remember that active scanning is not allowed in certain frequency bands in certain countries.

16.4 IEEE 802.11r for Fast Roaming

The emerging IEEE 802.11r is defining a *fast BSS transition* (FT), which is a BSS transition that establishes the states (e.g., those related with security and QoS) necessary for data connectivity before the reassociation rather than after the reassociation. The FT protocol provides a mechanism for a non-AP station to perform a BSS transition between APs in an IEEE 802.11i RSN or when QoS TS setup of IEEE 802.11e is required in the ESS. As shown in Figure 16.2, many frame exchanges are needed in order to set up the connectivity between a non-AP station and an AP. For a station with QoS traffic, this procedure might be a big burden since it will result in long delivery latency during the BSS transition. The mechanisms enabling fast scanning will help, but it can reduce only the time to receive probe responses.

We here present the 802.11r FT protocols based on IEEE 802.11r/D8.0 draft specification. As the standardization has not been finalized yet, the detailed protocols are subject to change. FT is intended to reduce the time duration that connectivity is lost between the station and the DS during a BSS transition. The FT protocols are part of the reassociation procedure, and apply only to station transitions between APs within the same *mobility domain* (MD) within the same ESS. The FT protocols require information to be exchanged during the initial association (and reassociation) between the non-AP station and AP. The initial exchange is referred to as the *fast BSS transition initial mobility domain association*. Subsequent reassociations with APs within the same MD utilize the FT protocols. Two FT protocols are defined as follows:

- *Fast BSS transition* is executed when a station hands off to a target AP and does not require a resource request prior to its handoff. A resource request can be conducted as part of the reassociation request/response exchange.
- *Fast BSS transition resource request* is executed when a station requires a resource request prior to its handoff. A resource request is conducted right after an authentication prior to the reassociation.

For a station to hand off from its current AP to a target AP by utilizing the FT protocols, the message exchanges are conducted using one of the following two methods:

- *Over-the-air:* The station communicates directly with the target AP using IEEE 802.11 *authentication* frames with the *authentication algorithm* set to "fast BSS transition."
- *Over-the-DS:* The station communicates with the target AP via the current AP. The communication between the station and the target AP is carried in newly defined *fast BSS transition* (FT) action frames between the station and the current AP, and between the current AP and target AP via an encapsulation method. The current AP converts between the two encapsulations

Accordingly, four different protocols are defined, namely, over-the-air FT, over-the-DS FT, over-the-air FT resource request, and over-the-DS FT resource request. As shown in Table 16.1, different protocols use a different set of frame exchanges for the authentication with a target AP and the resource request to the target AP. The FT protocols for both RSN and non-RSN are defined, but we will only consider the RSN case for the simplicity. Further details along with the frame exchange timings will be presented next.

16.4.1 FT Key Hierarchy

The FT key hierarchy is designed to allow a station to make fast BSS transitions between APs without the need to perform an IEEE 802.1X authentication at every AP within the MD. The fast BSS transition key hierarchy can be used with either IEEE 802.1X authentication or PSK authentication. A three-level key hierarchy is introduced to provide key separation between the key holders. A key holder is a

Table 16.1 Four Types of FT Protocols with Different Frame Exchanges for Authentication with a Target AP and Resource Request to the Target AP

	Authentication with Target AP	Resource Request to Target AP
Over-the-Air FT	802.11 Authentication Request/Response	Reassociation Request/Response
Over-the-DS FT	FT Request/Response	Reassociation Request/Response
Over-the-Air FT Resource Request	802.11 Authentication Request/Response	802.11 Authentication Confirm/ACK
Over-the-DS FT Resource Request	FT Request/Response	FT Confirm/ACK

component of RSNA key management residing in an authenticator or a supplicant. The FT key hierarchy is shown in Figure 16.11.

- *Pairwise master key, first level* (PMK-R0) is the first level of the FT key hierarchy. This key is derived as a function of the MSK or PSK and is stored by the PMK-R0 key holders, namely, the *PMK-R0 key holder in the authenticator* (R0KH) and the *PMK-R0 key holder in the supplicant* (S0KH).
- *Pairwise master key, second level* (PMK-R1) is the second level of the key hierarchy. This key is mutually derived by the S0KH and R0KH, and is distributed to PMK-R1 key holders, namely, the *PMK-R1 key holder in the authenticator* (R1KH) and the *PMK-R1 key holder in the supplicant* (S1KH). PMK-R1 keys are derived using PMK-R0 and the MAC addresses of the corresponding AP and the station and are delivered from R0KH to the R1KHs (i.e., other APs) within the same MD.
- *Pairwise transient key* (PTK) is the third level of the key hierarchy that defines the IEEE 802.11 and IEEE 802.1X protection keys as presented in Chapter 15, specifically in Sections 15.2.2 and 15.3. The PTK is mutually derived by the PMK-R1 key holders (i.e., R1KH and S1KH).

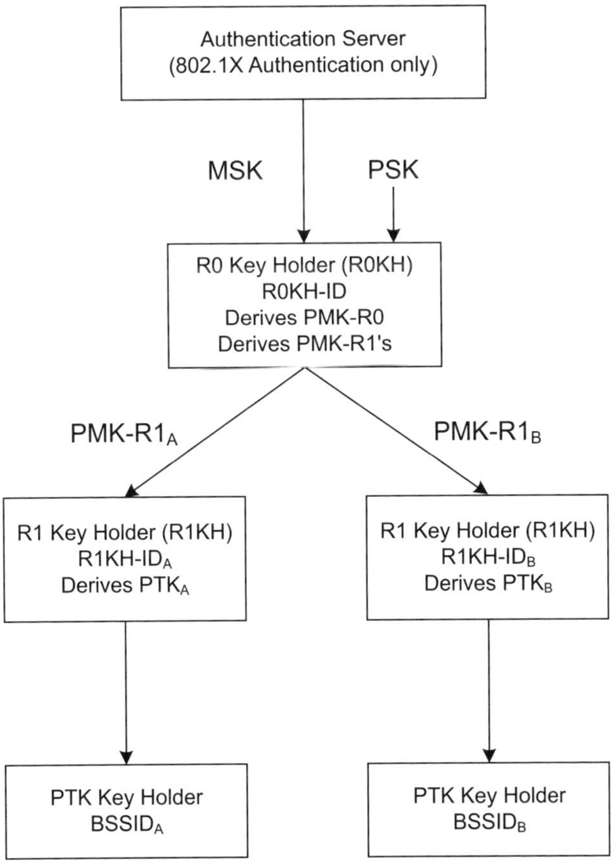

Figure 16.11 IEEE 802.11r FT key hierarchy. (*After:* [2].)

IEEE 802.11r defines a revised four-way handshake for the pairwise key distribution, referred to as *fast BSS transition four-way handshake*. This handshake confirms mutual possession of a PMK-R1 by two parties and distributes a GTK.

16.4.2 FT Initial MD Association

The first association or first reassociation within an MD is referred to as the *fast BSS transition initial mobility domain association*, which is established via the procedure illustrated in Figure 16.12. The notation for the EAPLO-key frames is defined in Section 15.3.2, where some new DataKDs, such as *mobility domain information element* (MDIE) and *fast BSS transition information element* (FTIE), are newly defined. The procedure in terms of the frame exchanges is basically the same as the

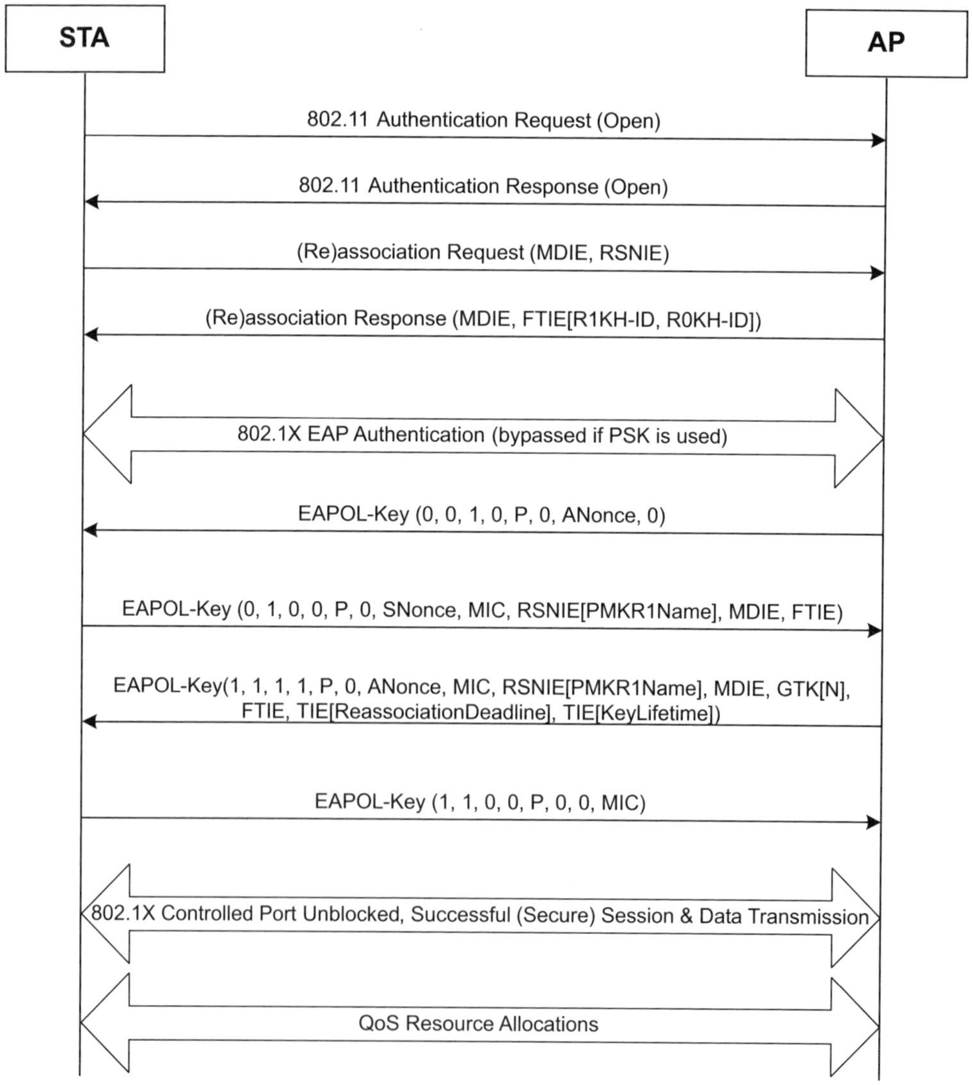

Figure 16.12 FT initial mobility domain association in an RSN. (*After:* [2].)

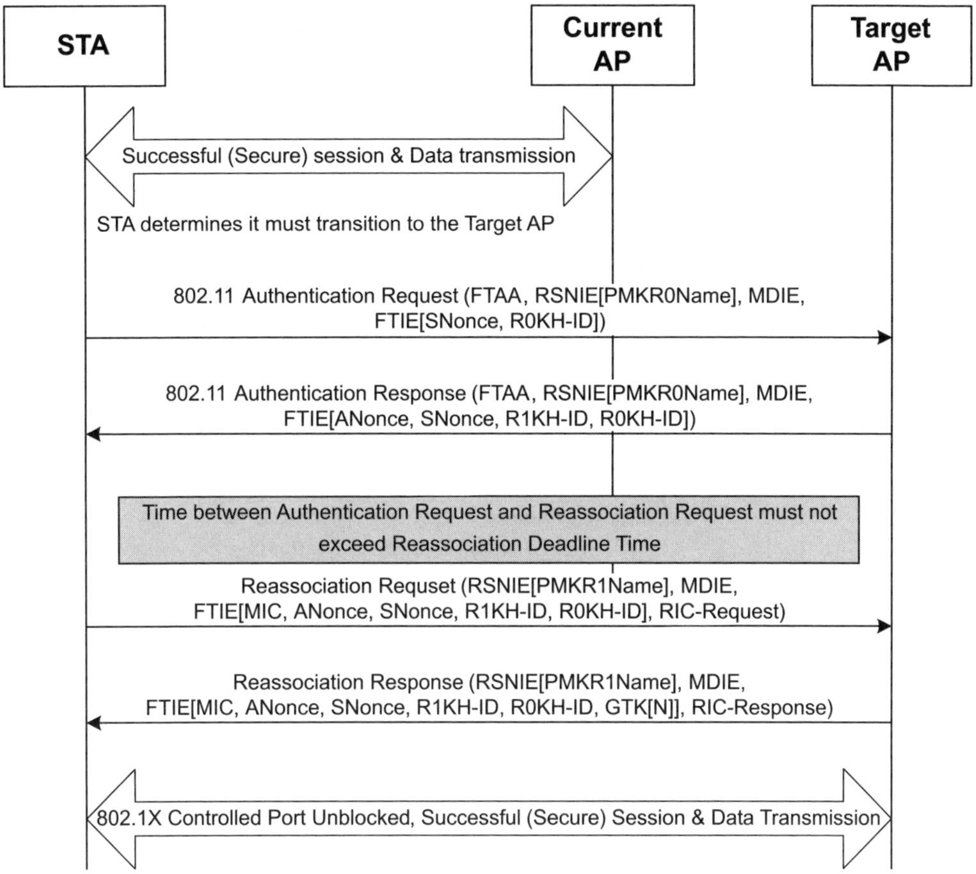

Figure 16.13 Fast BSS transition over-the-air in an RSN. (*After:* [2].)

one illustrated in Figure 16.2, but a new result out of the FT four-way handshake is the establishment of the FT key hierarchy.

16.4.3 FT Protocols

The FT protocol supports resource requests as part of the reassociation. The optional FT resource request protocol (see Section 16.4.4) supports resource requests prior to reassociation. IEEE 802.11 authentication frames, defined in Section 15.1.2, are used, where a new authentication algorithm, called the fast BSS transition authentication algorithm, is used for the FT protocol.

Over-the-Air FT

The station and the target AP use the FT authentication request and response exchange to specify the PMK-R1 SA and to provide values of SNonce and ANonce. This exchange enables a fresh PTK to be computed in advance of reassociation. The PTKSA is used to protect the subsequent reassociation transaction, including the optional *resource information container* (RIC)-request/response. Currently, the

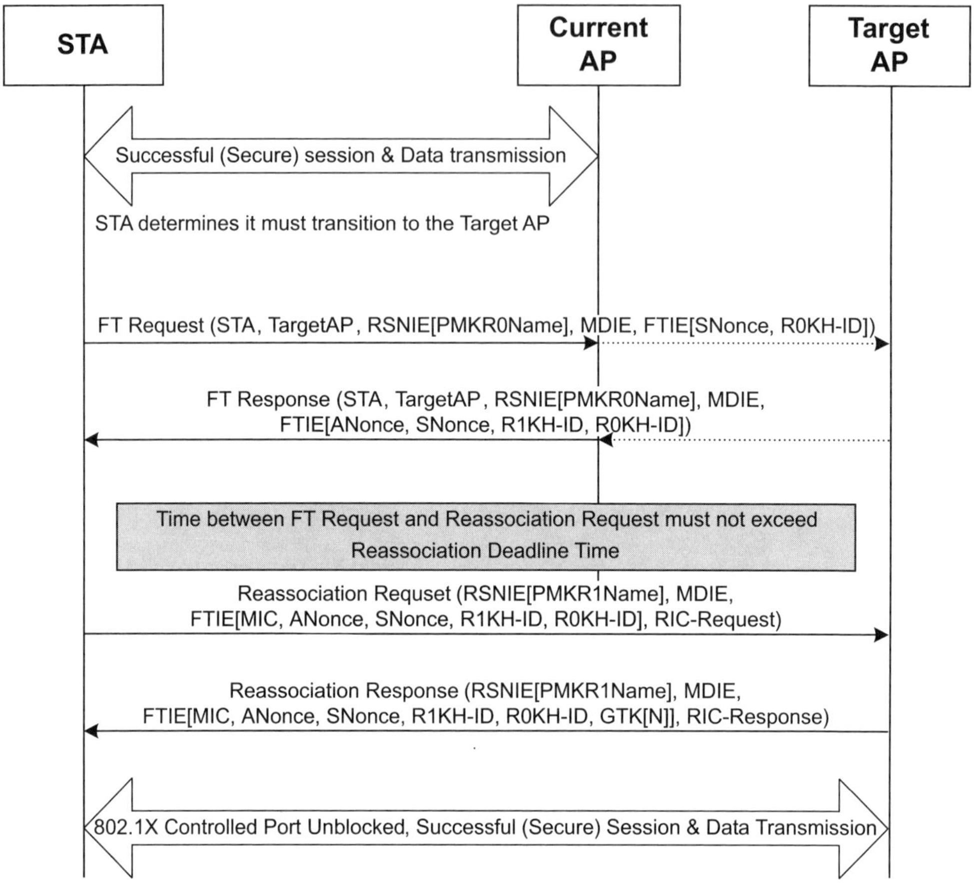

Figure 16.14 Fast BSS transition over-the-DS in an RSN. (*After:* [2].)

RIC-request/response can be used to set up the BlockAck mechanism between the station and the target AP. (See Figure 16.13.)

Over-the-DS FT

With the over-the-DS FT protocol (see Figure 16.14), the authentication with the target AP occurs through the current AP and the DS. The information exchange between the station and the current AP occurs via FT request and response frames.

16.4.4 FT Resource Request Protocols

The FT resource request protocol involves an additional message exchange after the authentication request/response, or FT request/response, in order to allow the station to request resources prior to reassociation. The additional message exchange for the FT resource request protocol is performed using the same method (i.e., over-the-air or over-the-DS) as was used for the authentication request/response or FT request/response.

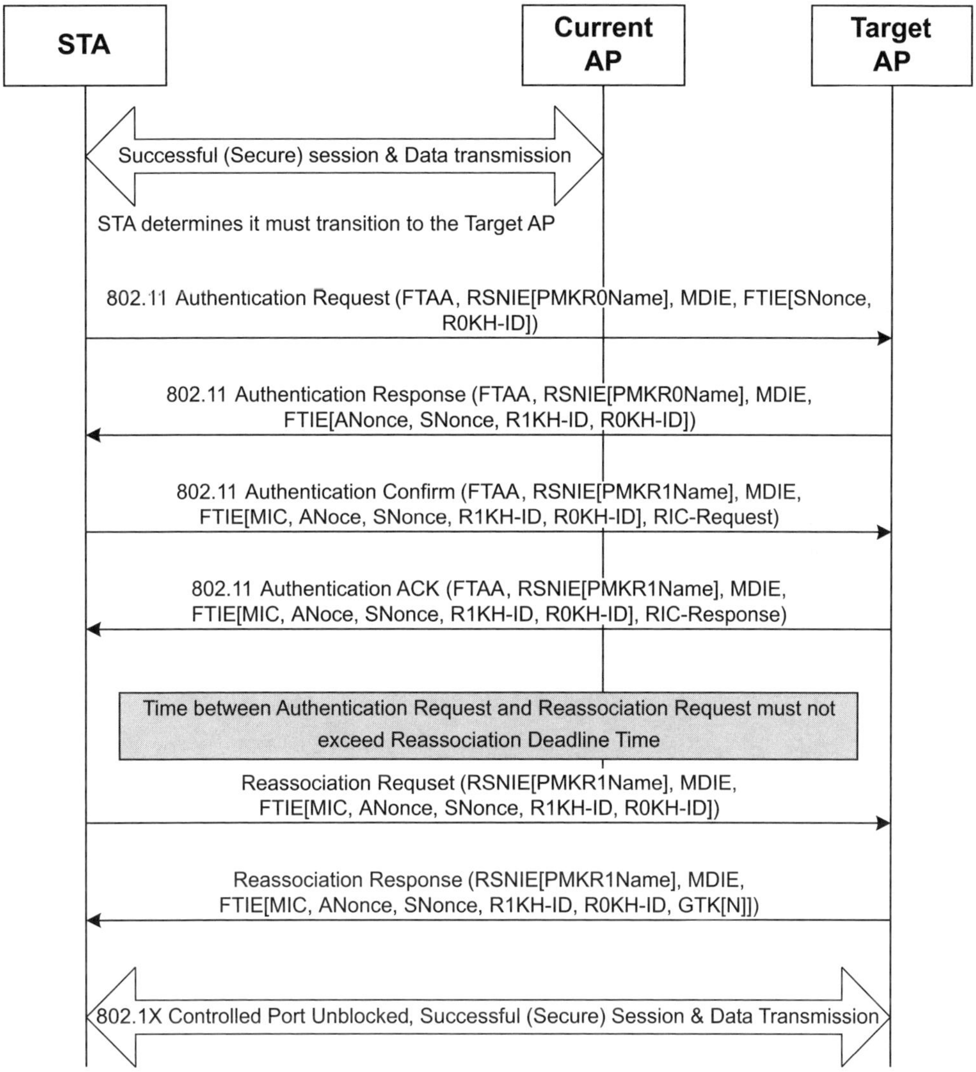

Figure 16.15 Fast BSS transition resource over-the-air in an RSN. (*After:* [2].)

Over-the-Air FT Resource Request

To perform an over-the-air FT resource request protocol to a target AP, after completing the authentication request/response exchange explained in Section 16.4.1, the station and target AP exchange the authentication confirm/ACK frames as shown in Figure 16.15. The authentication confirm and ACK frames basically convey RIC-request and RIC-response so that the resource request and setup can be made before the reassociation.

Over-the-DS FT Resource Request

To perform an over-the-DS FT resource request protocol with a target AP, after completing the FT request/response exchange presented in Section 16.4.3, the station and target AP (through the current AP and the DS) perform the exchange of FT confirm/ACK frames as shown in Figure 16.16. The FT confirm and ACK frames

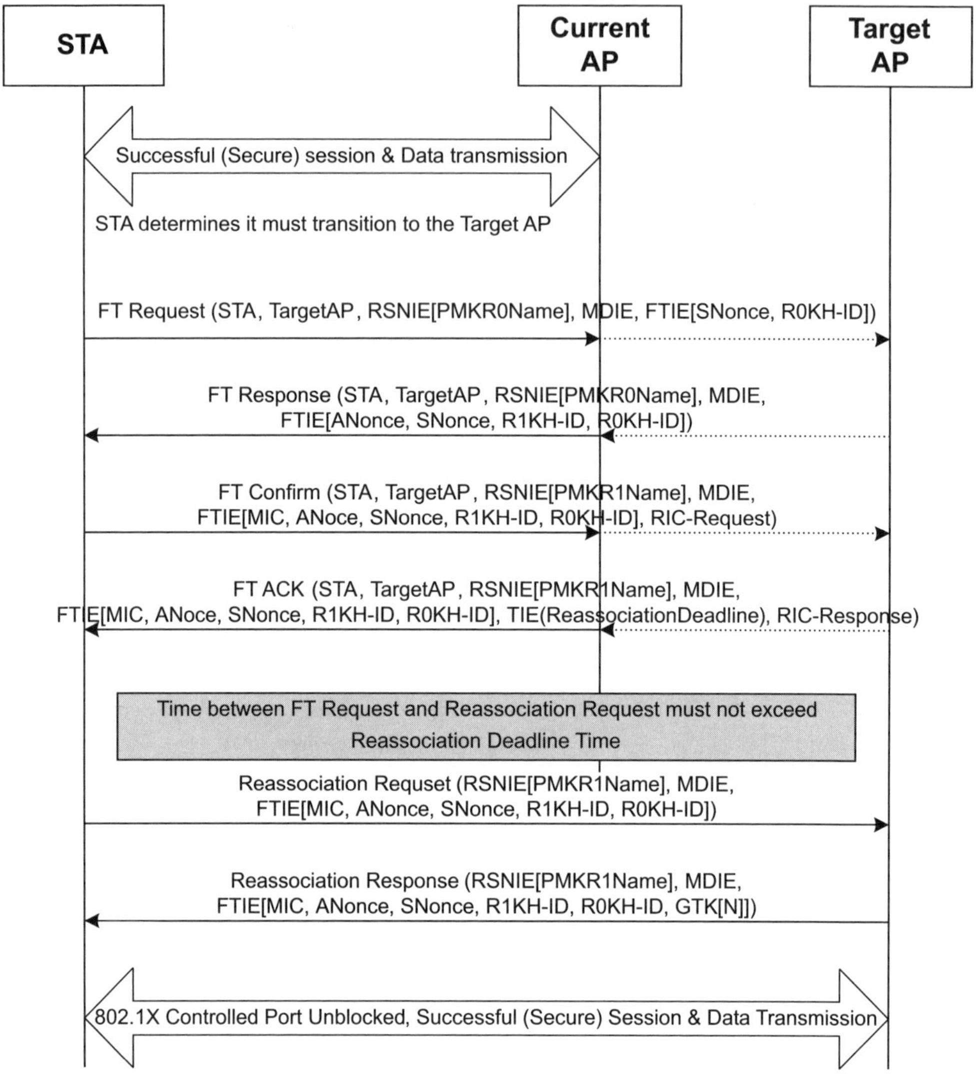

Figure 16.16 Fast BSS transition resource over-the-DS in an RSN. (*After:* [2].)

basically convey RIC-request and RIC-response so that the resource request and setup can be made before the reassociation.

References

[1] IEEE 802.11k/D9.0, Draft Supplement to Part 11: Wireless LAN Medium Access Control (MAC) and Physical Layer (PHY) specifications: Specification for Radio Resource Measurement, September 2007.

[2] IEEE 802.11r/D8.0 Draft Supplement to Part 11: Wireless LAN Medium Access Control (MAC) and Physical Layer (PHY) specifications: Fast ESS Transition, September 2007.

[3] IEEE 802.1X-2004, IEEE Standard for Local and Metropolitan Area Networks—Port-Based Network Access Control, 2004.

[4] IEEE 802.11e-2005, Amendment 8 to Part 11: Wireless LAN Medium Access Control (MAC) and Physical Layer (PHY) Specifications: Medium Access Control (MAC) Enhancements for Quality of Service (QoS), 2005.

[5] IEEE 802.11i-2004, Amendment 6 to Part 11: Wireless LAN Medium Access Control (MAC) and Physical Layer (PHY) Specifications: Medium Access Control (MAC) Security Enhancements, 2004.

[6] Velayos, H., and G. Karlsson, "Techniques to Reduce the IEEE 802.11b Handoff Time," *Proc. ICC'04*, Paris, France, June 2004, pp. 3844–3848.

[7] IEEE 802.11-2007, IEEE Standard for Local and Metropolitan Area Networks—Part 11: Wireless LAN Medium Access Control (MAC) and Physical Layer (PHY) Specifications, IEEE Std 802.11-2007, (Revision of IEEE Std 802.11-1999), June 12, 2007.

[8] Wong, D., and T.-J. Lim, "Soft Handoffs in CDMA Mobile Systems," *IEEE Personal Communications*, Vol. 4, No. 6, 1997, pp. 6–17.

[9] C. E., Perkins, "Mobile IP," *IEEE Communications Mag.*, May 2002, pp. 66–82.

[10] RFC 3220, IP mobility support for IPv4, January 2002.

[11] RFC 2131, Dynamic Host Configuration Protocol, March 1997.

[12] IEEE 802.1D-2004, IEEE Standard for Local and Metropolitan Area Networks—Media Access Control (MAC) Bridges (Incorporates IEEE 802.1t-2001 and IEEE 802.1w), 2004.

[13] IEEE 802.11F-2003, IEEE Trial-Use Recommended Practice for Multi-Vendor Access Point Interoperability Via an Inter-Access Point Protocol Across Distribution Systems Supporting IEEE 802.11 Operation, 2003.

[14] Kim, S., et al., "An Empirical Measurements-Based Analysis of Public WLAN Handoff Operations," *Proc. WILLOPAN'06*, New Delhi, India, January 8, 2006.

[15] IETF CAPWAP Charter, http://www.ietf.org/html.charters/capwap-charter.html.

[16] RFC 2865, Remote Authentication Dial In User Service (RADIUS), June 2000.

[17] IEEE 802.2-1998, IEEE Standard for Local and Metropolitan Area Networks—Part 2: Logical Link Control, 1998.

[18] Mishra, A., M. Shin, and W. Arbaugh, "Context Caching Using Neighbor Graphs for Fast Handoffs in a Wireless Network," *Proc. IEEE INFOCOM'04*, Hong Kong, March 2004.

[19] Mishra, A., M. Shin, and W. Arbaugh, "An Empirical Analysis of the IEEE 802.11 MAC Layer Handoff Process," *ACM Computer Communication Review*, Vol. 33, 2003, pp. 93–102.

[20] Shin, M., A. Mishra, and W. Arbaugh, "Improving the Latency of 802.11 Hand-Offs Using Neighbor Graphs," *Proc. ACM 2nd International Conference on Mobile Systems, Applications, and Services (MobiSys'04)*, Boston, MA, June 2004, pp. 70–83.

[21] IETF RFC 1157, "A Simple Network Management Protocol (SNMP)," May 1990.

[22] Jeong, M. R., et al., "Fast Active Scan Proposals," IEEE 802.11-03-0623-00, July 2003.

Selected Bibliography

Alimian, A, "Roaming Interval Measurements," IEEE 802.11-04-0378-r0, March 2004, http://www.drizzle.com/~aboba/IEEE/11-04-0378-00-roaming-intervals-measurements.ppt.

Alimian, A., and B. Aboba, "Analysis of Roaming Techniques," IEEE 802.11-04-0377-r1, March 2004, http://www.drizzle.com/~aboba/IEEE/11-04-0377-01-frfh-analysis-roaming-techniques.ppt.

Petroni, N., and W. A. Arbaugh, "An Empirical Analysis of the 4-Way Hand-Shake," IEEE 802.11-03-0563-00, July 2003, http://www.drizzle.com/~aboba/IEEE/11-03-0563-00-000i-tgi-4-way-handshake-timings.ppt.

Ramani, I., and S. Savage, "SyncScan: Practical Fast Handoff for 802.11 Infrastructure Networks," *Proc. IEEE INFOCOM'05*, Miami, FL, March 2005, pp. 675–684.

CHAPTER 17

Spectrum and Power Management

In 2003, the operation of IEEE 802.11a WLANs from 5.15 GHz to 5.725 GHz was confirmed by the harmonization and globalization of the 5-GHz frequency bands, adopted at the World Radiocommunication Conference (WRC 2003), provided some new regulations are implemented. The new regulations require the following four operations: (1) the system must be able to detect radar signals; (2) the system must be able to avoid interfering with radar operations; (3) the system must be able to uniformly spread its operation across all the usable channels; and (4) the system must be able to minimize the overall output power. While the 5-GHz bands, where the 802.11a operates, are unlicensed bands, there exist in fact primary users, which also use these bands. Those primary users include radar, satellite, and aeronautical and maritime navigation systems. These regulations were made in order to protect existing civil and military radars that already operate in the bands, as well as to minimize the hotspot that might show up in urban areas in radar images of earth resource satellite systems, which also use these bands [1].

The European Union had the same regulations even before 2003, and IEEE 802.11h-2003 was originally developed to satisfy regulatory requirements for the operation of IEEE 802.11a at the 5-GHz bands in Europe. The 802.11h defines spectrum and transmit power managements, including *dynamic frequency selection* (DFS) and *transmit power control* (TPC), which could be used to minimize the interference of the WLAN to the primary users. Now, according to the decision at WRC 2003, the same regulations are adopted in other regulatory bodies, including the United States and Korea. Therefore, IEEE 802.11h could be used for the operation of IEEE 802.11a at the 5-GHz bands in other countries to meet the regulations.

17.1 Regulatory Requirements

In this section, we first present the required TPC and DFS behaviors specified in (1) ETSI EN 301 893 [2] for the devices operating in 5.25–5.35-GHz and 5.47–5.725-GHz bands in Europe and (2) FCC CFR47 [3] for the devices operating in 5.25–5.35-GHz and 5.47–5.725-GHz bands in the United States. In the ETSI document, the term *radio local area network* (RLAN) is used to refer to the system operating in the 5-GHz bands, while the term *unlicensed national information infrastructure* (U-NII) is used to represent the system in the 5-GHz bands in the FCC document.

Tables 17.1 and 17.2 summarize the 5-GHz bands available for WLANs in Europe and the United States, respectively, along with their maximum transmission

Table 17.1 5 GHz Bands Available for WLANs in Europe

Bands (GHz)	Max. Tx Power	Remark
5.15~5.25	200 mW EIRP	Indoor only
5.25~5.35	200 mW EIRP	TPC & DFS
5.47~5.725	1 W EIRP	TPC & DFS

Table 17.2 5-GHz Bands Available for WLANs in the United States

Bands (GHz)	Max. Tx RF Power (with up to 6 dBi antenna gain)	Remark
5.15~5.25	40 mW	Indoor only
5.25~5.35	200 mW	TPC & DFS
5.47~5.725	200 mW	TPC & DFS
5.725~5.825	800 mW	ISM

powers. Note that TPC and DFS are not needed for 5.15–5.25-GHz and 5.725–5.825-GHz bands.

17.1.1 TPC Requirements

TPC is a technique to control the transmitter output power resulting in reduced interference to other systems. As presented next, the required TPC behaviors specified in ETSI EN 301 893 and FCC CFR47 are slightly different. It should be also noted that in both cases, the TPC mechanism, which enables a dynamic control of the transmit power, is not mandated; a fixed low transmit power level could be used while meeting the requirements.

TPC for RLAN in Europe

TPC should ensure a mitigation factor of at least 3 dB on the aggregate power from a large number of devices. This requires an RLAN device to have a TPC range with the lowest value of at least 6 dB below the values for the mean EIRP given in Table 17.1.

For devices with TPC capability, the output power, when configured to operate at the highest stated power level of the TPC range, will not exceed the levels given in Table 17.1. Moreover, the RF output power during a transmission burst when configured to operate at the lowest stated power level of the TPC range will not exceed the levels, which are 6 dB below the values from mean EIRP given in Table 17.1. For devices without TPC capability, the limits in Table 17.1 shall be reduced by 3 dB, and the devices are not required to have a capability to control the output power.

TPC for U-NII in the United States

A U-NII device is required to have the capability to operate at least 6 dB below the mean EIRP value of 30 dBm. A TPC mechanism is not required for systems with an EIRP of less than 27 dBm.

17.1.2 DFS Requirements

In this section, we present the required DFS behaviors based on ETSI EN 301 893. Similar behaviors are specified in FCC CFR47, where subtle differences in some operational parameters exist. In ETSI EN 301 893, the test radar signal patterns are also specified, which are missing in FCC CFR47. In this section, we use the term RLAN, as we refer to the ETSI document for the requirements.

An RLAN employs a DFS function to (1) detect interference from other systems and to avoid cochannel operation with these systems, particularly, radar systems; and (2) provide, on aggregate, a uniform loading of the spectrum across all devices. Within the context of the DFS operations, an RLAN device operates in either master mode (e.g., IEEE 802.11 AP) or slave mode (e.g., IEEE 802.11 non-AP station). RLAN devices operating in slave mode (i.e., slave devices) operate only in a network controlled by an RLAN device operating in master mode (i.e., a master device). Some RLAN devices, communicating in an ad hoc manner without being attached to an infrastructure, should employ DFS with the requirements applicable to a master.

DFS Requirements

The operational behaviors of DFS that are required for master and slave devices are as follows. First, master devices are required to operate as follows:

- The master device should use a *radar interference detection* function in order to detect radar signals.
- Before initiating a network on a channel, the master device should perform a *channel availability check* for 60 seconds to ensure that there is no radar operating on the channel. If no radars have been detected, the channel becomes an *available channel*, on which a network can be initiated.
- During normal operation, the master device should monitor the *operating channel* to ensure that there is no radar operating on the channel. This operation is referred to as the *in-service monitoring*.
- If the master device has detected a radar signal during *in-service monitoring*, the *operating channel* becomes unavailable. The master device shuts down the channel within the *channel move time* of 10 seconds by instructing all its associated slave devices to stop transmitting on this channel (which is to be unavailable soon).
- The master device should not resume any transmissions on this *unavailable channel* during the *nonoccupancy period* of 30 minutes after a radar signal was detected.

Second, slave devices are required to operate as follows:

- A slave device should not transmit before receiving an appropriate enabling signal from a master device.
- A slave device should stop all its transmissions whenever instructed by the master device with which it is associated. The device should not resume any

transmissions until it has again received an appropriate enabling signal from a master device.

- A slave device, which is required to perform the radar detection, will stop its own transmissions if it has detected radar signals.

Table 17.3 summarizes the applicability of DFS requirements for each of these operational modes. Table 17.4 summarizes the values of DFS parameters. We now explain some operations in more details.

Channel Availability Check

During the channel availability check of 60 seconds, the RLAN should be able to detect any of the radar test signals specified in Table 17.5 with a level above the interference detection threshold. As illustrated in Figure 17.1, the test signals contain a single burst of pulses. For a given radar test signal, *W* represents the pulse width, and *PRF* represents the pulse repetition frequency so that *1/PRF* represents the pulse repetition interval. Finally, *L* represents the number of pulses per radar scan, and is determined by [{Antenna Beamwidth (deg)} × {Pulse Repetition Frequency (pps)}] / {Scan Rate (deg/s)}.

The minimum interference detection threshold is determined based on the maximum EIRP emitted by a device: (1) for devices with a maximum EIRP of 200 mW to

Table 17.3 Applicability of DFS Requirements

Requirement	Operating mode		
	Master	Slave (without radar detection)	Slave (with radar detection)
Channel Availability Check	v	Not required	Not required
In-Service Monitoring	v	Not required	v
Channel Shutdown	v	v	v
Non-Occupancy Period	v	Not required	Not required
Uniform Spreading	v	Not required	Not required

Source: [2].

Table 17.4 DFS Parameter Values

Parameter	Value
Channel Availability Check Time	60 sec
Channel Move Time	10 sec
Channel Closing Transmission Time	260 msec
Non-Occupancy Period	30 min

Source: [2].

Table 17.5 Parameters of DFS Test Signals

Radar test signal	Pulse width *W* [µs]	Pulse repetition frequency *PRF* [pps]	Pulses per radio scan *L*	Detection probability with 30 % channel load
1 - Fixed	1	750	15	P_d > 60 %
2 - Variable	1, 2, 5	200, 300, 500, 800, 1000	10	P_d > 60 %
3 - Variable	10, 15	200, 300, 500, 800, 1000	15	P_d > 60 %
4 - Variable	1, 2, 5, 10, 15	1200, 1500, 1600	15	P_d > 60 %
5 - Variable	1, 2, 5, 10, 15	2300, 3000, 3500, 4000	25	P_d > 60 %
6 - Variable modulated	20, 30	2000, 3000, 4000	20	P_d > 60 %

Source: [2].

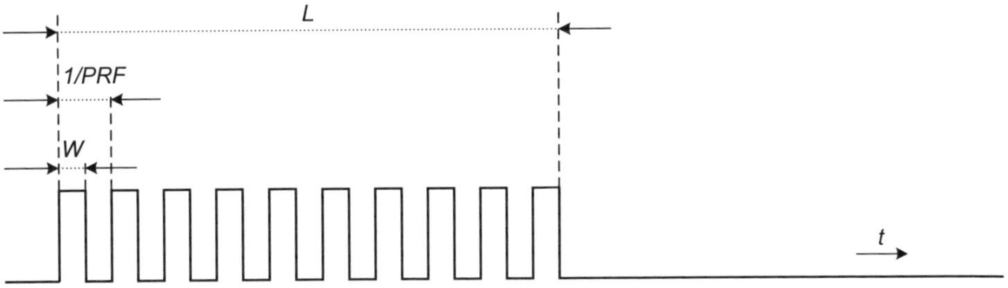

Figure 17.1 General radar test signal burst pattern per radar scan. The parameter values are found in Table 17.5. (*After:* [2].)

1W, the threshold is –64 dBm; and (2) for devices that operate with less than 200 mW EIRP, the threshold is –62 dBm. The detection threshold is the received power averaged over 1 µs referenced to a 0-dBi antenna. The detection probability for a given radar signal shall be greater than the value defined in Table 17.5. Available channels remain valid for up to 24 hours, after which a channel availability check should be performed again.

Channel Shutdown

The channel shutdown is defined as the process initiated by an RLAN device immediately after a radar signal has been detected on an operating channel. The master device should instruct all associated slave devices to stop transmitting on this channel within the channel move time of 10 seconds. Slave devices with a radar interference detection function should stop their own transmissions within the channel move time. The aggregate duration of all transmissions of the RLAN device on this channel during the channel move time is limited to the channel closing transmission time of 260 ms.[1] The aggregate duration of all transmissions does not include quiet periods in between transmissions.

Uniform Spreading

The uniform spreading is a mechanism to be used by the RLAN to provide, on aggregate, a uniform loading of the spectrum across all devices. This requires that a

1. For U-NII systems, this should be 200 ms [3].

RLAN device selects a channel out of the usable channels so that the probability of selecting a given channel should be the same for all channels. When implementing a frequency re-use plan across a planned network, the selection of the operating channel may be under control of the network. The probability of selecting each of the usable channels should be within 10 percent of the theoretical probability, where for "n" channels, the theoretical probability is $1/n$.

17.2 Introduction to IEEE 802.11h

IEEE 802.11h-2003 is an amendment of the baseline protocol and IEEE 802.11a PHY for "Spectrum and Transmit Power Management Extensions in the 5 GHz band in Europe." As the title indicates, the 802.11h was originally developed to extend the 802.11 operation in the 5-GHz bands in Europe by meeting the regulations discussed in Section 17.1. The 802.11h defines the DFS and TPC mechanisms on top of the 802.11-1999 MAC and the 802.11a PHY for these purposes. Note that, even though the 802.11h has been developed to satisfy the European regulatory requirements, it is also useful in other countries, especially, since other countries require similar regulations. Remind that TPC and DFS are not needed for 5.15–5.25-GHz and 5.725–5.825-GHz bands. Therefore, IEEE 802.11a stations, which are designed to operate only at the 5.15–5.25-GHz and 5.725–5.825-GHz bands, are not required to implement the TPC and DFS of IEEE 802.11h.

As will become clear later, the protocols defined in the 802.11h are more than what the regulatory bodies require. For example, TPC is not actually needed if the devices use a fixed low transmission power level. In Europe, a fixed power of less than or equal to the regulatory maximum power deducted by 3 dB can be used, and in the United States, a fixed power of under 27-dBm EIRP can be used. Accordingly, the protocols can be used for smart spectrum and power management for the optimization of the network performance, such as automatic frequency planning, reduction of energy consumption, range control, reduction of interference, and QoS enhancement, while meeting the requirements from the regulatory bodies. We first briefly present the TPC and DFS functions supported by IEEE 802.11h.

17.2.1 TPC Functions

As discussed in Section 17.1, radio regulations may require a WLAN operating in the 5-GHz band to use transmit power control, involving the specification of a regulatory maximum transmit power and a mitigation requirement for each allowed channel, in order to reduce interference with satellite services. The TPC service is used to satisfy this regulatory requirement. The 802.11h TPC service supports the following functions:

- Association of stations with an AP in a BSS based on the stations' power capability;
- Specification of regulatory and local maximum transmit power levels for the current channel;

- Selection of a transmit power for each transmission in a channel within constraints imposed by regulatory requirements;
- Adaptation of transmit power based on the estimation of path loss and link margin.

17.2.2 DFS Functions

As discussed in Section 17.1, radio regulations may require WLANs operating in the 5-GHz band to implement a mechanism to avoid co-channel operation with radar systems and to ensure uniform utilization of available channels. The DFS service is used to satisfy these regulatory requirements. The 802.11h DFS service supports the following functions:

- Association of stations with an AP in a BSS based on the stations' supported channels;
- Quieting the current channel so that it can be tested for the presence of radar with less interference from other stations;
- Testing channels for radar before using a channel and while operating in a channel;
- Discontinuing operations after detecting radar signal in the current channel in order to avoid interference with the radar system;
- Detecting radar in the current and other channels based on regulatory requirements;
- Requesting and reporting of measurements in the current and other channels;
- Selecting and advertising a new channel to enable the migration of a BSS or IBSS to the new channel after radar is detected.

17.2.3 Layer Management Model

Figure 17.2 illustrates the layer management model in the 802.11h to realize the DFS and TPC functions. Policy decisions, such as channel switching and transmit power control, reside in the SME while the associated protocols reside in the MLME.

Note that both DFS and TPC involve implementation-dependent algorithms. For example, a TPC algorithm is needed in order to determine the transmit power level of a frame transfer. Basically, the 802.11h defines the mechanisms/protocols to enable a right decision of the power level, not the implementation itself. It should be noted that there is virtually no change in terms of the channel access functions. That is, the 802.11 DCF/PCF or the 802.11e HCF are used to transmit the new management frames as part of the 802.11h.

17.3 Transmit Power Control (TPC)

For wide-area cellular systems, such as IS-95 *code-division multiple access* (CDMA) and the third generation *wideband CDMA* (WCDMA), TPC is critically important

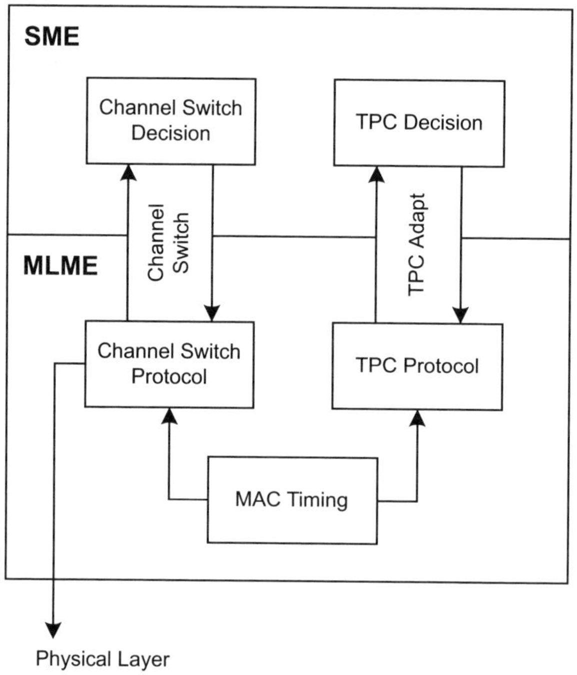

Figure 17.2 Layer management model in IEEE 802.11h. (*After:* [4].)

in order to (1) ameliorate the near-far problem, specifically, for CDMA uplink transmissions; (2) minimize the interference to/from other cells (i.e., cochannel interference); and (3) improve the system performance on fading channels by compensating fading dips [5, 6]. In comparison, most of today's 802.11 devices use a fixed transmit power for the frame transmissions, and TPC in 802.11 WLANs was not considered as critical to the success as in CDMA systems. However, the TPC in WLAN might be useful in many different ways: (1) to meet the regulatory requirements; (2) to control the range of a BSS; (3) to control the inter-BSS interference; and (4) to minimize the power consumption in order to reduce the battery drain.

17.3.1 Association Based on Power Capability

During a (re)association, a station provides an AP with its *minimum* and *maximum transmit power capability* for the current channel, using the *power capability element* in (re)association request frames. An AP might use the minimum and maximum transmit power capability of associated stations in order to determine the local maximum transmit power for its BSS. The local maximum transmit power specifies the maximum of the transmit power, which can be used within its BSS. The stations in the BSS are allowed to use the transmit power of their choice, which is smaller than or equal to the local maximum transmit power value. An AP may reject an association or reassociation of a station if the station's minimum or maximum transmit power capability is unacceptable (e.g., due to the local regulatory constraints).

17.3.2 Advertisement of Regulatory and Local Maximum

An AP in an infrastructure BSS or a station in an IBSS advertises the *regulatory maximum power level* and the *local maximum transmit power* for the current frequency channel in the beacon and probe response frames using the combination of a *country element* and a *power constraint element*. The stations in the BSS are allowed to use the transmit power of their choice, which is smaller than or equal to the local maximum value. The local maximum power should satisfy the following two conditions:

- It should be smaller than or equal to the regulatory maximum.
- It should be larger than or equal to the maximum of all the associated stations' minimum transmit powers.

The local maximum transmit power for the channel should also satisfy the mitigation requirements for the channel in the current regulatory domain. A conservative approach is to use the local maximum transmit power level equal to the regulatory maximum transmit power level minus the mitigation requirement (e.g., 3 dB in Europe). A lower local maximum transmit power level may be used for other purposes (e.g., range control and reduction of interference). The regulatory and local maximum transmit powers could be adapted during the run time of a BSS. However, the network stability should be considered in order to decide how often and how much these maximums are changed. The regulatory and local maximum transmit powers cannot be changed during the runtime of an IBSS.

17.3.3 Transmit Power Adaptation

IEEE 802.11h allows a station to select its transmit power level for each frame transmission as long as the power level is smaller than or equal to the local maximum transmit power. In order to determine the proper (or the best) transmit power level for a given frame, the transmitter station needs to know the link condition between the receiver station and itself. The 802.11h provides a transmit-power reporting mechanism to achieve the link condition. The transmit power adaptation procedure, illustrated in Figure 17.3, is based on the transmit-power reporting mechanism as detailed next.

Transmit-Power Reporting

For the estimation of the link condition between two stations, the 802.11h defines a transmit-power reporting mechanism, which works as follows:

- A station transmits an action management frame, referred to as the *TPC request frame*, to another station when it desires (i.e., in an event-driven manner). When to transmit this frame is implementation-dependent.
- Upon receiving the TPC request frame, the recipient station determines the *link margin* between the transmitter station and itself. The link margin is defined by the ratio of the received signal power measured during the reception of the corresponding TPC request frame to the minimum power level desired by the recipient station.

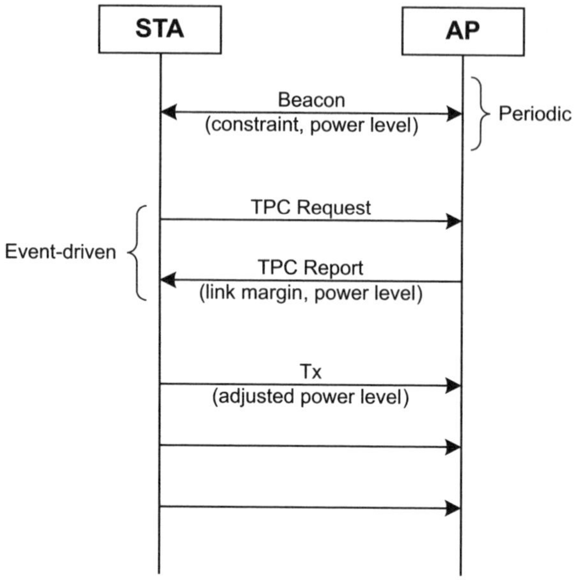

Figure 17.3 IEEE 802.11h transmit power adaptation procedure.

- The recipient station then responds with another action management frame, referred to as the *TPC report frame*, which includes a *TPC report element* that contains a *transmit power* field and a *link margin* field. The transmit power field simply contains the transmit power used to transmit the TPC report frame, and the link margin field contains the link margin measured during the reception of the corresponding TPC request frame.

Moreover, the AP in an infrastructure BSS or a station in an IBSS autonomously includes in any beacon frame a TPC report element with the transmit power field indicating the power level used for the beacon frame while the link margin field set to zero. This information can be used by the associated stations to periodically monitor the channel condition between the AP and themselves.

Link Margin–Based Power Estimation

Thanks to the transmit-power reporting mechanism, energy-efficient frame transmissions now become feasible in the 802.11h WLANs. Note that a station can utilize the received link margin and power level information in order to determine the best transmit power level in the future.

The best transmit power level could be directly determined from the link margin found in the TPC report frame. That is, the transmit power level could be estimated by the transmit power level (used for the TPC request frame) minus the link margin (found in the TPC report frame), where all the measures are in the units of decibels. While the link margin–based transmit power decision could be quite accurate, since the link margin is estimated by the receiver, this might be rather limited, since the link margin is valid only for the transmission rate used for the TPC request frame. In order to determine the best power level for each transmission rate, a station might want to transmit multiple TPC request frames, transmitted at different rates. Moreover, the best transmit power should be dependent on the frame length, and also on the goal for the transmit power adaptation. Therefore, relying on the receiver to determine the transmit power might not be always optimal.

Path Loss Estimation

Another way to determine the transmit power level for a given frame is based on the estimation of the link quality between itself and the receiver. With the 802.11h, a simple link-quality estimation scheme may work as follows. Whenever a station receives a frame containing the TPC report element, with the knowledge of the received signal strength via RSSI as well as the transmit power contained in the TPC report element, the station can estimate the link quality (in terms of path loss) from the transmitter station to itself by performing a simple subtraction. That is, the path loss (in dB) can be estimated by the transmit power level (indicated in the TPC report) minus the received power level (measured during the TPC report frame reception), where both are in the units of dBm. Then, the best transmit power level can be determined based on the estimated path loss.

The 802.11h does not specify how to implement the algorithm for the TPC operation, and, hence, it is an implementation-dependent issue as long as it satisfies any regulatory requirements. Based on the link quality estimation, the TPC mechanism could be used for an intelligent power management for various purposes. As most WLAN devices such as laptops and palmtops are battery powered, and extending the operation time of such devices is always desirable and important, applying TPC to save the battery energy is also an attractive idea [7–9]. Furthermore, in the multicell WLANs often found in office and public access environments, reducing the inter-BSS interference via TPC could be quite beneficial as well, since it results in better error performance in a given area [10, 11].

TPC Algorithm

We here briefly introduce a mechanism, called *MiSer* [7, 8], which utilizes both TPC and link adaptation intelligently in order to minimize the communication energy consumption. This mechanism can be used in combination with the PSM, which put a station into the doze state whenever there is no active traffic in order to minimize the energy consumption. TPC can enable an 802.11h station to use the best power level for each frame transmission in the transmit mode and is complementary with the PSM.

Due to the contentious nature of the 802.11 DCF, the effectiveness of MiSer relies on the condition that applying TPC to data transmissions will not aggravate the hidden station problem. Note that a station may become hidden from another station if a low transmit power is used, while they are not hidden from each other when a full transmit power is used. For this reason, MiSer exchanges RTS and CTS frames before each data transmission attempt to deal with the hidden station problem. More importantly, it is deployed in the format that the CTS frame is transmitted at a stronger power level than the RTS and ACK frames, and the data frame is transmitted at a lower power depending on the estimated path loss. This policy not only allows the data frames to be transmitted at lower power levels to save energy, but also ameliorates the potentially aggravated interference caused by TPC by transmitting the CTS frames at a stronger power level. MiSer uses a simple table-driven approach to determine the most energy-efficient combination of transmission rate and transmit power for each data frame.

The basic idea is that a station computes offline a rate-power combination table indexed by the data transmission status, where each entry of the table is the optimal rate-power combination in the sense of maximizing the energy efficiency under a given data transmission status. The data transmission status is characterized by the data payload length, the path loss between the transmitter and the receiver, and the number of unsuccessful transmission attempts for the data frame. The energy efficiency is defined as the ratio of the expected delivered data payload to the expected total energy consumption. At runtime, the station first estimates the path loss between itself and the receiver, and also updates the data transmission status. Then, based on the data transmission status, it selects the best transmission rate and transmit power for the current data transmission attempt via a simple table lookup.

17.4 Dynamic Frequency Selection (DFS)

In case of IEEE 802.11a WLAN operating at the 5-GHz bands, a BSS occupies a channel of 20 MHz. Today, there are 24 channels available for the 802.11a in the United States, while 19 channels are available in Europe. Out of them, 15 channels (i.e., 4 channels at 5.25–5.35 GHz and 11 channels at 5.47–5.725 GHz) require a DFS operation as discussed in Section 17.1.2. IEEE 802.11h defines a DFS protocol, which enables to switch the operational frequency channel of a BSS to another channel dynamically during the run time of the BSS. This should be differentiated from the measurement-based automatic channel selection, which is implemented in many APs today. That is, instead of a manual channel selection by the AP administrator, the AP might be configured to select the best channel based on its own measurement during the AP initialization phase.

The main purpose of the 802.11h DFS is to fulfill the regulatory requirement (i.e., if radar is detected in the current operating channel, the BSS should vacate the current channel). Just vacating the current channel upon the detection of radar is not desirable since the BSS should continue its operation. Accordingly, the DFS operation enables the entire BSS to switch to another channel after vacating the current channel. In addition to fulfilling the regulatory requirements, there might be many reasons why a BSS may want to change its operational frequency channel. One interesting example is when the current channel condition is too bad due to the interference from neighboring devices. In this context, DFS could be used to enhance the QoS of the WLAN.

In an infrastructure BSS, the DFS procedure is composed of the following steps, which basically occur in order.

- The AP selects a channel to initiate its BSS after checking the channel availability and condition.
- During the run time of its BSS, the AP continues to measure its current channel and also requests channel measurements to its associated stations.
- After measuring the channels as requested, stations report the measurement results to its AP.

- The AP decides whether to switch to another channel and which channel to switch to. If radar was detected during the measurement, a channel switch should occur.
- The AP announces the scheduled channel switch.
- Finally, a channel switch to a new channel occurs as scheduled.

More detailed operations are presented next.

17.4.1 Association Based on Supported Channels

During a (re)association, a station provides an AP with the list of channels in which it can operate, using a *supported channel element* in (re)association request frames. An AP should consider the supported channels of associated stations to select a new channel for the channel switch of its BSS. An AP may reject an association or reassociation of a station if the station's supported channels are unacceptable.

17.4.2 Quieting Channels for Testing

An AP in a BSS may schedule *quiet intervals* by transmitting one or more quiet *elements* in beacon frames and probe response frames. Only the station starting an IBSS, which is called a *DFS owner* as discussed further later, may specify a schedule of periodic quiet intervals, by transmitting one or more quiet elements in the first beacon frame establishing the IBSS. All stations in an IBSS should continue these periodic quiet interval schedules by including appropriate quiet elements in any transmitted beacon frames or probe response frames. Multiple independent quiet intervals may be also scheduled, to ensure that not all quiet intervals have the same timing relationship to TBTT, by including multiple quiet elements in beacon frames or probe response frames.

At the start of a quiet interval, the NAV is set by all the stations including the AP in the BSS or IBSS for the length of the quiet interval so that no transmission will occur during the quiet interval. Any frame transmission should be completed before the start of the quiet interval. Before starting transmission of a frame, a station should make sure that the frame exchange including the corresponding ACK will complete before the start of the next quiet interval. If it cannot complete it, the transmission should be deferred via a random backoff.

17.4.3 Measurement Request and Report

A station may measure one or more channels itself or may request other stations in the same BSS or IBSS to measure one or more channels on its behalf, either in a quiet interval or during normal operation. When requesting other stations to measure one or more channels, a station uses an action management frame, called a *measurement request frame*, containing one or more measurement request elements. Each measurement request element indicates: (1) the measurement type (one of three types as presented later); (2) the channel to be measured; (3) the measurement start time; and (4) the measurement duration.

The measurement request may be sent to an individual or group destination address, where the group address should be used carefully due to possible reply storms. Table 17.6 summarizes the allowed measurement requests for various cases. Note that a non-AP station cannot request another non-AP station to measure in the infrastructure BSS.

Upon the reception of a measurement request frame, a station starts the measurements at the times indicated by the measurement request elements. A station may ignore any group addressed measurement request frames. The measurement results will be reported to the requesting station in *measurement report elements* using an action management frame, called a *measurement report frame*. A station may also autonomously report measurements to another station in its BSS or IBSS using a measurement report frame. A station in an IBSS may also autonomously report measurements to other stations in the IBSS using the *channel map field* in the *IBSS DFS element* in a beacon or probe response frame. The channel map field indicates the status of each channel in terms of the basic type measurement results as discussed next.

DFS Measurement Types

There are three types of measurements, including *basic*, *clear channel assessment* (CCA), and *received power indication* (RPI) *histogram*. A measurement request frame can request to measure other station(s) using one or more types.

- The basic type determines whether each of another BSS, a non-802.11 OFDM signal, an unidentified signal, and a radar signal is detected in the measured channel.
- The CCA type measures the fractional duration of the channel busy period assessed by the PHY during the total measurement interval.
- The RPI histogram type measures the histogram of the quantized measures of the received energy power levels as seen at the antenna connector during the measurement interval as detailed next.

The *RPI histogram report* out of an RPI histogram type measurement contains the *RPI densities* observed in the channel for the eight RPI levels. To compute the RPI densities, the station measures the received power level on the specified channel, as detected at the antenna connector, as a function of time over the measurement

Table 17.6 Allowed Measurement Requests

Service set	Source of request	Destination of request	Type of measurement request allowed
BSS	AP	Non-AP Station	Individual or group
	Non-AP Station	AP	Individual only
	Non-AP Station	Non-AP Station	None
IBSS	Station	Station	Individual or group

Source: [12].

duration. The received power measurements are converted to a sequence of RPI values (i.e., RPI 0 to RPI 7), by quantizing the measurements according to Table 17.7. The RPI densities are then computed for each of the eight possible RPI values using [Duration receiving at RPI value] / [Measurement duration].

We know that the most critical measurement to meet the regulatory requirements is the radar detection as part of the basic type measurement. All others are complementary and can be used for a smart spectrum management purpose. For example, a station may use the RPI histogram to determine a new channel, to help avoid false radar detections, and to assess the general level of interference present on a channel.

17.4.4 Channel Switch in Infrastructure BSS

In an infrastructure BSS, it is the AP that determines when to switch and which channel to switch to. For this purpose, the AP should keep monitoring the status of the current and other frequency channels, and it may also request other stations to measure and report the channel status using the measurement types explained in Section 17.4.3. For the measurement, the AP may utilize the quiet intervals as explained in Section 17.3.2. Based on its own measurement as well as the reports from the associated stations, the AP continues to monitor the channel status so that the channel switching can be conducted at a proper instance. If radar is detected or the condition of the current channel is determined not acceptable, a channel switch may be decided by the AP, while the new channel should be supported by all the stations in the BSS.

If a channel switch is determined, it is announced to the stations in the BSS via an action management frame, called channel switch announcement frame. It could be also announced via periodic beacon frames and probe response frames. In order

Table 17.7 RPI Definitions for an RPI Histogram Report

RPI	Power observed at the antenna (dBm)
0	Power ≤ – 87
1	–87 < Power ≤ –82
2	–82 < Power ≤ –77
3	–77 < Power ≤ –72
4	–72 < Power ≤ –67
5	–67 < Power ≤ –62
6	–62 < Power ≤ –57
7	–57 ≤ Power

Source: [12].

to maximize the possibility to make all the stations informed including those in the PSM, the AP may announce the scheduled channel switch multiple times. The announced information includes the channel to switch to as well as the channel switching time. Along with the channel switch announcement, the AP can instruct the stations not to transmit any frames in the current channel until the scheduled channel switch occurs, especially, to meet the regulatory requirement. A channel switching occurs immediately before a TBTT, which is specified by the AP, so that a normal communication operation can be conducted for the following beacon interval at the new operational frequency channel. Note that the beacon frames are transmitted periodically.

17.4.5 Channel Switch in IBSS

An IBSS consists of multiple stations without an AP, and, hence, there is no central authority, like the AP in an infrastructure BSS, which can make the channel switching decisions. Basically, in an IBSS, the station initiating an IBSS assumes the *DFS owner*, and takes the responsibility of collecting the channel status as well as making the channel-switching decision. A beacon frame in an IBSS conveys the IBSS DFS element, which indicates the DFS owner. Each station transmits a beacon via the channel contention, and a beacon also conveys a channel map field within the IBSS DFS element. The channel map conveys the set of channels supported by the station and the basic type measurement report from that station.

If a station detects a radar signal in the current channel, the station broadcasts one or more measurement report frames indicating the presence of the radar. A DFS owner receiving such a measurement report frame selects and advertises a new operating channel. The DFS owner might make use of the information received in channel map fields as well as the measurements undertaken by other stations in the IBSS to select a new channel. The DFS owner selects a new channel that is supported by all the stations in the IBSS. The DFS owner then announces the scheduled channel switching via channel switch announcement frames, beacon frames, and probe response frames. The DFS owner might also announce that no frame transmission should occur until the scheduled channel switch time. A station that receives a valid channel switch announcement element repeats this element in all beacon frames and probe response frames that it transmits.

DFS Owner Recovery

If a station does not receive a valid channel switch announcement from the DFS owner within *DFS recovery time*, which is indicated in the IBSS DFS element of the beacon frames, since a radar notification was first transmitted by the station or received from another station, then it enters a *DFS owner recovery mode*. In the DFS owner recovery mode, the station assumes the role of DFS owner, and, hence, selects a new operating channel, and advertises the new channel by transmitting a channel switch announcement frame using the contention resolution algorithm defined for beacon transmissions in an IBSS (see Section 13.5.1). The station also includes the channel switch announcement element in all beacon frames and probe response frames until the scheduled channel switch time. A non-DFS owner station will not initiate a channel switch. If the station receives a valid channel switch announcement element, which is different from what it transmitted, from another station in the IBSS, the station leaves the DFS owner recovery mode prior to the channel switch, and then adopts the received channel switch information including a new DFS owner address.

There are many cases when the DFS owner recovery may be required. For example, the original DFS owner might have left the network (e.g., power off), or the original measurement report was not received by the original DFS owner. It should be noted that DFS owner recovery might temporarily make more than one DFS owner within the IBSS. However, because all the stations in an IBSS participate in sending beacon frames with the channel switch announcement over a number of beacon periods, multiple DFS owners should converge to one DFS owner. Unfortunately, if the channel switches scheduled by multiple DFS owners are to different channels, and the scheduled switches are too soon for the multiple DFS owners to converge, it is possible that an IBSS is partitioned into multiple IBSSs operating in different channels.

17.4.6 DFS Algorithm

The 802.11h does not specify how to implement the algorithm for the DFS operation, and, hence, it is an implementation-dependent issue as long as it satisfies any regulatory requirements. We here discuss the DFS algorithmic issues by considering an infrastructure BSS. Our discussion is based on the algorithm specified in [4]. A similar algorithm can be used for an IBSS DFS.

DFS Algorithm Example

In general, a DFS algorithm for an infrastructure-based 802.11h system consists of the following two phases: *startup* and *regular*. At startup, the AP performs a full DFS measurement on all frequency channels. Based on the measurement results, the AP selects a starting frequency channel that is unoccupied by the primary users for its BSS. Moreover, the AP attempts to avoid selecting the frequency channels that are already occupied by other secondary users (e.g., other 802.11a/h BSSs in operation) in order to maximize the performance of its BSS.

Once the algorithm chooses a starting channel, the AP begins broadcasting beacon frames so that other stations can detect its presence and associate themselves with the AP. The BSS then starts and enters the normal operation state of the regular phase. In the regular phase, the DFS algorithm can be described with the finite-state machine shown in Figure 17.4 [13]; it has four different states, namely, *normal operation*, *channel DFS test*, *full DFS test*, and *frequency change*.

The BSS remains in normal operation until either the measurement timer expires or the link quality degrades, which suggests possible interference. If the timer expires, the state changes to channel DFS test, during which the AP reassesses the current operating frequency channel. At this state, a quiet interval might be used to make a more accurate measurement. If no radar signal is detected and the channel status is good enough to continue operation, the state reverts to normal operation; otherwise, it changes to full DFS test, during which the AP performs a full DFS measurement on all frequency channels. During normal operation, if a radar signal is detected or the link quality degrades, state changes directly to full DFS test. In any

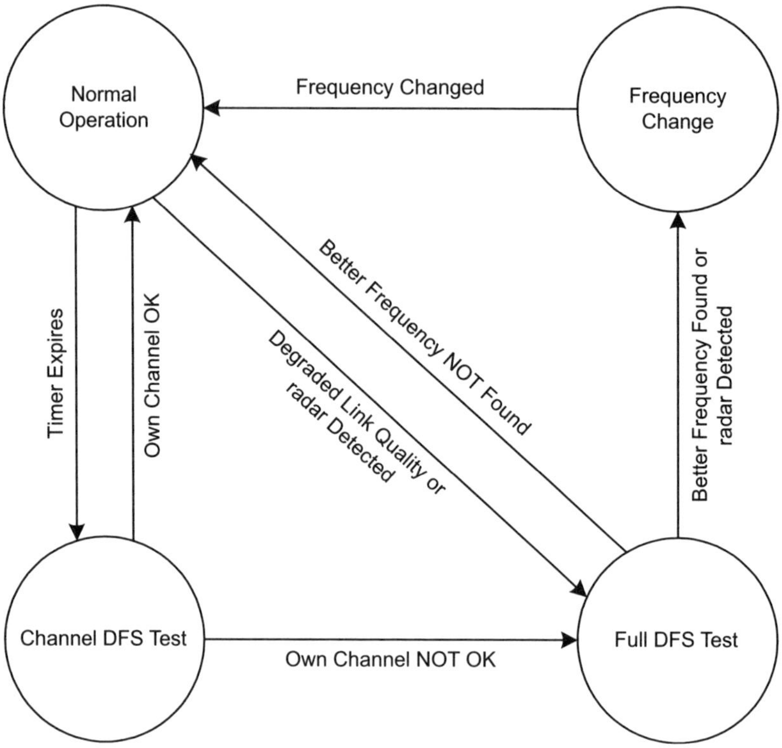

Figure 17.4 Finite-state machine for a DFS algorithm. (*After:* [13].)

state, if a radar signal is detected, all the normal data transmissions (i.e., data frame transmissions) should stop.

In full DFS test, the AP measures (or asks non-AP stations to measure) other frequency channels. Then, based on the measurement results, the AP makes a channel switching decision. If no radar signal was detected in the current frequency channel and the current channel is the best in terms of link quality, the BSS continues operating on the current channel by reverting to normal operation. Otherwise, the state changes to frequency change, in which the AP wakes up all the sleeping stations, if any, and also announces what the new frequency channel is and when operation in the new frequency channel will start. Finally, after the frequency channel switches successfully, the state returns to normal operation.

Radar Signal Detection

Another important function required for the DFS is the radar signal detection. The 802.11h does not specify how to implement the radar signal detection, and, hence, it is an implementation-dependent issue. That is, as long as the regulatory requirement, presented in Section 17.1.2, is satisfied, the actual implementation is up to the product designer. Figure 17.5, adopted from [1], illustrates an example of a typical implementation splitting the detection and monitoring functions between the PHY and MAC for radar detection. The basic method of detection is based on monitoring the RF energy (i.e., RSSI) at the receiver and using a threshold level detector to trigger events.

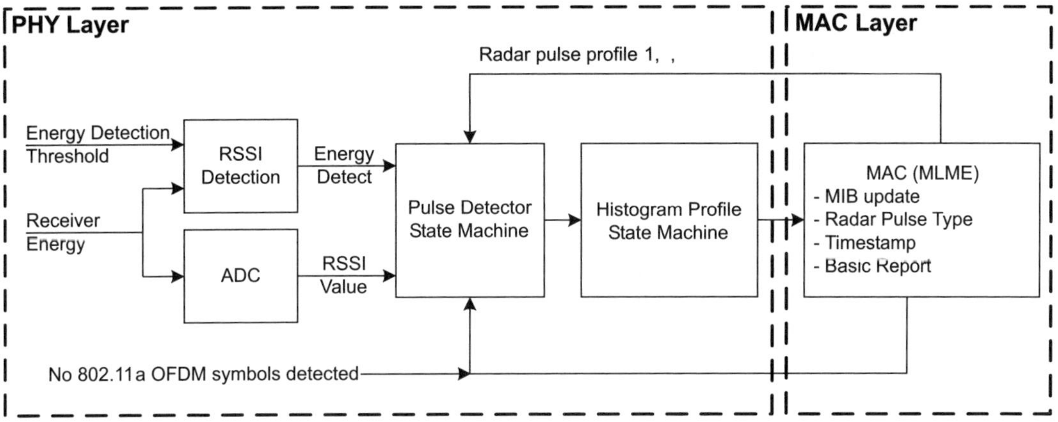

Figure 17.5 Radar pulse detection. (*After:* [1].)

From the example shown in the figure, if an IEEE 802.11a OFDM signal is not demodulated, then the signal energy is assumed to be from one of the following three cases: (1) noise in the channel; (2) an unknown interference source; or (3) other users of the frequency band, such as radars. A radar detection should be declared after detecting one of bursty and periodic signal patterns, as specified in Table 17.5.

References

[1] O'Hara, B., and A. Patrick, *IEEE 802.11 Handbook: A Designer's Companion*, 2nd ed., New York: IEEE Press, 2005.

[2] ETSI EN 301 893 v1.3.1, Broadband Radio Access Networks (BRAN); 5 GHz High Performance RLAN; Part 2: Harmonized EN Covering Essential Requirements of Article 3.2 of the R&TTE Directive, August 2005.

[3] FCC CFR47, Title 47 of the Code of Federal Regulations, Part 15: Radio Frequency Devices, Federal Communication Commission, September 20, 2007.

[4] Qiao, D., and S. Choi, "New 802.11h Mechanisms Can Reduce Power Consumption," *IT Professional*, Vol. 8, No. 2, March/April 2006, pp. 43–48.

[5] Vitervi, A. J., *CDMA: Principles of Spread Spectrum Communication*, Reading, MA: Addison Wesley Longman, 1995.

[6] Gilhousen, K. S., et al., "On the Capacity of a Cellular CDMA System," *IEEE Trans. on Vehicular Technology*, Vol. 40, No. 2, May 1991, pp. 303–312.

[7] Qiao, D., et al., "MiSer: an Optimal Low-Energy Transmission Strategy for IEEE 802.11a/h," *Proc. ACM 9th International Conference on Mobile Computing and Networking (MobiCom'03)*, San Diego, CA, September 14–19, 2003.

[8] Qiao, D., S. Choi, and K. G. Shin, "Interference Analysis and Transmit Power Control in IEEE 802.11a/h Wireless LANs," *IEEE/ACM Trans. on Networking*, Vol. 15, No. 5, October 2007, pp. 1007–1020.

[9] Qiao, D., et al., "Energy-Efficient PCF Operation of IEEE 802.11a WLAN Via Transmit Power Control," *Comput. Netw.*, Vol. 42, No. 1, May 2003, pp. 39–54.

[10] Kim, T. S., et al., "Improving Spatial Reuse Through Tuning Transmit Power, Carrier Sense Threshold, and Data Rate in Multihop Wireless Networks," *Proc. ACM 12th International Conference on Mobile Computing and Networking (MobiCom'06)*, Los Angeles, CA, September 24–29, 2006.

[11] Vanhatupa, T., M. Hannikainen, and T. D. Hamalainen, "Frequency Management Tool for Multi-Cell WLAN Performance Optimization," *Proc. IEEE 14th Workshop on Local and Metropolitan Area Networks (LANMAN'05)*, Chania, Crete, Greece, September 18–21, 2005.

[12] IEEE Std 802.11-2007, IEEE Standard for Local and Metropolitan Area Networks—Part 11: Wireless LAN Medium Access Control (MAC) and Physical Layer (PHY) Specifications, IEEE Std 802.11-2007, (Revision of IEEE Std 802.11-1999), June 12, 2007.

[13] IEEE 802.15-01/072, Liaison Statement on the Compatibility Between IEEE 802.11a and Radars in the Radio Location and Radio Navigation Service in the 5250–5350 MHz and 5470–5725 MHz Bands, IEEE, 2001.

CHAPTER 18

Ongoing Evolution of WiFi

Through Chapters 11 to 17, we have presented various functionalities of the 802.11 including PHY, baseline MAC, QoS extension, security mechanisms, mobility support, and spectrum/power management. As briefly presented in Section 1.3.2, this technology is still evolving. The summary of the ongoing standardization efforts within IEEE 802.11 WG [1] should be referred to Table 1.13.

Out of them, we have already discussed IEEE 802.11k for radio resource measurement and IEEE 802.11r for fast roaming in the context of the mobility enhancement in Chapter 16. In this chapter, we present IEEE 802.11n for higher throughput, IEEE 802.11s for mesh networking, and IEEE 802.11k for radio resource measurement (including the functions not related to the mobility). All the discussions in this chapter are based on the draft standard specifications, and, hence, the protocols are subject to change. However, it should be still useful to look at them in order to understand what they are for eventually.

18.1 IEEE 802.11n for Higher Throughput Support

Task Group N (TGn) was established in 2003 within the 802.11 WG in order to achieve a higher throughput by revising both the PHY and MAC of the 802.11. The group is basically targeting at a throughput of at least 100 Mbps measured at the MAC SAP. Since the 802.11a and 802.11g WLANs achieve about 25 Mbps maximum throughput at the MAC SAP in practice, this represents a WLAN that is at least four times faster.

We here present the 802.11n based on IEEE 802.11n/D3.0 [2]. IEEE 802.11n is built on top of IEEE 802.11-2007, including IEEE 802.11e MAC and 802.11 a/g PHYs. The 802.11n PHY increases the transmission rate by using multiple antennas or combining two frequency channels of 20 MHz. The PHY supports up to 600 Mbps by utilizing 4×4 antennas and a 40-MHz channel. It has been known that for the throughput maximization, increasing only the PHY transmission rate has inherent limitations due to the large protocol overhead in the MAC (e.g., backoff, IFSs, and ACK transmissions) [3]. It is the main reason why the maximum throughputs of the 802.11a and 802.11g are only about the half of the maximum PHY transmission rate (i.e., 54 Mbps). Various MAC mechanisms including frame aggregation are defined in order to enhance the protocol efficiency. In the context of the 802.11n, the term *high throughput* (HT) is used to refer to the 802.11n systems/stations, while the term non-HT is used to refer to the 802.11a/g systems /stations.

18.1.1 HT Control Field for Closed-Loop Link Adaptation

The MPDU format of IEEE 802.11n MAC is illustrated in Figure 18.1. Basically, a new field, called *HT control*, is appended immediately after the QoS control field. Moreover, the maximum frame body size is increased due to frame aggregation as detailed in Section 18.1.2.

The HT control field contains various kinds of information to support the HT operations. Specifically, the *link adaptation control* subfield can be used for the *closed-loop link adaptation*, in which a transmitter station requests the receiver station to feed back the recommended *modulation and coding scheme* (MCS) for the subsequent frame transmissions. A transmitter station can request an MCS feedback by setting an *MCS request* (MRQ) subfield in the *link adaptation control* field of its data frame, with the receiver station then indicating the recommended MCS in the *MCS feedback* (MFB) subfield in the link adaptation control field of its data frame. As discussed in Section 13.4.2, closed-loop link adaptation is not possible in the 802.11-2007. Since more than 150 transmission rates are available in the 802.11n PHY as discussed in Section 18.1.4, the support of this kind of close-loop link adaptation is very desirable.

The 802.11n also defines a means to extend the existing control frames by including the HT control field. A control frame, called the *control wrapper frame*, is newly defined. As shown in Figure 18.2, any control frame, conveyed within the *carried frame* field, can be wrapped within this new control frame along with the HT control field. Accordingly, the RTS/CTS exchange, wrapped by the control wrapper, can be used to request and feedback the recommended MCS.

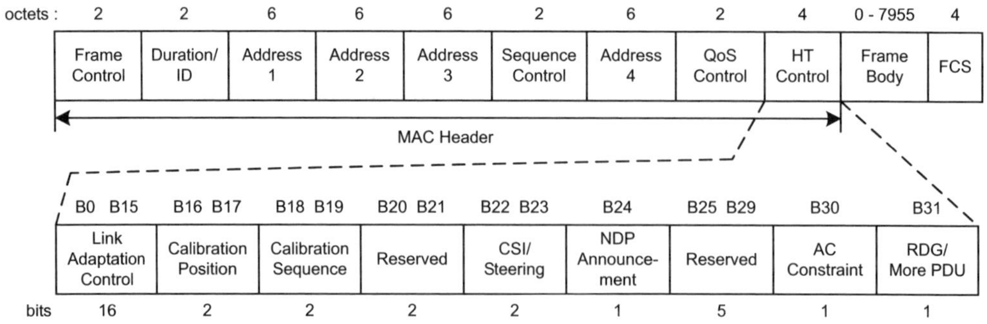

Figure 18.1 IEEE 802.11n MPDU format with HT control field. (*After:* [2].)

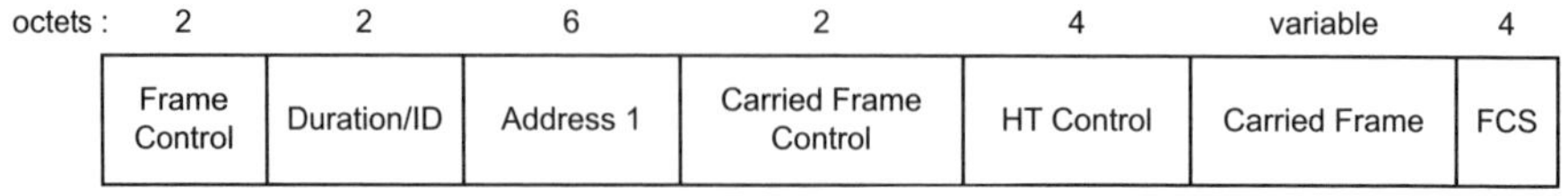

Figure 18.2 IEEE 802.11n control wrapper frame. (*After:* [2].)

18.1.2 Frame Aggregation

As discussed in Section 13.2.7, the throughput performance of the 802.11 WLAN depends on the frame length; the longer the frames, the higher the throughput performance. Accordingly, transmitting large frames is very desirable to maximize the throughput performance. However, there exist variable-size packets transferred in the Internet. Since the packet sizes depend on the applications, we cannot control the size of the packets. One way to address this is a technique called *frame aggregation*. This has been introduced in the literature [4].

The 802.11n defines two types of aggregation schemes, namely, *aggregate MSDU* (A-MSDU) and *aggregate MPDU* (A-MPDU), which are performed at the upper MAC and the lower MAC, respectively. The MAC sublayer could be conceptually divided into two entities, namely, the upper and lower MACs. The upper MAC takes care of the interaction with the LLC (e.g., the processing of an MSDU received from the LLC) and the lower MAC takes care of the interaction with the PHY (e.g., the forwarding of an MPDU to the PHY and the transmission of acknowledgments). Typically, the lower MAC involves more time critical operations, while the upper MAC involves less time critical ones.

A-MSDU

Under the A-MSDU scheme, multiple MSDUs are aggregated into a single A-MSDU, which is conveyed within a single MPDU (unless fragmented). The MSDU aggregation operation is conducted at the upper MAC. The maximum A-MSDU size is either 3,839 or 7,935 octets, depending on the receiver station's capability. Note that the maximum MSDU size per IEEE 802.11-2007 is 2,304 octets. Bit 7 in the QoS control field indicates the presence of the A-MSDU in the frame body of the MPDU. The transmission of A-MSDU frames is optional, while the reception is mandatory.

The A-MSDU format is illustrated in Figure 18.3. Each subframe consists of a subframe header and an MSDU from the higher layer. All the subframes should have the same transmitter and receiver addresses, since they are conveyed in a single MPDU. However, the subframes are allowed to have different source and/or destination addresses, and, hence, these addresses are indicated in the subframe header. The padding bits are needed to make each subframe length a multiple of 4 octets. In the 802.11 MAC, the unit for an acknowledgment is an MPDU, which includes its FCS. Accordingly, if any bit within an A-MSDU is erroneously received, the entire A-MSDU has to be retransmitted.

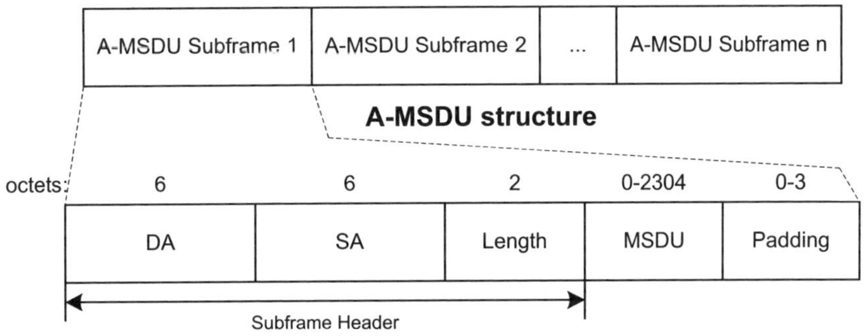

Figure 18.3 A-MSDU format. (*After:* [2].)

A-MPDU

Under the A-MPDU scheme, multiple MPDUs (possibly conveying A-MSDUs) are aggregated into a single A-MPDU, which is conveyed within a single PPDU at the PHY. The MPDU aggregation operation is conducted at the lower MAC. While the A-MSDU allows aggregating multiple MSDUs, the A-MPDU can aggregate various types of MPDUs (e.g., those conveying QoS data, ACK, BlockAck, and BlockAckReq). An A-MPDU is forwarded to the PHY, which receives it as a single PSDU, and, hence, the A-MPDU is transmitted within a single PPDU. The maximum A-MPDU size is 65,535 octets, where the actual maximum depends on the receiver station's capability. Each aggregated MPDU within an A-MPDU is limited to 4,095 octets; this maximum length applies to the MPDUs conveying A-MSDUs as well. An A-MPDU is indicated in the PLCP header of the PPDU (i.e., HT-SIG, as shown in later Figure 18.7).

The A-MPDU format is illustrated in Figure 18.4. Each subframe consists of an MPDU delimiter and an MPDU. All the subframes should have the same transmitter and receiver addresses, since they are conveyed in a single PPDU. Each subframe contains an *MPDU delimiter*, which includes the *MPDU length*, a *CRC-8*, and a *delimiter signature*. The purpose of the MPDU delimiter is to locate the MPDUs within the A-MPDU such that the structure of the A-MPDU can usually be recovered when one or more MPDU delimiters are received with errors. The CRC-8 protects the preceding 16 bits, and the delimiter signature is set to 0x4e, which may be used to detect an MPDU delimiter when the receiver MAC scans for a delimiter. The padding bits are needed to make each subframe length a multiple of 4 octets. The transmission of A-MPDU frames is optional, while the reception is mandatory.

In the 802.11 MAC, the unit for an acknowledgment is an MPDU, which includes its FCS. Accordingly, each subframe within an A-MPDU should be individually acknowledged. As multiple MPDUs are transmitted within a single PPDU, the usage of the BlockAck is needed for the A-MPDU scheme. Note that each subframe (i.e., an MPDU) can be individually retransmitted when BlockAck is used. To transmit a number of MSDUs over an erroneous channel, the A-MPDU might be a more efficient option than the A-MSDU, since each subframe is protected by an FCS. However, without errors, A-MPDU might be a more expensive option than A-MSDU due to larger protocol overhead (i.e., an MPDU delimiter, a MAC header, and an FCS for each subframe).

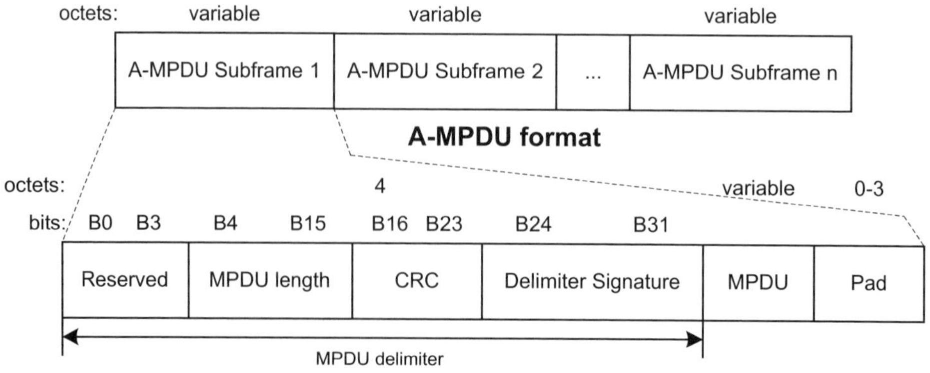

Figure 18.4 A-MPDU format. (*After:* [2].)

18.1.3 Other MAC Functions

Compressed BlockAck

As explained in Section 14.5.2, the block Ack has a potential to be more efficient than the normal ACK policy. However, as shown in Figure 14.25, the BlockAck frame defined in IEEE 802.11e includes a block Ack bitmap of 128 octets, and the efficiency of the BlockAck might be heavily compromised when the number of MPDUs acknowledged by a BlockAck is not that many. To overcome this potential problem, IEEE 802.11n defines a modified BlockAck frame, called *compressed BlockAck*, with a reduced bit map of 8 octets. Fragmentation is not allowed when the compressed BlockAck is used. Accordingly, a compressed BlockAck could acknowledge up to 64 nonfragmented MSDUs.

The support of block Ack is mandatory for the 802.11n MAC; note that the block Ack is optional per the 802.11e. Moreover, a BlockAck frame could be extended to include the bit maps for multiple TIDs. This extended BlockAck is referred to as *multi TID block Ack* (MTBA). In the following, the term BA is also used to represent the BlockAck frame.

Reverse Direction (RD) Protocol

The 802.11n defines an optional RD protocol, which allows a TXOP holder to grant part of its own TXOP to another station. The *RD grant* (RDG) is indicated in the HT control field of the MAC header. As shown in Figure 18.5, station 1 grants a TXOP to station 2 during its TXOP, and then also grants a TXOP to station 3 twice. The RD operation resembles the polling of the 802.11e HCCA, but is different from the HCCA polling in that station 1 does not need to be an AP. The RD protocol is developed to support interactive applications efficiently.

Power Save Multi-Poll (PSMP)

The 802.11n defines an optional PSMP protocol, which allows an AP to poll multiple stations by transmitting a single PSMP frame. Even if it is called a polling protocol, it is more like a dynamic TDMA in the sense that a PSMP schedules all the *downlink transmission times* (DTTs) and *uplink transmission times* (UTTs) in the subsequent PSMP sequence. Each of DTTs and UTTs is allocated a specific time interval (i.e., the start time and the duration), along with the corresponding station as the receiver or the transmitter. During a UTT, if the scheduled TXOP holder does not fully utilize the TXOP corresponding to the UTT, the residual TXOP is simply wasted. A potentially more efficient multipoll scheme has been also introduced in the literature [5].

Figure 18.6 illustrates a PSMP burst exchange sequence. Based on the 802.11e HCF, the AP first transmits a PSMP frame by announcing the schedules of DTTs and UTTs. Right after the PSMP, a downlink phase including two DTTs follows, where the first DTT is for station 1 and the second DTT is for station 2, respectively. After the downlink phase, an uplink phase follows with two UTTs, where the first UTT is for station 1 and the second UTT is for station 2, respectively. As shown in the figure, the AP can transmit another PSMP frame in order to continue the PSMP burst frame exchange after a SIFS interval.

The PSMP can reduce the overhead due to the backoff as well as the polling frame transmissions. Moreover, it can be particularly useful for power saving, as

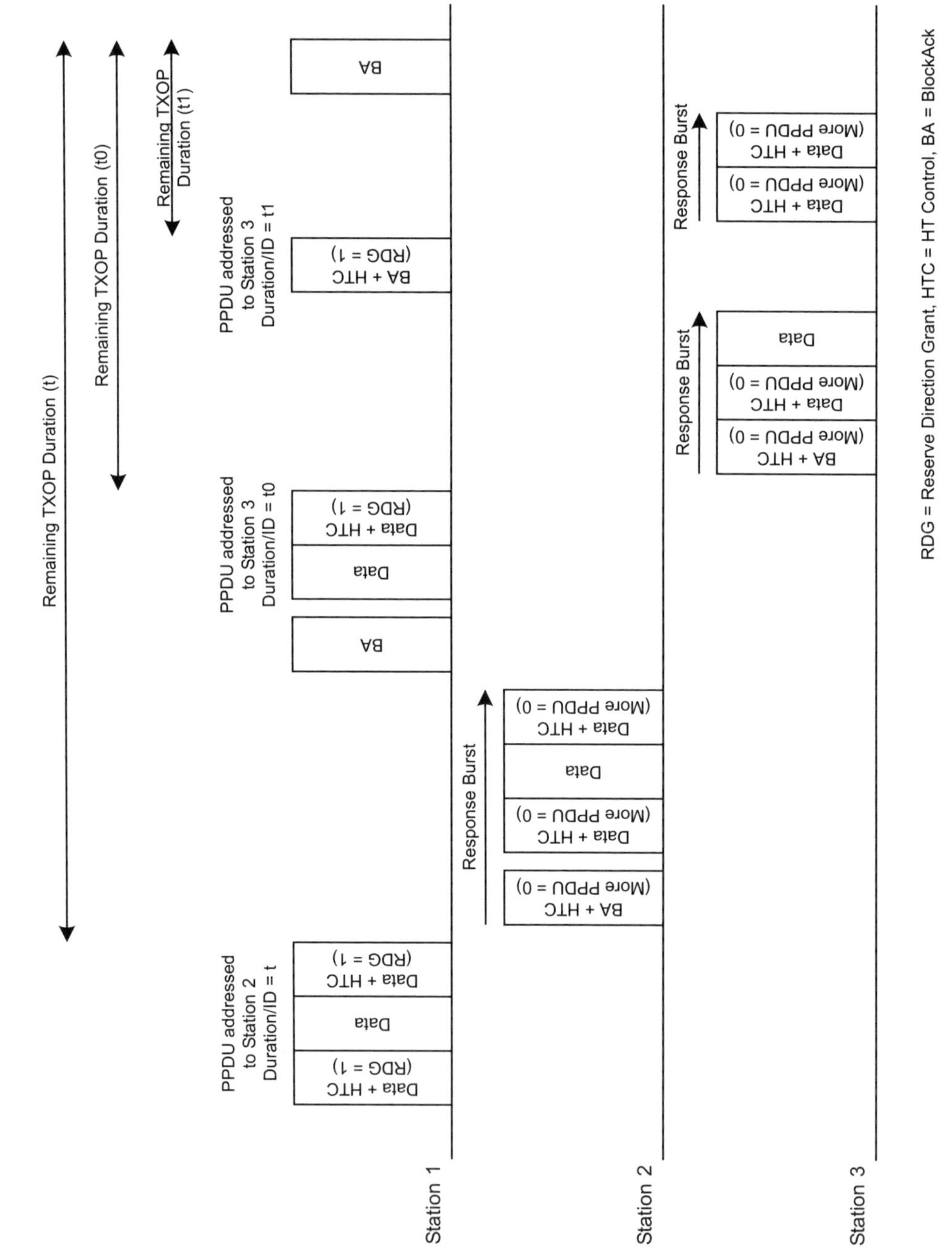

Figure 18.5 Example of RD exchange sequence. (*After:* [2].)

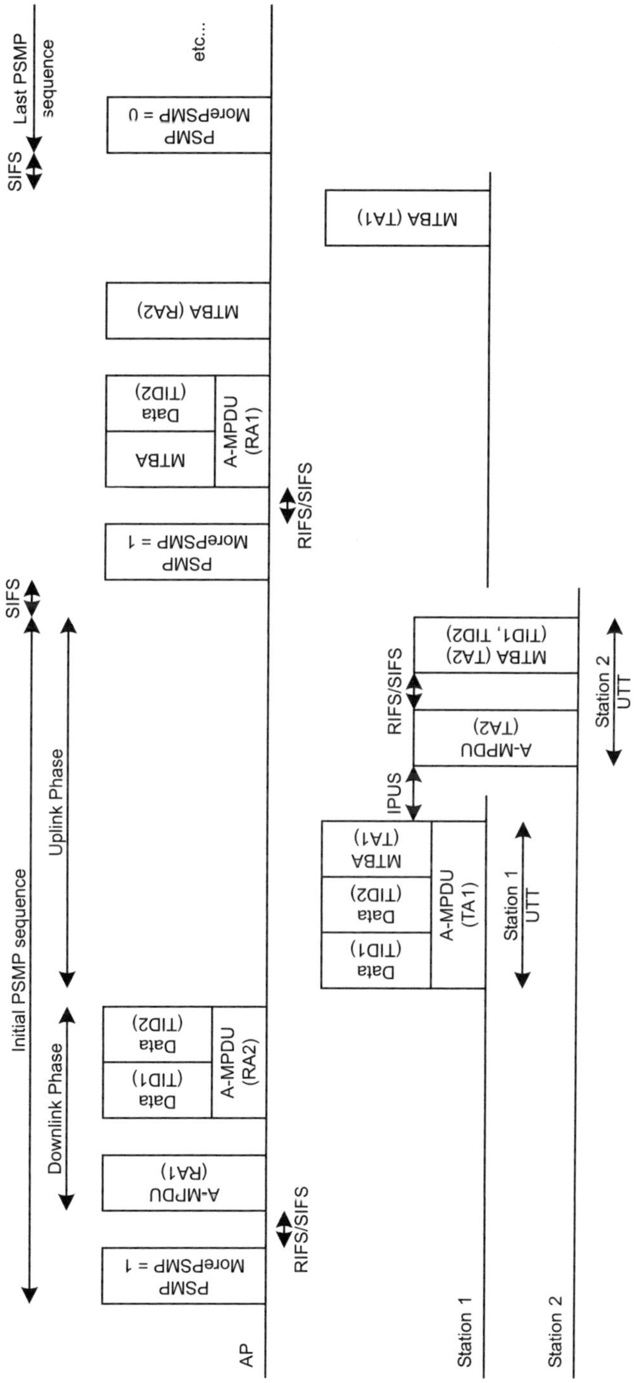

Figure 18.6 PSMP burst exchange sequence. (*After:* [2].)

the name indicates. Since each station knows when to receive and when to transmit frames in advance upon the reception of a PSMP frame, it could go to the doze state while it is not scheduled to receive or transmit. The PSMP allows a station to switch between the doze and active states in a smaller time scale than the 802.11e APSD can support.

18.1.4 HT PHY

IEEE 802.11n PHY is built upon the foundation of the OFDM system defined in IEEE 802.11a at 5 GHz and 802.11g at 2.4 GHz. The characteristics of the 802.11n PHY can be summarized as follows:

- Both 20- and 40-MHz channels are defined.
- Up to 4 spatial streams can be utilized by MIMO.
- The numbers of subcarriers are 56 and 114 for 20- and 40-MHz channels, where 4 and 6 pilot subcarriers are used, respectively. These are more efficient OFDM schemes than that in IEEE 802.11a/g in terms of the pilot percentage. Recall that 4 subcarriers out of 52 are used for the pilot in the 802.11a/g PHY, as discussed in Section 12.2.2.
- BPSK, QPSK, 16-QAM, and 64-QAM are used as in IEEE 802.11a/g.
- The convolutional codes with the rates of 1/2, 2/3, 3/4, and 5/6 are employed. Note that rate 5/6 is newly defined.
- *Low density parity check* (LDPC) codes can be optionally used instead of the default convolutional codes for stronger error protection, where the same set of code rates is defined.
- The two *guard intervals* (GIs)—mandatory 800-ns and optional 400-ns GIs—are defined.
- Both *space-time block code* (STBC) and *transmit beamforming* (TxBF) are optionally supported.

Modulation and Coding Schemes

IEEE 802.11n defines 306 different transmission rates, where 152 rates are for 20-MHz channel operations and 154 rates are for 40-MHz channel operations. Table 18.1 summarizes all the MCSs defined in IEEE 802.11n PHY.

- The first column represents the range of the MCS indices.
- The second column represents the number of spatial streams.
- The third column represents whether the *equal modulation* scheme is used for all the spatial streams (EQM) or *unequal modulation* schemes are used (UEQM). The UEQM schemes are utilized by STBC and TxBF.
- The fourth and fifth columns represent the triplet of the number of available transmission rates, the minimum transmission rate, and the maximum transmission rate for 20- and 40-MHz channels, respectively.

Note that a given MCS index is used for both 20- and 40-MHz channels. Moreover, the number of supported transmission rates is two times the number of MCS

Table 18.1 Summary of IEEE 802.11n PHY MCSs

MCS index range	Number of spatial streams	EQM vs. UEQM	20 MHz channel (Number of rates, Min. rate, Max. rate)	40 MHz channel (Number of rates, Min. rate, Max. rate)
0 - 7	1	—	(16, 6.5, 72.2)	(16, 13.5, 150)
8 - 15	2	EQM	(16, 13, 144.4)	(16, 27, 300)
16 - 23	3	EQM	(16, 19.5, 216.7)	(16, 40.5, 450)
24 - 31	4	EQM	(16, 26, 288.9)	(16, 54, 600)
32	1	—	—	(2, 6, 6.7)
33 - 38	2	UEQM	(12, 39, 108.3)	(12, 81, 225)
39 - 52	3	UEQM	(28, 52, 173.3)	(28, 108, 360)
53 - 76	4	UEQM	(48, 65, 238.3)	(48, 135, 495)

indices. The reason is that for the same MCS index, two GIs are defined, namely, 800 ns, which is the same as that of the 802.11a/g, and 400 ns. Apparently, a higher rate is supported with the shorter GI.

The mandatory MCSs are those for 20-MHz channel with MCS indices ranging from 0 to 7 (i.e., with a single spatial stream). Table 18.2 shows all the mandatory MCSs for 20-MHz channel, where Table 18.3 summarizes the symbols used in the MCS parameter table. Note that the maximum mandatory transmission rate for non-AP stations is 65 Mbps at the 20-MHz channel since 400-ns GI is optional.

For APs, the rates corresponding to MCS indices from 8 to 15 at 20 MHz are also mandatory. The maximum optional transmission rate is 600 Mbps, which corresponds to the MCS index = 31 (i.e., with 4 equally modulated spatial streams) for a 40-MHz channel. Table 18.4 shows the MCS parameters corresponding to MCS index from 24 to 31 for 40 MHz, including the maximum transmission rate. The MCS with index = 32 is available only for a 40-MHz channel and uses duplicated 20 MHz OFDM signals for the most reliable transmission.

PPDU Format

Figure 18.7 depicts both non-HT PPDU for IEEE 802.11a/g (i.e., that in Figure 12.2) and HT PPDU formats for IEEE 802.11n. Two formats are defined for HT PPDU (i.e., *HT mixed format* and *HT greenfield format*), where the mixed format can be used in an IEEE 802.11n WLAN with coexisting 802.11a/g stations, and the greenfield format can be used in a pure 802.11n WLAN. The elements in the PPDU formats are defined in Table 18.5. Note that the term *training field* (TF) refers to training symbols, which are in the PLCP preamble preceding the L-SIG in the case of IEEE 802.11a/g PHY. In the case of the 802.11n, some TFs (i.e., data and extension HT-LTFs) are located after the SIGNAL fields (i.e., L-SIG and HT-SIG in HT mixed

Table 18.2 Mandatory MCSs 0–7 for 20 MHz Channel; $N_{SS} = 1$, $N_{ES} = 1$, EQM

MCS index	Modulation	R	N_{BPSCS} (i_{SS})	N_{SD}	N_{SP}	N_{CBPS}	N_{DBPS}	Data rate (Mbps)	
								800 ns GI	400 ns GI See NOTE
0	BPSK	1/2	1	52	4	52	26	6.5	7.2
1	QPSK	1/2	2	52	4	104	52	13.0	14.4
2	QPSK	3/4	2	52	4	104	78	19.5	21.7
3	16-QAM	1/2	4	52	4	208	104	26.0	28.9
4	16-QAM	3/4	4	52	4	208	156	39.0	43.3
5	64-QAM	2/3	6	52	4	312	208	52.0	57.8
6	64-QAM	3/4	6	52	4	312	234	58.5	65.0
7	64-QAM	5/6	6	52	4	312	260	65.0	72.2
NOTE – Support of 400 nsec guard interval is optional for both transmission and reception.									

Source: [2].

Table 18.3 Symbols Used in MCS Parameter Tables

Symbol	Explanation
N_{SS}	Number of spatial streams
R	Coding rate
$N_{BPSCS}(i_{SS})$	Number of coded bits per single carrier for each spatial stream, $i_{SS} = 1,\ldots,N_{SS}$
N_{SD}	Number of data subcarriers
N_{SP}	Number of pilot subcarriers
N_{CBPS}	Number of coded bits per OFDM symbol
N_{DBPS}	Number of data bits per OFDM symbol
N_{ES}	Number of BCC encoders for the DATA field

Source: [2].

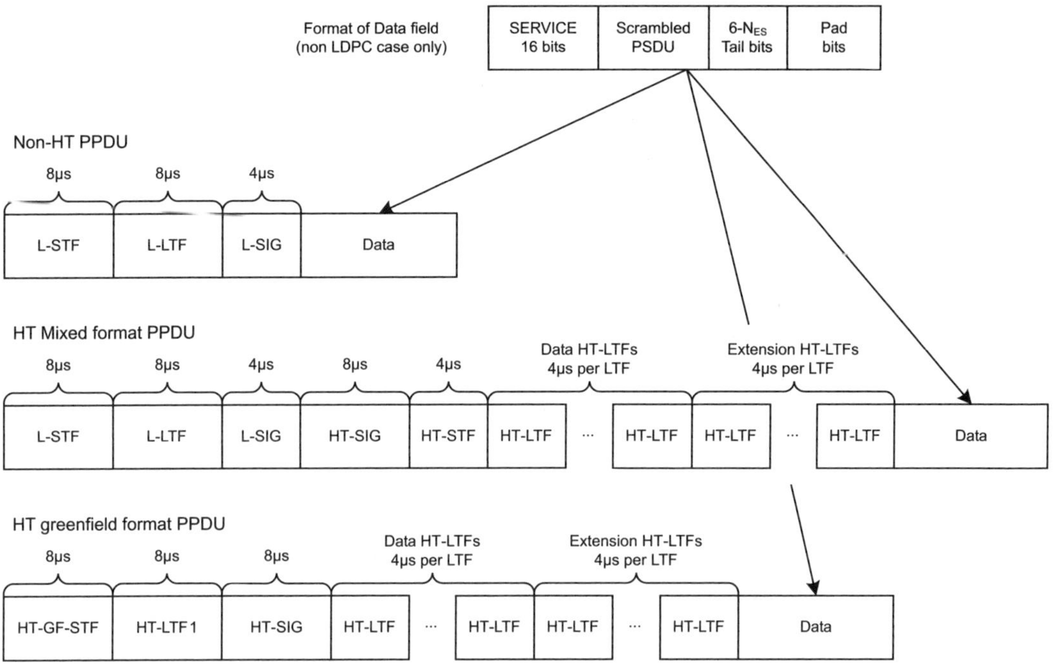

Figure 18.7 IEEE 802.11n PPDU format. (*After:* [2].)

Table 18.4 Optional MCSs 24~31 for a 40-MHz Channel, Where MCS 31 Is the Highest Rate MCS; $N_{SS} = 4$, EQM

MCS index	Modulation	R	N_{BPSCS} (i_{SS})	N_{SD}	N_{SP}	N_{CBPS}	N_{DBPS}	N_{ES}	Data rate (Mbps)	
									800 ns GI	400 ns GI
24	BPSK	1/2	1	108	6	432	216	1	54.0	60.0
25	QPSK	1/2	2	108	6	864	432	1	108.0	120.0
26	QPSK	3/4	2	108	6	864	648	1	162.0	180.0
27	16-QAM	1/2	4	108	6	1728	864	1	216.0	240.0
28	16-QAM	3/4	4	108	6	1728	1296	2	324.0	360.0
29	64-QAM	2/3	6	108	6	2592	1728	2	432.0	480.0
30	64-QAM	3/4	6	108	6	2592	1944	2	486.0	540.0
31	64-QAM	5/6	6	108	6	2592	2160	2	540.0	600.0

Source: [2].

Table 18.5 Elements in IEEE 802.11n PPDU

Element	Description
L-STF	Non-HT Short Training Field
L-LTF	Non-HT Long Training Field
L-SIG	Non-HT SIGNAL Field
HT-SIG	HT SIGNAL Field
HT-STF	HT Short Training Field
HT-GF-STF	HT greenfield Short Training Field
HT-LTF1	First HT Long Training Field (Data HT-LTF)
HT-LTFs	Additional HT Long Training Fields (Data HT-LTFs and Extension HT-LTFs)
Data	The data field include the PSDU (PHY Service Data Unit)

Source: [2].

format, and HT-SIG in HT greenfield format, respectively). In fact, the HT-SIG specifies how many HT-LTFs will follow.

For the HT mixed format PPDU, the first part of the PPDU is exactly the same as the non-HT PPDU so that non-HT stations can detect the PPDU and understand the L-SIG. The HT-STF appearing immediately after the HT-SIG is used to improve the AGC training in a MIMO system. Then, a number of HT-LTFs appear, where there are two types of HT-LTFs (i.e., *data* and *extension HT-LTFs*). The number of data HT-LTFs is 1, 2, or 4, depending on the number of *space time streams* being transmitted in the PPDU. The extension HT-LTFs may exist in *sounding PPDUs*, defined as part of beamforming, to provide additional reference to the receiver estimating the MIMO channels. The number of the extension HT-LTFs is 0, 1, 2, or 4. The maximum number of HT-LTFs is 5, which can occur when there are (1) four data HT-LTFs and one extension HT-LTF, or (2) one data HT-LTF and four extension HT-LTFs. Note that the combination of the training fields and the signal fields might occupy up to 52 μs, when there are 5 HT-LTFs, and this is well over the double of the non-HTs (i.e., 20 μs). In order to mitigate the increased PPDU overhead, it is more desirable to increase the PSDU size by aggregating frames at the MAC.

Transmission Operation

The transmitter block diagram used for the HT greenfield format PPDU and HT portion of the mixed format PPDU except for HT signal field is illustrated in Figure 18.8. We briefly summarize the transmission operations as follows:

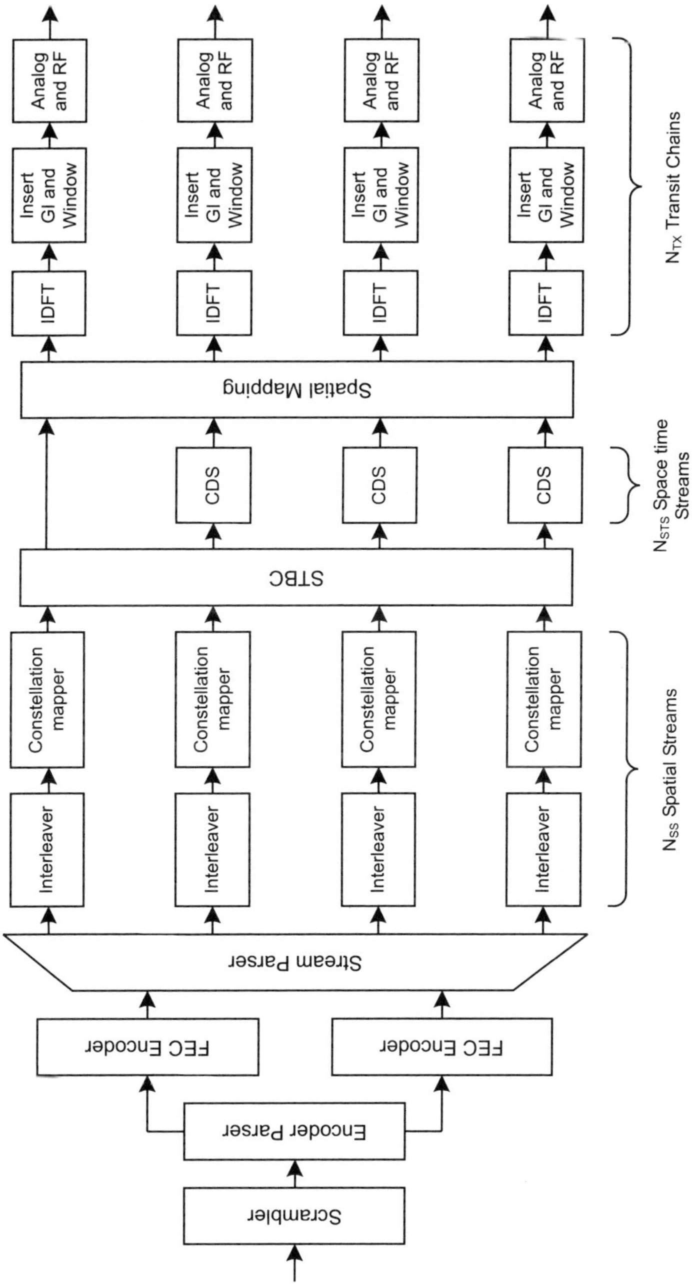

Figure 18.8 Transmitter block diagram assuming 4 antennas. (*After:* [2].)

- The scrambler is the same as that defined for IEEE 802.11a in Section 12.2.2.
- When the convolutional codes are used, there are 1 or 2 encoders. For MCSs supporting over 300 Mbps based on an 800-ns GI, two convolutional encoders are employed. When LDPC is used, there exists only a single encoder.
- The coded bit streams are divided into N_{SS} spatial streams by the *stream parser*, where $N_{SS} = 1 \sim 4$.
- The interleaving, which is also referred to as *frequency interleaving*, and constellation mapping are based on those defined for IEEE 802.11a in Section 12.2.2. The frequency interleavers are skipped when LDPC is employed.
- When an STBC is employed, N_{SS} spatial streams are mapped into N_{STS} space time streams, where $N_{STS} > N_{SS}$. The simplest STBC is 2×2 *Alamouti code* [6] with $N_{STS} = 2$ and $N_{SS} = 1$.
- After inserting the *cyclic shifts* (CSD) to prevent unintentional beamforming, the *spatial mapping* is conducted to map N_{STS} space time streams to N_{TX} transmit chains. Three types of mappings are defined, namely, (1) direct mapping, or $N_{STS} = N_{TX}$ via one-to-one mapping, (2) spatial expansion, used to produce the inputs to all transmit chains, such as for sounding PPDU, and (3) beamforming, as discussed next.
- Each of N_{TX} streams goes through IDFT, GI insertion, and analog/RF units for the transmission.

Tx Beamforming

The TxBF, which is also referred to as *beam steering*, is a technique in which the beamformer (i.e., the transmitter) utilizes the knowledge of the MIMO channel to generate a steering matrix Q_k that improves the received signal power at the beamformee (i.e., the receiver). The steering matrix Q_k, an $N_{TX} \times N_{STS}$ matrix, is used to rotate and/or scale the constellation mapper output vector (or the space time block encoder output vector, if STBC is employed) as part of the spatial mapping in Figure 18.8. In order to do TxBF, the MIMO channel has to be measured by the beamformer. The MIMO channel measurement takes place in every PPDU as a result of transmitting the HT-LTFs (possibly including extension HT-LTFs) as part of the PLCP preamble. This enables the computation of the spatial equalization at the receiver. The support of TxBF is optional for the 802.11n.

There are multiple methods of beamforming, depending on the way that the beamformer acquires the knowledge of the channel matrices H_k and on whether the beamformer generates Q_k or the beamformee provides feedback information for the beamformer to generate Q_k. Both implicit feedback beamforming, which relies on reciprocity in the TDD channel to estimate the channel, and explicit feedback beamforming, which relies on the measurement feedback from the beamformee, are supported. The explicit feedback beamforming is further divided into: (1) *channel state information* (CSI) matrices feedback, (2) noncompressed beamforming matrix feedback, and (3) compressed beamforming matrix feedback, depending on the specific format of the feedback information.

Figure 18.9 illustrates an example of the PPDU exchange used for unidirectional implicit transmit beamforming. Station 1 first transmits unsteered (i.e., nonbeamformed) PPDU to station 2 by indicating a training request (i.e., TRQ bit =

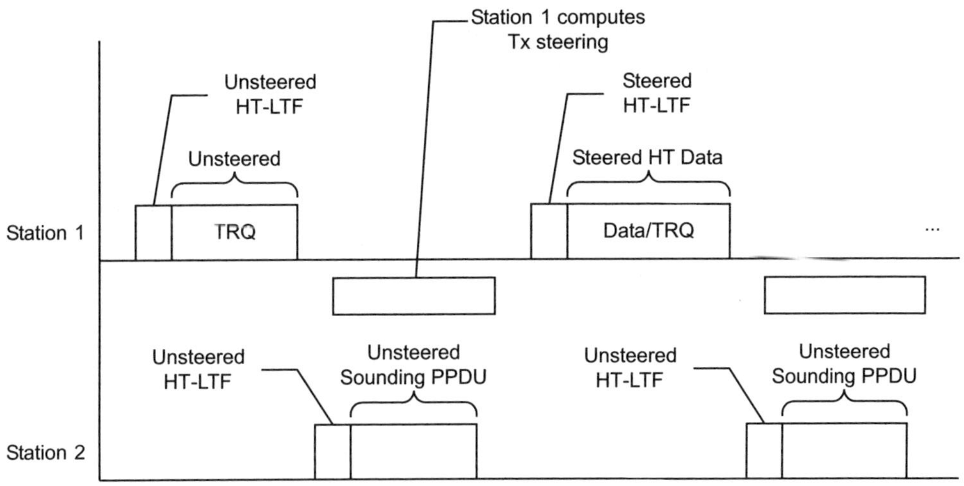

Figure 18.9 Example PPDU exchange for unidirectional implicit transmit beamforming. (*After:* [2].)

1 in the HT control field). Station 2 then transmits a sounding PPDU in response to the request. Station 1 measures the channel during the reception of the sounding PPDU and determines the steering matrices that will be used to send steered PPDUs to station 2 in the future.

18.2 IEEE 802.11s for Mesh Networking

Task Group S (TGs) was established in early 2004 within the 802.11 WG in order to develop the protocols that build interoperable wireless links and multihop paths between multiple APs (i.e., a *mesh network*). This wireless backhaul architecture can allow very flexible deployment of APs compared with the conventional backhaul construction, where APs are connected via wireline links.

We here briefly present the 802.11s based on IEEE 802.11s/D1.07 draft specification [7]. IEEE 802.11s is built on top of IEEE 802.11-2007 MAC, including the 802.11e. The 802.11s defines protocols for auto-configuring paths between APs over self-configuring multihop topologies to support both broadcast/multicast and unicast traffic in a mesh network.

18.2.1 WLAN Mesh Architecture

The *distribution system* (DS) of IEEE 802.11 could be constructed with the 802.11 wireless links or multihop paths between multiple APs. A network comprising these wireless links and wirelessly interconnected APs is referred to as a *mesh network*. It basically contains the following three types of entities:

- *Mesh points* (MPs) are entities that support the mesh services (i.e., the mesh formation as well as the operation of the mesh network, including the path selection and frame forwarding).
- *Mesh access point* (MAP) is an AP with the MP functionality, thus providing both the mesh services and the AP services.

- *Mesh portal* (MPP) is a portal with the MP functionality, thus interfacing the mesh network to other external networks.

Figure 18.10 illustrates an example of the 802.11 mesh network including MPs, MAPs, and MPP. While there is a single MPP in the figure, the architecture allows the existence of multiple MPPs. MPs, MAPs, and MPPs are interconnected via peer-to-peer mesh links, while each station and MAP pair are connected via downlink/uplink. The mesh services have nothing to do with nonmesh stations.[1] That is, IEEE 802.11s does not change any behavior of nonmesh stations. The MAP just looks like a normal AP to nonmesh stations. An MPP provides a MAC bridging functionality [8] between a mesh network and non-802.11 external networks.

18.2.2 Frame Formats

According to IEEE 802.11-2007, an MPDU contains either 3 or 4 address fields as discussed in Section 13.1.1. Frames within a mesh network are extended to contain 6 address fields as follows.

- Address 1 is the receiver address (RA)—the MAC address of the station receiving the frame in the link over which the frame is transmitted.
- Address 2 is the transmitter address (TA)—the MAC address of the station transmitting the frame in the link over which the frame is transmitted.

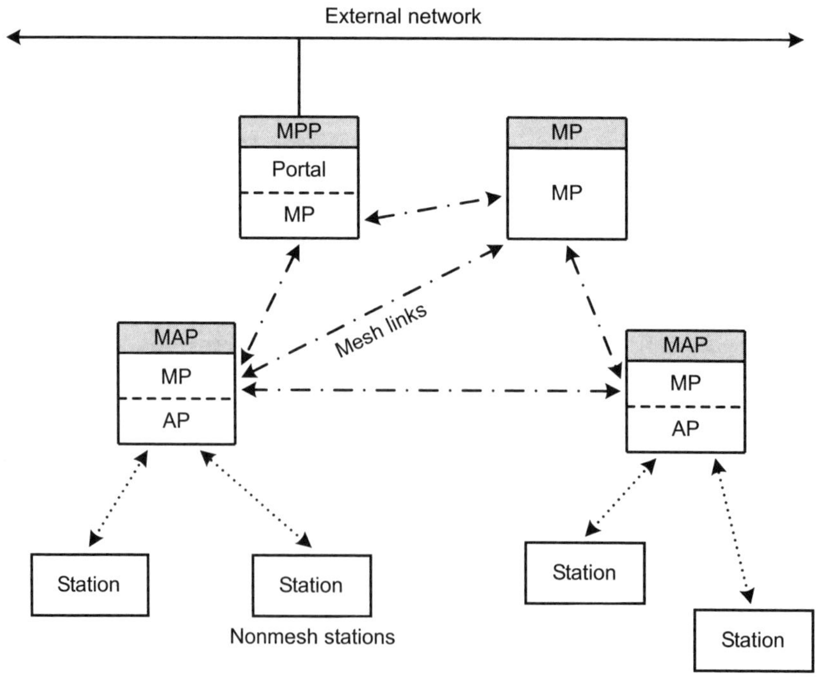

Figure 18.10 Example mesh including MPs, MAPs, and MPP.

1. We use the term *nonmesh* stations to refer to stations associated with an MAP. Note that the term *non-AP* station might also include MPs as well.

- Address 3 is the mesh DA—the MAC address of the destination station (i.e., destination MP) within the mesh network.
- Address 4 is the mesh SA—the MAC address of the source station (i.e., source MP) within the mesh network.
- Address 5 is the destination address (DA)—the MAC address of the destination station within the same subnet.
- Address 6 is the source address (SA)—the MAC address of the source station.

Different addresses can be easily understood using the example in Figure 18.11, where station 1 sends a frame to station 2, which is located outside the 802.11 network. A frame from station 1 traverses across (1) MAP1, which is the AP of station 1, (2) MP2, and (3) MPP3, to reach station 2. Figure 18.12 illustrates the addresses within the frame in each link. For the addressing scheme of the frame from station 1 to MAP1, please refer to the case with To DS = 1 and From DS = 0 in Table 13.3. Note that addresses 5 and 6 are not processed by intermediate MPs (i.e., MP2 in the figure). The intermediate MPs forward the frame based on addresses 3 and 4. The SA in the frame from MPP3 to station 2 is the MAC address of the source station (i.e., station 1). It should be clear why a frame within a mesh network is required to contain 6 addresses.

18.2.3 Routing Protocols

IEEE 802.11s provides extensible path selection and frame forwarding framework, where default path selection protocol[2] and link metric are defined for interoperability among devices from different vendors. Other path selection protocols and link metrics can be employed, where the support of such options is announced to the neighbors in the mesh network. Note that the 802.11s basically provides multihop routing at layer 2.

Airtime Link Metric

Typically, the end-to-end path metric is represented as the cumulative link metric across the end-to-end path. The most widely employed link metric value was simply

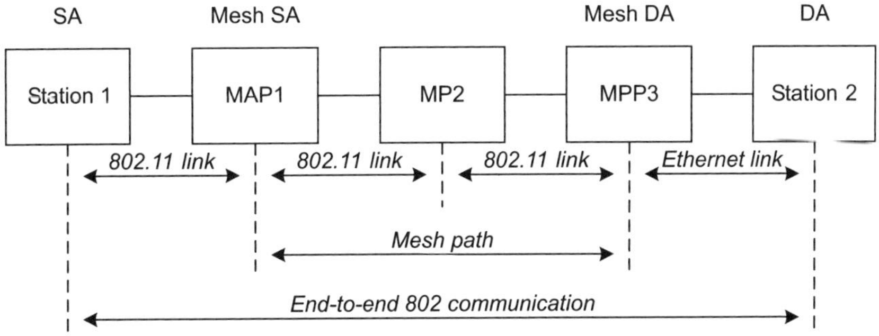

Figure 18.11 An example of the end-to-end 802 communication across a mesh network.

2. Path selection protocol is typically referred to as routing protocol in the literature.

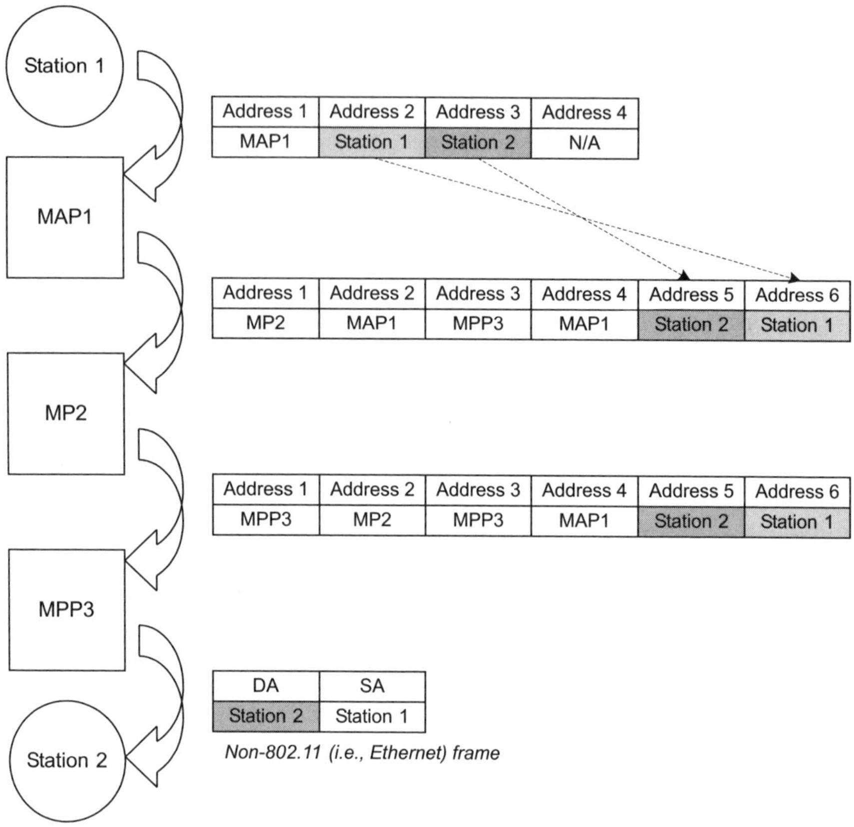

Figure 18.12 Address field usage for the end-to-end communication in Figure 18.11.

one, thus making the path metric the hop count. In mesh networks with various link conditions in terms of channel error and transmission rate, it has been known that the hop count is not a good path metric, and, in recent years, some link metrics have been developed for mesh networks [9, 10].

The 802.11s defines a default link metric, referred to as *airtime link metric*, which also reflects the wireless link characteristics. As the name stands for, the metric represents the amount of resources (i.e., airtime) to transmit a single frame over a mesh link. The metric for a given mesh link with estimated transmission rate r and frame error rate e_f for test frames of length B_t is given by

$$c = \left(O + \frac{B_t}{r}\right)\frac{1}{1-e_f}$$

where O represents the channel access overhead including the PLCP preamble/header, MAC header, IFSs, and ACK. Apparently, the lower the link metric value is, the better the link basically is. The path with the smallest cumulative airtime link metric is desired for an end-to-end communication.

Default HWMP

Now, the question is how to set up a path based on a selected link metric from a source to a destination. The 802.11s defines a default path selection, called *hybrid wireless mesh protocol* (HWMP), which combines on-demand path selection with proactive tree extension. Basically, MPs construct a tree topology, where a frame from an MP to another is forwarded basically following the path along the tree. This tree construction is done in a proactive manner. Therefore, the tree-based path selection enables MPs to communicate without time-consuming path resolution, while the end-to-end path might not be optimal.

On the other hand, the on-demand path selection, which is based on the popular *ad hoc on-demand distance vector* (AODV) protocol [11], triggers a path selection in a reactive manner (i.e., when an end-to-end path is unknown). Under the HWMP, a mesh tree-based path is available by default, and, hence, the on-demand path selection is enabled for the path optimization. Figure 18.13 illustrates an example of the on-demand path selection procedure.

The source MP (i.e., S) floods a broadcast frame, called *path request* (PREQ), to the mesh network. Upon receiving a PREQ, an intermediate MP rebroadcasts the PREQ with an updated link metric value only if the PREQ contains a better link metric value. A PREQ frame is updated to include a cumulative link metric value along the path. For the first arriving PREQ or when another PREQ with the best cumulative link metric arrives, the destination MP (i.e., D) responds with a unicast *path reply* (PREP) frame back to the source, and the PREP follows the reverse path, which the corresponding PREQ traversed along. An intermediate MP receiving a PREP creates a path (i.e., frame forwarding entry) to the destination MP. Upon receiving a PREP, the source MP creates a path to the destination MP.

Optional RA-OLSR

The 802.11s also defines an optional path selection protocol, called *radio aware optimized link state routing* (RA-OLSR), which combines OLSR [12] and fisheye state routing (FSR) protocol [13]. This protocol proactively maintains link-state for path selection.

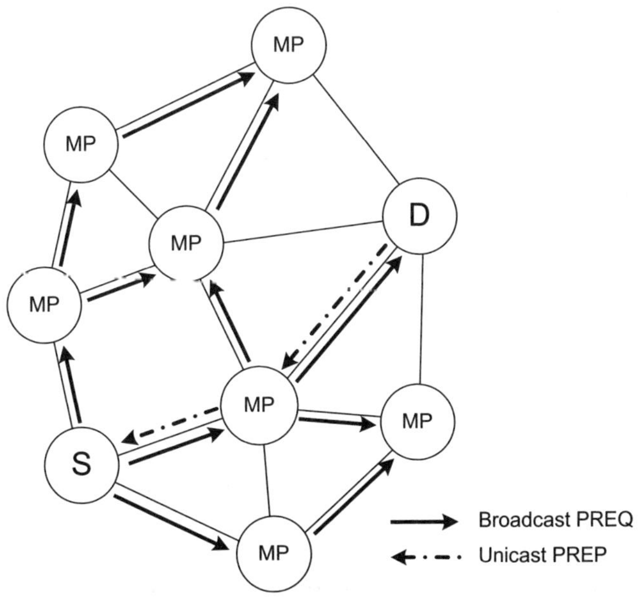

Figure 18.13 On-demand path selection procedure.

18.3 IEEE 802.11k for Radio Resource Measurements

IEEE 802.11k *radio resource measurement* (RRM) protocol is an emerging standard enabling stations to understand the radio environment in which they operate. The RRM enables stations to measure and gather data on radio link performance and on the radio environment. The measured data can be used for the optimization of the network performance. Some features of the 802.11k related with fast scanning, including AP channel report, neighbor report request/response, and measurement pilot, are introduced in Section 16.3.2. We here briefly introduce the complete list of measurements defined by the 802.11k. We refer to IEEE 802.11k/D9.0 draft specification [14], but as the standardization has not been finalized yet, the detailed protocols are subject to change.

The RRM enables 802.11 stations to understand the radio environment better by measuring it locally by themselves or requesting other stations to measure and report. The measurement request/response mechanism of the 802.11k is rooted in that in the 802.11h, which is presented in Sections 17.3.3 and 17.4.3. The 802.11k defines a set of new action frames with a new action category to enable a more versatile set of measurements. See Table 13.4 for the existing action categories. The 802.11k also defines an extensive set of MIB values so that the measured information can be also exposed to other network entities via SNMP [15].

18.3.1 Measurement Types

The request/response measurements defined in IEEE 802.11k/D9.0 include the following:

- The *beacon* request/report pair enables a station to request another station to report a set of beacons it receives on specified channel(s). Either active or passive scanning can be used for the measurement. Moreover, logged measurement results can be also reported without a further measurement.
- The *frame* request/report pair returns the information about all the traffic and a count of all the frames received by the measuring station. For each detected transmitter station, the address, the number of received frames, the average power level, and the BSS of this transmitter are reported.
- The *channel load* request/report pair returns the channel utilization (i.e., the fractional time during which the channel was assessed to be busy), as observed by the measuring station.
- The *noise histogram* request/report pair returns a power histogram measurement of non-802.11 noise power by sampling the channel when the channel was assessed to be idle.
- The *station statistics* request/report pair returns groups of values for *station counters* and for *BSS average access delay*: (1) the station counter group values include transmitted fragment counts, multicast transmitted frame counts, failed counts, retry counts, multiple retry counts, frame duplicate counts, RTS success counts, RTS failure counts, ACK failure counts, received fragment counts, multicast received frame counts, FCS error counts, and transmitted frame counts; and (2) BSS average access delay group values include AP aver-

age access delay, average access delay for each AC, associated station count, and channel utilization.

- The *location configuration information* request/report pair returns a requested location in terms of latitude, longitude, and altitude. The requested location might be the location of the requesting station or the location of the reporting station.
- The *neighbor report* request is sent to an AP that returns a neighbor report containing the information about known neighboring APs that are candidates for a BSS transition. The intended usage is discussed in Section 16.3.2.
- The *link measurement* request/report exchange provides measurements of the RF characteristics of a station-to-station link. This expands the TPC request/report pair of IEEE 802.11h presented in Section 17.3.3 by adding the information about the transmit/receive antennas and received signal strength measured during the link measurement request frame reception.
- The *transmit stream/category measurement* request/report pair enables a QoS station to inquire of a peer QoS station about the condition of an ongoing *traffic stream* (TS) link between them. The report includes many measurement results including: (1) the transmitted MSDU count, (2) the failed MSDU count due to the excessive retransmission attempts, (3) the discarded MSDU count due to the excessive retransmission attempts or queuing delay, and (4) the average transmit delay.

As a response-only mechanism, the *measurement pilot* frame, which is a compact action frame transmitted pseudo-periodically by an AP with a smaller interval compared with the beacon interval, is also defined. Further details about the measurement pilot frames are referred to Section 16.3.2.

References

[1] IEEE Working Group (WG), http://www.ieee802.org/11.

[2] IEEE 802.11n/D3.0, Draft Supplement to Part 11: Wireless LAN Medium Access Control (MAC) and Physical Layer (PHY) Specifications: Enhancements for Higher Throughput, September 2007.

[3] Xiao, Y., "Throughput and Delay Limits of IEEE 802.11," *IEEE Communications Letters*, Vol. 6, No. 8, August 2002.

[4] Kim, Y., et al., "Throughput Enhancement of IEEE 802.11 WLAN Via Frame Aggregation," *Proc. IEEE VTC'04-Fall*, Los Angeles, CA, September 26–29, 2004.

[5] Kim, S., et al., "MCCA: A High-Throughput MAC Strategy for Next-Generation WLANs," *IEEE Wireless Communications*, Vol. 15, No. 1, February 2008, pp. 32–39.

[6] Alamouti, S. M., "A Simple Transmit Diversity Technique for Wireless Communications," *IEEE Journal of Selected Areas in Communications*, Vol. 16, No. 8, October 1998, pp. 1451–1458.

[7] IEEE 802.11s/D1.07, Draft Supplement to Part 11: Wireless LAN Medium Access Control (MAC) and Physical Layer (PHY) Specifications: Mesh Networking, September 2007.

[8] IEEE 802.1D-2004, IEEE Standard for Local and Metropolitan Area Networks—Media Access Control (MAC) Bridges (Incorporates IEEE 802.1t-2001 and IEEE 802.1w), 2004.

[9] Couto, D. S. J. D, et al., "A High-Throughput Path Metric for Multi-Hop Wireless Networks," *Proc. ACM MobiCom'03*, San Diego, CA, September 2003, pp. 134–146.

[10] Draves, R., J. Padhye, and B. Zill, "Routing in Multi-Radio, Multi-Hop Wireless Mesh Networks," *Proc. ACM MobiCom'04*, Philadelphia, PA, September 2004, pp. 114–128.

[11] IETF RFC 3561, Ad Hoc On-Demand Distance Vector (AODV) Routing Protocol, 2003.

[12] IETF RFC 3626, Optimized Link State Routing (OLSR) Protocol, 2003.

[13] Pei, G., M. Gerla, and T. Chen, "Fisheye State Routing in Mobile Ad Hoc Networks," *Proc. ICDCS Workshop on Wireless Networks and Mobile Computing (WNMC'00)*, Taipei, Taiwan, April 10, 2000.

[14] IEEE 802.11k/D9.0, Draft Supplement to Part 11: Wireless LAN Medium Access Control (MAC) and Physical Layer (PHY) Specifications: Specification for Radio Resource Measurement, September 2007.

[15] IETF RFC 1157, A Simple Network Management Protocol (SNMP), May 1990.

Selected Bibliography

Akyildiz, I., X. Wang, and W. Wang, "Wireless Mesh Networks: A Survey," *Computer Networks Journal*, Vol. 47, 2005, pp. 445–487.

Ramachandran, K., et al., "On the Design and Implementation of Infrastructure Mesh Networks," *Proc. IEEE WiMesh'05*, Santa Clara, CA, September 26, 2005.

Sang, L., A. Arora, and H. Zhang, "On Exploiting Asymmetric Wireless Links Via One-Way Estimation," *Proc. ACM MobiHoc'07*, Montreal, Quebec, September 9–14, 2007.

Zhou, W., D. Zhang, and D. Qiao, "Comparative Study of Routing Metrics for Multi-Radio Multi-Channel Wireless Networks," *Proc. IEEE WCNC'06*, Las Vegas, NV, April 3–6, 2006.

Acronyms

1xRTT	1x radio transmission technology
3DES	triple DES
3GPP	Third Generation Partnership Project
3GPP2	Third Generation Partnership Project 2
AA	authenticator's MAC address
AAA	authentication, authorization, and accounting
AAD	additional authentication data
AAS	adaptive antenna system
AC	access category
ACID	HARQ channel identifier
ACK	acknowledgment
ACKCH	ACK channel
ACM	admission control mandatory
ACR	access control router
ADC	analog-to-digital converter
ADDBA	add block acknowledgment
ADDTS	add traffic stream
AES	advanced encryption standard
AGC	automatic gain control
AHARQ	asynchronous HARQ
AI_SN	HARQ identifier sequence number
AID	association identifier
AIFS	arbitration interframe space
AIFSN	arbitration interframe space number
AK	authorization key
AKM	authentication and key management
AKMP	authentication and key management protocol
AM	active mode
AMC	adaptive modulation and coding
A-MPDU	aggregate MPDU
AMPS	advanced mobile phone service
A-MSDU	aggregate MSDU
AN	access node

ANonce	authenticator nonce
AODV	ad hoc on-demand distance vector
AP	access point
APME	AP management entity
APSD	automatic power-save delivery
ARF	automatic rate fallback
ARP	address resolution protocol
ARQ	automatic repeat request
AS	application server; authentication server
ASN	access service network
ASN-GW	ASN-gateway
ASR	anchor switch reporting
AT	access terminal
ATIM	announcement traffic indication message
ATM	asynchronous transfer mode
AuC	authentication center
AWGN	additive white Gaussian noise
AWS	advanced wireless service
BAR	block acknowledgment request
BC	backoff counter
BCC	binary convolutional code
BE	best effort
BER	bit error rate
BF	beamforming
BGCF	breakout gateway control function
BGP	border gateway protocol
BISDN	broadband ISDN
BlockAck	block acknowledgment
BPF	bandpass filter
BPSK	binary phase-shift keying
BR	bandwidth request
BRO	bit reverse order
BS	base station
BSA	basic service area
BSC	base station controller
BSN	block sequence number
BSS	basic service set
BSSID	basic service set identification
BTC	block turbo coding
BTS	base transciever station
BW	bandwidth

BWA	broadband wireless access
B-WLL	broadband WLL
CAC	call admission control
CAP	controlled access phase
CAPWAP	control and provisioning of wireless access points
CB	customized browser
CBC	cipher-block chaning
CBC-MAC	cipher-block chaining message authentication code
CC	convolutional coding
CCA	clear channel assessment
CCK	complementary code keying
CCM	CTR with CBC-MAC
CCMP	counter mode with CBC-MAC protocol
CD	channel descriptor
CDD	cyclic delay diversity
CDMA	code division multiple access
CDR	call detail record
CEPT	European Conference of Postal and Telecommunications
CF	contention free
CFB	cipher feedback
CFM	confirmation
CFP	contention-free period
CFPRI	CFP repetition interval
CI	CRC indicator
CID	connection identifier
CII	CID inclusion indication
CINR	carrier-to-interference and noise ratio
CIR	carrier-to-interference ratio
CLPC	closed-loop power control
CMAC	CBC-based MAC
CMIP	Client MIP
CMS	contents management system
CoA	care-of-address
CoS	class of service
CP	cyclic prefix; contention period
CPS	common part sublayer; charge per sale
CPU	central processing unit
CQI	channel quality indicator
CQICH	CQI channel
CRC	cyclic redundancy check
CRF	charging rules function

CS	convergence sublayer; carrier sense
CSCF	call session control function
CSI	channel state information
CSM	collaborative spatial multiplexing
C-SM	cooperative-spatial multiplexing
CSMA	carrier sense multiple access
CSMA/CA	CSMA with collision avoidance
CSMA/CD	CSMA with collision detection
CSN	core or connectivity service network
CTC	convolutional turbo coding
CTR	counter
CTS	clear to send
CW	contention window
DA	destination address
DAC	digital-to-analog converter
DAMA	demand assigned multiple access
DBDM	dual band dual mode
DBPC	dynamic burst profile change
DBPSK	differential binary phase shift keying
DBTC	double binary turbo code
DCD	downlink channel descriptor
DCF	distributed coordination function
DD	delay diversity
DECT	digital enhanced cordless telecommunications
DELBA	delete Block Acknowledgment
DELTS	delete traffic stream
DES	data encryption standard
DFS	discrete Fourier series; dynamic frequency selection
DFT	discrete Fourier transform
DHCP	dynamic host configuration protocol
DiffServ	differentiated service
DIFS	distributed (coordination function) interframe space
DIUC	downlink interval usage code
DL	downlink
DLS	direct link setup
DMB	digital multimedia broadcasting
DMUX	demultiplexed
DNS	domain name service
DP	decision point
DPC	dirty paper coding
DPF	data path function

DQPSK	differential quadrature phase shift keying
DRM	digital rights mana
DS	distribution system
DSA	dynamic service addition
DSAP	destination service access point
DSC	dynamic service change
DSCP	differentiated service code point
DSD	dynamic service delete
DSSS	direct sequence spread spectrum
DSSS-OFDM	PHY using DSSS-OFDM modulation
DSTTD	double STTD
DTIM	delivery traffic indication message
DTPQ	delay threshold-based priority queuing
DTT	downlink transmission time
EAP	extensible authentication protocol (IETF RFC 3748-2004)
EAPOL	extensible authentication protocol over LANs (IEEE Std 802.1X-2004)
EAP-TLS	EAP-transport layer security
EAP-TTLS	EAP tunneled transport layer security
EC	encryption control
ECB	electronic codebook
ECINR	effective carrier to interference and noise ratio
ECRTP	enhanced compressed real-time transport protocol
ED	energy detection
EDCA	enhanced distributed channel access
EDCAF	enhanced distributed channel access function
EDCF	enhanced DCF
EDGE	enhanced data-rates for global evolution
EGC	equal-gain combining
EIFS	extended interframe space
EIK	EAP integrity key
EIRP	equivalent isotropically radiated power
EKS	encryption key sequence
EMS	elementary management system
EOSP	end of service period
EP	enforcement point
EQM	equal modulation
ERP	extended rate PHY
ERP-OFDM	PHYs using OFDM modulation
ERP-PBCC	PHYs using extended rate PBCC modulation
ertPS	extended rtPS

ERT-VR	extended real-time variable-rate
ESF	extended subheader field
ESS	extended service set
ETRI	electronics and telecommunications research institute
EV-DO	evolution-data only
EV-DV	evolution-data voice
FA	foreign agent; frequency assignment
FBSS	fast BS switching
FCC	Federal Communications Commission
FCH	frame control header
FCS	frame check sequence
FDD	frequency division duplex (duplexing)
FDM	frequency division multiplexing
FDMA	frequency division multiple access
FE	fast Ethernet
FEB	front-end board
FEC	forward error correction
FER	frame error ratio
FFR	fractional frequency reuse
FFT	fast Fourier transform
FGC	fractional guard channel
FH	frequency hopping
FHSS	frequency-hopping spread spectrum
FIFO	first in first out
FL	frame latency
FLI	frame latency indication
FRF	frequency reuse factor
FSN	fragment sequence number
FSS	frame-synchronous scrambler
FT	fast BSS transition
FTIE	fast BSS transition information element
FTP	file transfer protocol
FUSC	full usage subchannel
GC	guard channels
GCD	great common divisor
GE	gigabit Ethernet
GGSN	gateway GPRS support node
GI	guard interval
GM	grant management
GMH	generic MAC header
GMK	group master key

GNonce	group nonce
GPRS	general packet radio service
GPS	global positioning system
GRE	generic routing encapsulation
GSM	global system for mobile-communications
GTK	group temporal key
GTKSA	group temporal key security association
HA	home agent
HARQ	hybrid ARQ
HC	hybrid coordinator
HCCA	hybrid coordinate channel access; HCF controlled channel access
HCF	hybrid coordination function
HCS	header check sum
HE	horizontal encoding
HEMM	HCCA, EDCA mixed mode
HESM	horizontal encoding SM
H-FDD	half-duplex FDD
HIPERLAN	high performance radio LAN
HMAC	hash message authentication code
H-NSP	home network service provider
HO	handover
HOL	head-of-line
HR/DSSS	high rate direct sequence spread spectrum using the long preamble and header
HSDPA	high-speed downlink packet access
HSS	home subscriber server
HSUPA	high-speed uplink packet access
HT	header type; high throughput
HTTP	hypertext transfer protocol
HUMAN	high-speed unlicensed metropolitan area network
HWMP	hybrid wireless mesh protocol
IAPP	interaccess point protocol
IBSS	Independent BSS
IC	interference cancellation; interchange
ICI	intercell interference
ICIS	integrated customer information system
ICV	integrity check value
IDFT	inverse DFT
IE	information element
IETF	Internet Engineering Task Force

IFFT	inverse fast Fourier transform
IFS	interframe space
IM	instant messaging
IMS	IP multimedia subsystem
IMT-2000	international mobile telecommunications in the year 2000
IOT	interoperability test
IoTCH	interference over thermal channel
IP	Internet protocol
IPC	interprocessor communication
IR	incremental redundancy; infrared
IS-95	Interim Standard 95
ISDN	integrated services digital network
ISI	intersymbol interference
IS-IS	intermediate system to intermediate system
ISM	industrial scientific medical
ITS	intelligent transportation system
ITU	International Telecommunication Union
ITU-R	ITU-radiocommunication sector
ITU-T	ITU-telecommunication standardization sector
IUI	interuser interference
IV	initialization vector
KCK	EAPOL-Key confirmation key
KDE	key data encapsulation
KEK	key encryption key
KKT	Karush-Kuhn-Tucker
KMP	key management protocol
KT	Korea Telecom
LAN	local area network
LBS	location based service
LDPC	low density parity coding
LFSR	linear feedback shift register
LLC	logical link control
LLR	log-likelihood ratio
LMDS	local multipoint distribution service
LME	layer management entity
LNA	low-noise amplifier
LOS	line of sight
LPDU	link protocol data unit
LPF	lowpass filter
LRC	long retry count
LSB	least significant bit

LSDU	link service data unit
LTE	long-term evolution
LWDF	largest weighted delay first
MAC	medium access control; message authentication code
MAN	metropolitan area network
MANET	mobile ad hoc network
MAP	maximum a posteriori; mesh access point
MAPL	maximum allowable path loss
MBS	multicast-broadcast service
MCS	modulation and coding scheme
MCW	multicode word
MD	mobility domain
MDC	modification detection code
MDHO	macro diversity handoff
MDIE	mobility domain information element
MFB	MCS feedback
MGC	media gateway controller
MGCF	media gateway control function
MGW	media gateway
MIB	management information base
MIC	ministry of information and communications; message integrity code
MIMO	multiple input multiple output
MIP	Mobile IP
m-IP Channel	mobile Internet Protocol channel
ML	maximum-likelihood
MLME	MAC sublayer management entity
MMD	multimedia domain
MMDS	multichannel multipoint distribution service
MML	modified ML
MMPDU	MAC management protocol data unit
MMR	mobile multihop relay
MMS	multimedia messaging system
MMSE	minimum mean-squared error
MP	mesh point
MPDU	MAC protocol data unit
MPEG	moving picture experts group
MPP	mesh portal
MRC	maximal-ratio combining
MRQ	MCS request
MRT	maximal ratio transmission

MS	mobile station
MSB	most significant bit
MSC	mobile switching center
MSDU	MAC service data unit
MSI	maximum service interval
MSK	master session key
MTBA	multi TID block Ack
MTU	maximum transfer unit
N/A	not applicable
NAI	network access identifier
NAK	negative acknowledgment
NAP	network access provider
NAV	network allocation vector
NE	network element
NLOS	nonline of sight
NMS	network management system
NRM	network reference model
nrtPS	nonreal-time polling service
NRT-VR	nonreal-time variable rate
NSP	network service provider
NWG	network working group
OBSS	overlapping BSS
OFB	output feedback
OFDM	orthogonal frequency division multiplexing
OFDMA	orthogonal frequency division multiple access
OH-HO	optimized hard-handover
OLPC	open-loop power control
OMP	operation and maintenance platform
OSIC	ordered SIC
OSICH	other sector interference channel
OSPF	open shortest path first
OSS	operations support system
OUI	organizationally unique identifier
PAK	privacy authorization key
PAPR	peak-to-average power ratio
PBCC	packet binary convolutional code
PBR	piggyback request
PBRO	partial BRO
PC	point coordinator
PCB	power control bit
PCC	policy and charging control

PCF	packet control function; point coordination function
PCMCIA	personal computer memory card international association
PCRF	policy and charging rule function
PCS	personal communication service
PD	policy decision
PDA	personal digital assistants
PDN	packet data network
PDSN	packet data serving node
PDU	protocol data unit
PE	provider edge
PEAP	protected extensible authentication protocol
PEP	policy enforcement point
PF	proportional fairness
PF	policy function
PHS	payload header suppression
PHSF	PHS field
PHSI	PHS index
PHSM	PHS mask
PHSS	PHS size
PHSV	PHS valid
PHY	physical layer
PIFS	point (coordination function) interframe space
PIMS	personal information management system
PIN	personal identification number
PIS	personal information service
PKC	public-key cipher
PKI	public key infrastructure
PKM	privacy key management
PLCP	physical layer convergence procedure
PLME	physical layer management entity
PM	poll me
PMD	physical medium dependent
PMIP	Proxy MIP
PMK	pairwise master key
PMKID	pairwise master key identifier
PMKSA	pairwise master key security association
PMP	point-to-multipoint; portable multimedia player
PN	pseudo-noise (code sequence); packet number
PON	passive optical network
POP3	post office protocol version 3
PPDU	PLCP protocol data unit

PPP	point to point protocol
PRBS	pseudo-random binary sequence
PRF	pseudo-random function
PRN	pseudo-random number
PRNG	pseudo-random number generator
PS	power save (mode)
PSDN	packet-switched data network
PSDU	PLCP service data unit
PSH	packing subheaders
PSK	preshared key
PSM	power saving mode
PSMP	power save multi-poll
PSS	portable subscriber station
PSTN	public switched telephone network
PTK	pairwise transient key
PTKSA	pairwise transient key security association
PTT	push-to-talk
PTV	push-to-view
PU^{2}RC	per-user unitary rate control
PUSC	partial usage subchannel
QAM	quadrature amplitude modulation
QCS	quick connection setup
QoS	quality of service
QPSK	quadrature phase-shift keying
QRM-MLD	maximum likelihood detection with QR decomposition and M-algorithm
R0KH	PMK-R0 key holder in the authenticator
R1KH	PMK-R1 key holder in the authenticator
RA	receiver address or receiving station address
RADIUS	remote authentication dial-in user service (IETF RFC 2865-2000)
RA-OLSR	radio aware optimized link state routing
RAS	radio access station
RD	reverse direction
RDG	The RD grant
RF	radio frequency
RIC	resource information container
RLAN	radio local area network
RMB	RAS main block
RNC	radio network controller
RNG	ranging

RNP	radio network planning
ROHC	robust header compression
RP	reference point
RPI	receive power indicator
RR	resource reservation
RRA	radio resource agent
RRC	radio resource control
RRM	radio resource measurement
RSA	acronym for the three inventors, Rivest, Shamir, and Adelman
RSC	recursive systematic convolutional; receive sequence counter
RSN	robust security network
RSNA	robust security network association
RSNIE	RSN information element
RSP	response
RSS	really simple syndication; received signal strength
RSSI	received signal strength indicator
RTG	Rx/Tx transition gap
rtPS	real-time polling service
RTS	request to send
RT-VR	real-time variable-rate
RX	receive or receiver
S0KH	PMK-R0 key holder in the supplicant
S1KH	PMK-R1 key holder in the supplicant
SA	security association
SA	source address
SAIC	single antenna interference cancellation
SAID	security association identifier
SAP	service access point
S-APSD	scheduled automatic power-save delivery
SBC	subscriber-station basic capability
SC	single-carrier
SCW	single code word
SD	space diversity
SDH	synchronous digital hierarchy
SDMA	space division multiple access
SDU	service data unit
SETT-EDD	estimated transmission times earliest due date
SF	service flow
SFA	SF authentication
SFBC	space-frequency block code
SFD	start frame delimiter

SFID	service flow identifier or identification
SFM	service flow management
SFTP	secure file transfer protocol
SGSN	serving GPRS support nodes
SHARQ	synchronous HARQ
SI	slip indicator; service interval
SIC	successive interference cancellation
SIFS	short interframe space
SIM	subscriber identity module
SIMO	single-input multi-output
SINR	signal to interference and noise ratio
SIP	session initiation protocol
SIR	signal to interference ratio
SISO	single-input single-output
SLA	service level agreement
SM	spatial multiplexing
SME	station management entity
SMK	STSL master key
S-MML	sorted MML
SMS	short message service
SN	sequence number
SNAP	subnetwork access protocol
SND	sounding
SNMP	simple network management protocol
SNonce	supplicant nonce
SNR	signal-to-noise ratio
SOHO	small office home office
SOVA	soft-output Viterbi algorithm
SP	service period
SPA	supplicant's MAC address
SPID	subpacket identifier
SRG	shift register generator
SS	subscriber station
SSAP	source service access point
SSID	service set identifier
SSO	single sign-on
STA	station
STBC	space-time block coding
STC	space-time coding
STTC	space-time trellis code
STTD	space-time transmit diversity

SVD	single value decomposition
SYNC	synchronization
TA	transmitter address or transmitting station address
TB-CC	tail-biting convolutional coding
TBTT	target beacon transmission time
TC	traffic category
TCLAS	traffic classification
TD	transmit diversity; TXOP duration
TDD	time division duplex (duplexing)
TDES	triple DES
TDM	time division multiplex
TDMA	time division multiple access
T-DMB	terrestrial-DMB
TEK	traffic encryption key
TF	training field
TFTP	trivial file transfer protocol
TG	task group
TID	traffic identifier
TIM	traffic indication map
TK	temporal key
TKIP	temporal key integrity protocol
TLS	transport layer security
TLV	type length value
TMPTT	target measurement pilot transmission time
TPC	transmit power control
TPF	traffic plane function
TPS	triple play service
TRS	trunked radio system
TS	traffic stream
TSC	TKIP sequence counter
TSF	timing synchronization function
TSID	traffic stream identifier
TSN	transition security network
TSPEC	traffic specification
TSS/TP	test suite structure and test purposes
TTA	telecommunications technology association
TTAK	TKIP-mixed transmit address and key
TTG	Tx/Rx transition gap
TTP	trusted third party
TU	time unit
TWG	technical working group

TX	transmit or transmitter
TxAA	transmit antenna array
TxBF	transmit beamforming
TXOP	transmission opportunity
U-APSD	unscheduled automatic power-save delivery
UCC	user created contents
UCD	uplink channel descriptor
UDP	user datagram protocol
UE	user equipment
UEPS	urgency and efficiency-based packet scheduling
UEQM	unequal modulation
UGC	user generated content
UGS	unsolicited grant service
UI	user interface
UIUC	uplink interval usage code
UL	uplink
UMB	ultra mobile broadband
UMTS	universal mobile telecommunication system
U-NII	unlicensed national information infrastructure
UP	user priority
URL	uniform resource locator
USB	universal serial bus
UTT	uplink transmission time
VCC	voice call continuity
VE	vertical encoding
VESM	vertical encoding SM
VLAN	virtual LAN
VOD	video on demand
VoIP	voice over IP
VoWLAN	VoIP over WLAN
WAN	wide area network
WAVE	wireless access in vehicular environment
WCDMA	wideband CDMA
WDM	wavelength division multiplexing
WDS	wireless distribution system
WEP	wired equivalent privacy
WG	working group
WiBro	wireless broadband
WiFi	wireless fidelity
WiMAX	worldwide interoperability for microwave access
WIPI	wireless Internet platform for interoperability

WLAN	wireless LAN
WLL	wireless local loop
WMAN	wireless MAN
WMF	WiMAX forum
WMM	WiFi multimedia
WPA	WiFi protected access
WSM	WiMAX system manager
XID	exchange identifier
XOR	exclusive-OR
ZF	zero-forcing
ZT CC	zero tailing CC

About the Authors

Byeong Gi Lee received a B.S. and an M.E. from Seoul National University, Seoul, Korea, and Kyungpook National University, Daegu, Korea, both in electronics engineering, and a Ph.D. in electrical engineering from the University of California, Los Angeles. He was with the Electronics Engineering Department of ROK Naval Academy as an instructor and naval officer in active service from 1974 to 1979. He worked for Granger Associates, Santa Clara, California, as a senior engineer, where he was responsible for applications of digital signal processing to digital transmission from 1982 to 1984. He then worked for AT&T Bell Laboratories, North Andover, Massachusetts, as a member of technical staff responsible for optical transmission system development along with the related standards works from 1984 to 1986. He joined the faculty of Seoul National University in 1986 and served as the director of the Institute of New Media and Communications in 2000 and the vice chancellor for research affairs from 2000 to 2002.

Dr. Lee was the founding chair of the Joint Conference of Communications and Information (JCCI), the chair of the steering committee of the Asia Pacific Conference on Communications (APCC), and the chair of the founding committee of the Accreditation Board for Engineering Education of Korea (ABEEK). He served as the TPC chair of IEEE International Conference on Communications (ICC) 2005 and the president of Korea Society of Engineering Education (KSEE). He was the editor of the *IEEE Global Communications Newsletter*, an associate editor of the *IEEE Transactions on Circuits and Systems for Video Technology*, and the founding associate editor-in-chief and the second editor-in-chief of the *Journal of Communications and Networks* (JCN). He served for the IEEE Communications Society (ComSoc) as the director of Asia Pacific Region, as the director of membership programs development, as the director of magazines, as a member-at-large to the board of governors, and as the vice president for membership development. He served as a member of the Presidential Advisory Committee of Policy Planning and the Presidential Advisory Council on Science and Technology. He served as a vice president of the ABEEK, the president of Korea Information and Communication Society (KICS), and the first president of the Citizens' Coalition for Scientific Society (CCSS), a nongovernment organization for the advancement of science and technology in Korea. He currently serves as a commissioner of the Broadcasting and Communications Commission of the Korean Government and as the vice president for the member relations of the IEEE Communications Society.

Dr. Lee is a coauthor of *Broadband Telecommunication Technology*, first and second editions (Artech House, 1993 and 1996), *Scrambling Techniques for Digital*

Transmission (Springer-Verlag, 1994), *Scrambling Techniques for CDMA Communications* (Kluwer, 2001), and *Integrated Broadband Networks* (Artech House, 2002). He holds 13 U.S. patents with four more patents pending. His current fields of interest include broadband networks, wireless networks, communication systems, and signal processing. He received the 1984 Myril B. Reed Best Paper Award from the Midwest Symposium on Circuits and Systems, Exceptional Contribution Awards from AT&T Bell Laboratories, a Distinguished Achievement Award from KICS, the 2001 National Academy of Science (of Korea) Award, and the 2005 Kyung-am Academic Award. He is a member of the National Academy of Engineering of Korea, a member of Sigma Xi, and a fellow of the IEEE.

Sunghyun Choi is currently an associate professor at the School of Electrical Engineering, Seoul National University (SNU), Seoul, Korea. Before joining SNU in September 2002, he was with Philips Research USA, Briarcliff Manor, New York, as a senior member research staff and a project leader for three years. He received a B.S. (summa cum laude) and an M.S. in electrical engineering from Korea Advanced Institute of Science and Technology (KAIST) in 1992 and 1994, respectively, and received a Ph.D. at the Department of Electrical Engineering and Computer Science, University of Michigan, Ann Arbor, in 1999.

His current research interests are in the area of wireless/mobile networks with emphasis on the QoS guarantee and adaptation, resource management, wireless LAN/MAN/PAN, next generation mobile networks, data link layer protocols, and cross-layer approaches. He authored/coauthored more than 110 technical papers and book chapters in the areas of wireless/mobile networks and communications. He holds 14 U.S. patents, 9 European patents, and 7 Korean patents, and has tens of patents pending. He has served as a general cochair of COMSWARE 2008 and as a technical program cochair of ACM Multimedia 2007, IEEE WoWMoM 2007, and IEEE/Create-Net COMSWARE 2007. He was a cochair of the Cross-Layer Designs and Protocols Symposium in IEEE IWCMC 2006, the workshop cochair of WILLOPAN 2006, the general chair of ACM WMASH 2005, and a technical program cochair for ACM WMASH 2004. He is currently serving and has served on program and organization committees of numerous leading wireless and networking conferences, including IEEE INFOCOM, IEEE SECON, IEEE MASS, and IEEE WoWMoM. He is also serving on the editorial boards of the *IEEE Transactions on Moblie Computing, ACM SIGMOBILE Mobile Computing and Communications Review* (MC2R), and the *Journal of Communications and Networks* (JCN). He is serving and has served as a guest editor for the *IEEE Journal on Selected Areas in Communications* (JSAC), *IEEE Wireless Communications*, *Wireless Personal Communications*, and *Wireless Communications and Mobile Computing* (WCMC). He gave a tutorial on IEEE 802.11 in ACM MobiCom 2004 and IEEE ICC 2005. Since 2000, he is an active participant and contributor of IEEE 802.11 WLAN Working Group.

He has received a number of awards, including the Young Scientist Award (awarded by the president of Korea) in 2008; the IEEK/IEEE Joint Award for Young IT Engineer of the Year 2007 in 2007; and the Outstanding Research Award in 2008; and the Best Teaching Award in 2006, both from the College of Engineering, Seoul National University. Dr. Choi was a recipient of the Korea Foundation for

Advanced Studies (KFAS) Scholarship and the Korean Government Overseas Scholarship during 1997–1999 and 1994–1997, respectively. He also received a Bronze Prize at Samsung Humantech Paper Contest in 1997. He is a senior member of the IEEE, and a member of ACM, KICS, IEEK, and KIISE.

Index

B

C

E

N

Q

R

S

W

X

Z

Recent Titles in the Artech House Mobile Communications Series

John Walker, Series Editor

3G CDMA2000 Wireless System Engineering, Samuel C. Yang

3G Multimedia Network Services, Accounting, and User Profiles, Freddy Ghys, Marcel Mampaey, Michel Smouts, and Arto Vaaraniemi

802.11 WLANs and IP Networking: Security, QoS, and Mobility, Anand R. Prasad, Neeli R. Prasad

Advances in 3G Enhanced Technologies for Wireless Communications, Jiangzhou Wang and Tung-Sang Ng, editors

Advances in Mobile Information Systems, John Walker, editor

Advances in Mobile Radio Access Networks, Y. Jay Guo

Applied Satellite Navigation Using GPS, GALILEO, and Augmentation Systems, Ramjee Prasad and Marina Ruggieri

Broadband Wireless Access and Local Networks: Mobile WiMax and WiFi, Byeong Gi Lee and Sunghyun Choi

CDMA for Wireless Personal Communications, Ramjee Prasad

CDMA Mobile Radio Design, John B. Groe and Lawrence E. Larson

CDMA RF System Engineering, Samuel C. Yang

CDMA Systems Capacity Engineering, Kiseon Kim and Insoo Koo

CDMA Systems Engineering Handbook, Jhong S. Lee and Leonard E. Miller

Cell Planning for Wireless Communications, Manuel F. Cátedra and Jesús Pérez-Arriaga

Cellular Communications: Worldwide Market Development, Garry A. Garrard

Cellular Mobile Systems Engineering, Saleh Faruque

The Complete Wireless Communications Professional: A Guide for Engineers and Managers, William Webb

EDGE for Mobile Internet, Emmanuel Seurre, Patrick Savelli, and Pierre-Jean Pietri

Emerging Public Safety Wireless Communication Systems, Robert I. Desourdis, Jr., et al.

The Future of Wireless Communications, William Webb

Geographic Information Systems Demystified, Stephen R. Galati

GPRS for Mobile Internet, Emmanuel Seurre, Patrick Savelli, and Pierre-Jean Pietri

GPRS: Gateway to Third Generation Mobile Networks, Gunnar Heine and Holger Sagkob

GSM and Personal Communications Handbook, Siegmund M. Redl, Matthias K. Weber, and Malcolm W. Oliphant

GSM Networks: Protocols, Terminology, and Implementation, Gunnar Heine

GSM System Engineering, Asha Mehrotra

Handbook of Land-Mobile Radio System Coverage, Garry C. Hess

Handbook of Mobile Radio Networks, Sami Tabbane

High-Speed Wireless ATM and LANs, Benny Bing

Interference Analysis and Reduction for Wireless Systems, Peter Stavroulakis

Introduction to 3G Mobile Communications, Second Edition, Juha Korhonen

Introduction to Communication Systems Simulation, Maurice Schiff

Introduction to Digital Professional Mobile Radio, Hans-Peter A. Ketterling

Introduction to GPS: The Global Positioning System, Ahmed El-Rabbany

An Introduction to GSM, Siegmund M. Redl, Matthias K. Weber, and Malcolm W. Oliphant

Introduction to Mobile Communications Engineering, José M. Hernando and F. Pérez-Fontán

Introduction to Radio Propagation for Fixed and Mobile Communications, John Doble

Introduction to Wireless Local Loop, Second Edition: Broadband and Narrowband Systems, William Webb

IS-136 TDMA Technology, Economics, and Services, Lawrence Harte, Adrian Smith, and Charles A. Jacobs

Location Management and Routing in Mobile Wireless Networks, Amitava Mukherjee, Somprakash Bandyopadhyay, and Debashis Saha

Mobile Data Communications Systems, Peter Wong and David Britland

Mobile IP Technology for M-Business, Mark Norris

Mobile Satellite Communications, Shingo Ohmori, Hiromitsu Wakana, and Seiichiro Kawase

Mobile Telecommunications Standards: GSM, UMTS, TETRA, and ERMES, Rudi Bekkers

Mobile Telecommunications: Standards, Regulation, and Applications, Rudi Bekkers and Jan Smits

Multiantenna Digital Radio Transmission, Massimiliano "Max" Martone

Multiantenna Wireless Communications Systems, Sergio Barbarossa

Multipath Phenomena in Cellular Networks, Nathan Blaunstein and Jørgen Bach Andersen

Multiuser Detection in CDMA Mobile Terminals, Piero Castoldi

OFDMA for Broadband Wireless Access, Slawomir Pietrzyk

Personal Wireless Communication with DECT and PWT, John Phillips and Gerard Mac Namee

Practical Wireless Data Modem Design, Jonathon Y. C. Cheah

Prime Codes with Applications to CDMA Optical and Wireless Networks, Guu-Chang Yang and Wing C. Kwong

QoS in Integrated 3G Networks, Robert Lloyd-Evans

Radio Engineering for Wireless Communication and Sensor Applications, Antti V. Räisänen and Arto Lehto

Radio Propagation in Cellular Networks, Nathan Blaunstein

Radio Resource Management for Wireless Networks, Jens Zander and Seong-Lyun Kim

Radiowave Propagation and Antennas for Personal Communications, Third Edition, Kazimierz Siwiak and Yasaman Bahreini

RDS: The Radio Data System, Dietmar Kopitz and Bev Marks

Resource Allocation in Hierarchical Cellular Systems, Lauro Ortigoza-Guerrero and A. Hamid Aghvami

RF and Baseband Techniques for Software-Defined Radio, Peter B. Kenington

RF and Microwave Circuit Design for Wireless Communications, Lawrence E. Larson, editor

Sample Rate Conversion in Software Configurable Radios, Tim Hentschel

Signal Processing Applications in CDMA Communications, Hui Liu

Smart Antenna Engineering, Ahmed El Zooghby

Software Defined Radio for 3G, Paul Burns

Spread Spectrum CDMA Systems for Wireless Communications, Savo G. Glisic and Branka Vucetic

Technologies and Systems for Access and Transport Networks, Jan A. Audestad

Third Generation Wireless Systems, Volume 1: Post-Shannon Signal Architectures, George M. Calhoun

Traffic Analysis and Design of Wireless IP Networks, Toni Janevski

Transmission Systems Design Handbook for Wireless Networks, Harvey Lehpamer

UMTS and Mobile Computing, Alexander Joseph Huber and Josef Franz Huber

Understanding Cellular Radio, William Webb

Understanding Digital PCS: The TDMA Standard, Cameron Kelly Coursey

Understanding GPS: Principles and Applications, Second Edition, Elliott D. Kaplan and Christopher J. Hegarty, editors

Understanding WAP: Wireless Applications, Devices, and Services, Marcel van der Heijden and Marcus Taylor, editors

Universal Wireless Personal Communications, Ramjee Prasad

WCDMA: Towards IP Mobility and Mobile Internet, Tero Ojanperä and Ramjee Prasad, editors

Wireless Communications in Developing Countries: Cellular and Satellite Systems, Rachael E. Schwartz

Wireless Communications Evolution to 3G and Beyond, Saad Z. Asif

Wireless Intelligent Networking, Gerry Christensen, Paul G. Florack, and Robert Duncan

Wireless LAN Standards and Applications, Asunción Santamaría and Francisco J. López-Hernández, editors

Wireless Technician's Handbook, Second Edition, Andrew Miceli

For further information on these and other Artech House titles, including previously considered out-of-print books now available through our In-Print-Forever® (IPF®) program, contact:

Artech House
685 Canton Street
Norwood, MA 02062
Phone: 781-769-9750
Fax: 781-769-6334
e-mail: artech@artechhouse.com

Artech House
46 Gillingham Street
London SW1V 1AH UK
Phone: +44 (0)20 7596-8750
Fax: +44 (0)20 7630-0166
e-mail: artech-uk@artechhouse.com

Find us on the World Wide Web at: www.artechhouse.com